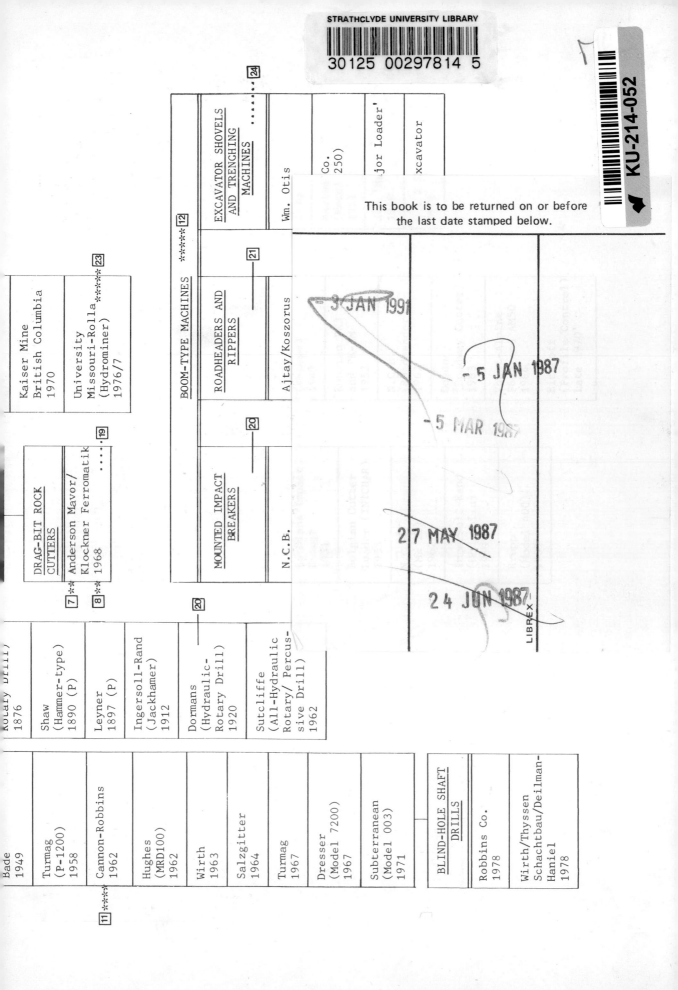

Rotary Drill)
1876

Bade
1949

Shaw
(Hammer-type)
1890 (P)

Turmag
(P-1200)
1958

[7] ** Anderson Mavor/
[8] ** Klockner Ferromatik ..... [19]
1968

Leyner
1897 (P)

[11] **** Cannon-Robbins
1962

DRAG-BIT ROCK
CUTTERS

Hughes
(MRD100)
1962

Ingersoll-Rand
(Jackhamer)
1912

Kaiser Mine
British Columbia
1970

Wirth
1963

Dormans
(Hydraulic-
Rotary Drill)
1920

University
Missouri-Rolla ***** [23]
(Hydrominer)
1976/7

Salzgitter
1964

[20]

Turmag
1967

Sutcliffe
(All-Hydraulic
Rotary/ Percus-
sive Drill)
1962

BOOM-TYPE MACHINES ***** [12]

Dresser
(Model 7200)
1967

MOUNTED IMPACT
BREAKERS [20]

ROADHEADERS AND
RIPPERS [21]

EXCAVATOR SHOVELS
AND TRENCHING
MACHINES .... [24]

Subterranean
(Model 003)
1971

N.C.B.

Ajtay/Koszorus

Wm. Otis

BLIND-HOLE SHAFT
DRILLS

Co.
250)

jor Loader'

Robbins Co.
1978

xcavator

Wirth/Thyssen
Schachtbau/Deilman-
Haniel
1978

Handbook
of
Mining and Tunnelling
Machinery

Handbook
of
Mining and Tunnelling
Machinery

Barbara Stack

A Wiley– Interscience Publication

JOHN WILEY & SONS

Chichester · New York · Brisbane · Toronto · Singapore

British Library Cataloguing in Publications Data:

Stack, Barbara
    Handbook of mining and tunnelling machinery.
    1. Mining machinery — History
    2. Tunneling machinery — History
    I. Title
    622'.028   TN345   80-41591

    ISBN 0 471 27937 4

Typeset in Great Britain by Photo-Graphics, Yarcombe, Devon
and printed by Pitman Press, Bath, Avon

To Miners and Tunnellers,
past, present, and future

Acknowledgements

I thank the many people, companies, and institutions, etc. listed below, who have contributed so generously in time and material towards the formation of this book. In particular I would like to thank H. M. Hughes, Head of Production Design Branch, National Coal Board, and Forrest S. Anderson of Anderson/Strathclyde, Limited, for their invaluable assistance with the section covering coal-mining.

Lastly, I would especially like to thank David B. Sugden (Consulting Engineer) for his patience, assistance, and encouragement during the writing of this book.

A.E.C. Inc., U.S.A. (Wm. J. Kogelmann)

Acker Drill Co. Inc., Scranton, Penn., U.S.A., (D. W. Wywonda)

Alfred Wirth & Co. (now Wirth Maschinen GmbH), West Germany

Allied Steel & Tractor Products, U.S.A.

Amax Coal Company, U.S.A.

American Embassy

American Institute of Mining, Metallurgical and Petroleum Engineers

American Society of Civil Engineers, New York, U.S.A.

Anderson Boyes Co. Ltd. (now Anderson Strathclyde Ltd), Scotland.

Anderson Mavor Ltd. (now Anderson Strathclyde Ltd), Scotland.

Anglo American Corp., Orange Free State, South Africa (R. N. Taylor)

Army Headquarters, Canberra, Australia

Army Information Office, London, U.K.

Army Information Office, Tas., Australia.

Army Museums Ogilby Trust, U.K.

Arthur Foster Construction Engineers Ltd. U.K.

Atlas Copco Australia Pty. Ltd. (J. M. Crook)

Atlas Copco Inc., Wayne, N.J., U.S.A.

Atlas Copco, Thun, Switzerland.

Atlas Copco, Sweden

Aubrey Watson Limited, Surrey, U.K.

Australian Coal Industry Res. Lab., Chatswood, Australia.

Australian Drilling Assoc. Ltd., Australia

Australian Institution of Mining and Metallurgy, Australia

Australian Tunnelling Association, Australia

Babcock Con. Equipment Ltd, Kent. U.K.

Bade & Company (now Bade & Theelen GmbH), W. Germany

Bergbau-Forschung GmbH (Dr. Ing. J. Henneke), W. Germany

Biblioteca Nazionale Centrale Vittorio, Rome, Italy

Bibliothèque Nationale, Paris, France

Binnie and Partners, London, U.K.

Birkenhead Town Hall, Town Clerk

Bituminous Coal Research Inc., U.S.A. (J. W. Igoe)

Bougard, J. F. Department of Transport, Paris, France

Bouygues, France

British Aircraft Corporation, Bristol, U.K.

British Channel Tunnel Company Limited

British Rail, Eastern Region, London, U.K.

British Railways Board, London, U.K.

British Standards Institution, London, U.K.

Bucyrus-Erie Company, South Milwaukee, U.S.A.

Bunt, E. F., Surrey, U.K. (Researcher)

Bureau of Reclamation, Region 7, Denver, Col., U.S.A.

Calweld, Santa Fe Springs, Cal., U.S.A. (W. A. Uski)

Calweld-Smith Inc., Newport Beach, Cal., U.S.A.

Canadian Institution of Mining & Metallurgy

Centre National de Documentation Scientifique et Technique

Chamber of Mines of South Africa Research Department, Johannesburg, South Africa

Chicago Metropolitan Sanitary District N. W., Intercept, U.S.A.

Coal Board, Queensland, Australia

Codelfa Construction Pty. Ltd.

Colorado School of Mines, U.S.A. (Dr. Fun-Den Wang and Mr. R. Miller)

Colorado School of Mines, U.S.A. (Prof. J. F. Abel)

Commissioner for Federal Districts, Mexico

CompAir Construction & Mining, U.K. (K. A. Tanner)

Consolidation Coal Co. (Messrs. Dahl and Petry), U.S.A.

Daiho Construction Company Limited, Japan

Demag AK (now Mannesmann Demag Bergwerktechnik), W. Germany

Denver Post Inc., Col., U.S.A.

Department for the Navy, London, U.K.

Department of Energy, Mines and Resources, Ottawa, U.S.A.

Department of Environment, U.K. (A. P. Moss)

Department of Film Production, Tas. Australia

Department of Mines, Hobart, Tas. Australia

Direction de L'Equipment, Paris, France

Dosco Overseas Engineering Ltd., London, U.K. (B. Reid)

Dravo Corp., Penn., U.S.A.

Dresser Industries Inc., Texas, U.S.A. (D. D. Nardo), (R. R. Durk)

Drilco Industrials, W.A., Australia

Dynadrill Co., Long Beach, Cal., U.S.A. (M. Emery)

Edmund Nuttall Ltd., London, U.K. (A. R. Biggart)

Eickhoff Maschinenfabrik (now Gebr. Eickhoff) Bochum, W. Germany

Electricity Supply Commission of S.A. (J. M. Barry)

Eton College, U.K.

Fairchild Incorporated, (P. L. McWhorter)

Foundation Com. Canada Ltd.

Fried, Krupp GmbH, W. Germany

Gardner B. H. Constructors (E. A. Horstketter)

Gardner Denver Ltd., Houston, U.S.A.

Gardner Denver Ltd., Tas. Australia

Gardner Engineering Corporation, Houston, Texas, U.S.A.

German Machinery Manufacturers, W. Germany

Germany, Federal Republic of, Consulate-General

Goldfields Diamond Drilling Company Pty. Ltd., Australia

Goodman Equipment Co. Chicago, U.S.A.

Gullick Dobson Ltd., Nottingham, U.K. (A. Purdy)

Guy F. Atkinson & Company, San Francisco, U.S.A.

Harding (Sir Harold), British Tunnelling Society, U.K.

Healy Tibbitts Construction Company, Cal. U.S.A.

Hecla Mining Company, Idaho, U.S.A. (D. Ferguson)

Hitachi Zosen, Osaka, Japan

Holman Bros. Ltd., Cornwall, U.K.

House of Commons Library, London, U.K.

Hughes Tool Company, Houston, Texas, U.S.A. (J. M. Glass Jr.)

Hurley Construction Company, Minneapolis, U.S.A.

Hydro-Electric Commission, Tas. Australia

Imperial War Museum, U.K.

Ingersoll-Rand (Aus.) Ltd., (J. H. Whitehead), Australia

Ingersoll-Rand Company, N.J., U.S.A. (J. W. Adams)

Institution of Civil Engineers, London, U.K.

Institution of Civil Engineers, Westminster, U.K.

Institution of Engineers, N.S.W., Australia

Institution of Mechanical Engineers, London, U.K.

Institution of Royal Engineers, Chatham, U.K.

Institut Nationale de L'Industrie Charbonnière de la Belgique, Belgium

International Nickel Company, Ontario, Canada

Japan Trade Centre, Melbourne, Australia

Japan Tunnelling Association, Japan

Jarva Incorporated, U.S.A. (E. W. Brickle)

JCB Sales Ltd., U.K.

Jeffrey Manufacturing Company, Ohio, U.S.A.

Jeffrey Mining Machinery Division, Dresser Industries, U.S.A.

John Connell, Mott & Anderson, Hatch, Jacobs, Melbourne, Australia (A. Neyland)

John Mowlem and Company Ltd., U.K.

Johns and Waygood, Victoria, Australia

Joint Coal Board, N.S.W. Australia (T. M. Clark)

Joy Manufacturing Company, Mining Machinery Division, U.S.A. (E. M. Warner)

Kawasaki Heavy Industries Ltd., Japan

Kinnear Moodie & Co. Ltd., U.K. (D. Scotney)

Kennametal Inc., Penn., U.S.A.

Kent Air Tool Co., U.S.A. (J. L. Smith)

Koken Boring Machine Company Limited, Japan

Komatsu Ltd., Japan

Koninklijke Vlaanse Ingenieursvereniging, Belgium

Kriegsarchiv., Vienna, Austria

Krupp & Co., N.S.W., Australia

Leeney, John, West Sussex, U.K.

Lee Norse Company, U.S.A. (T. Combs)

Library of the General Direction of the State Railways, Rome, Italy

Liverpool Town Hall, Town Clerk

London Transport, U.K.

Los Angeles Times, Los Angeles, Cal., U.S.A.

Lovat Tunnel Equipment Inc., Ontario, U.S.A.

M & H Tunnel Equipment Inc., Ontario, U.S.A.

M.U.R.L.A., Melbourne, Australia (F. Watson)

McGuire Shaft & Tunnel Corp., Arkansas, U.S.A.

Machinoexport Mosfilmowskaja, Moscow, U.S.S.R.

Manchester Geological and Mining Society, U.K.

Marcon International Ltd., U.K.

Marietta Manufacturing Company, W. Virginia, U.S.A.

Marion Power Shovel Company, Ind., U.S.A.

Markham & Co., U.K. (A. Armstrong)

Marshall Fowler Ltd., U.K.

Matthews Inc., (A.A.) Cal., U.S.A.

Maurer, W. C. (Dr.) Texas, U.S.A.

Melbourne and Metropolitan Board of Works, Melbourne, Australia

M.E.M.C.O. — Mining Machinery Equipment Company, U.S.A. (J. Tabor)

Metropolitan Water Board, London, U.K.

Metropolitan Water District of S. California, U.S.A.

Mid-Continent Coal and Coke Company, U.S.A.

Mini Tunnels International Ltd., U.K.

Mining Institution of Scotland, Inst. of Mining Engineers, Edinburgh, U.K.

Ministry of Defence, Library (Central and Army), U.K.

Ministry of Defence, Whitehall, London, U.K.

Mitchell Brothers Sons, London, U.K.

Mitsubishi Heavy Industries, Tokyo, Japan

Montabert, France

Morrison-Knudsen Company Inc., U.S.A.

Mott, Hay & Anderson, U.K. (J. V. Bartlett)

Museum of English Rural Life, University of Reading, U.K.

National Coal Board, U.K.

National Coal Research Advisory Committee, Melbourne, Australia

National Maritime Museum, U.K.

National Mine Service Company, U.S.A. (J. Karlovsky)

National Reference Library, Canberra, Australia

National Research and Development Corp., London, U.K.

National Water Well Association, Australia

New York Historical Society, U.S.A.

Nishimatsu Construction Company Ltd., Japan

Noyes Bros, Australia

O & K Orenstein & Koppel, West Germany

Oy Tampella Ab, Tamrock Industrial, Finland (J. Kuusento)

Osterreichisches Staatsarchiv, Austria

P.D.C. Construction Pty. Ltd. N.S.W., Australia

Patent Office, U.K.

Patent Office, U.S.A.

Patrick Harrison & Co. Ltd., U.S.A.

Paurat GmbH, Germany. (F. W. Paurat)

Perini Corp., U.S.A.

Perini News, Perini Corp., U.S.A.

Pipe Jacking Association, London, U.K.

Public Record Office, London, U.K.

Pugh, B. I. Ealing, London, U.K. (Researcher)

R.T.Z. Development Enterprises Ltd., U.K.

Racine Federated Inc., U.S.A. (R. Hayek)

Régie Autonome des Transports, France (J. Bougard)

Robbins Company, Kingston, Tas., Australia

Robbins Company, Seattle, U.S.A. (R. Robbins)

Robert L. Priestley Ltd, Gravesend, U.K. (R. Lewis)

Roberts-Union Corporation, Inc. Colorado, U.S.A. (C. F. Smith)

Rochester & Pittsburgh Coal Company, U.S.A. (W. E. Bullers)

Royal Aeronautical Society, London, U.K.

Royal Astronomical Society, London, U.K.

Royal Society, London, U.K.

Ruston-Bucyrus Limited, Lincoln, U.K.

S.E.L.I., Italy (C. Grandori)

S.E.T.I.S., Paris, France (M. Gervias)

Sato Kogyo Company Ltd., Japan (K. Miyata)

Schramm, Inc., U.S.A. (J. M. Deck)

Shea Company Inc., (J. F.), Utah, U.S.A. (E. G. Murphy)

Science Reference Library, Washington, U.S.A.

Sir Robert McAlpine & Sons Ltd, (J. Bland)

Smith Industries International Inc., Los Angeles, U.S.A.

Snowy River Council, Australia

Société Générale de Constructions Electriques et Méchaniques, France

Société Royale Belge des Ingenieurs, Belgium

Standards Association of Australia

State Railways, Library, Rome, Italy

Subterranean Tools Inc., (now Subterranean Equipment Company), Denver, U.S.A.

Sydney Morning Herald, Records Department, Australia

Tasmanian Government Railways, Australia

Tekken Kensetsu Company Ltd., Tokyo, Japan (H. Yamazaki)

Thomas, H. (Dr.) (Ex Hydro-Electric Commission, Tas., Australia)

Thyssen (G.B.) Ltd., U.K.

Thyssen Schachtbau GmbH, Germany (Dr. K. Wollers)

Transport Road Research Lab., Berkshire, U.K.

Tullock, I. G. (Ex Hydro-Electric Commission, Tas., Australia)

Tunnelling Equipment (London) Ltd., U.K. (R. R. Ross)

Turbo-Maschinen-Aktiengesellschaft, West Germany

Turmag Turbo-Maschinen AG, West Germany

U.S. Bureau of Mines, Washington, US.A.

U.S. Department of Commerce, Washington, U.S.A.

Underground Mining Machinery Ltd., U.K.

Union Carbide Corp., N.Y., U.S.A. (R. L. McNeill)

Union Industrielle Blanzy-Ouest, France

United States Department of the Interior, U.S.A.

United States Information Services, South Melbourne, Australia

University of Cambridge, U.K. (Prof. J. G. D. Clark)

University of California, U.S.A. (Prof. (Asst) W. Hood)

University of California, U.S.A. (Prof. G. W. Cook)

University of Edinburgh, U.K. (Prof. S. Piggott)

University of Missouri-Rolla, U.S.A. (Prof. C. R. Barker)

University of Missouri-Rolla, U.S.A. (Prof. D. A. Summers)

University of New South Wales, Australia (Prof. F. F. Roxborough)

University of New South Wales, Australia (Prof. J. P. Morgan)

University of Tasmania, Australia (Dr. M. S. Gregory)

Voest-Alpine AG, Australia and Austria

Wayss & Freytagg Aktiengellschaft, West Germany

Westfalia Lünen, West Germany

West Virginia Coal Mining Institute, U.S.A.

West Virginia University School of Mines, U.S.A.

White Pine Copper Company, Mich., U.S.A.

Whyte-Hall Pty. Ltd., N.S.W., Australia

Worthington Compressors, U.S.A. (R. G. Chambers)

Zachry Company, U.S.A.

Zokor International Limited, U.K.

Contents

II Excavation — Soft Ground

III Excavation — Coal-mining

Foreword

by Sir Harold Harding, founder Chairman of the British Tunnelling Society,
Consultant to the Channel Study Group 1958-1970

This is a splendid book and very readable. It has a fund of information, not only for those who burrow below ground — miners, tunnellers, geologists — but for other interested people. It is laced with lively accounts of examples of heroism mixed with pure skulduggery. There are useful and fascinating glimpses into history, with nice touches of wit and humour.

The author has succeeded in her stated intention of 'translating the archaic and sometimes almost incomprehensible language' used by the patent agents into the modern idiom. Her style is clear and pleasingly free of that other idiom — 'institutionese'. A thought which arises after reading the whole of the manuscript: who has benefited from the thousands of patents listed except the patent agents? A good deal of infringement seems to have occurred.

Several years ago I received a letter from a Mrs. Barbara Stack, in which she asked me a few questions about the Channel tunnel machines, for a book which she was writing on tunneling machinery. This surprised me as the letter came from Hobart on the south end of Tasmania. She could not get further away from the scene of the main 'action' without going to the Antarctic.

The correspondence continued, though my contribution was to utter words of encouragement. So it was a pleasure and a surprise to meet her when she came to England. She was far from the formidable matron whom I had visualized, being as small, lively, and active as the young Queen Victoria — happily married with four children who had completed their adolescence quite undamaged while this mammoth task was being undertaken. She told

me that she had started writing science fiction, which seemed a good apprenticeship for tackling the astonishing advances in recent years which would have seemed like science fiction to our forefathers.

In order to complete this work she had travelled 20,000 km in Australia, 45,000 km during travels overseas, and had spent over $4000 in postage of mail alone. She wrote, 'Of interest might be the fact that on several occasions both here in Australia and overseas I only gained entry to a mine or a tunnelling project because by some strange coincidence I bear the same name as the Patron Saint of Miners — St Barbara.'

United States 'sandhogs' boast that it is unlucky to allow a dame into a tunnel so it is strange that St. Barbara should be the Patron Saint of Miners. Perhaps it is because her father immured her in a tower before beheading her for becoming converted. She is also the Patron Saint of Armourers and Blacksmiths, who would be qualified to build machines, but her protection is especially sought against lightning, which might be helpful when blasting.

Mrs. Barbara Stack attended many conferences and had discussions with about 800 people ranging from managing directors of large organizations to the engineers on site in the workings and on the factory floor while visiting various mining and tunnelling installations.

There are three works which some of us regard as our gospels — H. S. Drinker (*Tunnelling, Explosive Compounds and Rock-drills*, John Wiley & Sons, New York, 1893), William Charles Copperthwaite (*Tunnel Shields and the use of Compressed Air in Sub-aqueous Works*, Archibald Constable & Co. Ltd.,

London, 1906) and B. H. M. Hewett and S. Johannesson (*Shield and Compressed Air Tunnelling*, McGraw-Hill, New York, 1922).

This missionary journey of St. Barbara's successor has been so fruitful that her work will rank in importance with the other three.

Harold Harding

Preface

At the beginning of 1973 I became interested in the history of tunnelling machines and sought information on the subject from the Tasmanian State Library and other Australian libraries through the Tasmanian State Library. It appeared that, although several books had been written on 'tunnelling', 'mining', or 'drilling' none was available on the history of the actual machines involved. I then decided to investigate the subject and, if sufficient information could be found, gather it together in the form of a book. With this end in view I began an extensive and exciting search by correspondence via companies, libraries, institutes of engineering, universities, and other associations throughout the world, gathering, sorting, and following obscure leads through various channels in an effort to obtain as much information on the subject as was procurable.

I found that the history of the rock-tunnelling machine was inextricably bound up with the evolution of other tunnelling and mining devices. Thus my area of research was gradually extended until it covered the entire range of associated tunnelling machines, including drilling machines — horizontal and vertical, raise drills and shaft-boring machines, shields, mechanized shields, and slurry machines, rock tunnelling machines, machines which developed in the coal-mines and which were later adapted to other uses, reaming machines, incline-boring machines and the various boom-type machines such as road-headers and mounted impact breakers, etc.

I found that I was able to trace the evolutionary development of these machines and their interrelationship from the earliest units introduced to the latest models now operating. In this latter connection I have been in regular contact over the past eight years with most major mining, tunnelling, and drilling machine companies throughout the world. I have also seen numerous machines of various types in operation and visited a large number of tunnelling projects, mining installations, and machine manufacturing companies in Australia, the U.K., U.S.A., and West Germany.

The particular manufacturers whose machines are featured in this book have not been selected because their product is necessarily deemed to be superior to that of their competitors, nor because they produce a large number of any type of machine, but rather because the company's engineers and/or designers have made some major contribution towards the development of the machine in question. Nor, with the exception of the hard rock-tunnelling machines, has there been any attempt to list all the manufacturers of any particular type of machine.

Behind the desire to list the various types of machine and their particular improvements was a personal wish to make the narrative both interesting and simple to understand. The hope was that a geologist, say, who had little mechanical knowledge of the drilling apparatus which produced his core sample, would, after reading the book, at least be on 'talking terms' with the drilling crew and better able to converse with some facility on the way the machine worked, its areas of application, its advantages and disadvantages, and, above all, its limitations.

Because the book was written during the transition period from the imperial to the metric system both figures have in most cases been given. However, while every care has been taken to ensure that these are accurate, some errors may have occurred — if so, I trust the reader will accept my profuse apologies.

To ensure the technical accuracy of the work the various sections have been sent to several experts or authorities for their comments. This has proved to be an enriching and valuable exercise as, in many cases, the MS. was returned with additional information, thus bringing that particular section up to date as well as filling any gaps.

Literally hundreds of patents covering early shields, mechanized shields, and tunnelling machines have been studied and their archaic and sometimes almost incomprehensible language translated into the modern idiom, I trust with accuracy.

Apart from such sources as papers, articles, etc. which are given in the references and bibliographies, a great majority of the material in the text was obtained through interviews with engineers, contractors, manufacturers, etc. and via personal correspondence which is not available to the general public.

I have deliberately broken with traditional practice so far as the format of a purely technical work is concerned and have included, for the information and interest of the reader, matters pertaining to the historical development of the machines, their mode of operation, and their performance under actual working conditions, etc.

It should be noted, however, that occasionally only one opinion, perhaps that of the project engineer, or the contractor, was available and, at such times I was constrained to view the performance of the machine through the eyes of that project engineer or contractor. Naturally, if other reports were available then these were included so that a more balanced view of the overall project or machine could be given. Indeed, relentless efforts were made during the research period to obtain as much information as possible on every aspect of the work.

While it is, of course, a pity when only one opinion of the performance of a machine is available, it has been considered that, in such cases, one view by a qualified person with intimate knowledge and experience is considerably better than none.

The reader's attention is also drawn to the fact that the technical data and ratings for particular machines were naturally, in most cases, supplied by the manufacturer of the machine in question.

Hobart, 1981   Barbara Stack

THE HYMN OF BREAKING STRAIN

RUDYARD KIPLING

The careful text-books measure
(Let all who build beware!)
The load, the shock, the pressure
Material can bear.

So, when the buckled girder
Lets down the grinding span,
The blame of loss, or murder
Is laid upon the man.
Not on the stuff – the man.

(Published first in *The Engineer*, 15 March 1935; courtesy of *The Engineer*)

Excavation — Rock

Introduction

Our trip through the earth's crust was but a repetition of my two former journeys between the inner and the outer worlds. This time, however, I imagine that we must have maintained a more nearly perpendicular course, for we accomplished the journey in a few minutes' less time than upon the occasion of my first journey through the five-hundred-mile crust. Just a trifle less than seventy-two hours after our departure into the sands of the Sahara we broke through the surface of Pellucidar....*[1]

Stone, Bronze and Iron Ages

We may grin with good-natured indulgence as we read Edgar Rice Burroughs's graphic description of life at the centre of the earth and the incredible speeds attained by his 'iron mole' the 'Prospector', yet despite this, one cannot help but notice the striking resemblance the modern tunnel borer has to this fantastic machine. Burroughs's tunnel borer was powered by a mighty engine which drove an equally large revolving drill. The mole itself was comprised of two skins,** an outer and an inner, and was articulated to allow for greater manoeuvrability around short-radius curves. Its slim steel cylinder was 30 m (100 ft) long. When the Robbins Company built their Model 181-122 (1968) for use in the White Pine Copper Mine in Michigan, it, too, was jointed. This was to allow the machine to negotiate curves of 30 m radius in both the horizontal and vertical planes.

Such speeds as those described by Edgar Rice Burroughs still belong, naturally, in the realms of science fiction. Nevertheless, modern technology is gradually whittling down the technical barriers, and tunnelling records, both in hard and soft ground, are constantly being broken. But in order to appreciate more fully man's latest achievements, it is necessary to view, if only briefly, the story of his beginnings — the important period of transition from *animal* to *man*.

The 'Ages' of man, i.e. the Stone Age, the Bronze Age, and the Iron Age are the names commonly given to the three specific phases of man's existence through which he has passed at some particular time. The terms have no definite chronological significance and may vary considerably from country to country. Indeed some people (e.g. certain primitive tribes in Africa and America) are only now emerging from the Stone Age.

During the Stone Age metal was unknown. Such tools or weapons as were shaped or used were made of wood, stone, bone, antlers, or ivory. In Europe, Asia, Egypt, and certain parts of Central America, the people passed through this phase about 2300 B.C. (Neolithic Age).

The term 'Bronze Age' describes that period of man's existence when he has learned the secrets of alloying copper with tin in the correct proportions to produce the desired hardening effect.

The people of Europe, Asia, Egypt, etc. entered the Early Bronze Age about 1700 B.C., the Middle Bronze Age about 1400 B.C. and

---

*David Innes's description of his journey to Pellucidar in Abner Perry's 'iron mole' the 'Prospector' from the book *Pellucidar* by Edgar Rice Burroughs, Grosset and Dunlap, 1923. (Courtesy Edgar Rice Burroughs Inc.)
**The Bade mechanized shield used in Vienna and São Paulo, etc. featured a double skin (see Chapter 12 on Pressurized plenum chamber machines).

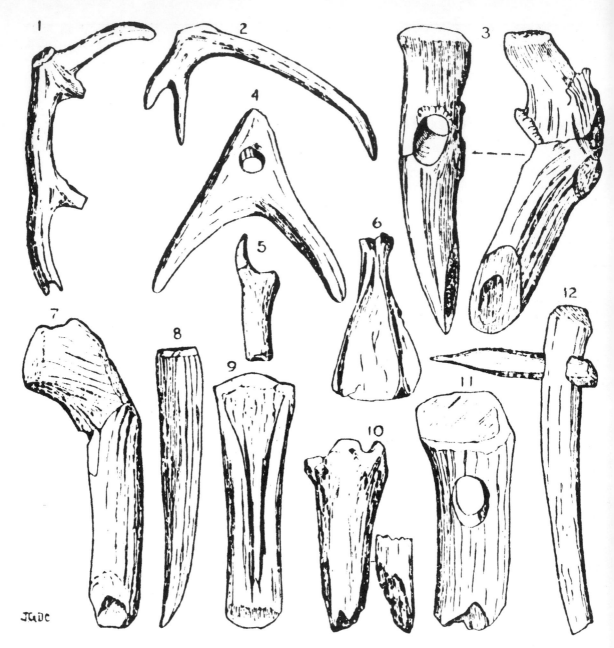

Figure 1. Implements of antler and bone from European Flint mines. (1) Antler pick; (2) rake; (3) axe-hammers; (4) rake; (5, 8-10) wedges and levers; (6) Scapulae — used as shovels; (7) antler crown hammer; (11) hammer; (12) two-piece pick. (Courtesy Prof. Grahame Clark and Prof. Stuart Piggott. From The age of the British flint mines, *Antiquity,* **7,** 1933)

the Late Bronze Age about 800 B.C. These people entered the beginning of the Iron Age (i.e. the period when man discovered iron and learned how to prepare and work it) about 450 B.C.

No doubt man's first implements or weapons were made of the nearest and handiest piece of wood or rock that chanced his way, but he soon noticed and took advantage of the more efficient weapons and tools supplied to him by a bountiful nature. The discarded deer antler or bone-wedge became a pick, a rake, or a hammer; an ox's shoulder-blade made a handy shovel; and pieces of stone an axe or knife.

As time went by man also discovered the difference between flint and ordinary rock. He found that though flint was hard it could nevertheless be flaked into pieces and shaped with another implement. Tools and weapons could be made and, if he struck his piece of flint with marcasite (an iron sulphide found in clay and chalk), he could also make fire. Modern man continued to use flint as almost the sole means of fire-making until fairly recently, by striking the flint against steel inside a tinderbox. The principle is of course retained today in the common cigarette or gas lighter.

When prehistoric man had exhausted local supplies of flint he began digging for this precious mineral underground. In fact mining for flints was the earliest of man's organized endeavours — coming as it did several millennia before agriculture — as witness the existence of Grime's Graves near Brandon in Suffolk, England, which are known to be more than 10,000 years old.

Man soon found that buried nodules were of a higher quality than the weathered surface pieces. This led him to dig ever deeper in his relentless search for suitable stone. In Europe, too, flint-miners ignored inferior material and sent their shafts down through Quaternary and Tertiary deposits of gravel and sand as far as 10 m (30 ft) or more in their efforts to reach the chalk which contained superior quality nodules of flint. As a result these flint-miners soon developed a high skill in the art of mining.

It is interesting to note how closely man's modern tools, i.e. the steel pick, hammer, shovel, awl, etc. resemble the shape of early bone implements and how little real change from these basic natural shapes there has actually been (see Figure 1).

The transition from stone to bronze, and bronze to iron was gradual, indeed the overlap continues today. Flint and bone were used extensively throughout the Bronze Age and the use of flint implements, hoes, etc. was not uncommon even during the earlier period of the Iron Age.

Of some interest is the fact that most iron objects found in Egypt and elsewhere were *weapons or ornaments* and not *tools.* These iron tools appeared at a much later stage when the Iron Age was well under way.[2,3]

References

1. Edgar Rice Burroughs, *Pellucidar,* Grosset and Dunlap, New York, 1923.
2. Grahame Clark and Stuart Piggott, The age of the British flint mines, *Antiquity,* **7**, 166-183, 1933.
3. Majorie and C.H.B. Quennell, *Everyday Life in the New Stone, Bronze and Early Iron Ages,* Batsford, Worc., revised 1945, pp. 12-15.

Bibliography

Chambers Encyclopaedia, Vol. V, new revised edn., Pergamon Press, London, 1966, p. 705.
Encyclopaedia Britannica, 1929, Vol. 2, p. 252.
Sir Ajril Fox, *Life and death in the Bronze Age,* Routledge and Kegan Paul, London, 1959.
H. Sandars, *The use of the deer-horn pick in the mining operations of the Ancients, Archaeologia,* 1 × 11 1911.

ONE

Drills

Horizontal Drills

Single- and Double-handed Hammer Drilling

Since man first emerged as a rationalizing, thinking animal, he has utilized the talents God bestowed upon him to devise easier and better methods of living and working.

For many hundreds of years all drilling work in mines, quarries, etc. had been done by hand. Two distinct methods have been employed by miners to carry out this particular type of work, namely either single- or double-handed drilling.

In single-handed drilling a light hammer weighing approximately 1.6 to 2.0 kg (3½ to 4½ lb) was used by one man. In one hand he held the drill or steel bit which he rotated slightly after each blow of the hammer, while in the other hand he carried the hammer which he pounded vigorously against the bit head.

In double-handed drilling one man held the hammer which could weigh from 4 to 4.5 kg (8 to 10 lb) while another man held the steel bit or drill bar. Sometimes when it was necessary to penetrate very hard rock or if very deep holes were required, it was considered essential to use two hammer-men. In such cases each wielded a 4.5 kg (10 lb) hammer against the head of a steel bit held by a third man.

Though man has continued to invent or devise better methods for easing his labours, paradoxically there have also always been those amongst us who have viewed each new device with scepticism and suspicion and who have earnestly questioned the value to mankind of each new innovation. Thus when enterprising men designed the first machines to ease the arduous labours of the miner, it was the miner himself who protested most vehemently at this intrusion into what he considered his private domain.

An excellent example of this antagonism is given by Gosta E. Sandstrom in his book *The History of Tunnelling.*[1] He tells in ballad form of the deeds of one John Henry, an American Negro who was said to have been employed in the Big Bend tunnel in the year 1870. Henry was only one of many such labourers who worked as hammer-men on the Chesapeake and Ohio Railroad. But in some respects Henry was different from his fellow workers. For Henry wielded not one but two 4.5 kg (10 lb) sledge-hammers. Moreover, Henry was a giant of a man and a veritable champion at his trade. When at work, he would stand about 1½ m (5 ft) from the man gripping the steel bar and, with a heavy hammer held in each hefty paw, he would swing these tools through the air, delivering a mighty blow to the bit-head with each downward stroke. To lesser men it seemed that when Henry was swinging his hammers they weighed nothing more to him than did featherweights.

Like any true champion, Henry was very particular about the care of his precious hammers. He meticulously maintained the suppleness of the long 1.2m (4 ft) switch handles with a regular application of tallow grease.

One day, Henry's ordered life was shattered by the appearance of a diabolic 'machine'. It was one of the new piston-type steam drills. Like many another hammer-man before him, Henry viewed their performance with contempt and swore that such puny devices could hardly match the drilling capabilities of a good hammer-man. Lesser men might fear replacement by the new-fangled contraptions, but he felt certain these ugly, new, noisy machines could never beat him, John Henry. His vociferous challenge was heard and at once accepted by the foreman in charge of his section. The foreman, sure of the superiority of the new mechnical device, offered to pay John Henry $100 if he could beat the drill. At Henry's request the foreman even procured two new hammers and handles for use at the contest.

On the big day Henry was ranged alongside the steam drill with his new hammers held expertly in his large work-worn hands. The long handles of the hammers were, of course, carefully tended with tallow grease of the best quality. With him was the shaker-boy, the person employed to hold the bit steady and rotate it regularly in the hole between blows in order to keep the tool free and work the loose debris out. Henry eyed the 2 m (6 ft) long steel bit and then commented grimly to his helper, the shaker-boy, that he had better pray he did not miss — for otherwise it could prove fatal for the lad.

Both Henry and the machine were set to work and a mighty contest it proved to be. Henry won the competition by drilling 4 m (14 ft) of rock in 35 minutes, against only 3 m (9 ft) penetrated by the new steam drill.

The foreman honoured his bet, but Henry's magnificent triumph was shortlived for he died that night. The tremendous strain had been too much, even for his great physique.

When his grieving workmates learned of his death, they refused to re-enter the tunnel, declaring that though Henry was dead, they could still hear the ring of his heavy sledge-hammers striking the steel bit. Eventually the exasperated foreman managed to persuade the men that what they were hearing was merely the seepage of water from the tunnel roof and not sounds emitted by some ghostly apparition. The men then reluctantly returned to their jobs.

Machine Drills Compared with Hand Labour[2]

Considering the heated controversy which rages today over the merits of various machine-bored tunnels as opposed to conventional methods, i.e. drilling, firing, and mucking, it is interesting to read that similar questions were asked in

Drinker's time on the relative economic advantages of machine labour (drills) and hand labour (hammer-work). The following copy of a letter from Walter Shanly, Civil Engineer, Montreal, Canada, dated 20 June 1877 addressed to Drinker and Drinker's comments thereon are typical of the major points of controversy which were discussed at that time. In his letter Shanly agrees with Drinker's opinion that it is uneconomical to use machines in 'short' tunnels of say less than 600 m (2000 ft) unless *time* was a primary consideration and *cost* a secondary one. However, Drinker later qualifies this view by observing that it would not necessarily apply to machines which were to be used and re-used in, say, mines. Today most experts on tunnelling would agree that when the entire cost of the machine is to be written off during the construction of one single tunnel it would be uneconomical to purchase a tunnel-borer for a tunnel of less than 4.6 km (15,000 ft) and certainly would not consider using one for a tunnel of say 3 km (10,000 ft) *unless,* that is, the machine was later to be used on another project. This of course would also apply for repetitive use in mines.

Office of Walter Shanly
Civil Engineer,
Montreal, Canada.
June 20 1877.

Henry S. Drinker, Esq.,
C.E. Philadelphia.

Dear Sir,

I have pleasure in responding to your request that I would state what may be the difference in cost of rock-drilling, and also in time gained, as between *machine*-work and *hand*-work, based on my experience in the Hoosac Tunnel.

I concur with you in thinking that in point of money economy there is nothing to be gained by applying machinery to the driving of *short* tunnels generally; but one cannot very well fix an arbitrary line at which 'short' shall end and 'long' begin. I observe that you incline to place the dividing line at 2000 feet; tunnels over that length to be classed as long and worked by machinery. Local conditions must largely govern in deciding on the use or non-use of power-drills. Favoring aspects of *power,* such as the proximity, abundance, and easy applicability of a good head of water, might render it judicious to drive a tunnel of much less length than your limit of 'short' by machine rather than hammer-work. Of

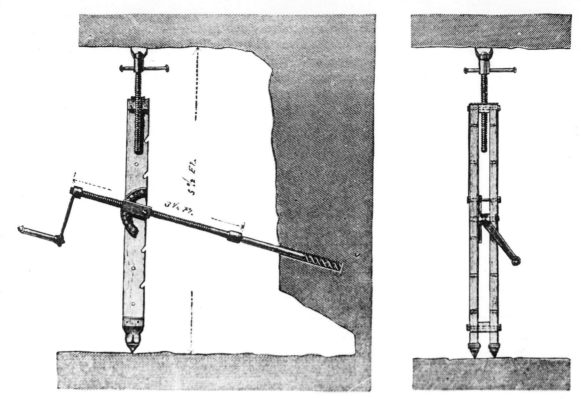

Figure 2. An example of one of the first rotary drills such as were used by Kranner and others in 1845

Figure 3. Couch's rock-drilling machine. First percussive drill ever made (1849)

course where *time* is the primary and *cost* a secondary consideration, mechanical appliances might have to be resorted to for almost any length of boring — 1000 feet, or less even — especially in cases where work would necessarily have to be carried forward on two faces only. In the Hoosac Tunnel I used both water and steam — i.e. applied both kinds of power to the compressing of the atmosphere, which third power, manufactured outside the tunnel, was sent to the drills inside under a pressure of from 60 to 64 pounds on the square inch. The rock pierced in the four and three quarter miles through Hoosac Mountain differed materially in structural character on the opposite slopes. The easterly half was a uniform mica-schist largely permeated with, and at intervals intersected by, veins or seams of pure quartz. It was a fairly good rock to drill, but somewhat tough of breaking. The westerly half of the work was in ground of varying character — passing from hard granite to granitoid gneiss, with much feldspathic and quartzose formation of exceeding hardness. The whole uniform in the one condition only of being extremely wearing upon tools and requiring deep drilling and heavy doses of nitro-glycerine to bring it out. At the easterly end of the tunnel, from which 'base' the best of the rock was to be attacked, I had water-power for compressing the atmosphere. The western workings were wholly dependent upon steam, causing, as you will perceive from the figures I give you below, a very marked difference in the expense of *manufactured power* as compared with that obtained through the agency of water at the other end, and which, in conjunction with the greater hardness of the rock, created a wide want of accord between the cost of drilling in the two divisions of the work.

In estimating the cost of machine-drilling, I have included mining labor, mechanical labor, blacksmith-work, materials consumed — such as iron, steel, oil, steam-coal, forge coal, etc. etc., and also allowed for interest on and depreciation in value of 'plant'.

The following footnote is given by Drinker as an explanation to the above letter.

This reference of Mr. Shanly's is to a letter from the author [Drinker] to Mr. Shanly, in which the author expressed the opinion that, as a general rule, machine-drilling would not pay in short tunnels — that the gain in time would not pay for the outlay for plant. Of course, in such a matter; no arbitrary limit can be laid down; as Mr. Shanly justly observes, local considerations will necessarily exercise a controlling influence; but, except where especially favorable circumstances occur, the author is willing to abide by the opinion that at the present rates of cost, and in the light of past experience, the use of machines in railroad-tunnels under 2000 feet in length will not be found to promote ultimate economy of work. This is largely owing to the fact that in every new railroad-tunnel of the present day [1882], the contractors and their hands have all to learn anew the art of drilling with machinery. When in the future the use of machine-drills becomes familiar to the great class of tunnelmen, so that so much time and money in each new tunnel need not be expended in taking costly lessons from new experience, then, probably, we shall find it economical to apply drills even in the shortest tunnels. Now the question is not, can Shanly, or Steele, or McFadden, or Sutro, drive short tunnels to good effect with power-drills; but, can the average contractor, who has never used them in former tunnels, do so? In February, 1876, the author [Drinker] asked Mr. Sutro what length of tunnel, in his opinion, ought to limit the introduction of machinery? His reply was 'That depends upon the hardness of the rock. I would not think of introducing machinery under any circumstances in a tunnel less than 1500 feet, that is accessible from both ends, and probably in a tunnel of that length it would hardly pay to go to the expense of erecting the proper compressing machinery, drill-carriages, etc. etc. It may be stated as an engineering proposition that in rock of average hardness, it would not pay to introduce machine-drills in a tunnel of less than 2000 feet.

These observations, of course, apply chiefly to railroad tunnels. In a mining region, where many successive short tunnels have to be driven, it may be found to pay well to use machinery on comparatively small pieces of work; for here the machinery is in constant use, and not laid aside on the completion of a single tunnel'.[2]

Early History

Because of the need to bore holes in rock for the placing of charges, the introduction of gunpowder for civilian purposes in mining led to the rapid development of the hand drill and subsequently to the invention of the first mechanical drill (Figures 2 to 9).

Kranner[2]

Kranner (Figure 2) used rotary drills for making stone water-piping in Prague in 1845 and later bored through limestone with them during the construction of the Karst Railroad in 1853-57.

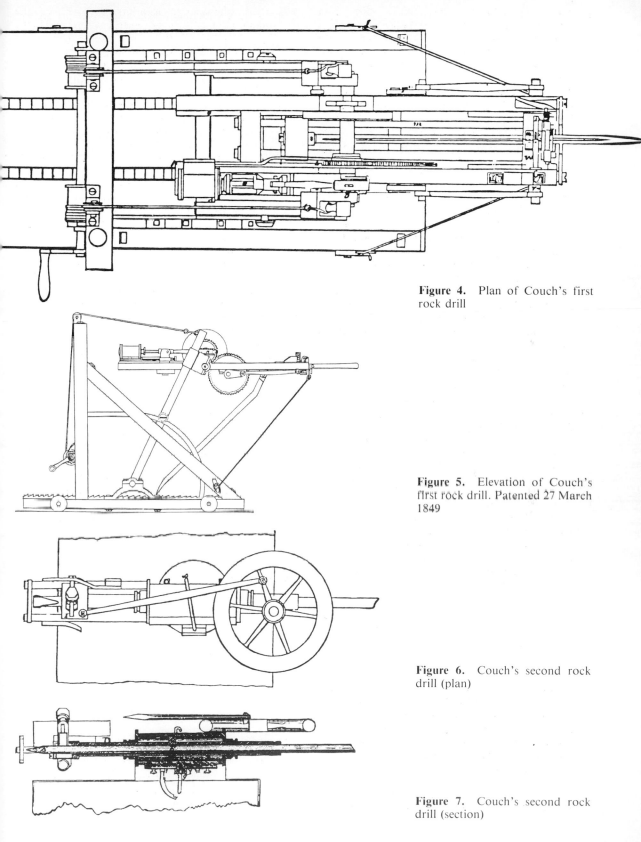

Figure 4. Plan of Couch's first rock drill

Figure 5. Elevation of Couch's first rock drill. Patented 27 March 1849

Figure 6. Couch's second rock drill (plan)

Figure 7. Couch's second rock drill (section)

Brunton — Compressed Air[2]

The first proposal for using compressed air to power drill-hammers (the air also serving to ventilate the face after use) was made by Brunton in 1844.*

J. J. Couch — First Percussion Drill[2]

In 1849 a patent for a rock drill (Figure 3) in which the drill was driven by steam-power and which acted independently of gravity was patented by J.J. Couch of Philadelphia on 27 March (U.S. Patent No. 6237). *This was the first machine percussion rock drill ever made.*

Couch was assisted in the construction of his drill by Joseph W. Fowle of Boston. During 1848 the two men carried out numerous experiments on a block of granite, using a working model of a drill built by Fowle for the purpose. However, apparently there was a clash of temperaments, for the following year Couch and Fowle separated and Fowle developed another quite different drill.

In Couch's machine the steam-engine drove a flywheel which, through a reducing gear chain, operated a slower crank that propelled a sliding frame back and forth. The drill bit which was supported in this frame was *gripped* on the *backward* stroke by a special gripper box or clutch and *thrown freely* against the rock on the *forward* stroke. Another interesting feature concerning the Couch drill was that rotation of the drill bar was achieved by means of a ratchet wheel which allowed the drill bar to pass freely through its centre. The wheel had a stud

projection which entered a groove cut in the drill bar. This stud caused the drill bar to rotate with the ratchet wheel.

J. W. Fowle — Direct Action Principle[2]

Though Couch is credited with designing the first percussion drill as distinct from a rotary borer, Fowle must be given credit for being the first engineer to apply the direct action principle in a drilling machine.

In his report to the Massachusetts Legislative Committee on the Burleigh claim in April 1874, Fowle said:

> My first idea of ever driving a rock-drill by direct action came about in this way: I was sitting in my office one day, after my business had failed, and happening to take up an old steam cylinder, I unconsciously put it in my mouth and blew the rod in and out, using it to drive in some tacks with which a few circulars were fastened to the wall. That was my first idea on the subject.

Fowle described his rock-drilling machine in the caveat as follows:

> 'My steam drilling-machine is distinguished from all others which have heretofore been contrived in the following particulars: in the first place, the drilling-tool is attached directly to the cross-head of the engine, or, in fact, to an elongation in a direct line of the piston-rod, and is impelled or driven forward by the entire power of the steam-engine, which has never before been attained, as in all other machines about one half of the power is expended in driving the machinery which connects the engine with the drill. In the second place, the mechanical means* for turning the drill as it comes back are much simpler and more effective than any other... heretofore devised. Thirdly, as the hole increases in depth, the entire engine with its frame is accurately fed forward by means of the momentum of the cross-head. Fourthly, in a machine of this description, where the momentum of the piston is availed of in direct action, the piston would be in danger of being driven through the head of the cylinder; but this difficulty I have obviated by a peculiar arrangement of the machinery, which moves the rocker-shaft and shifts the valves.... In the last place, I make the cutting edge of the drill I use in the shape of an S, which has decided advantages over any other form now in use, but this I do not claim as my invention.

---

*Steam[3,4] In 1712 the Newcomen steam-engine was put to work at a colliery near Dudley Castle. The application of this device, which constituted the first successful use of mechanical power underground, may be regarded as one of the forerunners of the Industrial Revolution.

Compressed air[3,4] According to Prof. Douglas Hay,[3] compressed air was used for power transmission at Goven Colliery, near Glasgow, in 1849 for a reciprocating engine to pump water and haul coal. Compressed air was also used for 'forcing water' in 1845. Later in a paper presented by Prof. Douglas Hay and N. E. Webster[4] at the Second World Power Conference in Berlin in 1930, it was said that: 'The first proposal to use compressed air underground was made in 1830 and in 1849 it was in regular use at a Scottish colliery for pumping and hauling below-ground'.

Electricity[3,4] According to Prof. Hay, electricity was first applied underground at Trafalgar Colliery when Sir Francis Brain used it to power a small pumping set in 1882.

---

*Fowle used a ratchet and pawl mechanism for rotating his drill.

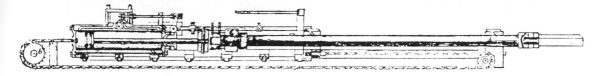

Figure 8. Fowle's first rock drill (1849)

Fowle's caveat was filed on the 9 May 1849. He was granted U.S. Patent No. 7972 in 1851.

Of Fowle's invention Drinker remarks:

It is interesting to note how completely this inventor (Fowle) comprehended the elements of the problem before him. He discarded light reciprocating parts, but instead of concentrating the weight in the piston-rod, as is now done, he attached to his cross-head a drill-bar (to which the tool was attached), which weighed over one hundred pounds. Having a drill-cylinder with light flanges, and fearing that he would break them by the piston striking the heads, and thus spoiling the cylinder, he secured the heads to each other by means of long rods passing from one to the other, and thus avoiding an attachment to the cylinder, and, as he reports, 'I thus saved my cylinder'. This principle has been incorporated into several modern drills. He used a flexible hose to conduct steam from the boiler to the machine. In 1850 or 1851, *he used compressed air* for driving his drill. He overcame the difficulty of making round holes with a machine-drill by making an S drill — a form which has since been used for the same purpose, though the Z and + bits are more commonly used because they are

more easily sharpened. When the Italian and French Commission came to this country to ascertain what they could find here that would be of service in the construction of the Mont Cenis Tunnel, they examined the models in the Patent Office, and finally adopted the Fowle type — i.e. they attached the drill to the piston-rod in such a way as to drive it directly by the piston.[2]

Cavé — First European Percussion Drill[2]

Cavé patented a percussion drill in France in 1851. The drill was designed to run with either steam, air, or 'electric currents', though the latter method of power was not fully explained by the inventor. Handles attached to the guiding rods of the drill rotated the drill tool. According to Bande,[5] 'The machine works slowly and is difficult to handle'. Apparently Cavé did actually run his drill with *compressed air* in 1851, using rubber (gutta-percha) conducting pipes. Though Cavé's drill is undoubtedly the first European percussion drill patented, it is preceded by the American drills. Ungainly as it was, it nevertheless embodied the

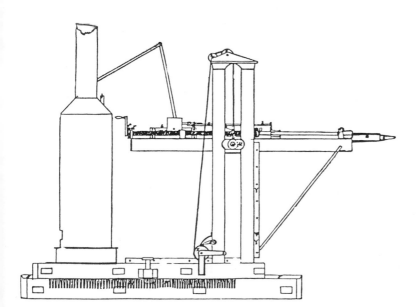

Figure 9. Fowle rock drill — elevation

13

essential principle of direct action, that is, the drill was moved by the direct action of the steam in the cylinder and not through a chain of gears or levers. Furthermore, the drill was attached directly to the piston rod. Drinker remarks sarcastically that: 'It can hardly be called a drilling-machine, but rather a very crude apparatus for making an experiment.'

As Fowle's caveat was filed in May 1849, about two months after the filing of Couch's original patent, Fowle's invention, as well as that of Couch, precedes, by about five years not only all German work in this direction (as the earliest work on Schumann's patent is recorded as 1854) but also Cavé's machine by more than two years (Cavé's machine was not patented until 15 October 1851).

T. Bartlett — Mont Cenis[2]

After Cavé, and following the failure of Maus's machine (see section on rock tunnelling machines), Thomas Bartlett of England invented a drill in 1854. This was proposed for use in the Mont Cenis tunnel and was patented in England on 23 August 1855, the Sardinian patent being taken out on 30 June 1855.

This drill, powered by steam and compressed air, was tested initially at Brighton in June 1854. In March and April of the following year it was again tried before an Italian Commission at St. Pierre d'Arena. However, it proved impracticable for the Mont Cenis project and was ultimately abandoned.

Sommeiller — Mont Cenis[2]

The Mont Cenis tunnel, lying between Modane in France and Bardonecchia in Italy is 12.8 km (7.98 miles) long with a horseshoe section 8 m × 7.5 m (26 ft 3 in × 24 ft 7 in). The material penetrated by the miners during the driving of the tunnel was mostly granitic.

In building the Mont Cenis tunnel, one of the biggest problems which faced the engineers at that time was the question of ventilation. This seemingly insurmountable difficulty was ultimately solved by the physicist, Colladon, of Geneva in 1852. Colladon's brilliant suggestion was that the local mountain streams should be harnessed as a source of power to compress air which in turn could be carried in pipes through the tunnel as the work progressed and thus blow the air clear at the working face. Colladon's scheme, which took approximately two years to perfect, was accepted, but still the question of some device to assist the miners in their work remained to be found. Without such a device and using conventional methods of hand drilling alone, contractors and workmen faced at least half a century of tunnelling work through the solid rock before they could hope to win through to the other side.

Maus's wonderful machine had been unsuccessful because the method he used for delivering motive force to the machine was not suitable for tunnels of great length, and Bartlett's drill had been rejected. It seemed likely, therefore, that the workers would be

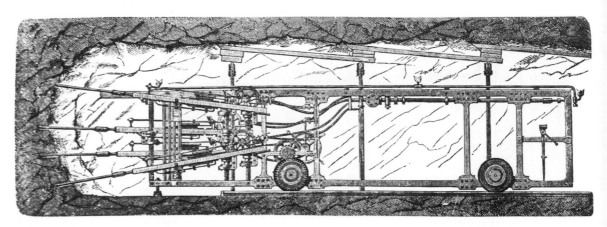

Figure 10. Sommeiller drills — mounted in heading (1857)

compelled to continue as their forefathers had before them, driving every inch by hand.

Eventually Sommeiller came to the rescue by inventing a drill (Figures 10 to 12). Patented first in Sardinia on 25 June 1857, this machine was partly Sommeiller's own idea and partly an improvement on Bartlett's drill (Maus, Bartlett, and Sommeiller had apparently worked together at one time on the Victor Emmanuel railway project. No doubt many ideas on the subject of drilling machines were exchanged by the three engineers at that time).

Sommeiller's new percussion machine was powered by compressed air and proved to be highly successful. It weighed approximately 200 kg (440 lb) and delivered 300 blows a minute (Sommeiller was also assisted by two other Italian engineers, Grandis and Grattoni, who, during the period 1850-60 continued developing the idea of using compressed air for powering drilling machines).

In 1861 machine drilling was commenced at Mont Cenis and finally in 1871 (10 years later) the tunnel was opened for traffic.

When the subject of drills is raised, men generally remember with gratitude the pioneering work performed by Charles Burleigh of America in conquering the mighty Hoosac. Yet Sommeiller's drill is, for the most part, mentioned only briefly and then, perhaps, forgotten. On the contrary it should be acknowledged that Sommeiller achieved for Europe what Burleigh later accomplished for America. By practical application he proved conclusively the soundness of engineering principles which hitherto had only been accepted in theory or, at best, had been attempted experimentally but had failed. Finally, it should be remembered that Sommeiller's drill preceded Burleigh's by at least five years and preceded the Dubois-François, McKean, and Ferroux drills by 7, 11 and 12 years respectively (Figure 22).[2]

Schumann[2]

Schumann's drill (Figure 13), though invented during the years 1854-55 was not tested in practice until 1856. The first model which was built in 1855 was a percussion machine driven by compressed air.

On 1 February, 1860 (Saxony), Schumann filed an improvement to his 1854 and 1857 patents. In 1862 Schumann made further improvements to his machine. However, these later modifications, though tested in practice, were not patented by Schumann. The drill was not very successful.

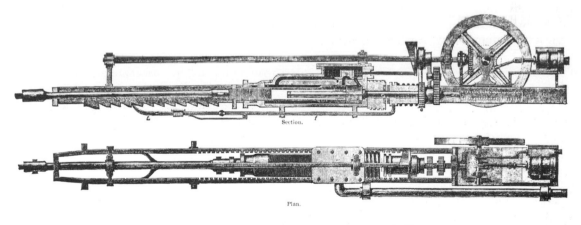

Section.

Plan.

Figure 11. Sommeiller rock drill — improved design

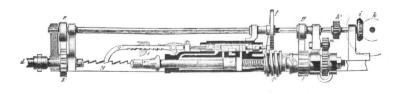

Figure 12. Sommeiller rock drill — third improvement

15

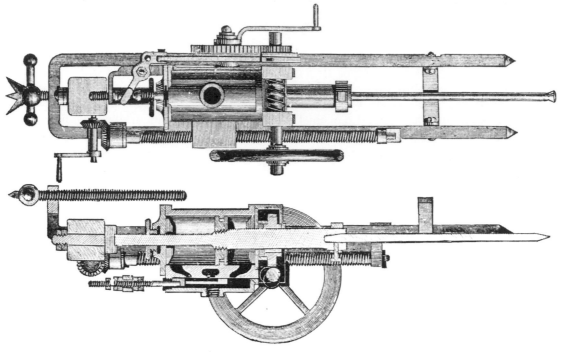

Figure 13. The Schumann drill (1855)

Schwarzkopff[2]

In 1857 (Patent No. 2,477 dated 5 November 1858) Schwarzkopff designed a drill which was rotated by steam power but utilized hand feed. The Schwarzkopff drill was later tried in the Mont Cenis tunnel but was not particularly successful.

George Low[2, 6]

In 1863, two years after Sommeiller's new drilling machine commenced work in the Mont Cenis tunnel, George Low of Dublin was granted a patent for a piston-type percussion drill. The first trial model was tested in 1864, but was found to be unsatisfactory. A second machine, built a short time later, also proved ineffective, and finally in 1865 a third machine (Figure 14) was built. This was tested in the Roundwood tunnel which was currently being excavated for the purpose of bringing water to the city of Dublin from the river Vartry. It was used successfully throughout the construction of this tunnel.

In his first machine the rotation of the piston and drill bit was effected by means of a vibrating lever, pawl, and ratchet (a flattened part of the drill bar passed through the ratchet so that it was forced to rotate with the ratchet.) In the later drill the rotation was produced by means of a spiral or 'twisted' bar which entered the rear end of the piston and operated both the valve and rotation of the piston. (So far as is known, this was the first record of a spiral bar having been used for producing rotation in a drilling machine.)

The machine was driven by compressed air and operated a chisel or 'jumper' for boring the holes. The chisel, which worked at high speed with a reciprocatory motion was forced to rotate slightly between blows. This rotary motion was necessary to keep the jumper free in the hole and ensure even wear of the drill bit. During operation of the machine a jet of water under high pressure was continually forced into the hole. Assisted by the reciprocatory action of the tool, this water kept the hole constantly clear of all loose material.

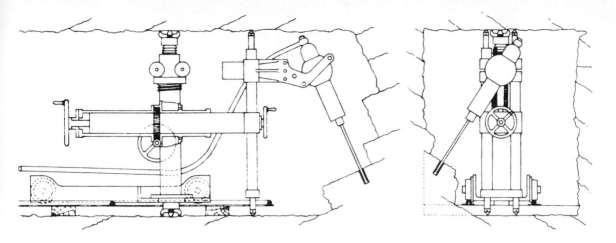

Figure 14. George Low drill. Patented 1863. Used in Roundwood tunnel, Dublin. (Courtesy Institution of Mechanical Engineers, London)

Another important innovation introduced by Low was the spherical trunnion on which the drill was mounted. It allowed universal movement so that the drill could be pointed in any direction. This was a considerable improvement on Sommeiller's drill which could only work in a horizontal direction.

Unfortunately, unlike the Mont Cenis tunnel, which excited worldwide interest, the Roundwood tunnel project was barely heard of outside the boundaries of Dublin and its immediate environs. Consequently, the important developmental work on drilling machines carried out by George Low was largely neglected and very soon forgotten. Later other men came to 'rediscover' many of the basic engineering principles initiated by Low.

Brooks, Gates, and Burleigh Drill[2]

The Brooks, Gates, and Burleigh drill (Figure 15) was patented on 6 March 1866. It was tried in the east heading of the Hoosac tunnel in June of that year. This drill had a hollow piston similar to the Couch model but, unlike the Couch, in which the tool was 'thrown' at the face, the tool of the Brooks, Gates, and Burleigh drill was attached to the piston rod with a central screw and was therefore driven in a similar manner to the Fowle drill.

About 40 of these machines were made for use in the Hoosac tunnel in 1866 but, because of the high cost of replacement parts, were ultimately superseded in November 1866 by the Burleigh drill. The Burleigh continued to be used throughout the entire construction of the Hoosac tunnel.

Charles Burleigh[2] (Figures 16-20)

Charles Burleigh worked as a machinist in the railroad shops of Fitchburg, Massachusetts, where the Couch/Fowle drill was built. Burleigh helped these two men build their machine. Having witnessed the poor success attained by the Brooks, Gates, and Burleigh machine, Burleigh decided to abandon the idea of developing his proposed new drill on the old Couch or hollow piston principle.

Instead he purchased the new Fowle patent and using this as a basic model built a machine (Figure 16) which incorporated many of his own improvements. He achieved this by dispensing with the crosshead and intermediate drill bar, attaching the drill directly to the piston rod and rotating the tool by rotating the piston. To operate his drill Burleigh also built a compressor which utilized water injected in a spray form into the cylinders, the purpose of this being to cool the air during the process of compression.

17

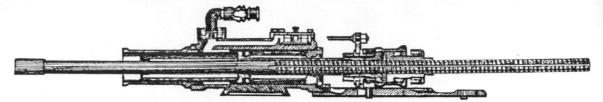

Figure 15. Brooks, Gates and Burleigh drill (1866)

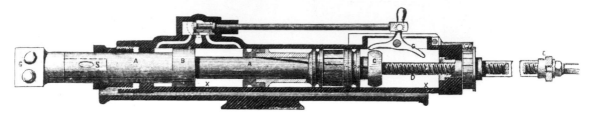

Figure 16. The Burleigh drill (1866)

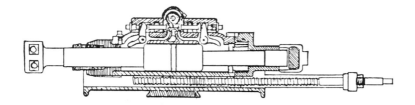

Figure 17. Improved Burleigh
rock drill

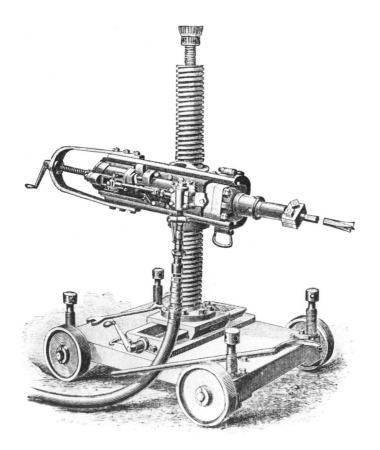

Figure 18. Burleigh mining drill
on carriage

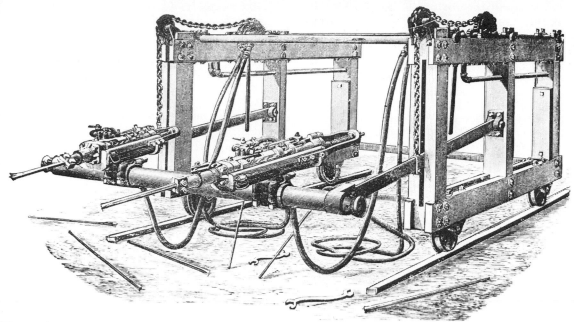

Figure 19. Burleigh tunnel carriage for mounting two drills

The Hoosac Tunnel[2]

Haupt, Gwynn, and Burleigh Drills

During the time Haupt was employed on the Hoosac he constructed several rock drills, the first of which was powered by steam, but after trials in the Hoosac this was eventually abandoned. Later Messrs. Gwyn and Haupt compared ideas and subsequently built a machine which was reported to have drilled on trial approximately 95 cm (3 ft) per hour in hard Rockport granite. The termination of Haupt's contract in 1861, however, temporarily halted these experiments. In 1864 Gwynn took out a patent for a hollow piston drill of the Couch type and in 1865 Haupt took out two succesive patents, the latter being for a drill which he exhibited at the Paris Exposition in 1867.

Machine drills were introduced into the east heading and west shaft of the Hoosac tunnel in June 1868,* and nitro-glycerine was first used in August 1868. During September of that year a linear advance of 16 m (51 ft) for five-sixths of the full month's work was achieved.

Figure 20. Burleigh stoping drill on stoping column

*There seems to be a slight discrepancy in Drinker's book[2] regarding the actual date machine drills were first reputed to have been used in the Hoosac Tunnel. On page 55 'Chronological table' Drinker states that both the Brooks, Gates, and Burleigh drill and the Burleigh drill were first tested in the Hoosac tunnel in 1866. Later, pages 315-34, 'The Hoosac Tunnel' Drinker states that machine drills were introduced into the east heading and the west shaft in June 1868.

19

At the end of 1868 Benjamin H. Latrobe[2] on tendering his resignation as consulting engineer, reported as follows:

> I think difficulties which have attended this great enterprise may be considered practically at an end. Vast as is the magnitude of the work, it has not, in fact, presented physical obstacles as formidable as those which have embarrassed many other undertakings of similar work.
>
> The 'demoralized rock' as it has been called, at the West end, was indeed a troublesome feature, but not so bad as the quicksand and slippery clays of some of the English tunnels. The water in the West shaft required only good pumps and ample power to keep it down. The central shaft has been dry and attended with no draw backs, except those which casualties that might have been avoided with ordinary care have produced; and, lastly, the East end had been so straightforward and really *comfortable* a piece of underground work as could be wished for. The real difficulties of the enterprise have been due chiefly to the other causes of an extraneous character to which, as it is hoped that they are now no longer in action, it is not necessary to refer more particularly. (p. 317)

The 'difficulties of an extraneous character' to which Latrobe was referring were heavy rainstorms which occurred on 3 and 5 October 1869. These storms caused a considerable amount of damage to the workings at the west end of the Hoosac tunnel. Apparently a stream, which had formerly crossed the railway line about 100 m (350 ft) west of the tunnel had been diverted into a newly excavated channel parallel with and about 46 to 60 m (150 to 200 ft) north of the railway line. The stream became abnormally swollen during the rainstorms and its banks were undermined by the fast-flowing torrent. In the process a dam of debris formed which backed water into the tunnel. The water behind this debris dam rose rapidly and within half an hour had risen 4.6 m (15 ft) above the brick arch at the mouth of the tunnel. The water then backed up through the arched tunnel and driftway filling the workings east of the west shaft as far as the advance heading.

Somehow all except one of the men at work there managed to escape. This unfortunate individual was trapped in the tunnel and drowned.

Contractors immediately set a large force of men to work at clearing away the mess. But so tightly packed had the debris become that, despite their efforts, the water was still 30 cm (1 ft) above the top of the brick arch at the western portal eight days later (i.e. 13 October).

Work on the Hoosac tunnel (which is 1048 m (1146 yds) long) was started in 1855 and completed in 1873. On 13 October 1875 (20 years later) the first passenger trains began to run through the tunnel. During 1870 the improved Burleigh drills gave highly satisfactory performances. Walter Shanly later commented that use of the Burleigh drills had saved at least two-thirds of the expense of drilling and that the expense of labour would probably have been fully three times the cost of machine-drilling. He estimated that an additional 12 years would need to have been added to the tunnel work time if the drilling had all been done by hand labour instead of machine.

Brandt Drill[2]

A. Von Brandt of Ebensee, Austria, was the inventor of the first hydraulic drill. Development of the Brandt rock drill which was patented in the U.S.A. on 5 February 1878 under No. 200,024 began in Germany in 1876. This drill, which was highly praised by German engineers of the day, was first tried in the Pfaffensprung tunnel of the Gotthard Railroad, and then subsequently in the Sonnstein tunnel, where special circumstances demanded an exceptionally rapid rate of advance. According to G.J. Specht, Engineer of the Sutro tunnel[2] the best results obtained with one Brandt drill was 3.81 m (12 ft 6 in) of hole in 2 hours 30 minutes, while average progress rates with one drill were reported to have been approximately 1.8 m (6 ft) of tunnel per 24 hours. (The Brandt drill was also used in the Arlberg tunnel in Austria in 1880[7].)

The Pfaffensprung tunnel of the Gotthard Railroad marked an important point in the history of the Brandt drill, as initially percussion drills (Frohlich system) were preferred for the work there. However, as the Brandt drill proved its worth time and again and most of the larger of the underground contractors began adopting it for contract work, the administrative officials of the

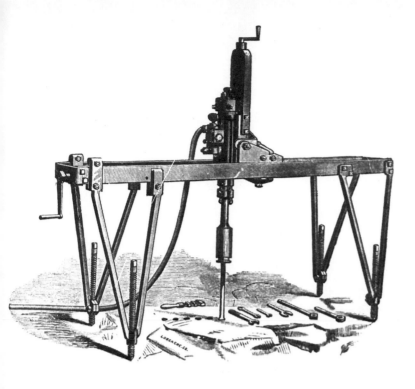

Figure 21. Woods improved drill and channelling frame

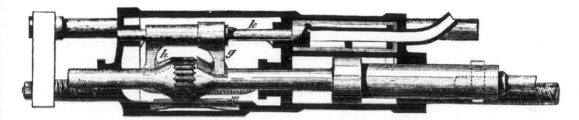

Figure 22. McKean drill

Gotthard Railroad finally decided to give the drill a trial. The Frohlich drill was used in the top heading and the Brandt drill in the bottom heading of the Pfaffensprung tunnel, both working in the same kind of rock (hard granite). The result of the trials proved conclusively the superiority of the Brandt drill, both in rates of progress and in drilling costs, and the Brandt drill was finally accepted. After the first important trials of the Brandt drill in the Pfaffensprung tunnel of the Gotthard Railroad, the drill was used in a great many mines in Austria and Germany. In a colliery in Altona, in Istria, the Brandt drill proved so successful that it even superseded the Darlington drill.

Distinctive features of the Brandt drill were two hydraulic engines, one on each side of the machine, which rotated the cylinder and forced the toothed-steel boring tool and tube against the rock, simultaneously rotating the bit. The boring tool or drill bit consisted of an annular steel ring provided with a series of hardened cutting teeth. The drill bit was screwed on the drill bar, which was tubular and made from a number of pipes connected with each other. The number of sections used depended upon the depth of the hole drilled which could be 4.5 m (15 ft) or more in length. After providing power for the rotation of the boring tool the waste water was also utilized to keep the cutting edges of the drill cool and for flushing out the

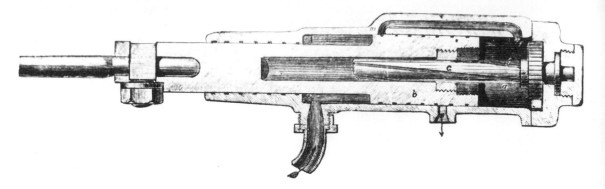

Figure 23. Darlington valveless borer

drill hole. (The waste water was fed to the drill hole via a flexible external pipe.) Lateral support for the machine was provided by means of a hydraulic jack consisting of a hydraulic cylinder and piston which pushed against the sides of the tunnel.

The principle of the Brandt drill was quite different from that employed by the diamond drill. The diamond drill (which is generally used as a core drill for producing rock cores and not for blast-hole drilling) worked with low pressure and at from 400 to 600 rpm, whereas the Brandt drill had a greater thrust and made only 5 to 8 rpm. In other words the Brandt drill, instead of boring by grinding a hard substance against it, gained headway by *crushing the rock.* So far as the author is aware this was the first application of the hydraulic rotary drill.[2]

Leyner, Ingersoll, Sergeant, and Rand Drills[8]

Though it is not proposed to discuss in detail each and every drill patented, nevertheless a few, such as the Ingersoll, Sergeant, Rand (Nutting and Githens), and Leyner drills merit a mention, if only because of the interesting life stories of the men involved (Figures 24 to 29).

Ingersoll-Rand Limited

Two great names — 'Ingersoll' and 'Rand' began independently approximately 100 years ago as the Ingersoll Rock Drill Company and the Rand Drill Company. They combined in 1905 to form the well-known firm of Ingersoll-Rand Limited.

Simon Ingersoll

Simon Ingersoll was born in Stanwich, Connecticut, on 3 March 1818 (the same year Brunel filed his patent for the first tunnelling shield). At the age of 12 Ingersoll's talents for invention were already evident. He made a small steamboat and powered it with a boiler constructed from a kitchen pot.

Educated in a country school, Simon Ingersoll married Sarah B. Smith when he turned 21. She bore him seven children but two died, leaving five surviving her at the time of her death in 1859. After her death Ingersoll remarried. His second wife, who bore him two additional children, was a Miss Frances Hoyt.

After his marriage to Sarah, the couple travelled to Astoria, Long Island, where Ingersoll became a truck gardener. Though little seems to be known of Ingersoll's work during this period, it seems highly likely that he continued to devise mechanical inventions. Evidence of this is indicated by the fact that when Ingersoll and his family returned to Connecticut in 1858 Ingersoll filed and was granted Patent No. 20,800 covering a rotating steam-engine.

In many respects the pattern of Ingersoll's life seemed to follow that set by a great many inventors of his time, whose efforts won for them a measure of fame but not fortune. His overwhelming urge to design and build, coupled with his frequently embarrassed pecuniary state, often forced Ingersoll to sell the rights to his latest invention in order to support his family and allow him to continue working on some new idea.

It seemed that all through his life Ingersoll was to be followed by adversity. Among his earlier inventions was a patent for a machine which automatically cut, shaped, and counted the plugs and wedges which were used by wooden sailing vessels. Unfortunately, these articles were no longer required after iron steamships superseded sail.

Though in many cases there is no doubt that Ingersoll's decision to sell his patent rights was influenced by his straitened pecuniary circumstances, this surely cannot always have been true. Perhaps the obvious fickleness of that first important market concerning his plug and wedge machine so impressed Ingersoll that, subsequently, he almost invariably sold the rights to his inventions shortly after perfecting them, rather than attempt to exploit these further himself.

Ingersoll also built a steam-carriage. When the trial model was ready, Ingersoll and his son drove it triumphantly into Stamford. However, instead of the acclamation he had expected, Ingersoll found himself severely reprimanded by Stamford's warden who, after giving the vehicle a jaundiced examination, proclaimed it extremely untrustworthy. He ordered Ingersoll to remove it immediately as he felt certain that the diabolic contraption would either explode or cause some dire catastrophe to occur. After further remonstrances to the effect that the vehicle was startling the animals, Ingersoll reluctantly assented to its removal. However, prior to his departure, Ingersoll defiantly made a speech prophesying that within the lifetime of many of the people there that day, machines as revolutionary as his steam-carriage, would be familiar sights on the streets. He added the thought, however, that these new vehicles would not necessarily employ the same motive force used by his steam-carriage.

In rapid succession Ingersoll designed and patented a great number of inventions during the 1860s. But with his creditors banging constantly on his door, Ingersoll was again forced to sell the rights to these patents, simply in order to make ends meet. Hard times followed and when he found that financial advances for proposed inventions were refused and no money was forthcoming for earlier ideas, Ingersoll packed his belongings and left Connecticut to return to Long Island (1870). Once more he became a truck farmer. Sufficient money from one of his latest assigned patents enabled Ingersoll to buy a stall in Fulton Street market, New York, where he was able to sell his garden products.

Ingersoll and John D. Minor

As there appears to be more than one version concerning Ingersoll's historic meeting with the Irish contractor John D. Minor, it is difficult to sort truth from legend.

One source states simply that Ingersoll met Minor in a chance encounter at his Fulton Street market stall. However, another authority depicts the meeting as having taken place aboard a public horse-drawn street car. It seems that the attention of a fellow passenger was attracted by the interesting device Ingersoll happened to be carrying at the time. Ingersoll obligingly displayed his latest invention, which had been designed specifically for a prospective buyer, to this receptive audience. Their animated conversation on the subject drew the attention of yet another passenger who introduced himself by quaintly demanding why Ingersoll did not apply his inventive energies to a more worthwhile project.

Stung by the implication Ingersoll demanded an explanation. Whereupon Minor answered suavely that what he had in mind was a *rock drill*. He then proceeded to qualify this statement by explaining that he had recently undertaken a rock excavation project in the city. This he said, was an arduous and time-consuming task which was costing him a great deal of money in labour. It required, Minor continued, at least three men. While one held the drill bit steady against the rock, the other two stood back a pace or so on either side and delivered alternate blows to the head of the bit, or steel bar, with sledgehammers. The process, Minor complained, was so slow that under average working conditions the three men could only manage one hole per day, ranging in depth from 8 to 10 ft. Minor expressed the opinion that a mechanical drill capable of drilling at a faster pace than that set by the men, would probably raise his meagre profit margin considerably. Minor then asked Ingersoll

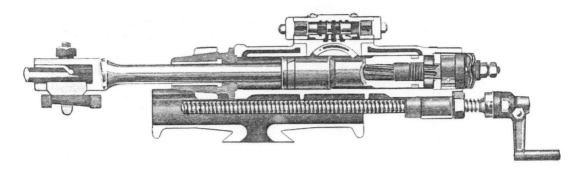

Figure 24. Longitudinal section of 'Sergeant' rock drill, showing valve mechanism. (Courtesy Ingersoll-Rand Ltd.)

Figure 25. 'Little Giant' rock drill. (Courtesy of Ingersoll-Rand Ltd.)

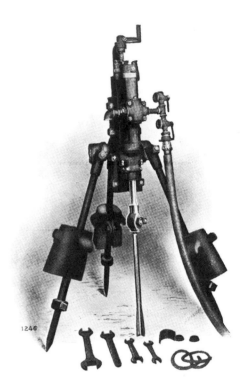

Figure 26. Ingersoll 'eclipse' drill. (Courtesy Ingersoll-Rand Ltd.)

whether *he* could make such a machine.

Such a challenge was one which undoubtedly would have appealed to Ingersoll. However, his prompt assent was coloured by his ever-present financial problems, for he told Minor he would be glad to undertake the work, provided Minor supplied the capital for the initial trial model. Minor immediately asked what amount Ingersoll considered was necessary and, after a quick approximate mental calculation,

Ingersoll suggested the figure of $50. Minor promptly handed over the sum stipulated. However, it seemed that Minor's ideas on such matters were far more practically inclined than were those of Ingersoll, for he added that should Ingersoll require further funds he must not hesitate to ask for them. Not unexpectedly, more money was required by Ingersoll from time to time and this was duly provided by Minor.

24

Figure 27. Sergeant 'Auxiliary Valve' drill. (Courtesy Ingersoll-Rand Ltd.)

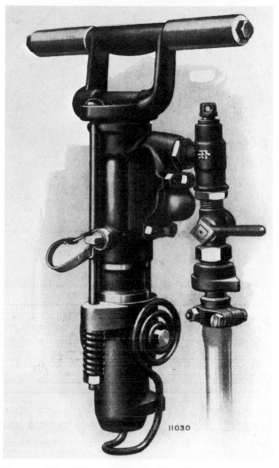

Figure 28. The original 'Jackhamer', Type BCR-430 drill. (Courtesy Ingersoll-Rand Ltd.)

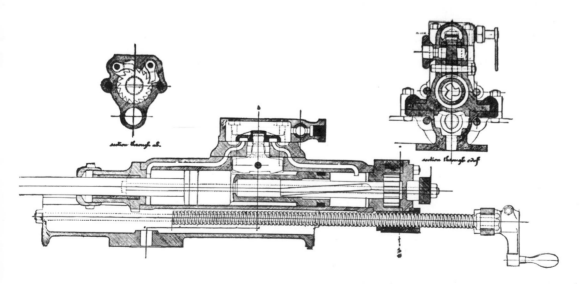

Figure 29. Rand 'Little Giant' drill (section)

A year of hard work coupled with many disappointments passed before Ingersoll was finally able to build his first full-scale working model. This he at last patented on 28 February 1871 under No. 112,254.

Ingersoll and Sergeant

Again in the matter of Ingersoll's meeting with Sergeant, several versions are to be found. One source suggests that having obtained $50 from Minor, Ingersoll arranged for and secured working space in a machine shop owned by a Spaniard, José Francisco de Navarro, and managed by Sergeant and Cullingworth. He then set to work conducting the experiments which eventually led to the construction of the first full-scale model. But another source postulates that the meeting took place *after* the completion of the first model.

In this latter version Minor and Ingersoll took the drill to a spot on Fourth Avenue, New York, where they proceeded to test it. Initially the machine performed faultlessly. It seemed that at last, after all the trials and tribulations of the past months, success was within their grasp. Encouraged, the two men left the scene determined to drink a toast in honour of Ingersoll's new achievement. But their elation was premature, for no sooner had they reached their destination than they were urgently recalled by a messenger with the dire news that the drill had broken down. They returned to find the front end fractured.

Minor and Ingersoll thereupon promptly took the drill to the nearest machine shop which happened to be the one managed by Sergeant and Cullingworth. Here Ingersoll met Navarro and Henry C. Sergeant.

Sergeant, being likewise an inventor in his own right, was quick to recognize the particular defect which had caused the breakdown of the Ingersoll drill. He noticed that the machine was prone to an occasional overthrow of the piston which then struck the solid end of the cylinder and fractured it.

Sergeant remedied the trouble by modifying the cylinder and fitting a removable front head which was spring-loaded to enable it to take a blow without fracturing. Later that year Ingersoll made several other improvements to his drill.

Of Ingersoll's drill Drinker says that it 'proved to be one of the best ever invented'. When this drill was first introduced to the public in 1872, it was the most compact of those then on the market and soon became popular with operators. The 5 in (127 mm) Ingersoll drill was used exclusively in the construction of the Musconetcong tunnel and reports from contractors indicate that it proved to be extremely efficient.

According to one source Ingersoll is said to have purchased the Fowle-Burleigh patents when the Ingersoll Rock Drill Company was first formed, but another suggests that Navarro bought the Burleigh patents which Ingersoll is accused of infringing at a later date when designing further improvements to drilling equipment. Whichever version is correct, there is no doubt that the Ingersoll drill was not a completely new invention, but merely an improvement on the Burleigh, as was the Burleigh an improvement on the Fowle, Couch, and Cavé drills. In other words they were all percussion machines.

Ingersoll's drill was carried in a guide through which it was propelled into the work by means of a feed screw which was operated by a crank handle and, a little later, also by an automatic pawl and ratchet system. This formed a very versatile mounting, permitting its use in almost any position. A distinctive feature of the Ingersoll drill was that it was mounted on a tripod consisting of two ordinary legs and one forked one. The legs were capable of telescopic adjustment and had provision for the temporary attachment of weights to enhance the stability of the whole structure.

Ingersoll and Navarro

In 1874 Ingersoll sold his drill patents to Navarro, whom he originally met at the firm of Sergeant & Cullingworth. Navarro founded the firm of Ingersoll Rock Drill Company with a capital investment of $45,000 out of a total of $50,000 required to start the company.

Though ironically named the *Ingersoll* Rock Drill Company, Ingersoll, by virtue of the sale of his patents to Navarro, forfeited all financial rights to the company. Despite this, Ingersoll's interest in drills remained, as is shown by the many patents associated with rock drills or their accessories, filed in his name.

Later in 1885 Navarro sold his shares in the Ingersoll Rock Drill Company to pay the creditors of a building project which he had initiated, called 'Spanish Flats'. Navarro had invested large sums in this venture which failed, partly as a result of estimation errors made by the architects on Navarro's copies of the building plans, and partly for other reasons concerning the financial state of the country as a whole at that time.

After selling his drill patents for a nominal figure to Navarro and realizing $400 for the sale of his market stall, Ingersoll returned to Stamford with the proceeds. Here he invested the money in a machine shop which traded under the name of Ingersoll, Betts, and Cox.

Until his death at the age of 76 on 24 July 1894, Ingersoll continued to design new models, a projectile gun for throwing life-lines and a steamboat anti-friction device, numbering amongst his many inventions.

Henry Clark Sergeant

Henry Clark Sergeant was born in Rochester, New York, on 2 November 1834.

Like most true artists, Sergeant was more interested in his work as a designer than in any financial gains which might have accrued from his many inventions. Nevertheless, though he never became excessively wealthy, he earned sufficient from his work to ensure a moderately steady and comfortable living.

In 1852, at the age of 18, Sergeant invented machinery which shaped the spokes, hubs, and felloes of wagon wheels. He was at this time working in a machine shop in Columbus, Ohio, and, not content with his initial success, followed it by securing for his employer large contracts for the production of these wagon-wheel components.

His talents must obviously have impressed his employer as two years later, the same year in fact that Sergeant filed his first patent for a steam boiler feed pump, he was taken in as a partner in the business. But the monotony of factory routine soon palled, Sergeant resigned, and entered the commercial world where he dabbled in a variety of ventures.

From this period onwards Sergeant's fertile and creative mind produced one device after another, the most notable being a governor for marine engines. It is said that the United States Navy had them fitted to their ships and that Sergeant, then only 23 years old, was paid the handsome sum of $10,000 for the use of his invention. Patents for steam pumps and boilers, water meters, brickmaking machines and gas regulators, etc. followed in rapid succession. Almost as fast as Sergeant invented new machines, he also changed his place of employment and abode. This restlessness seemed to be an inherent characteristic of Sergeant's temperament. An indication of this is to be found in the extraordinary fact that during the period covered by the years 1854 to 1869, Sergeant moved from city to city and from town to town where he worked for short periods. In all 26 different localities were embraced.

After a considerable period of travelling Sergeant eventually found himself in New York during the year 1868. He immediately acquired a machine shop and set up business. This undertaking prospered to such an extent that he subsequently entered into a partnership with Cullingworth, and together they acquired more spacious premises in the shop owned by Navarro. It was here in 1870 that the important meeting between Ingersoll and Sergeant, mentioned earlier, took place. Sergeant was on hand to effect the first basic improvements to Ingersoll's rock drill when he was asked to repair the damaged front head.

While Ingersoll's association with Navarro was of brief duration, Sergeant continued to benefit from the connection until Navarro left in 1885.

Four years after the Ingersoll-Sergeant-Navarro meeting took place, Navarro was induced to purchase Ingersoll's rock drill patents at the instigation of Sergeant. Sergeant becoming the first President of the newly formed company.

In some respects Ingersoll and Sergeant were

similar. Both invented a variety of machines and both showed a continued interest in rock drills from the time they first became acquainted with them. But while most of Sergeant's ventures prospered, Ingersoll ended his life as a pauper.

It was due to Sergeant's efforts that the drill was converted from steam to compressed air, and Sergeant's improved compressed air drills eventually superseded the earlier steam models. In addition Sergeant also made notable improvements to the compressor.

Sergeant's fascination with machines and in particular rock drills, is borne out by his activities during the years 1883-85 when, for a brief period, it seemed, he tried to get away from them.

At the beginning of that period, i.e. in the year 1883, Sergeant sold the financial rights which he owned in the Ingersoll Rock Drill Company and travelled to Colorado where he tried his hand at silver-mining. This venture was not successful and perhaps served, in some measure, to convince Sergeant that whatever else he did, he should at least stick to machinery. For after the loss of his money in Colorado he moved back to Bridgeport, Connecticut, in the East, where he again turned his attention to rock drills by founding the Sergeant Drill Company. However, the time spent in Colorado was not entirely wasted for it was during this two-year period that he designed and patented a new valve-motion drill (1884). The idea for this was originally conceived from a revolutionary steam pump designed and manufactured by A.S. Cameron who had occupied the same shop building on Second Avenue and 22nd Street, New York, as had Sergeant and Cullingworth during their stay there. It was known as the 'Eclipse' valve (Figure 26). Unlike the cam- or tappet-type valves the Eclipse valve was an independent valve. That is it had no mechanical connection with any other part of the drill. It was air-thrown and its action was controlled by the movement of the piston. The drill had a variable stroke and an uncushioned, heavy, powerful blow.

Under the new company founded by Sergeant, he produced a series of drills which were similar in many respects to the Ingersoll

models, but which embodied his own patented improvements. The two firms eventually merged, forming the Ingersoll-Sergeant Drill Company. For a short period after the merger, both types of drills were marketed. Later, however, a new rock drill was produced which combined the best features of both types of drill.

Sergeant lived to the age of 73. He died on 30 January 1907 in Westfield, New Jersey.

The Rand Brothers

Jasper Raymond Rand, father of the famous Rand brothers, Albert Tyler, Jasper Raymond, and Addison Crittendon, manufactured whips in Westfield, Massachusetts. Such was the reputation of the Westfield whip that at the height of the whip era, the district, which boasted some 30 whip factories, was producing the enormous quantity of 2 million whips per year.

In 1865 Jasper and Addison took over their father's whip business, but Albert Tyler declined to follow his brothers. He turned his attention instead to the manufacture and sale of black powder, which business he started in 1850. Later in 1855, Albert Rand merged his interests with another black powder merchant from Kingston called John Smith. Rand was elected President of the newly formed firm of Smith and Rand. In 1869 the Smith and Rand firm merged again with two more powder companies and the name of the firm was changed to Laflin and Rand. Three years later (1871) Alfred was joined by his brother Addison.

As miners were considered to be one of the largest consumers of black powder at that time, Albert Rand devoted his attention to wooing this market.

Most rock drilling in those days was done by hand. Only special projects such as important tunnels warranted the trial of uncertain and expensive equipment such as the new rock drills made by Burleigh. So far as the miners were concerned, the methods used by their grandfathers, and their grandfathers before them, were good enough. However, as can be imagined, the work was tedious and time-consuming. Because of the sheer physical effort involved in this type of labour, miners seldom drilled holes which exceeded 25 mm (1 inch) in diameter. Sometimes holes of a slightly larger diameter were started, but even these tended, after a period of drilling, to become narrower towards the bottom of the hole. In addition there was a natural tendency to halt the drilling some time before the desired depth was reached. These factors combined to produce an inefficient explosive charge which expended its energies near the surface rather than near the bottom of the hole where it was most needed. Albert Rand, realizing that these incorrectly drilled holes were costing his company good business, as it meant a reduced sale of black powder to the miners, gave the matter his full attention. He came to the conclusion that the only way to induce miners to drill correctly would be to provide them with a mechanical means of doing so. Though the Burleigh drill was available on the market it was still very much in its infancy and not completely accepted by operators. Albert Rand therefore decided to investigate the possibilities of his company building such a machine. He also apparently suggested that his younger brothers should take an active interest in the project.

The first important step in this direction came in 1872 when the company changed to that of the Rand and Waring Drill and Compressor Company. (Apart from the original patent dated 16 August 1872, covering a water-cooled air compressor, John B. Waring's name is also associated with several improvements to rock drills. These are filed under Patents Nos. 152,712, 156,003, and 169, 121, etc.) Later in 1879 the company again changed its name and became known as the Rand Drill Company.

The Rand brothers were always intensely interested in promoting the development of both compressors and rock drills but, so far as is known, none of them actually invented any of the Rand drills. (The original 'Little Giant' drill (Figures 25 and 29) was designed jointly by George E. Nutting and Joseph C. Githens — see list of drill patents). However, Addison Rand did hold patents for certain improvements to air compressors to which he devoted a considerable amount of his energies. Nevertheless, with the help of such men as

Frederick A. Halsey and Joseph C. Githens, Addison made many important contributions to the development of the early Rand drills. He may not have designed them, but his mechanical ability enabled him to suggest improvements which were accepted and incorporated into the drills by others.

C. H. Shaw — Hammer Drills

Almost side by side the Rand Drill Company and the Ingersoll-Sergeant Drill Company grew and prospered. Though there were minor refinements such as were incorporated in the 'Ingersoll Eclipse' drill which Rand tried unsuccessfully to emulate in the 'Rand Slugger' produced in 1880, all three drills, that is the Ingersoll, Sergeant, and Rand had one point in common, namely their basic method of construction. Fastened firmly to the shuttling piston, the drill bits of these machines had no option but to move simultaneously with it as it reciprocated in the cylinder. The chief drawback here was the high inertia of the piston assembly due to the heavy weight of the attached drill bit. This was most undesirable in a reciprocating component as the stresses set up demanded unnecessarily heavy mountings and powerful machines.

The first real improvement to this type of drill came from the hands of C.H. Shaw of Denver who, in 1890, built a drill intended for overhead work. Its distinctive feature was that the steel drill bit was not attached to the piston but was independently mounted and was struck hammer blows by the shuttling piston.

The drill was christened a 'stoper' because it was initially used for stoping work in the Colorado mines. Though mechanically successful it soon fell into disfavour with miners because of the dust it created and they aptly named it the 'widow-maker'.

Shaw also realized the problem of drilling uppers and very ingeniously equipped his machines with an air leg or a pneumatically operated feed leg which assisted the operator to hold the machine against the roof. But Shaw did not bother to patent this very important invention and so later other manufacturers used the idea.

Though nothing came of them, George Low

of England also filed a patent for a hammer-type rock drill in 1865 and, in 1884, Henry Sergeant was granted a U.S. patent for such a drill. The Sergeant drill was not successful because, as with Shaw's machine, a solid drill rod was used and it was found that the cuttings remained in the hole. This problem did not arise with the direct-action piston-type drills because when the drill rod is connected directly with the piston a type of reciprocating, pumping action occurs when it is operated and this pumps the cuttings out. Thus downward-sloping holes were drilled by injecting a stream of water into the hole next to the rod and a sludge would form. This was then mopped up or blown out. Holes which sloped upwards were dry-drilled as the cuttings fell by gravity or, if necessary, could be easily blown out. Probably for this reason Shaw's hammer drill was developed for overhead work and, as already mentioned, though it functioned well enough mechanically, it was ultimately rejected because of the dust it created.

John George Leyner

John George Leyner was born on 26 August 1860 on his parent's ranch in Boulder County, Colorado. His education was confined to that obtainable at a local public school. Like Ingersoll and Sergeant, Leyner's great love centred on machinery and his father's efforts to interest him in the farm failed. Gradually his talents drew him irrevocably towards engineering projects.

In 1879 he left his father's farm and for the following four years threshed grain for the farmers in the surrounding area. In 1884 a Jackson mining and milling company engaged him for two years as machinist, but Leyner was anxious to work on his own and, after a short period of time during which he was employed by a flour-milling company in Canfield, Colorado, Leyner broke away and at last was able to start his own machine and foundry shop in Longmont, Colorado, in 1886. However, aware that he lacked experience and training, Leyner disposed of his business after a relatively short period and bought a part share in a Denver machine shop which specialized in mining machinery repair work. It was here that

Leyner began serious work on his most important project — a rock drill. The first model, a piston-type drill, which appeared in 1893 was not significantly different from the Ingersoll, Sergeant, or Rand drills. But Leyner was not completely satisfied and eventually after another six years of work he produced a completely new model which was patented on 13 June 1899 under Patent No. 626,761.

Leyner's new drill followed the same basic principle as that initiated by Shaw. In other words the drill bit was driven into the rock by a series of hammer blows delivered to it by the reciprocating piston. In making this change, Leyner also succeeded in lightening his drill to one-quarter of the weight of his previous 27 kg (60 lb) piston-type drill.

Though these new hammer-type drills were obviously more efficient than the old piston drills, Leyner's machines suffered from one very serious disadvantage. While developing his revolutionary hammer-action design, Leyner had added yet another innovation which, no doubt at the time, must have seemed to him to be brilliantly conceived. He forced compressed air through a tubular cavity which ran down the centre of his drill and emerged through a hole in the drill bit. The idea was excellent as the air was intended to blow out the dust and debris which had collected at the bottom of the hole during the drilling process. Unfortunately, however, the dust created by this system of clearing was so excessive that miners stubbornly declined to use the Leyner drill after the initial trial. By that time Leyner had sold about 75 of his machines. These were returned to Leyner to his cost.

Leyner's ingenious answer to the dust problem was to substitute water for air, so that the cuttings emerged as mud. Unfortunately the remedy, so far as Leyner was concerned, came too late. His finances suffered a crippling blow and the initial teething problems experienced with the drill made the operators extremely reluctant to try the Leyner drill again. Leyner then turned his attention to other matters. He developed a drill-sharpening machine (Patent No. 917,777) and also made several important improvements to the air compressor. His principal contribution in this respect being covered by Patent No. 938,004 dated 26 October 1909.

In 1905 the Ingersoll-Sergeant Drill Company joined the Rand Drill Company and the firm of Ingersoll-Rand was formed. Not surprisingly, the new Ingersoll-Rand Company were quick to appreciate the improved characteristics of the Leyner jackhammer drill and, in 1911, Ingersoll-Rand negotiated for and won the entire Leyner business, comprising various types of mining machinery together with all of Leyner's drill patents. In 1912 the first of the new Ingersoll-Rand 'Jackhamer' drills (Figure 28) made their appearance in Phillipsburg, New Jersey.

Leyner then turned his attention to farm machinery and on 29 January 1918 he was granted a patent covering a caterpillar-type farm tractor. Encouraged by the initial success attained in this new field of engineering, Leyner formed a manufacturing company. Two prototypes of Leyner's new tractor had just been completed when Leyner was involved in a serious motor-car accident, which claimed his life. He died on 5 August 1920.

Although at first the world was slow to accept the new Leyner drill, improved though it was, its advantages over the older type of piston drill were so obvious that in time these became obsolete. This was especially so after 1914 when Leyner's original patent covering his jackhammer drill expired and manufacturers everywhere were free to use his ideas.

It is, of course, always easy to criticize with hindsight, yet had Leyner conducted sufficient trials at the actual workface it seems extremely unlikely that the inherent disadvantages of the air-cleaning system would have escaped his attention. In this event Leyner would no doubt have rectified the fault *before* the drill was marketed and thus saved himself considerable embarrassment and financial loss.

The Gardner-Denver Mnf. Company[9] (now Gardner-Denver/Cooper Industries)

Early steam-engines were powerful but erratic machines which, once they started to gain speed, would continue to operate at peak power regardless of load. As a result they tended to run uncontrollably and very quickly wore out or, in extreme cases, even exploded.

Robert W. Gardner, aware of these problems, set to work to find some means of taming the monster and in 1859 he devised a fly-ball governor. The invention of this device which revolutionized the steam industry also marked the beginning of the Gardner Governor Company.

In 1880 the company added steam pumps to its production list and in 1883 their first spring governor was patented and put on the market.

During the difficult days of the 1894 depression, Gardner turned his eyes towards Texas where fortunes were being made in the newly discovered oilfields there. Steam water pumps were adeptly converted into 'mud pumps' and a flourishing market was found for both pumps and governors.

The Colorado mines next drew Gardner's attention and a new Gardner product was put on the market — an air compressor.

In the meantime a D.S. Waugh of Denver (who had been associated with C.H. Shaw at the time he built his stoper drills) decided to improve upon and patent some of the outstanding features of the Shaw drill. In 1890 the newly founded Denver Rock Drill Manufacturing Company began producing the improved Shaw-type drills. They were powered by compressed air.

In 1920 a successful bid by the Denver Rock Drill Company for a large government road contract, led to the production of the mobile drill rig which was the forerunner of the modern giant crawler-type drill rig currently in use today.

Because pneumatic drills and compressors are inextricably linked, the two, by this time prosperous, companies, decided to amalgamate and in 1927 the Gardner-Denver Company was formed.

Outlined below are some significant contributions made by Gardner-Denver towards the development of the percussive rock drill.

1937. G.C. Pearson and Irving Carpenter were granted a joint patent dated 13 July 1937 covering a rock drill guide shell feed mechanism, with dual ratchet racks. This patent was assigned to Gardner-Denver. Later other inventors patented their own version of a feed mechanism which was basically similar in concept to the Pearson-Carpenter design. These

Figure 30. Gardner-Denver 'Universal III' three-boom rubber-tyred Jumbo of type introduced to the mining industry about 1964. (Courtesy Gardner-Denver Co.)

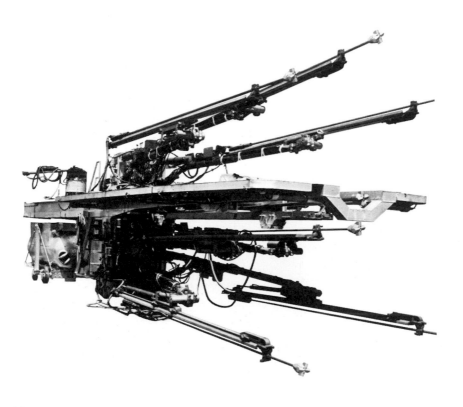

Figure 31. Gardner-Denver early rail-mounted jumbo. (Courtesy Gardner-Denver Co.)

Figure 32. Gardner-Denver air motor propelled rubber-tyred mounted ring drill Jumbo with 1.82 m (6 ft) drillsteel change screw feed and independent rotation. Manufactured in U.S.A. and introduced to Mt. Isa Mines Ltd., Australia in 1975. (Courtesy Gardner-Denver Co.)

Figure 33. Gardner-Denver current fan drill jumbo: Used at Meramec Iron Mine, Sullivan, Miss., U.S.A. (Courtesy Gardner-Denver Co.)

patents were, in turn assigned to various manufacturers such as the Cleveland Rock Drill Co. and the Worthington Pump and Machinery Corporation.

1950. This year saw the introduction by Gardner-Denver of carburized hollow drill rods, shanks, and couplings incorporating special thread designs which allowed the joints to be loosened, yet still effectively transmitted the energy of the hammer down through the string. Developmental work on this was carried out by F.R. Anderson and B.C. Essig who were granted a patent covering both the carburization process and the thread design system on 27 August 1957.

These important contributions (coupled with the independent development of tungsten carbide insert bits — see section on drill steels) made possible the use of longer drill strings. These had hitherto been limited to about 10 m (33 ft) in length.

1960. Prior to the 1960s most drills were fitted with valve chests which channelled air via ports to the front and rear chambers of the drill. Because gaskets cannot be used on a drill, manufacturers needed to adhere to very close tolerances in order to produce highly finished flat air-sealing surfaces on the cylinders. In addition, the projecting valve chests were frequently subjected to mechanical abuse.

On 28 June 1960, M.C. Huffman, G.C. Pearson, and H.C. Gustafson of Gardner-Denver were granted joint patent rights for an 'underground mole drill'. The significant feature of this drill was that the automatic valve was placed in the cylinder behind the hammer. This new streamlined design eliminated the need for a valve chest with all its inherent problems.

1963. The fact that an independently rotated percussive drill would be more effective than one where the bit was rotated by the percussive mechanism (i.e. the rifle-bar system) had been recognized by design engineers for many years. However, various attempts by engineers to design such a drill met with failure for one reason or another.

One of the main problems was the need to provide torsional elasticity in the rotation system which was subjected to severe shock waves from the percussive action of the hammer and bit. During the 1890s D. S. Waugh made several independently rotated drills, but these were not successful for the above reason.

On 26 March 1963, M.C. Huffman of Gardner-Denver filed Patent No. 3,082,741, covering a design for an independently rotated drill. Huffman had found that if the rotary mechanism was driven by a long drive shaft, the intermittent shock waves, produced by the percussive action of the hammer and bit, were absorbed by the shaft which acted as a type of spring.

Later the separate development by Gardner-Denver of a special spring-shank obviated the need for a long torsional shaft. The new spring-shank was an elongated version of a conventional shank that had been tapered and thinned between the connection points so that, in effect, it acted as a type of spring. Thus instead of a great proportion of the energy produced by the hammer being dissipated by the shank, as was the case with the original component, the blow energy was transmitted directly through the new shank and into the drill rods where it was required. This increased the overall efficiency of the drill (Figures 30 to 33).

Rotary/Percussion Drills

Since the introduction of the Ingersoll-Rand 'Jackhamer' drill in 1912, percussion drills made by Ingersoll-Rand and other major manufacturers such as Gardner-Denver, Atlas-Copco, etc. have been used extensively throughout the world. These drills were used for hand drilling, as pavement breakers, or mounted on carriages or jumbos in tunnels, for blast-hole work in mines and quarries, and for a hundred other purposes where it was necessary to penetrate a hard material or some difficult rock strata for excavation, etc.

While minor differences such as the size and length of the piston or perhaps the particular type of valve action used were evident between one manufacturer's drill and another, they were, nevertheless, basically similar in concept. They were generally powered by compressed air or steam. These energy sources were fed in turn first to the back and then to the front end of a cylinder. By this means a piston within the cylinder (which separated the rear chamber from the front chamber) was forced to reciprocate back and forth so that it struck a tool held in a chuck. The tool, or drill bit, in turn impacted against the rock or other material being attacked. Entry into the chambers was automatically controlled by a valve.

The air or steam was exhausted from these cylinders by the opening at the correct moment of a port in the front and rear chambers of the cylinder respectively.

In order to keep the tool from jamming in the hole, the drill bit was made to rotate slightly on the return stroke by means of a rifle-bar and ratchet mechanism. A rifle bar is a spirally fluted or grooved shaft which passes along the centre of the piston. A 'rifle nut' seated (either by screwing or pressing) into the head of the piston, guides the movement of the piston on the bar. The head of the rifle bar is fitted with a ratchet mechanism which forces the piston to rotate in an anticlockwise direction as it moves along the flutes of the bar on its return stroke. On the forward stroke the head of the rifle bar rotates freely in the ratchet ring so enabling the piston to move toward the drill steel without rotating. At the same time that the piston head moves on the rifle bar, the striking end of the piston travels back and forth along splines in the chuck. The chuck in turn guides the drill steel. Thus each time the piston is forced to rotate on its return stroke by the splines on the rifle bar, the chuck and therefore the drill steel are also made to rotate. When too high a thrust was applied to the feeding mechanism of drills using a rifle bar rotation system, they tended to jam. This also periodically happened when drilling through badly fractured strata and, if these conditions persisted, the drill would eventually stall and the drill string would become stuck in the hole. Conversely the application of too little thrust would cause the drill bit to rotate too fast or spin freely, and penetration rates would be seriously affected. Because of these problems the independent air-driven rotation motor was introduced. These motors which were built on to either the front or the back-head of the machine, rotated the drill bit through a train of gears, etc.

Although the need to rotate the drill steel through the piston and rifle bar was obviated by the introduction of the independent rotation mechanism, it was still found necessary to lubricate the shuttling piston efficiently. For this reason some manuacturers continued to incorporate a rifle bar system in the drill which served to lubricate the piston but which did not act to rotate the drill steel. Other manufacturers overcame this problem by providing one or more transverse grooves on the piston and then forced the piston to rotate as it reciprocated by splines along the striking end of the piston which registered in the upper section of the chuck. The piston was then rotated by the chuck at the same time that the chuck turned the drill steel.

Because of their weight independent rotation mechanisms are not generally mounted on hand drills.

An important aspect concerning rock drills is lubrication. Three methods are used: hand-oiling systems, oiling systems integral with the drill itself, and airline lubrication.

The disadvantages associated with hand-oiling arrangements are obvious. Oil is injected into the airline connection on the machine and is fed to the various moving parts of the drill by the air. Unless the operator inserts the oil regularly, when it is required, the moving parts will wear rapidly. In most cases the operator feeds oil into the system when he observes that the drill is no longer operating efficiently. Unfortunately, this only becomes noticeable when the drill has been working dry for some time.

Airline lubrication systems are considered very efficient provided they are positioned about 3 m (10 ft) or less from the drilling machine. At this distance the oil is finely atomized and thus penetrates all the moving parts properly. Sometimes, however, operators fit the lubricator on the other end of the airpipe away from the machine. Over long distances the atomized oil particles form into drops which are deposited on the inner surface of the pipeline. These are carried along the line by the air, and because it is no longer atomized the oil does not lubricate the drill efficiently. A greater quantity of oil is thus needed to do the job properly, most of which is lost via the exhaust pipe and is therefore wasted.

The Air-leg or Feed-leg Drill

The first air-leg or pneumatically operated feed-leg was introduced by C.H. Shaw of Denver. Shaw did not patent his feed-leg invention because he apparently did not fully appreciate its value at that time. Later the idea was copied and used by other manufacturers.

During the early 1940s the air-leg drill made its appearance in Europe. The first of these was produced in Germany in 1938. These drills were a great improvement on the light, hand-held drills. While still retaining the flexibility and versatility of the hand-held drill, the addition of the air-leg enabled manufacturers to build a heavier and therefore more powerful drill, the additional weight being supported by the air-leg.

The air-leg itself consists simply of a single cylinder and piston which is operated by compressed air. The cylinder is attached to the drill with a swivel connection and the piston rod is attached to a spike which is thrust into the floor. There is another type of air-leg in which the piston rod is attached directly to the drill. In this case the piston rod is tubular to carry the air from a control valve near the connection to the drill down to the piston head. This type of air-leg is popular for stoping work because it enables the operator to climb on to the handle of the air-leg in order to shut off his machine when he has drilled a particularly high hole.

The purpose of the air-leg is twofold. It provides support and forward thrust for the drill. Modern air-leg or feed-leg drills have double-acting legs which may be retracted or extended by means of controls grouped conveniently with the main controls of the drill.

The Stoper Drill

An interesting variation of the air-leg is the 'stoper' drill. The air-leg on the stoper is attached rigidly to the drill and directly in line with it. This is to provide thrust to enable the operator to drill vertically upwards.

The word 'stope' means 'to excavate or to remove the contents of a vein' of coal or ore, etc. and originally the 'stoper' was so named because the first of these (invented by Shaw) was tested in the Colorado Mine stopes. It was designed to drill vertical or nearly vertical holes in the roof of the stope for blasting purposes. However, since the advent of modern coal-getting machines, this type of drill is now used mainly for roof-bolting rather than as a 'stoper'.

Drill Steels[10]

In 1897 J.G. Leyner obtained a patent for a 'hammer-type' drill which was based on principles initiated by Shaw.[8] However, Leyner added an important innovation. In order to channel air and then later water through the drill steel so that it emerged in a stream at the bit end where it was needed, he drilled a longitudinal hole through the drill rod so that in effect it was hollow. However, because rolled hollow steel was not available at that time, he was constrained to drill the holes in much the same manner that rifle barrels for guns were pierced — a laborious and difficult procedure.

Another problem faced by Leyner was the quality of the steel which was then very much inferior to modern steels. This resulted in a high rate of failure of shank ends, drill rods and bit ends, etc. The problem was aggravated by the longitudinal hole which further weakened the rod. Recent developments in cored alloy steel has improved the situation, but to offset this the drills themselves are now capable of exerting much harder hammer-strokes which again leads to the search for better and stronger materials to withstand these greater shocks imposed upon them.

Just as manufacturers and designers have striven over the years to improve the efficiency and performance of the percussion drilling machine so, too, has developmental work been undertaken on ways and means of improving the drill steel.

In the words of F.R. Anderson[10] (Gardner-Denver Company) 'percussion drill steel' is defined as '... the energy transmission line connecting the rock drill tool to the rock-cutting bit, transmitting impact pulses, feed forces, rotation, direction control and providing the chip removal medium'.

Despite the fact that a percussive drill is capable of penetrating hard rock more efficiently than a rotary drill, for many years —

indeed until fairly recently — percussive drills were mainly used for shallow holes, the depth not exceeding about 10 m (33 ft). Holes were driven by means of one-piece drill rods or steels, forged by (hopefully) experienced blacksmiths, men who fully understood the mysteries of tempering, so that the machined shank and bit ends of the steel were properly heat-treated, to enable them to withstand the rough handling to which they would be subjected.

Drilling of the hole was commenced with a comparatively short rod but, as the hole grew in depth, it became necessary to fit succeedingly longer drill steels to the drilling machine, until the required depth was eventually reached. If holes deeper than 10 m or so (30 ft) were required, then it was customary to use the rotary drill fitted with a diamond-impregnated drilling bit — a more expensive and slower method of making holes.

In percussive drilling, depending of course upon the type of strata being penetrated, wear on the drill steels or rods was considerable and this necessitated continual transportation of large quantities of these components from the job site to the machine shop and back again.

The first important innovation introduced in the early 1930s was the detachable bit. While this did reduce the cost of bit replacements, problems with the connection point arose. Irrespective of whether tapered or straight threaded joints were used, this constituted a weak point. There was the risk that the joint might become uncoupled during operation, energy was inevitably lost, and there was also a tendency for the steel or the bit to break at the coupling point.

During the succeeding years and particularly after the mid 1950s and early 1960s the wearing properties of drill tools underwent dramatic changes with the introduction of tungsten carbide insert bits. These were either inserted at the ends of one-piece hollow drill steels or were available set in detachable tools which could be coupled to the drill steel. Improved methods of hardening the surface of steel have recently led to the production of better couplings, shanks, and drill rods, etc. and this in turn has encouraged the use of percussive drills for deeper holes than was previously considered

practical. The incidence of breakage in this area, the problems associated with energy loss, and the susceptibility of the joints to becoming uncoupled during operation, has become less acute due to the advent of improved designs in couplings. In addition various angles and pitch depths have been tried on the male and female threads of couplings, rods, and bits, etc. in an effort to produce a better joint and thus reduce energy loss and the tendency of the coupling to become undone. Since the early 1920s when extension rods were first tried, it was discovered that if the joint was tightened too much, the operator experienced great difficulty in undoing it when this became necessary. Conversely, if the thread was so designed that the joint remained 'loose' there was the danger that the sections or the bit might become uncoupled in the hole.

In his paper, 'Percussion drill steel', F.R. Anderson[10] discusses the subject and gives several examples of the changing trends in thread design ranging from those used in 1917 to those at present in vogue.

Mountings and Control

As the drill itself has developed and become more efficient, so too, have the mountings and controls, particularly in the heavier types of machine which are rig- or jumbo-mounted. This has come about because of the increased demand for accurately blasted tunnels which need little or no trimming or filling after the explosives have done their work.

The advent of hydraulics hastened this development, and drill carriages were fitted with hydraulic positioning rams which greatly increased their stability and thus their accuracy. Column-mounted drills gave way to complicated rig-mounted units which could be moved into a variety of positions either beside the boom or below it by such devices as 'roll-over' mechanisms.*[11] In addition the drill was given extended reach by the incorporation of hydraulic telescopic booms. These allowed the drill more

*The 'roll-over' mechanism was developed to enable the drill operator to reach points which were regarded as inaccessible without the repositioning of the entire rig. These inaccessibilities were usually caused by mechanical obstructions.

movement than that provided by the normal feed mechanism.

Holman Brothers Limited of Cornwall, England, introduced a drill rig which was fitted with an automatic means of keeping the drill carriage parallel with the tunnel axis. The rig, designated the 'Holman Autopanto' was equipped with a 'linear actuator' for roll-over and the entire mechanism was operated by means of a hydraulic servo system. When it was necessary to drill holes at an angle, an override of the automatic parallel motion system could be brought into effect. On completion of the angle hole, manipulation of the appropriate control automatically returned the mechanism to parallel motion without the need for resetting.

Another step towards complete automation was the fitting of mechanisms which enabled the drill to collar a hole (start it) and then continue drilling (with the correct feed pressure) until the drill had reached the end of its travel, after which it was retracted and switched off.

Some large Holman drill rigs were fitted with the above mechanisms and delivered to Romania for use in tunnelling work. The particular drill rigs in question were not, however, fitted with the automatic parallel motion features, although according to the manufacturer, these could have been added if required.

A sophisticated trial rig, capable of automatically drilling the entire 'round' sans operator, was fitted to a 'Holman Autopanto' and displayed at the International Mining Equipment Exhibition in London in 1969. The whole sequence was regulated by a metal (or plastic) 'punched card' template, operated by a 'pneumatic fluidic circuit'.[11] In his article on the subject J. Hodge (Director of Holman Brothers Limited) commented that the rig described above was, of course, still in the experimental stages and that while it could operate successfully under ideal conditions, it had yet to be proved in the hostile environments of tunnels or mines where the operator was frequently faced with changing and difficult rock formations.

Nevertheless, the mere fact that such a rig has been built is an indication of what the industry may expect in the not too distant future.

Development of the All-Hydraulic Rotary/Percussive Drill [12-14]

As was noted earlier, the first successful percussive drills were designed by Sommeiller, Grandis, and Grattoni and were used in the Alpine railroad tunnel under Mont Cenis during the period 1861-72. A. Von Brandt of Germany began developmental work on the first hydraulic rotary drill in 1876 and it was used in the Pfaffensprung tunnel of the Gotthard Railroad. However, this latter type of motive force, i.e. hydraulic power, was not used as extensively as compressed air or steam for powering drills. In fact even the steam drill was gradually superseded by drills powered by compressed air which, for a time, was used almost exclusively as the power source for percussive drills (Figure 36).

Apart from some developmental work which was carried out on hydraulic drills during the early 1920s by Dormans of Stafford, England, very little, if any, research work was done in this field during the period between the 1870s and the early 1950s. (The main problem associated with the Dorman hydraulic drifter concerned the 'hose'. Apparently the hydraulic fluid was transmitted in lengths of steel pipe connected by ball joints which proved troublesome.)

During the 1950s and early 1960s the introduction of new mining machines and roof supports powered by hydraulics gave impetus to the development of hydraulic rotary drills which initially were used only for the softer types of material such as coal, potash, rock salt, or gypsum. When drilling through harder types of material the air-powered percussive drill was still found to be the most effective tool.

Developmental work was also taken one step nearer the all-hydraulic drill during the 1960s with the introduction of a hydraulic motor for providing rotary power for the conventional percussive drill, but the percussive energy was still imparted to the drill bit via a piston driven by compressed air. Thus a large corner of the drill market was dominated by manufacturers of percussive drills, and companies producing hydraulic rotary drills felt the need to enter into this field in order to expand their interests.

Figure 34. Bretby two-boom hydraulic percussive drill rig at Cloud Hill quarry, Breedon-on-the-Hill, U.K. 1962

Figure 35. Dynaboart hydraulic jack hammer at Swadlincote surface trials Centre, N.C.B. (M.R.D.E.) 1974

In 1961 at the request of the N.C.B. two hydraulic rotary/percussive drifters (Figure 34) were manufactured by Richard Sutcliffe Ltd., Horbury, Yorkshire. They were delivered to the N.C.B. (M.R.D.E.) in December 1961. These drills, each of which weighed 250 kg (550 lb) were then mounted on a motorized Bretby two-boom hydraulic percussive drill rig which weighed 18 t and travelled on rails. The two independent booms carrying the drills were mounted on a bridge structure which had a central aperture large enough to allow conventional loading-out equipment to pass through. It was designed for 4.88 m (16 ft) by 3.66 m (12 ft) arched roadways and could not therefore be used underground owing to the diminution of tunnelling in British coal-mines at that time. The drills at Breedon performed fairly well between March and November 1962, but were inclined to be somewhat temperamental. Further developmental work was continued by Sutcliffe and the N.C.B. on hydraulic machines and a 50 kg (110 lb) drill was subsequently built.

In 1970 as a result of collaborative work carried out between the N.C.B. and Victor Products (Wallsend) Ltd., of Northumberland on the one hand and the N.C.B. and Reyrolle Hydraulics Ltd./Boart & Hard Metal Products Ltd. on the other, light drills weighing only 26 kg (57 lb) were produced.

Both companies now market these light-weight hydraulic percussive drills,* the second drill (Figure 35) being known as the 'Dynabort hydraulic jackhammer'.

According to H.M. Hughes of the N.C.B. (M.R.D.E.), machines from both these firms 'gave notable demonstrations' in the N.C.B. coal-mines in 1976, 'drilling hard rock without clutter where compressed air was not available'. Hydraulic pressure in these machines is limited to 11 MPa (1600 lbf/in²) for manual reaction, but can be raised to 17 MPa (2500 lbf/in²) when they are mechanically reacted.

*The drill rods in these hydraulic jackhammers are indexed between blows while the rod is jumping. The drill rod is not continuously rotated as is the case with rotary/percussive drifters therefore the hydraulic jackhammers produce no torque at the handles and can be manually operated. The torque of the rotative motor is reacted by the cradle in the case of rotary/percussive drifters.

General Development of the All-Hydraulic Drill [12, 13, 14]

In the meantime the advent of the original Sutcliffe drills encouraged developmental work in this field in other parts of the world. Reports of the manufacture of hydraulic drills came from Russia and Sweden, and various prominent American and Japanese manufacturers who, according to rumours circulating at that time, were said to be busily designing and developing their own versions of the high-pressure hydraulic drill.

During the early 1970s a variety of machines were built by different manufacturers including Gardner-Denver (who based the design of their prototype units on the Bouyoucos-General Dynamics patents), Ingersoll-Rand of America, Atlas-Copco of Sweden, and Krupp of Germany, etc. Three different models of the all-hydraulic rotary-percussive drill were produced by Montabert of Lyons, France. Roger Montabert of Lyons began developmental work on hydraulic drills during the early 1960s. In March 1970 the first all-hydraulic three-boom jumbo was delivered to the Largentière mine (Pennorayo group) and was later sold to a French mining contractor. The machine is still operating after drilling some 600,000 m (about 2 million ft) in eight years. During that period it was only out of service for two months for overhaul and maintenance purposes. So successful was this drill that Montabert began mass producing hydraulic drifters the same year (1970), a few years earlier than most other manufacturers whose hydraulic drills were only in the developmental stage at that time.

According to J.H. Clark,[15] the field population of hydraulic drills as at January 1978 was approximately 1500. Of these 680 were produced by Montabert.

There are currently over a dozen manufacturers producing or developing high-energy, all-hydraulic, rotary-percussive drills. These include Atlas Copco (COP 1038 HD), Alimak (AD 101), Bohler (HS338 and HS352), Ingersoll-Rand (Hard III), Krupp (HB 51 and HB 101, 102 and 103), Le Roi-Dresser (LHD 155), Montabert (H-40, H-60, H-45, H-70 and H-100. H-40 and H-60 are first-generation drills which are now obsolete. H-45 and H-70 belong to the second-generation Montabert drills), Salzgitter (HH 5001), Conrad-Stork, Secoma (RPH 35 and RPH 400), SIG (HBM 100 and HBM 110), Tampella-Tamrock (HE 425 and HL 538), Joy (JH-2), Gardner-Denver, R.H.L./Boart (90 MFR and 150 MFR). The machines being developed by R.H.L. (Reyolle Hydraulics Ltd.) and Boart & Hard Metal Products S.A. Ltd.* use a rifle-bar system in their drills, similar to those previously used on most compressed-air percussion units. The intention being to produce a light hydraulic percussion drill suitable for manual use.

The all-hydraulic rotary-percussive drill is basically similar in concept to the original compressed-air drill. That is hydraulic fluid (instead of air) is fed first into the front and then the rear chamber of a cylinder. This causes a piston therein to reciprocate back and forth so that it impacts against a tool which, in turn, strikes the rock or other material being attacked. Rotary power for the bit is provided by a hydraulic motor via a series of gears, etc.

The number or frequency of blows delivered by the piston and the rotation of the drill bit is regulated by the entry and discharge through the respective ports of the hydraulic fluid. The exhausting of the fluid from the front and back chambers is either through a 'kicker port valve' (i.e. a port which is opened or closed by the movement of the piston in the cylinder), through a valve (operated either by the piston or by the hydraulic fluid) which closes and opens by seating on a flat surface, or perhaps by a sliding or rotating disc, depending upon the preference of the manufacturer and/or designer of the unit concerned.

In his paper, 'Excavation methods for long highway tunnels and ventilation shafts in the Swiss Alps',[16] Rudolf E. Pfister (Director, Electrowatt Engineering Services Limited, Switzerland) enumerates the specific advantages of the all-hydraulic drilling hammers (Montabert Type H-60) which were used for the portal sections of the Seelisberg highway tunnel. These include a marked reduction in power consumption, the elimination of compressed-air pipes, less noise,

better visibility due to the absence of fog caused by moist compressed air being exhausted into the tunnel environment from the drill, and the greater adaptability of the drill to the peculiar conditions of the material being attacked, by the regulation of the rotary or percussive operating power of the drill. Pfister also mentioned that drilling performance was improved by some 30-40 per cent.

Richard L. Bullock in his article, 'An update on hydraulic drilling performance,'[13] also draws attention to the fact that an added but not always recognized advantage gained by the use of hydraulic drills was that because there was no fog, there was also an absence of dust particles which usually became suspended in the air when compressed-air drills were being used.

An interesting facet of the new all-hydraulic drill due to increased penetration rates is its apparently voracious appetite (particularly when drilling in softer materials) for flushing water, which must be delivered at a higher pressure and in larger volume than was previously necessary with the compressed-air drill. If water was not supplied in sufficient quantity and at the correct pressure the drill bit tended to jam in the hole. Generally speaking, the majority of the new breed of drills, with perhaps one or two exceptions (such as the Ingersoll-Rand Hard III), use accumulators either in the rear chamber alone or in both the rear and front chambers. Front- and rear-chamber accumulators are similar, although used for different purposes.

The accumulator in the rear chamber is designed to impart extra energy to the piston on the forward impacting stroke. It consists basically of a nitrogen-filled compartment or chamber which is separated from a compartment containing hydraulic fluid* by a flexible diaphragm and operates as follows.

When the hydraulic fluid is pumped into the rear chamber it also fills the accumulator oil reservoir which compresses the nitrogen gas in the compartment at the back of the rear chamber. As soon as the piston is thrust forward by the fluid in the main rear chamber

---

*These are the same drills as the Dynaboart hydraulic jack-hammer units mentioned earlier.

*In some models the nitrogen gas is compressed directly by the oil in the main rear chamber and not via an oil reservoir as described above.

the nitrogen gas expands and pumps the oil in the reservoir into the main rear chamber, thus transferring the extra energy stored by the compressed nitrogen gas to the piston and ultimately to the drill tool.

So far as the front accumulator is concerned this is put there to cushion the drill against any high peak pressures which may be caused by an erratic action of the piston when it is thrown forward. In air drills, such shocks or high-kinetic energy waves are automatically absorbed by the compression of the air in the front chamber. However, when hydraulic oil is used this is not possible as there is very little cushioning effect in a liquid — thus an alternative method was felt by the majority of manufacturers to be necessary, and this has been provided by the installation of an accumulator.

Vertical Drills

Oil and Shaft Digging — Early History[17-19]

The story of oil and the history of drilling are so intimately interwoven that it is difficult to investigate the one without coming across numerous references to the other.

Deep-hole drilling was reputedly invented by the Chinese at least 100 years before the coming of Christ. They drilled for water, oil, and natural gas by means of coupled bamboo rods which were lifted, dropped, and rotated repetitively. The oil and gas was then transported by the Chinese to their temples through bamboo pipes where it was used as a source of heat and light. According to Drinker,[2] the art of deep-hole drilling was brought to Europe by Jobard. (The author has been unable to trace any other reference to 'Jobard'.)

Henning Huthmann,[2] Rector of Ilfeld, proposed the first 'boring machine', a drop drill, in 1683. The drill was raised by a rope drawn by two men and then dropped. This drill, according to Calver (1763), would sink a hole 1½ in (38 mm) deep and a hand's breadth wide and long with each stroke.

In 1721 Barthels[2] of Tellersfeld invented a machine for drilling or boring shafts (this in the

opinion of H.S. Drinker, was probably a drop drill), and in 1803 a machine which worked 'quicker than a miner' was reported to have been made by Gainschrigg at Salzburg. In 1838 I.M.* and John A. Singer[2] experimented with a large drop drill on Section 54 of the Illinois and Michigan Canal about 50 km (30 miles) south of Chicago. The trials proving successful they patented the drill in May 1839. After this about 10 or 12 similar machines were built and used on the canal until the suspension of the work in 1841-42. These machines were subsequently put to use in the Mount Washington Cut near Hinsdale for the Western Railroad of Massachusetts. (Steam-power was employed to lift the drill bar from the hole, but it descended on the downward stroke by means of gravity.) From that time until the end of the nineteenth century two distinct methods of deep-hole drilling developed. One was by means of a series of rotating rigid rods attached to a drill bit which cut as its end revolved against the rock. The other used a rope which was also attached to a heavy drill bit which cut into the rock each time it was dropped and struck the bottom of the hole. These methods, i.e. rotary drilling and percussion drilling with minor variations, are described below.

Earth Augers — Soft and Cohesive Soils[19]

These comprised two types, namely the *spiral auger* and the *pod auger.*

Pod Augers

The most common type, the pod auger, consisted of two concave metal plates which were attached to each other by a hinge. When the two plates were closed they formed a pod-shaped cavity which was open at its lower end. The top end of one of the concave metal plates was attached to the end of a rigid rod, while the second plate (attached to the other plate only by its hinge) was left free in order that it could be opened. In operation the rod would be plunged into a hole and forced downward until the pod had filled with soil, then it would be lifted, and

*Later Isaac M. Singer became famous for his invention of the sewing machine.

42

Figure 36. Rand drills — sinking a shaft using columns

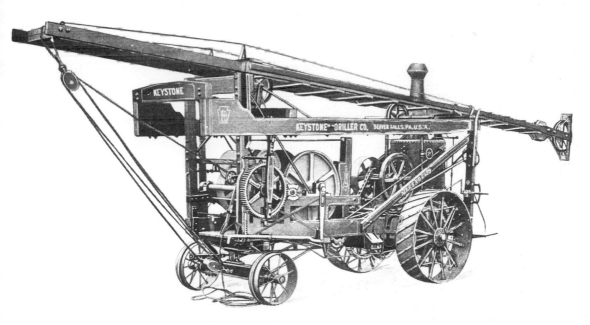

Figure 37. Keystone Driller Co. Beaver Falls, Pa. Cable tool drilling rig, 1910 Model No. 5 (Missouri Special for 304 m (1000 ft), powered by 10.4 kW (14 hp), boiler 40 × 72 with 99 cross-tubes, combination friction and cog hoist, 10.3 m (34 ft) high extension derrick, 76.2 cm (30 in) diameter crown Pulley, and 55.8 cm (22 in) diameter spudding sheaves. (Courtesy Goldfields Diamond Drilling Co. Pty. Ltd.)

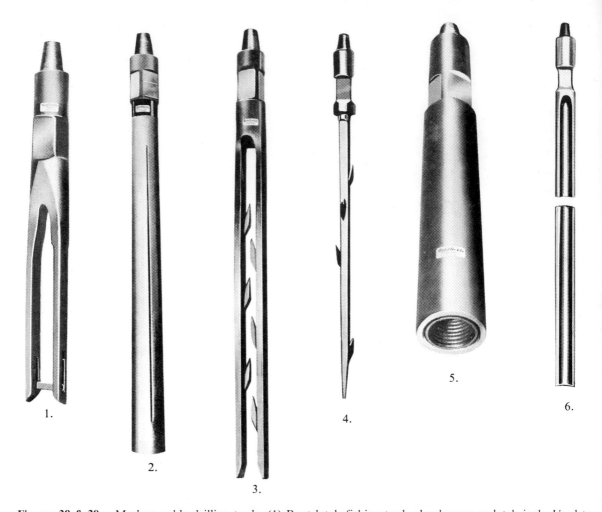

1.

2.

3.

4.

5.

6.

Figures 38 & 39. Modern cable drilling tools. (1) Boot latch fishing tool, also known as latch jack. Used to recover bailers, sand pumps and broken jars. The latch near the bottom is hinged and the tool may also be used with fishing jars; (2) Horn socket. Used to recover lost tools. Consists of long tapered barrel slotted down one side; (3) Two-pronged rope grab. Used to remove broken cable from the hole; (4) Single-prong rope spear. Also used to remove broken cable; (5) Combination fishing tool. Used to catch a rope socket pin, a tool joint pin, or any cylindrical object left in the hole. Fitted with grooved steel slips for gripping; (6) Spud. Used to drill around and loosen stuck tools or to straighten bits or other tools which have been pushed over into the wall in such a way that a fishing socket cannot be placed over them; (7) Side cutting bits (enlarging bits). The tool is machined off-centre, thus throwing the cutting edge over into the wall of the well. It passes through the casing and enlarges the hole below the casing to allow the advancing of the casing without under-reaming; (8) Drilling jars. Usually incorporated in a tool string to prevent and correct trouble. Although they do not add to the direct efficiency of the drilling tools, they greatly assist in dislodging them should they become stuck in the hole. Constructed with two telescoping reins; (9) Sinker bars or drill stems. Provides weight for the tool string; (10) Drive or chop pumps. Used in place of the chisel bit when drilling through gravel or stony formations. A clack valve is fitted to retain the cuttings inside the tool and the shoe is angled to give a better penetrating edge for drilling in loose stones; (11) Sand pumps or bailers. Used for removing cuttings, mud, and sand from hole. Operated by a separate line of smaller diameter than the drilling cable (sand line) and may be used with a clack valve or a dart valve; (12) Clack valve. Allows close contact of the bailer with the bottom of the hole; (13) Dart valve. Allows the bailer to be emptied at ground level without being up-ended; (14) Vacuum sand pump — Morahan type. Piston-type bailer. Upward movement of the plunger causes displacement of fluid so that a suction action is created. Used with a clack valve; (15) Modern Goldfields Drilling Company's percussion cable tool drilling unit. Has an 11 m (36 ft) telescopic mast capable of carrying approximately 583 kg (1290 lb) of tool. It is powered by a 19-26 kW (25-35 hp) petrol or diesel engine. May be converted to a rotary drilling unit with a conversion kit. (Courtesy Goldfields Diamond Drilling Co. Pty. Ltd.)

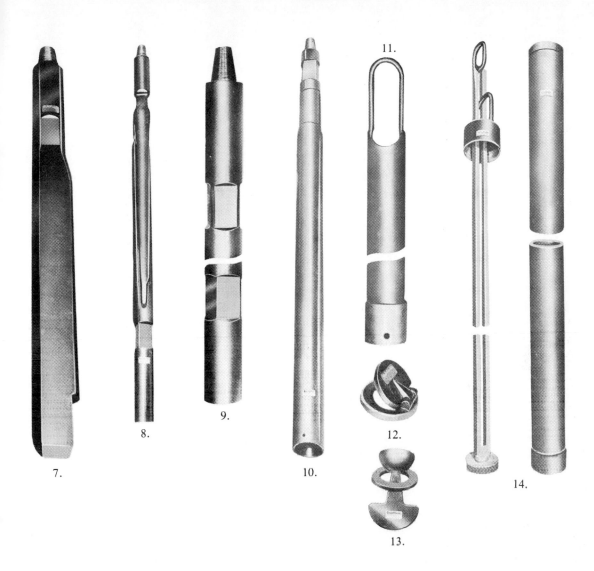

7.

8.

9.

10.

11.

12.

13.

14.

15.

45

such soil and debris as were trapped within the pod was cleared by opening the door of the pod.

Spiral Augers[19]

Spiral augers are similar in design to those used by carpenters. The spiral auger is screwed to one or more socket-jointed rods which are lowered into the hole and rotated by means of horizontal levers clamped to the surface rod.

Drop Drill and Rod — Hard Ground[19]

The drop drill and rod consists of a chisel bit which is attached to long lengths of rigid rod by screw-joints. Because of the excessive wear caused by the weight of the rods as they descend into the hole on their downward stroke, and to prevent rod buckling a pair of sliding links is inserted between the bit and the first rod. Thus when the drill is plunged downward the bit strikes the ground ahead of the rods, and the shock of impact is lessened by the action of the links closing as the rods continue to descend until the slack between the links has been taken up.

Drop Drill and Rope (Later to be known as Churn or Cable Tool Drills)[19]

These were used mainly for prospecting and for oil and water drilling during the nineteenth century. This method is similar to that of the drop drill and rods system, except that the rigid metal rods are replaced by wire rope (Figures 37 to 39). The use of a rope instead of rigid rods allows the bit to be raised to the surface and lowered again much faster than is possible when individual metal rods have to be removed or added in the process. This advantage becomes more noticeable as the hole grows deeper.

Drive Pipes[19]

Drive pipes were mainly used for exploratory testing to determine the depth and nature of soil covering rocks, etc. These consisted of open-ended iron pipes ranging in size from 13 mm (½ in) to 152 mm (6 in) in diameter. At the lowest

end of the pipe an annular steel shoe is attached, while the top end carries a drive head designed to withstand hammer blows. As the hole deepens additional pipes are attached at the top.

Cleaning[19]

The normal method of cleaning the hole was by 'bailer'. This device consisted of a hollow tube of from 1.2 to 1.8 m (4 to 6 ft) in length. At the lower end of the tube is a valve which allows the water and cuttings to enter on the downward thrust, but which closes and retains the contents on the upward stroke. The tube is raised and lowered a number of times until filled, after which it is drawn to the surface and emptied. This method, however, was not found effective for very deep holes. In such cases a hollow pipe carrying a drill bit would be lowered into the hole. Water would then be pumped down the hollow pipes until it emerged through holes in the bit, thus forcing the spoil to the surface through the annular space between the rod and the shaft wall. This latter method of cleaning led indirectly to the development of the *rotary flush drill*.

The Hydraulic Rotary Drill or Rotary Flush Drilling[19]

Often the presence of layers of swelling clays or running sands would prove to be an insurmountable barrier for operators using the old percussion drill bits. Such areas would perforce have to be abandoned in favour of more amenable strata. But, hand in hand with the increased demand for oil, came the demand for faster and more efficient drilling methods. It was only a matter of time before someone devised a better method of penetrating these thick clayey beds of mud. This came in the form of the hydraulic rotary drill which was used initially in Louisiana and Texas. In both these areas operators using conventional percussion drills had been forced to admit defeat in the face of thick beds of running sands.

The new hydraulic drills were installed and they pushed through the wet sands with ease. These drills used water which was forced down hollow drill pipes under high pressure. It emerged in strong jets through two small holes set just above the wings of the 'fish-tail' drill bit. As the bit rotated, the jetting water scoured the mud and clay from the bottom of the hole and forced it back up the shaft on the outside of the drill pipes.

After piercing the mud veins successfully, *the same type of drill was also tried on rock formations.* At first plain water was used, but inevitably it became apparent that sludge or mud was forming. To the amazement of the operators, instead of this proving to be a nuisance they found, on the contrary, that the mud was actually beneficial. Penetration rates increased and the condition of the shaft wall improved. This occurred when the mud-laden water was being returned to the surface via the annular hole outside the drill pipes. As the thick fluid moved upwards, the vibration of the rotating drill pipes caused a certain amount of the mud to be plastered into fissures and cracks along its route. In addition it was found that the borings were removed more efficiently in the mud than in plain water.

This in turn led to the development of an entirely new science in oil-well drilling, i.e. the control and use of special muds and chemicals in drilling fluids.

Some Difficulties Encountered with Early Drilling Methods[17, 18]

Though rods were considered to be extremely effective as they allowed the driller to administer a sharp blow at intervals during the turning or screwing motion, this advantage was outweighed, so far as deep holes were concerned, by the extreme inconvenience of raising them from great depths when it became necessary to change from boring to cleaning.

If the hole was not cleaned regularly the accumulated debris formed a barrier between the bit and the solid rock surface. Continued pounding only served to reduce this debris to dust without noticeably advancing the depth of the hole. As a result, many valuable work hours were lost by drillers using this method in deep-hole drilling.

Both rope and rod drills were subject to jamming by pieces of rock which became

dislodged from the walls of the hole by the passage of the drill bit. Generally, when this happened the drill could be released by jarring. Periodically, the bit or reamer would become so firmly wedged that jarring was ineffective and though smaller pieces of rock, which were perhaps the cause of the problem, could sometimes be hammered into the sides of the wall by the upward action of the jarring tool, this did not necessarily apply to the larger fragments. In such cases the disheartening decision to abandon the well often had to be made, and one can readily imagine the frustration and disappointment of the workers when this happened after the hole had been sunk to depths of several hundred feet or more.

The drilling jar consists of a telescoping sliding link. The stroke of the link being about 17 or 20 cm (6 or 8 in). The drilling jar is usually incorporated in a tool string to prevent and correct trouble and is added to the string when the hole is about 9 m (30 ft) deep. It is inserted between the rope cable (the top pin joint being connected to the rope socket) and the top of the drill stem. Generally speaking, the drill jar is not used and remains closed during the drilling operation unless the drill bit becomes stuck or wedged in the hole due to a pebble or piece of rock, etc. falling in from the walls or from the surface. If this happens the bit is dislodged by jerking the cable upwards and then dropping it again. This causes the links in the drill jar to open and close and tends to jerk the drill bit free while the pebble is hammered out of the way into the walls of the shaft.

Reamers and drill bits also frequently become unscrewed from their auger-stem sockets, or sand-pumps would break free from their ropes and plunge to the bottom of the hole. Unless the operator happened to remove the drill from the hole for purposes of cleaning, etc. shortly after the displacement occurred he was very often left unaware that the bit was no longer attached to the stem and would continue to pound on top of it with the heavy auger stem. This action would naturally minimize his chances of recovering the bit or reamer intact.

Another cause of trouble (i.e. dropped ropes, drill tools, etc. necessitating 'fishing') was the use of rope or cable with the incorrect 'lay'.

47

Normally ropes are made with a right-hand 'lay' or twist, but proper drilling ropes or cables are spun with a left-hand lay, and are known as 'hawser laid' manila cable. This ensures that the cable at the end of its stroke, when it is taut and unwinds, will turn clockwise, or to the right. When the cable is on the downward stroke it turns anti-clockwise or to the left, and when it is being drawn up it turns clockwise. Thus the drill bit is turned constantly in the hole during the drilling operation by the action of the rope or cable automatically twisting and untwisting. However, when a hole is commenced the operator usually has to turn the rope for a short distance by hand to ensure that the drill bit remains free in the hole. As the screw-joints on a drill string are all right-handed, purchase of the wrong kind of rope or cable may cause these joints to unscrew with disastrous results.

When the bit strikes the bottom of the hole the cable, if it is the correct length, should become taut, and an experienced operator learns to judge, by the vibration of the cable at this point, whether or not the correct amount of cable has been released. If too much cable is released into the hole penetration rates are adversely affected. Normally the links of the jarring bit remain closed during the drilling operation, but if the operator is letting out more cable than is necessary this may cause the links to open.

During the period when these drilling methods were extensively used, operators cunningly learned to devise a variety of both tools and procedures for either retrieving lost implements or extracting those which had become jammed.

Mud Veins[17, 18]

Before the advent of the hydraulic rotary system of drilling, the biggest nightmare of the driller was the fear of encountering a mud vein. This is a layer of mud or clay usually several inches thick. When mud or clay was encountered near the surface it did not usually create a very severe problem, but when it was met with at depth, it often possessed alarmingly cohesive properties. Should it flow into the well

and surround the bit, it could set very quickly into a consistency almost similar to that of the surrounding rock, thus jamming the bit firmly in the hole.

A. Beeby-Thompson in his book *Oil-field Exploration and Development,*[18] describes this clay as being bright blue in colour. He comments that various authorities such as Pascoe, Arnold, and Eldridge had noted the presence of this type of clay in close proximity to oil regions in Burma and Assam, the Eastside Coalinga oilfield and the Kern river oil district respectively. Pascoe considered that this particular type of clay could even be related to the oil genetically, and ancient Rumanian shaft-diggers in seeking sites for well-sinking, regarded areas exhibiting this clay as being highly suitable for their activities.

Many of the difficulties outlined above which were experienced by late nineteenth and early twentieth century shaft-diggers disappeared with the introduction of the rotary flush drill.

Rotary Drilling with Compressed Air

During the early 1940s some small rotary drills were fitted with air compressors and air, in place of water or mud, was tried as a medium for removing the cuttings.* This was found to be an extremely efficient method of getting rid of the debris, especially if ground conditions were favourable. The advantages of this method were that penetration rates increased, the bit life of the drill was prolonged, and, most significantly, it could be used in dry areas, where providing a continuous supply of water in the large quantities needed for a rotary drill constituted a major problem.

The system used is basically the same as that required for a mud- or water-circulating system except that the mud-pump assembly is replaced

---

*According to J. M. Glass, Jr., of the Hughes Tool Company, a rotary blast-hole rig using compressed air as a flushing medium was developed jointly by the Hughes Tool Company and the Joy Manufacturing Company during 1948/49. This is reputed to be the first application of compressed air flushing for a rotary blasthole rig. However, in his letter to the author, Mr. Glass comments that as far as he was aware, compressed air was used on percussive wagon drills during the 1920s and 1930s and natural gas was used in the oilfields as a flushing medium during the early 1930s.

by the air compressor, an air receiver is fitted for directing the air down the hollow pipes, and a disposal system is added to get rid of the cuttings when they reach the surface. Sometimes a small amount of water is injected into the air stream. This is to control the dust and also to help keep the drill bit cool at the bottom of the hole.

The air is forced down the drill stem and emerges through openings in the bit at the bottom of the hole, where it collects the cuttings and carries them to the surface through the annular space between the drill pipe and the hole wall.

Air is only suitable as a circulating medium in rock or very hard clay. It is not considered suitable for vuggy* or unconsolidated ground. If the ground is broken and full of fissures, circulation may be lost. If the air is stopped the hole walls are no longer supported, and when drilling in loose or unconsolidated ground pieces of the wall tend to fall or cave in against the drill stem causing it to become jammed, etc.

Experience with the rotary air method showed that frequently comparatively 'dry holes' became wet as they got deeper.** This presented a problem when drilling with air as mud rings or mud collars then formed and these caused the drill string to stick. Operators learnt to overcome this difficulty by again injecting water with the air to keep the hole clear. In addition various foaming agents have been successfully used. A mixture recommended by the National Water Well Association of Australia in their *Drillers Training and Reference Manual* for producing a stiff foam consists of '12 lb bentonite, ¾ lb soda ash, ½ gallon of foaming agent and 40 gallons of water'. The mixture is added to the air stream.

Reverse Circulation Drilling[20]

During the late 1940s and early 1950s a method of drilling known as 'reverse circulation' was introduced.

---

*Vug — Cornish miner's name for a cavity in the rock; 'vuggy' — adjective.

**Another major problem associated with deep holes is that ingress of water creates a considerable 'static head' which may effectively prevent the escape of exhaust air.

A rotary table and rotating drill string and bit are used but the flushing fluid is not *pumped into the hole through the hollow drill pipes.* Instead water is allowed to flow down to the bottom via the annular space between the drill pipe and the hole wall. Then, together with the cuttings, the water is *pumped (or sucked) up* the hollow drill stem by a powerful pump and deposited in a settling pit or tank near the drilling rig. The tank or pit is divided into two compartments by a partition of wood or some suitable material. The cuttings and water flow into the first compartment which holds the debris while the water is allowed to pour over the divider into the second compartment. From this second compartment the water is redirected back to the hole by means of a trench or a pipe.

The pump is typically capable of handling from 1100 to 2300 litres/min (250 to 500 gal/min) or more, as well as the cuttings, which could include some large pebbles (sometimes up to 15 cm (6 in) or so in diameter). Centrifugal suction pumps are the type most commonly used for this purpose.

This method of drilling is most suited to sandy or gravelly formations. Mostly it is used to bore large-diameter high-capacity wells, but it has also been used to excavate shafts and feed-holes leading to underground storage caverns.

Mud is not generally used, the idea being to keep the hole clean so that the aquifer is not sealed off from the hole. However, the introduction of drilling mud to cope with particular sections of extremely porous ground may sometimes be necessary in order to stop a fluid circulation loss.

The hole walls are supported by the hydrostatic pressure of the water, and casing is not inserted until the well has been drilled to the bottom. Then again, if extremely porous ground which will not seal with mud is penetrated, it is sometimes necessary to insert casing through the problem area.

When natural clays which act as a sealant are encountered, the drilling operation must be slowed and water circulated until the clay is eliminated and the hole is 'clean', when normal drilling may proceed.

A large supply of water is necessary for

drilling by reverse circulation and in his article 'Specialised rotary drilling methods' R. Lauman[21] comments that a 1100 litres/min (250 gal/min) source should be available.

Once work starts it is customary to continue drilling on a 24 hour basis until the hole is bottomed, when it is pumped clean and then cased. The method is extremely fast and very often the bottom is reached within 24 hours or so. If, for some reason, it becomes necessary to halt the drilling operation, then a constant watch must be maintained to ensure that the water level does not drop during this period. Otherwise the walls of the hole, which are supported by the water, may cave in causing damage to drilling equipment and even the loss of the hole.

Compressed air injected into the drilling fluid has also been found to be advantageous. Its use has speeded penetration rates and allowed the operator to attain greater drilling depths than was previously possible using water alone. Drilling rates of approximaely 60 cm/min (2 ft/min) have, according to Lauman, been achieved with compressed air and water.

The compressed air is forced into the outgoing stream of water at the bottom or about half-way up the hole, via an airline which is coupled to (or set inside) the hollow drill pipes. This makes the rising mixture much less dense than the incoming fluid and has the effect of increasing its velocity so that cuttings, etc. are more readily carried to the surface, and problems associated with clogging, etc. are minimized.

Closed-Circuit Reverse Circulation Drilling[20]

Another interesting variation of the reverse circulation drilling method has been the introduction of what is commonly known as the 'closed-circuit reverse circulation' system.

In this a double hollow tube drill rod is used. In other words the drill string consists of drill rods that have an inner hollow pipe which is in turn surrounded by an outer load-bearing pipe in such a way that an annular space exists between the inner and outer tubes of the drill rod.

The drilling fluid, be it air or water, is directed down between the double skins to the bottom of the hole where it picks up the cuttings made by the drill bit and carries them to the surface via the hollow centre of the inner tube.

The main advantage of this method is that because the hole walls are supported by the external surface of the outer pipe, neither mud nor casing is required until the hole is bottomed. Moreover, fluid loss through porous or badly fractured areas is greatly minimized and soil samples are less susceptible to contamination from disturbed areas which may have been encountered *en route*.

Double-skinned or 'dual pipes' range in size from about 7.62 cm (3 in) o.d. to approximately 15.2 cm (6 in) o.d.

Apart from drill rigs which have been specifically designed for use with dual pipes, the system may be adapted for use with a conventional rotary drill rig (having either kelly

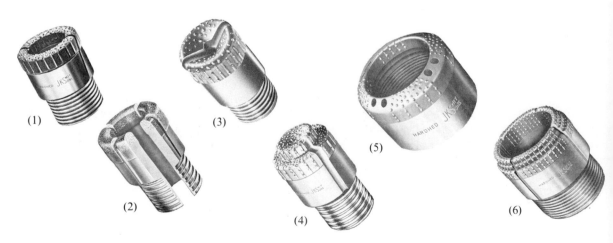

(1) (2) (3) (4) (5) (6)

or top-head drive) so that either hammer drilling, rotary drilling, continuous coring, or jet lifting (cuttings carried to the surface by air inserted at high pressures) may be carried out.

High Pressure Water Jetting[20]

This system is of particular value when penetrating sandy soil and consists in pumping about 300 to 400 litres (70 to 90 gals) of fluid per minute at a pressure of approximately 0.34 MPa (490 lb/in²) down the hollow pipes of the drill stem, so that it emerges in a strong jet at the bottom through the holes of the drill bit. The ground is thus excavated by the scouring action of the high pressure stream of fluid (usually water or a mixture of water and drilling mud) and by the percussive action of the bit. The cuttings are then carried to the surface by the fluid via the annular hole between the drill pipe and the walls of the shaft.

Core Drilling and the Diamond Drill

The early Egyptians are reputed to have used tubular drills for boring short holes during the construction of the Pyramids, and some sources believe that the bits in these hand drills were set with gemstones. However, other sources postulate that abrasive powders were probably used and that the first diamond-core bit was made by a Swiss engineer, Jean Rudolphe Leschot, while he was residing in Paris. Leschot and Pihet (a mechanic) made a hand-operated drill which was worked by two men and used for drilling blast-holes. An annular bit studded with black diamonds

(carbons) measuring 30 mm (1.2 in) i.d. and 40-42 mm (1.6-1.7 inches) o.d. was used in the drill. In 1864 Leschot used an annular bit set with black diamonds for blast-hole work in the Mont Cenis tunnel.

Important developmental work concerning the application of the diamond drill to both tunnelling and drilling (particularly for prospecting work) was made by C. J. Appleby and by Major Beaumont (See Figures 192 and 193). (Major Beaumont with T. English designed the first Channel tunnel machine.)[22]

In his paper, which was presented at the Institution of Mechanical Engineers in 1875, Beaumont described his prospecting machine thus:

> The first Prospecting machine was made by Messrs. Appleby of London; and in Figures 1 and 2, Plates 8 and 9, is shown the largest form of prospecting machine, as made by Messrs. Ormerod Grierson and Co. of Manchester, who are now constructing the prospecting machinery for the company. The principle of the two is the same, but the latter machine is far more powerful

Figure 40. Diamond drilling bits. (1) Surface set coring bit; (2) Impregnated coring bit (the impregnated Ferret bit is set with small evenly graded diamond particles moulded in the matrix — the diamond particles are distributed throughout the depth of the matrix so that the bit may be consumed without resetting); (3) Concave bit (used for blast-hole work, etc.); (4) Pilot bit (used in conjunction with a reamer, etc.); (5) Large series bit; (6) Step bit; (7) Casing bit; (8) Casing shoe bit; (9) Reaming shells; (10) Tapered reaming shell; (11) Thin-walled coring bit (specially suitable for taking core samples in reinforced concrete work). (Courtesy J. K. Smit & Sons (Aus.) Pty. Ltd.)

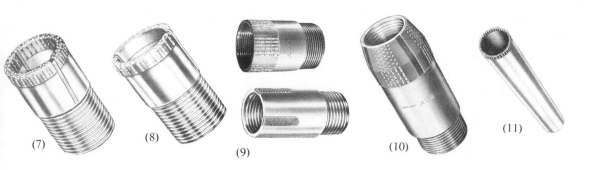

(7)　(8)　(9)　(10)　(11)

than the former one, and is capable of dealing with the heavy boring bars and large crowns required to cut cores of 3 to 6 in. diam. and several feet in length, such as those of which portions are exhibited. The lifting gear for taking the boring rods out of the hole is also more powerful, and an arrangement has been made by which the quill and turning gear can be lifted clear of the line of the hole, and so allow the boring rods to be more readily manipulated.

The prospecting machine consists of two vertical girders, carrying between them a hollow quill made of sufficient size to grasp and turn the boring bars. This quill has a rotating motion given to it by means of an inclined shaft driven through bevel gear, power being transmitted by means of a belt from the flywheel of a portable engine. The pressure necessary to cut the rock, as previously described, is given by the weight of the rods, reduced where necessary by counterbalance weights. For holes of a moderate depth, say up to 1200 ft. or 1300 ft. an 11 in. cylinder engine with 40 lb. steam pressure is sufficient.

The boring rods, are withdrawn whenever it is desired to know the nature of the strata being passed through; and as the rods are not so large as the hole, a core tube has to be used at the bottom of them, to receive the core cut, and the length of this tube limits the boring that can be done at one operation. In practice the core tube is about 10 ft. long. The diamonds are kept cool, and the debris continuously removed, by a stream of water pumped down through the hollow boring bars, which after passing under the crown rises up to the surface through the space between the rods and the sides of the hole.

Amongst various other projects described in his article Beaumont mentions the drilling of a borehole at Widdrington in Northumberland which was commenced on 9 February and was completed to a depth of 1565 ft (477 m) in 265 days.

Because of its versatility, since there was no restriction to the angle at which it bored, the advent of the diamond drill constituted a significant development in the history of the drill in general. Holes could be driven up, or down, or horizontally, whereas in the first four methods described, the drill could only operate vertically downwards. Moreover, the diamond drill could either make an annular hole which was particularly suitable for the removal of a core for exploratory or prospecting work, or a solid bit could be attached to the rods for use in blast-hole work, etc.

In the case of core samples the drill string consisted of open-ended pipes joined by screw-joints and having an annular soft steel bit set with from six to eight or more diamonds (according to its diameter) attached to the lower end. The diamonds (which originally were hand-set) were carefully placed in the bit so that each stone projected very slightly from the surface of the steel bit. Two types of diamonds were used. These were 'carbons' and 'borts'.

Carbons have no cleavage planes and are opaque, dark in colour, and are tougher than the brilliants, so they were used for hard rock. Borts or 'congos' are imperfect brilliants (diamond fragments or dust) which were originally used only for the softer rocks. However, because these stones were relatively cheap in comparison with carbons, and also because a method was found whereby they could be mechanically set, as they are at the present time, thus further reducing the price of borts-impregnated bits, this latter type of stone became more extensively used for general drilling purposes and has now entirely superseded carbons-set bits except in some special cases such as when drilling through reinforced concrete, etc.

Between the diamond bit and the first rod a core lifter and core barrel were inserted. The core barrel fitted neatly in the hole, and during the early stages of the development of the drill was occasionally grooved to allow the passage of water and debris. It served two main purposes, namely that of protecting the core and guiding and keeping the shaft on course. The core lifter within the barrel gripped the core and allowed it to be broken away from the hole. The core lifter also prevented the core from falling out when the barrel was raised. (In addition, core barrels may have inner tubes which support the core and keep water off it. During the development of the drill, a variety of other devices were introduced by manufacturers which improved core recovery and presentation.)

If no core was needed then a solid bit was attached to the lower end of the pipe instead of the annular bit. This allowed the drilling to proceed at a much faster rate without the need for extracting the core. When the hole required cleaning this was done by pumping water down the hollow pipes. (Later, water was pumped

Figure 41. Independent rotation drifter introduced by Gardner-Denver in 1959. Photograph taken at Blue Metal Industries quarry in Prospect, N.S.W., Australia. (Courtesy Gardner-Denver Co.)

Figure 42. Gardner-Denver independent rotation 8.9 cm (3½ in) bore drifter — six-boom shaft jumbo. Used by Cementation Aust. Ltd. for work at Mt. Isa, and COBAR, Australia (hydraulic booms and drill positioner — 10 ft screw feeds). (Courtesy Gardner-Denver Co.)

53

continuously during the drilling operation.) The water forced the cuttings up through the annular space left between the pipes and the walls of the shaft.

The drill was operated by means of a string of pipes. The speed of the bit could be varied from say 300 to 800 rpm or more, depending upon the diameter of the bit and the type of rock being attacked.

It should be recognized that the above paragraphs merely outline the history of diamond drilling during its early stages of development. While the basic principles involved remain the same, many of the methods and/or equipment used have undergone significant changes which have improved the art of diamond drilling.

The wire line core barrel for instance (see below) and the use of soluble oil, which latter product has greatly increased bit life and assisted with the general lubrication of the drill string, are amongst some of the more outstanding innovations introduced fairly recently. Another has been the marriage between diamond drilling and other techniques, particularly down-hole percussive units which has widened the scope of the vertical drill. Using these two methods, faster, cheaper holes may be drilled through non-core surface rocks and then they may be continued by diamond core drilling where required. 'All-purpose' drill rigs have been developed which handle this marriage admirably.

As an aid to mobility a frame measuring approximately 8 ft (wide) by 20 ft (long) (2.4 × 6.1 m) was specially constructed. This frame was fitted with hydraulic jacks at each corner to assist the operator in lifting the drill, and mast, on or off a truck on his own.

The Wire Line Core Barrel[23]

The process of raising and lowering a long string of drill rods to recover the core every metre or so is a very time-consuming operation. Because of this the wire line core barrel was conceived. It was first used by oil-well prospectors in the petroleum industry.

Significant developmental work in this area was carried out by the E. J. Longyear Company over several years. This work culminated in the introduction in 1954 of their BX size hole (6 cm (2 in)) wire line core barrel.

Basically the wire line core barrel consists of an outer barrel which is attached at its upper end to the drill string and has at its lower end the diamond drilling bit. Within the outer barrel is a retrievable inner barrel which holds the core. The inner barrel is fitted at its upper end with a spear head.

When the core is to be retrieved a cable fitted with a lifting dog is dropped down inside the hollow pipes. The lifting dog attached to the end of the cable grabs the inner core barrel by the spear head and raises it and the core sample to the surface while the outer core barrel and the drill string, etc. remain in place. So that no time is lost during the inspection and removal of the core, etc. another spare inner core barrel is immediately dropped down the hole into the outer barrel and drilling is recommenced.

Driving and Underreaming Casing[23]

Before drilling with a diamond bit may commence, the hole must be driven through to bedrock and the walls of the hole protected with casing. To do this the casing, which is normally about 3 m (10 ft) in length, is driven straight into the hole. The first pipe down the hole is protected at its lower end with a coupling or, if the ground being penetrated is particularly hard, with a steel drive shoe. The top of the casing, which receives the blows of the drive hammer, is protected from damage by the insertion of a special tee piece or drive head. Sometimes the casing may encounter a boulder or even very hard ground through which it cannot be driven. When this happens the casing is withdrawn approximately 20 cm (8 in) and a reamer which has a solid concave diamond pilot bit at its end is put down the hole. Immediately above the pilot bit are three tapered reamers which are set with diamonds on their outer faces. When downward pressure is applied to the pilot bit the reamers are forced to expand outwards in the hole. In this manner the hole diameter is enlarged to a size just slightly greater than that of the casing. As the reaming operation proceeds, the casing is rotated in the hole so that it follows down immediately behind the reaming bit. When the obstruction

Figure 44. Ingersoll-Rand DM-6 Drillmaster (rotary machine). Speed 0-100 rpm torque to 11.9 kNm (8750 lb ft). Hole size diameter 25-28 cm (9⅞-11 in). Depth to 47 m (155 ft). (Courtesy Ingersoll-Rand Ltd.)

Figure 43. Marion blast-hole drill, Model M-4, 31 cm (12¼ in) diameter hole. (Courtesy Marion Power Shovel Co. Inc.)

has been penetrated the reamer is raised slightly, thus relieving the pressure on the pilot bit. This allows the reaming lugs or cutters on the bit to retract and the reamer is then pulled out of the hole.

The above method of underreaming was used farily extensively in Canada, but when particularly hard boulders were encountered explosives were used to break them. In Australia the general practice has been to drill a larger diameter hole and then place casing, sometimes fitted with a 'shoe' set with diamonds or hard metal, into it. This shoe enabled the drill to drive the casing through any caving ground which might have fallen into the hole.

Sandvik Coromant O.D. Drilling Method[24]

Glacial type overburden in Scandinavia was deemed difficult to drill through for two reasons; the hole walls tended to collapse thus necessitating the use of casing tubes, and there was a high risk that boulders would be encountered.

Such were the problems faced in 1955 by Skanska Cement Gjuteriet AB, a large civil construction company in Sweden which was excavating a channel.

To overcome these problems Sandvik Coromant developed the O.D. drilling method (O.D. = overburden drilling method) and Atlas Copco built a special pneumatic rock drill capable of producing the high torque required for the O.D. method.

Equipment for the O.D. method consists of a shank adapter, an inner string of drill rods, couplings, casing tubes, and a tungsten carbide bit casing ring. The tungsten carbide tipped ring leads the casing tubes which are connected to the shank adapter. As both the inner drill string and the casing tubes are connected to the rock drill, rotary and percussive forces are simultaneously transmitted through these components to the tungsten carbide tipped ring bit and the drill bit at the bottom of the hole. Thus, using this method, it is possible to drill with the casing tubes and the drill steels simultaneously. Alternatively, it is also possible to drill with either the pilot drill bit alone, or with the casing tube bit alone. The former

method (i.e. with pilot bit) is useful when drilling through bedrock, while the latter method (i.e. with casing tubes) is frequently used for obtaining undisturbed soil samples. This is achieved with the aid of a soil sampler which is dropped down through the tubes to the ground ahead of the casing tubes.

The ODEX Underreaming Method[24, 25]

While the O.D. system overcame a number of problems and featured distinct advantages over conventional drilling methods, it had one major disadvantage. This was the requirement for expensive high-quality casing tubes capable of withstanding the high torque and percussive stresses imposed on them. There was, therefore, a need for a drilling method which could overcome the problems in a similar way to the O.D. method, yet which did not require the use of expensive casing tubes. The casing tubes could then be more economically left in the hole if this was desired.

An answer was found in the development of the ODEX drilling method which was first marketed by Atlas Copco and Sandvik early in 1970.

The manufacturers recommend its use in varying types of overburden strata namely sand, gravel, clay, and those with a high occurence of boulders.

Description of ODEX equipment

ODEX equipment may be used with either top hammer or down-the-hole hammer drilling equipment. Basically the ODEX drilling unit (Figure 45) consists of the following components: a pilot drill, a reamer and guide, and casing. (A special bit tube, is used for down-the-hole hammer work.)

The reamer is a circular member with an eccentrically located bearing hole which swings about an eccentric bearing shaft. Attached to the eccentric bearing shaft is the pilot drill. The pilot drill is fitted with a standard type rock-drilling bit. A threaded shaft projects upwards from the top end of the eccentric bearing shaft and screws into the 'guide'.

The guide consists of a strainer part at its lower end and a spiral guide part above. The

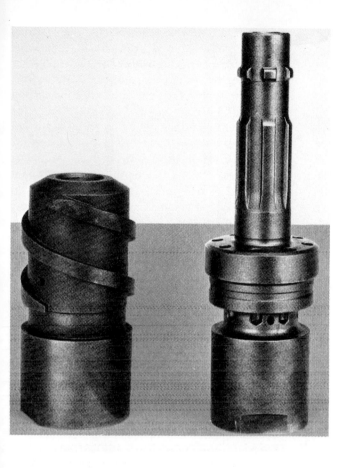

Figure 45. Components for ODEX underreamer drill. (Top left) guides for eccentric bits; (left) top hammer drill; (right) down-the-hole drill. (Courtesy Sandvik Coromant-Atlas Copco)

spiral lands on this part of the guide bear against the inner walls of the tube casing and steer the bit in the casing tube. The strainer part of the guide limits the size of the cuttings which are allowed to move upwards. Their upward transportation is facilitated by the provision of two flushing holes which clean out the area.

The flushing agent enters through the inner string of extension rods and emerges through the various flushing holes in the pilot bit, the reamer, and the guide. Cuttings are then carried upwards past the strainer part, through the spaces between the spiral lands on the guide, and up the annular hole between the drill string and the casing. They emerge through an elbow connection in the driving cap situated above a threaded adapter sleeve. (Conventional flushing agents such as air, water, or a liquid foaming mixture may be used.)

When the bit is being rotated in drilling, the reamer swings to its 'felled out' position which allows it to cut at a diameter sufficient to pass the casing (i.e. slightly larger than the casing tube). The reamer is held in this position while drilling by the driving contact between inclined mating surfaces on the pilot bit and reamer. On rotating in the opposite direction the reamer swings on the eccentric until it is fully retracted and comes to rest against a stop. With the reamer returned to the 'felled in' position the drill string and ODEX reaming equipment may be withdrawn up through the casing tubes. The reamer is fitted with two cemented carbide cutting inserts.

Top hammer equipment

In top hammer drilling (Figures 46 and 47) impact and rotation are transmitted to the drill string in the conventional way, but a portion of the impact energy is transferred via a 'shank adapter' to a 'driving cap' positioned above the casing tube. (Hole cuttings are discharged through a rubber hose connected to the driving cap.)

The string of casing tubes is driven down by means of impacts from above while the drill string is rotated anti-clockwise, or to the left. The impact mechanism and rotation unit form an integral part of the feed component.

When drilling commences the casing tubes drop into the hole by their own weight but gradually, as the hole deepens and increased frictional forces are encountered, it becomes necessary to drive the casing down by means of the shank adapter and the driving cap. In top hammer drilling the guide is screwed to the extension rods by means of a short extension rod. The rods of the drill string are jointed with a wing coupling which is in contact with the casing tube at roughly every other joint.

Down-the-hole equipment

In down-the-hole drilling (Figure 48) (see also section on down-the-hole percussion drills), similar ODEX components are used except that a special leading 'bit tube' is fitted to the casing and the pilot bit is threaded to the right. (Rotation of the drill string is to the right and, of course, the mechanism by which the ODEX reamer opens to its 'felled out' position is of the opposite hand to comply with this opposite direction of rotation.) The guide is equipped with a shaft for attachment to the down-the-hole drill and a connection thread for the pilot bit. An external shoulder on the guide transfers the impact energy of the down-the-hole drill to a corresponding shoulder on the inner surface of the bit tube. (The driving cap of the top hammer version is thus replaced by a flushing discharge which leads away the cuttings.)

Drill tubes are used instead of extension rods and only one wing coupling is required. It is jointed between the first drill tube and the top sub of the drill. As in top-hammer drilling the flushing agent (which in this case is the exhaust air from the down-the-hole drill) is forced out through channels in the guide, the reamer, and the pilot bit. The cuttings are carried upwards past the strainer part of the guide and emerge through an elbow which is fitted to the 'discharge head'.

(*Note*: For an ODEX 76 mm drill the inner diameter for the casing is 77 mm (3.03 in), the diameter of the reamed hole is 96 mm (3.78 in) and the diameter of the pilot bit hole is 70 mm (2.75 in). These sizes vary correspondingly for the larger-size drills, e.g. the ODEX 127 has a casing hole diameter of 128 mm (5 in),

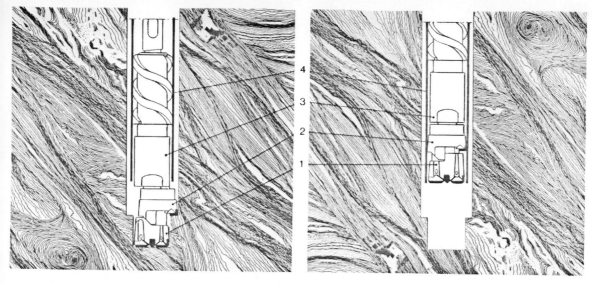

A. Drilling

1. Pilot bit   3. Guide
2. Reamer   4. Casing tube

B. Taking up the drilling equipment

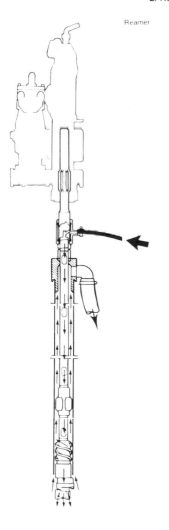

Reamer

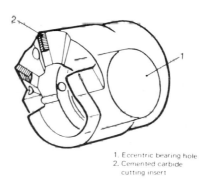

1. Eccentric bearing hole
2. Cemented carbide
cutting insert

Figure 46. ODEX top-hammer overburden drilling with eccentric underreamer. (Courtesy Copco-Sandvik Coromant.)

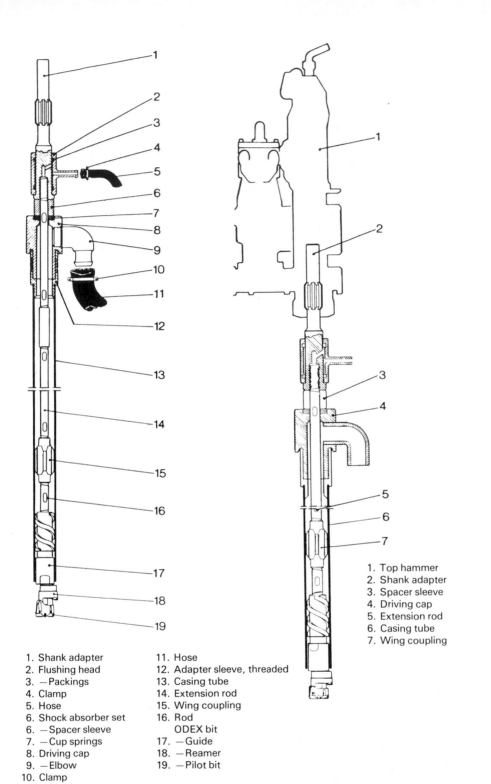

1. Shank adapter
2. Flushing head
3. —Packings
4. Clamp
5. Hose
6. Shock absorber set
6. —Spacer sleeve
7. —Cup springs
8. Driving cap
9. —Elbow
10. Clamp
11. Hose
12. Adapter sleeve, threaded
13. Casing tube
14. Extension rod
15. Wing coupling
16. Rod
  ODEX bit
17. —Guide
18. —Reamer
19. —Pilot bit

1. Top hammer
2. Shank adapter
3. Spacer sleeve
4. Driving cap
5. Extension rod
6. Casing tube
7. Wing coupling

1. Bit part
2. Eccentric part
3. Threaded part
4. Guide part
5. Strainer part
6. Reamer

Figure 47. ODEX top-hammer overburden drilling with eccentric underreamer (Courtesy Atlas Copco-Sandvik Coromant.)

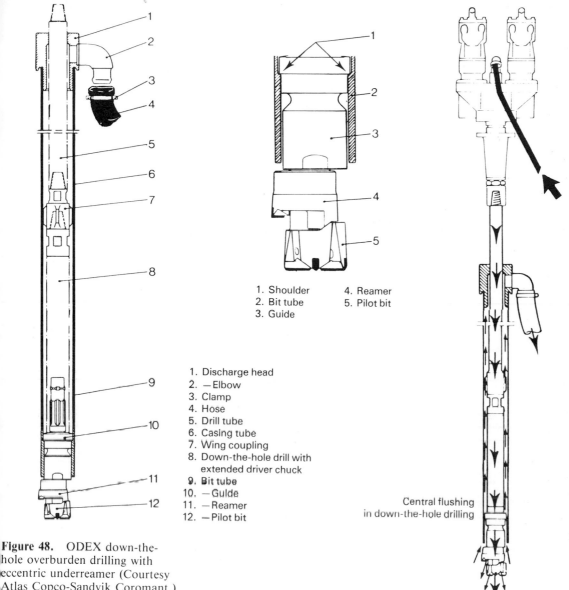

1. Shoulder   4. Reamer
2. Bit tube   5. Pilot bit
3. Guide

1. Discharge head
2. — Elbow
3. Clamp
4. Hose
5. Drill tube
6. Casing tube
7. Wing coupling
8. Down-the-hole drill with
    extended driver chuck
9. Bit tube
10. — Guide
11. — Reamer
12. — Pilot bit

Central flushing
in down-the-hole drilling

Figure 48. ODEX down-the-hole overburden drilling with eccentric underreamer (Courtesy Atlas Copco-Sandvik Coromant.)

a reamed hole diameter of 162 mm (6.4 in) and a pilot hole diameter of 110 mm (4.3 in).)

General developmental history of ODEX equipment

According to R. Sanden (Technical Representative of Sandvik Coromant Australasia) it was found, during the early developmental stages of the equipment, that in most cases it was essential for the hole cuttings to be finely ground so that they could be lifted up through the casing tubes. It was for this reason that the guide on ODEX

bits was designed to screen the cuttings in the column so that only the smallest cutting particles were allowed to pass. Generally speaking, foam flushing was found to be the most efficacious method of removing the cuttings. The foam separated the cuttings and lifted them out of the hole. It also lubricated and sealed the hole walls, thus making it possible to drill deeper than when only air or water flushing was used. Casing tubes also tended to slide more easily down into the hole and there was a marked reduction of wear on the equipment.

Sanden also comments that it is important to use the specific casing tube dimensions recommended by the manufacturers. If the wrong size is used flushing problems might be experienced.

During the initial developmental period of ODEX very thin-walled diamond drilling casing tubes were tested, but these proved unsuitable. A slightly heavier casing tube was then tried and found to be satisfactory. However, Sanden cautions that these are still not as robust as the old O.D.-type casing tubes which were strong enough to withstand fairly rough treatment without being damaged. Care was therefore required when feeding the new ODEX casing tubes to ensure satisfactory service.

Experience with ODEX equipment had indicated that the main components subject to wear were the eccentric reamer inserts and the steel around these inserts. Therefore, future developmental work on ODEX equipment will be aimed at attempts to 'protect the eccentric reamer bit from steel wear by using tungsten carbide buttons'.[24] Sanden also suggests that it may be possible in the future to use a plastic-type material casing tube and, in the case of water-well drilling, casing tubes with ready built-in screens may become available.

Acker Drill Company Casing Underreamer
(Figures 49 and 50)

An overburden casing underreamer produced by the Acker Drill Company of Scranton, Pennsylvania, U.S.A. features a pilot drill fitted with carbide cutter blades. About two-thirds up the pilot drill are twin underreamer blades. Normally, through soft ground, these blades reside in a retracted position, but when boulders or difficult ground is encountered the downward pressure on the drill string is increased, causing the underreamer blades to expand and thus cut the necessary clearance for the casing to follow.

An adapter is provided between the 'drill spindle' and casing to rotate the casing which is protected by a carbide casing shoe. A clutch fitted on top of the spindle allows the operator

Figure 49. Acker casing underreamer fitted with tri-cone roller rock bit. (Courtesy Acker Drill Co. Inc.)

62

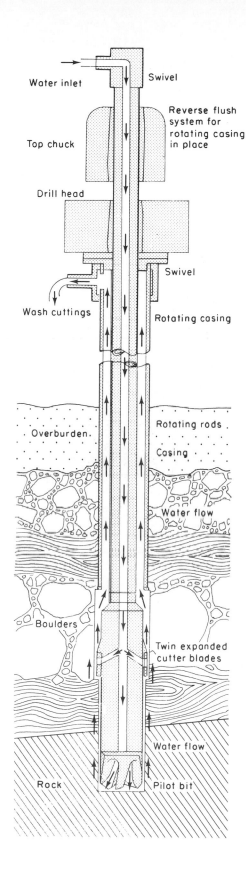

Water inlet

Swivel

Reverse flush system for rotating casing in place

Top chuck

Drill head

Swivel

Wash cuttings

Rotating casing

Overburden.

Rotating rods.

Casing

Water flow

Boulders

Twin expanded cutter blades

Water flow

Rock

Pilot bit

to vary the distance between the back of the casing and the underreamer to maintain the flow of the flushing medium.

The internal drill rod and the casing underreamer assembly and pilot bit are rotated by a separate top drive manual or hydraulic chuck.

The flushing medium enters through an inlet at the top swivel, passes down the centre of the drill string, and emerges through flush holes in the pilot bit. The flushing medium, together with cuttings from the casing underreamer and the carbide cutter blades of the pilot bit, is then forced upwards past the underreamer. It flows in the annular space between the drill string and the interior walls of the casing, emerging through the side swivel casing adapter assembly above the surface.

According to the manufacturers, the drill may be fitted with tri-cone roller rock bits or down-the-hole hammer components.

The Calyx Core Drill[26]

When the diamond drill was originally developed the price of diamonds was relatively low, but as time passed these stones became more and more expensive until eventually the stage was reached where the price of the diamond bit alone equalled or even exceeded the cost of the entire drilling equipment. The use of borts alleviated the situation to some extent, particularly after mechanized setting was introduced, but they were only of use at that time in softer formations and the carbon stone was still found to be necessary for very hard rock.

Towards the end of the nineteenth century an Australian named Francis H. Davis developed a drilling machine which was somewhat similar in concept to the diamond drill, except that it used a tubular bit fitted with forged steel cutters. However, although it penetrated soft ground fairly well it could not cope with very hard rock.

This drill was taken by Davis to the United States in 1893, and was patented in the U.S.A. under No. 642,587 on 12 October 1898. In 1900 Davis found an unusual answer to his problem

Figure 50. Acker casing underreamer for overburden drilling. (Courtesy Acker Drill Co. Inc.)

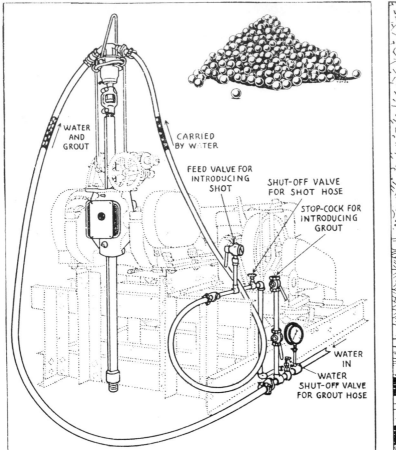

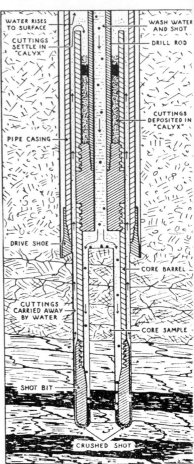

Figure 51. Illustration of 'Calyx' method of delivering grout, shot, and wash-water to the drill rod. (Courtesy Ingersoll-Rand Ltd.)

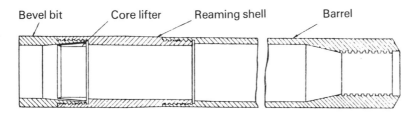

Equipped with bevel bit, reaming shell, and core lifter

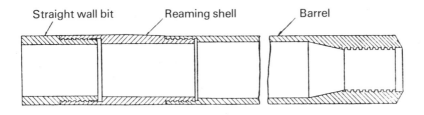

Equipped with straight wall bit, reaming shell and no core lifter

Figure 52. Single tube core barrels for diamond core drills. (Courtesy Ingersoll-Rand Ltd.)

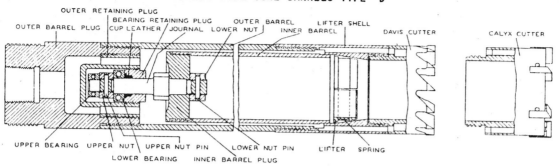

DOUBLE TUBE CORE BARRELS—TYPE "D"

Figure 53. Double tube core barrel for use with small-diameter 'Calyx' core drills. (Courtesy Ingersoll-Rand Ltd.)

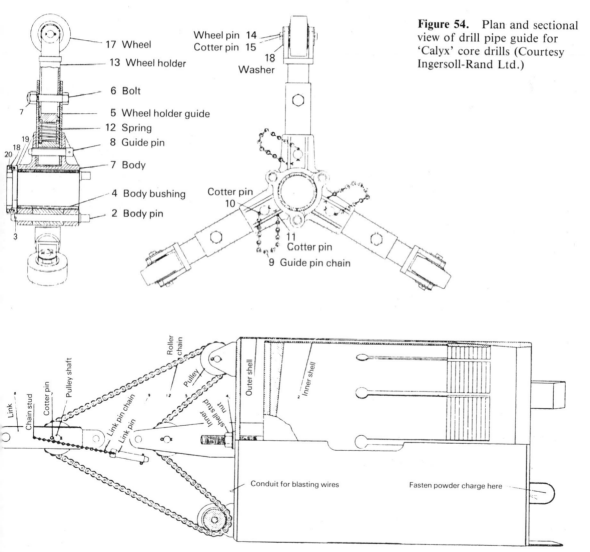

Figure 54. Plan and sectional view of drill pipe guide for 'Calyx' core drills (Courtesy Ingersoll-Rand Ltd.)

17 Wheel
13 Wheel holder
6 Bolt
5 Wheel holder guide
12 Spring
8 Guide pin
7 Body
4 Body bushing
2 Body pin

Wheel pin 14
Cotter pin 15
18 Washer
Cotter pin 10
11 Cotter pin
9 Guide pin chain

Figure 55. 'Calyx' core drill core lifter for removing cores from 90 cm and over. (Courtesy Ingersoll-Rand Ltd.)

65

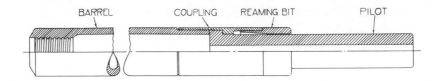

Figure 56. Diamond core drill reaming barrel. (Courtesy Ingersoll-Rand Ltd.)

of drilling through hard rock. He inserted chilled shot with the flushing water. The chilled shot travelled down the hollow rods to the bottom of the shaft. Here it was crushed into small abrasive particles by the rotary action of the tubular bit. These particles of abrasive metal were found to wear away the hard rock almost as effectively as had the diamonds, and the Calyxite core drill was born. To develop his drill Davis formed the Davis Calyx Drill Company. The patent concerning the latter method of drilling was filed by Davis on 8 August 1901 under U.S. Patent No. 694,534 and was granted on the 4 March 1902.

Later the name of the drill was changed to 'Calyx'. Apparently this name was adopted because, like a flower which traps moisture in its calyx, the Calyx core drill traps the sludge which forms during the drilling process in the annular space which is located in the barrel above the drill bit (Figure 51). This area is known as the 'calyx'. The flushing water forces the cuttings and sludge up through the annular space around the bit until it reaches the top of the calyx. At this point the slightly decreased velocity of the water allows the sludge, etc. to settle in the calyx, while the water continues its journey to the top of the hole. This trapped sludge serves as a reverse record of the material being penetrated, and each time the core is raised and removed, the sludge is also extracted. If for some reason the core were to become lost, then the sludge could be used as a record of the type of strata being penetrated.

In 1905 the Ingersoll-Rand Company bought the patent rights of the Calyx drill when it took over the Davis Calyx Drill Company. During the succeeding years the drill was manufactured by the Ingersoll-Rand Company which also continued with its developmental work (Figures 52 to 56).

The advent of the Calyx drill made possible the relatively economical boring of cores ranging in diameter from 6.3 cm to approximately 183 cm (2.4 in to 6 ft). It was widely used in many parts of the world for a variety of purposes which included pavement testing, exploratory holes for dam sites,* etc. The slate quarries of Pennsylvania were amongst the first to make use of this novel type of core drill.

When slate is extracted from a quarry it is normally removed in blocks. These blocks are parted from the main bed by means of channels which are drilled into the slate with percussion drills. However, the disadvantage of this is that the percussion drill tends to shatter the slate so that out of each block a fair proportion of material is lost. To overcome this problem it was decided to use the wire saw method which had apparently been tried with success on the Continent.

Basically a wire saw consists of an endless length of steel cable measuring approximately 6.3 mm (¼ in) in diameter. The wire is pulled tight across the slate by means of two tension posts which are installed at each end of the block being cut. These posts are fitted with sheaves. The wire is then run over the pulleys and across the slate while abrasive sand is fed into the groove. To erect the tension posts it was necessary to provide a hole at each end of the block and, because the conventional method of making these holes (i.e. by drilling and blasting) would have had the same shattering effect on the slate as the percussion drills, thus defeating the purpose of using the wire saw, a Calyx drill was tried. The success of this type of drill, where even the cores extracted for the posts were found to contain good usable material, led to its extensive adoption in the majority of Pennsylvania's slate quarries (Figures 57 to 62).

*The Calyx drill was considered to be extremely valuable because it allowed for a personal 'down-the-hole' inspection of dam foundations, etc.

Figure 57. (top left) Class 'PT' 'Calyx' drill taking a sample core from a concrete highway. (Courtesy Ingersoll-Rand Ltd.)

Figure 58. (top right) Long small cores made by a 'Calyx' drill in Cuba. The machine is owned by the Dept. of Public Works of the Cuban government. (Courtesy Ingersoll-Rand Ltd.)

Figure 59. (bottom left) 'Calyx' core drill shot bit (Courtesy Ingersoll-Rand Ltd.)

Figure 60. (bottom right) 'Calyx' bit for drilling in soft and moderately hard materials in which 'Calyx' or crushed steel would not be suitable. Tooth cutters are used. (Courtesy Ingersoll-Rand Ltd).

Figure 61. Tension post holes being drilled with 910 mm (36 in) 'Calyx' core drills for a wire-saw in the Phoenix slate quarry near Wind Gap, Pa., U.S.A. (Courtesy Ingersoll-Rand Ltd.)

Figure 62. 36-in sample cores being taken from the Norris dam as a final check of the grouting operations. (Courtesy Ingersoll-Rand Ltd.)

Basically the Calyx drill consists of:

(a) *a drilling or driving head* (the chuck or driving sleeve of which is rotated by an electric motor — usually about 30kW (40 hp) — via a series of bevel gears and chain drive, etc.);

(b) *the drill rods, couplings, drill rod guides, starting barrel, coring barrel, etc.*;

(c) *the hoisting gear* (i.e. winch, derrick, wire ropes, etc.);

(d) *the core lifter.*

Before the core drill can be used it is necessary to go down through any overburden until bedrock is reached. (For example, when a 91 cm (3 ft) diameter manway was sunk at the Phillipsburg, New Jersey, plant of the Ingersoll-Rand Company, the overburden penetrated before bedrock was reached was

approximately 2 m (7 ft). A 94 cm (3 ft 1 in) i.d. cylindrical form of concrete was then made on this ledge. The concrete extended up through the overburden to about 30 cm (1 ft) above the surface ground. This not only served as casing to prevent the collapse of the overburden into the hole but was used for starting the Calyx drill.)

Once bedrock is reached, drilling of the rock core may be commenced, the core barrel at the bottom of the shaft being rotated by the motor at the surface, via the string of drill rods which extends down the hole. As the hole gets deeper and the drill string becomes longer, there is a tendency for the rods to vibrate and whip about in the hole. To control this whipping action it is necessary to insert drill rod guides or spiders. The drill rods pass through the hub of the spider and support is provided by three arms which extend radially against the walls of the shaft. These arms are spring-loaded and each arm has a rubber wheel at the end.

The chilled shot used may vary in size depending upon the diameter of the core being drilled, but generally is about 2.4 mm (0.09 in) in diameter. It is made by atomizing molten iron or steel and then suddenly chilling it. The resulting material becomes hard enough to scratch glass. Under pressure of the shot bit this material breaks into small sharply edged particles which by rotation and contact mill away the rock beneath the bit and, to a somewhat lesser extent, the bit itself.

Operators of the Calyx shot drill learnt with experience how much shot to feed per minute. This tended to vary according to the diameter of the coring bit and the type of rock being cut. If too much shot was fed down the hole the coring bit would rotate as though it were mounted on ball-bearings, while too little shot could cause the bit to wear away more rapidly than was normal, without making any appreciable progress down the hole.

After a core is cut (usually about 150 to 180 cm (5 to 6 ft) long) the drill string and core barrel are lifted out of the hole and the core lifter is lowered down the hole.

A core lifter is considered necessary for all large-diameter cores. However, a double tube Calyx core barrel was available for use when coring small-diameter cores. These barrels came equipped with their own core lifters.

The Core Lifter

The core lifter (see Figure 55) which is normally about 180 cm (6 ft) long consists of an inner and outer shell. The outer shell is tapered on its inside, while the inner shell has slots cut into it from the bottom to approximately half-way up, leaving a series of flexible fingers surrounding the core. The inner shell is capable of sliding inside the outer shell, but when the two shells are being lowered into the hole they remain attached. When the inner shell is resting around the core near the bottom of the hole, it is detached by the drill operator from its clevis so that only the outside shell remains supported by the hoist. The outer shell is then raised and, as

it travels upwards the wedge ring at its lower end makes contact with the flexible tapered fingers of the inner shell. This causes the fingers on the inner shell to grip the core tightly so that it may be raised.

The core is broken free by the firing of a small charge of explosives attached to the bottom of the core lifter, or by means of wedges driven into the slot between the core lifter and the shaft wall.

In an attempt to preserve some of the groove at the bottom of the hole to facilitate the commencement of the following drilling cycle, the outer shell of the core lifter is fitted with three 23 cm (9 in) long legs. These are designed to keep the explosive charge away from the bottom of the groove during firing.

If the bottom of the groove is damaged by the explosive charge, then it is necessary to lay sandbags at the bottom of the shaft before the next core is drilled. This is to contain the shot. Sandbags are also used at the commencement of the drilling operation until an adequate groove is formed that will contain the shot.

The following patents were granted to F. H. Davis and his associates:

Australia
(before federation of the States)
Victoria 18,905/02 — boring holes. N.S.W. 11,705/1902, N.S.W. 13,488/1902, W.A. 3725/1902.

After federation
13,265/1908 — shot feeding, etc.

U.S. Patent Specifications
642,587 (F. H. Davis) — apparatus for boring. 12 October 1898. 693,117 (F. H. Davis and C. A. Terry) — bit. Filed 22 September 1900. 710,438 (F. H. Davis and C.A. Terry) — rotary drilling machine. Filed 6 June 1901. 694,534 (F. H. Davis) — process of boring holes in rock, etc. Filed 8 August 1901. 790,330 (C. A. Terry/Davis Calyx Drill Co.) — core drill. Filed 10 June 1904. 790,331 (C. A. Terry/Davis Calyx Drill Co.) — core-drill apparatus. Filed 25 August 1904. 902,564 (F. H. Davis) — shot-feeding means for core-drilling apparatus. Filed 26 April 1907. 694,535 (F. H. Davis) — apparatus for boring holes in rock. Filed 7 September 1901.

J.B. Newsom Shot Drill[27, 28]

Early History

For many years the Idaho Maryland Mines Corporation of California had used conventional methods (i.e. blasting and mucking) to sink vertical shafts in its mine. However, because of the inherent disadvantages of these methods, namely shattered uneven walls which frequently required support, etc. the problem was discussed with J. B. Newsom, consulting (mining) engineer, of California. Newsom had had previous experience using large-diameter core drills, but felt that although these drills overcame some of the problems associated with conventional methods of shaft sinking, they had other disadvantages which were equally important and are discussed later in this section. As a result Newsom proposed that he should design and supervise the building of an entirely new drill with the assistance of the Idaho Maryland Mines Corporation. This corporation agreed to make available an experimental site at Grass Valley, Cal., away from the main surface plant where Newsom could try his drill without interference. The proposed shaft would be close enough to the mine workings on the 152 m (500 ft), 229 m (750 ft), and 335 m (1100 ft) levels to permit connections to be made between them when the shaft was finished.

The drill was built and work was begun in 1934. As was natural with new equipment, Newsom encountered many difficulties and experienced a great number of frustrating delays before he was satisfied that he had overcome the majority of the problems associated with his prototype unit. By 1936 the Newsom drill had successfully completed boring a 1.52 (5 ft) diameter shaft which was some 335 m (1100 ft) deep. Connection was then made with the other mine works. All in all the shaft had taken some 21 months to drill.

In 1937 Pickands Mather & Co. of Duluth, Minn., decided that an additional ventilation shaft connecting the surface with their underground workings was needed for their Zenith iron ore mine, which was located in the Lake Superior ranges. The success of the Idaho

experiment prompted them to enter into an agreement with Newsom and the Idaho Maryland Mines Corporation for the drilling of a 366 m (1200 ft) deep, 1.68 m (5 ft 6 in) diameter hole from the top of the mine to the fourteenth level. As a result of the experience gained by Newsom on the previous job, the core-drilling equipment used in Minnesota was completely redesigned in the Duluth office of Pickands and it carried many improvements which undoubtedly contributed towards the excellent progress made on the new project.

Work on the Minnesota shaft was commenced on 1 April 1938 and by August the hole was 212 m (700 ft) deep. It was finally completed in October 1938, i.e. a little under seven months from the time of commencement and, at completion, was some 366 m (1200 ft) deep.

The Drill

For a long time Newsom had felt that the most significant disadvantage of large-diameter core drills, similar to the Calyx drill, was that such drills were operated from the surface. In other words the large core bit measuring perhaps 1 m (3 ft) or more in diameter was rotated by the drive and head via a long string of drill rods which often measured several hundred metres in length. Apart from the problem of drill whip, the larger the diameter of the core the greater was its weight which also increased with the length of the core. This factor dictated that the cores be kept relatively short which in turn meant frequent raising and lowering of the equipment (i.e. the core barrel, drill rods, and drill rod guides) each time the core was lifted from the hole. After the core was extracted from the core lifter all the equipment had to be replaced down the hole, a long and time-consuming operation.

It was to overcome these disadvantages that Newsom designed his new drill (Figure 63), which in essence consisted of a conventional type coring bit that used chilled shot in much the same manner as did the Calyx drill. However, instead of the power being provided at the collar of the hole, Newsom built a special cabin which carried an operator and the power plant. The cabin was lowered into the

70

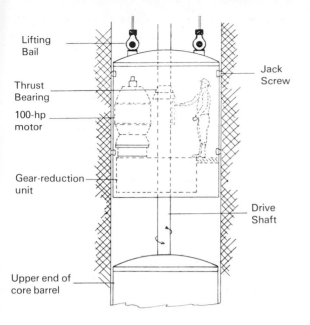

Lifting Bail

Jack Screw

Thrust Bearing

100-hp motor

Gear-reduction unit

Drive Shaft

Upper end of core barrel

Figure 63. Drawing depicting the Newsom core drill

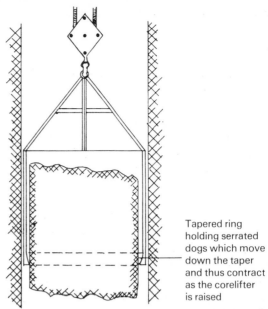

Tapered ring holding serrated dogs which move down the taper and thus contract as the corelifter is raised

Figure 64. Drawing depicting the Newsom core lifter

hole and was positioned immediately above the core barrel to which it was attached by a drive shaft. Eight screw jacks which thrust against the walls of the shaft prevented the cabin from turning in the bore during operation. The core barrel was rotated by a 75 kW (100 hp) motor located in the cabin via a series of reduction gears which turned the central drive shaft. Chilled shot was dropped on to the top of the core barrel from where it found its way by gravity to the bottom of the hole.

A movable thrust bearing which operated against a collar on the main drive shaft controlled the downward thrust of the shoe. The thrust bearing could be raised or lowered on the spline of the shaft by the operator in the cabin. This was done by means of a hand operated winch.

The drilling cycle consisted of lowering the drilling cabin and core barrel to the bottom of the hole. Then the operator was lowered into the cabin. The operator braced the cabin against the walls with the screw jacks, connected the power cable, and brought the air hose into the cabin. Drilling was commenced after the power had been switched on by the surface operator. This continued until the core barrel was full unless the operator was forced to

stop for any other reason such as a mechanical failure or a rock slip, etc. When drilling had been completed the power cable was disconnected and, together with the air hose, was pulled up to the surface. The jacks were then released and the operator hoisted up followed by the drilling machine itself. When the hole had been pumped dry and cleaned, the operator descended again and broke the core by driving in wedges between it and the shaft wall. When the core had been broken free the operator was lifted to the surface and the core puller lowered until it rested around the core. The core puller (Figure 64) consisted of a number of serrated dogs arranged around the inner circumference of a tapered ring which was suspended from a clevis. When the ring was lowered over the core the dogs would slide up in the ring to the widest part of the taper, but when the ring was raised the dogs would move down the taper and thus contract around the core. The core would then be raised to the surface.

Newsom suggested that a number of these manoeuvres could be eliminated if the core barrel was fitted with its own core lifter, and the author understands that towards the end of the Grass Valley project in California, Newsom

71

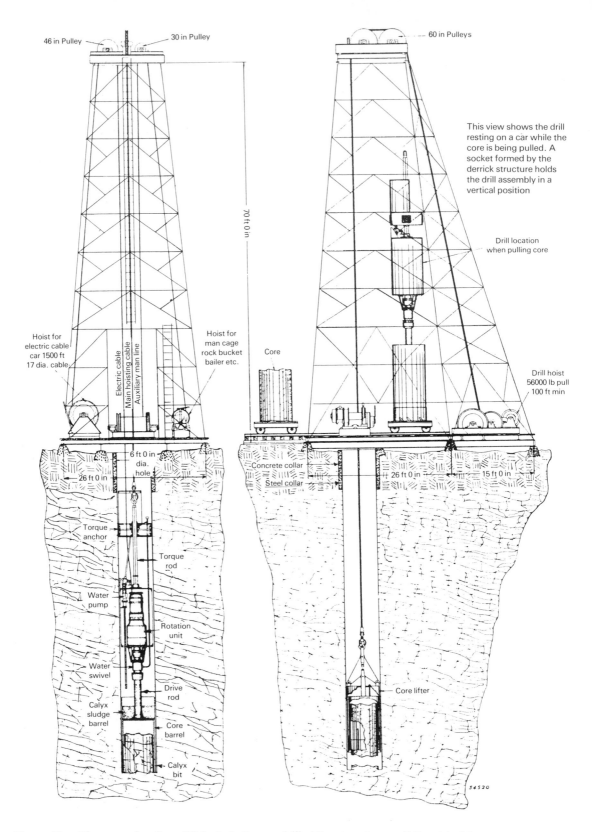

Figure 65. Diagram of rodless 'Calyx' shaft core drill. (Courtesy Ingersoll-Rand Ltd.)

72

did in fact have a core puller built inside the core barrel. However, although it worked fairly well, he felt that several modifications would have improved its performance. In addition he expressed the view that if the hoist were provided with sufficient lifting capacity the core could probably be pulled free without the necessity for either an explosive charge or wedges.

The Rodless 'Calyx' Core Drill

No doubt also recognizing the problems associated with large-diameter deep-hole core drilling, the Ingersoll-Rand Company introduced their 'Rodless Calyx bore unit' (Figure 65). As will be seen in the diagram, the power unit follows the core barrel down the hole (sans operator) and the bit is rotated by this unit via a drive rod which connects the two sections together. The motor is prevented from turning in the bore by a torque rod mounted inside an expanding torque anchor which braced against the walls of the shaft. Power is supplied by cables from the surface.

Rolling Cone Cutter Core Drill[29]

In 1953 the Hughes Tool Company was asked to design and build a 193 cm (76 in) diameter rolling cone cutter core drill (Figure 66) for the Zeni Corporation of Morgantown, West Virginia, which specialized in shaft-sinking.

When the prototype (Model X-153) was ready it was put to work drilling a mine shaft through fairly hard sandstone.

The X-153 unit was similar in concept to Newsom's drill in that the operators were down the hole with the drill. The operator's platform was located immediately above the core barrel and was connected to the core barrel by the drive shaft. The drive shaft was rotated by hydraulic motors located on the operator's platform. The entire drill was suspended by hoist and was prevented from rotating in the bore by two horizontally placed anchor jacks which thrust against the walls of the shaft just above the operator's platform. Another vertical jack situated immediately beneath the horizontal jacks provided downward thrust on the core barrel and the cutters. The lower end

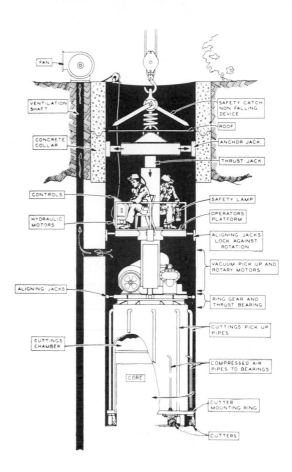

Figure 66. Zeni down-the-hole coring machine. (Courtesy Hughes Tool Co.)

of the core barrel was fitted with six rolling (individual-cone) cutters mounted on a conventional tri-cone type bit bearing.

No mention is made in the article on the drill written by J. H. Allen[29] of how the core was removed after it had been cut, but presumably this was raised in the normal way with a separate core lifter after the drilling machine had been hoisted to the surface.

Penetration rates of up to 1 m/hour (3 ft/hour) were obtained with the prototype unit on the first shaft. By 1959 the machine had bored some 14 mine shafts of varying depths up to 152 m (500 ft). Later a new machine (Figure 67) was developed which enlarged the full face of the hole to 193 cm (6 ft 4 in) after a 31 cm (1 ft) pilot hole had first been drilled through to the level below. The cuttings from the reamer bit were either flushed or blown down the pilot

73

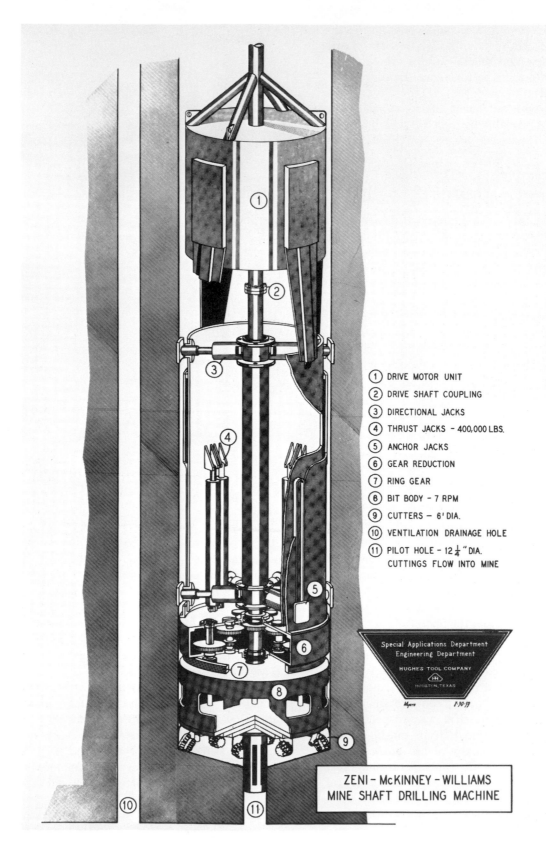

1. DRIVE MOTOR UNIT
2. DRIVE SHAFT COUPLING
3. DIRECTIONAL JACKS
4. THRUST JACKS – 400,000 LBS.
5. ANCHOR JACKS
6. GEAR REDUCTION
7. RING GEAR
8. BIT BODY – 7 RPM
9. CUTTERS – 6' DIA.
10. VENTILATION DRAINAGE HOLE
11. PILOT HOLE – $12\frac{1}{4}$" DIA.
    CUTTINGS FLOW INTO MINE

Special Applications Department
Engineering Department

HUGHES TOOL COMPANY

HOUSTON, TEXAS

Myers   7-30-59

ZENI – McKINNEY – WILLIAMS
MINE SHAFT DRILLING MACHINE

Figure 68. The 'Calyx' shot drill has been superseded by the core barrel, shown above fitted with tungsten carbide insert drag cutters (Courtesy Hughes Tool Co.)

hole to the intersecting lower level. This machine eliminated the need for the extraction of the core and was to some extent the fore-runner of the modern raise drill which is now quite widely used for this type of work (see section on raise drills).

After the advent of the first large-diameter rolling cutter core-drilling machine described above, a few large-diameter core drills were produced with carbide insert drag bit cutters attached to the core barrel. These units were mainly designed for use in the softer formations. Where large-diameter holes have been required in harder strata, full-bore drilling machines fitted with either rolling or disc cutters have been used (Figure 68 to 70).

Ingersoll-Rand have now ceased to manufacture the Calyx core drill and it is of interest to note that nearly all exploratory drilling work nowadays is accomplished with small-diameter core drills.

Figure 67. Zeni-McKinney-Williams full-bore down-the-hole machine (Courtesy Hughes Tool Co.)

Down-the-hole Percussion Drills

After the introduction of the rotary flush drill the previously extensive use of the vertical percussion drill (both rod and cable tool types) declined. The need for vertical percussion drilling had not, of course, disappeared entirely, as there still remained areas of application where this type of action gave penetration rates superior to the rotary drill.

While use of the rotary flush drill spread, developmental work on percussive drills continued and subsequently resulted in the successful introduction of 'down-the-hole' hammer drills, which widened the scope of application for vertical percussive units.

In 1952 the first down-the-hole percussion drill[30] (Model M628) designed by Andre Stenuick, with a drilling diameter of 10 cm (4 in) was produced in Belgium. Later, arrangements were made to sell these drills in England through the Halifax Tool Company.

The successful application of the Halco-Stenuick units encouraged other drill manufacturers to produce their own versions of this type of drill and in 1954 the Belgian

Figure 69. In the harder materials full-bore bits fitted with toothed roller or carbide insert cutters are now generally used for large diameter holes (Courtesy Hughes Tool Co.)

Figure 70. Hughes Tool Company drilling bit fitted with milled tooth rolling cutters. (Courtesy Hughes Tool Co.)

machine was followed by the introduction of Ingersoll-Rand's first unit (Model DHD400) which was capable of drilling 17 cm (6½ in) diameter holes.

Down-hole percussive drills (Figures 71 and 72) consist primarily of a string of drill rods attached to a percussion drill bit at the bottom of the hole. However, instead of the mechanical energy being transmitted to the bit from the surface of the hole via the entire string of rods, the power unit is located 'down the hole' immediately above the bit, thus eliminating energy loss through friction of the drill string against the shaft walls and through the threaded connections of the drill rods. A rotary rig is used to operate the drill because rotation of the bit is effected through the drill pipes but,

in effect, the drill uses basically the same components as a conventional 'compressed-air pavement breaker' or 'percussion drill'. That is, a large piston or hammer *within the drill pipe* is reciprocated up and down by means of compressed air so that it impacts against the head of the tool at the bottom of the hole. Use is made of the exhausted air to keep the hole clean and remove the debris.

Advantages

Penetration rates remain constant irrespective of whether the unit is operating at a depth of 20 or 100 m (65 or 330 ft). This is because of the absence of drill string losses.

Down-hole drills are also a great deal quieter

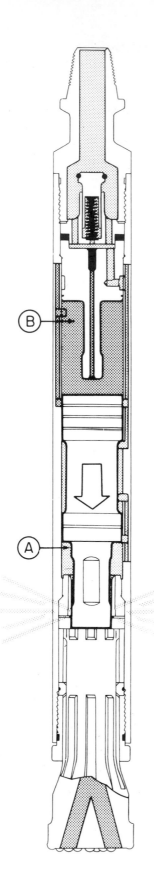

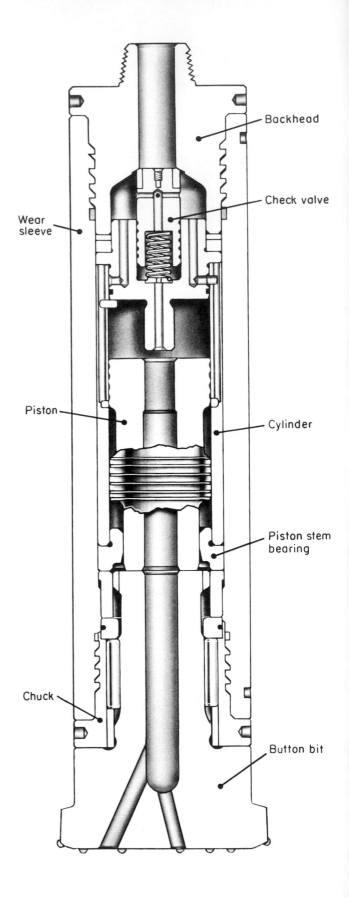

Backhead

Check valve

Wear
sleeve

Piston

Cylinder

Piston stem
bearing

Chuck

Button bit

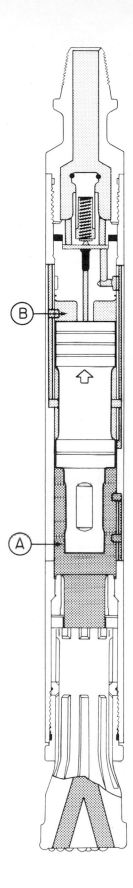

than conventional percussive drills because the noise of the piston striking the bit (i.e. steel on steel) is absorbed by the surrounding ground of the hole.

Because the down-hole drill is a rigid member behind the bit there is very little tendency for the bit to 'run-off' or deviate as is the case with conventional percussive units and an extremely straight hole is therefore drilled with these machines.

Disadvantages

Disadvantages associated with this type of drill are that either a rockfall or the formation of a mud collar above the drill bit could cause the string to become jammed to the point where withdrawal of the string becomes extremely difficult or, occasionally, impossible. In the latter event the operator would be forced to abandon it.

In addition, because the machine is taken down the hole instead of being positioned on

Figure 71. Sectioned diagram of Ingersoll-Rand down-the-hole percussive drill. (Courtesy Ingersoll-Rand Ltd.)

Figure 72. Sectioned diagram of Ingersoll-Rand down-the-hole percussive drill. (1) The drilling cycle commences when air under pressure opens the check valve and enters the drill. As the bit is fed on to the rock, the shank moves upwards until it seats under the stem-bearing washer. This pushes the piston upward, uncovering the bottom ports, and allowing live air to enter chamber A; (2) the air, acting on the undersurface of the piston flange, rapidly accelerates the piston upward. Continued piston travel cuts off the air supply to chamber A, but expansion and inertia cause the piston to continue its movement towards the top of the stroke; (3) The ports supplying chamber B are now opened, admitting live air above the piston. Increasing pressure stops its upward travel and stores the energy for the power stroke. In chamber A, the neck of the piston has unseated in the stem bearing and this air is exhausted out the face of the bit; (4) At this stage the energy generated by the piston mass and the power of expanding compressed air is released, forcing the piston down on to the striking face of the bit. When the top of the piston pulls away from the air distributor stem, the air in chamber B is vented down the centre of the piston out the face of the bit. The exhaust air carries the broken rock out of the hole. (Courtesy Ingersoll-Rand Ltd.)

79

the surface, where no limitation is imposed on its size or shape, the down-the-hole drill is subject to several restricting factors, all of which affect its ultimate performance or drilling speed.

An annular space must be provided around the outside of the cylinder within which the piston operates, to allow room for the cuttings to be carried to the surface by the exhaust air. Space is also occupied by the cylinder walls which obviously need to be fairly robust if they are to withstand the stresses imposed upon them. The ratio of piston diameter to drill bit diameter decreases as the hole becomes smaller thus, other factors remaining constant, the power of the hammer is reduced relative to bit size as the hole decreases in diameter. For this reason very few manufacturers build down-the-hole drills measuring less than about 75 mm (3 in) in diameter and the general tendency has been to increase both the diameter of the drill and the fluid pressure used, the 100 to 150 mm (4 to 6 in) diameter range being the most popular size.

As the exhaust air is also used to remove the cuttings from the hole, a significant fringe benefit derived from increasing the air pressure has been improved cuttings removal which, in turn, has resulted in better bit wear (i.e. less abrasion of the bit by grinding, etc.). However, J. Hodge*[11] warns that excessive velocities may cause hammer erosion.

New Developments

Ingersoll-Rand who have made significant contributions towards the development of this type of drill have recently introduced a 61 cm (2 ft) diameter down-the-hole drill for work on the Alaska Pipeline project, where holes were needed for supports for the elevated pipeline. Originally the Alyeska Pipeline Service Company tried several machines of various types, including rotary and screw-type auger drills. However, the frozen river silt which at times may develop the characteristics of tough

concrete as well as the bedrock encountered under the silt, mostly defied their efforts and penetration rates were generally extremely poor. As a result the company began looking for an alternative method of drilling. Ingersoll-Rand were approached and an order for about six 62 cm (2 ft) diameter down-the-hole percussion drills was placed. This order constituted something of a risk in that so far as was known the largest down-the-hole drill previously supplied by any manufacturer was a unit produced by the Mission Manufacturing Company (Model A100-10) which used 38.7 cm (15¼ in) solid head button bits. However, the new drills proved to be so successful that Ingersoll-Rand was given an order for another 8 machines, making a total of 14 units altogether. The Alaskan units are reputed to have penetrated the permafrost and rock at the rate of 61 cm/min (2 ft/min) compared with less than 30 cm/min (1 ft/min) obtained by the rotary and other units tried there.

Since then Ingersoll-Rand have produced a new 76 cm (2½ ft) diameter percussion drill, the IR DHD-130 model, and various other contractors, etc. are now, despite the high cost of these units compared with the conventional rotary machine, gradually beginning to appreciate their advantages. The South African Chamber of Mines, for instance, has ordered a 76 cm diameter unit for mine rescue work, and three machines were recently delivered to the New Jersey Drilling Company (U.S.A.). According to an article which appeared in *Construction* entitled 'A rock drill with twice the speed', Johnson Brothers Drilling Company, Mableton (U.S.A.) reported that they were penetrating limestone formations at the rate of 3 to 7.6 m/hour (10 to 25 ft/hour) with the new drill. Previous penetration rates through this type of formation using a rotary drill had been as low as 60 cm/hour (2 ft/hr). The units were being used to drill foundation holes for an office building block in Knoxville, Tenn. Before the purchase of the 76 cm (30 in) diameter units the Johnson Brothers Drilling Company considered using shot core drills or one of Ingersoll-Rand's magnum drills (which is a cluster of smaller down-the-hole drills). These units were, however, rejected in favour of the larger-diameter percussion drill, since

---

*During the mid-1960s Holman Brothers Limited of Cornwall, England, carried out extensive research work on the design of D.T.H. drills with the result that these drills were then able to operate at much greater pressures than had hitherto been possible.

estimated drilling rates ranging from 90 cm/day (3 ft/day) to a possible 30 cm/hour (1 ft/hour) did not compare favourably with reported drilling rates already achieved with the new large-diameter percussion drill.

Various other buyers have since shown their interest and, if present trends continue there may well be a steady and growing market for this type of drill for specific purposes. According to the manufacturers, the 76 cm (30 in) diameter drill should be supplied with at least 75 cu m/min (2700 cu ft/min) of air at 9 kg/cm^2 (125 lb/in^2), if maximum efficiency is to be obtained. Other advantages claimed by the manufacturers are:

(a) that because only 10 to 15 rpm at low torque levels are required for rotation of the button bit, the mounting is not called upon to provide large amounts of torque reaction and can therefore be of lighter construction than those used for other drilling methods;
(b) that minimal hole deviation is encountered from the action of the hammer which impacts directly against the button bit at the rate of 700 blows per minute;
(c) that as there is only one moving part, namely the piston, the drill should prove to be reliable and maintenance costs, etc. should be minimal.

Ingersoll-Rand also suggest that the drill should be suitable for pile-driving if the unit were mounted on a crane, as its own weight of 4.53 t would provide ample down-feed force.

Down-the-hole drills were used primarily for drilling from the surface. Their successful application in this field led to the development of a down-the-hole, or in-the-hole as it is also called, drill, which was specially adapted for operation in the confined spaces of a mine.

Apart from raises, drain and ventilation holes, etc. the adaptation of in-the-hole drills for underground work has also made possible the economic development of long blast-hole stopes.

This method of mining (i.e. long blast-hole stope or raise work) is, of course, by no means new — the International Nickel Company of Canada Limited (I.N.C.O.), Ontario Division, began using long blast-hole layouts with the new heavy duty percussive drifter or fan drills which were introduced during the late 1930s, and 10 cm (4 in) diameter in-the-hole percussive drills were tested by I.N.C.O. during the early 1960s. However, at that time the drills were not considered to be particularly successful as, apart from problems associated with hole deviation, the modern button bits had not yet been developed and the conventional tungsten carbide cross-bits proved uneconomical as they tended to wear too rapidly, requiring constant sharpening.

In 1971, using conventional rotary-type drills with roller-cone carbide-insert bits, I.N.C.O. drilled some large-diameter trial holes ranging in size from 17 to 28 cm (6¾ to 11 in).

These large holes proved successful and in 1972 two in-the-hole surface drills were obtained from different manufacturers. The units were tested at the Clarabelle open pit in rock with compressive strengths of up to 40,000 lb/in^2 (280 MPa). Eight vertical holes with nominal diameters of 6 in (15 cm) were drilled — penetration rates being about 10 to 15 ft/hour (3 to 5 m/hour). There was less than 1 per cent hole deviation.

I.N.C.O. then had one of the surface drills converted for use underground and its successful operation encouraged other drill manufacturers to enter this field.

While differences naturally exist between one manufacturer's unit and those of its competitors, the basic requirements for an underground drill remain the same. Below is given a description of a typical I-R CMM/DHD blast-hole jumbo drill powered by compressed air.

Its main components, i.e. the drill tower, rotary head, power pack, jacks, and control equipment are mounted on a base structure which uses crawler tracks for propulsion. The drill tower, which may be tilted either to right or left of its longitudinal axis for drilling purposes, may also be hydraulically lowered for tramming, or raised for drilling work.

The crawler tracks are hydraulically powered and independently driven, and the drill is levelled by means of four hydraulic jacks, fitted with foot-pads, which extend to the floor beneath the machine. In addition a roof jack with a swivel foot-pad is mounted on the tower to ensure that correct alignment is maintained during the drilling operation.

Down-the-hole Rotary Drills

Loss of power resulting from friction between the drill string and the walls of the hole is as much a problem to the operator and manufacturer of a rotary drill as to that of a percussive one. This is particularly true in deep holes and, in this respect (i.e. in deep holes), K. Smyth[20] of Drilco Industrials (Western Australia) comments that of the total power available at the surface, probably only as little as 20 per cent may actually be used in advancing the hole.

The answer to this problem has, of course, been obvious to engineers for many years — simply put the motor down the hole. However, this is easier said than done when the hole diameter is as small as 100 or 150 cm (4 or 6 in).

So far as is known, the first patent (No. 142,992) for a single-stage, hydraulic, turbine-type down-hole motor was granted to Christopher G. Cross of Chicago, Illinois on 23 September 1873. It was filed by Cross on 14 June 1873. However, there appears to be no record of the tool being tested in practice.

Apart from the above patent no further work appears to have been carried out in this field during the succeeding 40 or 50 years until about 1924 when both multi-stage and single-stage turbo-drills were simultaneously developed in the U.S.A. and Russia. The same year Matvey A. Capelivshicoff of Baku, Russia, was granted two patents covering a 'Well drilling tool — with planetary speed reducer'. This was the first 'geared' single-stage turbine. This particular design was not apparently developed further, possibly due to difficulties involved in perfecting it mechanically. Subsequently turbo-drills were reportedly being used in Russia and, so far as is known, are probably still being used there.

In America, development turned towards use of the positive displacement motor for bit rotation. Using this principle, developmental work on the Dyna-drill (Figures 73 and 74) was initiated by the Smith Tool Division of Smith International in 1955. The tool, which was in effect powered by a multi-stage Moyno pump operated in a reverse application, was first built in 1956. It was initially tested for straight-hole drilling, but short bearing life, and the failure of the drill bits when subjected to high rotational speeds, made the tool uneconomical. Subsequently further tests with the drill indicated a new application for it, namely as a directional drilling tool and, because of the urgent need for such a tool in the industry, almost all developmental work up to 1968 was concentrated on achieving this.

Latterly, however, additional research and development has led towards the production of a down-the-hole rotary drill which has been successfully tested in straight-hole drilling applications. Furthermore, the Dyna-drill has also been successfully operated with air, instead of a liquid drilling fluid, which is particularly valuable in those applications where use of a liquid is not practical.

Both hydraulic and turbine motors are presently being manufactured, while the possible use of electricity as an alternative power source is also currently being investigated.

Friction loss in this type of unit is minimal as only the bit itself rotates, the drill string serving merely as a conveyor line for the hydraulic fluid. Again, as with the percussive down-the-hole drill, the fluid, after doing its work of powering the drill, is discharged into the bottom of the hole where it serves to collect the cuttings and carry them to the surface in the conventional way via the annular space between the drill pipes and the hole wall. The fluid is then channelled through a series of settling tanks until all cuttings are removed and it is then recycled.

In 1964 the Dyna-Drill Division of Smith International was established as a company and this has now become an independent entity. Dyna-drills are used throughout the world and are particularly popular in Australia.

One of the most important attributes of this tool and probably its greatest value to the industry, has been its ability to drive an angled or curved hole sans hole wedges. By either fitting a 'bent' sub or alternatively substituting a 'connecting-rod housing' which incorporates 'kick-off' shoes in place of the conventional housing component for connecting the rods, the drill may be made to cut holes with angles of up to approximately 8° per 30 m (98 ft). While it is, of course, possible to exceed the

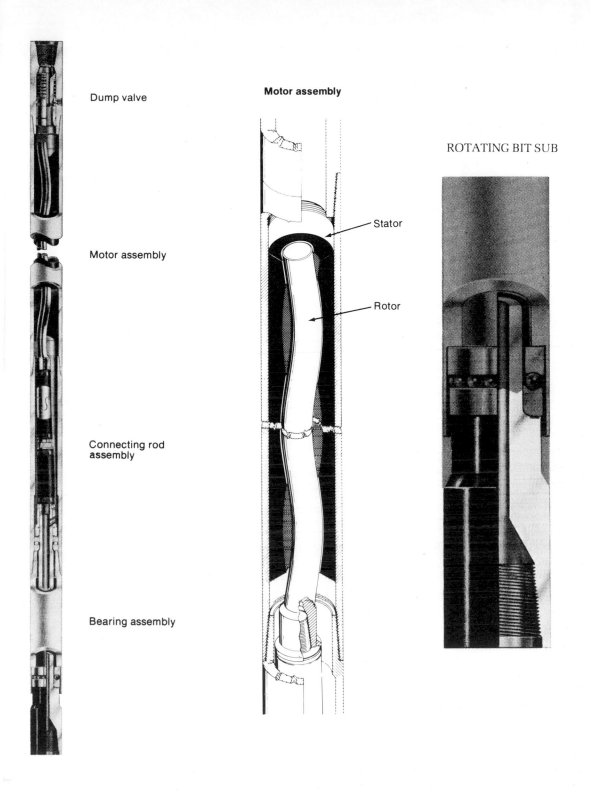

Dump valve

Motor assembly

ROTATING BIT SUB

Motor assembly

Stator

Rotor

Connecting rod
assembly

Bearing assembly

Figure 73. Dyna-Drill: showing motor assembly and rotating bit sub.
(Courtesy Dyna-Drill — Div. of Smith International Inc.)

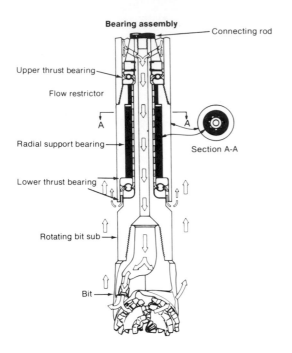

Bearing assembly

Connecting rod

Upper thrust bearing

Flow restrictor

A

A

Section A-A

Radial support bearing

Lower thrust bearing

Rotating bit sub

Bit

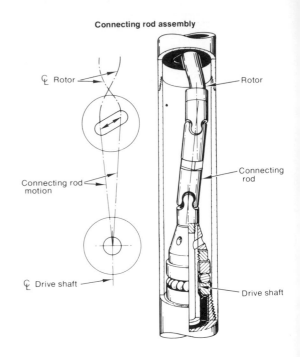

Connecting rod assembly

Rotor

℄ Rotor

Connecting rod motion

Connecting rod

℄ Drive shaft

Rotor

Drive shaft

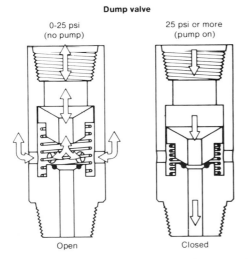

Dump valve

0-25 psi (no pump)

25 psi or more (pump on)

Open

Closed

Figure 74. Dyna-Drill: showing bearing assembly, dump valve, and connecting rod assembly. (Courtesy Dyna-Drill — Div. of Smith International Inc.)

a specially designed motor consisting of only two parts, a rotor and a stator. The stator is a moulded rubber-like spiral passageway having an 'obround' cross-section. The rotor is a solid steel spiral shaft of round cross-section.

Rotation occurs when fluid under pressure is forced into the cavities between rotor and stator. The upper end of the rotor is unattached and exposed to the flow of drilling fluid. The lower end of the rotor is attached to a flexible connecting rod which in turn is connected to the drive shaft. This connecting rod converts eccentric rotation of the rotor to concentric rotation of the drive shaft. The lower end of the drive shaft is designed to accommodate the drill bit.

As fluid will not flow easily through the tool unless the motor is operating, a bypass or dump valve assembly is supplied which will allow the drill pipe or rods to fill or drain while they are being lowered into or pulled out of the hole.

Briefly, the dump valve assembly consists of a sliding piston, a sleeve seat, a coil spring, and external parts. When there is no fluid circulation the spring holds the piston in the 'up' position. This opens the external ports and allows fluid entry and exit through the sides of the valve body. The piston is activated by fluid velocity which pushes down the piston and holds it against a seat, thus closing the external

'kick-off' rate angle of 8° per 30 m (98 ft), Smyth warns that to do so would be to court possible trouble in the form of a damaged drill string. For special occasions when very rapid angle changes are required or when formation hardness prevents a conventional bent assembly from being effective, a combination of two bent assemblies in common alignment will multiply the lateral force effect.

Simply, the Dyna-drill is described as

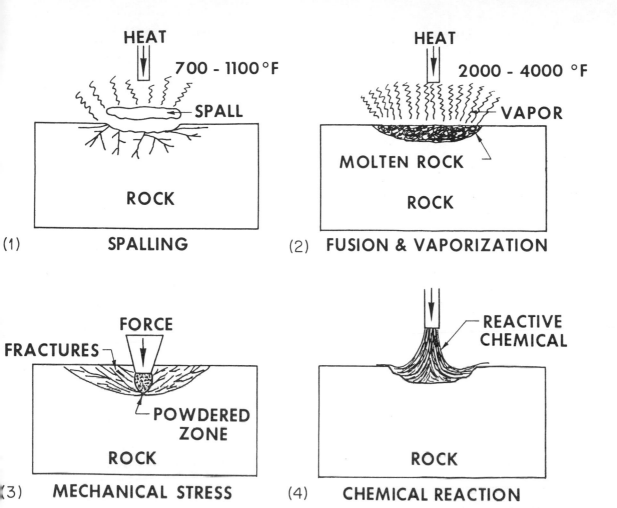

Figure 75. (1) Spalling; (2) fusion and vaporization; (3) mechanical stress; (4) chemical reaction. (Courtesy Dr. William C. Maurer from his papers — see References)

ports. The ports remain closed until pumping is stopped when the pressure on the spring is released, and it is allowed to return the piston to the 'up' or 'open port' position, thus allowing fluid to bypass the motor assembly.

Dyna-Drills range in size from the 44.45 mm (1¾ in) 'Micro-Slim' tool to the 'Hevi-Drill' which is 244.48 mm (9⅝ in) in diameter, powering 311.15 mm to 444.5 mm (12¼ in to 17½ in) diameter bits.

Such is the versatility of this tool that it has been used successfully in a number of diverse applications which include making horizontal holes beneath roads or railway lines, and for surface-to-surface holes beneath lakes and other awkward areas, for carrying power and telephone lines, or for water or gas pipes, etc.

It should be noted, however, that the efficiency and performance of these drills is affected by the fact that the power in small units is limited to about 4 kW (5 hp), whereas a conventional diamond drill of comparable size would transmit about 20 kW (30 hp) to the drill bit. As hole depth increases so is the power of the diamond drill increased to compensate for losses through the rotating string.

In addition, it is inconvenient to have a core barrel (or anything other than a bit) in front of the Dyna-drill. Therefore the hole must be drilled non-core, which in turn calls for more power.

Additional Drilling Machines and Methods[31-33]

The processes whereby rock can be excavated or drilled may be divided roughly into five basic categories:

(1) impact;
(2) abrasion;
(3) thermally induced spalling;
(4) fusion and vaporization;
(5) chemical reaction.

Apart from the so-called conventional percussive or rotary-type drills already described (and which are not included in this section) men have, over the years, devised various types of machines or techniques which have employed one or other of the above methods to drill holes in rocks (Figure 75).

Some of these drilling machines or techniques are almost as old as the first of the 'conventional' percussive or rotary drills, while others have only recently been introduced and, indeed, are still undergoing laboratory tests.

In his book *Novel Drilling Techniques,*[31] and in his papers,[32, 33] Dr. William C. Maurer covers this subject fairly extensively. It is not, therefore, the author's intention to give a detailed description of each and every novel or unusual drill which has been invented. Rather the purpose of this section is to outline briefly some of the more significant developments in this field so that the reader may be made aware of their existence.

The main aim behind the development of most new drilling concepts is an attempt to increase penetration rates while keeping costs as low as possible. Apart from the actual cutting or excavating mechanism, one of the main barriers to the achievement of higher penetration rates in deep-hole drilling is the great amount of time needed to extract drill pipes from the hole, so that the cutters may be changed.

According to Maurer, conventional roller and drag bits used in rotary drilling become blunt after approximately 10 to 20 hours of operation, while diamond bits will become dull after 50 to 200 hours. In deep holes it could take up to 15 hours to extract the drill string and replace the bit — making this an extremely costly operation.

The Pellet Drill

It was an attempt to devise a drill with a cutting edge which could be replaced at the bottom of the hole that led the Carter Oil Company to try the pellet drill (Figure 76).

This drill could, perhaps, be referred to as a distant cousin of the shot drill described earlier, except that instead of the shot or pellets being used as an abrasive, they are impacted against the rock.

At the bottom of the drill string the pipe is divided into two sections. These two sections form the cutting tool. The pipe diameter of the first or upper section of the cutting tool is constricted into a venturi jet and an opening allows fluid and pellets to enter at this point. The second or lower section of the cutting tool reverts to normal pipe size (which corresponds to the drill string). Thus, in effect the cutting tool is somewhat similar in design to a venturi pump.

Steel pellets measuring 3.2 cm (1.2 in) in diameter are inserted via the drill pipe so that

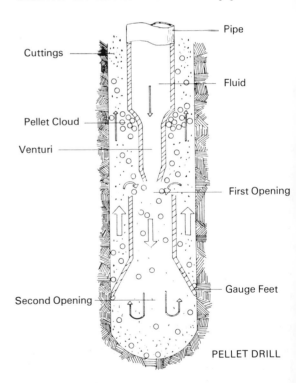

Figure 76. Pellet drill (J. E. Eckel, F. H. Deily and L. W. Ledgerwood, Development and testing of jet pump pellet impact drill bits, *Trans. A.I.M.E.,* **207,** 1-9, 1956.) (Courtesy Dr. William C. Maurer from his papers — see References)

they fall to the bottom of the hole. Drilling fluid is then forced down the drill pipe through the venturi, past the first opening, and out the lower end. The pellets and cuttings are picked up by the fluid and carried towards the surface via the annular space around the drill pipes. However, as the pellets, fluid, and cuttings pass the venturi opening in the pipe, some of the fluid (together with a number of pellets) is drawn into the downward-flowing stream through the venturi opening. The remainder of the fluid and cuttings continue on their way to the surface but the pellets, because they are heavier, collect in groups just above the venturi opening where the fluid velocity of the upward stream is reduced. The pellets remain suspended there for a while but gradually drop back to the venturi opening where they re-enter the circulation. When drawn into the downward stream through the venturi opening the pellets increase in speed as the fluid forces them through the second and lower section of pipe. Thus, by the time they emerge at the bottom of the hole and impact against the rock, they are travelling at a speed of approximately 23 m/sec (75 ft/sec).

Two cutting feet are fitted to the bottom of the drill to maintain gauge and the entire string is rotated slowly during operation.

During the study period many different designs of pellet drilling tools were tested, and drilling rates of about 30 m/hour (98 ft/hour) were achieved in some sandstone and limestone formations. For these trial projects in the sandstone and limestone formations the drill was designed to mechanically cut gauge while the pellets excavated the centre of the hole. Unfortunately, however, the pellets also tended to wear away the gauge cutters and for this and various other reasons further developmental work on pellet drills was discontinued.

Continuous Penetrator-type Drills

Various types of penetrator drills have been devised. The main advantage of these (apart from the fact that the cutting tool does not need changing) is that no drilling fluid is necessary. This latter fact is of considerable importance in porous ground or unconsolidated material where loss of circulation may be a real problem.

And it is for such strata that the penetrator drill was designed.

Basically it consists of a solid conical drill tool or moil point which is either hammered into the ground or is pushed in by means of weights (i.e. doughnuts or drill collars). Some of these penetrator drills have been fitted with hydraulic mechanisms which anchor the tool to the hole wall and then hydraulically thrust down the penetrator tip. A crushed rock zone is formed around the tool as the penetrator is advanced downwards.

In view of the fact that greatly increased force is needed to advance the penetrator in larger-diameter holes, Maurer suggests the possibility of using a smaller-diameter penetrator and then subsequently reaming the hole to enlarge it to the required size. If necessary a circulating fluid could also be used with the reaming tool in order to clear the hole of cuttings.

Implosion Drills

According to Maurer, N. P. Ostrovskii of Russia suggested the use of an implosion drill. He proposed that sealed vacuum capsules be pumped through the drill pipes to the bottom of the hole where the capsules would be ruptured by impacting them against the rock.

Some work carried out by Rayleigh has indicated that severe pressure waves are created when a vacuum bubble is collapsed in an incompressible liquid. His work showed that the deeper the hole, the denser the fluid therein, and the bigger the capsule size the more powerful the effect of the implosion. However, Maurer personally carried out a number of tests using 1 cm diameter capsules in Indiana limestone and Berea sandstone wells having a hydrostatic pressure of up to 800 kg/cm² (11,000 lb/in²). Maurer found that although in all cases the capsules were broken into small pieces by the implosion, the rocks remained intact.

Explosion Drills

Some explosive drills (Figure 78) have been tried by the Russians who, according to Maurer, drilled approximately 3000 m (9900 ft) of hole by this means.

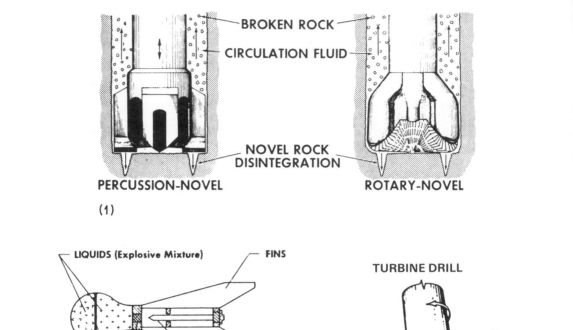

Figure 77. (1) Novel rock disintegration (percussion and rotary); (2) Russian explosive capsule (J. Raynal, Discussion of paper by L. W. Ledgewood, Efforts to develop improved oilwell drilling methods, *J. Pet. Tech.,* **219,** 63-67, Dec. 1960). (3) Turbine drill (G. E. Cannon, Development of a high-speed low-torque drilling device, presented at A.I.M.E. Meeting, Dallas Texas, 6-9 Oct. 1957). (Courtesy Dr. William C. Maurer from his papers and *Novel Drilling Techniques,* Pergamon Press, 1968.)

The method used is similar to that proposed by Ostrovskii for his implosion drill. That is, explosive capsules are pumped to the bottom of the hole via the drill string and are detonated when they emerge and strike the rock.

Each capsule (Figure 77) carries two liquids which are separated by a thin diaphragm. The liquids are mixed together when the dividing diaphragm is deliberately broken by forcing the capsule through a narrowed section of the drill near the bottom. These two liquids when mixed form an explosive compound.

Immediately above the explosive liquids the capsule is fitted with fins which remain together as the capsule passes down the drill pipe and through the constricted channel at the bottom. However, as soon as the fins emerge from the bottom of the drill, they open out or separate and this releases the percussion pin which in turn fires the detonator when the capsule is impacted against the rock.

To protect the drill from the explosive forces the nozzle end is held about 20 to 40 cm (8 to 16 in) away from the bottom of the hole. Care must also be taken to space the capsules accurately otherwise there is the risk that a descending capsule may be accidentally exploded by the shock wave from a preceding capsule. Maurer suggests that they be spaced at least 1.5 sec apart.

Because of the cushioning effect of uncleared debris at the bottom of the hole, penetration rates are also affected by the detonation rates which must be adjusted to allow for adequate

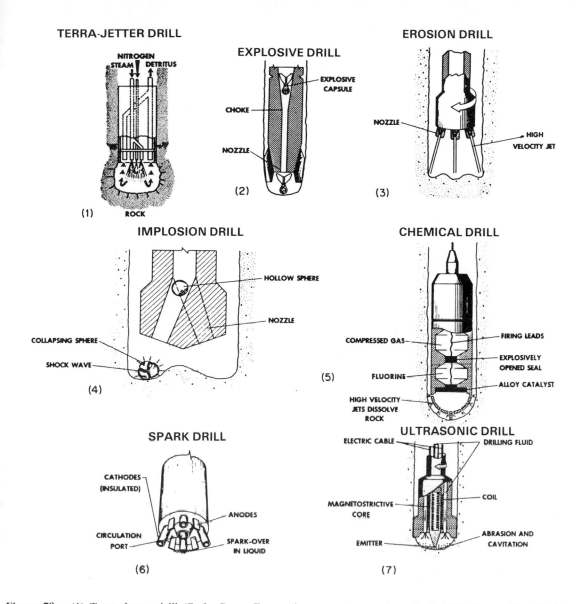

Figure 78. (1) Terra-Jetter drill (S. L. Ross, Excavating apparatus and method, U.S. Patent No. 3152651, 1964); (2) Explosive drill (N. P. Ostrovskii, *Deep-hole Drilling with Explosives,* Gostoptekhia 'dat Moscow, 1960, trans. by Consultants Bureau Enterprises Inc., New York); (3) Erosion drill (N. P. Ostrovskii, ibid.); (4) Implosion drill (N. P. Ostrovskii, ibid.); (5) Chemical drill (L. W. Ledgerwood, Efforts to develop improved oil-well drilling methods, *J. Pet. Tech.,* pp. 61-74, April 1960); (6) Spark drill (N. I. Titkov, M. A. Varzanov, I. I. Slezinger, O. P. Petrova, and G. I. Borisov, Drilling with electrical discharges in liquids, *Neft. Khoz. USSR,* 5-10, 1957, English trans. by Associated Technical Services Inc. New Jersey); (7) Ultrasonic drill (E. A. Neppiras, Ultrasonic machining and forming, *Ultrasonic,* **2,** 167-173, Dec. 1964). (Courtesy Dr. W. C. Maurer from his papers (see References) and *Novel Drilling Techniques,* Pergamon Press, 1968)

clearing between explosions according to the type of material being drilled.

Other explosive drills tried include one where the two liquids plus a chemical initiator were carried in reservoirs in the drill and were allowed to flow in regulated amounts into a detonation chamber at the bottom of the drill. The initiator then caused them to explode. But various factors associated with the flushing fluid which tended to dilute the explosive liquids, and with hole-clearing, etc. made these drills inefficient.

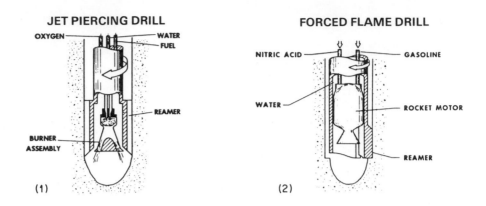

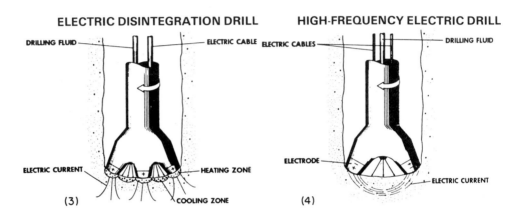

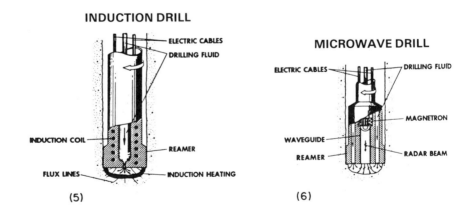

Figure 79. (1) Jet piercing drill (J. J. Calaman and H. C. Rolseth, Technical advances expand use of jet piercing process in taconite industry, *Mining Research,* Pergamon Press, p. 473, 1962); (2) Forced flame drill; (3) Electric disintegration drill ((a) E. Sarapuu, Electrical fracturing and crushing of taconite, *Proc. 7th Symposium on Rock Mechanics,* Penn. State University, pp. 314-324, July 1965. (b) E. Sarapuu, Electrical disintegration drilling, *Rock Mechanics,* Pergamon Press, pp. 173-184, 1963; (4) High-frequency electric drill; (5) Induction drill; (6) Microwave drill. (Courtesy Dr. W. C. Maurer from his papers (see References) and *Novel Drilling Techniques,* Pergamon Press, 1968.)

Spark Drills (Figures 78 and 80)

Several types of spark drills have been designed and have been laboratory tested. Most of these operate by the discharge of a high-voltage spark underwater. This spark in turn creates pressure waves which impact against the rock and break it. Maurer states that a spark drill was used to drill a 2.1 cm (0.8 in) hole in dolomite and penetration rates of about 0.67 cm/min (0.3 in/min) were achieved. The compressive strength of the dolomite was approximately 225 kg/cm^2 (3200 lb/in^2).

Jet Drills (Figure 78 — erosion drill (3))

A very promising area of research is the developmental work at present being carried out on high-pressure water jet drills.

Several types of rock including sandstone, limestone, marble, and granite, etc. have been drilled by forcing water at varying pressures up to approximately 490 MPa (70,000 lb/in^2) through 1-5 mm (0.04-0.2 in) diameter nozzles.

Ultrasonic Drills (Figure 78)

A tool at the bottom of the drill is vibrated by means of ultrasonic emissions from a magneto-strictive core which is activated when a high-frequency current is passed through coils surrounding it.

Pressure waves caused by the vibrator form vacuum bubbles in the water. These tend to move outwards towards the surrounding rock where they implode and break the rock. In addition abrasive particles are inserted in the hole and these are projected against the rock by the vibrator when it strikes them. Thus two distinct forces act on the rock surface, namely the imploding vacuum bubbles and the impacting abrasive particles.

These drills are still being tested in the laboratory.

Thermal or Jet-piercing Drill (Figure 79)

These drills, which were first tried in 1946, operate by applying heat produced by the flame from the ignited mixture of oxygen and fuel oil. The flame, which travels at a velocity of some

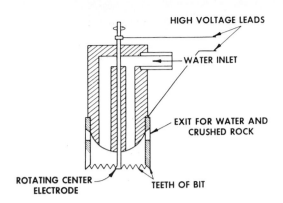

Figure 80. Russian radial spark drill. (Courtesy Dr. W. C. Maurer from his papers (see References) and *Novel Drilling Techniques,* Pergamon Press, 1968.)

1800 m/sec (5900 ft/sec), is directed by nozzles on to the rock at the bottom of the hole. Temperatures attained are as high as 2400°C. However, only a small proportion of the applied heat (approximately half) is effectively absorbed by the rock which then expands and spalls. The rest of the heat is either dissipated by conduction through the walls of the hole, or it rises to the surface in hot gases and is therefore wasted.

During operation the burner and nozzle components of the drill are kept cool by water and some of the heat produced by the flame jets is lost in vaporized water which also rises to the surface.

Forced Flame Drills (Figure 79)

A mixture of nitric acid and fuel oil instead of oxygen and fuel oil has also been tried. According to Maurer, nitric acid and fuel oil produce a much hotter flame than oxygen and fuel oil and increased penetration rates of up to four times that obtained with the oxygen/fuel oil mixture have been achieved.

As some types of rock spall more easily than others the drilling rates vary considerably according to which type of rock is being drilled. These rates range from 0 to 3.7 m/hour (12 ft/hour) for limestone, shale, dolomite, and slaty taconite respectively, and from 5.5 to 12.2 m/hour (18 to 40 ft/hour) for syemite, conglomerate, granite, sandstone, magnetic taconite, quartzite, and jasper, respectively.

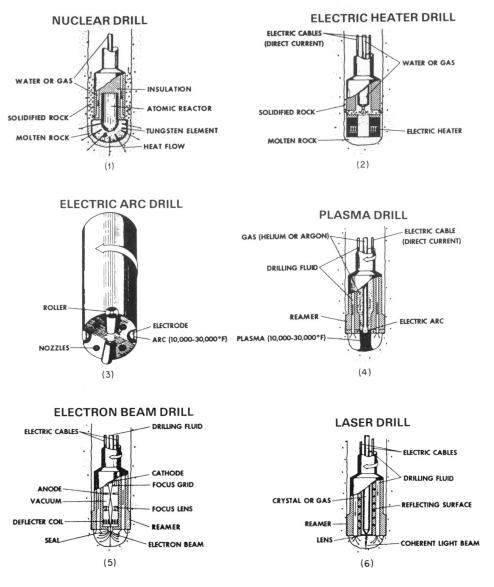

Figure 81. (1) Nuclear drill; (2) electric heater drill; (3) electric arc drill; (4) plasma drill; (5) electron beam drill; (6) laser drill. (Courtesy Dr. W. C. Maurer from his papers (see References) and *Novel Drilling Techniques,* Pergamon Press, 1968.)

Additional Thermal-type Drills

Other thermal-type drills which have been tried or suggested are as follows: electric disintegration drills (Figure 79) — high-voltage, low-frequency electric current is transmitted to the rock which then heats along the path of the current, the thermal stresses induced in the rock cause it to spall; terra-jetter drills (Figure 78) — the rock is heated by steam then cooled by liquid nitrogen, this causes the rock to break; high-voltage, high-frequency electric drills, (Figure 79) — the rock spalls or is broken by dielectric and resistance heating which takes place when high-voltage electrodes are connected to the rock and high-frequency currents are passed through it; microwave drills (Figure 79) — heating by electromagnetic waves causes spalling; induction drills (Figure 79) — heating produced by a magnetic field causes spalling.

92

Fusion or Vaporization Drills (Figure 81)

Several types of fusion or vaporization drills have been designed and tested practically, either in the laboratory or in small holes. These are listed below. (The nuclear drill has not been tried.)

Electric heater drill
Electric arc drill
Plasma drill
Nuclear drill
Laser drill
Electron beam drill

While the compressive strength or abrasive property of the rock poses no barrier, nevertheless, because of the great amount of energy needed to produce fusion or vaporization, the use of this type of drill has so far been strictly limited to relatively small holes or to some specific commercial applications such as diamond dies. By spreading the heat produced by these drills over a larger area some of the fusion or vaporization drills have also been used for spalling or breaking rock.

Basically the drills operate by producing high temperatures of between 1000°C and 2000°C which is concentrated on to a small area of the rock, causing it to melt or vaporize.

In his book,[31] Maurer describes in some detail the various power sources used to produce these high temperatures.

Chemical Drills (Figure 78)

Maurer states that chemical drills have been tested in the laboratory where they have successfully drilled sandstone, limestone, and granite. However, he comments that because of the high cost of chemicals and the inherent dangers involved in handling these substances (fluorine, etc.) which are highly reactive, this type of drill does not appear to be suitable for rock drilling on a large scale.

References

1. Gösta E. Sandström, *The History of Tunnelling,* Barrie and Rockliff (Barrie Books Ltd), London, 1963.
2. H.S. Drinker, *Tunneling, Explosive Compounds and Rockdrills,* John Wiley, New York, 1893.
3. Prof. Douglas Hay, *Historical Review of Coal Mining,* Fleetway Press, London, 1925.
4. Prof. Douglas Hay and N. E. Webster, A comparison from the standpoints of economics and of practical working between compressed air and electricity for use in collieries. *Trans. Inst. of Mining Engineers, U.K.,* **79,** 1930.
5. Bande, *Berg u. Hutt. Zeitung,* 1864, p. 148. (See Drinker (2), page 294).
6. George Low, Description of a rock boring machine, *Proc. Inst. Mech. Eng.,* London, 1865, pp. 179-200.
7. H. M. Hughes, The hydraulic hammer in coal mining, Paper C27/74, *Proc. Inst. Mech. Eng.,* London, 1974, pp. 83-7.
8. George Koether, Men, machines and a company, *Compressed Air Magazine,* Jan.-Nov. 1971. **76,** Nos. 1, 3, 5, 7, 9, 11, Washington.
9. Gardner-Denver Co. 1964, *Rotary or Percussion Drilling,* Gardner-Denver Co., Dallas, U.S.A.
10. F. R. Anderson, *Percussion Drill Steel: Operating, Maintenance Effects on Service Life,* Gardner-Denver Co., Dallas, U.S.A.
11. J. Hodge, Some recent developments in percussive rock drilling for mining, quarrying and tunnelling, *Proc. Inst. Mech. Eng.,* London, 1974, pp. 19-27.
12. Hydraulic percussive drills, *Mining Magazine,* Sept. 1976, London, pp. 194-205.
13. Richard L. Bullock, An update on hydraulic drilling performance, *Proc. R.E.T.C.,* Las Vegas, 1976, pp. 627-47.
14. Richard L. Bullock, Industry-wide trend towards all-hydraulically powered rock drills, *Mining Congress Journal,* Oct. 1974, Washington, D.C., **Vol. 60,** No. 10, p. 54.
15. J. H. Clark, *Canadian Experience Using Hydraulic Rock Drilling Equipment,* C.I.M., A.I.M.E., Vancouver, April 1978.
16. Rudolf E. Pfister, Excavation methods for long highway tunnels and ventilation shafts in the Swiss Alps, *Proc. R.E.T.C.,* Las Vegas, 1976, pp. 377-97.
17. C. Isler, *Well Boring for Water, Brine and Oil,* 3d edn., E. & F. N. Spon, London, 1921?
18. A. Beeby-Thompson. *Oil-field Exploration and Development,* Vols. 1 and 2, The Technical Press, London, 1950.
19. *Encyclopaedia Britannica,* 1929, Vol. 3, pp. 706, 792, 902; Vol. 7, pp. 662-3.
20. *Drillers Training and Reference Manual,* National Water Well Association of Australia, N.S.W., Australia.
21. R. Lauman, Specialised rotary drilling methods, *Drillers Training and Reference Manual,* National Water Well Association of Australia, N.S.W., Australia.
22. F. E. B. Beaumont, On rock boring by the diamond drill, and recent applications of the process, *Proc. Inst. Mech. Eng.,* London, 1875, pp. 92-125.
23. J. D. Cumming, *Diamond Drill Handbook,* Diamond Products Ltd., Toronto, 1956 (J. K. Smit & Sons).

24. R. Sanden, *Sandvik Coromant Eccentric Drilling Method O.D.E.X.,* Sandvik Australia Pty. Ltd., Australia.

25. The ODEX method. Overburden drilling with the eccentric method, *Atlas Copco Mining and Constructions Methods,* MCT, Printed Matter No. 15490, Atlas Copco M.C.T. A.B., Stockholm, 1975.

26. H. L. Pearce, Calyx shaft core drilling at Australian Iron & Steel Ltd. *Collieries,* Collieries Australian Iron and Steel Ltd., pp. 10-14.

27. J. B. Newsom, Shaft boring found inexpensive and safe, *Engineering & Mining Journal,* Sept. 1936, New York.

28. J. B. Newsom and W. D. Haselton, Borehole at the Zenith Mine, Ely, Minnesota, *Proc. American Inst. of Mining and Met. Eng.,* May 1939, New York, pp. 1-14.

29. James H. Allen (Hugh B. Williams Mnf. Co.), Super drills for boring shafts and tunnels, *Mining World/World Mining,* Dec. 1959, San Francisco.

30. Abatage en masse en Carrières: Par le service d'inspection du groupement des poudres et explosifs pour les Carrières. (Ref. Andre Stenuick), *Explosives Magazine,* **6,** No. 1, Jan.-Mar. 1953, pp. 25-9, Brussels.

31. Dr. William C. Maurer, *Novel Drilling Techniques,* Pergamon Press, London, 1968.

32. Dr. William C. Maurer, Novel rock disintegration techniques, *Proc. 15th Annual ASME Symposium on Resource Recovery,* Albuquerque, N.M., March 1975, pp. 57-75.

33. Dr. William C. Maurer, Drilling R & D underway in the United States, *Proc. 1977 Drilling Technology Conference, International Assoc. of Drilling Contractors,* 16-18 March, Houston, U.S.A., pp. 1-21.

Bibliography

Assessory Equipment — Diamond Core & Calyx Drills — Form 2493-A, Ingersoll-Rand, U.S.A.

Calyx Core Drilling — Form 4042, Ingersoll-Rand, U.S.A.

Calyx Core Drills — Form 4047, Ingersoll-Rand, U.S.A.

N. A. Creet and W. J. Taylor, Big hole blasthole drilling at the International Nickel Company of Canada, Limited, Ontario Division, *Underground Mining,* Canada.

Downhole drilling motors and accessory equipment, *Dyna-Drill 1978-1979 Catalog,* Dyna-Drill Division of Smith International, Long Beach, Calif.

Dyna-drill Handbook, 2nd edn., Dyna-Drill, Division of Smith International, Long Beach, Calif.

Dyna-Drill Mining/Industrial Catalog, Dyna-Drill Division of Smith International, Long Beach, Calif.

Joseph Gies, *Adventure Underground,* Robert Hale, London, 1962.

Goldfields Cable Tool Drilling Cat, Diamond Drilling Co. Pty. Ltd., Moorabbin, Australia.

G. R. Green, Big hole blasthole at INCO Limited, Ontario Division, American Mining Congress Convention, Denver, U.S.A., 1976.

B. M. Hewett and S. Johannesson, *Shield & Compressed Air Tunnelling,* McGraw-Hill, New York, 1922.

Ingersoll-Rand, Largehole blasthole stoping drilling equipment and application, Industrial Minera, Mexico S.A., Seminar, 1975.

Ingersoll-Sergeant Rock Drills. Cat. No. 45, Ingersoll-Rand Company, U.S.A.

'Jackhammers' — Hand and Mounted Types etc. — Form 4046, Ingersoll-Rand Company, U.S.A.

Wal. Kaempffert, *A Popular History of American Invention,* Vols. 1 and 2, Charles Scribner's Sons, New York, 1924.

New advance scored in boring holes of large diameter, *Engineering & Mining Journal,* Sept. 1936, McGraw-Hill, New York.

K. McGregor, *The Drilling of Rocks,* C.R. Books, London, 1967.

A rock drill with twice the speed, *Business Week — Construction,* 12 July 1976, McGraw-Hill, New York.

Rock Drilling Data, Gardner-Denver Company, Dallas, Tex.

Diamond Drill Bits, J. K. Smit and Sons (Aus.) Pty. Ltd., Moorabbin, Australia.

Water and Oil Drilling Machines and Methods, Goldfields Diamond Drilling Co. Pty. Ltd., Moorabbin, Australia.

Correspondence

Ingersoll-Rand Ltd., Australia and U.S.A.

Dyna-Drill (Division of Smith International Inc.), U.S.A.

Gardner-Denver Co., Australia and U.S.A.

Holman Brothers Ltd. (Comp. Air Con. Mining), U.K.

National Coal Board, U.K.

Atlas Copco, Australia and Sweden.

Acker Drill Co., U.S.A.

J. K. Smit & Sons (Aus.) Pty. Ltd., Moorabbin, Australia.

Hydro Electric Commission, Drilling Dept., Tasmania, Australia.

TWO

Raise Drills and Shaft-Boring Machines

The vertical 'entry' or opening which descends into a mine is generally known as a 'shaft'. *Within the area of the ore body* vertical openings which connect crosscuts, drives, levels, or other similar developments, are known as raises or winzes.

A raise (rise in the U.K.) is an opening or shaft which is worked upwards, whereas a winze is an opening or shaft worked downwards. However, depending upon which level is used as a reference, the same connecting opening may be referred to as a raise or a winze.

As winzing is generally considered to be both expensive and laborious (all cuttings, etc. must perforce be *lifted* or shovelled out of the hole as it is drilled or blasted), most openings between levels and drives, etc. are usually worked upwards. As a result it is now common for vertical openings or shafts within the ore body to be referred to generally as 'raises' or 'rises'.

These openings or raises are required for a number of reasons. They are necessary for dropping the ore from one level to another (ore passes), service ways for either men or material, ventilation shafts, etc. The raises may be vertical or on an incline and may connect two or more levels, or, indeed, run from the bottom of the mine to the top, connecting all levels.

In order to drive a raise by conventional methods, holes are drilled in the roof or 'back' of the raise. These are then loaded and fired. In most cases where the ground is unstable the shaft walls need timbering or shoring. As the raise proceeds further it becomes necessary to build platforms to enable the miners to reach the working area or 'back'. But before each new round of explosives is fired the platform must be removed.

Though raise driving is cheaper and less laborious than winzing, it has long been considered to be a particularly dangerous part of the mine development work. Ground near the surface which is weathered or fractured is obviously unstable and hazardous, while ground lying at considerable depth below the surface is generally subject to great stresses. The immediate effect of blasting is to relieve these stresses, thereby frequently causing dangerous rock bursts or scaling. Workmen are

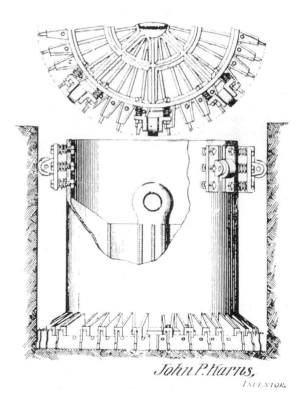

John P. Karns,
INVENTOR.

Figure 82. J. P. Karns early sink shaft drill: patented in U.S.A. on 24 December 1907 under No. 874,848

Figure 83. Alimak raise climber: Model STH-5E electric-driven Alimak. Also shown behind the climber is the Alitrolley or service climber (Courtesy Alimak Australia Pty. Ltd.)

Figure 84. Alimak raise climber: Model STH-5L, air-motor-driven Alimak. (Courtesy Alimak Australia Pty. Ltd.)

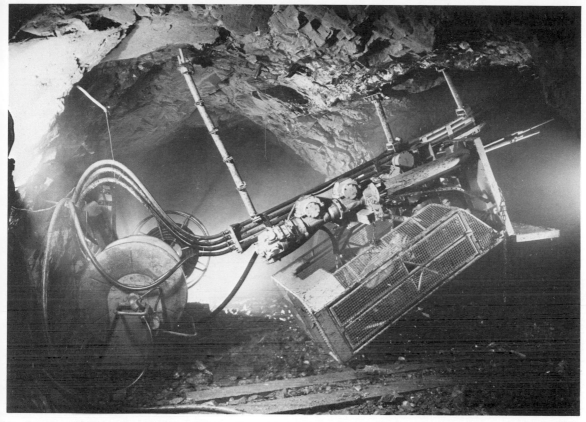

Figure 85. Alimak raise climber: Model STH-5L, air-motor-driven Alimak. (Courtesy Alimak Australia Pty. Ltd.)

then compelled to re-enter the confined space of the raise shaft in order to continue the work. The danger caused by loose and fractured rock to which these workmen are exposed need not be further emphasized. This is particularly applicable to the man or men responsible for barring loose any obvious rock debris which might constitute a danger to those doing the drilling work.

One method of evading these hazards was devised by Linden Alimark (Figures 83-85) who designed a climbing trolley or staging which was equipped with a head-cover. Thus protected, the miners were able to proceed at a faster rate than was previously possible and in comparative safety. The increased need in mines for raise drives produced a special type of worker who developed high skills driving both conventional raises and those where an Alimak-type cage was used. However, as time progressed the services of these special miners became increasingly difficult to procure. It

grew apparent that less and less men were willing to undertake this highly skilled type of labour.

Bade — Pioneer of the Modern Raise Drill

The first major breakthrough in this area came in 1949 from the design board of a German engineer, Herr Bade. Bade conceived of a brilliant method of raise driving which he hoped would eliminate the necessity for the presence of workmen within the shaft during the dangerous period of raise driving. Basically this equipment consisted of a rope winch, a borer, and a control point (Figure 86).

Before his equipment could be put into operation, however, it was necessary to bore a pilot hole between the levels to be connected. A rope winch was erected on the upper level and the borer unit was installed on the lower level.

97

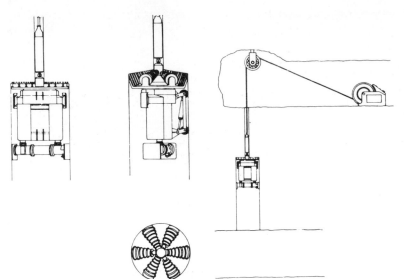

Figure 86. Bade raise drill: Model ABS 2, 1.52 m (5 ft) diameter. (Courtesy Bade & Theelen GmbH.)

A rope was thereafter attached to the winch and passed down through the pilot hole where it was connected to the borer unit below. Bade's machine was then able to propel itself up the hole by means of its hydraulic propulsion jacks, the rope serving merely to support the borer unit during that stage of the cutting operation when the jacks, having been extended to their fullest length, were being retracted and repositioned, ready for the next upward thrust.*

The tool carriers on the head of Bade's units were fitted with steel cutters carrying tungsten carbide blades. These cutters were radially arranged so that circular grooves were cut by them in the rockface. The ridges remaining between the grooves were dealt with by breakers fitted behind the cutting blades. These breakers came into operation as soon as the cutters reached a certain depth. Power was supplied to the unit by means of a trailing cable from the lower level.

During the early stages, Bade encountered many problems. These were gradually overcome until the unit was operating satisfactorily. They were used successfully for many

years in certain iron ore mines and coal-mines in Germany.

Though there were, of course, many disadvantages still associated with this machine, nevertheless it should be recognized that it was the forerunner of subsequent large-diameter shaft drills designed and built in Germany.

The Cannon/Robbins Raise Drill[1, 2]

Developmental work on the Robbins raise drill was initiated during the late 1950s by Robert E. Cannon,[3] Consulting Mining Engineer with Security Engineering Division, Dresser Industries Limited.

The first Cannon machine which evolved quite independently of its German cousins was the product of a joint development programme by James S. Robbins and Associates (who built the machine) and Security Engineering Division of Dresser Industries (who supplied the bits and reamers). It embodied in its operation basic principles initiated by Bade, i.e. it drilled a small pilot hole which was then enlarged by back-reaming. However, its mode of operation (which is described later) was significantly different from the German machines. It incorporated a completely independent unit, free of cables or winches and it did not rely for upward movement on propulsion jacks which thrust against the walls of the shaft. This meant

---

*According to J. Whitehead of Ingersoll-Rand a Lawrence raise borer using the Alkirk patent pilot-pull system with pilot mechanism was tried in 1968. During the repositioning the borer was held by a cable through the pilot hole, but the system was not satisfactory, the machine eventually falling out of the hole (see caption to Figure 87).

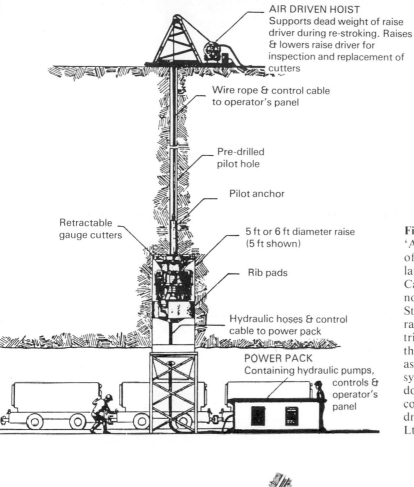

AIR DRIVEN HOIST
Supports dead weight of raise driver during re-stroking. Raises & lowers raise driver for inspection and replacement of cutters

Wire rope & control cable to operator's panel

Pre-drilled pilot hole

Pilot anchor

Retractable gauge cutters

5 ft or 6 ft diameter raise (5 ft shown)

Rib pads

Hydraulic hoses & control cable to power pack

POWER PACK
Containing hydraulic pumps, controls & operator's panel

Figure 87. Early Ingersoll-Rand 'Alkirk raise driver'. A prototype of this machine was built in the late 1960s and tried in the Calhoun mine at Coleville in the northern part of Washington State. Its similarity to the Bade raise drill is of interest. Initial trials with the machine showed the existence of serious problems associated with the cable-pull system and the design was abandoned in favour of the more conventional 'cannon-type' raise drill. (Courtesy Ingersoll-Rand Ltd.)

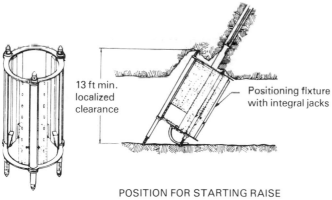

13 ft min. localized clearance

Positioning fixture with integral jacks

POSITION FOR STARTING RAISE

RAISE DRIVER
POSITIONING FIXTURE

Figure 88. Alkirk raise driver: positioning fixture (Courtesy Ingersoll-Rand Ltd.)

that if badly fractured or poor ground was encountered, the overall efficiency of the machine was not affected. Thrust for Cannon's machine was supplied to the reamer head through the drill stem from the unit stationed on the *upper level*.

Apart from other important advantages over conventional methods, Cannon's new machine meant that all personnel connected with the work of driving could, during the process of drilling or reaming, be stationed on the level above the raise, away from the danger area.

99

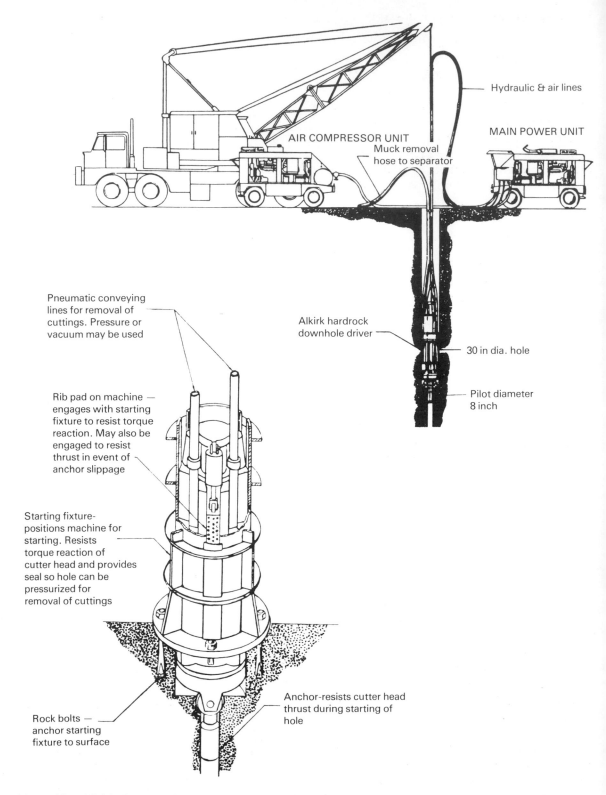

Hydraulic & air lines

AIR COMPRESSOR UNIT

MAIN POWER UNIT

Muck removal hose to separator

Pneumatic conveying lines for removal of cuttings. Pressure or vacuum may be used

Alkirk hardrock downhole driver

30 in dia. hole

Pilot diameter 8 inch

Rib pad on machine — engages with starting fixture to resist torque reaction. May also be engaged to resist thrust in event of anchor slippage

Starting fixture-positions machine for starting. Resists torque reaction of cutter head and provides seal so hole can be pressurized for removal of cuttings

Anchor-resists cutter head thrust during starting of hole

Rock bolts — anchor starting fixture to surface

Figure 89. Alkirk down-the-hole driver. This machine was designed at about the same time the Alkirk raise driver was produced. However, probably as a result of the unsatisfactory performance of the Alkirk raise driver, the 'down-the-hole' unit was not built. (Courtesy Ingersoll-Rand Ltd.)

100

Development of the Cannon Raise Drill

Cannon's drill evolved as a result of a fire which had started in an old open stope in the Homer-Wauseca mine near Iron River, Mich., owned by the M. A. Hanna Company. J. D. McAuliffe, Mine Superintendent, on finding that the area was inaccessible instructed that a hole be drilled from the level above into the back of the stope where the fire was burning, in order to force waste material down the hole and into the burning area, and thus extinguish the fire.

A soft-coal auger machine was used for the drilling operation which was completed satisfactorily and the fire was duly killed. However, the idea behind the 'drilling manoeuvre' germinated and led to further developments which culminated in the trials of the 31R.

At first a conventional auger drill, such as was used by McAuliffe, was mounted on a skid. Rotary power was supplied by a two-speed a.c. electric motor. Downward thrust was provided by air-powered chains. It was soon discovered, however, that the device, designed as it was for work in soft coal, lacked the torque and thrust which was necessary if it was to make full use of hard rock rotary drilling tools. Various steps were taken to overcome these problems. The modifications effected enabled the mine to drill 165 mm (6½ in) diameter holes with the unit fairly satisfactorily. Further modifications were made, including the installation of a Schramm rotary drilling machine which was hydraulically powered. The Schramm unit which was usually carried on a truck was skid-mounted in two units for the drilling machine. One unit carried the electrically powered hydraulic pump, the other carried the derrick with the rotary and chain thrust machinery. Though a considerable improvement on the original auger machine, the modified unit was not completely satisfactory, still not enough torque and thrust could be applied to the reamer to get the most out of the tools. An erratic rotary action developed as a result of attempts to utilize maximum available thrust. Moreover, another difficulty arose — that of inadequate anchorage. The machine frequently shifted position when rock bolts used to anchor the unit to the floor worked themselves loose. This happened in all cases except where the unit was housed on uniformly textured hard rock. It was in the process of solving this particular difficulty of anchorage that a new and unique engineering principle was tried. A 4½ in (114 mm) diameter hole was drilled from an upper level down to a lower level. When the initial drilling had been completed the rock bit was removed from the drill stem which was left in place in the hole. A reaming bit was then turned upside down and attached to the drill stem at the lower level. Rotation was recommenced, the thrust mechanism was put in reverse, and the drill string and reaming bit were slowly drawn back up the original hole, thus pulling the machine hard against the floor rather than pulling on the bolts as previously.

Because the debris naturally fell to the bottom level, the problem of cuttings removal was automatically solved. Additional fringe benefits derived from this new method of raise driving were improved shaft wall surfaces and extended cutter life.

A rental-purchase contract for production of a machine capable of carrying out the above functions was signed by the respective companies involved in the project and work was started on the 31R 1101* early in June 1962 (Figure 90). By September of that year the prototype had been completed and was delivered to the Hanna Mining Company's Homer-Wauseca mine in Iron River, Michigan, for its initial trials. Using a 6¾ in (17 cm) bit, a pilot hole was drilled which was later back reamed with a 40 in (100 cm), six-cutter raise head. By 1964 nine 40 in (100 cm) diameter raises

---

*The original 31 R was tested in the Homer-Wauseca mine in upper Michigan by the M. A. Hannah Company who later purchased the unit from the Robbins Company. It was used by the M. A. Hannah Company until 1969 when it was sold to Patrick Harrison & Company Limited. The drill was sent back to the Robbins Company that year and modified. Changes were made to the main frame and gearbox and a 56 kW (75 hp) motor was fitted to up-grade the unit to the same capabilities as that of a standard 41R machine. Since then Patrick Harrison & Company Limited have used the 31R at seven mines to bore raises of from 1.20 to 1.83 m (4 to 6 ft) diameter. In 1975 the 31R1101 raise drill was installed at the Sunshine Mining Company mine near Kellogg, Idaho, where the Patrick Harrison Company bored a series of 1.83 m (6 ft) diameter raises, 68 m (220 ft) in length.

Figure 90. Robbins 31R (1133-1) raise drill. (Courtesy The Robbins Company.)

totalling approximately 1800 ft (550 m) had been completed by this machine.

Initial problems were experienced with the hardened steel mill-tooth-type cutters used on the prototype unit, as these required frequent changing. However, after Security Engineering replaced these with tungsten carbide insert cutters which dealt more effectively with the difficult ground conditions present there, this problem was largely overcome.

Operation

Before a raise drill can be used, a suitable site is needed to house the machine and other sundry equipment necessary for its operation. This equipment would supply power, compressed air, and water, etc. Naturally the site may vary in size depending upon which type of raise drill is used but, roughly speaking, could range from an area 5.5 m (18 ft) long by 4.9 m (16 ft) wide by 4 m (14 ft) high for say a 41R* Robbins raise drill, to an area 5.5 m long by 4 m (14 ft) wide by 5.8 m (19 ft) high for a 61R Robbins raise drill.

After a suitable area is excavated a concrete pad is poured on which the drill is mounted. The size of the pad would again vary according to the size of raise drill, etc. but would measure about 2 m × 3 m × 30 cm (6 ft × 9 ft × 12 in) for a 41R model. Drilling commences with a pilot hole which is bored downwards using a tungsten carbide tri-cone drill bit on the end of a rotating drill string. The size of the pilot hole may also vary slightly but, generally, would

---

*The 41R drill has now been superseded by the Robbins 32R which is a more compact unit requiring only 3 m (10 ft) of height (an important consideration in a mine) and 4 m of width.

measure about 23 cm (9 in) in diameter.* When the desired intersection or opening is reached, the drill bit is removed and is replaced with a reamer head carrying cutters similar to those used on horizontal hard-rock tunnelling machines. The hole is then enlarged by back-reaming.

Initial trials carried out with the prototype 31R machine were successful and proved that the idea was viable. Consequently, another machine was constructed by the Robbins Company in 1963. This was the 41R.

Once the Robbins Company had shown the way, other well-known drill manufacturers were quick to follow with designs of their own. But most of these machines embodied basic design principles initiated by Robert E. Cannon.

International Nickel Company, Ontario Division[4, 5]

During the period when Robbins were testing their first machines in 1962-63, the International Nickel Company, Ontario Division (INCO), were commencing stope development at their Creighton mine. At that time raise driving by the International Nickel Company was carried out using either conventional methods or with Alimak raise climbing trolleys. However, there were several disadvantages associated with these methods, not the least of which was the continued shortage of skilled miners able to carry out this type of developmental work. In addition, unless ground conditions were good, the use of Alimak trolleys was not practicable as there was no way of introducing a mechanical lift in an area requiring heavy timbering for shoring purposes. These and other factors induced

---

*The diameter of both the pilot hole and the drill rods increases according to the size of the raise drill being used. For instance, for a Robbins raise drill:

Model	Nominal diameter of reamed hole	Diameter of drill rods	Diameter of pilot hole
61R	183 cm (6 ft)	25.4 cm (10 in)	27.9 cm (11 in)
84R	244 cm (8 ft)	28.6 cm (11¼ in)	31.1 cm (12¼ in)

INCO to investigate alternative methods of raise driving. Naturally, word of so revolutionary a machine as the new Robbins raise drill spread quickly throughout the mining industry and excited great interest. Amongst those who gave it serious consideration were executives of the International Nickel Company[4, 5]. But INCO realized that however promising the new drill may have sounded, initial capital outlay would be high for what was a *virtually untried product*. On the credit side, however, were such factors as a complete redundancy of explosives with its implicit dangers, speedier raise driving, a possible lowering of raising costs, smoother circular walls less liable to later deterioration, and, of course, the fact that their skilled miners, whose services were so difficult to obtain, could be redirected towards other important developmental work elsewhere in the mine.

A decision was ultimately made to give the new borer a trial and a Robbins 41R machine was ordered. It was delivered in July 1964.

The slow rate of progress experienced during the first 18 months of operation of the 41R was mainly attributed to the design of this unit which has been built to cope with rock of much lower compressive strengths than were being encountered in the Creighton mine. This and other factors which emerged during the initial trial period showed the necessity for major modifications if improvement in boring rates and general efficiency was to be obtained. These modifications included the following:

1. Change of drive box from a rigid type to a floating design.
2. Conversion of the drill pipe from Taper-Loc, with acorn and bolt connections, to heavy-duty threaded pipe with Drilco DI 22 threads.
3. Removal of the universal joint at the reaming head.
4. Reinforcing of the reaming head and later complete redesign.
5. Change of reaming cutters from tooth or disc type to tungsten carbide insert or 'button' type.[4]

Though the particularly hard rock encountered in Creighton's No. 8 shaft restricted use of the reamer to holes of 90 cm (3 ft) diameter, it was found that for most purposes 120 cm (4 ft) diameter holes could be bored in other areas with the newly redesigned 41R unit.

Development of Larger Raise Drills

But still INCO were not satisfied. Certainly the new 41R machine had performed satisfactorily and allowed them to proceed with vital new developments. Nevertheless there remained an urgent need in these and other areas for numerous raises, preferably of a larger diameter than 4 ft. These additional raises were required for purposes of temporary ventilation, ore passes, manways, etc. INCO felt that 1.5 to 1.8 m (5 to 6 ft) diameter raises would probably meet their requirements. They therefore approached the Robbins Company who agreed to produce a machine capable of boring these larger diameters and, in October 1967, the first 61R was delivered to the Stobie mine. This machine also proved satisfactory and plans were put in hand to use mechanical raise drills for all raises over 23 m (75 ft) in length. In effect this meant that most INCO mines with a production capacity of say between 3000 and 3500 t per day would have at least one raise borer permanently in operation, whereas those of their mines producing more than 3500 t per day could even warrant the installation of two such units. By 1975 INCO had purchased eight additional 61R drills.

Though the 61R was originally designed with 1.80 m (6 ft) diameter holes and raise lengths of 250 m (800 ft) in mind, these machines have, in favourable ground conditions, successfully bored holes of 2.1 to 2.4 m (7 to 8 ft) diameter, some as long as 450 m (1500 ft).

Mining operations of the Ontario Division of INCO are located in the Sudbury Basin district. Hard Pre-Cambrian rock dominates this area and, in order to enable INCO's 61Rs to cope with this type of ground many modifications were found to be necessary. These modifications included changes in the hydraulic system which increased the machine's thrust capacity from 1700 kN to 2200 kN (380,000 lb to 500,000 lb) and changes to the main thrust bearing, the bottom drive assembly, and the main frame. (This enabled the machine to withstand the increased forces produced by the

now more powerful hydraulic system.) After modification the 61Rs successfully bored 2.1 m (7 ft) diameter raises of up to 180 m (600 ft) deep. Further general improvements to the design of reamer heads and cutters enabled INCO in 1972 to bore a 2.10 m (7 ft) by 160 m (522 ft) ore pass at 85° with a 61R carrying 14 cutters. This work was executed in extremely hard Sudbury Granite Breccia at their Copper Cliff North Mine. Penetration rates in this difficult ground material averaged from 30 to 38 cm/hour (12 to 15 in/hour). However, the newly designed 'flat head' back-reamed the entire raise without cutter change, and all bearings and seals were in good condition at the completion of the raise. It was estimated that the gauge (outside) cutters travelled approximately 1.8 million metres (6 million feet) during the 450 hours of reaming. At the end of this mammoth journey the carbides of the gauge cutters had worn flush with the matrix. (Note: the introduction of the 'flat head' type of cutter head in place of the original 'christmas tree' type constituted an important development in cutterhead design.)

Ingersoll-Rand Raise Drill

Despite the success of the above machines INCO felt that even more powerful equipment was needed to bore 2.10 m (7 ft) diameter raises of greater depth. By that time Ingersoll-Rand were producing large diameter raise drills and in 1973 INCO installed an Ingersoll-Rand raise drill at their Copper Cliff North Mine for trial purposes. This machine though weightier and less portable than the 61R had a thrust capacity of 3300 kN (750,000 lb). Its design appearance was somewhat similar to the 61R and the machine was found to be satisfactory. Subsequently the drill was used to back-ream raises of 2.10 m (7 ft) to depths of 300 m (1000 ft). (Note: the Raise Drill Division of Ingersoll-Rand was taken over by The Robbins Company on the 20 November 1979.)

Comparative Costs of Conventional versus Bored Raises

In their report 'Raise Boring at the International Nickel Company of Canada,

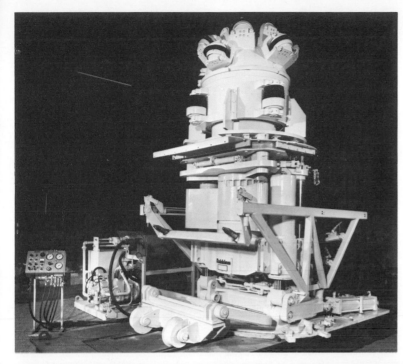

Figure 91. Robbins raise drill, Model 52R-1199: used by Buffelsfontein Gold Mine Co. Ltd. (Courtesy The Robbins Company.)

Figure 92. The Robbins Company raise drill, Model 61R-1163 in the factory. (Courtesy The Robbins Company.)

Figure 93. The Robbins Company 61R raise drill ready for transportation. (Courtesy The Robbins Company.)

Limited, Ontario Division',[5] Messrs. W. J. Taylor (Superintendent of Drilling) and J. R. England (raise borer specialist), mentioned that in their opinion disadvantages such as the increased 'set-up' and 'tear-down' time which is required for the more powerful type of machine was favourably balanced by the increased reaming capacity of these machines. In effect this allowed for the elimination of one 'set-up' station per 300 m (1000 ft).

Set out here are two tables (Tables 1 and 2) given by Messrs. Taylor and England in their report. These cover 'Typical time distribution' and 'Comparative costs of conventional vs.

Table 1. Typical time distribution: 500 ft— 7ft dia. raise

Set up	6 shifts
Pilot hold (6-8 ft per hour)	14 shifts
Connect head	2 shifts
Reaming (12-15 in per hour)	75 shifts
Misc. delays	2 shifts
Pull rods, tear down	3 shifts
Total	102 shifts

Source: Taylor and England.[5]

Table 2. Comparative costs: conventional vs. bored raise

	7 ft × 9 ft cribbed raise	5 ft dia. bored raise[a]
	$	$
Labour cost	86.61	20.18
Timber supports, etc.	30.68	—
Drill steel, repairs, explosives	13.79	6.07
Borer supplies	—	3.72
Borer repairs	—	11.83
Pilot bits	—	6.00
Reaming head and cutters	—	30.00
Stripping raise cribbing	13.10	—
Raise borer station	—	22.97
Total cost per foot raise	147.66	100.77

[a]For ore pass or ventilation purposes.
Source: Taylor and England.[5]

bored raise' obtained by INCO, up to the year 1973.

Further Developments — The Robbins Co.

In 1970-71 the Mt. Isa mines, Queensland, Australia, required 2.4 m (8 ft) diameter holes

106

Figure 94. Robbins 61R raise drill in operation underground. (Courtesy The Robbins Company.)

Figure 95. Robbins raise drill Model 81R: used at Mt. Isa Mines Ltd., Australia and various projects in the United States. (Courtesy The Robbins Company.)

of varying lengths up to 600 m (2000 ft). Robbins designed their 81R machine for this purpose. It consisted of two 61R drive systems coupled together and mounted on a large frame. Two of these machines were built, one for the Mt. Isa mines and one for use by the Hecla Mining Company of America. The Mt. Isa machine was affectionately christened the 'Wombat'. With a total thrust capacity of 5600 kN (1,250,000 lb), these two machines were recognized as being the largest and the most powerful raise borers built at that time in the world. Robbins also built the 71R (1972) for 2.1 m (7 ft) by 300 m (1000 ft) raises (this model is similar to the 61R, but incorporates a third cylinder for the necessary additional thrust required) and the 83R (1973). (This machine

was built for Anglo-America of Africa and it has bored raises of 3 m (10 ft) diameter.) Since then an Ingersoll-Rand raise drill RBM-211 with an RBH-8ER20 raise boring head has been used in the Monterey Coal Company No. 1 mine near Carlinville, Ill. to bore two shafts with diameters of 6.17 m (20 ft 3 in).[6] The initial 27 to 30 m (90 to 100 ft) were sunk using conventional methods, then the shafts were machine-bored another 64 m and 61.5 m (210 and 202 ft) to 93 m (305 ft).

Removal of Cuttings from Pilot Hole

During the years which followed the installation of the first raise drill at INCO, many trials were made to discover the best

107

Figure 96. Robbins raise drill 84R: used at Mt. Isa Mines Ltd., Australia. (Courtesy The Robbins Company.)

method of removing cuttings from the pilot hole. Water, dry air, and drilling mud were all tried with varying results, depending upon the ground conditions at each particular location. Where conditions were favourable, dry air appeared to have been preferred by many operators as it tended to prolong the life of the pilot hole drill bit. The dust problem created by the dry air method of cutting removal was successfully overcome by the introduction into the discharge line of water in the form of a fine spray. In some cases, however, due to the poor nature of the terrain, it was found necessary to use drilling mud to consolidate the walls of the pilot hole. But, in most raises, the use of drilling mud was considered to be both costly and inconvenient in the underground situation.

Up-hole Pilot Hole Drilling as Opposed to Down-hole Pilot Hole Drilling.
Experiences at Hecla Mining Company's Lucky Friday Mine[7]

Inevitably the merits or otherwise of drilling a pilot hole up and then reaming it down instead of drilling down and then reaming up, as had hitherto been the practice, was questioned by operators. This matter was seriously considered by engineers of the Hecla Mining Company when it was found that raises were needed at the Company's Lucky Friday mine between those levels under high production and lower developmental levels.

Not only would raise boring from these production levels have seriously hampered current work programmes but it would also

Figure 97. Robbins raise drill 85R: used by the Anglo-American Corp. of South Africa. (Courtesy The Robbins Company.)

have necessitated the removal of sill timber which, of course, would ultimately have had to have been replaced after the raise was completed and before further production work on that level could proceed.

On the other hand up-drilling could be carried out from the developmental level below, with little or no inconvenience.

'RD-1' Drilling Machine (Nichols Universal Drilling Company)[7]

In 1964 five 'up-holes' were drilled, the work being carried out by the Nichols Universal Drilling Company, who used one of their own units which had been manufactured and assembled to their specifications. This was the 'Nichols RD-1' drilling machine.*

(At that time the Nichols Drilling Company was a subsidiary of the J. R. Simplat Company. Later the Nichols Universal Drilling Company was taken over by the Security Division of Dresser Industries. Since then the name 'Security' has been dropped and their machines are now marketed as 'Dresser'.)

Hecla Mining Co.) at Mullan, Idaho. Four (4) 32″ diameter raises were up-reamed at this mine. This same machine completed 32″ diameter raises for Sunshine Mining at Kellogg, Idaho.

'The *second* raise drill RD-2 was built in 1965 which was commissioned in July 1965 at the American Smelting and Refining Co.'s Galena mine at Wallace, Idaho. Three (3) 48″ diameter up-reamed raises were completed.

'The *third* raise drill RD-3 was built in August 1965 which was displayed at the American Mining Congress Show in Las Vegas. In December 1965, the Security Division of Dresser Industries took over the Nichols Universal Drilling Company. RD-3 was commissioned in January 1966 and began boring 72″ diameter up-reamed raises for the Anaconda Co. at Butte, Montana.'

Quote from letter received from W. T. Folwell, Application Engineer, Mining Services and Equipment Division, Dresser.

---

*'Nichols Universal Drilling Company produced their *first* raise drill RD-1 in 1964 which was commissioned in November, 1964 at the Lucky Friday Mine (owned by the

Figure 98. Raise boring machine reaming head dressed with Ingersoll-Rand tungsten-carbide button ring cutters. (Courtesy Ingersoll-Rand Ltd.)

After the completion of five 'up' raises totalling approximately 940 ft (286 m) in length, R. S. Hendricks (Senior Industrial Engineer of Hecla Mining Company) expressed the opinion in his report, 'Hecla Mining Company Raise Boring Experience',[7] that up-hole drilling was, generally speaking, an inferior method of driving raises and should only be resorted to where special circumstances warranted it. He based his conclusion on the following points:

1. The headroom requirements, and site preparation costs are higher in the case of up-drilling since 5-6 feet of space must be provided above the drill for a hopper to catch drill cuttings.
2. The hole collaring operation is more difficult in the case of up-hole drilling since a long length of drill rod projecting above the machine must be supported by timber or other means. In the case of down-hole drilling, a bushing inserted in the work deck supports the rod string a few inches above the bit.
3. The reaming of an up-hole is more expensive from the standpoint of cutter costs. It is relatively difficult to keep the rock face clean and considerable regrinding of cuttings takes place. Tests were conducted which indicated that reamer cutter costs for a down-reamed 32-inch hole would be 50% higher than an up-reamed hole of the same size. The relative difference can be expected to increase as the hole size is made larger.
4. The down-reaming operation is slower from a penetration standpoint, and therefore, more expensive in labor cost/foot, again because of the unavoidable regrinding of cuttings.[7]

Some Difficulties Encountered by Raise Drill Operators

Just as operators of early shaft drilling equipment were confronted by various mishaps and problems, so, too, were operators of the modern raise drill.

Pilot Holes

In the Tynagh mine,[8] which lies about 140 km (90 miles) to the west of Dublin, Ireland, it was found after the pilot hole was drilled that the hole was approximately 90 cm (3 ft) away from the predetermined target area. The raise borer was moved to a new location, 4.6 m (15 ft) from the first hole, and drilling recommenced. Despite positive steps taken to overcome the difficulty such as alteration of the borer alignment and modification of the stabilizer system, the second hole also broke through 1 m (3 ft) off target. It was therefore assumed that the peculiar formation of the ground strata in that region was responsible for the deviations (i.e. the drill bit was being deflected by the various layers of rock which were inclined at an angle). The raise borer was therefore reinstalled at the original location and the first pilot hole was back-reamed.

At the Kaiser Steel Corporation's Sunnyside No. 1 mine, Utah,[9] it was found necessary to bore a hole to dewater a 4.9 m (16 ft) diameter main airshaft in order to prevent freezing during the winter. The first 6.7 m (22 ft) of the pilot hole was lined with casing so that fill and overburden lying in the vicinity of the borehole would not be eroded away by water during the drilling process. At the 140 m (460 ft) mark the valves in the water line were blown out by the back-pressure and the machine had to be stopped. Some while elapsed before the faulty valves were eventually replaced. By this time it was found that the cuttings had settled solidly to the bottom, binding the drill pipe, so that it was impossible to force water through in order to dislodge them due to the hydrostatic head keeping them held in position.* As the drill pipes could not be rotated it was necessary to raise the drill string. The raise drill, however, was only able to perform this operation to the extent where 12 drill pipe sections or 18 m (60 ft) were removed before it finally jammed, so that it would not move up or down, or rotate. At this stage helpful experts came forward with so many ways and means of loosening the jammed string that it took 30 days to try them all out. After weeks of protracted experiment they were able to raise the next 30 m (100 ft) by a complicated rig consisting of two 450 t hydraulic longwall chocks and a specially reinforced box girder placed under the wrench flats of the drill pipe and over the chocks. After 30 m (100 ft) had been raised like this the raise borer was again able to take over and remove the remaining pipes. They survived this exasperating struggle only to discover that the 6.7 m (22 ft) of casing originally inserted at the head had slipped nearly 6.1 m (20 ft) further down the shaft and a cavity had developed under the raise drill. New casing was installed and set in a mixture of coke breeze and cement after which a plug was lowered 23 m (75 ft) down the shaft and this section was filled with grout and redrilled. No doubt feeling rather satisfied, they recommenced drilling at the 140 m (460 ft) level. At 146 m (480 ft), on investigating a roughness in the drilling, they discovered that the drill was grinding away the original 6.7 m (22 ft) section of lining which had now slid all the way to the bottom. Having ground away this lining, the drilling was completed in peace.

Reamer Heads

Though there have not been many problems associated with this side of raise drilling, occasionally the reamer head has parted from the rest of the drill string and dropped to the bottom of the hole. Mostly this can be attributed to human error. However, when raise-drilling work was being carried out in the Tynagh mine (Dublin), the reamer was dropped on no less than four occasions. Two of these were due to rod failure (in the one case a stabilizer* and, in another a saver sub-

---

*Similar problems were experienced by early shaft drillers when they encountered a mud vein and the mud compacted around the drill bit (See section on early drilling methods.)

*When the stabilizer broke, to the surprise and dismay of the operator the resultant release of torque caused the remainder of the drill string to unscrew and this, together with the reamer head, plummeted to the bottom of the hole.

assembly). The third drop was due to the shearing off of the bolts which attached the 2.1 m (7 ft) bolt-on stage, and the last drop occurred when the weight of the heavy reamer head bent and pulled down a support beam during the final stages of breakthrough.

On another occasion in the U.S.A., 305 m (1000 ft) of drill string and the reamer head were lost down a 610 m (2000 ft) hole because the operator did not ensure that the drill string was correctly threaded to the subassembly after pipe removal during the back-reaming operation. Amid stunned silence the Mine Manager looked down and remarked, casually, 'Yup, she's gone all right'.

Williams-Hughes Raise Drill

In 1960, about the same time that Cannon and the Robbins Company began developing their first raise drill, the Cleveland-Cliffs Iron Co. approached the Hughes Tool Company in order to discuss the feasibility of using large rotary-type equipment in its Mather mine.

Large-diameter rotary drills in use at that time were, however, cumbersome machines not suited to the confined spaces usually encountered underground. Moreover, most of them were only capable of drilling in soft sedimentary or metamorphic rock strata, not in the hard, highly abrasive greywacke rock formations which predominated in the Mather mine.

After investigating the various types of boring machines available (including tunnel-boring machines) and their specific areas of application, it was finally decided to concentrate on a unit which could be used for drilling raises.

A machine for this purpose was then designed and built by the Hugh B. Williams Manufacturing Company (a subsidiary of the Hughes Tool Company) with Hughes supplying the cutting bits. The machine was delivered to the Mather mine in 1962.

Initially, a pilot hole was drilled from the lower to the upper level using a 32 cm (13 in) diameter steel tri-cone bit. This hole was then back-reamed downwards by the machine to a diameter of 122 cm (4 ft).

In order to direct the cuttings towards the pilot hole the face of the conically shaped reamer head was fitted with replaceable rubber wiper blades held in place by metal brackets. A hopper with a spout attached was fitted to the top of the drill tower in such a position that it caught and directed the cuttings away from the drilling equipment and into a truck. (A similar disposal system was also used by the German companies of TURMAG and Wirth for deflecting the muck away from their raise- and shaft-boring equipment which, like the Williams-Hughes unit was stationed on the lower level.)

Water was used as a flushing medium and to keep the cutters cool, after a trial with air as a flushing medium proved unsatisfactory due to the dust created by the latter. Machine specifications were as follows:

Drive: Rig was driven at 10, 20, 38, and 60 rpm by a 74.6 kW (100 hp) electric motor
Thrust: 890 kN (200,000 lb)
Torque: 7192 kgm (52,000 lb-ft)

Note: As the rotary column frame was limited to a height of 3 m, six (three mounted on each side of the frame) triple-telescoping, double-acting, hydraulic thrust cylinders were installed. These cylinders provided a maximum thrust stroke of 6 ft (1.8 m).

The material penetrated was mainly hard greywacke interspersed with quartz, and the original steel shaft cutters fitted to the reamer wore away rapidly (i.e. two sets in 10.4 m of raise). These were then replaced with tungsten carbide insert cutters which proved satisfactory. As a result, the hard steel tri-cone pilot hole bits were also changed to tungsten carbide for subsequent raises.

The drill string stabilizers for the first pilot hole proved to be inadequate and the hole tended to deviate from the vertical. This in turn caused problems with the reamer because the drill string guiding it naturally followed the line of the pilot hole. As a result the reamer head frequently stuck on one side of the hole. It was freed by using the machine's rotational power, but the additional strain imposed on these components ultimately led to their failure. The

Figure 99. Wirth raise-boring machine. (Courtesy Wirth GmbH.)

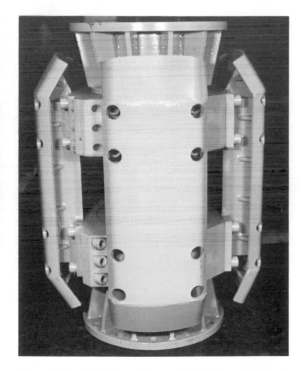

Figure 100. Wirth stabilizer which follows the 'hole opener' during the pilot hole drilling. (Courtesy Wirth GmbH.)

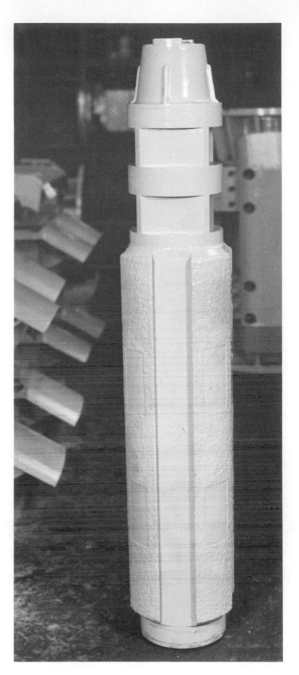

Figure 101. Wirth stabilizer which precedes the reaming head and assists in maintaining hole accuracy during the final stages of the raise-boring operation. (Courtesy Wirth GmbH.)

problem was overcome by using adequate reamer-stabilizers.

During the boring of six subsequent raises the machine was modified and improved several times so that by the time the sixth hole was being drilled the machine's downtime (changed from 56 to 12 min/m (17.1 to 3.7 min/ft) of raise) and consequently the cost of raise boring, had been dramatically decreased.

113

Figure 102. Wirth 2.40m (7 ft 10.5 in) diameter reaming head for their raise-boring machine. (Courtesy Wirth GmbH.)

Wirth Raise-Boring Machines

It is of interest to note that in 1963 the Wirth Company of Germany introduced an electro-hydraulic raise drill which followed the German system, that is the pilot hole was drilled from the lower level to the upper level and the hole was then reamed *downwards*. Since then the Wirth Raise Boring Machines have undergone considerable development and the latest models follow the Cannon principle in that the pilot hole is drilled down and the hole is then back-reamed upwards (Figures 99 to 102).

In keeping with other major manufacturers of raise drills, Wirth Maschinen und Bohrgerate Fabrik GmbH also now produce a box-hole drill.

Blind-hole Raise Drills[15]

The first blind-hole or 'box-hole' raise borer was introduced by the Calweld-Smith company in about 1967. Designated VTB (vertical thrust borer) (Figure 102a), the unit was planned for use in the New Mexico uranium fields.

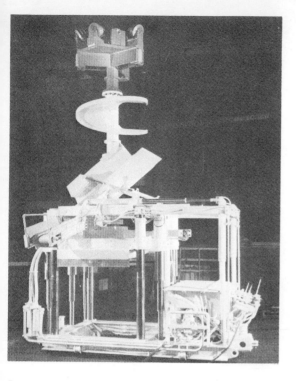

Figure 102(a). Calweld-Smith blind-hole raise drill, Model VTB. (Courtesy Calweld-Smith Inc.)

Rotational and thrust forces for the cutter head were provided via the drill string from the power unit located on the mine floor at the foot of the raise. The hole was bored in a single pass, no pilot hole being necessary. Facilities were also provided for the deflection, collection and loading of the cuttings.

In 1974 an experimental blind-hold raise drill built by Calweld-Smith and designated Model BH-80 was installed in the South African Orange Free State gold mines.

The following year (1975) the Anglo American Corporation purchased a Subterranean box-hole drill Model UR-60 for use in the South African gold mines. The first of these machines was produced in 1973, the same year Subterranean Tools Inc. became a wholly owned subsidiary of Kennametal Inc. (Later, in 1978, Kennametal Inc. ceased marketing rock-cutting machines and a new company 'Subterranean Equipment Co.' was formed to service Subterranean Tools Inc. customers).

The UR-60 was capable of boring a nominal raise diameter of 1.5 m (5 ft) and three of these

units were built. The smaller UR-36, capable of cutting 0.9 m (3 ft) diameter raises was produced in 1975.

An important lesson learned from the use of these early machines was the unsuitability of the original flanged pipe connections and the rotating stabilizers which constantly failed. Threaded pipe replaced the flanged pipe connections and non-rotating stabilizers were introduced.

In the case of machines using rotating drill strings the non-rotating stabilizer consists of arms attached to the drill string via roller bearings which enable the string to turn while the stabilizer remains static. So far as the Subterranean machine is concerned, use is also made of rotating centralizers which assist in maintaining gauge immediately behind the cutter head.

In 1973, at about the same time as the BH80 was introduced, the Robbins Company were commissioned to build a blind-hole raise drill designated Model 51R. This machine successfully drilled a number of holes and served as the prototype model for the 52R.

Six blind-hole raise drills (Model 52R) (Figures 102b and 102c) made by the Robbins Company and first introduced in 1974 were delivered to various gold-mining companies in South Africa. These machines differed from the Calweld-Smith and Subterranean machines in that, although thrust was still provided via the drill string, torque was supplied at the cutter head, thus obviating the necessity for the pipes to rotate. Large diameter pipes continuously supported along the length of the drill string by stabilizer fins were necessary for this type of drill in order to provide rigidity and prevent buckling. This arrangement facilitated high thrusts through the drill string with the reduced possibility of buckling. When boring upwards, the cuttings were allowed to drop down between adjacent sets of stabilizers onto a muck-collecting apron fitted above the derrick assembly. A 1.52 m (60 in) diameter adapter ring was bolted to the top of the muck collector and during the drilling operation the muck collector was raised until the ring entered the bored raise. This ensured that the work area below was kept completely protected from rocks and dust.

Figure 102(b). Robbins 52R blindhole raise drill being assembled in the factory. (Courtesy The Robbins Company.)

A 'head stabilizing stinger' or pilot drill (Figure 102d) with a button tri-cone bit mounted on the end of the stinger (somewhat similar to the pilot drill used on the Alkirk hard rock tunneller built by the Lawrence Mnf. Co. — see section on 'Modern rock tunnelling machines: Lawrence Mnf. Co.') was fitted to the centre of the domed cutter head. Extra stabilization was provided by four rollers mounted on the skirt below the cutter head.

Either disc (for rapid and economic excavation) or tungsten carbide button (for hard-ground conditions) cutters could be fitted to the cutter head and flushing water and spray nozzles provided dust suppression, cooled the cutters and tri-cone bit and cleared rock chips and cuttings from the cutter saddles.

Three stabilizer fins were longitudinally mounted on each drill pipe section. These fins were enclosed U-shaped pieces and acted to stabilize the drill pipes and also provide complete enclosed protection for the various hoses and cables feeding hydraulic and electric power to the rotational drive at the cutter head.

The derrick assembly consisted of a base plate, main frame, a crosshead, two columns, two thrust cylinders and an upper worktable or headframe. The headframe was machined to allow the non-rotating drill pipe to pass through with a sliding fit.

After each new pipe had been connected to the string after passing through the derrick crosshead, the three stabilizer fins were manually attached, being spaced at 120°. Once attached the fins interlocked to provide continuous stabilization throughout the total length of the hole.

A hydraulically operated clamp fitted to the headframe of the derrick supported the drill string column when a pipe was being added. As a safety precaution the pipe clamp could not be opened when in a closed position without manually opening a special safety locking mechanism.

Figure 102(c). Robbins 52R blindhole raise drill being installed underground. A cuttings collector which collects and deflects the cuttings is fitted to the unit above the derrick. During the drilling operation an adaptor ring is fitted to the collector and the collector and ring are raised until the ring enters the bored raise. This effectively seals the work area below from dust and debris. (Courtesy The Robbins Company.)

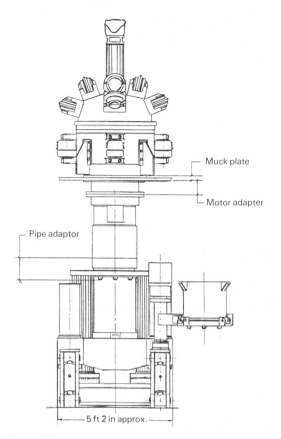

Muck plate

Motor adapter

Pipe adaptor

5 ft 2 in approx.

Figure 102(d). Specification drawing showing 'head stabilizing stinger' or pilot drill. Front view (shown extended by one pipe length). (Courtesy The Robbins Company.)

The derrick assembly was in turn mounted on a swivel base which was mounted on a sub-base. This allowed the blind-hole borer to be rotated to any desired drilling angle.

Initially the heads of the Robbins machines were powered hydraulically, but this system was not found to be successful because of the power loss experienced as the hole became deeper. (Hydraulic fluid was pumped to the cutter-head motor through hoses from the power unit below.) As a result later models were fitted with an electric drive. The first of these models was introduced towards the end of 1979.

Since the introduction of the first Calweld-Smith unit in 1967, blind-hole raise drills have undergone considerable developmental improvement, becoming smaller and thus more manoeuvrable underground. The site erection system has also been improved.

According to C. Smith,[15] experience in the South African mines has indicated that the optimum boring depths for these machines was between 15 and 90 metres (50 and 300 ft). Below 15 metres set-up time and transport made their use uneconomical, while the machines have not proved practicable beyond a distance of 90 metres.

In a telex message received by the author in November 1980 the Robbins Company reported that at that time there were only six blind-hole raise drills with rotating drill strings in operation, while there were 14 machines with

117

Figure 103. Koken segment erector blind-hole raise drill. (Courtesy Koken Boring Machine Company Ltd.)

non-rotating drill strings in the field and a further four were to be commissioned within the following six months. This seemed to indicate a preference in the field at that time for the non-rotating type of machine.

Segment Erector Blind-hole Raise Drill[10]

In the Matsumine black ore mine in the northern part of Akita Prefecture Japan, difficult strata involving soft ground, heavy earth pressures, clayey collapsible rock, and water problems, led to the development of a novel segment erector blind-hole raise drill.

This machine was built when it became apparent that conventional methods such as raise boring, cage raising, stage blasting and blind-hole raise drilling were not only unsuitable and inefficient but extremely dangerous methods if used in such difficult ground conditions.

The development of the unit was a joint undertaking by the Dowa Mining Company Limited and the Koken Boring Machine Company Limited.

Basically the unit consists of a boring

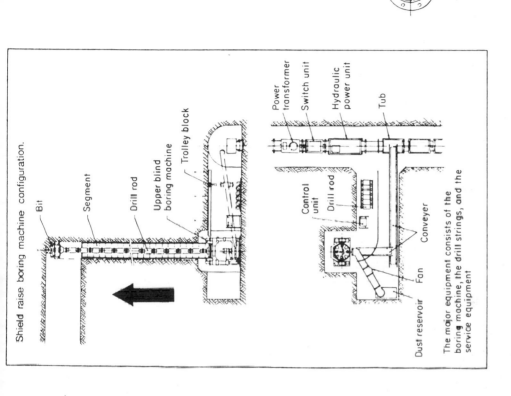

〈Roller bit〉

〈Drill pipe〉

Shield raise boring machine configuration.

Bit
Segment
Drill rod
Upper blind boring machine
Trolley block

Power transformer
Switch unit
Hydraulic power unit
Tub

Control unit
Drill rod
Fan
Conveyer
Dust reservoir

The major equipment consists of the boring machine, the drill strings, and the service equipment

Description of the Equipment

1 Turn table
2 Segment frame
3 Drill pipe
4 Drill pipe positioner
5 Drill head

6 Base
7 Thrust cylinder
8 Column
9 Segment jacking cylinder
10 Segment holder

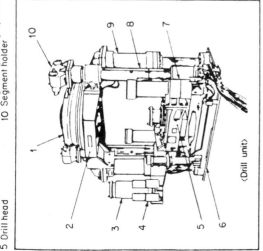

〈Drill unit〉

Figure 104. Segment erector raise drill — description of equipment. (Courtesy Koken Boring Machine Company Ltd.)

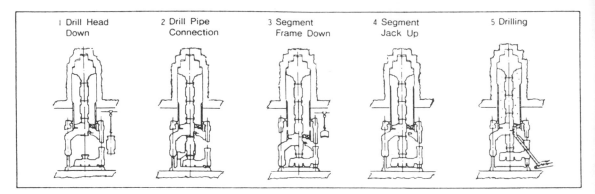

Figure 105. Segment erector raise drill — operation procedure. (Courtesy Koken Boring Machine Company Ltd.)

machine, the drill string, and the segment erector components (Figures 103 to 105). A conventional Christmas tree shape consisting of four tungsten carbide tip insertion kerf roller cutters headed by a tri-cone bit form the cutter-head section of the blind-hole drill apparatus. Beneath the cutter head is a turntable thrust ring on which the shaft-lining segment rings are assembled. Cuttings from the cutter head are directed by a scraper inside the segment erector frame to a discharge chute. Beneath the segment erector assembly is the drill head frame which provides revolution and thrust to the drill pipes. Torque is provided by two oil hydraulic motors (cam rotor type) which have a low profile to decrease machine height.

Two columns, each fitted with a jacking cylinder, are positioned on either side of the drilling machine. At the head of each column is a segment holder which supports the segment as it is assembled. A spring on the segment holder enters automatically into a hole in the segment to support it as it is pushed upwards.

Five segments or blocks complete a ring of shaft lining. In order to reduce friction as the segments are moved up the shaft, each segment is lined on its outer surface with steel plate. Square holes are provided at the appropriate positions to facilitate assembly from the external side of the segment.

A hinged bar-like step is fitted to the inside of a segment so that a climbing ladder is made available to workmen as the lining is erected. The drill pipes which are conveyed by trolley hoist are gripped and positioned by the drill

pipe positioner which automatically centres them.

The machine body may be reduced in height to 1394 mm (4 ft 7 in) from rail surface for transportation underground if the segment jack assembly, segment holder, column upper half, and upper frame are removed. Rail wheels (rail gauge 762 mm (2 ft 6 in)) are fitted to the base of the machine body.

When the shaft is completed the cutter head and drill string are withdrawn down the shaft by decreasing the cutter diameter. According to Haruo Ohshika, Manager, Project Engineering, of the Koken Boring Machine Company Limited, this is accomplished by manually moving the cutters inward. The head is also withdrawn in this manner for maintenance purposes, if required.

Operation procedure is as follows: at the end of the boring stroke the chute and conveyor which discharges the cuttings are disconnected and removed, the drill pipe holder is closed and the drill head is lowered. A new drill pipe is brought in by the drill pipe positioner and connected. The drill pipe holder is opened and the segment frame is lowered for the connection of another segment piece. The segment is jacked up. After a complete ring of segments has been jacked up, the chute and conveyor are reconnected and a new boring cycle commences.

Boring depth is limited to 30 m (100 ft), and within this limit the 406 mm (16 in) diameter drill pipes are strong enough to accept the maximum thrust load imposed on them without requiring the use of a stabilizer.

The drill pipes are the same diameter as the flange and can thus pass through the centre hole of the segment frame without a gap. The drill pipes are connected with a pair of tapered pins for easy assembly.

A New Generation of Raise-boring Machines

Following the development of the 121R for 3.6 m (12 ft) diameter holes, 914 m (3000 ft) deep, built for the Western Mining Corporation, Western Australia, the Robbins Company were commissioned in 1978 to build two drills, Model 121BR for the S.A. Healy Company for use on the Chicago Drainage Project and Model 80BR for use by the Ball-Healy-Horn (Joint Venture). These machines, that is the 121BR and the 80BR, were in effect somewhat of a cross between a conventional big hole drill and a raise drill and were designed to drill 2.5 m (8 ft) blind holes using the 'closed circuit reverse circulation method' of drilling. In the case of the 121BR, when the blind hole was completed a crew would be lowered into the hole to excavate a starting area for a 3.6 m (12 ft) expanding reamer head and the pilot hole would be back-reamed in the normal way, the muck falling to the bottom of the hole. Later this muck would be removed by a loader and skip arrangement. If these machines and new method of drilling prove successful the entire concept of shaft sinking may be changed. (For further information on the Robbins 121BR machine and expanding reamer see section on 'Robbins blind-shaft borer: 121BR blind-hole/raise drill reverse circulation, air lift system'.)

Large-diameter Shaft Drills — The German Experience

The TURMAG Shaft Drill[11]

Serious developmental work on large-diameter bore-hole drills was undertaken in Germany in 1954. In 1959 Turbo-Machinen-Aktiengesellschaft Nüsse & Grafer (TURMAG) produced their revolutionary large-diameter hole drill Model P-1200. As a result of this in 1962 the company was presented with the Blue Ribbon Mining Award for their contribution towards the development of large hole drills.

A significant feature of the TURMAG drill was that a pilot hole of 193 mm (7½ in) was first drilled upwards. This hole was later down-reamed to a diameter of appoximately 800 mm (2 ft 7 in) by a unit powered by compressed air.

In 1964 Salzgitter Maschinen AG used a multi-stage reamer head to enlarge a pilot hole to a diameter of 4.80 m (15 ft 9 in). Again the pilot hole was drilled from the lower level to the upper one and then later reamed downwards. This machine was also air powered. According to Dr. K. Wollers[12] of Thyssen Schachtbau GmbH, rotational torque was applied to the reamer head of the Salzgitter unit from the upper level through a drill string while thrust was applied from the lower level by means of a chain and winch system. However, apparently this method of reaming was not particularly successful, mainly because the reamer head tended to deviate from line. The machine was finally abandoned.

The Salzgitter machine was followed by another TURMAG drilling system which was produced in 1967, namely Model TE-500. It comprised a power unit and a reamer head with the drill string as the connecting element. The cone-shaped reamer head of this machine was fitted with disc-type cutters of different diameters. These cutters were arranged in pairs and each pair was individually driven by means of a series of reduction gears which controlled their speed of rotation. On the first shaft the pneumatically operated power unit P1200 was used. In 1974 this was replaced by the electro-hydraulic unit E-6000 — 132 kW (180 hp). Torque was applied to the reamer head through the drill string from the power unit positioned on the lower level, while thrust was primarily applied by the dead weight of the reamer head, as the thrust of the P1200 was only 12 t.

As the diameter of the cutter head increased towards the upper end of its cone so, correspondingly, the diameter of the cutting discs was increased in size so that when the cutter head was 5.40 m (17 ft 9 in) in diameter, each of the disc cutters at that point were 1.20 m (3 ft 11 in) in diameter.

In this manner TURMAG strove to counteract the torque resistance which

Figure 106. Wirth shaft-boring machine, Model GSB-V450/500 used in the 'Emil Mayrisch' and Walsum colliery shafts. (Courtesy Wirth GmbH.)

Figure 107. (right) Wirth shaft-boring machine. (Courtesy Wirth GmbH.)

increased with the diameter of the bore, and thus obviate the necessity to install more powerful motors and heavier drill pipes when boring large-diameter shafts.

In addition, to reduce wear and tear on the cutters, TURMAG adjusted the diameters of the cutters so that the peripheral speed of each pair was kept constant in relation to the neighbouring pair of cutters.

Immediately above the cutter head a ring-beam platform was erected which could be raised or lowered on a winch. A man-cage was also provided to give quick access to the platform from the top level. The platform was fitted with hydraulic clamping jacks which thrust against the walls of the shaft. As the cutter head moved downwards the shaft was simultaneously lined by workmen on the platform. A hydraulic lifting mechanism was also fitted to the platform which was capable of supporting the cutter head. This support was found to be necessary when cutters needed

122

Figure 108. Cutter head for Wirth's new shaft-drilling machine — diameter 6.0 m (19 ft 8 in). (Courtesy Wirth GmbH.)

changing, the reaming operation was about to commence, or when soft formations were penetrated. In the latter event if the shaft had been bored with the full weight of the head, the torque required would have been too high.

According to Messrs. Karl-Heinz Brümmer and Karl Wollers in their paper 'Experience with shaft boring and mine developments in German coal mines',[12] which was presented at the Las Vegas 1976 Rapid Excavation and Tunnelling Conference, two shafts in coalmines were at that date drilled with this unit. They measured 4.40 and 5.40 m (14 ft 5 in and 17 ft 9 in) in diameter and were about 100 m (330 ft) in depth. The pilot holes measured 1.6 m (5 ft 3 in) in diameter.

Further developmental work in this direction was carried out by Wirth & Co. KG.

Wirth Shaft-Boring Machines

In March 1971 work was commenced in the Emil Mayrisch colliery on the pilot hole for a shaft. The work was undertaken by a consortium of shaft-sinking and tunnelling contractors, Messrs. Deilmann-Haniel GmbH of Dortmund and Thyssen Schachtbau GmbH of Mülheim/Ruhr. A 1.20 m (3 ft 11in) diameter pilot hole was first drilled with a Wirth raise borer and this hole was then enlarged to 4.50 m (14 ft 9 in) with a Wirth GSB-V-450/500 (Figures 106 and 107) shaft-boring unit. The shaft borer's cylinders were capable of exerting a maximum thrust of 2940 kN (660,000 lb) and its cutter head was dressed with disc cutters.

The shaft machine was kept on course by means of a laser guidance system and alignment of the unit was carried out from the operator's control panel in the machine.

As the borer descended the shaft was lined from a working platform at the rear of the unit. The distance of this platform from the edge of the cutter head was approximately 8 m (26 ft). However, if conditions such as fractured or faulty rock strata were met special arrangements could be made to install the lining down the shaft to within 3 m (10 ft) of the cutter

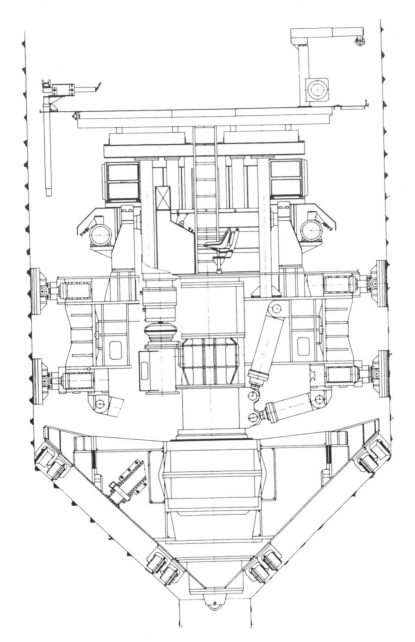

Figure 109. Diagram of Wirth's new shaft-boring machine, Model SB VI-500/600. (Courtesy Wirth GmbH.)

head. The 240 m (790 ft) deep shaft was holed through on 10 August 1971. Later the same machine was used to bore the Bergbau Ag. Niederrhein Walsum colliery shaft. Work on this project was commenced in January 1972 and was completed in September 1972.

Average penetration rates through the Carboniferous formations were approximately 2 m (6 ft) per day and 11.4 m (37 ft) per day for the Emil Mayrisch and the Walsum shafts respectively. Shafts sunk by this machine since 1972 are given in Table 3.

Based on six years' experience with the prototype GSB V-450/500, a new machine, (Model SB VI-500/650) which incorporated various modifications, was manufactured by Wirth. In August 1977 it was installed at the Ibbenbüren coal-mine. By December 1977 it had successfully drilled a 525 m (1700 ft) deep shaft, averaging approximately 10 m (33 ft) per drilling day.

Later the unit was moved to the Göttelborn mine of the Saarbergwerke AG where it was used to deepen the No. 3 'day shaft' an

Table 3

Year	Mine	Depth		Average drilling rate	
		m	(ft)	m/day	(ft/day)
1973	Zollverein	227	(745)	9.02	(20)
1974	Sterkrade	196	(643)	13.52	(44)
1974/75	Carl Alexander	228	(748)	8.00	(26)
1975	Minister Stein	302	(990)	9.64	(32)
1976	Emil Mayrisch	226	(741)	6.60	(22)
1976/77	Rossenray	212	(696)	10.30	(34)

additional 440 m (1440 ft). Penetration rates down the 6.50 m (21 ft) diameter shaft averaged about 13.83 m (45 ft) per day. During this period (i.e. from March 1978, when the deepening work was commenced, to May 1978, when it was completed) a new record of 30.10 m (99 ft) sunk in a single day, was established. According to Dr. Wollers, the high penetration rates of the shaft-boring machine were matched by the fast installation of the new steel ring lining. This lining system was designed and produced by Thyssen Bergbautechnik.

Wirth/Thyssen Schachtbau/Deilman-Haniel Blind-hole Shaft Borer

The same year (i.e. 1978) Thyssen Schachtbau GmbH and Dellmann-Haniel GmbH used machine Model GSB V-450/500 for boring a trial blind-hole shaft (without a pilot hole) at the Gneisenau coal-mine. To accomplish this the unit was equipped with a hydraulic muck removal system. The cuttings were hydraulically removed from the shaft face by centrifugal pumps which forced the mixture of slurry and cuttings through pipes to the head of the blind shaft.

According to Dr. Wollers, the information gained from this test has been of considerable value and will undoubtedly influence the design of the new shaft-boring machine proposed by the Thyssen Schachtbau Company. In particular, it is expected that the hydraulic muck removal system will be redesigned to overcome certain difficulties experienced with the trial machine. Apart from this, the new unit will probably differ in many respects from conventional shaft-reaming machines so far produced.

Robbins Blind-shaft Borer[13]

In 1976 the Robbins Company of Seattle received a contract from the United States Bureau of Mines for the construction of a 7.44 m (24 ft 5 in) diameter blind-shaft boring machine, which was to be used by the Cementation Company to bore a mine shaft for U.S. Steel.

The borer, designated Model 241SB-184, (Figures 110 and 111) was completed in May 1978 and delivered in August. It was capable of transportation by truck and, according to the manufacturers, was designed to bore at a maximum of 12.3 m (40 ft)/day. This complex machine is designed to bore vertical holes downwards and remove the muck upwards without the benefit of a pilot hole.

Basically the machine consists of an 11.6 m (38 ft) long central column or cylinder. This column is carried at the bottom by the cutter-head support and extends to the hydraulic swivel at the top of the machine.

Four decks which are used for muck transfer, machine control, and shaft lining are mounted on this central column. The muck-measuring pockets, chutes, and stationary sections of the muck-collecting carousel are also supported by the centre column.

Cutter head

The cutter-head support, which houses the non-rotating inner face of the 86 in (218 cm) i.d. (98 in (248 cm) o.d.) Torrington double-rowed tapered roller-bearing, is the principal load-bearing structure of the machine, providing both torque reaction from the rotating cutter head to the stationary gripper ring and the

BLIND SHAFT BORER
ROTATING COMPONENTS

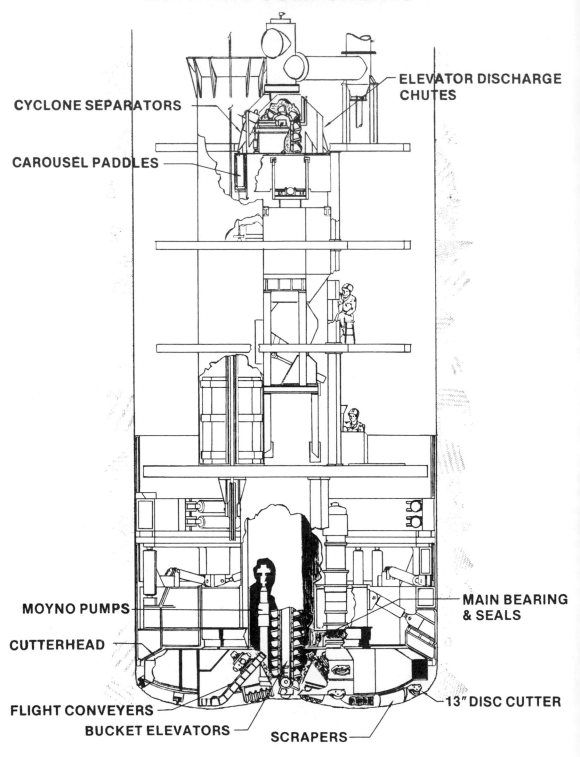

CYCLONE SEPARATORS

CAROUSEL PADDLES

ELEVATOR DISCHARGE CHUTES

MOYNO PUMPS

CUTTERHEAD

MAIN BEARING & SEALS

FLIGHT CONVEYERS

BUCKET ELEVATORS

SCRAPERS

13" DISC CUTTER

Figure 110. Robbins blind shaft borer: rotating components (Courtesy The Robbins Company.)

126

BLIND SHAFT BORER
NON-ROTATING COMPONENTS

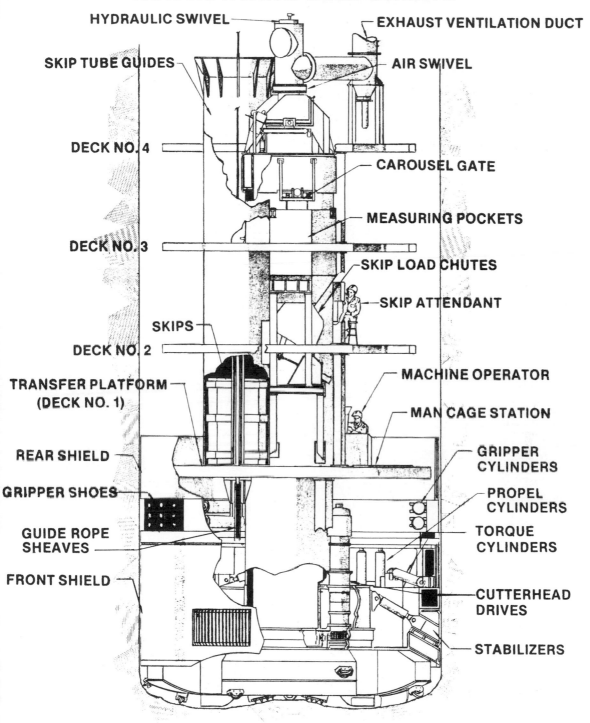

HYDRAULIC SWIVEL

EXHAUST VENTILATION DUCT

SKIP TUBE GUIDES

AIR SWIVEL

DECK NO. 4

CAROUSEL GATE

MEASURING POCKETS

DECK NO. 3

SKIP LOAD CHUTES

SKIP ATTENDANT

SKIPS

DECK NO. 2

MACHINE OPERATOR

TRANSFER PLATFORM
(DECK NO. 1)

MAN CAGE STATION

REAR SHIELD

GRIPPER
CYLINDERS

GRIPPER SHOES

PROPEL
CYLINDERS

GUIDE ROPE
SHEAVES

TORQUE
CYLINDERS

FRONT SHIELD

CUTTERHEAD
DRIVES

STABILIZERS

Figure 111. Robbins blind shaft borer: non-rotating components. (Courtesy The Robbins Company.)

integral with the central circular section and two are removable outrigger plates. These removable sections allow the cutter head to be reduced in overall weight and size to permit it to be hoisted through limited clearances in the shaft.

Fifty-six disc cutters are mounted on the cutter head. At the centre of the cutter head, twelve 33 cm (13 in) cutters are arranged in three groups in the form of a cross (each group consisting of a cluster of four discs — two twin disc cutters). The remaining forty-four 33 cm (13 in) cutters are dispersed singly over the balance of the head.

Six independent 93 kW (125 hp) electric motors totalling 560 kW (750 hp) provide rotational power to the cutter head. The output drive pinions of these motors (each driven through a two-speed gearbox) mesh with a common internal ring gear on the cutter head.

The six electric motors and their gearboxes, as well as four 37 kW (50 hp) double-ended electro-hydraulic power units and one 22 kW (30 hp) unit, are mounted on the cutter-head support deck. These latter units provide the power for all the machine hydraulics: motors, cylinders, conveyors, elevators, and pumps.

Oil for the various hydraulically driven auxiliary functions which rotate with the cutter head, such as the flight conveyors, bucket elevators, and slurry pumps, etc. must pass through the rotary swivel at the top of the central column before being fed to the hydraulic motor in question.

Figure 112. Robbins blind shaft borer, Model 241SB-184. (Courtesy The Robbins Company.)

thrust required for penetration. In addition, the cutter head support structure absorbs the thrust of the stabilizers on the shaft wall, providing directional stability for the machine.

The head consists of a central circular section with four radial extensions each carrying six roller cutters, including the gauge cutters, and a scraper. These extensions are arranged at 90° intervals around the periphery. Two are

Propel System

Nine hydraulic propel cylinders (acting in groups of three — the groups being arranged at 120° intervals around the central section), mounted between the gripper ring and the cutter-head support are capable of producing a total thrust of 6361 kN (1,430,000 lb).

During the propel cycle stabilizer shoes are extended against the wall, but not gripped, to provide stability. Gradual steering corrections may be effected by deliberately creating an imbalance between the stabilizer shoes.

The machine has a boring stroke of 76 cm (30 in).

Muck-Removal System

Scraper blades on the cutter head move the cuttings towards the lowest point of the cut face where they are collected by two horizontally positioned single-chain flight conveyors and swept up ramps. The cuttings are deposited from each of the two ramps into independent hydraulically driven bucket elevators. Polyethylene buckets mounted on these chain-type elevators lift the muck vertically about 14.3 m (47 ft) and drop it into an annular collection hopper or carousel. Paddles which extend down from the rotating structure in the centre into the non-rotating collection carousel, move the spoil around to either of two openings in the base of the trough, so that it falls into one of two 2.8 cu m (100 cu ft) measuring pockets.

Hydraulically controlled gates on the measuring pockets allow the spoil to drop into a skip capable of taking the full pocket load. The system is so arranged that when the measuring pocket is full, the gate under the carousel closes and a gate beneath the measuring pocket opens to allow the spoil to drop into a waiting skip. The skip carries the material to the top of the shaft. While the skip is being filled (approximately eight seconds) the paddles in the carousel move the material into the other measuring pocket. The hydraulically controlled gates on the carousel and measuring pockets are fitted with an interlock system which ensures the timely discharge into or from the measuring pockets.

Ground Support and Gripper System

Temporary ground support is provided by a full shield which extends from immediately behind the cutter head to 60 cm (2 ft) above the muck transfer deck. The stabilizer shoes and the grippers are operated through windows in the shield.

The gripper system uses three shoes, forming a ring mounted around the periphery of the machine. At each joint the shoes are connected by a pair of hydraulic cylinders, tangential to the circumference, which expand the ring against the shaft wall. Alternating protrusions and recesses are provided on the outer surface of the gripper ring. These protrusions extend through openings in the shield and contact the shaft wall.

Dewatering and Air-ducting Systems

Two 380 litre/min (100 gal/min) Moyno slurry pumps which also rotate with the head are available for dewatering the hole bottom. Their output may be directed through a cyclone or, if desired, dropped directly into a settling tank. Two 380 litre/min clear-water pumps are also available for pumping excess water to the surface.

To keep the manned areas free from dust and methane gas, fresh air is ducted into the working section via a 90 cm (36 in) diameter pipe at the rate of up to 850 cu m/min (30,000 cu ft/min). The air passes down to the face of the machine through an annular clearance allowed between the upper cutter-head support deck and the shaft wall. From the hole bottom the air is drawn up through the machine central column and into a 76 cm (30 in) diameter duct running to the surface.

The following safety interlock features are incorporated in the machine:

(a) The muck-loading gate cannot open unless a skip is present.
(b) The upper gate under the carousel closes when the storage pocket is full.
(c) The skip cannot leave until the lower gate is closed.
(d) If both measuring pockets are full, machine 'shutdown' is activated.
(e) If gripper pressure drops below 12.4 MPa (1800 lb/in^2) head rotation is interrupted.
(f) Head rotation is also halted if low lubrication oil pressure is indicated.
(g) If a gas level of 2 per cent is recorded on any of the three continuous methane monitors located at strategic positions, power at the machine is automatically interrupted.

Other safety interlock features incorporated include stage separation, ground check circuits, short and overload protection, low air pressure, etc.

In his paper entitled, 'Blind shaft construction new equipment update',[14] James E. Friant

Figure 113. Robbins 121BR Blind-hole raise drill — pipe-loading sequence — pipe section placed in loading arm. (Courtesy The Robbins Company.)

Figure 114. Pipe-loading sequence — table wrench secures the string. (Courtesy The Robbins Company.)

121BR Blind Hold/Raise Drill Reverse Circulation, Air Lift System[14]

(the Robbins Company, Seattle), reported that during the first period of boring a number of hydraulic, electrical, and mechanical faults had developed. In addition, handling problems had arisen because of wet muck which created difficulties during delivery to the bucket elevators and discharge from the buckets at the top of the machine.

After some minor modifications, the conveyors, on the other hand, had handled the wet muck satisfactorily. A temporary solution to the bucket problem had been to add more water and thus decrease the viscosity of the muck. Friant commented further that while this has assisted in keeping the muck moving, attempts to dry the shaft had nevertheless not been entirely abandoned at that stage.

Other innovations introduced by the Robbins Company and described in James E. Friant's paper included the 121BR blind-hole/raise drill reverse circulation, air lift system, an expandable reamer and the 1211 shaft reamer machine.

In 1978 the Robbins Company were commissioned to build a raise drill (Model 121R) for use by the Western Mining Corporation, W. Australia. Like the Robbins 81R raise drill, this unit was designed specifically as a surface rig for aboveground operation. Made for 3.6 m (12 ft) diameter raises to depths of up to 900 m (3000 ft) the unit uses 32.7 cm (12⅞ in) diameter drill rod. According to the manufacturers, the machine is capable of producing a maximum thrust of 9000 kN (2,000,000 lb) and 700 kNm (500,000 lb ft) of torque.

This machine formed the basis for the design of Model 121BR (blind raise) for the S.A. Healy Company for use on the Chicago Drainage Project and Model 80BR for use by the Ball-Healy-Horn (Joint Venture).

The 121BR machine was used to bore 2.5 m (8 ft 3 in) shafts in Chicago limestone using the closed circuit reverse circulation drilling system

Figure 115. Pipe-loading sequence — pipe section installed. (Courtesy The Robbins Company.)

Figure 116. Pipe-loading sequence — operation restored. (Courtesy The Robbins Company.)

(see section on Vertical Drills: Closed-circuit reverse circulation drilling), except that in place of conventional big-hole drilling equipment, raise drill equipment was specially designed and built for the purpose by the Robbins Company. The difference between this type of machine and conventional big-hole drilling equipment is the replacement of a kelly drive system by an in-line travelling chuck and drive train hydraulically powered for thrust in the same manner as a conventional raise drill derrick.

Two beneficial side effects from this revolutionary new type of drill have been better control of torque input and bit loads in rotary blind drilling, providing added protection to the pipe connections. This refinement is not possible with normal fixed-gear drive systems used on conventional drilling rigs.

By the end of 1979 about eight of the shafts on the Chicago Drainage Project had been completed. These are approximately 91 m (300 ft) in depth.

After the initial 2.7 m (9 ft) diameter pilot shaft had been drilled using the reverse circulation method, two of the shafts were back-reamed to a diameter of 3.66 m (12 ft).

A domed head fitted with disc cutters leads the string of double tube pipes into the hole. Cuttings are moved to the mouth of a slightly off-centre pickup by scrapers attached to the cutter head. Two hydraulic motors combine to supply 336 kW (450 hp) to the floating drive head of the 121BR. The head is mounted on an underground raise drill type of derrick. Use of relatively short pipe sections (3.5 m (11.5 ft)) has permitted the production of a powerful but nevertheless lighter and more compact machine than a conventional drill unit normally used for this type of work on the surface. Each pipe section consists of two concentric tubes (i.e. an inner tube of 25.4 cm (10 in) i.d. and an outer tube of 41 cm (16 in) o.d.).

No kelly is required as the head drives the drill pipes directly. A table wrench (Figure 114)

131

holds the drill string when new pipe sections are added.

A rotary swivel at the drive head discharges slurry when the drill is operated in the large-diameter blind-drilling mode and introduces bailing fluid when the machine is being used for drilling a pilot hole with direct circulation.

The unit's lifting capability of 5560 kN (1,250,000 lb) represents maximum thrust available for when the machine is being used for back-reaming. Thrust, in the blind-drilling mode, is provided by steel weights (doughnuts) placed on the mandrel behind the head. Control on bit pressure is accomplished by increasing or decreasing the lift on the drill string.

During operation the shaft is maintained nearly full of water at all times. High-pressure air from a diffuser built into the drive head is injected into the bailing fluid via the annulus between the two pipes. The air emerges a few feet above the pick-up point and enters the central column, reducing the density of the medium in the column. This condition in turn creates a hydrostatic pressure differential which causes an upward surge of the medium in the central column. The discharge from the machine flows into a large vertical tank where the air is allowed to escape and then into an inclined trough. Heavy material is moved up the trough incline by an auger or screw conveyor and fed into a discharge chute, while the water drops back to the lower end of the trough. The trough water overflow is then pumped through a centrifugal separator and back to the shaft for recirculation. The 'paste' or 'thick water' which emerges at the bottom of the separator is channelled to a waste basin.

Friant reports that during the boring of the first shaft the drill head was equipped with carbide button cutters mounted on standard 30 cm (12 in) cutter bearings. Drill weights totalled 1780 kN (400,000 lb). Maximum penetration achieved was 0.46 m (1.5 ft) per hour and this figure was only reached after the central cutters had been replaced by disc-type cutters. Torque and thrust were varied up to 1300 kN (300,000

Figure 117. Robbins expandable reamer after completion of the first shaft in Chicago limestone. (Courtesy The Robbins Company.)

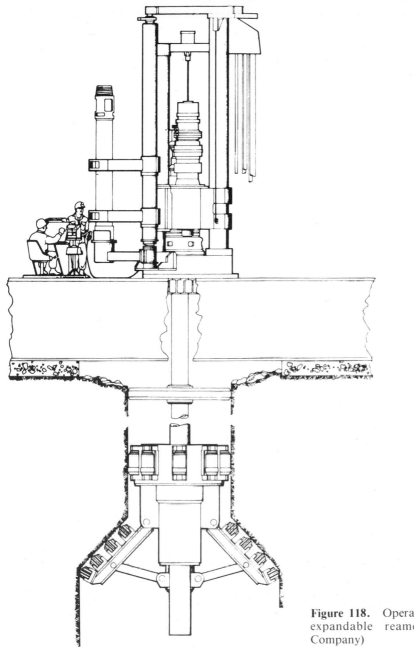

Figure 118. Operational sketch of the 121BR and expandable reamer. (Courtesy The Robbins Company)

lb) of bit thrust and 407 kNm (300,000 ft lb) of torque. Before commencing the second shaft the head was equipped with disc cutters only and 61 cm (2 ft)/hour was achieved with only 890 kN (200,000 lb) of bit force and 190 kNm (140,000 ft lb) of torque at 6 rpm. Subsequent shafts were drilled at rates exceeding 1 m/hour (3 ft/hour).

During the boring of the second shaft after a start-up period of 3.5 min to establish a flow, flow water (after separation of the muck) was measured at about 5000 litres/min (1100 gal/min). Lifting velocity was therefore calculated at 82 m/min (270 ft/min), slurry load by weight being calculated at 3.6 per cent.

Friant further reported that the chips resembled tunnel borer cuttings rather than normal drill cuttings. Indeed one chip found actually measured 18 × 9 × 3.3 cm (7 × 3.5 × 1.3 in). It weighed 0.68 kg (1.5 lb). Despite the fact that such unusually large chips were being lifted, no blocking was experienced either at the bit gauge or in the string itself.

Bit pressure and torque were increased slightly to 1100 kN (250,000 lb) and 2000 kNm (150,000 ft lb) respectively, during the boring of the third shaft, resulting in an average penetration rate of 91 cm/hour (3 ft/hour). Fears that the cuttings from the disc cutters would prove too large to be moved to the pick-up point, or that they would not float well proved groundless. In fact, as Friant commented rather the reverse was true as the hubs and cutter retaining ring bearings in the pick-up region were battered by chips because of the excessive turbulence. Friant suggests that gradually, as improvements are made to the prototype unit, penetration rates of from 1.52 to 1.8 m/hour (5 to 6 ft/hour) could well be expected.

The change from carbide button cutters to disc cutters brought about a marked cost reduction worth noting, namely from U.S.$7.60 to U.S.$11.40 per cu m (for carbide button cutters) to U.S.$1.57-U.S.$2.66 per cu m (for disc cutters).

Expandable Reamer[14]

After the first shaft had been bored, it was enlarged conventionally at the bottom to accommodate an expandable reamer (Figures 117 and 118) equipped with 39 cm (15.5 in) disc cutters.

A quill shaft was inserted into the standard double-wall pipe and the expandable reamer was fitted to the quill. The reamer was then expanded to the required diameter of 3.66 m (12 ft) by rotating the quill shaft which spread the reamer arms.

The reamer was pulled up the shaft and rotated by the 121BR raise drill. Cuttings from the reamer fell back into the shaft, and in this respect Friant comments that 'the cuttings did not "catch up" to the reamer'.

While reaming performance was satisfactory and cutter wear minimal, it was found necessary to make some modifications to improve the expansion and contraction actions of the reamer.

1211SR Shaft Reamer[14]

Another prototype shaft-reaming machine used on the Chicago Water Storage project and described in Friant's paper was the 1211SR shaft reamer, also built by the Robbins Company.

This machine was specifically designed to enlarge an existing 1.82 m (6 ft) diameter vertical shaft to a diameter of 3.66 m (12 ft) in relatively competent rock. By 1979 three shafts, 91 m (300 ft) deep, had been completed in the Chicago limestone.

The unit functions in much the same manner as a conventional rock T.B.M., in that grippers are extended to anchor the machine against the shaft walls while the cutter head and support are thrust downward into the shaft face by the propel cylinders. However, there are significant and interesting differences. The 1211SR is not manned but is remotely controlled from the surface.

The fully shielded machine consists of two sections, the main front components include the cutter head and cutter-head support system. At the rear the shield is split into two semi-cylinders which are connected by double-acting hydraulic pistons so that, in effect, the rear shield is also the gripper subassembly in which shield and grippers are combined.

Sixteen standard 30 cm (12 in) disc cutters are fitted to the cutter head.

An extension of the cutter head (i.e. a guide cylinder) (Figure 119) fits into the 1.82 m (6 ft) pilot hole ahead of the cutters and keeps the machine on course.

Gravity and scrapers combine to move the cuttings through a window in the guide cylinder into a 3.8 cu m (5 cu yd) capacity bucket suspended beneath the machine. Bucket size corresponds to the amount of cuttings collected during a 30 cm (12 in) boring cycle. While the machine is being repositioned for the next boring phase, the bucket is hoisted through the centre of the machine to the surface and emptied.

During the boring cycle the rear shield gripper assembly holds the unit in the shaft while the propel cylinders thrust the cutter head and cutter-head support down into the hole. Torque and propel forces are reacted through the gripper pads. At the end of the boring

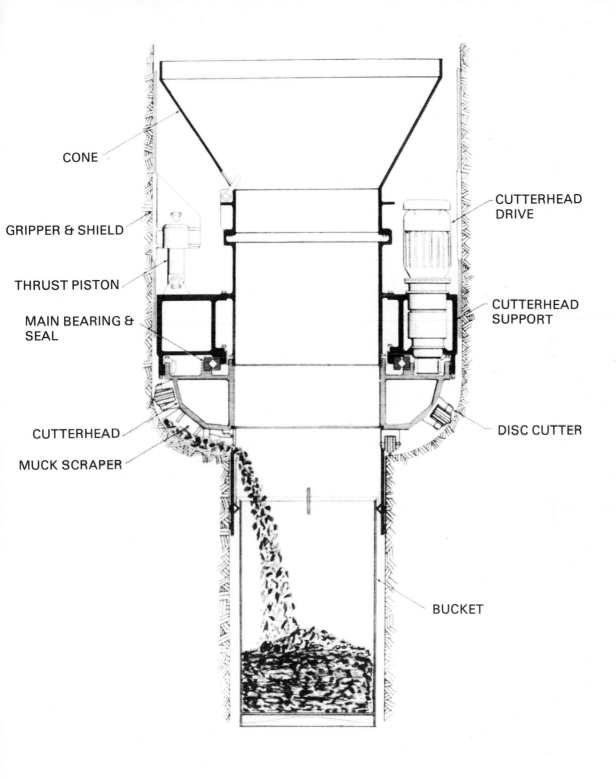

CONE

CUTTERHEAD DRIVE

GRIPPER & SHIELD

THRUST PISTON

MAIN BEARING & SEAL

CUTTERHEAD SUPPORT

CUTTERHEAD

MUCK SCRAPER

DISC CUTTER

BUCKET

Figure 119. Robbins shaft reamer Model 1211-194 — operational sketch. (Courtesy The Robbins Company.)

Figure 120. Robbins shaft reamer, Model 1211SR — 'resting' over a weekend. (Courtesy The Robbins Company.)

Vibration was apparently caused by the clearance of the nose guide in the pilot hole which provided a much less stable head than those on tunnel-boring machines. Most T.B.M.s feature stabilizer shoes which are positioned immediately behind the cutter head.

On the credit side the machine was relatively light (45 t (U.S. 50 tons)) and simple to set up. It was also easy to operate. Friant predicted that when its teething problems had been overcome it would 'provide an economical method of down reaming a blind hole, and removing the muck concurrently'.

References

1. Richard J. Robbins, The status of tunnel and raise boring, paper presented at Annual Meeting of the American Inst. of Min. Met. and Pet. Engs. Inc., New York, Feb. 1964.
2. Richard J. Robbins, Raise and shaft drilling, a continuing development, *Journal of the South African Inst. of Min. and Met.,* Sept. 1973.
3. Robert E. Cannon, New raise drilling techniques and equipment, *Mining World,* Feb. 1963, San Francisco, U.S.A.
4. R. M. Brown, *Raise Boring at the International Nickel Company Ontario Division,* Canadian Inst. of Min. and Met. Mont., April 1969.
5. W. J. Taylor and J. R. England, Raise Boring at the International Nickel Company of Canada, Limited, Ontario Division.
6. Shaft Bored at Monterey, No. 1, *American Mining Congress Journal,* Nov. 1978, **6,** 11, 13, Washington, D.C.
7. R. S. Hendricks, Hecla Mining Company, Raise Boring Experience, 4 Sept. 1969, Idaho, U.S.A.
8. R. A. J. T. Oram and D. J. Bedford. Raise-boring practice and the boring and equipping of an ore-hoisting shaft at Tynagh Mine, Ireland, extract from *Trans./Section A of the Inst. of Min. and Met.,* **82,** 1973.
9. W. L. Wright, *Kaiser's Sunnyside mines go all out for raise boring* (adapted from a paper presented at the Rocky Mountain Coal Mining Inst. Meeting, June-July at Snowmass-at-Aspen. Colo.) Reprinted from *Coal Age* Jan. 1971, McGraw-Hill, New York, U.S.A.
10. T. Jinno, Y. Kotake, H. Ohshika and K. Yamatani. Development of blind raise boring machine used for collapsible formation, paper presented at International Conference on Mining Machinery, July 1979, Brisbane, Australia.
11. Kurt Trosken, *The latest experience in drilling large holes with roller bits in the Ruhr coal mines,* trans. from the German *Gluckauf* **97,** No. 14, 1961. Essen, W. Germany.

stroke the machine rests on the shaft bottom while the rear grippers or shield halves are released and pulled down the shaft by the retraction of the propel cylinders.

The cutter head is powered by two 93 kW (125 hp) motors through gearbox assemblies meshed to a common bull gear. Maximum torque is 298 kNm (220,000 ft lb) and maximum thrust is 1668 kN (375,000 lb).

Friant reports that during the excavation of the three shafts the best performance attained was an advance of 17 m (55 ft) in 9.5 hours, which included 60 regrip cycles.

Some problems connected with water, humidity, and vibration were experienced. To overcome the first two, plans were under way to remove all control equipment from below ground to obviate maintenance problems associated with the underground environment.

12. Karl-Heinz Bruemmer and Karl Wollers, Experience with shaft boring and new developments in German coal mines. *Proc. R.E.T.C.,* Las Vegas, 1976.
13. Blind Shaft Borer. The Robbins Company Model 241SB-184, 27 Sept. 1978, the Robbins Co. Seattle, Washington. (Unpublished)
14. James E. Friant, Blind Shaft Construction New Equipment Update, the Robbins Co., Seattle, Washington. (Unpublished)
15. Colin F. Smith, Blind hole raise boring, *Mining Congress Journal,* **66,** No. 6, June 1980. Washington, D.C.

Bibliography

Arne J. Andelin, *Raise Boring at the Mather Mine,* Cleveland-Cliffs Iron Co. U.S.A.

H. R. Hammond, Raise boring at Bluebell, Riondel, B.C. Canada 41R-1108, Paper presented at B.C. Mining Assoc. Meeting, May 1968, Canada.

Lok W. Home, An appraisal of the role of the mechanized tunnel borer in mine applications, Paper presented at Northwest Mining Assoc. Conf., Dec. 1971, Spokane, Washington, D.C.

M. Mellish and R. Crisp, Raise boring at Rhokana, *Mining Magazine,* **122,** No. 6, June 1970, London.

J. W. Wilson and P. C. Graham, Raise-boring experiences in the gold mines of the Anglo-American Corporation Group, extract from *Trans/Section A of the Inst. of Min. and Met.,* **81,** 1972, U.K.

Dr. John W. Wilson, Shaft sinking technology & the future needs of the mining industry, *Proc. R.E.T.C.,* Las Vegas, 1976, A.I.M.E., New York.

Correspondence

The Robbins Company, Seattle, U.S.A.

Ingersoll-Rand Ltd., Australia and U.S.A.

International Nickel Company, Canada

Hecla Mining Company, Idaho, U.S.A.

Alfred Wirth & Co. KG (now Wirth Maschinen und Bohrgerate Fabrik GmbH) Erkelenz, West Germany

Thyssen Schachtbau GmbH, Mulheim (Ruhr), West Germany

THREE

Early Rock Tunnelling Machines

The difference between a full face tunnel-boring machine and a drill is a question of magnitude. For in effect a full face machine is merely one for which a single drill hole constitutes the entire tunnel.

While a solid steel drill — practically the diameter of the hole being bored and tipped with a harder material — is accepted naturally in the case of a comparatively small hole, the mind boggles at the prospect of a directly scaled-up version for cutting holes, or rather tunnels, 12, 4, or even 2 m in diameter. This, of course, is a perfectly natural reaction in man — being the size he is. Such a machine would be unmanageable, even when broken down into components, let alone in its fully assembled form. We can, however, quite easily imagine another world on which men standing some 80 m (260 ft) in their stockinged feet, could use a rotary drill scaled up 40 or so times and operating a solid steel bit to drill 3 m (10 ft) diameter holes in hillsides — such holes we would call tunnels.

On the other hand the idea of fabricating a 100 mm diameter (4 in) drill bit out of 10 mm (0.4 in) steel shafting, 10 mm × 5 mm (0.4 in × 0.2 in) steel beams and 2 mm (0.08 in) steel sheet may sound ridiculous. Yet, in effect, that was the kind of thing man was attempting to do when he constructed his first full face tunnelling machines from parts which, at that time, he felt were heavy enough. But as history shows, these machines repeatedly failed because, in fact, the materials were too light and flimsy for the task set them.

It is not surprising to find, therefore, that though in most cases the engineering principles involved in the design of the old full face tunnel-boring machines were more or less sound, the finished article was incapable of withstanding the strains imposed upon it.

The gradual realization of this fact then prompted engineers to tackle the problem with cunning and an ingenious combination of levers and gears, etc. so that to some extent they offset this disability. Thus the first successful full bore tunnelling machines were made.

Because steel-making technology was, at that time, also in its infancy, these machines were largely confined to materials such as chalk, clay, soft sandstone, or soft coals. Later the advent of new alloys encouraged further development in this field and the first successful rock machines made their appearance.

During the period 1846-1930 close to 100 rock or hard-ground tunnelling machines of various types were designed and patented, but the actual physical manifestation did not in all cases reach the light of day.

Many of the ingenious devices and/or engineering principles depicted or attempted by those early engineers have either been ignored or forgotten. Others can be seen incorporated in the modern tunnelling machine of today. For this reason descriptions and, where possible, pictures or drawings of patents have been included in this book. These patents cover as many as could be traced of rock or hard-ground tunnelling machines designed during that period, whether or not the actual machine was ever built.

In all cases where information was available the author has stated whether the machine in question was actually built and/or tested. However, although the machine might very well have been constructed, if no record of such construction was available at the time of writing then only a description of the patent has been given.

1846-1930 Machines

Henri-Joseph Maus — Inventor of the first (Percussive) Rock Tunnelling Machine

The Mont Cenis Tunnel[1-6]

From the earliest times travellers wishing to cross the chain of mountains (known as the Cottian and Graian Alps) which divides France and Italy were forced to climb 2101 m (6893 ft) in order to use the Mont Cenis pass.

In 1803-10 a carriage road was built across this pass by Napoleon who used it for military purposes. Despite this improvement the route was unsatisfactory for many reasons, the main one being that the road was often closed by falls of ice and snow for long periods during the winter months.

With the advent of the locomotive many people, and especially those living in the kingdom of Piedmont, thought longingly of a railway line between the two countries.

As early as 1832 Giuseppe Francesco Médail, a native of Bardonecchia, dreamed of such a connection, the ultimate convenience and economic advantages of which were easy to envisage.

Over the succeeding years Médail, who had an intimate knowledge of the local terrain, drew up a relief map of the mountain region lying between Modane and Bardonecchia. He was assisted in this work by Trafora. The map, the culmination of two years of hard work involving a considerable amount of personal financial expenditure, was presented to General Racchia with a report on 30 August 1839. In his memorandum Médail set out what he considered to be the most favourable route for a tunnel through the Alps. General Racchia put Médail's report before King Carlo Alberto of Sardinia. In May 1840 and again on 20 June 1841 Médail submitted two further reports which were examined by the Sardinian government.

A Commission was appointed to examine Médail's project, but Médail left Suse on 5 November 1844, before any concrete action could be taken to bring about the realization of his dream. Médail's work served to focus public attention on the subject and the question of a rail tunnel link through the Alps became a matter of national importance which was discussed freely in the press. In 1843 Brunel submitted a project for a railway from Turin to Genoa, and in 1845 King Carlo Alberto sanctioned the building of this railway at the expense of the Sardinian government. At about the same time he learned of the accomplishments of a certain Belgian engineer, Henri-Joseph Maus. Maus was asked by the Sardinian Home Minister for Public Works, Des Ambrois, to direct the laying of the Turin-Genoa line and also to review the proposed future rail plans for the kingdom of Piedmont. His salary was set at the figure of 20,000 lire per annum.

Henri-Joseph Maus

Maus was born in Namur on 22 October 1808. Evidence of Maus's brilliance showed in his early school-work during the time he attended the public secondary school of Namur. In 1827, at the age of 19 years, Maus was appointed Supervisor of Mines for the Associate de Luxembourg. In this capacity he carried out certain research work on rock salt, which occupation also incidentally improved his geological knowledge. By 1833 Maus was managing a large colliery in the district of Liège. But the advent of the railways, an exciting new innovation in those days, attracted the young Belgian engineer, and Maus became involved with the construction of the inclined plains of Ans and the funicular railway up them on the line from Liège to Aix-la-Chapelle.*

Soon after Maus's arrival in Piedmont in 1845 he established workshops for the Turin-Genoa railway line. At the same time he turned his attention to the important question of a railway link between Modane and Bardonecchia. This link was, of course, merely a section of the proposed main line which would run from Chambéry in France to Turin in Italy, passing on its way through the mighty

---

*As a result of the success of this venture the Loire Railway Administration Council consulted Maus regarding the application of his system of inclined planes and funicular traction to their railways.

Alpine chain which formed the border between these two countries.

Maus was fortunate in having at that time the assistance of Angelo Sismonda, Professor of Geology at the University of Turin. Together these two men studied the problem thoroughly and came to the conclusion that Médail's choice of a route which lay directly under Mt. Fréjus and some 20 km (12 miles) to the south-west of Mont Cenis was, after all, the best.

The difficulties associated with such a project seemed insurmountable. Though gunpowder was at that time being used fairly extensively, all drilling was performed by hand (i.e. by hammer-men). Moreover the problem of an adequate ventilation system within the confines of the tunnel itself remained unsolved. When the Giovi tunnel (3260 m (10,700 ft) long) was constructed, 14 intermediate shafts were dug between the two main entrances. These shafts assisted in ventilating the tunnel and removing noxious fumes. But due to the geological structure of Mt. Fréjus this was not possible with the Mont Cenis tunnel.

Maus planned to enter the mountain at a point some 1363 m (4500 ft) above sea level at Bardonecchia which lay near the valley of Rochemolle in Piedmont, Italy, and then travel downwards in a gradual slope of 1.88 per cent so as to emerge at a height of 1150 m (3770 ft) above sea level on the opposite side of the mountain close to Modane, which lay near the Arc valley in France. The tunnel which Maus estimated would be approximately 12,290 m (40,300 ft) long would be situated near the base of the mountain in order to reduce the steepness of the approaches. Nevertheless the difference in levels between the two valleys necessitated the use of a funicular traction system as was used at Ans. In this case five gradients of 35 per cent would need to be linked.

So far as the excavation of the tunnel itself was concerned, Maus decided to eliminate the use of gunpowder entirely. He planned, instead, to build a tunnelling machine.

A place was made for him in the Valdocco arms factory and, with money granted by the Sardinian government for the purpose, Maus designed and by 1846 had built a prototype of his rock-cutting machine — the first rock-tunnelling machine ever made.

The Machine

The Maus machine (Figure 121) consisted of a metal framework which carried percussion drills or chisels. However, there appears to be some confusion over the number of drills used by Maus on his prototype unit, many authors preferring to skirt around the problem by not actually stating the number, while others used the figure of 116 given in Maus's own report, which seems a reasonable thing to do. The drawing accompanying Maus's report, dated 1849, makes a centre block of 80 chisels very obvious, while the probable location of a further 36 drills is not particularly clear; due to the fact that the side elevation shown, appears to be partly in section and so leaves the design of the outside vertical rows of chisels to the imagination of the reader. The only indication of such rows is seen in the plan drawing where each horizontal row consists of a centre section of 16 chisels with one on either end being clearly separate from this block, and being the top members of a row of 18 chisels down each side. This interpretation would give a total number of 116 chisels as stated by Maus himself.

The central block appears to consist of 80 chisels (five rows of 16) which actually oscillate back and forth horizontally while drilling, and so cut the rock into five superimposed blocks separated by the grooves so cut. The purpose of the outside vertical rows of chisels being presumably to sever the lateral section of these blocks on their ends. This would leave four horizontal superimposed blocks about 2 m × 0.5 m (6 ft 6 in × 1 ft 6 in) attached to the mother rock only by their posterior faces. These blocks were then removed by driving wedges into the grooves between them, causing them to snap off along this posterior anchoring plane. (Maus suggested in his report that these neatly cut blocks could later be put to good use as sleeper supports under the rail lines.)

A system of cams operated by shafts and gears obtained their power from the main pulley-driven shaft. These cams drew the chisels back against the action of powerful coaxial springs and then released them suddenly, to be flung against the rockface by the reaction of these same springs. The cams also imparted a slight rotary movement to the drill shaft each

Elevation

Plan

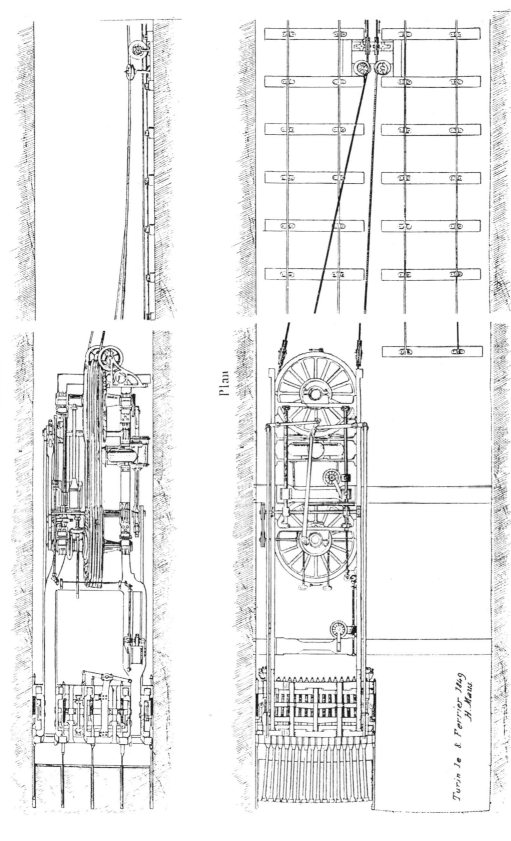

Turin le 8 Ferrier 1849
H. Maus

Figure 121. Henri-Joseph Maus: first rock-tunnelling machine was designed and built by Maus in 1846. In 1848 and 1849 Maus submitted two reports — the culmination of four years' work on the Mt. Cenis project. The sketch illustrated was attached to the 1849 report

time the springs were compressed. This caused the bit to turn in the hole thus preventing jamming. During the drilling operation a small jet of water under high pressure was forced into the hole and along the drill bit. The water cleared the hole or groove and also kept the tempered metal cool. One hundred and fifty blows per minute were delivered by the drills.

The entire frame was moved forward during operation by a manually manipulated crank handle, the speed being determined by the hardness of the rock.

Maus suggested that half the tunnel face could be cut by his machine at one time. When the frame had been extended to its fullest length into the face, it would be retracted and the entire machine barred sideways. It would then be set to work on the remaining uncut section of the face, thus leaving the area immediately in front of the prepared rockface clear for the workmen who would be moving in to dislodge and remove the rock.

Power would be transmitted to the rock-cutting machine in the tunnel by a series of endless cables and pulleys, driven by large turbine wheels stationed outside the entrances. Maus stated that these turbine wheels should be capable of handling 2 or 3 m (6 or 10 ft) of water per second and should be strong enough to withstand the strains imposed on it by the action of the various cables and pulleys. Maus also considered it essential that the turbine wheels should be safe to use and easy to repair. Power for the water wheels would be provided by the Alpine streams which would probably need to be diverted for the purpose. (These could also be used later for agriculture.)

Maus estimated that this machine would excavate a 4.40 m (wide) × 2.20 m (high) (14 ft 5 in × 7 ft 2 in) gallery which could later be enlarged using conventional methods (i.e. gunpowder) or, alternatively, two machines could be worked side by side. Maus also estimated that the 44 mm (1¾ in) diameter endless cable would travel at a speed of approximately 12 m/sec (39 ft/sec) over a series of pulleys spaced 10 m (33 ft) apart along the length of the tunnel. Ventilation would be provided by fans installed at the base of some of the pulleys, the actual number being determined by necessity.

The prototype machine was submitted to practical tests which involved numerous blocks of rock of various kinds, during the two years it was installed at Valdocco. Because the construction material available at that time was too soft to take the strains imposed on it, Maus was obliged to make several modifications to his machine in order to overcome this problem. The machine drew admiring crowds, and eminent scientists and engineers came from all quarters of the globe to view the amazing 'mountain slicer' as it was called. The King of Sardinia, Carlo Alberto, frequently called at the Valdocco factory accompanied by a great retinue of army personnel.

Bolstered by the success of these initial experiments, Maus calculated in his 1849 report that at the rate of 1, 2, or even 3 cm/min (0.4, 0.8 or 1.2 in/min) the machine should be able to advance 7.20 m/day (24 ft/day) (which would be reduced to 5 m (16 ft) when the tools needed changing). Working from both ends of the tunnel he anticipated a progress of 10 m (33 ft) a day, so that it would take four years to complete the project. Allowing another year for delays, breakdowns, etc. he thought five years would be the maximum time needed for the entire excavation.

Maus also planned to excavate a service gallery alongside the main tunnel so that machinery and transport would not interfere with the other works.

Maus expected to build improved models of his machine for the actual work in the tunnel.

During the period Maus was in Turin he submitted four reports to the Sardinian government. These were dated 8 August 1845, 26 March 1846, 29 June 1848, and 8 February 1849. The 1848 and 1849 submissions were lengthy reports concerning the work he had accomplished and setting out his proposed project for the excavation of the tunnel.

However, due to major political events which took place during the years 1848-49* public interest in Maus's project waned, and the initial euphoria and excitement generated by this brilliant invention became dulled.

---

*Complete political union with Sardinia was granted in 1848 when the vice-regal government of Piedmont was suppressed.

Instead engineers and scholars began expressing grave doubts that such a device could work when tested practically in the tunnel. They argued that it was all very well under the ideal conditions set up at Valdocco but, due to its heavy weight and cumbersome size, the machine would not be manœuvred easily within the narrow confined space of a tunnel. The method of power transmission in particular was strongly condemned by such men as Bella (engineer and senator of the kingdom) who pointed out that the further one tunnelled into the mountain the longer the traction rope would grow and consequently the greater the resistances due to the tightness of the cable, the friction of the pulleys, etc. According to Albert Duluc, Maus himself admitted that a considerable power loss would occur between the turbine wheel and the drilling machine, estimating that if 75 kW (100 hp) were transmitted from the beginning, probably only 22 kW (29 hp) would remain at the extremity, and even this would be cut in half or greatly diminished if it were to operate the ventilators on the pulleys. Despite this, Maus still retained faith in his device, preferring it to the greater disadvantages of gunpowder.

A Commission of Enquiry was set up by the Sardinian government to review Maus's project. Among other eminent personalities selected to serve on the Commission were Menabrea (Colonel of Engineering) and Paleocapa (Professor of Mechanics in the Artillery). Paleocapa (a Venetian engineer) strongly supported Maus's project despite the general criticisms. Paleocapa was obviously influenced by his personal determination to see the accomplishment of a tunnel link, whatever the means. He expressed the opinion that such difficulties as might become apparent would most probably be resolved on site, and that if modifications were necessary to the machine or ancillary equipment these could only be tackled efficiently during the operation of the machine in the tunnel.

Paleocapa felt that even assuming Maus had grossly underestimated the cost of the project any additional expenses involved would be fully justified, in view of the great importance of the rail link to the two countries concerned.

The Commission completed its work in November 1849 with a recommendation that an excavating machine be built (for which an amount of 720,000 lire should be granted) and that all necessary preparations for the start of work on the northern side be commenced.

Unfortunately, however, the political and military developments which took place during 1848 and 1849 had depleted the coffers of the treasury and Parliament voted that the project be temporarily shelved. So far as Maus's project was concerned this delay proved fatal, as it allowed time for research in other fields to develop.

During the succeeding years those most intimately concerned with the project of piercing the Alps were continually on the look-out for easier and, if possible, better ways of tunnelling than had so far been proposed. One of these was compressed air.

The idea of using compressed air to ventilate a tunnel was not new. Many had thought of it. But the general opinion was that great losses of pressure would be incurred if compressed air were transmitted long distances. At that time experiments with compressed air were simultaneously being undertaken by a number of people in different parts of the world. Among them was Colladon, the physicist from Geneva, who conducted numerous practical tests using metal conduit pipes of varying diameters and lengths. He succeeded in proving that the expected resistances were not as great as had hitherto been anticipated. Having overcome this obstacle, Colladon applied for a Sardinian patent to cover his new system of excavating tunnels. He suggested that a drilling machine operated by compressed air (either similar to the one built by Maus or one that simply drilled holes for mines) be used in the proposed Mont Cenis tunnel.

By 1853 Maus realized that his project was doomed and on 23 November that year sought permission to return to Brussels. Before he left, Maus indicated his willingness to remain at the call of the Sardinian government should they decide at any time to proceed with the excavation, using his machine. He even agreed to waive acceptance of a salary until such time as the machine had proved itself in practice. When Maus left Piedmont in October 1854 the honorary title of Inspector of Sardinian

Engineering was conferred upon him and he was given the insignia of Commander of Saint Maurice and Lazare. Maus's contribution to the development of the Piedmont railways and the construction of the Giovi tunnel in particular, during his stay in Turin, is undisputed. And though his machine was found to be unsuitable for the great Apennine tunnel due to an inefficient power supply system, it must be remembered that he did build the first rock-tunnelling machine. One, moreover, which worked successfully for two years in the arms factory at Valdocco, cutting rocks of various types. Just as Brunel is acknowledged as the father of the shield, so too must Maus be acknowledged as the father of the rock-tunnelling machine. His machine, built by 1846 precedes even Couch's percussion drill which was only ready for practical testing in 1848 — two years after Maus had completed his prototype unit.

After his return to Brussels Maus was appointed Inspector-General of Highways and later, Director of Mines. In 1872 he represented Belgium on the International Commission of Metre. Maus died in Brussels at the age of 85 on 11 July 1893.

The Mont Cenis Problem — Its Final Solution

In 1854 Thomas Bartlett took out an English patent for his steam-powered compressed-air drill. Bartlett's drill consisted of a piston with a long piston rod terminating in the drill bit. This piston was housed in a pneumatic cylinder and was caused to reciprocate by the increase and decrease of air pressure in front of the piston. In this way up to 200 or 300 blows were delivered to the rock per minute. The fluctuating air pressure was provided by mechanically linking a steam-driven piston and a pneumatic piston which forced the air back and forth to the drill.

Bartlett's drill which was patented in Sardinia in 1855 was examined by a Sardinian-appointed Commission with a view to using it in the Mont Cenis tunnel. However, though it performed fairly satisfactorily during the excavation work carried out on the Aix-San Giovanni to Moriana railway line, it could not be used in a tunnel because of the air pollution caused by the engine which produced the steam.

Some time prior to this, three engineers, Grandis, Grattoni, and Sommeiller had carried out research on a hydraulic pneumatic traction system. This research work was done on the inclined planes of Giovi. A 'column compressor' which had been patented in 1853 was used to produce the compressed air. (Air was compressed by using the energy provided by a column of water — hence the name.) The House of John Cockerill of Seraing in Belgium built a similar compressor at Sampierdarena in Coscia. Various experiments were carried out at Sampierdarena during the months of March and April 1857 by the Italian Commission which consisted of Menabrea, Des Ambrois, Giulio, Sella, and Ruva. Not only was the column compressor tested thoroughly but Colladon's method of transmission of compressed air in conduit pipes, several hundred metres long, was also carefully examined. In addition the new drilling machine which had been recently developed by Sommeiller, Grandis, and Grattoni and which was powered by compressed air was vigorously tested.

The success of these tests led to the final acceptance by the Sardinian government of this system for the excavation of the Mont Cenis tunnel.* Sommeiller's drill was used successfully throughout the excavation. (For further information on the Sommeiller drill and on the Schwarzkopff rotary drill which was also tried in the Mont Cenis tunnel, see Chapter 1 — Horizontal drills.)

E. Talbot — 1853

On 13 March 1847 Charles Wilson of Springfield, Hampden County, Mass., filed a patent (No. 5,012) (Figure 122) for a device designed to cut, turn, or split stone by means of a *revolving cutter*. When Ebenezer Talbot of Windsor, Hartford County, Connecticut, saw this device, it occurred to him that a similar component could, if driven by a machine, be used for boring horizontal tunnels through rock.

---

*The 12.8 km (8 mile) long tunnel was eventually constructed 27.3 km (17 miles) west of the Mont Cenis pass below the Col de Fréjus. This was about one kilometre (⅝ mile) east of the point originally proposed by Maus.

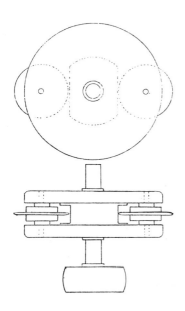

Figure 122. C. Wilson's patent (No. 5,012 USA) for a stone dressing device using roller cutters, dated 13 March, 1847

Talbot filed a patent for his machine on 7 June 1853 under No. 9,774 (U.S.A.) (Figure 123) and promptly set to work building the unit. His machine consisted of a framework which was mounted and slid on rails laid at the bottom of the tunnel. The machine was advanced towards the face by screw jacks attached to the main framework of the machine. These jacks thrust against the tunnel walls.

The revolving cutter head [a] or wheel carried a central framework on which two rocker shafts were mounted. These each carried a sector [b] bearing two arbors fitted with a rolling disc cutter [c] apiece. The axis of the arbor in every case being set at an angle of approximately 45° to the axis of the rocker shaft on which the sector pivoted.

The angle of each arbor, and thus each disc cutter, was adjustable to enable the operator to set the cutting edges of the discs so that each traversed a different path on the face.

While the cutter wheel rotated, a reciprocating motion was also imparted, by a connecting rod, to each sector. This enabled the cutters to move from the centre of the wheel to

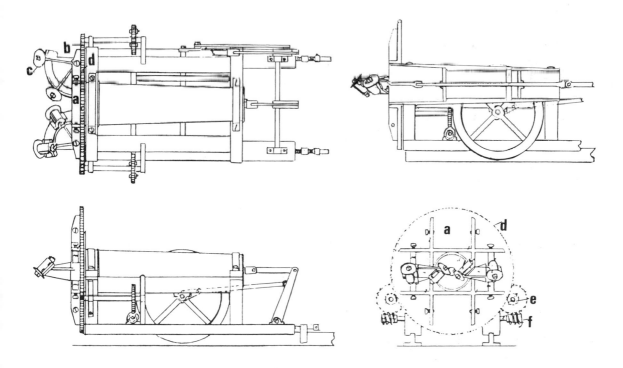

Figure 123. E. Talbot tunnelling machine. Patented 7 June, 1853 under No. 9,774 (U.S.A.)

the periphery and back again, thus covering the entire face.

The cutter wheel was peripherally driven through a circumferentially mounted annular rack [d] on the wheel, a pinion [e], and two transversely positioned worm gears [f] attached to the sides of the main frame near the bottom.

According to Drinker,[7] Talbot's machine was built and tried near Harlem, New York, in 1853. It was adapted to cut out a core 17 ft (5.18 m) in diameter, but was a failure. Nevertheless the reader is invited to compare the cutting action of the Talbot machine with the rock-tunnelling machine built by Union Industrielle Blanzy-Ouest of Paris, France, during the early 1970s.

C. Wilson — 1856/7

Although Wilson's design for a T.B.M. was not patented until 18 March 1856, his machine was actually built in 1851 in South Boston especially for the Hoosac tunnel and was tested there early in 1853 before Talbot's machine was built and tried. On 23 March 1853, A. F. Edwards, then Chief Engineer in charge of the Hoosac tunnel project submitted a report[7] on Wilson's machine to the Joint Special Committee of the Legislature of Massachusetts. It stated, *inter alia,* that:

> ...among the various improvements in machinery for drilling stone and making excavations, no machine at the present day should command the attention of railroad men more than Wilson's patent stone-cutting machine for tunnel excavation in rock; the first working model, of 100 tons weight, is now at the Hoosac Mountain, experimenting in various ways, and perfecting the principles of the same. The result of its workings in the natural rock, under every disadvantage, in the different experiments, has been from 14 to 24 inches forward per hour, on a full circle of 24 feet diameter; exceeding the expectations of its most sanguine friends, and bidding fair to revolutionize the whole system of railroad-building. At the view, by the committee of the Massachusetts Legislature, of the working of the model, under many disadvantages, the machine cut the rock at the rate of 1⅓ lineal feet per hour, with the thermometer at only two degrees above zero, on that morning, although exposed in the open air, with all its cast iron fixtures, which should be of wrought-iron. The machine has never worked less than 14 inches per hour. To show with what progress this tunnel could be worked, I will say

the machine will cut 12 inches per hour, 33⅓ per cent less than was actually witnessed by the committee of the Legislature. In two hours the machine will cut, at that rate, 2 feet forward; the ring of the tunnel (of which the diameter is 24 feet, interior diameter of the core 22 feet — making, in quantity cut, 2¾ cubic yards, thus leaving a core of 22 feet diameter = 14 cubic yards) to be taken out by blasting, the same to be transported out under the machine, after it is run back, which I propose to rest for two hours, that the machine and engine, which is to be attached to the machine for its working and its locomotion, when perfected, may be examined, wiped, oiled, and any cutter or nut adjusted. During this two hours, the core will be blasted by means of cast-iron shells, so encased in wood as to fit exactly a portion of the ring; the same will be fired by electric battery...

Edwards then goes on to calculate the time and number of workmen necessary to carry out the mucking procedure and concludes by saying:

> ...the whole tunnel excavation will advance 2 feet on one face in four hours = 12 feet in 24 hours, and with two machines, one at each end, at the same rate, will progress 24 feet per day — thus completing the entire excavation of the whole tunnel in 1005 days, and present a hammered faced surface upon each and every side.

Edwards, however, like Maus, was counting his chickens before they were hatched. The Wilson machine was subjected to numerous trials during the initial excavation period of the Hoosac tunnel, but the total distance actually cut by the machine amounted to a mere 3 m (10 ft), after which it was abandoned.

Charles Wilson filed two patents for tunnelling machines, these were Nos. 14,483 (U.S.A.) and 17,650 (U.S.A.) dated 18 March 1856 and 23 June 1857 respectively.

Although the machine in Patent No. 14,483 has been depicted by some authors as being 'the machine' built and tested in the Hoosac tunnel, the author is of the opinion that what in fact was actually used was one resembling that depicted in Patent No. 17,650. This contention is borne out both by Edwards's description of Wilson's machine and by Latrobe's sharply worded and sarcastic report on the various units proposed for use in the Hoosac tunnel.[7]

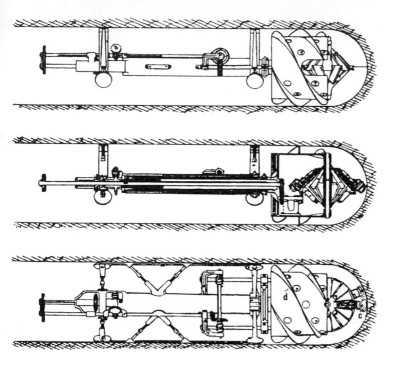

Figure 124. C. Wilson's T.B.M. This machine was tried in the Hoosac tunnel, but proved unsatisfactory. Pat. 18 March, 1856 under No. 14,483 (U.S.A.)

The 1856 Machine (Patent No. 14,483)

The cutter head of Wilson's 1856 Patent (Figures 124 and 125) consisted of a series of tapered plates [a], which were set at an angle of approximately 45° with the centre line of the shaft on which they rotated, and which were joined together so that they formed two cones united at their bases. On each plate a stock or arm [b] was mounted which carried a removable rolling disc cutter [c] set at an angle of about 45° to the shaft of the cutter wheel.

The cutter wheel (Figure 125) was mounted with its axis across the open end of a revolving cylindrical drum, and both the drum and the cutter head were fed forward into the face by screw jacks. Thus, in operation, the cutter wheel cut a circular groove in the plane of the axis of the tunnel in the face, while the turning of the drum steadily advanced the groove clockwise around the face so that after one revolution of the drum a cut had been taken off the entire hemispherical heading.

Any debris cut by the discs which fell into the drum was passed out through various openings in the drum skin and then moved to the rear by scraper blades [d] mounted in a helical pattern on the outer skin of the drum.

The 1857 Machine (Patent No. 17,650)

Wilson's second patent dated 23 June 1857 (Figures 126 to 128) described a horizontal upper frame [j] which slid longitudinally upon a wheeled carriage running on rails. At the front end of the machine was a large-diameter, short hollow cylinder [h]. Around the front edge of this cylinder a series of rolling disc cutters were mounted. These discs which were in groups of two or three were set at the same angle of inclination, but arranged so that one was slightly in advance of the other. However, while all the discs of a particular group were set at the same angle of inclination, alternate groups were set at opposite angles of inclination to the general plane of the wheel, so that when the wheel rotated, an annular groove was cut out of the tunnel face. Two or three additional rolling disc cutters [k] were centrally mounted on the wheel to enable the machine to cut out a core. These cutters projected laterally from the axis of the main drive shaft. Immediately behind the cutters Wilson arranged a helix or screw [l] which fitted the bore of the hole made in the face by the central disc cutters. This screw was mounted around the outer circumference of the main drive shaft and was designed to direct the

147

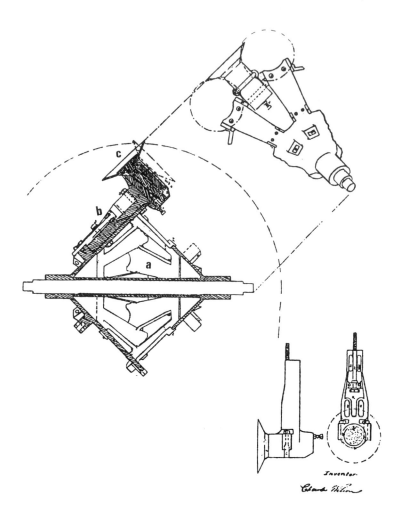

Figure 125. Cutter head of Wilson's 1856 machine

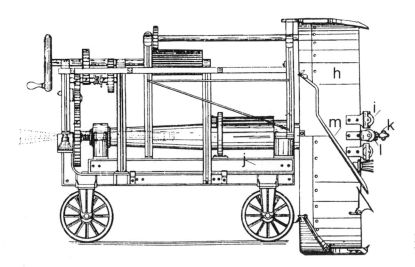

Figure 126. Side elevation of Wilson's improved T.B.M.

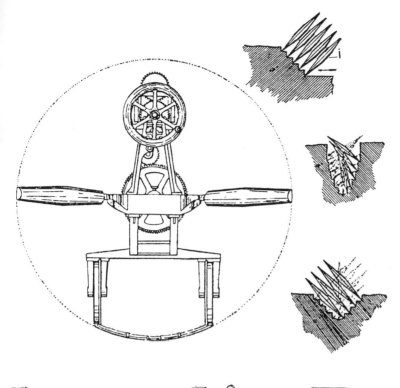

Figure 127. Rear end elevation of Wilson's improved T.B.M. and cutters showing the angle at which these will enter the face

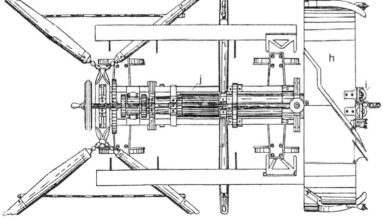

Figure 128. Wilson's improved T.B.M. Patented in U.S.A. on 23 June 1857 under No. 17,650. Top or plan view of machine

debris cut by the central discs towards the rear. A series of inclined planes or scrapers [m] were also attached to the periphery of the cutting wheel. These lifted the spoil and directed it into a hopper which carried it via a chute into a waiting truck.

Wilson's plan was, therefore, to bore a single annular groove at the periphery of the tunnel face and also to cut out a central hole. The machine would then be run back and the face would be broken by a charge of gunpowder.

B. H. Latrobe (later appointed Consulting Engineer for the Hoosac tunnel project) in his report dated 1 October 1862 to the Commissioners of the Hoosac tunnel was obviously disillusioned with these machines as he wrote:

The novel and ingenious machinery for driving the tunnel, either by an annular groove or a cylinder bore in the centre of the section, I could entertain no confidence in, from the first suggestion, as they require the machines to do too

149

much and the powder too little of the work; thus contradicting the fundamental principles upon which all labor-saving machinery is formed.[7]

According to Drinker a third machine was constructed in New York at about that time (1853). It was designed to cut out a 2.43 m (8 ft) diameter core.

Herman Haupt (of Haupt & Company) who during that period worked as a contractor in the Hoosac tunnel, spent about $25,000 in developmental work in an effort to make the machine workable but, despite this, it was never tested in practice in the tunnel. Another machine, similar to this, was also apparently tried in California where it was reputed to have cut a 1.83 m (6 ft) hole in hard rock at the rate of 58 cm (23 in) in 1 hour and 45 minutes (i.e. at a rate of about 8 m (26 ft) per day). But nothing more was heard of this machine and presumably it failed as did the rest. Unfortunately, Drinker does not give the names of the inventors of these two latter machines.

I. Merrill — 1856

Another machine design (Figure 129) which was patented but, so far as the author could ascertain, was not built, was that conceived by Ira Merrill of Shelburne Falls, Franklin County, Massachusetts. This patent described a machine which, although crude, was similar in concept to that built by Henri Maus in 1846. It consisted of a main frame [s] constructed of

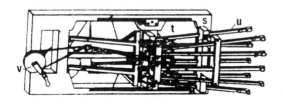

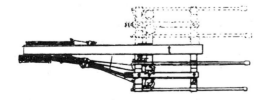

Figure 129. I. Merrill's T.B.M. Patented 22 April 1856 under No. 14,755 (U.S.A.)

150

heavy timber upon which was mounted a secondary frame [t] which supported a series of chisels [u] set in vertical rows. By means of a crank wheel [v] which operated two slide rods, the secondary frame and thus the chisels, could be alternately raised and lowered at each revolution of the wheel. Powerful springs mounted on shafts to which the chisels were attached were drawn back by the action of cams. As the cams passed pawls on the shaft the springs were suddenly released, thus flinging the chisels forward against the face. The cams were so arranged on their respective shafts that only one chisel could strike the rock at a time. Thus, as the camshafts were rotated the chisels struck the face a rapid series of successive blows which, Merrill claimed, would form a set of vertical grooves in the stonework.

When the chisels had penetrated the face as far as they were able to go (i.e. the length the chisels extended beyond the main frame), the machine was drawn back, the chisels were removed, and the entire secondary frame was slid sideways to the opposite side of the main frame. After the chisels had been replaced the machine would be ready to commence cutting another set of grooves. In the meantime the first section of the face could be broken down by workmen with wedges, etc. The blocks so removed would then be useful for building or other purposes. Merrill suggested that by employing a similar arrangement of camshafts and fixtures, etc. the same machine could be used to cut horizontal grooves across the top or bottom of the face.

Herbert Newton Penrice — 1856-76

Herbert Newton Penrice, a Captain in the Royal Engineers stationed at Newcastle upon Tyne, county of Northumberland, filed two patents for tunnelling machines.

The first (Figures 130 and 131) dated 26 September 1856 (U.K. 760) described a massive unit powered by steam or compressed air.

The cutter head consisted of a number of radially arranged arms [a] on which were mounted rows of chisels or picks [b]. It was attached to a large rod or plunger [c] which in turn was attached to a piston. The piston rod and thus the cutter head were driven forward

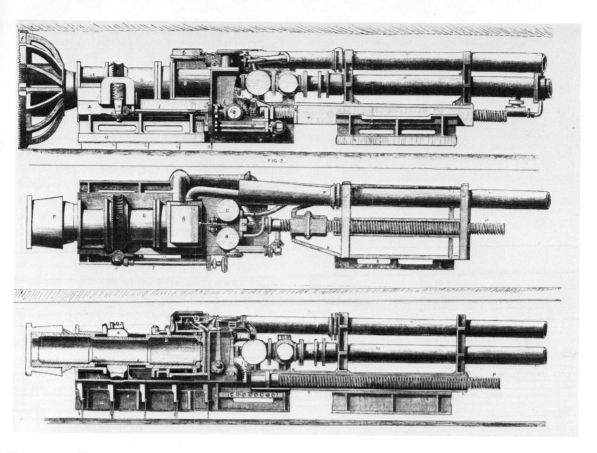

Figure 130. H.N. Penrice tunnelling machine depicted in *The Engineer* 17 June 1859. (Courtesy *The Engineer*)

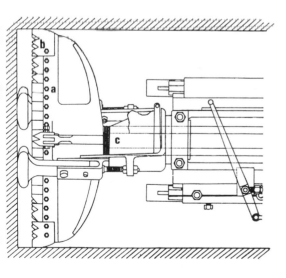

Figure 131. Cutter head of Penrice's machine patented in London on 29 March 1856. Herbert Newton Penrice filed two further improvements which were patented under Nos. 2,999 and 794 and dated 26 August 1875 and 25 Feb. 1876 respectively

against the rock by the direct pressure of the steam or air as it entered the main cylinder chamber housing the piston. In addition the head was made to revolve slightly between strokes by a worm gear and ratchet which were actuated by the return stroke of the piston rod. As the rock at the face was chipped away by the rapidly repeated percussive strokes of the cutters, the whole machine was gradually advanced by a large longitudinal screw mounted on a separate fixed back carriage. The front end of the screw was held in a bearing on the bed of the front carriage and was driven via a series of gears, etc. by the engine. Provision was made for the disconnection of this main advance mechanism from the automatic drive system so that it could be worked manually if desired.

The back carriage was supported on sledges or sleighs (or Penrice suggested that it could be run on rails) and during operation was

151

anchored by pins or bolts driven into the walls of the tunnel. Penrice suggested that the holes for these bolts could be made by two small drills mounted transversely on the back carriage.

The steam cylinder and cutter head, etc. were mounted on a frame which slid on a lower carriage or bed, carrying a rack. Pawls on the upper carriage prevented the machine from backsliding during the boring phase of the cycle. These pawls could be lifted out of the rack by a lever when necessary. The lower carriage was also mounted on sleighs or rails.

A telescopic pipe supplied the machine with steam or air and another telescopic pipe carried away the exhaust steam, etc. During operation the machine was advanced against the face until the telescopic pipes were at their fullest extent. The machine was then stopped so that new pipes could be inserted. Its advance against the face was regulated by two stops or 'bumpers' which pressed against the face and which were mounted on the cutter-head frame so that they revolved with it.

Beneath the machine was a long sliding bar or rod which carried a series of blades or scrapers. These scrapers were given a reciprocatory motion via a series of bevel gears and pinions, etc. so that they moved backwards and forwards. As the bar moved towards the cutter head the scrapers folded back and lay flat so that they passed over the broken rock lying on the floor of the tunnel. When the rod returned the blades automatically assumed a rigid upright position so that they scraped or raked out the debris. (This principle is also used on many modern conveyors.) Penrice suggested that although he had only taken the scraping mechanism half-way along the machine, it could, of course, be extended to cover the full length of the machine. However, he felt that this was unnecessary as the muck could very easily be removed from the halfway point by a workman wielding a long-handled rake.

A drawing of the machine described in Penrice's 1856 patent appeared in *The Engineer* on 17 June 1859.

In his second patent dated 25 February 1876 (U.K. 794) Penrice describes a percussive tunnelling machine which was also powered by steam, air, or hydraulics, but which cut a small-diameter annular groove rather than the entire face. A distinctive feature about this machine was that in addition to the circumferentially arranged main chisels which were carried on the cutter head, Penrice provided three additional chisels which were mounted on arms. These chisels, although connected with the main cutter head, rotated independently of it. The auxiliary drills penetrated the rock simultaneously with the chisels on the cutting head and bored two lateral holes and one upper hole which were fired after the machine had been withdrawn. The idea was to force the rock between these holes and the main groove to break into the groove of the main bore.

Penrice expressed the opinion that the 'direct explosion of the dynamite or other explosive used suffices in most cases to shatter the small core'. Penrice also made arrangements for a jet of water to enter the area of the main bore and also each of the auxiliary holes during the drilling operation so that the cuttings were flushed out.

According to an article entitled 'Merseyside's first mole' (Mott, Hay and Anderson),[17] Captain Penrice is reputed to have designed and built his machines so that they could be used to undermine the walls of Sebastopol during the Crimean War, but the siege was terminated before the units could be tested there in practice.

Cooke and Hunter — 1865-67

Revolutionary pioneering work was carried out by William Fothergill Cooke of Aberia, Caernarvon, North Wales, and George Hunter of Maentwrog, Merionethshire, Wales, in the development of two distinct types of machine. These machines constituted the forerunners of the Stanley and thus of the Mariette through the McKinlay, and also of the modern rotary drum-type continuous miners which first appeared during the mid-1960s (see section on continuous miners).

The Cooke and Hunter machines were apparently developed and tested in quarries in Wales during the late 1860s and the early 1870s.

The first patent (Figures 132 and 133) covering a tunnelling or slate-quarrying machine was filed by these two gentlemen in

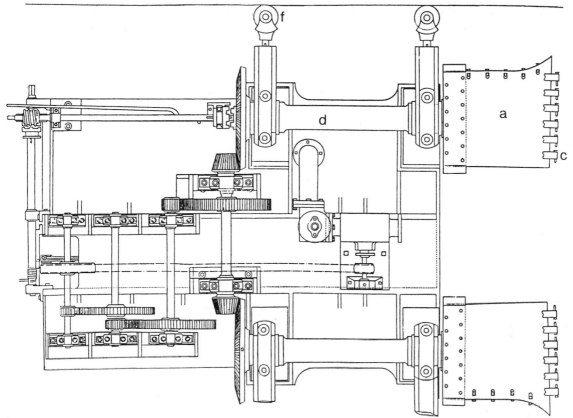

Figure 132. Cooke and Hunter trepanner T.B.M. Pat. 3,297 (plan view)

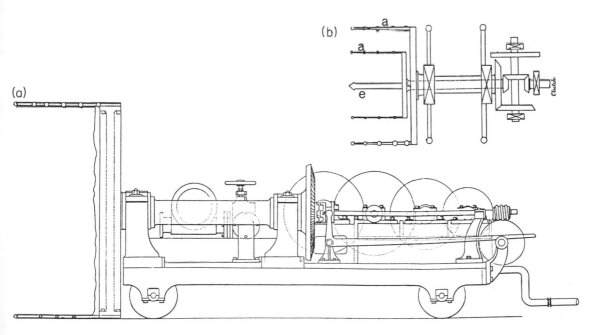

Figure 133. Trepanner-type tunnelling machine designed by W. F. Cooke and G. Hunter. U.K. Pat. No. 3,297 dated 20 Dec. 1865. (a) Side elevation; (b) showing two revolving cutters, one within the other

1865 under No. 3,297. It described a robustly constructed carriage on wheels which carried at its front two revolving cutter heads, each formed of two segments of hollow cylinders. The cutter heads [a] were fitted with a number of projecting teeth [c] on their forward ends. Additional teeth also projected from the sides of the cylinder segments.

The two cutter heads were mounted on chucks on the ends of two parallel shafts [d] which extended forward beyond the front end of the machine. When in operation the shafts, and thus the cutters, counter-rotated.

The purpose of the side teeth on the cutter heads was to act as reamers to allow the rear of the cutter blade to penetrate the face without restriction as the machine advanced.

The forward-most end of each cutter drive shaft was fitted with either a central drill [e] or a cutter tooth which projected into the face. The purpose of this drill was to bore a hole in the centre of the core made by the cutter head to receive a blasting charge.

The two heads were so positioned that they cut two annular grooves either next to each other or slightly intersecting each other. The machine was then run back to allow the remaining core to be broken by wedges if the centre material was of good-quality slate, or by blasting if the core was not required.

The axles carrying the cutters were hollow and if the shaft was not fitted with a drill, then strong jets of water could be flushed through the hollow shaft into the cutting to clear away the debris. Additional flexible pipes for water flushing were also provided around the cutter head. These pipes could be directed at either the top or bottom of the cut as required by the operator.

During operation the machine was braced against both the sides and top by steadying rollers [f]. The rollers which were either hydraulically or spring-operated, were also used to steer the machine.

Cooke and Hunter also suggested that the cutter heads could be so arranged that one cylinder revolved within the other, the shaft of the inner cutter passing through the centre of the outer cutter. The two cutters would then be turned in opposite directions, the inner cutter revolving faster than the outer cutter. In this configuration the cutters could also be so arranged that the entire surface was removed from the hole rather than just the outer circumference.

The following year — 1866, Cooke and

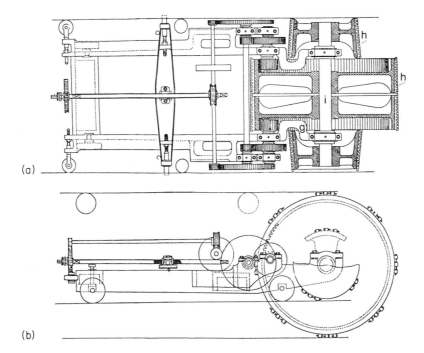

(a)

(b)

Figure 134. Rotary drum type tunnelling machine designed by W.F. Cooke and G. Hunter. U.K. Pat. No. 433 dated 12 Feb. 1866. (a) Plan view; (b) side elevation

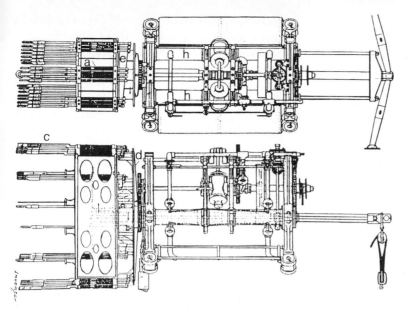

Figure 135. T. Lindsley tunnelling machine. Patented 12 June, 1866, No. 55,514 (U.S.A.)

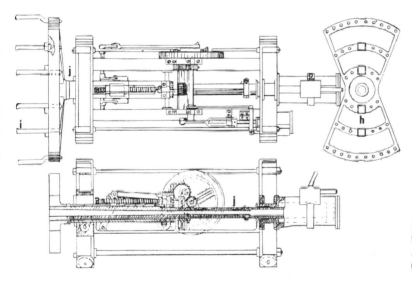

Figure 136. E. M. Troth's tunnelling machine. Patented 2 July, 1867 under No. 66,422 (U.S.A.)

Hunter filed another patent, No. 433, (Figure 134) covering a machine with a very different cutting action.

The main structural assembly remained the same, but the cutter-head mechanism consisted of three rotating drums [h] fitted with a series of cutting tools. These drums rotated on a transverse axle or shaft [i] mounted on arms [g].

Thus, instead of a trepanning action, the teeth on the cutter-head drums ripped the material away from the face in much the same

manner as did the rotary-drum cutter-head type machines first produced by the Jeffrey Manufacturing Company in 1965. There was, however, no provision for raising or lowering the cutter head after sumping.

While Cooke was testing the trepanning machine in the slate quarries, he found that it took about three times as long to remove the central core by hand through the middle of the machine as it did for the machine to cut the actual groove. This meant in practice that the machine took about three hours to cut a groove

155

and the workmen nine hours to remove the core, while the machine lay idle. To overcome this problem of machine downtime, Cooke and Hunter designed a turntable which allowed the operator to move the machine back from the face along the rail and then transfer it to an adjacent parallel rail track, thus enabling the machine to cut another groove while the first core was being broken away. After cutting the second groove the machine was then run back and moved on to a third track so that, in all, two more annular grooves were cut by the machine while the first core was being taken out. By this means a larger quantity of slate was removed in a given time than could be accomplished when the machine was employed for cutting one tunnel only.

Apart from the aforementioned tunnelling or slate-quarrying machines, Cooke and Hunter also patented several improvements relating to stone-cutting apparatus.

T. Lindsley — 1866

Patent No. 55,514 (Figure 135) dated 12 June 1866 granted by the United States Patent Office to Thales Lindsley of Rock Island, Illinois, described a percussion machine which, by means of a series of spring-loaded [a] drills [c] activated by cams [e] and powered by compressed-air cylinders, cut concentric circles in the rockface. The drills were mounted on a main cutter-head plate [d] which revolved slowly during the operation of the drills. The cutter-head plate, and thus the drills, were fed forward into the face by two compressed-air cylinders [h] mounted on the main frame immediately behind the cutter head.

E. M. Troth — 1867

Like the Lindsley patent, that granted to Edward M. Troth of New York on 2 July 1867 (No. 66,422) (Figure 136) covered a percussive tunnelling machine featuring a drill wheel [h] fitted with drills [i] which was mounted on the forward end of the main central shaft [j] of a carriage on wheels. At the rear of the carriage was the compressed-air engine which powered the unit.

However, whereas the drills on the Lindsley unit were spring-loaded and activated by cams powered by compressed-air cylinders, Troth proposed using the direct action of the cylinders to impart a reciprocatory motion to the main shaft carrying the drill wheel. At the same time the shaft, and thus the drill head, were also rotated by a worm wheel through which the shaft slid longitudinally when reciprocating.

R. C. M. Lovell — 1867

Richard C. M. Lovell of Kentucky proposed a machine (Patent No. 67,323) (Figure 137) which could be used as a type of 'shortwall' coal-mining unit or for heading work. It would run on a straight or curved track.

The machine consisted of a carriage which moved horizontally and automatically along tracks laid beside the face by means of the engagement of a spur wheel on a shaft beneath the carriage with a rack-bar [a] which lay between the tracks. At the same time two reciprocating chisels [c] cut a groove along the bottom of the face so as to form an undercut ready for blasting.

The chisels and also the movement of the machine on the rack-bar were activated by the oscillating action of the pistons of two main air cylinders [d] which were joined by a plate [e] (or 'working beam') at the rear of the machine. Lovell proposed that power for the machine's cylinders be provided by compressed air supplied from the mine's compressor units.

The entire carriage could also be fed into the face for a distance of approximately 2 in (50 mm) by means of a hand-operated crank attached to the main drive shaft. At the end of a cut the machine was moved back across the face to the beginning in order to commence the next cut.

T. F. Henley — 1870[8]

On 26 August 1870 Thomas Frederick Henley of Pimlico, Middlesex, England, filed a patent (U.K. 2,349) (Figure 138) for a percussive tunnelling machine. While percussive machines were, by this time, no novelty, there were one or two unusual features of Henley's design which are of interest.

156

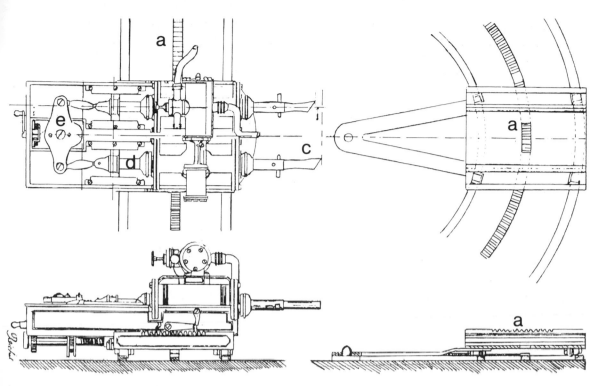

Figure 137. R. C. M. Lovell's tunnelling machine. Patented 30 July, 1867 under No. 67,323 (U.S.A.)

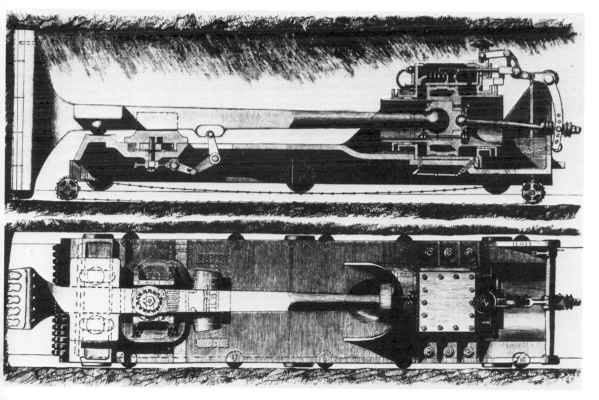

Figure 138. Henley's tunnelling machine — Pat. No. 2,349 dated 26 Aug. 1870 (U.K.). (Courtesy, *Engineering,* 13 Jan. 1871.)

Henley's machine consisted basically of a large ram, the head of which was adapted to hold a series of chisels or rock-cutting tools. The ram was mounted on a main frame or base-plate which ran on six wheels. At the rear end was an engine which delivered compressed air or steam to a piston which impacted against the rounded rear end of the ram shaft. The top or forward end of the piston which struck the ram was, in turn, concave. This entire section was enclosed by a gland which was constructed in three pieces, i.e. two halves of gun metal fitting the sphere and an outer ring of wrought iron holding these halves together. This enabled the ram shaft to be moved in an arc of a circle on the main centre point at the back end of the ram. A pair of lateral projections or lugs extended from the ram shaft a short distance from the head. The undersides of these projections were hollowed and they bore upon a pair of spheres which rolled on a table on the base-plate. The spheres carried almost the entire weight of the ram shaft and head and were so placed that when the ram was at half-stroke they were almost directly under the centre of gravity of the combined head and ram shaft, the idea being to relieve the piston of any unnecessary wear and tear during operation.

The two spheres were 43 cm (1 ft 5 in) apart laterally from centre to centre, forming a wide enough base to keep the head from canting.

Immediately behind the lateral projections was a carriage which guided the ram shaft. By means of a series of worm gears, pinion and ratchet, etc. the carriage was capable of lateral movement. This was controlled by pawls which acted during the return stroke of the ram. Thus the head was reciprocated back and forth by the piston, while at the same time it was moved slowly to one side of the tunnel by the carriage and then back again. It could therefore cut a tunnel of rectangular section.

The chisels on the head were so arranged that they stopped short 10 cm (4 in) from the bottom of the head on each side. Because of this the gallery would be cut with a step or ledge 10 cm high × 10 cm wide (4 in × 4 in) at each of the lower corners. Along this ledge ran the wheels, thus leaving room beneath the machine for the removal of debris. Henley then proposed to remove the debris by means of an endless chain of scrapers which would sweep the cuttings from the face to the rear of the machine.

E. A. Cowper — 1871

Patent No. 1,612 (Figure 139) filed by Edward Alfred Cowper of Westminster, Middlesex, England, on 20 June described a percussive machine which featured a series of 'jumpers' (chisels) set in horizontal rows on a number of frames. Each chisel was individually actuated by a separate piston which was powered by either steam or compressed air. According to Cowper's patent it was proposed that the tools be given a reciprocating movement only. They did not rotate on their own axes. However, Cowper suggested that as the chisels moved back and forth the frames could be laterally vibrated (either by manual labour or by power

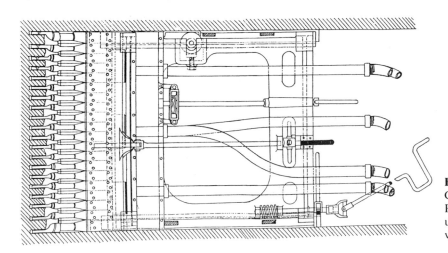

Figure 139. Edward Alfred Cowper tunnelling machine. Patented in U.K. 20 June 187 under No. 1,612. Sectional ele vation

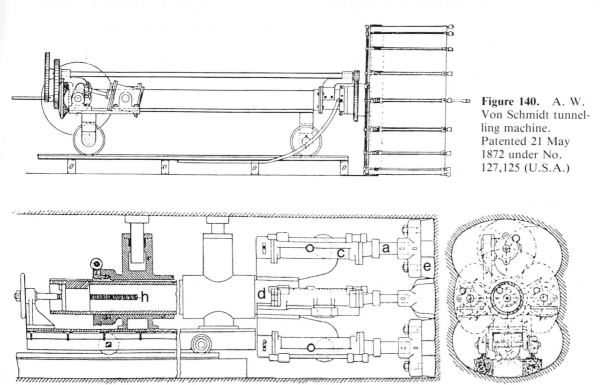

Figure 140. A. W. Von Schmidt tunnelling machine. Patented 21 May 1872 under No. 127,125 (U.S.A.)

Figure 141. Frederick Bernard Doering's T.B.M. Patented 27 Sept. 1881 under No. 4,160 (London)

through the medium of two wire ropes or other such mechanical means) so that grooves would be cut in the face. The entire apparatus was designed to advance slowly by means of a screw rack and pinion during operation.

A. W. Von Schmidt — 1872[8]

Allexey W. Von Schmidt's Patent (U.S. No. 127, 125) (Figure 140) dated 21 May described a machine which consisted of a circular cutting head mounted on a main base frame. A series of diamond-tipped drill tools was carried by the cutting wheel along its periphery. In addition there was also a diamond drill in the centre of the wheel which was connected directly to the main shaft and which bored a hole for blasting purposes. The rear of each drill shaft terminated in a pinion.

This set of circumferentially arranged pinions was driven by one large central gear. At the same time as the drills revolved on their axes the entire wheel and, of course, the central drill, were turned by means of the revolving main

shaft which was powered by two engines at the rear of the machine. By this means Schmidt proposed that an annular groove would be made in the face and the core would afterwards be shattered by blasting.

F. B. Doering — 1881

On 27 September 1881, Frederick Bernard Doering of Trefriw, North Wales, filed a patent (U.K. No. 4,160) (Figure 141) covering a rotary/percussive rock-tunnelling machine.

To enable the machine to bore either an oval or circular tunnel Doering proposed to use four heads which were, in effect, large percussive drills [a], each drill being fitted with its own piston, cylinder [c], and twist or rifle bar. These drills were mounted around a central drill which was rigid and was attached directly to the rotating main shaft [d]. The four outer drills were pivoted so that they could be turned outwards when required, to increase the size of the gallery in case this became reduced through wear of the outer gauge chisels.

159

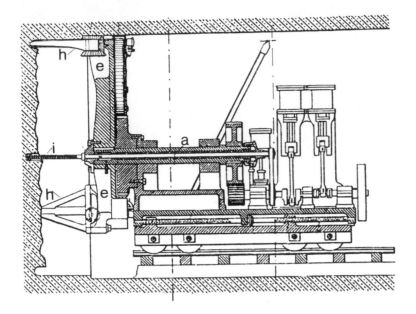

Figure 142. H. S. Craven tunnelling machine. Patented 28 Oct. 1884 under No. 307,379 (U.S.A.)

The whole head, consisting of all the drilling machines including the rigid central drill, was rotated by means of a worm gear.

According to Doering, by correctly arranging the position and size of individual drilling machines and by causing the whole cutter head to rotate completely or only oscillate through a small arc, it would be possible to cut either circular or oval tunnels.

Each drill was fitted with a crosshead which carried a number of chisels [e]. A percussive motion was imparted to the gangs of chisels on each drill head by the reciprocatory action of the pistons in the cylinders. In addition to the percussive movement each crosshead was also turned by the rifle bar on the back stroke of the piston.

The cutting head was advanced by means of a screw on the main shaft [h].

While there appears to be no positive indication that Doering's machine was ever built and tested in practice, the wording in the specifications accompanying the patent intimate that Doering did at least test some of the components described by him during the design phase.

H. S. Craven — 1884

U.S. Patent No. 307,379 (Figure 142) granted to Henry S. Craven on 28 October related to a rotary tunnelling machine which consisted of a central rotating shaft [a] on a base frame. The frame was in turn mounted on a carriage on wheels which ran on tracks on the tunnel floor. During the boring operation the carriage was fastened securely in place in the tunnel and the base frame carrying the shaft and cutter head, was slid forward into the face.

Three radial arms [e] projected from the forward end or head of the shaft, each bearing a drill [h] or chisel which rotated on its axis. In addition a central axial drill [i] was provided. While each drill revolved, the entire head turned so that a groove or chase was cut in the face. After the annular groove had been cut, Craven proposed that the core be broken by a charge of explosive placed in the central drill hole.

R. Dalzell — 1885

The Dalzell Patent No. U.S. 332,592 (Figure 143) dated 15 December described a revolving or oscillating central tubular shaft [a] mounted on a sturdy base frame on wheels. The forward end of the shaft carried a series of radiating arms [d]. At the outer end of each arm was a drill [e] which was reciprocated by compressed air or steam supplied to it by flexible tubing which was connected to the hollow central shaft. By this means Dalzell proposed that an

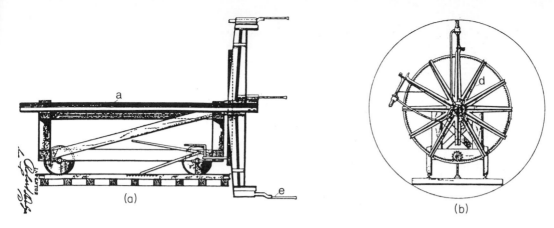

Figure 143. R. Dalzell T.B.M. (a) longitudinal section; (b) front-end view of machine. Patented 15 Dec. 1885 under No. 332,592 (U.S.A.)

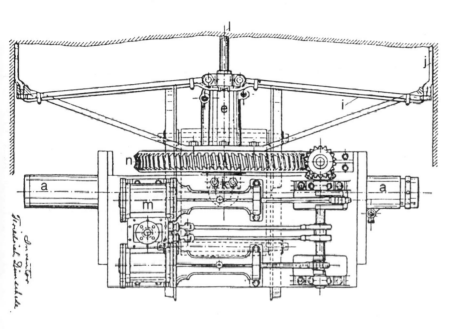

Figure 144. F. Dünschede T.B.M. Patented 31 Oct. 1893 under No. 507,891. (U.S.A.)

annular groove could be cut in the face and the core thus formed removed later by blasting.

F. Dünschede — 1892

By 1892 several trepanner-type percussive or rotary tunnelling machines had been designed. However, the trepanner machine patented by Friedrich Dünschede of Essenberg, Germany, on 5 May 1892 under No. 66,876 was different from its predecessors because Dünschede used water for the first time as the motive power instead of steam or compressed air. (He was granted U.S. Patent No. 507,891 (Figure 144) on 31 October 1893.)

A horizontal mounting bar [a] was held in place across the tunnel by a hydraulic ram located at one end. The engine and other ancillary parts were mounted on this bar.

Fixed to the centre of the bar and projecting forward along the axis of the tunnel was a hydraulic piston [e]. Over this piston fitted the hydraulic cylinder which formed the hub of the cutter head. This head consisted of two radial arms [i] with braces. These arms carried drills [j] at their ends and, as the hub was rotated by a system of gears, cut an annular groove in the face at the periphery. The hydraulic piston at the centre served both as a pivot [k] for the rotating head and as the means of applying

161

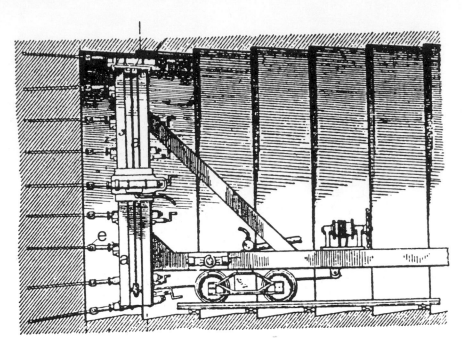

Figure 145. H. Byrne tunnelling machine. Pat. No. 545,675, dated 3 Sept. 1895. (U.S.A.)

forward thrust. In addition a centre drill [l] was fixed in the end of the cylinder so that a central hole, suitable for blasting out the core, was drilled as the head advanced.

The gears which rotated the head were powered by two hydraulic cylinders [m] positioned transversely on the machine immediately behind the large crosshead gear wheel [n].

During the boring operation Dünschede proposed that the water, after having served its purpose in the various cylinders, be directed to the face via a canal which extended from the hollow space of the hub or central cylinder into the hollow portion of the central drill. A lateral continuation of the canal was connected to conduit pipes leading to the peripheral tool holders, the water being discharged through small pipes provided on each of the outer tool holders.

The author was unable to ascertain whether the Dünschede machine was ever built and tested in Germany.

H. Byrne — 1895

Harry Byrne of Chicago designed a percussive machine (Patent No. 545,675 (Figures 145 and and 146) dated 3 September) which cut a series

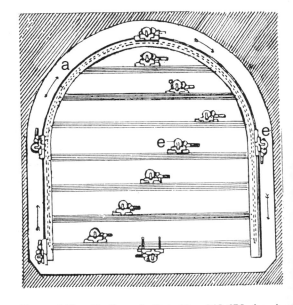

Figure 146. H. Byrne's Pat. No. 545,675 showing apparatus at work in the tunnel

of horizontal channels within an arch-shaped profile. He suggested that the rock remaining between the channels could be either drilled or blasted away.

His machine consisted of a carriage on wheels at the head of which was a massive upright arch-shaped framework [a]. Four or more percussive drilling machines [e] powered

162

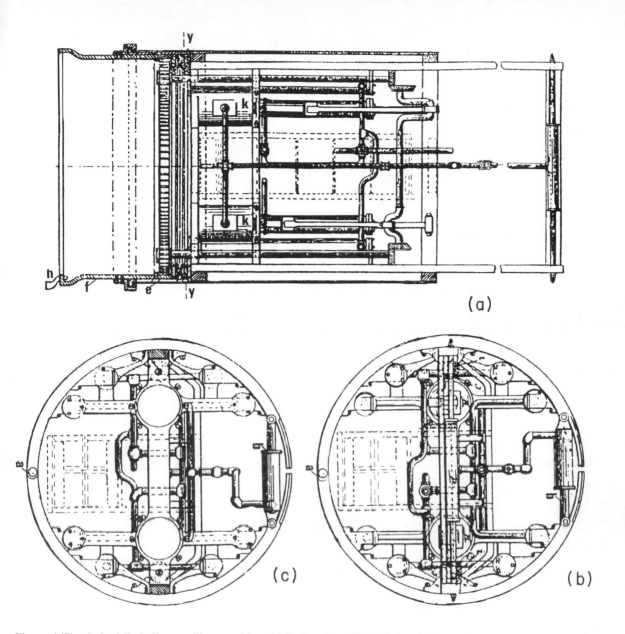

Figure 147. J. L. Mitchell tunnelling machine. U.S. Pat. No. 537,899 dated 23 April 1895. (a) Longitudinal horizontal section; (b) Rear view; (c) vertical transverse section on line y-y of (a)

by either steam or compressed air were mounted on horizontal slideways within the arched profile. Another four machines were mounted in a recess or cavity around the outer surface of the framework. A flexible wire-rope connection, which extended around guide sheaves and to an operating capstan, linked the drills and enabled them to be drawn along their tracks so that a series of channels were cut in the face.

The profile was cut with the four machines mounted on the outer surface of the framework. Two of these were adapted to have a curvilinear reciprocating travelling movement upon the semicircular top portion of the frame so as to form the crown of the tunnel, while the other two, one on each side of the arch, travelled in a vertical plane to form the sides of the tunnel.

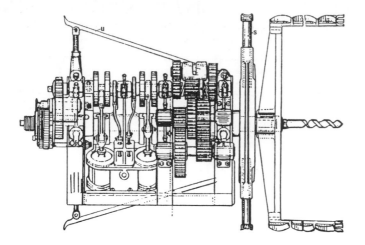

Figure 148. Improved design of Mitchell's machine. U.S. Pat. No. 565,494 dated 11 Aug. 1896

J. L. Mitchell — 1895-6

By the end of the nineteenth century a number of trepanner-type machines had been designed and built and were thus no longer considered a novelty. Having proved the basic design concept to be practical, engineers then began submitting patents covering 'improvements' to this type of machine, and during the late 1800s and early 1900s a number of patents relating to these machines were granted.

On 23 April 1895 under No. 537,899 (Figure 147) and on 11 August 1896 under No. 565,494, (Figure 148) Jonas L. Mitchell of Illinois was granted two U.S. patents covering improvements to the trepanner-type machine.

The first of these (dated 1895) described a hydraulically powered machine which was supported in a bed frame consisting of four curved metal bars (two at the front and two at the rear of the machine). Each pair of bars was joined by a hinge [a] at the bottom or floor and at the top by hydraulic jacks [b] which allowed the gap between the upper ends to be expanded or contracted when necessary. The two ring-like frame supports were connected together by several longitudinal girts or bars. Sheet metal was then secured to the exterior surfaces of the base frame along the longitudinal bars and served both as grippers to be expanded (by means of the hydraulic jacks at the top) against the tunnel wall during the boring operation, and also as a protective shield for the working parts of the machine.

Within the base frame were carrier guides along which the cutter-head support frame ran. A large-diameter tube [e] (the size of the tunnel) was mounted on the forward end of the cutter-head support frame and into this was fitted another tube [f]. This latter tube was provided with sockets for receiving cutters [h] around the circumference of its forward end. The rear end of the inner tube or cutter head formed a bearing flange and was fitted with an internal ring gear driven by two pinions.

During operation, thrust from the rotating cutter head was received upon balls or rollers which ran between the bearing flange and a plate.

Power for the rotating cutter head was supplied by hydraulic cylinders [k] via a series of bevelled pinions, wheels, and crank shafts, etc. These cylinders, together with those supplying forward thrust to the cutter head were positioned near the gripper sheeting, above and below the horizontal, central, longitudinal plane of the machine. This particular arrangement, according to Mitchell, gave thrust to the head in line with the components (i.e. the cutters, etc.) which attacked the face. The concept of a hydraulic instead of a screw feed was claimed by Mitchell to be a new contribution towards the development of tunnelling machines (see F. Dünschede, 1892).

The following year Mitchell filed a patent (No. 565,494) which covered improvements to Reginald Stanley's patents first issued in the U.S. on 12 November 1889 and reissued on 9

164

May 1893 (see Chapter 18 on coal-mining — Great Britain — Stanley heading machine).

These improvements related to various modifications to the anchorage and advance systems originally adopted by Stanley.

In Stanley's 1889 patent (which, incidentally, was practically identical with the 1888 patent (No. 1,763) granted to Stanley in the U.K.) the main frame of the machine (after the completion of the cutting cycle) was advanced as follows. A pair of anchoring arms [u] were attached to the rear of the cutter shaft and these were secured against the tunnel wall. The cutter shaft was thus prevented from rotating so that the main frame could be drawn forward by the engines until it rested once more against the rear of the cutter head.

In U.S. Patent No. 565,494 dated 11 August 1896, Mitchell suggested that hydraulic jacks [s] be installed on the cutter shaft immediately behind the cutter head and in advance of the main frame. When it became necessary to reposition the main frame, these jacks could be extended against the floor and ceiling, thus

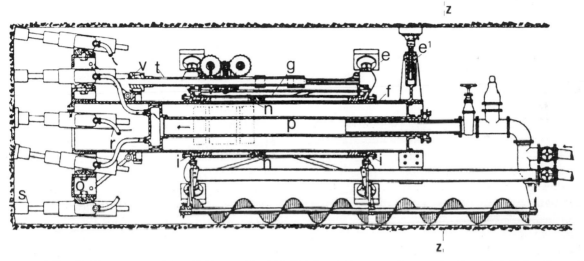

Figure 149. P. Unanue T.B.M. U.S. Patent No. 732,326 dated 30 June 1903. (application filed 23 Dec. 1901). Longitudinal central section of machine in elevation

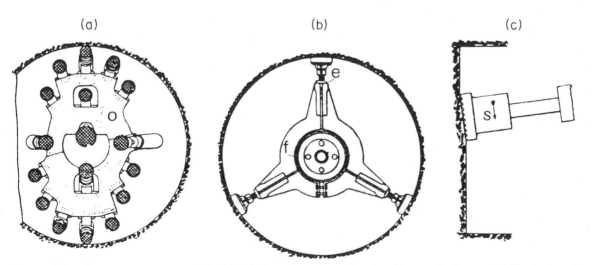

Figure 150. Unanue T.B.M. (No. 732326). (a) Face view of ram-head and its attached rams; (b) Vertical section on the line z-z on Figure 149; (c) enlarged view of one cutter

165

securing the cutter shaft so that the main frame could be advanced while the cutter head remained anchored.

P. Unanue — 1903

Pedro Unanue, a Spaniard residing in Mexico, was granted a U.S. Patent (No. 732,326) (Figures 149 and 150) on 30 June 1903 (application filed 22 December 1901) for an improvement in tunnelling machines.

While there is no positive evidence that Unanue built and tested his machines, the wording of the patent itself implies that the 'improvements' suggested therein were the result of previous trials on earlier models. In addition it seems unlikely that Unanue would have gone to the trouble and expense of procuring a foreign patent for his 'improvements' if earlier experiments had not been encouraging.

The Unanue patent described a percussive machine which consisted of a cylindrical casing [g] which could be anchored to the tunnel wall by two sets of adjustable radial screw jacks [e]. Within the casing was a tube [f] which extended through glands or stuffing boxes [i] beyond the forward and rear ends of the casing. A piston ring [n] was fitted to the exterior of the tube, between the tube and the casing, and it travelled with the tube back and forth inside the casing. The tube was attached at its forward end to the ram head [o] and thus provided thrust for this component. At the rear of the tube (where it extended beyond the casing) was another set of radial screw jacks [e¹]. Their purpose was to anchor the tube when it became necessary to reposition the casing in readiness for the next cutting cycle. During the boring or cutting operation these jacks were kept in a retracted position. Unanue proposed that the piston be reciprocated by oil pressure.

Within the tube was a second tube [p] which conveyed air or steam to the ram head. The air was conducted to a head at the forward end of this inner air-carrying tube. From the head of the air tube flexible tubes [r] carried the air to each of a number of rams mounted on the main ram head. These rams were, in effect, conventional percussive rock drills fitted with ram heads [s] instead of chisels or drills. The individual percussive rams were positioned at different radial distances from the centre in overlapping peripheral paths which ensured that the entire face was attacked during the rotation of the ram head.

Unanue commented that in previous machines the percussive ram cutters were arranged to '...strike in a plane parallel to the surface of the material acted upon or in a plane parallel to the plane of revolution of the ram head, and consequently the cutters after the first blow would not strike squarely, but would impinge on the uncut material and consequently tend to produce a bending movement on the rammer-rods'.

Unanue suggested that this difficulty could be overcome by setting the rams to strike the face at a slightly inclined angle which corresponded to the revolutionary direction of travel of the ram head.

The ram head was made to revolve by means of a slide bar [t] which travelled (in conjunction with the tube and piston within the casing) between guideways fitted to the exterior of the main casing. This slide bar was rotated by a steam or air motor mounted on standards connected to the casing. The slide bar rotated a pinion [v] which meshed with the main rotary gear driving the ram head.

J. P. Karns — 1903-18[9, 10]

To some extent commensurate with the contribution made by Reginald Stanley of England towards the development of the coal-heading machine, was that made by John Prue Karns of America towards the development of rock-tunnelling and shaft-cutting machines during the early 1900s.

It was during the year 1890 when Karns was Construction Superintendent for the Revenue tunnel at Ouray, Colorado, that he designed his first tunnelling machine. Unassisted, he built the unit in the tunnel's blacksmith shop using 61 cm (24 in) diameter water pipe for the band, common steel for the cutters, and wrought iron for the main structure. The cutters extended 38 mm (1½ in) beyond the band, thus giving the machine an overall cutting diameter of 68 cm (27 in).

The tunnelling machine was powered by a

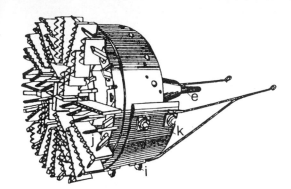

Figure 151. Cutter head of John Prue Karns T.B.M. U.S. Patent No. 744,763 dated 24 Nov. 1903 (application filed 16 Sept. 1902)

conventional No. 13 Slugger drilling machine.

When it was tested the machine cut 60 cm (2 ft) of tunnel in 42 minutes. Despite the encouragement of this initial trial, Karns was unable to obtain financial backing for the continued development of his machine. It was thus ultimately dismantled and dumped at the bottom of the Old Wheel of Fortune shaft at Ouray. However, during the succeeding years Karns continued to study the various problems associated with tunnelling machines. He commenced building his second machine in 1899 while working on the Climax property near Victor. The final designs for this model were filed on 16 September 1902 and he was granted Patent No. 744,763 (Figure 151) on 24 November 1903.

The machine consisted basically of a central percussive drill shaft [e] powered by a steam- or air-driven drill machine motor. The circular cutter head consisted of a pair of concentrically arranged rings united by a number of radiating spokes. A circular web, capable of holding ball-bearings was attached to the outer ring. Each ball-bearing was individually held in place in one of a series of sockets which fitted into the web in a continuous circular row around the perimeter of the head. These bearings permitted free rotative and forward movement of the drill head during the operation of the machine. The drill head rested on a pair of curved plates [k] or shoes which were also provided with ball-bearings [i] to permit the forward movement of the entire machine during the repositioning cycle.

The shoes themselves formed the lower section of the drill-head support frame which consisted of two vertical columns connected by a horizontal bar about three-quarters of the way down and, across the top, by a curved plate which was mounted on two screw jacks. When the machine was repositioned in readiness for the next cut this curved upper plate was extended to the tunnel ceiling by the screw jacks to provide anchorage during the drilling operation.

Radially and angularly arranged holders [j] mounted on the radiating spokes and rings carried chisel-type cutters which were simultaneously reciprocated against the face. Although Karns does not describe this in his patent, the drill head was also apparently rotated or indexed slightly, presumably by the drill motor piston (to which the main shaft was attached) during its backward stroke.

The 32 cm (52 in) diameter machine was tested in the India tunnel in 1904 where it cut some 23 m (75 ft) of tunnel.

As usual, unfavourable predictions were made by various critics of the day regarding the future prospects of the machine but, despite these, Karns managed to secure sufficient financial backing to continue with his developmental work. By 1907 two larger machines of 1.8 m and 2.4 m (6 and 8 ft) diameter had been built by the American Machine Company of Cleveland, O. The 1.8 m (6 ft) diameter machine was tested in the Cumberland tunnel of the India Gold Mining Company 16 km (10 miles) from Boulder, Colo., on 16 July 1908. On that day before an audience of eminent mining and machinery men it was run continuously for 17½ minutes during which time it cut 7 cm (2¾ in) from the face which consisted of a tough mixture of quartz and spar devoid of seams or rock joints. It was powered by a conventional compressed-air percussive drill machine motor operating at an average pressure of 0.69 MPa (100 lb/in²). After each forward stroke the head was rotated 13 cm (5 in) to the right on the backward stroke by rifled cams.

Between the years 1903 and 1918 Karns was granted 16 patents, all of these relating to shaft-sinking or tunnelling machines or various components used in the construction thereof.

It is obvious that the method employed for

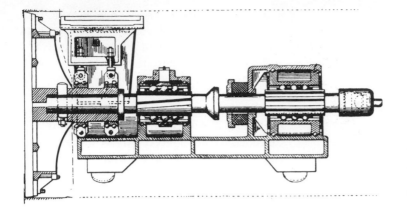

Figure 152. Karn U.S. Pat. No. 957,687 dated 10 May 1910 (application filed 25 May 1909). Longitudinal section, partly in elevation of T.B.M.

indexing the drill head of his 1903 unit was not satisfactory, as in the 1907 patent dated 26 March (No. 848,107) Karns introduced a rifled main shaft. This method of rotating the head was further developed in Patent No. 875,664 dated 31 December 1907 when ball-bearings running in a rifled groove were used to rotate the head, as depicted in Figure 152.

Throughout all his patents it is apparent that Karns was a firm believer in the efficacy of the ball-bearing as an anti-friction device, as these were used whenever or wherever it proved practicable. All the machines designed or built by Karns were percussive units.

W. A. Lathrop — 1906

Strictly speaking, William Arthur Lathrop's patent (No. 816,923) (Figure 153) dated 3 April belongs to the section entitled 'Excavation — Coal Mining' as it described a heading machine which he proposed for use in bituminous coal-mines, rather than for cutting rock in tunnels. However, it is included in this section not merely because it was designed to cut fairly hard material other than clay, etc. but also to give the reader the opportunity to compare the basic design of this type of unit with other rock machines which were built or proposed at that time.

Lathrop's machine was similar in concept to some of the continuous miners which made their debut during the late 1940s and early 1950s. Its action is also reminiscent of the bucket dredge or wheel excavator.

The cutter head consisted of a large framework carrying two transversely oriented drums or cylinders [e], one above the other. The top drum was slightly larger in diameter than the bottom one.

A series of sprocket chains [k] fitted with cutting picks [m] ran around these drums travelling up over the forward portion of the drums so that any cut coal or debris would be thrown back upwards and over the two drums into a hopper [o] which extended across behind the upper drum and parallel to it. The hopper had two screw conveyors [s] leading from its outer ends towards the centre where an opening was provided to enable the coal to fall through into a chute and thence on to another screw conveyor [r]. This latter conveyor carried the coal up and towards the rear for delivery into coal carts, etc. The rear end of this conveyor was fitted with a double-pivoted spout [p] to enable the operator to direct the coal into either of two mine cars.

Lathrop further provided hydraulic jacks [j] which lifted the upper section of the framework. This enabled the operator to adjust the inclination of the frame and consequently the vertical height of the cut. Lathrop proposed using an electric motor to power his unit.

Forward thrust for the cutters was provided by two jacks [t] at the rear of the machine which reacted against the tunnel walls. Two further jacks were provided at the sides of the unit so as to enable the operator to reposition the machine when necessary.

August M. Anderson — 1906

August M. Anderson filed two patents (Figures 154 and 155) covering 'excavators' on 17 April

168

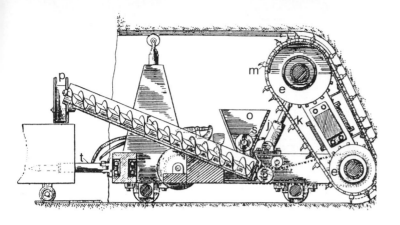

Figure 153. W. A. Lathrop tunnelling machine. U.S. Pat. No. 816,923 dated 3 April 1906 (application filed 31 Aug. 1903)

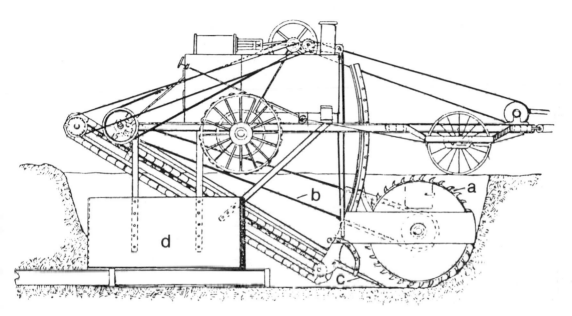

Figure 154. A. M. Anderson excavator. Patented in the U.S. on 17 April 1906 under No. 818,214

1906. These were Nos. 818,214 and 818,215 (U.S.A.) respectively.

In both the aforementioned patents the cutting wheel was supported upon a shaft mounted transversely between two flat metal bars running parallel with the axis of the trench. Behind the excavating wheel was a plough.

Power was transmitted to the excavating wheel [a] through a chain [b] which rotated the wheel so that its teeth dug downward into the earth in advance of the machine. The loosened material was scooped backwards and up over the plough [c] on to the front end of the main

bucket conveyor. From this point it was lifted and carried by an auxiliary conveyor which deposited it at the side of the machine into skips or wagons.

Though the basic outline of these two machines vary slightly, in the main they follow a similar design pattern. Anderson, however, provided protective guards [d] in Patent No. 818,214 which were not included in Patent No. 818,215. These guards consisted of rectangular plates held vertically near the rear of the machine. They were designed to act as retainers in order to prevent the side walls of the trench from collapsing.

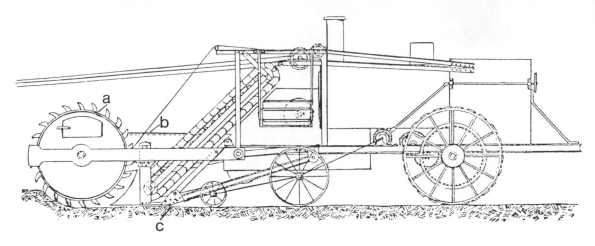

Figure 155. A. M. Anderson excavator. Patented on 17 April, under No. 818,215

The above patents covered machines designed primarily for digging ditches or trenches in open-cut work, and they were the forerunners of the modern wheel-type excavator developed during the mid-1930s and also of the ripper-type continuous miners of the 1950s.

An Anderson excavator approximately 3 m long was apparently worked in a tunnel in Detroit in 1914 sans shield. Later another Anderson excavator was used in a Cleveland West Side water supply tunnel in 1915 with a shield. The machine was, however, withdrawn before construction work on the tunnel had been completed. According to *Engineering News* dated 18 January 1917, this was because unsuitable ground had been encountered. But later it was found that practically the same advance rates were achieved by hand excavation. Moreover, stoppage problems encountered due to mechanical failure were avoided, as were the high costs of machine maintenance and operation.

A. D. Lee & F. J. E. Nelson — 1907

Alva D. Lee and Francis J. E. Nelson were granted a U.S. Patent (No. 874,603) (Figures 156 and 157) on 24 December for a rock-tunnelling machine.

Messrs. Lee and Nelson's patent described a machine which cut a wide annular groove in the rock by means of a circular cutter-head plate [a], the diameter of the proposed tunnel.

The plate carried a series of groups of freely revolvable cutters [e], so positioned on the plate that when it turned the entire face of the wide peripheral groove was attacked. The cutters were cylindrical roller cutters with the cutting edges forming ridges on the surface of the roller. Within each group, individual cutters had their ridges running in different directions, some spirally clockwise, some spirally anti-clockwise, some diagonally and some simply around at right angles to the axis. Two diamond gauge cutters [i] on the periphery of the wheel ensured that the bore diameter remained constant. Messrs. Lee and Nelson proposed that a spray of water [m] be directed over the face as it was being cut. A hole behind each group of roller cutters was designed to allow the water and cuttings to drop through to the bottom of the bore, where it was collected and then pumped to the rear through a pipe running beneath the rails along which the main machine carriage travelled.

As the core was cut it was supported by a long cylindrical member [p] which extended to the rear of the machine. Lee and Nelson proposed that this core be broken off and removed as desired when it reached a suitable length.

S. A. Knowles and W. E. Carr — 1907

Patent No. 875,082 (Figure 158) dated 31 December granted to Silas A. Knowles and Walter E. Carr, described a percussive machine powered by compressed air or steam.

170

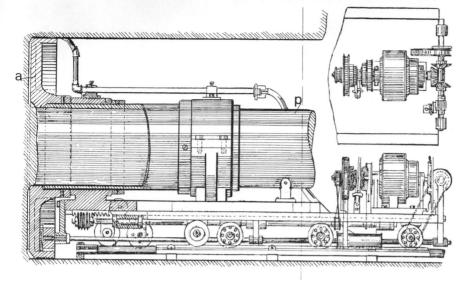

Figure 156. A. D. Lee and F. J. E. Nelson, Jr. T.B.M. Patented on 24 Dec. 1907 under No. 874,603) (U.S.A.)

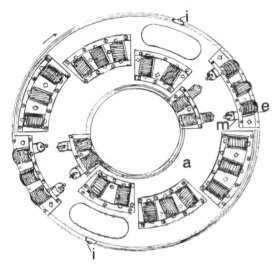

Figure 157. Front elevation of Lee and Nelson T.B.M. revolvable cutter plate

Although Knowles and Carr mentioned that the head could be made any desired shape or size, i.e. large enough to cut, say, a full rectangular tunnel or perhaps, of a smaller size, for bench and heading work, preference was given in the patent to a narrow cutter head, the full height of the tunnel.

In this case the head was adapted to move across the face from wall to wall. This was achieved by using a curved gear-tooth rack [m] to swing the rear end of the guide arm [g] on the pivot on the feed frame. The cutter head was fed forward into the face by a feed screw [k]

which moved the feed frame and guide arm (and thus the cutter head [i] to which the latter was attached) towards the face.

The head was reciprocated by a conventional drilling motor mounted on the guide arm. It was connected to the hub [h] of the cutter head by a key [p] which extended down through both the hub and the drilling motor plunger.

W. J. Hammond — 1908

U.S. Patent No. 885,044 (Figure 159) dated 21 April granted to William J. Hammond described a percussive machine having a revolvable head or arm [m] on which were mounted a number of individually pneumatically powered hammers [e]. These hammers consisted of a series of teeth or chisels [i] placed closely together to form rectangularly shaped hammers.

Hammond proposed that the whole cutter-head frame should be made to rotate by means of the main shaft [p] at approximately one revolution per minute.

G. A. Fowler — 1908-11

Using the Fowler patents, a full-size tunnelling machine, carrying a narrow rectangular swing-block cutter head with 41 percussive rock drills, was built by the International Tunnelling and Machine Company in 1909 at the Davis Iron Works in Denver, U.S.A., where it was tested

171

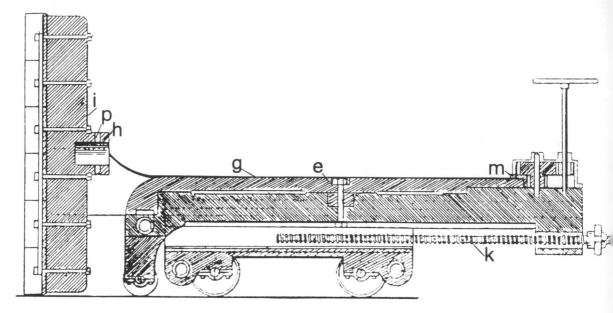

Figure 158. S. A. Knowles and W. E. Carr T.B.M. U.S. Pat. No. 875,082, dated 31 Dec. 1907

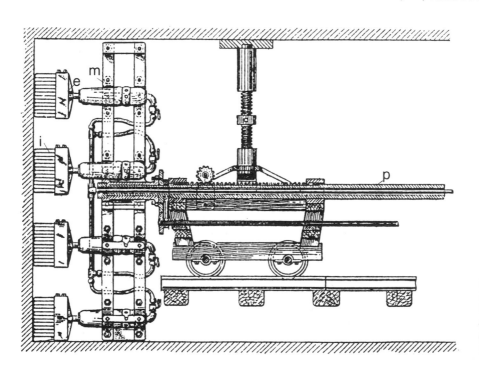

Figure 159. W. J. Hammond Jr. T.B.M. U.S. Pat. No. 885,044, dated 21 April 1908 (application filed 15 Nov. 1906)

on large blocks of concrete. However, as the machine proved unsatisfactory it was not developed any further.

George Allen Fowler filed two patents covering improvements to tunnelling machines, one on 30 July 1907 and another on 22 May 1909. These were granted on 23 June 1908 (No. 891,473) (Figure 160), and 4 July 1911 (No. 996,842) (Figure 161) respectively.

The first patent (No. 891,473) described a a machine with a narrow, rectangular swing-drill-block head [e] the full height of the tunnel. The swing drill block was fitted with a series of percussive air-powered drills [i] arranged in

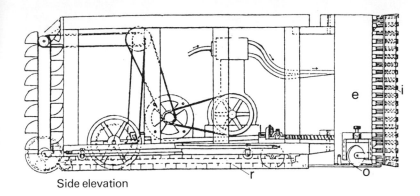

Side elevation

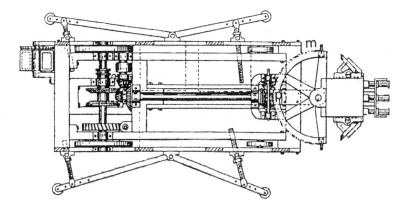

Figure 160. Side elevation and plan view of G. A. Fowler T.B.M. U.S. Patent No. 891,473, dated 23 June 1908 (application filed 30 July 1907)

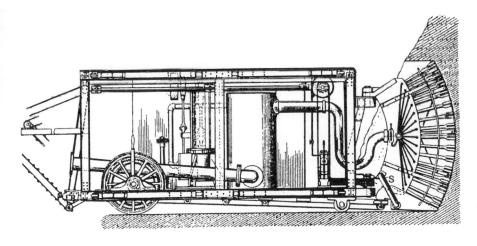

Figure 161. G. A. Fowler's 1911 T.B.M. U.S. Pat. No. 996,842 dated 4 July (application filed 22 May 1909)

groups of three. Air was fed to the front and rear chambers of each group of three drills in such a manner that the groups were made to strike alternate blows at the face.

While the various drill groups were striking alternate successive blows at the face, the entire block was made to swing in an arc from one side of the face to the other by means of a large quadrant gear [m] (to which the drill-head block was attached) and a worm [p] actuated by the rotating main shaft.

During the drilling operation the face was sprayed with water and the cut debris fell into a muck pan which was connected by a stud to the

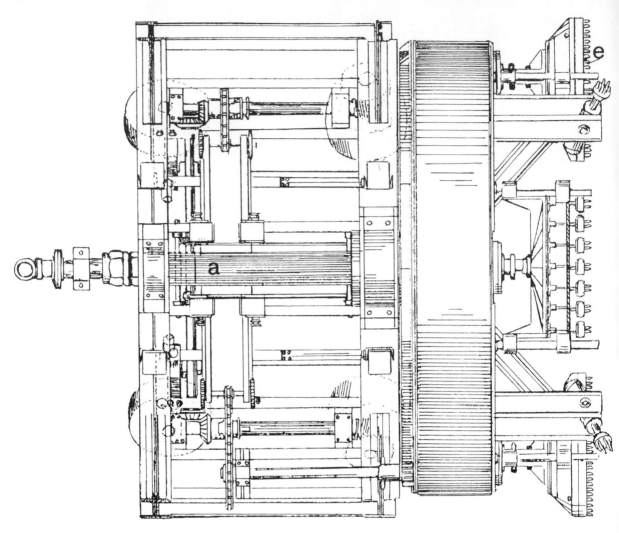

Figure 162. E. F. Terry Tunnelling Machine. U.S. Pat. No. 917,974 dated 13 April 1909 (application filed 23 Oct. 1905)

lowest drill plunger on the head. Thus, as the plunger was reciprocated, the muck pan [o] was also made to reciprocate. This caused the muck thereon to be thrown from its rear end on to the conveyor apron [r] running along beneath the machine. The apron then carried the muck to the rear and deposited it into the boot of a bucket conveyor which in turn lifted it up and dumped it into a rail-car.

Apart from some modifications, Fowler's second patent (No. 996,842) dated 1911 described a similar type of machine. These modifications included arrangements whereby the exhaust air [s] from the drill block could be used to blow the rock cuttings rearwards and

deposit them upon an endless conveyor. In addition, the frame carrying the drill block was adapted so that it could be raised or lowered at its rear end, to enable the operator to drill any desired grade of tunnel, either straight or curved.

O. S. Proctor — 1908 and E. F. Terry — 1909[11-13]

Olin S. Proctor of Denver, U.S.A., was a qualified dentist. His first tunnelling projects, with holes no larger in diameter than that of a knitting needle, were perforce executed in the teeth of his squirming patients. But Dr. Proctor soon tired of this and sought bigger and better

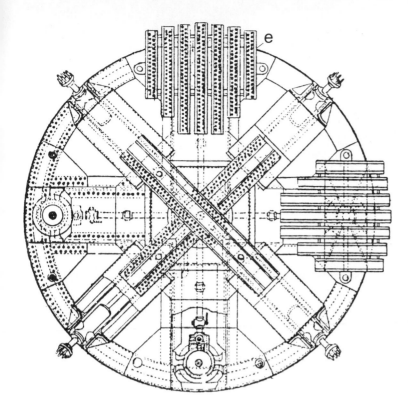

e

Figure 163. Front view of Terry T.B.M., partly in section. Pat. No. 917,974

areas, preferably ones that remained stationary while he worked.

To pay his way through dental college, Dr. Proctor worked in the mines at Cripple Creek and it was there that he first learned of the refractory characteristics of rock as a substance.

The rough design of his first tunnelling machine gained the interest of one of the largest contracting firms in the country, the Terry and Tench Company. This was probably because one of its directors, Edward F. Terry, was intensely interested in the subject and, in fact, had already filed an application for a patent covering his own tunnelling machine design on 23 October 1905 (Figures 162 and 163). (Terry's patent was finally granted on 13 April 1909.)

The tool head of Terry's machine was mounted on a main central hollow rotating shaft [a]. The head carried nine gangs [e] of drills composed of a central gang, and eight sets of side and corner gangs. Each of the side gangs, the corner gangs, and the central gang of drills was powered by compressed air supplied by three cylinders on the main tool head.

Screw jacks were fitted to each of the carriage wheels. By raising or lowering these jacks Terry proposed to alter the side and vertical cant of the machine with respect to the tunnel axis for steering purposes. Terry proposed to remove the debris from the face with a series of scraper blades pivotally secured at their upper ends to a longitudinal bar which was suspended beneath the machine by four arms. By means of the eccentrics on a shaft connected to the main engine, the arms, and thus the bar and scraper blades suspended below the machine, were made to reciprocate. When the bar was swung backwards the blades assumed a vertical position so that the debris could be moved rearwards where it was pushed up a small incline and then allowed to fall on to a conveyor belt at the rear of the tunnelling machine.

In view of his own intense interest in the subject, it is easy to understand that Terry might easily be persuaded to 'blow into his partner's ear' and convince him of the benefits of building a tunnelling machine for Proctor.

The first unit which took two years to

175

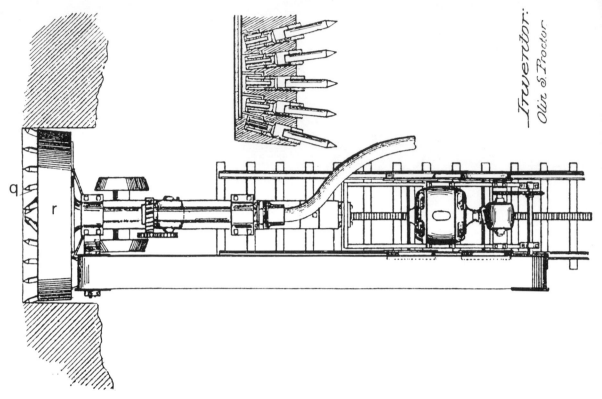

Inventor:
Olin S. Proctor

Figure 164. O. S. Proctor T.B.M. U.S. Pat. No. 900,950, dated 13 Oct. 1908

complete was constructed in a large machine shop belonging to the company. It cost about U.S. $17,000. It was tested at the new Grand Central Terminal Works on Manhattan Island where it worked well enough to demonstrate its potential. The $17,000 package of steel and iron was then returned to the company's yards and dumped, and draughtsmen commenced working on the plans for another machine.

While still retaining some of the essential features of the original model, the new machine was nevertheless radically different and more compact. Moreover, during its construction other changes were made until ultimately the machine which emerged bore very little resemblance to the first drawings made of it.

The patent covering this machine (No. 900,950) (Figure 164) was filed on 26 February 1907 under the name Olin S. Proctor and was assigned to the newly formed Terry, Tench, and Proctor Tunnelling Machine Company. It was granted on 13 October 1908 and to some extent resembles the actual machine built by the

company. (Terry's patent was also assigned to the new company.)

The head of Proctor's second machine, when it was finally built, carried 15 percussive drills [q]. These drills were mounted at various angles on four curved arms forming part of a massive cast steel cutter head [r] which revolved. In appearance the cutter head was somewhat like the blades of a propeller. The drills were individually powered by compressed air fed to them at 0.76 MPa (110 lb/in^2) pressure through the hollow main horizontal shaft.

The main shaft was rotated by a 25 hp motor through a powerful worm gear and a 112 cm (40 in) diameter worm wheel. Two 20 cm (8 in) diameter horizontal hydraulic jacks provided thrust for the main cutter shaft and head.

A scoop attached to one of the cutter-head arms lifted the muck and deposited it into a hopper. The hopper dropped it on to a conveyor which carried it to the rear and delivered it into a dump car.

A pair of 40 t hydraulic jacks with a 20 cm (8

in) stroke thrust against the tunnel roof at the front and rear to provide anchorage for the machine during the cutting phase. The machine was 5.5 m (18 ft) long and weighed some 23 t. It was run by one operator.

An interesting refinement was that the percussive drills on the head did not operate until they were thrust against the face with a force of some 317 kg (700 lb). This pressure pressed them back against the striking head of the plunger with sufficient force to move the plunger back in the air cylinder until the exhaust ports were opened. The opening of the ports actuated the drills which were capable of delivering 500 to 800 blows per minute. When the machine was penetrating soft rock the drills, because of the above-mentioned arrangement, ceased to function as percussive tools but acted instead as fixed drag teeth or chisels on a rotary head. (The writer assumes that this refinement was a later development, as there does not appear to be any reference to it in the patent filed by Proctor.)

While the second Proctor machine was being built, a 1 m square (12 sq ft) block of concrete, embedded with large pieces of the various types of rock found around New York, was being prepared. By the time the unit was ready to be tested the block had had a chance to cure thoroughly and it was therefore extremely hard.

After the machine had successfully bored its way through most of the concrete block, it was moved to a hill on the banks of the Harlem Ship Canal. This area had already gained the dubious reputation of possessing the hardest rock in the vicinity of New York, the material in question was a conglomerate of marble and granite. In fact, so hard was this material that the cutting tools lasted only three to four minutes against its surface. It soon became obvious that a better tool was needed and experiments to find one were put in hand.

When the Simplon tunnel was constructed the rock encountered there was found to be so hard that work was all but halted. Eventually the Simplon problem was solved when a steel containing vanadium and tungsten was made. It was tough enough to handle the Simplon rock and has since been referred to as 'Simplon steel'.

A two-pronged cutting tool made from Simplon steel was used for the Proctor machine and this appeared satisfactory. But at that stage one of the air compressors went out of action after a discharged engineer dropped a nut into it. This left one dilapidated old compressor so 'wheezy' it could only provide sufficient air to run the tunnelling machine for a few minutes at a time, after which it had to be stopped to allow it to get its breath back. Under these harrowing conditions the trials continued until eventually the company was satisfied that the unit was ready for service in Pennsylvania.

According to an article which appeared in the publication *Technical World,* Vol. 16, dated 1912, plans were in hand at that time to build two 4.8 m (16 ft) diameter units with cutter heads capable of carrying some 48 percussive drills each, and 8 smaller units of 2.4 m (8 ft) diameter. How far this work proceeded is not known as, according to the New York Historical Society, the Terry, Tench, and Proctor Tunnelling Company ceased to be listed after 1913, the company apparently continuing as the Terry, Tench company (steel contractors).

In their book *Modern Tunnelling,*[14] Messrs. Brunton and Davis mention that the App tunnelling machine was developed by O. App of the Terry, Tench, and Proctor Tunnelling Machine Company, and in this respect the reader is invited to compare the drawings of these two machines, i.e. the Proctor unit and the App unit (Figure 164 — see also Figure 183) which is described towards the end of this chapter, and note their similarity.

Russel B. Sigafoos — 1908

In effect the Sigafoos machine (Figure 165) (U.S. Pat. No. 901,392 dated 20 Oct.[10]) was a horizontal stamp mill. It had a rotary head carrying 10 hammers or cutter heads with flat-faced hardened steel cutting teeth [e]. The hammers struck the face in succession.

The rear ends of the outer and inner circumferential rows of cutter-head stems were provided with collars [m] which bore and rotated against the cam edges of the interior and exterior cams of the cam drum [r]. As the stem collars rotated with the drum, the stems were drawn backwards on the rear stroke of

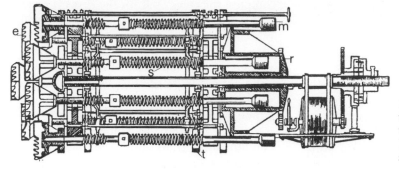

Figure 165. R. B. Sigafoos tunnelling machine. Patented in the U.S. on 20 Oct. 1908 under No. 901,392 (vertical longitudinal sectional view)

Figure 166. Hole made by Sigafoos T.B.M. in 1910 3.2 km (2 miles) east of Georgetown, Colorado. Compressive strength of rock approximately 200 mPa (29,000 lb/in²)

their reciprocal movement. The cams were wedge-shaped, thus when the bevelled edges of the stem collars dropped off the abrupt shoulder of the cam, the hammers were thrown forward against the face by the energy of the compressed coiled spring which was mounted on each stem. When the stems were drawn back by the cam on the return stroke the springs [s] were again compressed ready for the following stroke.

The entire machine, which was electrically driven, rotated in the bore on wheels [t] which could be given a helical pitch. Thus, by its own rotation and the particular pitch selected for the wheels, the machine could be made to advance or withdraw from the face. How the rotation of the machine was achieved is not clearly explained by Sigafoos in his patent.

Several experimental machines were built. Amongst these was a 2.4 m (8 ft) model constructed by the American Rotary Tunnelling Company. This unit was apparently tested near Georgetown, Colorado, and the accompanying photograph (Figure 166) is reputed to be the hole bored by this machine. Unfortunately, however, further developmental work on the machine was halted due to lack of funds.

J. Retallack — 1908

According to U.S. Patent No. 906,741 (Figures 167 and 168) dated 15 December, half of Joseph Retallack's patent was assigned to John H. Redfield of Denver, Colorado. The machine had a rotary head carrying a number of percussive drills. The drills were so

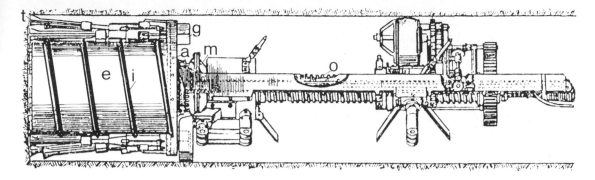

Figure 167. J. Retallack T.B.M. U.S. Patent No. 906,741 dated 15 Dec. 1908

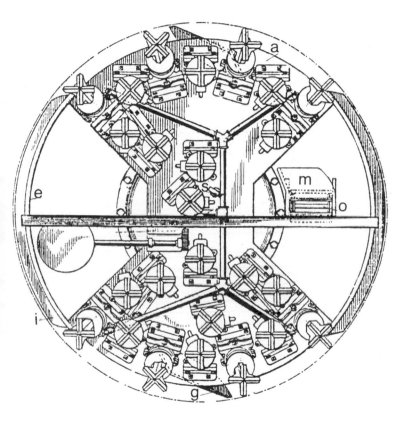

Figure 168. J. Retallack T.B.M. Face view of drill head and drills. U.S. Pat. No. 906,741

positioned on the head as to form the shape of two quadrants, thus leaving space on either side of the two drill groups for the passage of a workman in case repairs to the head became necessary. The head was mounted on a central shaft with a screw feed.

Connecting the two quadrant-like members of the head [a] were two arcuate plates [e] that were curved to conform to the contour of the periphery of the head. Helical ribs [i] which extended out to the tunnel walls were fixed around these plates. The helical ribs were designed to carry the cuttings from the face to a number of lifting buckets [g] at the rear of the head. The buckets carried the cuttings up and deposited them into a hopper [m] which directed them on to an endless conveyor [o] running along the side of the machine towards the rear. One or more rods extended forward from the head towards the face. Each rod was fitted at its forward end with a roller [s] which bore on the tunnel face. This rod and roller

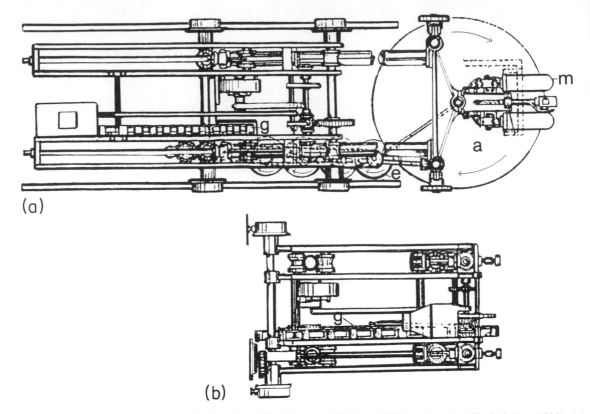

(a)

(b)

Figure 169. C. A. Case T.B.M. U.S. Pat. No. 910,500, dated 26 Jan. 1909 (application filed 13 Jan. 1908). (a) plan view; (b) end view

limited the forward feed of the head, thus preventing the percussive drills [t] from being crowded excessively against the face and also ensuring that the drills remained in the best position for the most effective percussive stroke at all times. The rear end of the rod was in view of the operator and by its position he could determine how close the head was to the actual face.

In 1922, according to Messrs. Brunton and Davis,[14] a 2.4 m (8 ft) diameter Retallack machine was in the course of construction at the Vulcan Iron Works in Denver. It was designed to carry 28 percussive drills symetrically arranged on the revolving head, each drill having 7.94 cm (3⅛ in) bits with cruciform cutting faces 15 cm (6 in) in diameter.

Fitted around the head was a 1.8 m (6 ft) long steel tube 2.26 m (7 ft 5 in) in diameter (18 cm (7 in) less than the bore), carrying 76 cm (3 in) high flanges arranged as a worm conveyor. This served to carry the cuttings to the rear of the

tube where the cuttings were deposited into buckets at their lowest point of travel, and then emptied at their highest point by a trip straight on to a rubber belt conveyor for transport to a dump car at the rear of the machine. No further news of the success or failure of this machine appears to be available.

Charles A. Case — 1909

The Case patent (Figure 169) (U.S. Pat. No. 910,500 dated 26 Jan.) described a machine consisting of a 'power truck' and a 'hammer truck'. The hammer truck carried the head or hammer frame bearing a series of hammers which were reciprocated against the face by means of springs and cams, etc. driven by a motor.

Unusual features included Case's proposal alternately to heat and cool the face, before attacking it with the hammers, in order to render the rock brittle. His patent describes a

180

method whereby electric arcs could be used for heating the rock, but he suggests that petroleum lamps, or gas jets, etc. would be just as effective.

Arrangements were made whereby the carbon electrodes [m] could be vertically raised and lowered. This was alternately to heat and cool the rock during the drilling operation. A water nozzle for additional cooling was also provided, or refrigerated air, liquid air, or other refrigerants could be employed. Another unusual feature connected with the Case patent was his proposal for a disc-type conveyor system.

One large horizontally positioned revolving conveyor disc [a] mounted beneath the hammer frame collected the debris and, as the disc revolved the material was moved on to a series of smaller revolving conveyor discs [e] by a rigid scraper bar. Each of these smaller discs and their scraper bars then successively moved the material on to the adjacent rear disc until it eventually reached a series of lifting buckets [g] which carried the material up to the rear of the machine and deposited it into a rail truck.

William R. Collins — 1910

The Collins patent (Figures 170 and 171) (U.S. Pat. No. 973,107 dated 18 Oct.) described a machine with a rotary head [a] mounted on a central tubular shaft [e] about which the head rotated. The head carried a series of percussive drills [i] on eight arms [n] which radiated from the hub. The head was designed to cut an annular groove, leaving a small core which was received by the central tubular shaft and then later broken off. The Collins machine was similar in concept to the Lee and Nelson machine except that percussive drills were fitted to the rotary head instead of roller cutters.

George R. Bennett — 1910

The Bennet machine (U.S. Pat. No. 958,952 dated 24 May)[10] consisted of a battery of pneumatically powered percussive drills arranged in rows on a head. It was electrically driven. The patent filed by Bennett described two batteries of seven drills each, but according to an article by W. L. Saunders[10] the machine

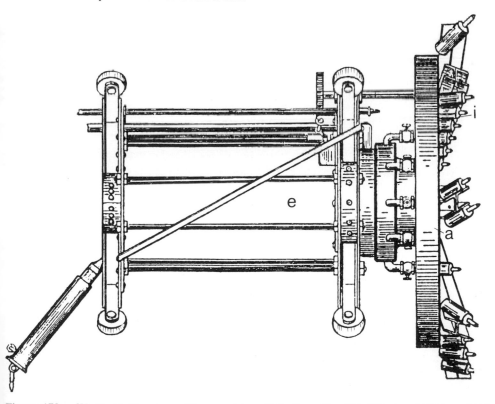

Figure 170. W. R. Collins tunnelling machine. U.S. Pat. No. 973,107, dated 18 Oct. 1910

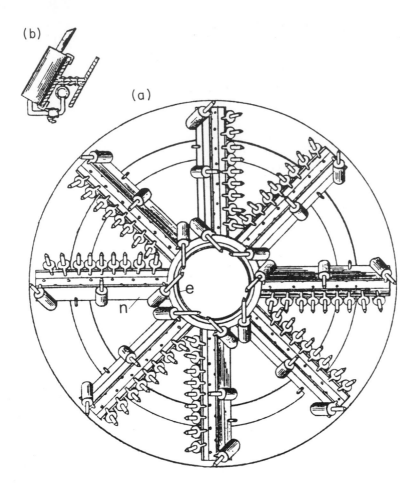

(b)

(a)

Figure 171. (a) Front view of Collins T.B.M. cutter head; (b) detailed section through one of the cutter-head arms

constructed carried 48 hammer drills with 10 cm (4 in) bits which drilled to a depth of 60 cm (2 ft). The drills, which were set eight in a row horizontally, were then withdrawn and reset both laterally and vertically until eight successive sets of holes had been drilled. This covered a rectangular area 2.4 m (8 ft) by 1.6 m (5 ft 4 in) having 384 holes. The repositioning of the head after each drilling had been completed was handled automatically by the machine. The drills were horizontally and vertically positioned so that 20 cm (8 in) was left between drills on the horizontal rows and 10 cm (4 in) on the vertical rows.

The batteries of vertical drills struck the face alternately. As each set of drills completed penetrating the face the required distance of 60 cm (2 ft), it was withdrawn and the power automatically shut off until all eight sets of

drills had completed their cuts. The drill head was then repositioned 10 cm (4 in) to the side and another series of holes was drilled. When these holes had been drilled the drill head dropped 30 cm (1 ft) and the drilling cycle was recommenced.

In effect the Bennett machine was a refined and improved version of the Maus and Merrill units described earlier in this section except that, instead of using cams and springs, it was pneumatically driven. The hammers were drawn back and mechanically held until they reached a point where they were released by a trip mechanism, when the expansion of air in the cylinder behind the hammer provided the energy for the delivery of the blow.

According to Bancroft,[15] Bennett claimed that: 'Although the machine is operated automatically in every detail, its workings may

182

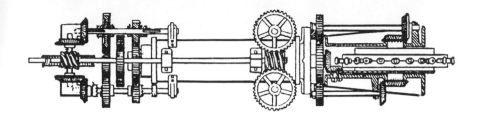

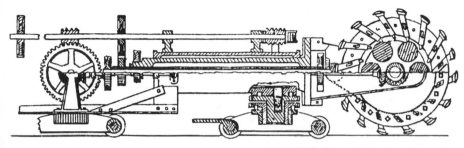

Figure 172. A. G. and E. G. Seberg T.B.M. U.S. Pat. No. 976,703, dated 22 Nov. 1910

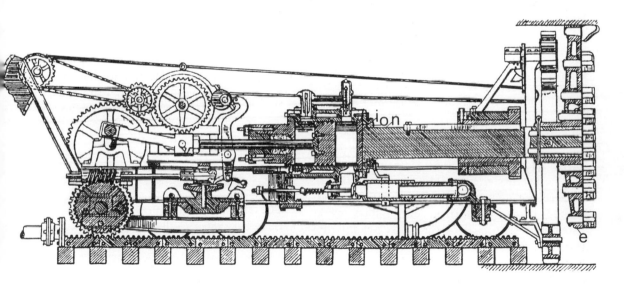

Figure 173. L. F. Sleade tunnelling machine. Longitudinal central section in elevation. U.S. Pat. No. 945,623, dated 4 Jan. 1910

also be directed by the machine operator; the drills work automatically, yet no single drill is dependent upon the others. The machine is constructed with interchangeable parts, and the most serious break-down would not delay the work over ten minutes.'

Tests with the Bennett machine were apparently made in a rock heading near Golden, Colorado, but the outcome of these trials is not known.

A. G. & E. G. Seberg — 1910

The Seberg machine (Figure 172) (U.S. Pat. No. 976,703 dated 22 Nov.), a forerunner of the modern rotary-drum type continuous miner (first produced by the Jeffrey Mnf. Co. in 1965) is similar in concept to the Cooke and Hunter 1866 patent, except that a single cutter head instead of three was used. It was driven by a central shaft through a worm.

183

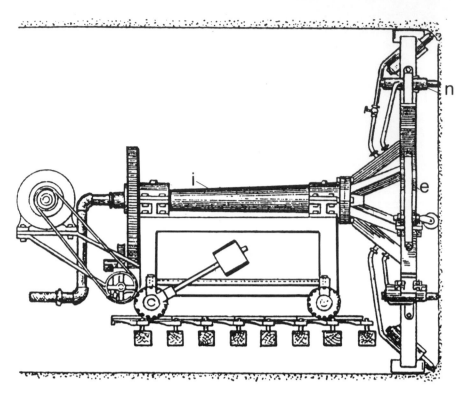

Figure 174. F. M. Iler tunnelling machine. U.S. Pat. No. 986,293, dated 7 Mar. 1911

Louis Franklin Sleade — 1910

The Sleade patent (Figure 173) (U.S. Pat. No. 945,623 dated 4 Jan.) described a machine which was similar in concept and fell into the same class as the Penrice, the Henley, the Karns, and the Knowles and Carr units, in that a single percussive head [e] was used to attack the face. Sleade proposed using an eccentrically shaped head. Cruciform cutter bits were distributed irregularly on the cutter-head plate. The head was mounted on a central shaft [n] which extended rearward, terminating at its rear end in an annular flange [i] held in a bearing cap [o] which was secured to the forward head of an internal combustion engine cylinder. Thus the head was given a forward movement by the internal combustion engine while it was also being rotated by an electric motor through a series of gears, etc. Arrangements were made whereby the electric motor and the internal combustion engine were connected in order to impart a return movement to the cutter head.

Sleade proposed that the material which was cut by the cutters and fell to the bottom of the bore would be scooped up by a series of buckets positioned immediately behind the cutter head. The buckets would lift the material and drop it via a chute on to a belt conveyor which extended from the top forward end of the machine to its rear.

Franklin M. Iler — 1911

The Iler patent (Figures 174 and 175) (U.S. Pat. No. 986,293 dated 7 Mar.) is similar to several already mentioned. The cutter head [e] of the machine described carried a number of pneumatic percussive drills [n] mounted on radially arranged arms [p]. The head was revolved by a central hollow main shaft [i] supported by a carriage on wheels.

Robert Temple — 1911

By means of two compressed-air cylinders [e], percussive power was supplied to the horseshoe-shaped cutter head [n] of Temple's machine (Figure 176) (U.S. Pat. No. 1,001,903 dated 29 Aug.). The head was fitted with a number of fixed cutters or chisels. While the cutter head was being reciprocated it was also

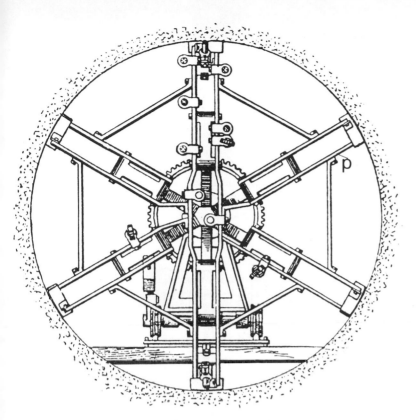

Figure 175. F. M. Iler T.B.M. Front view of cutter head. U.S. Pat. No. 986,293

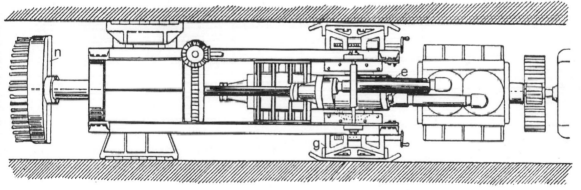

Figure 176. Plan view of R. Temple tunnelling machine. U.S. Pat. No. 1,001,903, dated 29 Aug. 1911

simultaneously being moved transversely in a circular path across the face. Hydraulic grippers [g] which extended sideways against the tunnel walls provided anchorage for the machine.

John Nels Back — 1911

The Back patent (Figure 177) (U.S. Pat. No. 1,011,712 dated 12 Dec.) describes two rectangular frames, one within the other, mounted

on wheels. The inner frame [n] could be slid longitudinally within the outer frame [p] between guides on the outer frame.

The cutter head consisted of a bucket [e] fitted with teeth which moved in a transverse direction along the front of the frame, pulled by a cable. It was held against the face by the pressure of the inner rectangular frame in its forward position. As the bucket was moved across the face the teeth cut the material which fell into the bucket. When the bucket reached

185

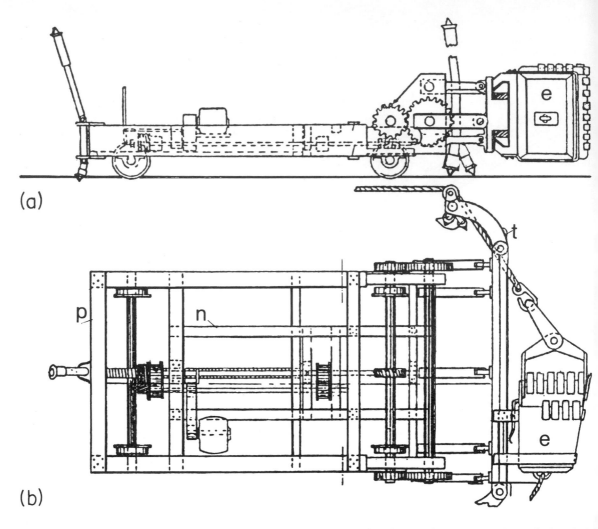

(a)

(b)

Figure 177. J. N. Back tunnelling machine. (a) Side elevation; (b) plan view. U.S. Pat. No. 1,011,712, dated 12 Dec. 1911

the end of its transverse run a trip mechanism [t] discharged the contents on to the ground. The movement of the cable on the winding drum at the rear of the machine was then reversed, and the bucket was returned to the opposite side of the face. Back suggested that the material dropped to the floor by the bucket could then be dragged *by the cable* to the rear where it could be disposed of!

The machine was electrically powered.

Edward O'Toole — 1911

O'Toole's patent (Figure 178) (U.S. Pat. No. 1,011,994 dated 19 Dec.) covered a tunnelling

machine which he envisaged would primarily be used in a coal-mine, but he commented that his invention was not, in its broader aspect, limited to such specific application.

The machine itself was unusual enough to warrant some description.

A cutter head [e] and conveyor pipe [n] were mounted in the frame of a carriage which was provided with traction and supporting wheels [w] beneath and above the carriage. The traction and supporting wheels were each fitted with heavy springs [s] which thrust the wheels firmly against the floor and roof of the tunnel to hold the machine in position during the cutting operation.

186

The revolvable cutter head was funnel-shaped. Around the rim of the funnel (i.e. the periphery of the head) were several cutter teeth [t] which cut into the face as the head rotated. The central opening of the cutter head formed a discharge orifice through which the cut material was carried to the conveyor pipe. The inner face of the funnel-shaped cutter head was provided with spirally arranged conveyor blades [b], so positioned that, as the head rotated, they presented their concave sides. The material was then moved through the conveyor pipe to the rear of the machine by suction.

The cutter head was also pivotally mounted so that it described a reciprocal or windscreen-wiper motion in a horizontal plane. The motion was effected through a series of gears driven by the machine motor and a chain [c] which was secured to opposite ends of the carriage frame. As the head reached the limits of its oscillation

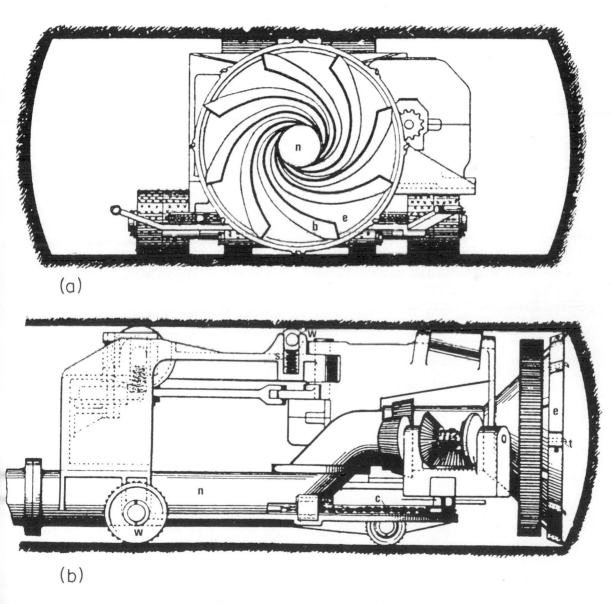

(a)

(b)

Figure 178. E. O'Toole tunnelling machine. U.S. Pat. No. 1,011,994, dated 19 Dec. 1911. (a) Front view of cutter head; (b) side elevation

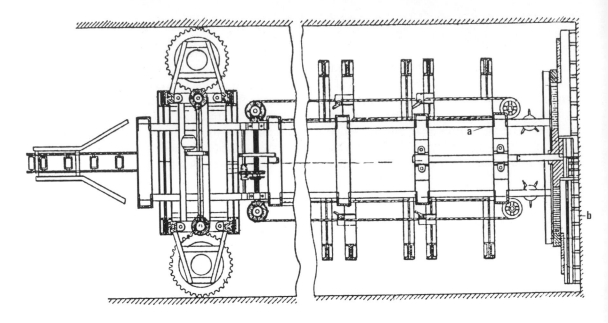

Figure 179. Plan view of longitudinal frame and cutter head of A. F. Walther tunnelling machine. U.S. Pat. No. 1,026,335, dated 14 May 1912 (application filed 22 May 1911.)

a trip mechanism engaged a clutch which automatically changed its direction of travel. A lever mechanism interposed between the oscillating cutter head and the traction wheels of the carriage effected the turning of the traction wheels and a corresponding advance of the entire carriage.

Adolph Fred Walther — 1912

In designing his tunnelling machine (Figure 179) (U.S. Pat. No. 1,026,335 dated 14 May) Walther gave the main emphasis to the tunnel support system during the working and operating cycle of the machine. His patent described longitudinal bars supported along their length by several pairs of octagonally shaped bracing frames. Mounted on the outer faces of each of the pairs of octagonal braces were eight jacks which extended against the floor, walls, and roof of the tunnel. (In all, the patent drawings depicted approximately 56 jacks being used.) The top of the longitudinal frame [a] was provided with grooved rollers mounted in bearings on the octagonal frames. These allowed the longitudinal frame to be advanced or retracted from the face. A rotary cutter head

[b] was mounted on a central shaft [c] at the forward end of the longitudinal frame. It was revolved by the motor through a series of gears, etc. The cutter head carried radial arms on which the cutters were fitted.

Lebbeus H. Rogers — 1912

A unique design for a tunnelling machine (Figure 180) (U.S. Pat. No. 1,039,809 dated 1 Oct.) was submitted by L. H. Rogers of Manhattan. His patent described a machine which had a dome-shaped cutter head [u] carrying a large number of grinding wheels [o]. The head was rotated by a central hollow shaft [d] to the rear end of which was fitted a plunger [q]. The shaft and its plunger moved longitudinally within a cylinder [x]. By introducing air to the chamber behind the plunger, the revolving central shaft, and thus the head, were moved forward towards the face.

As the head carrying the grinding wheels rotated, air was also fed to the motors of the individual grinding wheels causing them to rotate about their axes. During the grinding operation water was sprayed at the face of the

188

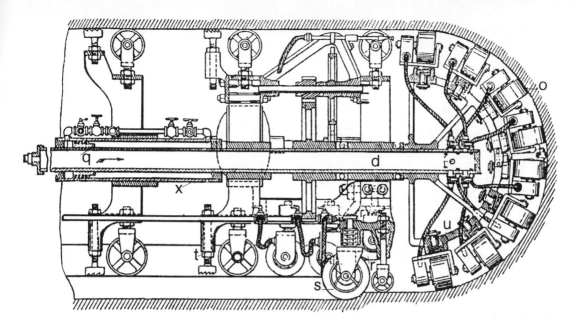

Figure 180. L. H. Rogers tunnelling machine. Partial longitudinal section in elevation. U.S. Pat. No. 1039,809, dated 1 Oct. 1912 (application filed 3 Jan. 1911.)

tunnel, serving to wash the grindings therefrom. In addition the face was swept clean by several brushes fitted to the head. After a cylindrical portion of tunnel had been cut by the grinding wheels on the cutter head, additional grinding wheels [r] positioned immediately behind the head on the main frame of the machine came into action and cut channels to form the lower corners of the tunnel; while a large wheel [s] which was forced downwards by a heavy spring, cut a gutter in the floor of the tunnel. The slurry mixture from the face could then be channelled away in this gutter towards the rear.

While the machine was in operation jacks [t] were extended against the floor and the roof to anchor the unit.

William F. Wittich — 1912

The Wittich patent (Figure 181) (U.S. Pat. No. 1,043,185 dated 5 Nov.) described a tunnelling machine with a doughnut-shaped head [e] which revolved on a central hollow shaft [n]. The cutter head consisted of a number of chisels or drill bits [q] mounted on a plate. Each drill bit was fitted with a spring which served to

retract the bit after it had been flung forward by a cam on a cam wheel [s] which conformed to the shape of the cutter head. The cam wheel was mounted behind the drills and, to some extent, resembled a multiple-blade propellor in that it was provided with a series of cam surfaces. It was mounted concentrically with the head and was fixed to the central shaft. An independent drive was supplied to the head through a ring gear and pinion. This allowed the relative speeds of the head and the cam wheel to be varied. Thus the head could be rotated and advanced slowly while the cam wheel revolved at its normal speed.

Using the cams to provide the forward force and the springs to retract the drills instead of vice versa was an unusual practice.

Two hydraulic cylinders positioned on either side of the main shaft supplied power for the feed apparatus which advanced or retracted the cutter head.

Wittich used a complicated conveyor system to dispose of the cuttings'. It appears that the muck, which was dropped to the bottom of the bore, was lifted by a screw conveyor in a spout positioned immediately behind a bulkhead [u] which separated the head from the remainder

189

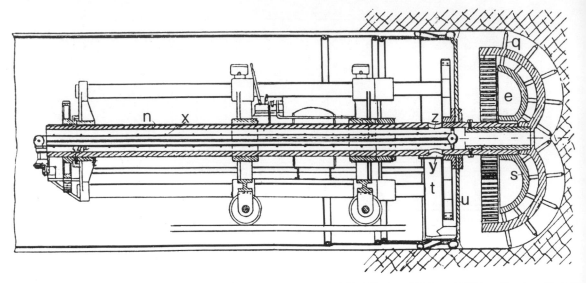

Figure 181. W. F. Wittich tunnelling machine. U.S. Pat. No. 1,043,185, dated 5 Nov. 1912 (application filed 27 Nov. 1908.)

of the machine. The hollow main shaft which rotated the cam wheel passed through this diaphragm. Within the hollow shaft was a conveyor belt [x]. According to the patent, Wittich proposed lifting the muck by the screw conveyor and depositing it in a trough [y]. The trough was designed to direct the material through a pair of openings [z] in the shaft onto the belt conveyor inside, which carried it through the hollow main shaft to the rear of the machine.

Oliver O. App — 1917-18

According to D. W. Brunton and J. A. Davis,[14] O. O. App was associated with the Terry, Tench, and Proctor Tunnelling Machine Company during the time it functioned (see section on Proctor and Terry tunnelling machines).

Apart from three patents covering tunnelling machines, App was also granted a number of patents relating to fluid pressure-operated tools, rock breakers, and power-actuated implements.

The first tunnelling machine patent filed by App on 10 September 1910 was granted seven years later on 20 March 1917 under U.S. Patent No. 1,219,419 (Figure 182). It referred to a machine with two cutting heads [a] which described a planetary motion in that, apart

from their individual rotation about their respective axes, they also revolved about a common axis, namely, the main central shaft [b] of the machine.

The main shaft was supported in a robust octagonal-shaped frame [d]. An outer supporting frame [c] mounted on a pair of trunnions [e] on the inner frame carried four hydraulic jacks [f]. This construction in effect provided a universal joint which allowed the axis of the machine to be swung in any direction with respect to the outer supporting frame. Two sets of jacks were depicted on the drawings which accompanied the patent.

The cutter heads carried 22 tools and App suggested that these could be rigid fixed tools which acted as cutters or, alternatively, they could be made either to rotate or reciprocate, depending upon the type of material being attacked. App also commented that although his patent embodied two cutter heads, it was within the scope of his invention to use either a single head or several heads, as the case may be. App proposed disposing of the muck by means of a bucket or drag conveyor positioned beneath the machine and running from the forward end to the rear. App also suggested using a diaphragm or muck shield behind the cutter head to protect the main structure and works from muck which might be flung rearwards by the cutters.

190

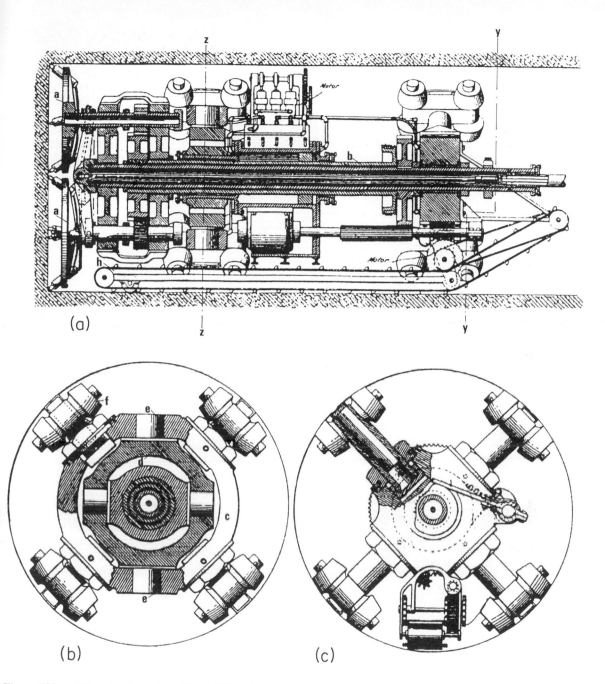

(a)

(b)   (c)

Figure 182. Oliver O. App. Pat. No. 1,219,419 (U.S.), dated 20 March 1917 (application filed 10 Sept. 1910). (a) Vertical axial section of machine; (b) transverse section on line z-z of (a); transverse section on line y-y of (a)

The second patent No. 1,283,618 (Figures 183 and 184) which was filed on 15 March 1917, was granted on 5 November 1918. This machine was basically similar in concept to that patented by Proctor in 1908, in that it described a tunnelling machine with a cutter head [n] mounted on a central shaft. The cutter head carried four radially arranged arms [o] on each of which were mounted four pneumatically powered percussive drills [p]. App suggested that the head could be either rotatable, oscillatory, or fixed to the frame. In each

191

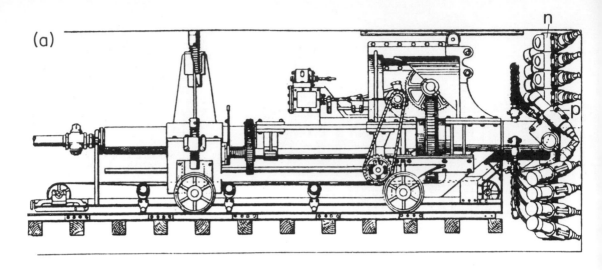

(a)

n

p

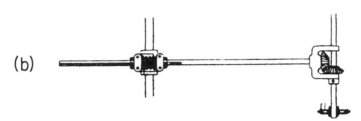

(b)

Figure 183. App's 1918 T.B.M. U.S. Pat. No. 1,283,618, dated 5 Nov. 1918 (application filed 15 March 1917). (a) Side elevation of machine; (b) detail of gearing mechanism for advancing the frame which carries the cutter head

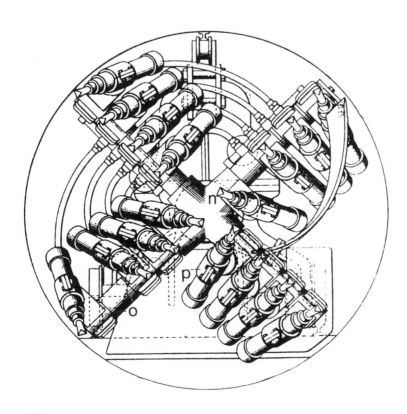

Figure 184. Front view of App's 1918 machine showing cutter head

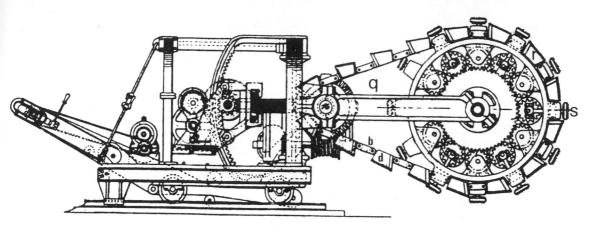

Figure 185. Side elevation of App's 1919 T.B.M. U.S. Pat. No. 1,290,479, dated 7 Jan. (application filed 20 Feb. 1911)

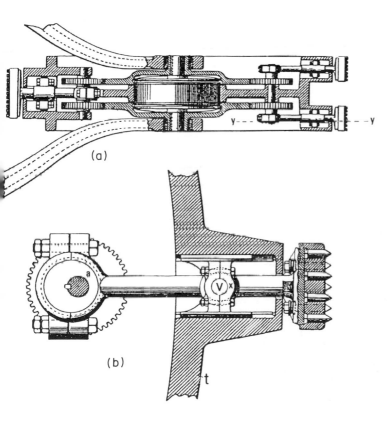

(a)

(b)

Figure 186. (a) Lateral arms of App's 1919 machine; (b) detailed section taken on line y-y of (a)

instance the tools should be mounted at such an angle to the work that when the face was struck by the hammer, the impact of the blow would chip away the material.

Apparently a machine similar to that described in Patent No. 1,283,618 was constructed by App, but the outcome of its trials, if any, are not known.

His third patent, No. 1,290,479 (Figures 185 and 186) which was filed on 20 February 1911 and granted on 7 January 1919 described a machine which was radically different from the previous two. It was, in effect, a trenching or excavating machine with an arm or jib [q] which extended forward beyond the main carriage of the machine.

193

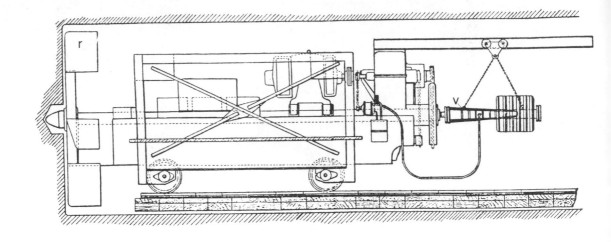

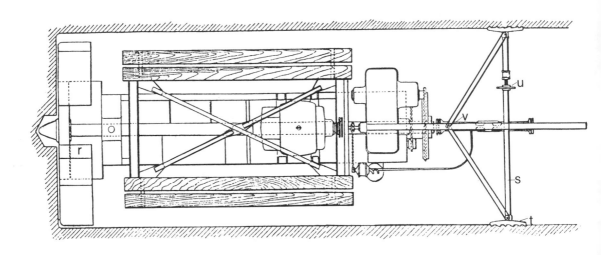

Figure 187. (Top) Side elevation of Douglas Whitaker tunnelling machine. U.K. Pat. No. 159,342, dated 3 March 1921 (application filed 3 Sept. 1920.)

Figure 188. (Middle) Plan view of Whitaker's 1921 machine

Figure 189. (Bottom) Whitaker T.B.M. used on Main Drainage Scheme, Manchester. Photograph supplied by Richard Trigg — former Managing Director of Edmund Nuttall Ltd.)

A large wheel [r] was mounted at the forward end of the jib. This wheel carried a number of cutter tools [s], each with a series of projecting points or teeth on its head. The tools were radially arranged around the wheel, and the rear end of the tool shaft was enlarged to form a crosshead [x] which terminated in a recess formed within a tooth of the sprocket wheel [t]. A pin [v] which was journalled in each crosshead passed through an eye which formed the outer extremity of a connecting rod. Each of these rods had an eccentric [a] at its inner extremity which converted the rotary motion of the shaft on which the eccentric was mounted into a reciprocatory movement. Thus each tool on the wheel described a recriprocatory movement as it was revolved around the wheel. A chain [b] carrying buckets [d] ran around the large front wheel and a smaller wheel at the rear of the jib. The buckets were so arranged that, as they moved on to the large front wheel, they took up alternate positions with the cutter tools.

Douglas Whitaker — 1921

Douglas Whitaker filed his patent (Figures 187 and 188) for a rock-tunnelling machine on 4 December 1919, it was accepted in 1921.

The patent (U.S. Pat. No. 159,342 dated 3 Mar.)[16] which was used by Sir William Arrol and Company, Limited, as the basic design for the units they built, described a machine with the rotary cutter head [r] and mechanism for driving it, supported by a carriage on wheels.

Anchorage and thrust for the machine was provided by an adjustable transversely disposed stretcher bar [s], with shoes or pressure plates [t], which thrust against the side walls of the tunnel. The stretcher bar was actuated by a screw and nut device, operated by a handwheel [u].

The head could be fed forward into the face by either a screw and nut device or a hydraulic ram [v]. (In the machines constructed, hydraulic rams with a 90 cm (3 ft) stroke were used.)

Four radially arranged cutter arms fitted with replaceable cutters positioned at different radii on the arms so that the entire face was covered, revolved about a central shaft. Four muck buckets with renewable cutters were also positioned at different radii on the arms. The material collected by the buckets was carried to the rear of the machine by a spiral conveyor, after which it was delivered on to a belt conveyor which discharged it into dump trucks. The spiral conveyor operated in a steel trough and was fed from a semicircular hopper at the cutter end.

During the early 1920s three Whitaker machines following the above specifications were built.

A 2.1 m (7 ft) diameter version was constructed in 1922 and was used to drive a water tunnel in Folkestone. The same year a 3.6 m (12 ft) diameter machine was built for further Channel tunnel trials at Dover, while in 1926 a 4.5 m (15 ft) diameter unit was used by the Edmund Nuttall Company to tunnel through Red Sandstone for the Manchester Main Drainage Scheme (Figure 189).

The 3.6 m machine after driving some 150 m (480 ft) was left protected at the mouth of the tunnel covered by a wooden shed. Later, during the Second World War, a landslide damaged the shed and partially buried the machine.

In August 1924 the Metropolitan Water, Sewerage, and Drainage Board of Sydney, Australia, purchased two Arrol-Whittaker tunnelling machines (similar to those which had been used at Folkestone and Manchester) for work on the pressure tunnel between Potts Hill and Waterloo. They were delivered in June 1925 and commenced work at Shaft No. 9, Watson Avenue, Ashfield, and Shaft 11, Weston Street, Summer Hill. However, according to the Sydney Water, Sewerage, and Drainage Board the strata in which the machines were to be used, namely, sandstone and shale, were entirely different from those which had been successfully tunnelled by this type of machine in England. One machine cut about 6 m (20 ft) in shale with great difficulty, and the other cut hardly any distance at all in the sandstone. They were equipped with steel cutters which wore away rapidly under the abrasive action of the shale and especially the sandstone.

The board came to the conclusion that the machines were not capable of effectively tunnelling through such strata. The 23 t machines were abandoned and finally disposed of as scrap in 1956.

195

References

1. *Il Traforo del Fréjus* (The Perforation of Fréjus), a cura del Collegio Ingegneri ferroviari italiani — a celebrazione del 1 centenario dell'aviazione della 1 strada-ferrata per l'Europa, Rome, 1971, pp. 47-48.
2. Louis Figuier, *Les Nouvellas conquêtes de la science: grands tunnels et railways métropolitains,* Paris, n.d.
3. Enea Bignami, *Cenisio e Fréjus,* Florence, Italy, 1871.
4. Ferruccio Pisano, Gli studi, le invenzioni ed i lavori per la realizzazione del traforo — Sta. in: *Il traforo del Fréjus,* Rome, 1971, pp. 101-9.
5. Albert Duluc, *Le Mont Cenis — sa route, son tunnel. Contribution a l'histoire des grandes voies de communications,* Hermann, Paris, 1952, pp. 38-44.
6. Henri Maus, *Report,* Turin, 1849.
7. H. S. Drinker, *Tunneling, Explosive Compounds and Rockdrills,* Wiley, New York, 1893.
8. Tunnelling Machines, *The Builder,* 17 June 1871, London.
9. R. L. Herrick, The Karns tunnelling machine, *Mines and Minerals,* October, 1908.
10. W. L. Saunders, Driving headings in rock tunnels, *Trans. American Inst. of Mining Engineers,* Bulletin No. 28, 1909, New York.
11. A machine for boring tunnels, *Engineering News,* **60,** No. 21, 19 Nov., 1908. McGraw-Hill, New York.
12. John Tyssowski, Trial of a tunnel-boring machine, *Engineering and Mining Journal,* 1909, pp. 126-46, McGraw-Hill, New York.
13. Phillip R. Walton, Great augers to bore holes in mountains, *Technical World,* **16,** Chicago, 1912.
14. D. W. Brunton and J. A. Davis, *Modern Tunnelling,* Wiley, New York, 1922.
15. G. J. Bancroft, A history of the tunnel boring machine, *Mining Science,* 13 Aug., 1908, Denver, U.S.A.
16. The Whitaker rotary tunnelling machine, *Engineering,* 26 Jan. 1923, London.
17. Mott, Hay and Anderson, 'Merseyside's First Mole', March 1973.

Bibliography

George J. Bancroft, A History of the tunnel boring machine. *Mining Science,* July 23, pp. 65-8; 6 August, pp. 106-8; 13 August, pp. 125-7; 20 August, pp. 145-6; 27 August, pp. 165-6; 1908, Denver, U.S.A.

Profs. George T. Bator and Niles E. Grosvenor, History of tunnel boring, *The Mines Magazine,* March 1971, Golden, U.S.A.

I. R. Muirhead and L. G. Glossop, Hard rock tunnelling machines, Paper presented at meeting of Inst. of Mining and Metallurgy, 10 Jan. 1968. A.I.M.E., New York.

Correspondence

Sir Harold Harding, U.K.
A. P. Moss, Department of the Environment, U.K.
Metropolitan Water Sewerage and Drainage Board, Sydney, Australia.

Patents

C. Wilson	14,483/1856, 5,012/1847, 17,650/1857, U.S.A.
E. Talbot	9,774/1853, U.S.A.
I. Merrill	14,755/1856, U.S.A.
H. N. Penrice	760/1856, 794/1876, 2,999/1875, U.K.
F. E. B. Beaumont	1,904/1864, 4,166/1875, U.K.
T. Lindsley	55,514/1866, U.S.A.
R. C. M. Lovell	67,323/1867, U.S.A.
E. M. Troth	66,422/1867, U.S.A.
J. D. Brunton	1,424/1876, 1,784/1866, 80,056/1868, 2,544/1876, 1,618/1869, 4,113/1881, U.K.
T. F. Henley	2,349/1870, U.K.
W. F. Cooke and G. Hunter	3,297/1865, 433/1866, 2,580/1867, 2,192/1866, 1,202/1870, U.K.
F. B. Doering	4,160/1881, U.K.
E. A. Cowper	1,612/1871, U.K.
A. W. Von Schmidt	127,125/1872, U.S.A.
T. English	4,347/1880, 5,317/1881, 307,278/1884, 1,482/1881, 4,160/1881, U.K.
H. S. Craven	307, 379/1884, U.S.A.
R. Dalzell	332,592/1885, U.S.A.
F. Dünschede	507,891/1893, U.S.A.
J. L. Mitchell	537,899/1895, U.S.A.
H. Byrne	545,675/1895, U.S.A.
A. Bailey	640,621/1900, U.S.A.
P. Unanue	732,326/1903, U.S.A.
C. T. Drake	747,869/1903, U.S.A.
W. A. Lathrop	816,923/1906, U.S.A.
A.D. Lee and F. J. E. Nelson	874,603/1907, U.S.A.
S. A. Knowles and W. E. Carr	875,082/1907, U.S.A.
W. J. Hammond	885,044/1908, U.S.A.
G. A. Fowler	891,473/1908, 996,842/1911, U.S.A.
E. F. Terry	917,974/1909, U.S.A.
O. S. Proctor	900,950/1908, U.S.A.
J. Retallack	906,741/1908, U.S.A.
C. A. Case	910,500/1909, U.S.A.
W. R. Collins	973,107/1910, U.S.A.
G. R. Bennett	958,952/1910, U.S.A.
L. F. Sleade	945,623/1910, U.S.A.

R. B. Sigafoos	901,392/1908, U.S.A.	F. M. Iler	986,293/1911, U.S.A.
J. P. Karns	744,763/1903, 848,107/1907,	R. Temple	1,001,903/1911, U.S.A.
	867,511/1907, 874,848/1907,	J. N. Back	1,011,712/1911, U.S.A.
	875,664/1907, 879,822/1908,	E. O'Toole	1,011,994/1911, U.S.A.
	888,137/1908, 889,136/1908,	A. F. Walther	1,026,335/1912, U.S.A.
	892,849/1908, 906,496/1908,	L. H. Rogers	1,039,809/1912, U.S.A.
	957,687/1910, 977,955/1910,	W. F. Wittich	1,043,185/1912, U.S.A.
	1,030,663/1912, 1,023,654/	C. O. App	1,219,419/1917, 1,283,618/
	1912, 1,156,147/1915,		1918, 1,290,479/1919, U.S.A.
	1,271,321/1918, U.S.A.	D. Whitaker	159,342/1919, U.K.

FOUR

Channel and Mersey Railway Tunnel Machines

Beaumont, Brunton, and English and their Combined Contribution Towards the Development of the First Channel Tunnel Machines

F. E. B. Beaumont

Military Career

Lieutenant-Colonel (Hon. Colonel) Frederick Edward Blackett Beaumont, M.P. R.E. was born on 22 October 1833 and, contrary to usual Beaumont family tradition which favoured Eton College, was educated at Harrow.

Beaumont entered the Royal Military Academy, Woolwich, as a gentleman cadet on 29 May 1848 and, at the age of 18 years, received his commission as a second lieutenant. In 1854, 1859, 1866, 1872 and 1877, he was promoted to the ranks of lieutenant, second captain, captain, major, and lieutenant-colonel respectively. He retired from the Royal Engineers on 27 October 1877 and was granted the honorary rank of colonel.

After being commissioned into the Royal Engineers he followed the usual programme for all young officers entering that corps by attending a 15 months' course (July 1852-September 1853) in field engineering and forti-fication at the Royal Engineers Establishment, Chatham (now known as the Royal School of Military Engineering, situated in Brompton Barracks, Chatham, Kent).

On completion of his initial training as a young officer, he was given normal military duties with a company of the Royal Sappers and Miners at Woolwich. (The Royal Sappers and Miners were incorporated into the Corps of Royal Engineers in 1856. Prior to that date the Royal Engineers had been solely an officer corps, but had provided officers for duty with the Royal Sappers and Miners.) It was probably during this period that Beaumont's thoughts first turned towards ways and means of easing the miner's lot by mechanizing the operations.

During 1854 and 1855 he was detached for duty at the Laboratory, Woolwich (which later became known as the Royal Arsenal, Woolwich) and, although there is no record of the nature of his duties there, it has been suggested by Lieutenant-Colonel J. E. South, librarian of the Institute of Royal Engineers, that these probably included the testing of artillery projectiles on defensive constructions for the information of the Inspector-General of Fortifications at the War Office.

From June 1855 to August 1856 he was one of a number of R.E. officers attached to the Turkish forces in the Crimea — known as the Turkish contingent of Royal Engineers. During this phase of his army career Beaumont was awarded the Turkish medal.

Later, after his return from a tour of duty in India, Beaumont served with the 7th Company, Royal Engineers, in Canada, from about 1859 to 1862, and during this period he managed to attach himself to the Federal Army balloonists in the American Civil War. On his return from Canada in 1862 to Chatham and during the period 1863 to 1864 Beaumont was employed with his company on the Dover defences, where his interest in mining the Channel chalk no doubt developed. In March 1867 he was posted for special duty in connection with the organization of the Paris Exhibition of 1867. Here he was involved with the installation of prime movers, boilers, and engines, specially adapted to the requirements of that Exhibition.

From 1872 until he retired from military service in October 1877, Beaumont was given long periods of leave each year which usually extended from February to September or October. His military employment during the remainder of that time was as follows:

October 1872-February 1873 — Portsmouth District
October 1873-March 1874   — Woolwich District
October 1874-February 1875 — Home District
(London)

Lieutenant-Colonel South suggests that these unusual extended periods of leave were granted to Beaumont to enable him to fulfil his obligations as a Member of Parliament as well as his duties as a serving officer.

The Beaumont/Adams Revolver

As may be seen from the list of Beaumont's patents at the end of this section, he began designing and devising various things as early as 1853. It appears from the pattern of his life that many of his inventions were strongly motivated by his activities and environment at the time. His first patent of significance covered improvements to fire-arms — in particular the lock mechanism of the revolver. (This Patent was filed shortly before he left for Turkey for war service there.) The story behind this particular invention begins with Robert Adams, a British designer and manufacturer of fire-arms who, for several years, had been working assiduously at improving the design of his revolver so that it would become acceptable to the Army. At that time Samuel Colt of America had set up a London factory and was busily supplying the Army and Navy with weapons. These consisted of a great quantity of Colt Navy revolvers and pistols (the total order amounting to over 23,000 arms) which were being used for Crimean service.

It was at this stage that Adams and Lieutenant Beaumont became acquainted. By inserting a 'short sear' Beaumont managed to adapt the long sear on Adam's original gun so that the hammer and trigger mechanism were combined. This adaptation permitted the operator to 'thumb cock' the weapon. Both the

original Adams revolver and the Colt were, of course, self-cocking weapons, but neither could be 'thumb-cocked'.[1]

The Army promptly ordered 2000 of these improved guns. Under a royalty contract which ensured him a small percentage of the value of each gun sold, Beaumont assigned his patent rights to Adams. Adams then broke away from his existing partnership with John Deane and John Dean, Jn., and founded the London Armoury Company (Ltd.), which company thereafter sold the Beaumont-Adams revolvers. As a result of the success attained by Adams, Colt left England and returned to America where there was a very large demand for Colt revolvers.

Beaumont's 1864 Percussive T.B.M.

In 1864 Beaumont filed an application for his first tunnelling machine patent (Figures 190 and 191). Prior to the filing of this patent a machine was actually built and tested for a short time in 1863 in the driftway of the Vartry Waterworks (City of Dublin) in the Wicklow hills. (This was also where George Low's drill was tested — see section on horizontal drills.)

Beaumont's 1864 patent (No. 1,904) dated 30 July described a percussion machine in which a series of chisels or jumpers, evenly fitted around the periphery of a strong disc or wheel, were made to strike the face rapidly and simultaneously a successive number of blows. The disc or cutter wheel was mounted on a strong shaft which was supported in bearings on a base plate or carriage.

The shaft was capable of sliding longitudinally and it could also rotate. The longitudinal motion was provided by a cylinder and piston powered by compressed air or hydraulics. Thus by these means a percussive and rotary action was imparted to the disc which then cut an annular groove in the face. The core was to be broken by blasting. (Beaumont suggested that the rotary motion could be given to the disc 'in any convenient manner, it may be by having a groove on the axis into which there enter a stud capable of being slowly transversed around the centre of motion').

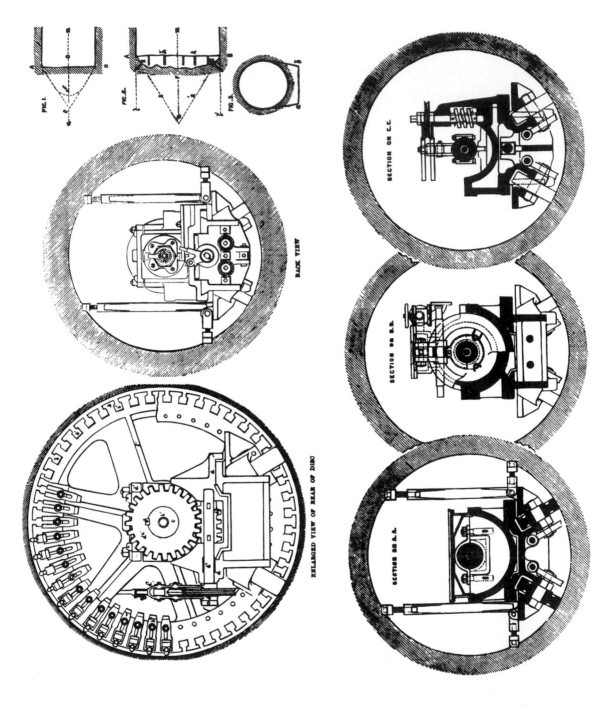

Figure 191. Captain Beaumont's tunnelling machine. Patented on 30 July 1864 under No. 1904. (Courtesy *The Engineer*, 26 April 1867.)

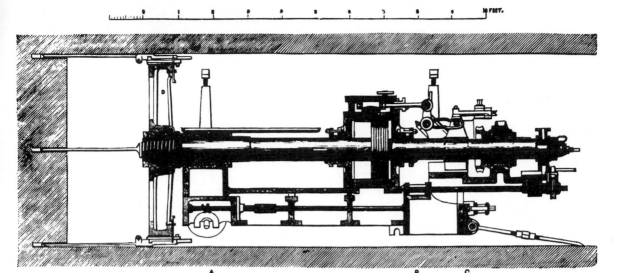

Figure 190. Captain Beaumont's tunnelling machine — 1864 — percussion: Exhibited at Paris Exhibition 1867 (Courtesy *The Engineer,* 26 April 1867.)

A model of this machine with modifications was exhibited at the Paris Exhibition which Beaumont was so conveniently attending at that time and a report of it appeared in *The Engineer* dated 26 April 1867.[2] This report is worth quoting almost in its entirety. Not only because it criticizes Beaumont's invention but because it is indicative of the views of the engineering fraternity of those times.

As originally patented the machine was constructed somewhat differently in detail from that now exhibited. The most important difference consists in the removal of the worm wheel and endless screw, which produce the slow motion of rotation of the cutter block or large disc carrying the ring of jumpers from the front of the actuating cylinder for compressed air to the rear of the same. By this a good deal of simplification has been effected, and some of the more vital parts of the machine got further to the rear — a not unimportant matter when the whole is drawn back a yard or two upon occasion of firing a blast. As now produced there is, in fact, almost nothing in advance of the air cylinder, except the central stem and cutter block, and the forward portions of the upper and under frames which carry the whole. The machine exhibited has been constructed by Messrs. Bryan Donkin and Sons, of London, who are also the constructors of the horizontal engine and air pump for the compression of the air which transfer the motive

power to the machine. The inventor proposes to work at an air pressure of only about 30 lb. per square inch, upon the ground that the higher the amount of condensation the more of the original power of the steam or fuel will be uselessly employed, or wasted in the development of heat in the process. The air pump produced is a double-acting horizontal one, the piston rod being coupled on direct to that of the steam cylinder. The valves are clasp or hinged valves, rectangular in form, and faced with india-rubber; and, as these prevent the piston from coming home against the cylinder end at either half-stroke, there must be a rather considerable loss of useful effect due to this, which seems to us an imperfect construction, even at so low a tension as 15 lb to 30 lb per square inch. We do not know whether this form of air pump is attributable to Captain Beaumont; but, in any case, should it be possible to make any exact experiments as to what the exhibited machine may be capable of performing, it would be but fair to make due allowance for any faults in the condensation apparatus.

We fear, however, that it will be impracticable to work the machine upon the slender foundations upon which it has alone been feasible to place it against any block of sufficient magnitude and hardness to prove what its real powers may be as regards useful effect. It is indeed one of the many machines whose functions are percussive, the movements of which may be exhibited, but whose *work* can only be shown in and upon its proper element and place. It will be in the recollection of some of our readers that this

201

machine in its first or original form was tried for a very short time in the drift way or tunnel of the Vartry Waterworks, of the city of Dublin, in the Wicklow hills, and that the results were not favourable.

It is, however, but fair to Captain Beaumont to state that the machine there employed was almost a first essay, and had not his improvements in arrangement here visible, independently of the unfortunate conditions that must have existed for a *full* trial of a novel machine, in the case of a tunnel which was under contract with limitations as to time, and so precluding that indispensable sort of patient meeting of small difficulties and remedying them as they arise. We are ourselves in a condition to state, from personal knowledge, that the rock perforated in the Vartry tunnel is of a quality that we believe no rock-cutting machine whatever could be expected to answer well in, if at all. The formation is a dense hard quartzose Silurian rock in laminae, and with joints bent and contorted in every conceivable way, passing from hard crystalline quartz in seams and irregular nodules and patches, into rock so Magnesian as to be almost a serpentine, and this often within a few feet, the very type in fact of heterogeneity in rock. Hardness, however great, with comparative uniformity, may be dealt with; but we entertain great doubt whether any rock-*cutting* machine, which in the strict sense, this machine of Captain Beaumont's is, can be made to work in rock whose characteristic is the absence of uniformity for even a yard together.

In the general design of this machine we deem the inventor clearly right in two different respects.

First, it proposes to act by *percussion*. No machine that shall be designed to cut out a cylindrical plug of rock, for that is really the function aimed at, by *grooving* or *planing* out an annular channel by means of edged tools acting like those of a slotting or boring machine on metals, can ever answer at all in even moderately hard, and especially siliceous rock, nor for long in a satisfactory way even in stone as soft and uniform as Tertiary Limestone. Even coal cutting upon the principle of the planing machine will in the end prove a failure as compared with striking tools.

In this respect, then, Capt. Beaumont's machine has escaped the fate that overtook some earlier ones, bearing a general resemblance to it as to functions, but proposed to work by grooving or scraping.

It is scarcely worth while to go into the physics of the relative actions of percussion and of grooving or planing, i.e. cutting tools in the strict sense, but it is easily shown that in anything like hard or gulley rock the hardest and best steel tools that can be formed suffer much more from being *ground* away by the rock than does the rock from being eaten into by the tool — eaten into mainly as to the barb — for an edge is out of the question here.

In the percussive tool, the jumper, the rock is *fractured* so is the edge of the steel jumper, but the tougher body suffers less just as the velocity of the blow is greater, and the stroke therefore more effective; whereas in the cutting tool and small rate of movement the relative suffering is almost as the relative hardness, modified by molecular constitutions in rock and tool;...

...Having thus pointed out that, in principle, Captain Beaumont is right, both as to the way in which he proposes to act upon the rock percussively in forming his cylindrical groove, and as to the value of the groove when got, it remains for us to say a few words as to the practical aspects of the machine itself as it presents itself to us.

The mass of the outer block, as we have called it, or circular disc that holds the jumpers with its large central wrought iron shaft, is necessarily very great. The rate at which it is intended to make it beat 150 or more strokes per minute with the short strokes of only about two or three inches, though at first may not prove full of difficulty, must we fear, prove so ere long. The reaction upon the upper and lower frames of the machine must bear a very sensible relation to the action delivered into and by the cutter block and jumpers, for the weight of the latter is a large fraction of the total weight of the machine. The severe repercussions passed through this ponderous mass of central shaft, disc, sockets, wedges, and jumpers must transmit their reactions primarily to the piston, cylinder, and valve gear, and as strokes made thus short and sharp are in fact little different from vibration, and are at once resolved into vibrating jars, we apprehend rapid deterioration of the fit and adjustments of both the piston and cylinder, and of the air valve faces. Even assuming the rock absolutely homogeneous, and also the steel of all the jumpers, and that those all wear alike (a thing not to be anticipated), still the laws of elastic compression and extension in solid column enable us to see that very severe work will be exercised upon the wrought iron neck of the central shaft close to its shoulder and junction with the circular disc.

But should the rock be uneven and possess hard places, or the jumpers cut faster at one side of a diameter than at the other, then also very severe transversal strains will be visited upon the wrought iron at this same region.

Beaumont/Appleby Patents on the 'Application of the Diamond Drill to Tunnelling'

Another significant contribution was made in 1868 (Patent No. 1,682) and in 1872 (Patent No. 392) (Figures 192 and 193) by Beaumont and C. J. Appleby towards tunnelling by the adaptation of the diamond drill.

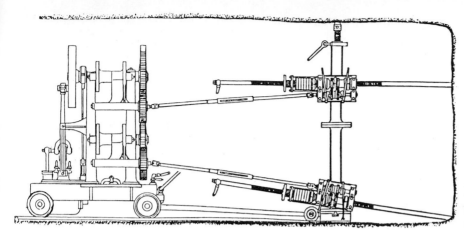

Figure 192. Beaumont and Appleby's adaptation of the diamond drill to tunnelling (side elevation)

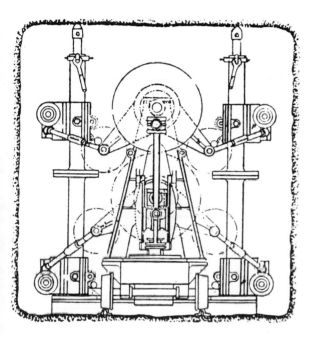

Figure 193. Beaumont and Appleby's adaptation of the diamond drill to tunnelling (end elevation). British Pat. No. 1,682/1868

Basically the Beaumont/Appleby diamond drill consisted of a 'couple of standards carrying drills... which are all set in motion by one engine driven by compressed air'. These drills and their application to prospecting, tunnel driving, shaft-sinking and subaqueous operations were described by Major Beaumont, R.E., M.P. of London, in his paper[3] which was presented at the Institution of Mechanical Engineers in 1875. Of interest in the light of the critical report of Beaumont's first T.B.M.

made by *The Engineer* in their issue dated 26 April 1867, are Beaumont's comments regarding tunnelling machines in general which were made in this paper:

In boring into rocks, the hardest known steel is of no use except percussion is used, because, if it is attempted to give a cutting or rubbing action to a steel tool, it is impossible to construct it so that it can work for more than a comparatively short time. Mechanics having only steel at their disposal have been driven therefore to employ percussion in designing machinery for making holes in rocks. It is well known how comparatively difficult it is to give a definite percussive motion and to control it; and how, unless an altogether disproportionate amount of strength be given to the parts, it is impossible to construct a machine to do a great amount of work in striking blows, and yet to be durable. Moreover the work done by the hammer, and its source of power, ought not to be far apart; or in other words, a series of blows represents work done in a shape in which it is very difficult to transmit it to a distance.

The use of the diamond in this case breaks, as it were, the bottom of Columbus' egg, and many of the difficulties attending machinery for dealing with rocks vanish at once. As an illustration, what a wonderful improvement would be the use of an auger in boring holes in wood, if the only means of drilling that material were hammering a nail in and pulling it out again, or cutting a hole with a gouge and mallet.[*3]

From 1869 to February 1872 Beaumont was employed on engineer duties in the

---

*Some of the Beaumont/Appleby drills were, the author understands, used in the Mont Cenis and the St. Gotthard tunnels.

Sheerness district which probably involved him in the maintenance of buildings and defences of the garrison and naval base. During this period Beaumont filed a number of patents for various items amongst which were three covering improvements to drills (i.e. 1,919/1873, 1,149/1874 and 829/1875).

Beaumont's 1875 Rotary T.B.M.

On 2 December 1875 under No. 4,166 (Figure 194) Beaumont filed his second patent for 'Improvements in tunnelling machinery'. This time the patent described a 'rotary' tunnelling machine which Beaumont envisaged would be used for driving tunnels or galleries in soft rock, shales, or strata such as chalk.

A crown or cutter head consisting of two or more radial arms was mounted on the end of a horizontal revolving shaft. The head was conically shaped and sockets in the form of steps were provided on the arms for the reception of the steel cutters.

The spearhead of the crown consisted of a conventional (1875)-type rock drill which cut a central cylindrical recess. On either side of the head, sockets and their respective cutters were so arranged that when the head was advanced against the face and rotated, each cutter scraped away the material in the form of a circular ledge or step slightly in advance of the following outside cutter.

By this arrangement Beaumont stated 'all the debris produced by the cutters will readily fall to the bottom of the tunnel or gallery whence they can be removed by means of an endless band, or a screw or creeper, such as is used in flour mills, delivering the debris into trucks behind the carriage, or by workmen who can shovel it away when it has fallen clear of the cutters'.

It was proposed by Beaumont that the shaft which carried the boring head be mounted on a carriage running on rails on the bottom of the gallery. Also mounted on the carriage would be the motor engine or set of engines which would impart a slow rotary motion to the boring shaft by suitable speed reduction gears.

Beaumont's Patents

The patents granted to Beaumont are given in Table 4.

Brunton[4, 5]

Little is known of the private life of the civil engineer from Middlesex, John Dickinson

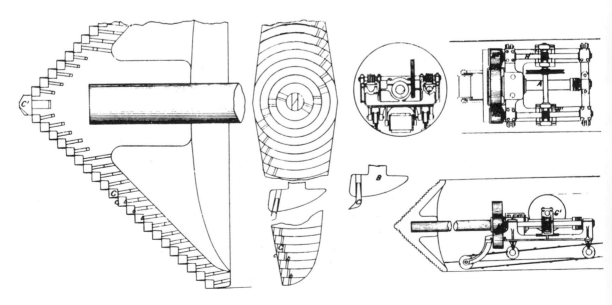

Figure 194. Frederick Edward Blackett Beaumont tunnelling machine. Patented on 2 Dec. 1875 under No. 4,166 (U.K.). (Courtesy *The Engineer*, 8 Dec. 1916.)

Table 4

No.	Date	Description
1,134	1853	Constructing dwelling houses or other buildings, peculiar shaped bricks and tiles to be used for that purpose.
1,378	1853	Bricks or tiles.
374	1855	Improvements to revolver.
867	1857	Lamps and apparatus used in coal-mines
1,904	1864	Tunnelling machine
1,682	1868	With C. J. Appleby. Application of diamond drill to tunnelling
392	1872	With C. J. Appleby. Rock drilling, tunnelling etc.
1,919	1873	Rock drilling
392	1874	Steering vessels
4,402	1874	Small trunk engine for valve and rotation
2,424	1874	Steering apparatus
1,149	1874	Improvement to drills
3,079	1875	Indicating, etc. speed of carriages
829	1875	Improvement to drills
4,196	1875	With W. Pilkington. Roller skates
4,166	1875	T.B.M.
7	1876	Elastic fluid motor engines
2,010	1876	Steam-steering apparatus
1,061	1876	With W. Pilkington. Roller skates
1,390	1876	With Jos. Foster. Percussive rock drills
3,277	1876	With W. H. Ashwell and F. J. Bolton. Making and utilizing slag castings
1,833	1877	With T. Barningham and C. Thompson. Tramways and rolling stock
3,500	1878	Stampers worked by fluid pressure
5,824	1882	With E. C. Wickers. Tramway cars
20,455	1890	Pneumatic guns
22,931	1896	Velocipede, etc. saddles
22,932	1896	Velocipede, etc. brakes
4,030	1897	Velocipedes
12,280	1897	Velocipedes and brakes

Note: During the period 1897-99 Beaumont filed 11 additional patents, all of these relating to bicycles

Brunton, except that he was acquainted with Brunel and in his day was apparently well respected as a man of ability in the field of engineering. He was (in 1844) the first to propose the use of compressed air to power drill hammers (the air also serving to ventilate the face after use). His inventions, many of which relate to aspects of tunnelling in one form or another, are numerous and date from about 1866 to 1905.

Brunton filed seven patents in England relating directly to designs for or improvements to machines which he suggested could be used for 'sinking shafts or pits or driving tunnels or galleries'. The first of these was dated 5 July 1866. This was Patent No. 1,784 (Figure 195) and it described a tunnelling (or shaft-sinking) machine which consisted of a giant central screw mounted on a carriage on wheels.

The cutter head had three or more radially arranged arms attached to the forward end of the machine. Each arm carried a circular disc of steel. The periphery of each disc was formed into a cutting edge which Brunton proposed keeping sharp by allowing it to come into occasional or continuous contact with a

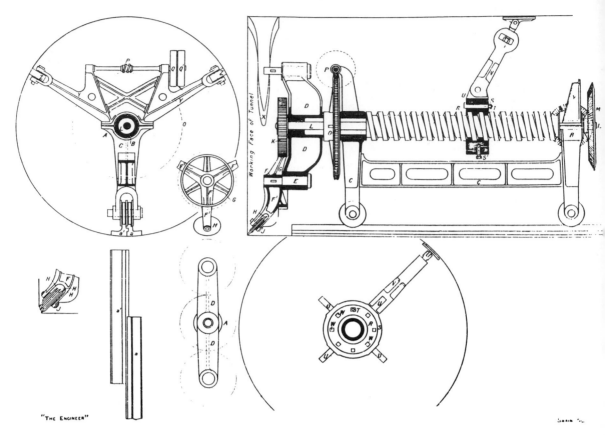

Figure 195. The Brunton tunnelling machine — 1866. Pat. No. 1784 dated 5 July. (Courtesy *The Engineer,* 8 Dec. 1916.)

whetstone or grindstone fixed on the machine. The discs were set at right angles to their planes on pivots or journals on which they rotated freely. Brunton proposed that the inclination angle of each disc when in its place on the arm, should be set according to the requirements of the material being penetrated. In addition to the orbital motion of the cutters around the central axis, the cutters were also given a spiral motion so that the material was pared away from the face in a spiral pattern. Of interest was the spiral scraper blade attached to one of the cutter arms. This was similar in design to those used by Reginald Stanley on his machines some 19 years later.

On 21 July 1868 under Patent No. 80,056 which was filed in America, Brunton suggested various improvements to his earlier U.K. patent. These mainly concerned the cutter head.

In place of the three radially arranged arms which each carried a disc cutter, Brunton suggested the use of two 'chucks' or 'face-plates' each carrying six disc cutters. These face-plates were to be mounted on arbors which were carried at opposite ends of a crosshead which, in turn, revolved about the central shaft. The arbors about which the face-plate and discs rotated were eccentrically positioned. By means of a worm wheel and gear and the eccentrically positioned arbors it was possible to move the chuck a small distance outward or inward. This was designed to compensate for any wear of the cutters.

A model of a machine similar to the one just described was exhibited at the Conversazione of the Institution of Civil Engineers in Paris in May 1868 (Figure 197).

In subsequent patents Brunton proposed various improvements to his machines. These

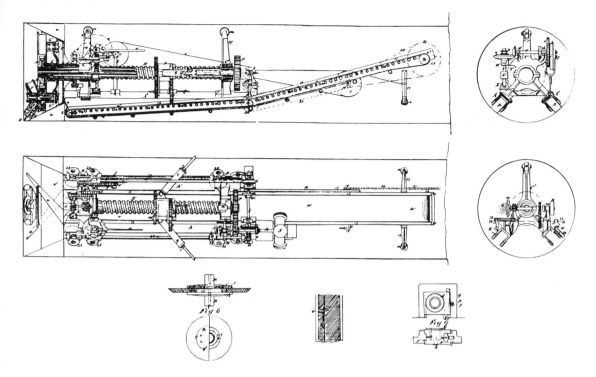

Figure 196. Brunton's tunnelling machine. Patented on 20 June 1876 under No. 2,544 (U.K.)

were mainly concerned with the cutters and the muck removal arrangements. The main structure, that is the central screw-threaded shaft mounted on a carriage, remained unchanged. For example Patent No. 2,544 (1876) (Figure 196) refers to 'the construction of the working parts, ...the mode of imparting motion to the travelling wheels, ...the cutters, the mode of collecting and removing the debris produced, and a means for excluding dust from the working parts'.

Brunton suggested that the debris be

> removed from the front of the machine when at work by a rotating drum having attached thereto a number of scoops so arranged that as the drum rotates the scoops raise the debris from the bottom of the tunnel and discharge it over a plate into a shute with which it is provided for the purpose of conducting it on to an endless travelling apron for conveying it to the rear of the machine. The endless apron is supported upon rollers carried in a suitable frame, and is actuated by means of chains and chain wheels, belts, or gearing, which impart motion to carrying rollers provided with teeth or

projections engaging with bars of wood fixed to the back side of the endless apron.

He suggested that any debris which fell inside the 'endless apron' could be removed by means of an Archimedean screw which operated within a trough placed there to collect such debris.

In his letter dated 23 November 1907 addressed to Forrest Brunton, Brunton* points out that his machine was 'regarded as *the* machine for the Channel tunnel'.

Before the advent of the Beaumont/English machine, which will be described later, current thinking amongst the engineering fraternity of that time appeared definitely to favour Brunton's machine as being the one which

---

*There appear to be two errors in this letter (see illustration) (Figure 203) firstly the initial 'T. D.' should be J. D. — (which may be a typing error as the letter depicted is a copy of a copy). Secondly, Brunton says the scheme for the Channel tunnel was vetoed in '1871'. In fact the boring on the English side was stopped by injunction at the instance of the Board of Trade on 23 July 1882.

Table 5

No.	Date	Description
1,784	1866	Machinery or apparatus for sinking shafts or pits and drilling tunnels and galleries
302	1868	Machinery or apparatus for cutting, dressing, planing, turning and shaping stone
2,198	1868	Tools and machinery or apparatus for cutting slate and other rock
1,618	1869	Machinery for tunnelling, shaft sinking, and stone dressing
1,673	1870	Ventilation of tunnels
702	1875	Dressing, etc. of stone
2,544	1876	Excavating tunnels and sinking shafts
1,424	1876	Tunnelling machinery
1,254	1877	Cutting, dressing, planing, and shaping stone
4,591	1877	Cutting stone
548	1878	Cutting rock; dressing, etc. of stone
3,045	1879	Boring, cutting, dressing, and shaping rock and stone
4,754	1879	Safety-lamps; pyrometers
1,651	1881	Excavating tunnels, levels, or galleries
4,113	1881	Tunnelling, shaft-sinking, and excavating
505	1882	Dressing, turning, and moulding stone, etc.
2,803	1888	Artificial stone
10,824	1889	Mixing materials for concrete etc.
13,762	1892	Drying peat, etc.
6,770	1893	Drying peat, etc.
1,276	1896	Organs etc.
545	1896	Improvements in the manufacture of peat fuel and in the machinery employed therein
13,522	1905	Wire ropes etc.

would ultimately be used for the proposed Channel tunnel. On this Drinker[6] remarks:

> Brunton's tunnelling-machine for boring through chalk, proposed for use in the Channel Tunnel, is the latest production.... The Brunton machine, which appears to give good promise, will, if it succeeds, be the first machine of the circular-cutting type that has stood a practical test, and even it is not intended for use in hard rock with alternate blasting, but simply in soft chalk where it is expected to bore its way like an auger. It has not as yet had any extensive practical test in a large tunnel though many successful trials with it have been reported. It has been hoped that it might be applied in the construction of the long-proposed Channel Tunnel, should the latter ever be constructed.

British Patents Granted to Brunton

The patents granted to Brunton are given in Table 5.

English[7, 8]

Lieutenant-Colonel Thomas English was born in 1843. He married Clara Jane (daughter of Lieutenant-General H. J. Savage, R.E.) in 1867. He was educated at Kensington Grammar School and later enrolled at the Royal Military Academy. He obtained his commission in the Corps of Royal Engineers in 1862.

In the Corps, English's duties at the War Office during the period 1867-84 were mainly concerned with the design and construction of armour-plated coast defences. He designed a method whereby large armour plates could be cross-rolled and this invention was adopted by the government. In addition he also designed a special spherical nut for armour bolts.

From 1878 to 1884 he was engaged in a number of pursuits which included an invention for soft metal caps for shells (1878). However, as often occurs, the importance of this

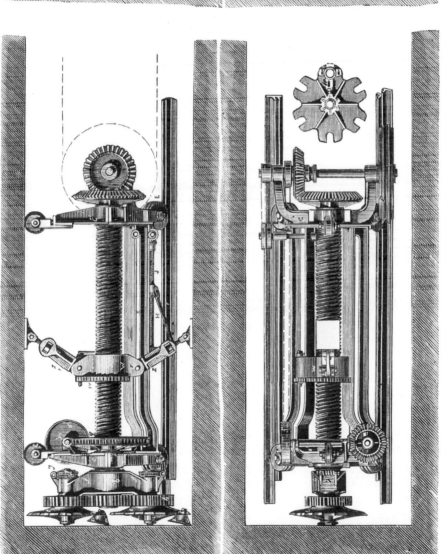

Figure 197. J. D. Brunton rock tunnelling machine exhibited at the Conversazione of the Institute of Civil Engineers (Paris). (Courtesy *Engineering* 28 May 1869.)

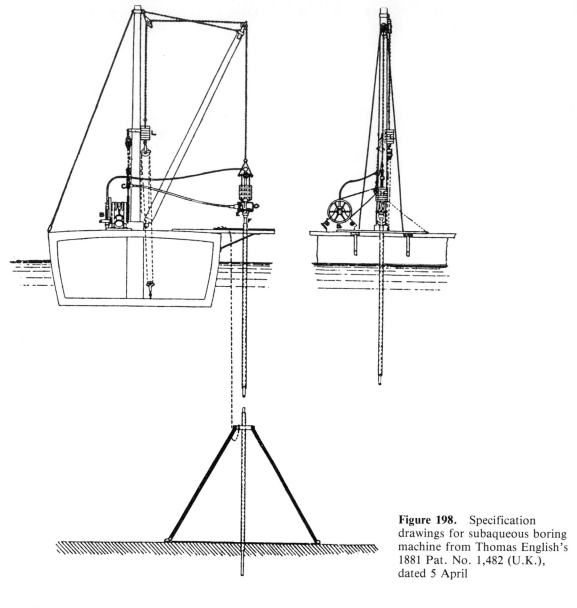

Figure 198. Specification drawings for subaqueous boring machine from Thomas English's 1881 Pat. No. 1,482 (U.K.), dated 5 April

particular invention was not recognized by the British government until it had been developed and used effectively by foreign countries.

On 25 October 1880 under Patent No. 4,347, English filed his application for a patent covering improvements to Beaumont's patent dated 2 December 1875 (No. 4,166). The following year (1881) English filed Patent No. 1,482 (Figure 198) dated 5 April covering 'Apparatus for subaqueous boring' and later that year (i.e. 5 December) he filed Patent No. 5,317. This latter patent is described more fully in the section entitled 'reaming machines'.

In 1882 English was seconded to the Dean and Chapter of St. Paul's Cathedral so that he could direct the work of raising the 'great Paul' bell into its correct position.

The year 1884 saw him involved with the design and construction of rolling-stock for use on the Suakin-Berber railway. After his retirement from the Army and from his post as Superintendent of the Royal Carriage Department, Woolwich Arsenal, which position English held from 1887 to 1889, he entered civilian life by accepting the position as General Manager of Palmer's Ordinance Works at

Jarrow. His inventive talents were soon recognized and he was made a director of the company's shipyard. In this capacity he designed and superintended the construction of some 21 ships. Included in the group were the first of the British destroyers to be launched, namely the *Janus,* the *Lightning,* and the *Porcupine.*

He left this company in 1896 and joined the European Petroleum Company. This led him to carry out investigations of oil properties in Turkey, Romania, and Galicia.

During the First World War English re-enlisted as Assistant Inspector of Steel for the Navy. At that time (1915-17) his son, Captain Douglas English was serving with the 21st London Regiment, while his grandson, D. R. B. English, was with a naval division, thus three generations of the same family were simultaneously on active service.

After the war and until his death at Whitby on the 20 June 1935, at the age of 92, English concerned himself with mathematical research work.

The Subaqueous Drilling Machine

English's patent covering 'Apparatus for subaqueous boring' which was filed under No. 1,482 on 5 April 1881, is described below in English's own words:

My invention relates to apparatus for boring under water, arranged in such a manner as to permit boring operations to be continued notwithstanding tidal or other currents or changes of the water level, or movements of the barge or other vessel from which the apparatus is worked. For this purpose I employ a rotating boring tube, having at its end suitable cutters, which when hard rock has to be bored may be arranged as diamond rock drills; and I steady this tube in a vertical or a more or less inclined attitude by a frame resting on the bottom, the tube itself extending some distance above the surface of the water. At the upper end of the tube I provide a bearing for it in a framing, in which there is also a bearing for a short shaft at or about right angles to the tube carrying a bevel wheel, which gears with a bevel wheel on the tube. The framing also carries a weight, the effect of which may be increased or diminished as required for the nature of the work by means of a counterweight connected to the

Table 6

No.	Date	Description
96	1864	Construction of motive-power engines
3,844	1868	Construction of tubes or cylinders capable of resisting great internal pressure
1,247	1869	Screw bolts, bolt-holes, nuts and washers for armour-plated structures
3,211	1871	Screw-bolts and nuts
1,465	1871	With G. Wilson. Bending armour plates
4,347	1880	Improvements to Beaumont's T.B.M.
1,482	1881	Subaqueous drilling machine
2,557	1881	With D. Grieg. Variable expansion gear for link-motion reversing engines
5,317	1881	Reaming machine
1,427	1882	Expansion gear for link-motion engines
556	1883	Variable expansion gear
3,869	1883	Bogie trucks for locomotives, etc.
8,817	1884	Armour plates
2,209	1886	Bogie trucks for locomotives
14,621	1886	Maintaining heat and preventing condensation in cylinders of steam engines
10,738	1889	Making fish-hooks
	1890	15 patents concerning heavy artillery
	1892	1 patent concerning heavy artillery
3,186	1897	Nautical course and position finder

frame of the drill tube by a rope or chain passing over pullies on a jib or derrick mounted on a barge or other suitable floating vessel moored in the required position. On board this vessel I provide an engine or other suitable motor, and I connect a revolving shaft of this motor to that of the bevel wheel, which works the drill by a flexible twisted wire shaft, such as is frequently employed for working drills in various positions and attitudes. I also connect from pumps on the barge or floating vessel a flexible hose, conducting water under pressure to the interior of the drill tube. The steadying frame, which may be tripod or open pyramidal form, being placed on the bottom. The boring tube is lowered by means of the jib or derrick chain into guides provided in the steadying frame, and its counterweight is adjusted to suit the nature of the ground or rock to be bored; it is then by means of the flexible shaft and gearing caused to revolve while water is forced through it to scour out the borings, the tube descending as the hole becomes deepened. When it has descended a certain distance the gearing and its frame are detached, an additional length of tube is added, and the boring is continued.

Drilling machines of this type fitted with diamond drill bits were used for deepening the river Clyde in 1886.

British Patents Granted to T. English

The patents granted to English are given in Table 6.

The Beaumont/English Channel Tunnel Machine (English's 1880 Patent No. 4,347 Covering Improvements to Beaumont's 1875 Patent)

In the preface of the specifications accompanying Patent No. 4,347 (Figure 199) English says:

> In the Specification of a Patent granted to F. E. B. Beaumont on the 2nd day of December 1875, No. 4166, was described machinery for cutting cylindrical tunnels in soft rock by cutters arranged stepwise on a head made to revolve and advance. My Invention relates to improvements in machinery of this class, having chiefly for their object means for mounting firmly the revolving head, and of effecting its step by step advance as the excavation proceeds. The main framing of a tunnelling machine, according to my Invention, consists of two parts. The under frame or bed is of trough-like form, fitting the lower part of the circular tunnel, and having at its upper edges suitable guides parallel to the axis of the tunnel, along which the upper part of the framing is fitted to slide longitudinally.

The upper frame carried the working machinery which consisted of the engine (driven by compressed air), the boring head (the axis of which was mounted in bearings), and the gearing. That part of the axis or shaft of the boring head which extended backwards was

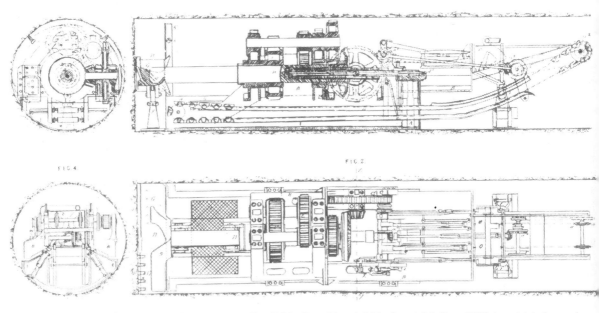

Figure 199. Drawings attached to Thomas English's Pat. No. 4,347, dated 25 Oct. 1880 in which he makes several improvements to Beaumont's Pat. No. 4,166, dated 2 Dec. 1875 (U.K.)

212

tubular and fitted with a piston, the rod of which projected backwards and was attached at its end to the under frame.

The compressed-air engine operated a hydraulic pump, the stroke of which was adjustable. This forced water into the hollow cylinder of the boring head through a channel provided within the piston rod. Thus the head was made to advance by means of hydraulic power, as it was revolved by compressed-air power delivered through a series of gears from a compressed-air engine.

When the upper frame and head had advanced as far as was possible (i.e. as far as the length of the guides permitted) the gearing by which it was revolved was disconnected from the engine, the cylinder relieved of hydraulic pressure, and the entire machine was raised slightly by means of four hydraulic jacks which projected down from the upper framing and bore against the invert of the tunnel.

Hydraulic pressure was then admitted into the annular space around the piston rod, and while the upper frame remained stationary the under frame or bed was advanced against the face. After which the four hydraulic jacks were retracted and the machine was repositioned in readiness for the next cut.

English suggested that instead of hydraulic pressure, screw jacks could be used for raising the machine and to force the grippers against the tunnel walls. The head could likewise be advanced and rotated by means of a screw on the axis of the head which engaged with a fixed nut on the under frame.

English further proposed that the boring head be constructed of two strong radial arms which projected in opposite directions from the axis, each arm being perforated with cylindrical holes for the reception of tool holders which were to be clamped in position by keys. Any of these could then be withdrawn without the necessity of retracting the boring head. The cutting tools were to be arranged to cut concentric annular grooves in the face, leaving between them narrow ridges of material which could easily be broken away or which would probably crumble away during operation of the machine.

In the specifications English suggested that inclined worms (Archimedean screws) to carry

the excavated material backward and upward, be used for delivering the muck into the trucks. However, on the drawing which accompanied these specifications an endless chain of buckets was depicted. This latter method of spoil removal was the one ultimately adopted for the Channel tunnel machine.*

This then was the design for the machine (Figure 200) which was ultimately built for the Channel tunnel project and which, for the purposes of this book, will be referred to as the 'Beaumont/English' tunnel boring machine (T.B.M.).

The actual Channel tunnel machine had a boring head consisting of two strong arms which projected from a boss on the hollow boring shaft. A series of chisel-edged cutting tools were arranged side by side along each arm. The rear section of each arm was hollowed slightly to form a trough, the mouth of which sloped backwards so that the debris cut from the face would be directed into the buckets on the conveyor.

A platform was provided beside the head so that a workman could stand there and ensure that the material so cut would be properly deposited into the conveyor buckets.

The significance of this particular machine cannot be overemphasized when it is remembered that this was the first *successful* rock machine (albeit chalk). Moreover, the basic engineering principle which ensured its success is still widely used today, namely

---

*According to an article[9] which appeared in *The Engineer* dated 8 December 1916, the excavated material from the machine used in the experimental heading of the Channel tunnel was conveyed to the shaft in trucks which ran along rails. At the time, as was the case with Brunel when he built the Thames tunnel, numerous 'self-styled' experts converged on the scene bringing with them their various suggestions and inventions which they felt would improve the work of excavation. Amongst these was a certain T. R. Crompton who suggested that the tunnelling machine should be operated entirely by hydraulic power and that the material cut from the face be crushed and mixed with the water from the machine (after it had done its work there). This mixture of debris and water could then be pumped out of the tunnel to the surface as a *slurry* (see also section of slurry machines). He calculated, from personal experience with this method which he had apparently used in his own business, that the ratio of water to cut material would need to be about 6 : 1. This method was seriously considered by the engineers involved with the project. Unfortunately, however, before it could be tested in practice the tunnel works were stopped by order of the Board of Trade.

Figure 200. Drawing of Beaumont/English tunnelling machine: used both sides of the Channel. (Courtesy *The Engineer* 8 Dec. 1916)

Figure 201. Tunnel bored by Beaumont/English machine in 1881

English's proposal to build the unit in two parts so that by hydraulic power and compressed air the powerful boring head could be slowly and evenly advanced against the face, while the under frame provided a steady and solid support.*

---

*In his 1875 patent Beaumont proposed that the 'advance of the boring head as the work proceeds may either be effected by advancing the carriage on which it is supported, or by sliding forward the shaft of the boring head by any known means'.

Channel Tunnel[10-12]

Work on a 2.50 m (8 ft) diameter shaft at Sangette (France) was commenced on 1 June 1878 by the French Channel Tunnel Company. The diameter of this shaft, which was completed to a depth of 86 m (280 ft) on 22 December 1880 was later (at a depth of 57 m (190 ft)) increased to 3.25 m (10 ft 8 in). The first shaft was used for pumping and another larger shaft of 5.40 m (17 ft 9 in) diameter was then begun.

214

Channel Tunnel Works.

Experimental Boring (7ft. dia.) in an easterly direction from No. 2 Shaft, near the western end of Shakespeare Tunnel, Dover.

Lengths of heading bored from week to week from April to July, 1882, by the Beaumont Boring Machine. (Progress not recorded previous to Apl. 13, began work on April 4th)

Week ending. 1882.	Length of Heading bored during week.	Remarks.
April 20th	173 feet	Delayed by mechanical defects and by visitors.
" 27	143 "	Delayed by visitors, by fixing tubbing and by moist chalk adhering to buckets.
May 4	131 "	Hydraulic jacks gave way.
" 18 (Fortnight)	144 "	Delayed by repairs, renewals and visitors.
" 25	95 "	No delays.
June 1	76 "	Not sufficient hands one day; hydraulic jacks gave way on another.
" 8	104 "	No delays.
" 15	110 "	Delayed by visitors.
" 22	109 "	No delays.
" 29	112 "	No delays.
July 6	106 "	Delayed by visitors.
July 13	87 "	Hydraulic jacks gave way on 7th & 11th; delayed by visitors on 8th July.
July 20	98 "	Delayed by visitors and by Board of Trade inspection on 15th July.
14 weeks	1,488 feet.	106.3 feet = average progress per week.

Figure 202. Weekly progress report on Beaumont machine used for experimental boring at Dover, April to July 1882. (Courtesy British Railways Board, Channel Tunnel Department.)

Of interest was the method used to line this shaft. The shaft was allowed to fill with water to a level convenient for the workmen, who then used a floating floor or platform designed by Drou, the second engineer, to carry out the work of lining the shaft with cement. The height of the working platform in the shaft could thus be easily controlled simply by opening or closing a tap. The main shaft was completed on 4 February 1882.

The same method of lining a shaft from a floating platform was used during the late 1970s by Deilmann-Haniel GmbH, Deutsche Tiefbohr AG (Deutag) and Gutehoffnung-shutte Sterkrade AG, contractors of a joint venture during the construction of a 410 m (1340 ft) shaft for the Sophia-Jacoba Colliery Company. During construction bentonite was inserted between the lining and the shaft walls and the centre was filled with water. After the lining had reached the bottom of the shaft a concrete mixture was pumped into the annular space between the lining and the shaft wall which displaced the bentonite. Subsequently the water in the central column was pumped out.[13]

By about the beginning of 1881 when the British were at last ready to proceed with the experimental headings for the Channel tunnel, the Beamont/English machine (Figure 200) had been built and, in order to test its capabilities, No. 1 Shaft was sunk near the western end of Abbotscliffe tunnel on the South Eastern Railway between Folkestone and Dover. The shaft was 22.6 m (74 ft) deep and when this was ready the Beaumont/English machine was installed there and a 2.13 m (7 ft) diameter heading, 804 m (2600 ft) long, was driven with this unit. The machine was powered by a compressed-air engine.*

---

*The engine supplying this motive power was apparently designed by Beaumont who, for the past four years had been doing considerable work in this field. For his experimental work he used an engine which was run in the Royal Arsenal at Woolwich. Early in 1880 trials were carried out with a prototype compressed-air locomotive on the Met. Railway and, judging from the reports of these trials in *The Times,* (dated 27 May 1880 and 15 March 1881) the locomotive performed satisfactorily.[14]

According to the Science Museum in London, the head of the T.B.M. used in the Channel tunnel project was rotated through a reducing bevel-and-spur gear in the ratio of about 1 : 50 by a two-cylinder engine operated by compressed air which was delivered at 25 lb/in^2 (0.17 MPa).

While the British were busy testing the Beaumont/English T.B.M., the French completed their main shaft and, by February 1882, were ready to commence work on the tunnel. The French promoters of the Channel tunnel, like their English counterparts, also wished to avoid the use of explosives and they accordingly examined the possibility of using a tunnelling machine. Two types of machine were given serious consideration, namely the Brunton machine which had been known of for some time, and the Beaumont/English unit, which appeared to be working satisfactorily in England on the other side of the Channel.

An agreement was signed between the French Channel Tunnel Company and Messrs. Beaumont and Company and also between the Channel Company and the company who had acquired the right to exploit the Brunton patent in France.[10] A machine from each of these companies was then installed in the main shaft and members of the Society of Mineral Industry went down to see them in operation.

According to the agreement, the Beaumont/English T.B.M. was to cut a test gallery to the east and, simultaneously, the Brunton unit was to cut another heading in the same direction 42.2 m (138 ft) above the Beaumont/English machine.

In his report covering the period 1 June 1878 to 18 March 1883, C. Breton (Chief Works Engineer for the French side of the project) stated that after several tests the Brunton machine was found to be unsatisfactory, due principally to the weakness of several main components, which eventually failed and forced the contractors to dismantle the machine.

The Beaumont/English machine began work on 19 July 1882 and, after some initial teething problems had been overcome, and the working crew had had time to accustom themselves to the intricacies of operating it, daily advances of 24.80 m (81 ft) were achieved with the machine.

During 53 days (i.e. from 13 January 1883 to 18 March 1883, when work was finally halted) the machine advanced a total of 810.3 m (2659 ft) — a record tunnel-driving achievement. In all, from the time the machine commenced work on 19 July 1882 until operations were ceased on 18 March 1883, 1839.6 m (6036 ft) of tunnel were driven. Apart from an irruption of water totalling approximately 2 million litres (500,000 gal) in 24 hours which was experienced in the gallery, no other major problems occurred.

So far as the English side was concerned the Beaumont/English machine bored a 7 ft (2.13 m) diameter tunnel 1985 yd (1815 m) making the total length of the heading 2024 yd (1850 m).

Work on the English side was stopped in 1882 due to political pressure.

Beaumont

There were, of course, many who were disappointed at the turn of events, not the least being Beaumont who, apart from his contribution towards the development of the machine itself, had become personally involved

I had a model of my Tunng. Machine, which I lent to a man who represented a Contractor in Sydney. I have not had it retd. to me - nor do I know what has become of it - Philip has made enquiries without result.

My Machine was put to work in a grey chalk quarry for the purpose of exhibiting its capabilities, & it was regarded as the Machine for the Channel tunnel - But this, as you know, was vetoed by the Govt. in 1871 - so my machine was never put to work in the Channel.

Beaumont had the help of Sir Edward Watkin, & by that means got his machine adopted, & put to work - I, unfortunately was associated with the other party connected with the Channel tunnel project - They were very supine & did nothing. Since then the Tube Railways have had in use a machine which has succeeded very well - & if the Ch. Tunnel be ever carried out - this machine will do it.

The Scheme is now, however, again blocked - & I at all events, shall not live to see it proposed afresh nor do I think it probable that my machine would be used, altho' it has many advantages over any other.

So my machine may be regarded as dead & done with. For hard rock it will not compete with the existing drills.

I was very unfortunate in the trials of it.

First was a slate quarry in N. Wales, but the wear of the cutters was so considerable that it failed. 2dly in a mine in Westmorland. They wanted a drift in a hardish clay - in which a tunnel had been successfully driven by hand.

It was quite capable of this - but the ground wd. not hold up - & so, after almost losing the machine by the downfall of the ground - that had to be given up.

3dly The late - (can't recall the name) had a tunnel to carry through in chalk in Kent - & resolved to use the machine - I had one of 9 ft. diam. made for him, & it was put to work, but here again, the ground gave way - & it ended in an open cutting instead of a tunnel.

The same machine was carried to South Wales, when a tunnel was to be driven in hard rock. Here the Cutters failed - the wear was too great, so the machine was withdrawn, & has by this time, I expect been sent to the scrap heap.

(Signed) T.D. Brunton

[handwritten margin note:] This refers to the new Shield & air-lock process.

[handwritten margin note:] This was Walker

Figure 203. Copy of letter from T. D. Brunton [sic] 41 Cambridge Road, Southend to Forrest Brunton, dated 23 Nov. 1907, by whom side notes were made ('T. D.' should probably read 'J. D.'). (Courtesy Sir Harold Harding, U.K.)

217

in the actual excavation works on both sides of the Channel.

Opposition to the scheme was focused primarily around the threat the tunnel appeared to pose towards England's insular security and, in addition, there was grave doubt that so long a tunnel could be adequately ventilated. On both these issues Beaumont and his supporters wrote numerous letters and articles[15] which were published in the various periodicals and newspapers of the day. But opposition to the scheme continued to grow. After the works were temporarily stopped by an injunction of the Board of Trade on 23 July 1882, the final death knell was sounded by the joint committee of both Houses of Parliament which met on 10 July 1883. By a vote of six to four, the joint committee decided that Parliamentary sanction should not be given to a submarine communication between England and France.

Brunton

Brunton, too, was disappointed, as he had hoped that *his* machine would be *the one* chosen for the prestigious work of excavating the Channel tunnel, and it is plain from his letter (see photograph of T. D. Brunton's letter) (Figure 203) that he hoped his machine might yet be used, despite the initial rejection by both the French and English promoters.*

English

English, for his part, was embittered because most of the credit for the invention of the Channel tunnel machine appeared to be directed towards Beaumont. On at least two occasions English was moved to express his

displeasure at this publically. The first occasion concerned an article in *The Engineer*. English wrote a stern letter to *The Engineer* which is quoted below:

> Sir,
>   Referring to a passage in your leading article of the 18th inst., on the subject of the "Channel Tunnel", to the effect that "it is well known that Colonel Beaumont, R.E.... is the inventor of the machinery used to bore it," I beg to forward, for your information, a copy of the specification of my patent No. 4347, dated 25th October, 1880, under which the boring machines are constructed.
>   THOS. ENGLISH, R.E.
>   Hawley, near Dartford.[17]

As a result of English's protest, *The Engineer* published another article on 15 June 1883 headed 'English's boring machine'.[18] This article commenced with the words: 'At page 456 we illustrate the boring machine used in the Channel Tunnel. The engraving is copied from a photograph placed at our disposal by Mr. Thomas English, the patentee of the invention....'

The second protest occurred at a meeting of the Institution of Civil Engineers when the subject of the Mersey railways was being discussed. Someone apparently mentioned the 'Beaumont' tunnelling machine. This incensed Major English who immediately rose and stated 'that the tunnelling machine alluded to as the Beaumont was one patented by him, and was worked by Colonel Beaumont in the Mersey Tunnel, under a licence from him, which had since been withdrawn. The Tunnel-Driving Company, Limited now held his sole licence to use it.'[19]

Despite these protests, numerous authors have referred (and still continue to refer) to the machine as the 'Beaumont' tunnelling machine or, in one or two cases as either 'Beaumont and an unknown inventor's' or 'Beaumont and a co-inventor's' tunnelling machine. The reason for this is easy to understand. While English was certainly responsible for the major modifications which enabled an otherwise

---

*In a paper entitled 'the Channel tunnel',[16] J. Clarke Hawkshaw mentions the Brunton machine in connection with some comments he was making regarding earlier geological and engineering investigations which were undertaken to ascertain where and how a tunnel could be made beneath the Channel. He says, *inter alia*: 'The thickness of the chalk was ascertained by borings 500 feet deep, made through it on the two coasts; and a machine, made by Mr. Brunton, for excavating chalk was tested, and was found to work as rapidly and as efficiently as the machine lately used between Folkestone and Dover, of which so much has been said. All the geological work was done in the years 1865-7; the machine was tested in 1870. Trials were made with Mr. Brunton's machine in the grey chalk at Snodland on many occasions. On the 8th

September, 1870, it excavated a heading 7 feet in diameter at the rate of 44 inches per hour; on the 20th January, 1871, the rate was 45 inches an hour, and on 25th February, 1871, it was 49 inches an hour.'

ineffective machine to function properly*, Beaumont, for his part, was responsible for the design of the compressed-air engine which provided the motive power and also for the operation and development of the various machines *after* they were installed in the tunnels. The author believes it was because of this personal involvement and dedication to the works themselves, that people then, and later, tended to associate 'the machine' with Beaumont. Today the Beaumont/English machine lies abandoned at the head of the Channel tunnel No. 2 shaft.

The First Mersey Tunnel[20-22]

For hundreds of years the Mersey river between Birkenhead and Liverpool was crossed by small boat, or via a public ferry service. But the service generally was a haphazard affair, regulated for the most part by the weather, which could delay the ferry crossing for many hours.

As the population of these two cities grew, it became apparent that an alternative method of crossing was needed. Mark Isambard Brunel who, at that time (the 1840s), was involved with the design and layout of Birkenhead's dock system, considered the two alternative suggestions put forward, namely a tunnel or a bridge. Finally Brunel recommended a road tunnel. But, like the Channel tunnel, there were many delays and procrastinations before anything definite was decided upon and action taken.

Eventually the decision to build a railway tunnel was made, the line to link up with the L. & N.W. and G.W. Joint Railways on the Cheshire side. No link-up with existing lines on the Lancashire side was initially planned, instead it was decided that the line would terminate there.

The Mersey Railway Company was formed to carry out this work although finance was difficult to obtain, investors being reluctant to gamble on a scheme fraught with so many

difficulties.* However, sufficient money was collected to enable the commencement of work in 1880 on the two pumping shafts. Unfortunately, the contractor got into financial difficulties before the job was completed and the Mersey Railway Company approached Major Isaac. An agreement was forthwith entered into with this gentleman (who was a wealthy trader) whereby Isaac, in return for certain remunerations, would undertake to settle all outstanding debts and complete the job. (Remuneration included promise of payment to Isaac of all residue share capital and debentures, which remained unissued after the completion of the project.)

As Major Isaac was a trader and not an engineer, he promptly sublet the work to the contractors John Waddell and Sons of Edinburgh and, under the guidance of two competent engineers (James Brunlees and Charles Douglas Fox) who were appointed by the company, the work was finally completed.

Owing to various problems associated with the levels of the drainage headings and those of the main tunnel, it became necessary to divert a portion of the drainage heading on the Liverpool side so that a 'loop heading', as it was afterwards called, was formed.

By October 1881 the two shafts had been sunk and a start had been made on the drainage tunnels from each side of the river. Initially this work, which progressed at the average rate of approximately 10 m (33 ft) per week was carried out by hand. In March 1883 a Beaumont/English machine was installed on the Birkenhead side and average penetration rates immediately leapt to 15 m (51 ft) per week with a maximum of 31 m (102 ft) being attained in one week.

The machine was pneumatically operated and the cutting head revolved at the rate of one and a half revolutions per minute.

During the driving of the loop section, penetration rates increased to 49 m (54 yd) per week after some modifications had been carried out on the cutters of the T.B.M.

---

*Beaumont's earlier machine (i.e. that patented in 1864) was a percussive unit, and while it failed to cut through the hard rock in the Wicklow mountains (for the Dublin water tunnel) it did cut softer freestone very effectively in Paris in 1867, when it was exhibited there.

*Other subaqueous tunnels, i.e. the Severn tunnel, the Hudson river tunnel and the Detroit river tunnel, which at that time were under way, had all met with severe irruptions of water, these problems ultimately leading to their temporary abandonment.

The machine was later used to drive the entire 2060 m (2250 yd) long ventilation tunnel from Birkenhead to Liverpool, and penetration rates of 59 m (64 yd) per week were recorded in this section.

All the machined tunnels were 2.24 m (7 ft 4 in) in diameter, and the wall surfaces were said to be 'as smooth and as circular as a gun barrel'. Excavation of the double-line main tunnel was carried out by hand labour as it was feared that heavy blasting would disturb the rock.* In all some 244,000 cu m (320,000 cu yd) of material was taken from the main tunnel, the crown of which was only 9.14 m (30 ft) below the bed of the river at the tunnel's highest point.

The Beaumont/English machine was operated by a crew of seven, and while the driving of the ventilation and drainage tunnels constituted an outstanding engineering feat because they were bored with the first commercially successful rock-tunnelling machine, the ultimate fate of most of the seven men who manned the unit was not so happy. Nearly all of them died of silicosis shortly after the railway was opened to the public because of the great quantity of rock dust produced by the machine.

After some 90 years the tunnels bored with the Beaumont/English machine are still in good condition and the tool marks made by the cutters remain clearly visible.

Summary

English appears to have been a quiet, extremely intelligent engineer whose brilliant work and personality were to some extent overshadowed by the flamboyant and extravert temperament of his fellow officer/engineer, Beaumont.

While English's contribution towards the development of the first successful rock machine has today been largely forgotten, Beaumont's name shines as a beacon and lives on in the minds of tunnelling folk all over the world.

When Beaumont died on 20 August 1899 an obituary by Major-General E. C. Sim and a letter by Major-General E. Renouard James were published in *The Royal Engineers Journal** dated 2 October 1899 and 1 December 1899, respectively. Because the opinions of these two men, colleagues of Beaumont, who obviously knew him well, give an insight into the character of Beaumont, the contents of these two documents are quoted in full.[23, 24]

Obituary

Away in the depth of the country I took up my *R.E. Journal* last week and saw announced the death of my old friend Fred. Beaumont, who at one time promised to be a very distinguished officer of the Corps, and also a prominent member of the Parliamentary world.

Unfortunately, disastrous speculations put an end to Beaumont's career.

My first real acquaintance with him took place at Gravesend in 1862, when he landed in command of a company of Sappers from Canada in the *Himalaya,* after the *Trent* affair, and I brought the remains of my old company from Eastern Australia in the *Lincelles,* a sailing ship which took 168 days to perform the passage from Fremantle to the Thames. We marched into Chatham together, and served side by side for some months at Brompton Barracks, when he was full of stories from the United States Army; then on the Potomac under General McLennan, whom he accompanied in balloons and in some skirmishes — I am afraid without leave. He told me in his lovely American drawl, which he never lost in after days, that the only two words of command he ever heard were, 'Form Lump to the Nor'-East', and then 'Split and Squander'.

I next met Beaumont at Castle Hill Fort, Dover, now called Fort Burgoyne, I believe, which he more or less completed with his company of Sappers in 1863-4. He had previously marched his men by road from Chatham to Dover to 'do them good', as he said. Soon after that he got into Parliament as a Liberal, I think, for South Durham, and he remained in Parliament somewhere about 10 years — until I fancy his means were straitened.

But he was always an inventive genius. First a trigger for a revolver, which he patented, and on which he received a royalty until the company burst and could not pay. Then balloons, diamond rock borers, electric engines, and motors, and when I last saw him a short time ago he was patenting some improved gear for bicycles, which he rode habitually. His father, who had been in a Hussar regiment in the early part of the century, was also an inventor, and ran some patent pottery

---

*Triassic sandstone.

*Courtesy of the Institution of Royal Engineers.

works in the midland counties when I saw him at Chatham in 1862-3.

Beaumont saw service in the Indian Mutiny, and it was said of him that he was so clever and rather more previous than his Colonel that he let off a mine some minutes sooner than ordered, much to the C.R.E.'s disgust, who would never recommend him for a brevet, which I believe he had really earned, for he was the 'coolest of the cool', under fire and elsewhere.

He was one of the best looking men in the Corps in the sixties, and in great demand in society on account of his experiences and amusing stories.

He had travelled in many parts of the world, pushing some of his inventions and making others.

He was a most devoted son to an invalid mother.

I regret his loss as a former intimate friend, and as a positive genius, which, if he had only been able to apply practically and commercially, might have made him a millionaire. He only just missed being a gigantic success. Peace be to his memory.

E.C.S.

Letter

12th September, 1899.

To the Editor of THE ROYAL ENGINEERS JOURNAL.

Sir, — Major-General E.C. Sim's characteristic and genial style is so typical of a man we are all endeared to that we always welcome any article from his hand, and his interesting and kindly notice in the *R.E. Journal* of October, 1899, of an officer with whom probably only a few near his own standing were personally acquainted — the late Lieut.-Col. F.E. Beaumont — must have been appreciated by the Corps at large. General Sim, in a few graphic touches, brought Fred Beaumont back at once to the memories of those who knew him as one who failed to become great through his own unbusinesslike qualities. Nothing could be truer; but in leading us to the due appreciation of the mechanical genius — it was nothing less — of the man, I scarcely think the bright style of the writer does full justice. Two of Beaumont's inventions may be placed in the list of those which in our time — and that is saying in *all* time — have left their mark in the world; and on these I beg to offer a few remarks.

The revolver improvement was patented about the time of the breaking out of the Crimean War in 1854. The first, and almost the only, revolver before then was the Colt, the breech-block of which was turned after each discharge by hand, the arm being re-cocked in like manner; and, of course, the aim was withdrawn for some time to do this. Under the new patent the revolution and re-cocking were effected instantaneously by the pressure of the trigger finger, the aim being maintained. That was all! The company which brought out this improvement fell into difficulties, and our very unbusinesslike brother officer was soon deprived of all pecuniary interest in it. The invention became public property, and is now applied universally to repeating arms; and these are in use all over the world. The facility with which an intoxicated ruffian can empty his six-shooter into one's body is certainly not an unmixed blessing, and the Corps may not, perhaps, care to be reminded that we owe this to the genius of a brother officer.

The second invention of note was, however, beneficial to humanity in a high degree. It was the diamond boring apparatus, by means of which — or by some modification of it — the boring of tunnels through the hardest rock has been brought within practical limits of time and cost. By the rapid revolution of a number of drills armed with diamond points, and actuated to move simultaneously, the working face of the tunnel is pierced; and by the continuous revolution of each of the borers in a circle masses of rock are rapidly cut clear, and can then be easily removed by blasting or otherwise. The Mont Cenis, the St. Gotthard, and the Blackwall tunnels — which rank among the more remarkable engineering achievements of recent years — one and all owe a debt to the boring-shield as improved; and should the Channel tunnel ever come to maturity, it can only do so by the same aid. Poor Beaumont benefitted very little in a pecuniary sense by his conception; but 'honor virtutis praemium' and we may, at least, claim what honour we can for our own brother officer.

Beaumont's other inventions were legion, and nothing was too small for him. I remember being much amused at his appearance in Hyde Park on a bicycle with folding wings, by the aid of which, when running down-wind, he claimed to be able to gain at least two miles per hour in speed. I need not say he did not make money of this, more than he had done with his many similar clever toys.

General Sim has remarked on Beaumont's coolness, and I can corroborate what he tells us by relating a personal experience of the latter's conduct in 1852 at a fire in the street outside Brompton Barracks. Before the fire engines arrived he, without taking off his mess dress, mounted to the upper floor of a house and rescued some individuals at great personal risk to himself. He was complimented for this in the next day's garrison orders, and surely the act was of the character of those which are thought worthy of high reward when performed before the enemy in the field of war.

Figure 204. According to a note found in the archives of Marshall-Fowler Ltd., this machine is reputed to have worked on the pilot shaft for the Channel tunnel in 1881. (Courtesy Marshall-Fowler Ltd.)

Beaumont had many faults, but was his own enemy rather than the enemy of others. Let us, at any rate, uphold his memory in regard to what he did for the honour and the credit of the Corps.

I am, etc.,

E. RENOUARD JAMES,
Major-Gen., late R.E.

Other Channel Tunnel Machines

The Fowler Machine

According to Marshall-Fowler Limited of Gainsborough, England, a Fowler machine (Figure 204) is reputed to have been used on the pilot shaft of the Channel tunnel in 1881. A photograph of this machine accompanied by a brief note was found recently in the company's archives. The note reads:

> This boring machine, made by Fowlers in Leeds, actually worked on the pilot shaft for the Channel Tunnel in 1881. The main shaft for this original tunnel was driven at the foot of Shakespeare Cliff, Dover, the work backed by the Submarine Continental Railway Company. It is said that a

Fowler Steam-engine lies along the pilot tunnel the entrance to which was filled in when, for military reasons, the Government refused to sanction further work. (In those days there was still fear of a French Army sneaking into Dover.)

As a result of the receipt of this information, an extensive search was undertaken by the author in various papers and periodicals of the time, but to date no further evidence or mention of this particular machine has been found. In addition the British Railways Board of London was approached in the hope that some data on the unit might have been available in the railway archives.

The Public Records Office (British Transport Historical Records) which holds old railway records and in particular data from the minutes of the South-Eastern Railway Company replied as follows:

> I have examined the South Eastern Railway Minutes 1879-1885 (SER 1/47-49), and whilst there is considerable reference to Beaumont/English transactions, and the activities of the Submarine Continental Railway (merely a

subsidiary of the SER), there is no mention of either Brunton or Fowler machines. The 1881 experimental boring (for which periodic progress reports are given) appears to have been conducted with the Beaumont/English machine.

Whilst I am not suggesting Fowlers did not build an experimental machine intended for use in the tunnel, the impression given in the minutes is that the Beaumont/English machine proposal was the only one considered, so that *selection* is not a relevant factor. In this case one could conceive that tests of the Fowler machine took place elsewhere and that because these were found unsatisfactory, the firm did not put in a tender to compete in the test bore proper. If this was so then record of such experiments would not necessarily be revealed in the SER records. If indeed the Fowler experiments were backed by the SCRC it is unfortunate their records do not mention the fact. We have no SCRC records here, information concerning their activities appearing in the context of their relationship to the SER.

Mr. P. A. Keen, Director, Channel tunnel, British Railways Board, who advised the author of the findings of the Public Record Office comments:

I have also discussed the matter with the Public Relations Consultants to Channel Tunnel Investments (the successors to the old Channel Tunnel Company). Unfortunately many of the original records of the Channel Tunnel Company were destroyed by enemy action in the 1940 London Blitz. Enquiries have also been made in the Public Relations Office British Channel Tunnel Company and Mott, Hay and Anderson (Consulting Civil Engineers). Available records indicated that trial boring tests in 1881 were carried out with 2 Beaumont/English machines, one on the British side (initially at Abbotscliff and later at Shakespeare) and on the French coast at Sangatte. However, no information is available from any of these sources on the Brunton* or Fowler machines.

It would seem that there are 3 possibilities:

(i) The Beaumont machine may have been constructed by Fowlers;
(ii) Fowlers may have constructed machines for driving shafts;
(iii) Fowlers may have designed a machine to be used for Tunnel boring, initial tests showed

that it was not satisfactory, and as such Fowlers did not take part in the test bores.

(Incidentally the Marshall-Fowler archives would appear to be incorrect in one or two respects. Although strategic factors brought about the abandonment of the 1881 project it was much later that this heading was cut in connection with coal mining operations and flooded. The 1881 borings had revealed the existence of coal and this led to the subsequent development of the Kent coalfields, several of which remain in active operation although the Shakespeare Colliery later closed.)

The Whitaker Machine[25]

After numerous vain attempts by Sir Edward Watkin and others to revive the issue of the Channel tunnel during the ensuing years, the matter was again raised after the First World War by Sir Percy Tempest, Chairman of the South Eastern and Chatham Railways and the Channel Tunnel Co. Ltd. (Sir Percy was successor to Sir Edward Watkin).

During the war the Royal Engineers had carried out rigorous tests on a Whitaker machine at the War Office testing ground in Claygate. Apparently this machine was inspected by Sir Percy and by Sir Douglas Fox and partners, and as a result of the favourable recommendations made by Sir Douglas Fox and partners, Sir Percy persuaded the directors of the Channel Tunnel Company to purchase one. The argument put forward being that possession of such a machine would 'strengthen the hand of the Channel Tunnel Company'.[12]

The original price quoted by Whitaker was £6150, but during the manufacture an increase was made to the horse-power (120 to 126) and in addition its structural strength was improved. As a result Whitaker requested that the price be raised to £8150. Tempest would not agree to this and finally it was decided that the matter be reconsidered by the directors of the Channel Tunnel Company after the unit had been tested in practice. As the machine performed satisfactorily a further sum of £1000 was subsequently paid to Whitaker, thus making the total purchase price £7150.

On 18 June 1922 the 3.6 m (12 ft) diameter machine was put to work boring a trial heading in the cliff face of the Lower Chalk at Folkestone Warren (Figures 205 and 206).

---

*Mention of the trials carried out with the Brunton machine may be found in the French paper 'Rapport sur les traveaux exécutes au tunnel sous-marin depuis l'année 1878, jusqu'au 18 mars 1883,' (Report on work carried out on the undersea tunnel from 1878 to 18 March 1883) which was prepared by C. Breton, Chief Works Engineer for the French Channel Tunnel Company.[10]

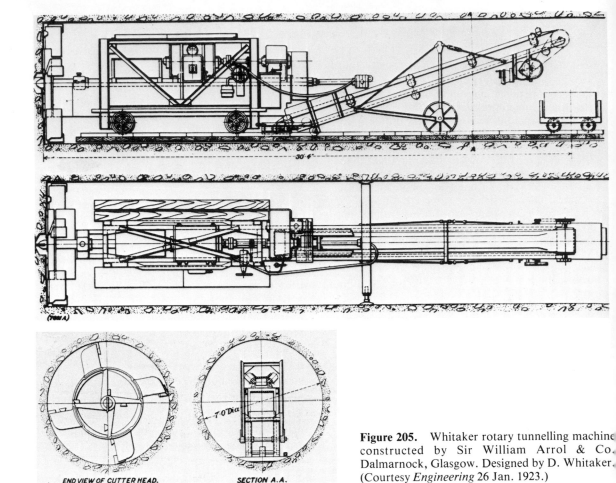

END VIEW OF CUTTER HEAD.   **SECTION A.A.**

Figure 205. Whitaker rotary tunnelling machine constructed by Sir William Arrol & Co, Dalmarnock, Glasgow. Designed by D. Whitaker. (Courtesy *Engineering* 26 Jan. 1923.)

Average driving rates of approximately 2.7 m (9 ft) per hour were achieved. Work on the trial heading was terminated on 26 September 1923; after 146 m (480 ft) of tunnel had been bored.

Estimated working costs for the Whitaker of £16 5s. 1d. per eight-hour shift were made up as follows:

20% Depreciation	£6. 4. 0.
6% Interest	1. 17. 0.
75 BHP Electric power @ 1½d per unit — a unit = 1 H.P.	2. 14. 0.
1 Driver	15. 0.
1 Fitter	15. 0.
2 Labourers in front of Excavator 12/-	1. 4. 0.
1 Conveyor driver	15. 0.
2 Labourers bringing sleepers 12/-	1. 4. 0.
Oil and waste	2. 6.
£200 per year upkeep	14. 7.
Total cost of Excavating and delivering material into wagons	£16. 5. 1.

After completion of the trials Sir Percy put another Channel tunnel scheme before Parliament. But when the matter was debated upon on 30 June 1930 the motion was defeated by 179 to 172, a matter of only 7 votes.

The machine was abandoned and subsequently a scrap metal dealer offered the Channel Tunnel Company £6 for the unit. As a result K. W. Adams was requested by the company to inspect the machine and submit a report. Adams found that the Whitaker unit had become half-buried in the chalk from a landslide, while a shed which had originally been built over the machine to protect it had been partially removed by vandals. He expressed the opinion that the machine's value as scrap far exceeded the amount offered by the dealer and, in any case, there was a possibility

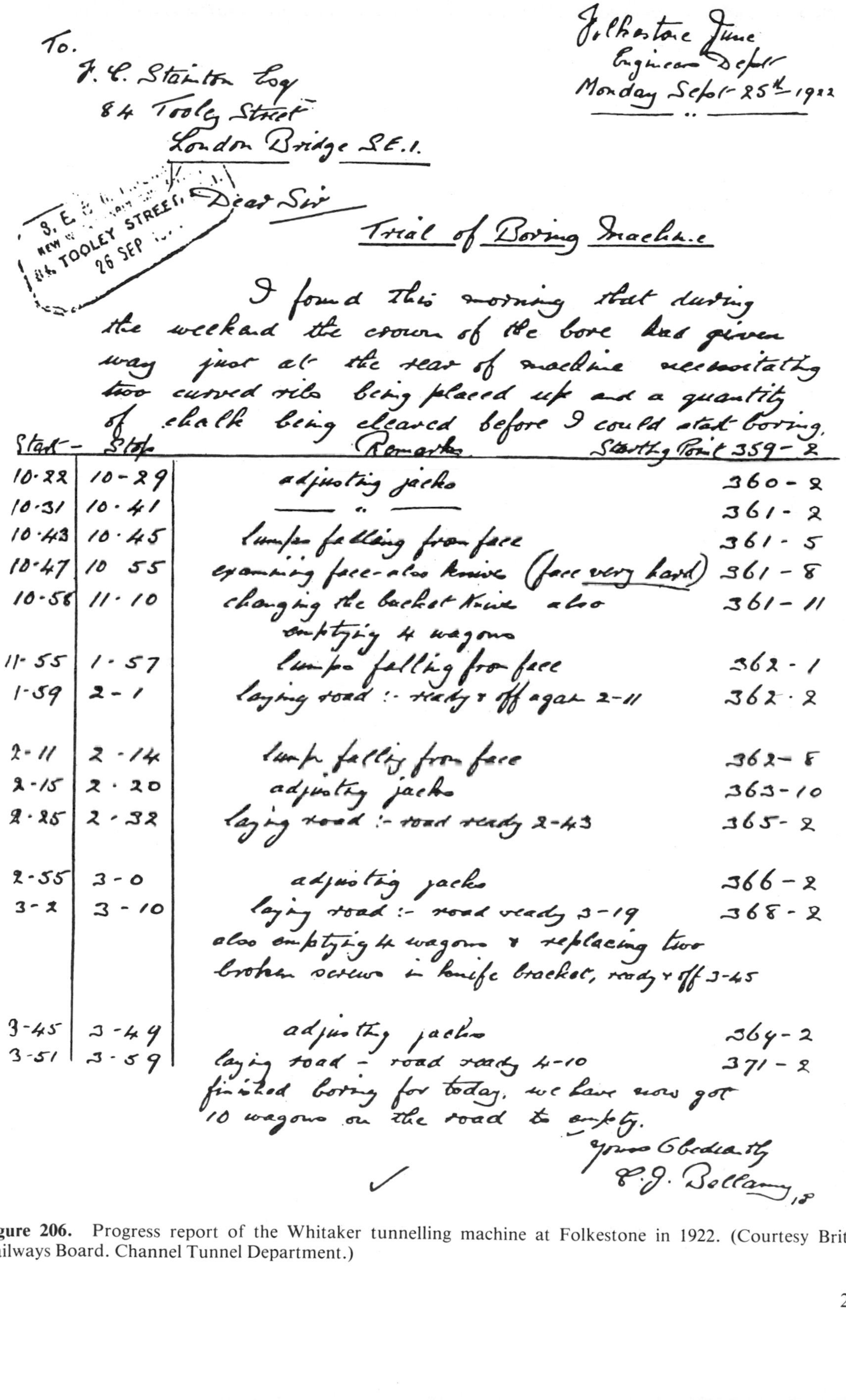

Folkestone Junc
Engineers Dept
Monday Sept 25th 1922

To.
F. C. Stainton Esq
84 Tooley Street
London Bridge S.E.1.

S. E. & C. R. ... NEW W... 84 TOOLEY STREET ... 26 SEP ...

Dear Sir

Trial of Boring Machine

I found this morning that during the weekend the crown of the bore had given way just at the rear of machine necessitating two curved ribs being placed up and a quantity of chalk being cleared before I could start boring.

Start	Stop	Remarks	Starting Point 359-2
10·22	10-29	adjusting jacks	360-2
10·31	10·41	"	361-2
10·43	10·45	lumps falling from face	361-5
10·47	10 55	examining face - also knives (face very hard)	361-8
10-58	11·10	changing the bucket knive also emptying 4 wagons	361-11
11-55	1·57	lumps falling from face	362-1
1-59	2-1	laying road :- ready & off again 2-11	362·2
2-11	2-14	lumps falling from face	362-8
2-15	2·20	adjusting jacks	363-10
2·25	2·32	laying road :- road ready 2-43	365-2
2-55	3-0	adjusting jacks	366-2
3-2	3-10	laying road :- road ready 3-19 also emptying 4 wagons & replacing two broken screws in knife bracket, ready & off 3-45	368-2
3-45	3-49	adjusting jacks	369-2
3-51	3-59	laying road - road ready 4-10	371-2

finished boring for today, we have now got 10 wagons on the road to empty.

Yours Obediently
E. J. Bellamy, 18

Figure 206. Progress report of the Whitaker tunnelling machine at Folkestone in 1922. (Courtesy British Railways Board. Channel Tunnel Department.)

that the unit could be removed and preserved for historical purposes.

According to A. P. Moss (Department of the Environment) who visited the site a few years ago, the Whitaker machine still lies partially buried at the entrance to the trial tunnel which is above the Martello tunnel at Folkestone Warren.

Modern Channel Tunnel Machines

During the late 1950s and early 1960s extensive work was undertaken by the Channel tunnel study group to determine again the feasibility of a Channel tunnel. A paper on the subject entitled 'The work of the Channel tunnel study group 1958-60' by Professor John McGarva Bruckshaw (Prof. of Applied Geophysics at the Imperial College of Science and Technology); Jean Goguel, (Ingénieur Général des Mines. Head of the French Geological Survey); Sir Harold John Boyer Harding (Consulting Engineer), and René Malcor (Ingénieur en Chef des Ponts et Chaussées, delegate of the group) was given to and published by the Institution of Civil Engineers in February 1961.

The paper described the group's work which included, *inter alia,* planning and survey, geology, and a traffic and economic study.

Although a preliminary figure of £468 million was given for the project it was anticipated that this would probably escalate to £846 million by the time the work was completed, due to inflationary trends.

Both the French and British governments gave the project their official blessing towards the end of 1973.

Construction work, which was scheduled to begin in 1975 and be completed in 1980, was to be undertaken by an international group headed by the British Channel Tunnel Company and La Société Française Du Tunnel Sous La Manche, Paris. (However, when the agreement was drawn up, provision was made to allow either party to withdraw at any time, if they so desired.)

Three tunnels were planned, namely two 'main tunnels' (each carrying a single railway line) which would run parallel with one another, and a third 'service tunnel' which would be situated beneath and between the two main tunnels. The tunnel route which would extend from Cheriton near Folkestone in South-east England to Frethun near Calais in France would have a length of 32 miles (51.5 km), 23 miles (37 km) of which would be beneath the sea. It was anticipated that the journey would take approximately three and a half hours, which was about the same time it took to fly from London to Paris. At present it takes about eight hours to make the crossing by boat and rail.

Both sides elected to use a tunnelling machine for the excavation work and orders for these were given to Robert L. Priestley Limited of England and the Robbins Company of Seattle, U.S.A., by the British and French cross-Channel contractors respectively.

The Priestley Machine

In 1973 Robert L. Priestley Limited were asked by the cross-Channel contractors to build a unit for the work which was to be carried out on the British side of the proposed service tunnel running from Dover (England) to Sangatte (France). Project managers for this work were R.T.Z. Development Enterprises Limited.

The 5.27 m (17 ft) diameter machine (Figure 207) designed for the work was a conventional fully mechanized shield of cylindrical shape, equipped with a thrust reaction ring. Its cutter head (powered by a 560 kW motor) could be rotated in either direction and was fitted with either twin action drag picks or disc cutters. Push rams capable of exerting a total forward thrust of some 762 t advanced the machine.

The segment erector at the rear of the machine was also designed and built by Priestley. Its erection system was so arranged as to permit preliminary pre-erection work on segment erection to be carried out continuously. This considerably cut down actual erection time during the appropriate part of the cycle. The 100 t erector (46 m (15 ft) long) was capable of handling individual segments weighing approximately 4 t. Five segments each 1.25 m (4 ft) long and the key segment were needed to complete a ring of tunnel lining. Because the erector's advance along the tunnel was independent of the main machine, it was possible to carry out tunnel-

226

Figure 207. Priestley Channel tunnel machine for use on the British side of the service tunnel. (Courtesy Robert L. Priestley Ltd.)

lining erection and excavation work simultaneously. While the machine was advanced by the use of thrust rams reacting against an expanding anchor ring, which locked against the tunnel walls, the erector was advanced by means of hydraulic grippers, which thrust it forward from the rails along which it travelled.

Although this project was shelved at the beginning of 1975[26] for political reasons just as manufacture was completed, the Priestley machine was nevertheless put to work excavating some 300 m (1000 ft) of tunnel. During this initial drive the machine exceeded expected boring rates by excavating at the rate of up to 10 m/hour (33 ft/hour), leading to the view that the unit would have been able to achieve tunnel completion well within the prescribed time had the project been proceeded with.

The Robbins Machine

The Robbins machine, Model 165-162 (Figure 208), ordered by the French Company SITUMER (which was the new company created to make investigative preparations for the Channel tunnel and to prepare and receive tenders before contractors were involved in the project) was delivered in 1974. However, although it was assembled on site it was, unfortunately, not put down the shaft and tested in the Chalk.

Because of the stringent specifications laid down by both the French and British contractors, the Robbins and Priestley units were basically similar in concept. Each machine had, however, distinctive characteristics peculiar to the manufacturer.

The Robbins machine featured a flat cutter head with large muck openings. Above the

227

Figure 208. Robbins Channel tunnel machine (Model 162-165) for use on the French side of the service tunnel. (Courtesy The Robbins Co.)

muck openings were six robustly constructed radial arms with deep scoops. These arms were fitted with mountings for either disc cutters or drag teeth.

The unit was a full-length, two-piece, cylindrical, telescopic, rotary shield which was articulated for steering purposes.

The gripper system operated somewhat on the principle of conventional drum brakes used on motor vehicles; they expanded circumferentially out of the shield skin against the tunnel wall. These grippers provided thrust reaction for the unit's propulsion jacks. In addition a system which thrust off the tunnel lining was also provided.

The evenly spaced steering jacks on the front shield enabled precise control of line and grade.

An unusual rotary segment erector which could travel axially on a long gantry beam was provided for installing cast-iron segments either behind the machine or inside its tail shield. This arrangement (like that of the Priestley Channel machine) allowed for the erection of segments at a different rate from that of the machine.

Machine specifications for the 165-162 were as follows:

Diameter	4.95 m (16 ft 3 in)
Cutter head	615 kW (825 hp)
Thrust	471,000 kg (1,040,000 lb)
Torque	109,000 k.m (788,000 lb/ft)
Weight	245 t (270 ton)
Cutters	30 discs and 5 drag bits (centre cutter)

References

1. A. W. F. Taylerson, *Revolving Arms,* Herbert Jenkins, London, 1967, p. vi.
2. Rock cutting machinery at the Paris Exhibition, *The Engineer,* 26 April 1867. London.

3. Major Beaumont, R. E., On rock boring by the diamond drill, etc., *Proc. Inst. Mech. Eng.,* London, **26,** 92-125, 1875.
4. Brunton's tunnelling machine, *Engineering,* 28 May, 1869.
5. Brunton's tunnelling machine, *The Colliery Guardian,* April 1874.
6. H. S. Drinker, *Tunneling, Explosive Compounds and Rockdrills,* Wiley, New York, 1893.
7. Lieutenant-Colonel Thomas English, obituary, *The Times,* 28 June 1935.
8. Lieutenant-Colonel Thomas English, obituary, *The Engineer,* July 1935, London.
9. The Channel Tunnel and other projects, *The Engineer,* 8 Dec. 1916, London.
10. C. Breton, Report on work carried out on the undersea tunnel from 1878 to 1883, unpublished.
11. K. W. Adams, The story of the Channel Tunnel, unpublished.
12. K. W. Adams, Folkestone Warren: Channel Tunnel trial boring machine, unpublished.
13. *Technik und Betrieb. Glückauf,* **114,** Maschinelles Teufen eines Wetterschachtes nach dem Lufthebebohrverfahren, No. 11, 1978.
14. *The Times,* 27 May 1880, 1 Feb. 1881, 4 March 1881, 15 March 1881, 27 April 1881, 27 May 1881, 23 July 1881, 15 Aug. 1881, 22 Oct. 1881, 27 Oct. 1881, 21 Nov. 1881, 21 Feb. 1882.
15. F. Beaumont, The proposed Channel Tunnel, *The Nineteenth Century,* Feb. 1882, 288-333, New York.
16. J. Clarke Hawkshaw, The Channel Tunnel, paper read before the Mechanical Section of the British Assoc. for the Advancement of Science, Southampton, Aug. 1882.
17. *The Engineer,* 25 May 1883, 398, London.
18. English's boring machine, *The Engineer,* June 1883, London.
19. *Minutes of Proceedings, Institution of Civil Engineers,* London, **86,** 116, 1885.
20. G. W. Parkin, *The Mersey Railway* Oakwood Press, McWinnville, OR, U.S.A., 1968.
21. Mott, Hay and Anderson, *Merseyside's First Mole,* March 1973, Mott, Hay & Anderson, Croydon, U.K.
22. Railway tunnel under the Mersey, *Encyclopaedia Britannica,* Vol. 22, p. 562, 1929.
23. E. C. Sim, The late Lieutenant-Colonel Frederick E. B. Beaumont R.E., obituary notice, *The Royal Engineers Journal,* 2 Oct. 1889,Chatham, p. 216.
24. E. Renouard James, The late Col. F. E B. Beaumont, correspondence, *The Royal Engineers Journal,* Dec. 1899, Chatham, p. 258.
25. The Whitaker rotary tunnelling machine, *Engineering,* 26 Jan. 1923, London, p. 103.
26. *The Mercury* Hobart, Tasmania, 13 Sept. 1973, 22 Jan. 1975.

Bibliography

J. M. Bruckshaw, J. Goguel, H. J. B. Harding and R. Malcor, *The Work of The Channel Tunnel Study Group,* British Section of the Société des Ingénieurs Civils de France, March 1961.

The Channel Tunnel, *The Engineer* 18 May 1883, 383-4.

Correspondence on the Mersey Railway, *Minutes of Proceedings, Institution of Civil Engineers,* **86,** 1886.

A. W. F. Taylerson, R. A. N. Andrews, and J. Frith, *The Revolver, 1818-1865,* Herbert Jenkins, London, 1968.

Correspondence

Science Reference Library, London, U.K.
Department of the Environment, U.K.
British Railways Board, U.K.
Marshall-Fowler Limited, U.K.
Mott, Hay, & Anderson, U.K.
Institution of Royal Engineers, U.K.
Science Museum, U.K.
The Robbins Company, Seattle, U.S.A.
R. L. Priestley Limited, U.K.
Sir Harold Harding, U.K.
County Borough of Birkenhead, Central Library, U.K.
Merseyside Passenger Transport Executive, U.K.
Institution of Civil Engineers, U.K.

FIVE

Modern Rock Tunnelling Machines

By the end of the 1920s, after the repeated failures which had attended the introduction of most rock-tunnelling machines, interest in their development tended to wane and tunnelling projects in hard rock were tackled using time-honoured conventional methods, namely drilling, blasting, and mucking. For the next 20 years this situation continued and very few, if any, patents for rock machines were submitted by engineers, nor were any units built.

America

The Robbins Company

James S. Robbins

In 1952 F. K. Mittry from the Mittry Construction Company of Los Angeles, approached the Goodman Company (manufacturers of mining equipment) and enquired whether they could design and build a large jumbo-type structure on which a Goodman coal cutter could be mounted. Mittry wanted the machine to cut a ring or annular groove as the periphery of the cross-section of a 7.6 m (25 ft) diameter diversion tunnel in soft Pierre shale at Oahe dam, South Dakota. His intention was to blast out the central core and use the jumbo structure to aid in installing tunnel rib supports (ring beams) close to the face. The president of the Goodman Company, Bill Goodman, suggested he employ James S. Robbins, the company's Consulting Engineer, to design and supervise the construction of the device.

Born on 12 August 1907, James S. Robbins attended Michigan Technological University and earned the degree of Bachelor of Science in mining engineering in 1927.

After this he was employed as a miner and a mining engineer at various mines. This included the position of General Manager of Arctic Circle Exploration Inc. during the period 1935 to 1946.

In 1947 he launched out on his own as a consulting engineer until 1948 when he formed the Mechanical Miner Company which was located on South Michigan Ave., Chicago. During this period Robbins and his small staff of engineers developed a continuous coal borer which was used by various South Illinois coal companies including the Marietta Boat and Barge Company for whom he was consultant. At that time he also acted as consultant for the Chicago Wilmington and Franklin Coal Company and became involved in the development of the 'Marietta miner' (see also Chapter 19 on coal mining — America — the Marietta miner).

About four continuous coal-miners with cutting heights varying from about 91 cm to 274 cm (3 ft to 9 ft) evolved as a result of the work carried out by the Mechanical Miner Company. (The width of these machines varied, but was usually about twice the cutting height of the unit concerned.)

Twelve or thirteen patents covering these units, as well as the name 'Mechanical Miner Company', were then sold to the Goodman Manufacturing Company. Because of these arrangements, Robbins was at that time restricted by contract from developing competitive multi-head coal-miners and he worked as a consultant for the Goodman Company for about five years, devoting about 50 per cent of his time to their work during the first year and,

Figure 209. Model 910-101 T.B.M. designed by James S. Robbins & Associates in 1953. The 8 m (26 ft 3 in) diameter machine was used at Oahe Dam in South Dakota. (Courtesy The Robbins Co.)

during the following four years decreasing this to 40 per cent, 30 per cent, 20 per cent and 10 per cent for each year respectively.

Development of the miner was continued by Goodman who marketed the unit as the 'Goodman Miner'. About 200 of these units were sold during the ensuing 20 years.

A complex arrangement then developed whereby Robbins received royalties from Goodman sales and Goodman financed the patenting of developments made by Robbins in the tunnel boring machine field in exchange for assignment of these developments and inventions.

In 1952 the company, James S. Robbins and Associates, was formed and Robbins paid Goodman a royalty for the use of his own patents which Goodman had paid to acquire through the use of the Goodman Company's patent attorney.

When Mittry, at Goodman's suggestion, approached Robbins about the jumbo and trepanning machine required, Robbins, after careful study of the problem, managed to persuade Mittry that it would be more practical

to design and build a full-face tunnel-boring machine with counter-rotating inner and outer heads, rather than use the coal cutter and blasting design. Robbin's plan was to attach the counter-rotating cutter heads and their support and drive systems to the front of the large jumbo, thereby retaining that part of Mittry's concept. By doing this ring beams could be placed even closer to the face than would have been possible using the original concept. A Goodman engineer, Mr. Bud, devised a method of handling the ring beams in four segments and rotating them into place on a ring beam jig. (This method is still used today around the world.)

The first history-making tunnel borer dubbed the 'Mittry mole' (Figure 209) was the result of a collaboration of Robbins, Mittry, and Bud. All patents and designs, and details of construction, etc. were developed by Robbins and his small team of two or three engineers and draughtsmen, except for the Bud 'ring beam jig'. Mittry paid the bills. Apart from the work done by Bud, the Goodman Company was not involved in the project. The machine, which

231

was designated Model 910-101, was built in various fabrication and machine shops in the Chicago area. It was put to work in 1953.

The unit was 27.4 m (90 ft) long, and the 7.8 m (25 ft 9 in) diameter cutting component consisted of two counter-rotating heads — an inner section with three radially arranged cutting arms, and an outer section, fitted with six cutting arms. The outer section rotated in the opposite direction from the inner head. The head was powered by two 149 kW (200 hp) wound-rotor motors.

A row of radially arranged fixed tungsten carbide drag bit cutters was mounted on each arm together with a parallel row of freely rolling disc cutters (made from mild steel) which were set slightly behind the fixed cutters. The leading edge of each disc cutter was so positioned that it followed the centre of the ridge left between two adjacent kerfs.

Of interest is the fact that while some 362 fixed cutters were consumed only 6 disc cutters were replaced during the driving of four of the tunnels and then only because of seal and bearing failure, rather than because of actual cutter wear.

A 19 kW (25 hp) motor powering hydraulically operated 'feet' moved the unit forward on tracks.

Specifications for the project demanded that the tunnel be kept at a 95 per cent humidity level to prevent deterioration of the walls. This high humidity affected the electrical equipment during the initial stages, but after redesigning this section the problem was overcome. In addition, lateral steering difficulties were experienced and the steering was changed to the forward end. However, while this cured the lateral steering problems, the shoe under the forward end then tended to dig in when negotiating soft ground and, at times, this was quite troublesome.

While the machine was not designed for use in particularly hard material and, as such, could not strictly speaking be classed as a 'rock' machine, it nevertheless did cut through the soft Pierre shale so successfully that Mittry later sold it to Oahe Constructions at a considerable profit. At the time it broke existing world records by advancing 61 ft (18.5 m) in an eight-hour shift and 161 ft (49 m) in a day.

Figure 210. Second machine designed by James S. Robbins & Associates for the Oahe dam project in 1955. (Courtesy The Robbins Co.)

Figure 211. Sixth T.B.M. Model 131-106 designed by James S. Robbins & Associates. This unit was used in the Humber river sewer tunnel in Toronto in 1965 and was the first successful application in medium hard-rock strata. (Courtesy The Robbins Co.)

The Oahe dam project involved four or five major contracts for upstream and downstream sections, which included seven diversion tunnels and five power tunnels. All of these tunnels were in the 7.6 m to 9.1 m (25 to 30 ft) diameter range. Oahe Constructions commissioned another T.B.M. (Model 930-102) (Figure 210) from Robbins. This was to be built by about 1954 with some improvements.

Following these two machines a series of three smaller diameter units (Models 101-103, 102-104, and 103-105) were built by Robbins for use in sewer tunnels in Pennsylvania and Chicago.

Interbedded shale and limestone in Pittsburgh and hard limestone in Chicago with compressive strengths ranging from 35 to 83 MPa (5000 to 12,000 lb/in^2) and 124 to 183 MPa (19,800 to 20,500 lb/in^2) respectively, soon pinpointed the inherent weaknesses of these machines which were used by Perini, Dravo, and Healy. In addition, dust became a major problem in the harder ground.

Ducting was installed right up to the face in an attempt to alleviate this nuisance, but although the amount of dust in the air was reduced somewhat, it was not eliminated entirely.

There was a high breakage rate of the tungsten carbide drag bits, the drive shafts were too flexible, the machines lacked rigidity and strength, and their chain conveyors and high-pressure hydraulic systems tended to fail. None of these machines bored more than 30 m (100 ft).

Up to this time the machines had carried both fixed and disc cutters. The fixed cutters were designed to cut the rock, while the disc cutters were required to deal merely with the outstanding ridges of material left between the kerfs or grooves. The disc concept had evolved from the McKinlay (see 'McKinlay entry driver') which Robbins senior had serviced during his association with the Chicago Wilmington and Franklin Coal Company. The McKinlays carried 'wedging wheels' which followed the cutting blades and broke off the concentric rings of coal left by the cutters. In 1956 a new 3.28 m (10 ft 9 in) diameter machine (Model 131-106) (Figure 211) carrying both fixed and disc roller cutters was built for the Foundation Company of Canada for use in the Humber River sewer tunnel project in Toronto, Ontario. It was put to work, and during its initial trials experimentation showed that when the fixed cutters (which had proved constantly troublesome due to their weakness under shock loading) were removed, the machine actually bored at the same rate as it had done previously; moreover, the discs cut and wore better in the harder material than did the fixed

233

Figure 212. Third Oahe dam T.B.M. Model 351-107 built by Pacific Car & Foundry Co. for James S. Robbins & Associates in 1960. The 9 m (29 ft 6 in) diameter unit was used by Morrison-Knudsen, Kiewit, Johnson. (Courtesy The Robbins Co.)

drag teeth. It successfully bored its way through 4510 m (14,800 ft) of sandstone, shale, and crystalline limestone, with compressive strengths in the 55 to 186 MPa (800 to 27,000 lb/in²) range.

The Toronto machine had a single rotating head on which the cutters were mounted. Four open-bottomed removable buckets positioned around the periphery of the head picked up the cut material and deposited it on to a belt conveyor at the top. Gripper shoes which expanded against the tunnel walls provided both torque and thrust reaction. Forward thrust was provided by hydraulic cylinders reacting against these gripper shoes, and vertical and lateral steering was catered for by steering jacks

mounted near the head of the machine.

Although this machine was more robust and powerful than its predecessors, success was not, of course, immediate, but rather the result of several modifications which were effected on site as the work progressed.

In 1956 Richard J. Robbins, son of James S. Robbins, graduated from the Michigan Technological University, receiving his B.Sc. degree in mechanical engineering. He joined his father's workforce as a design engineer in 1958. His initiation into the game was in the field, modifying one of the earlier machines in an attempt to make it function properly.

In 1957 when the first successful hard-rock machine built by the company was being tested in practice, young Robbins was doing military service. The same year Morrison-Knudsen obtained the Oahe dam upstream section contract and they commissioned James S. Robbins and Associates to build them a machine. This unit (Model 351-107) (Figure 212) was about two-thirds built when James S. Robbins was killed on 7 December 1958, while flying his own light twin-engined plane from Denver to Seattle. The tunnelling machine was completed by the remainder of the team of about 12 or 13 designers and draughtsmen at James S. Robbins and Associates and it went to work early in 1959.

Torque reaction for the 9 m (29 ft 6 in) diameter single rotating head of this unit was carried by the dead weight of the machine, instead of by the grippers as previously.

Richard J. Robbins took over the management of the firm but, at that time, it was hard going as he had had as yet very little experience in this field. To cap matters his chief engineer left for service elsewhere.

The last contract for the Oahe dam went to Johnson, Drake & Piper, Winston, Green, American Pipe & Foley. A fourth machine was thus needed for the dam but, at that time, the owner of Foley Bros. (who had lost confidence in the Robbins Company since the death of J. S. Robbins) decided to give the order to American Hoist Pacific as they had a design group in Seattle who were experienced in heavy construction machinery. It was designated the 'Prairie miner' and it successfully mined the downstream section of the power tunnels, proving that large-diameter tunnelling machines could be used through material such as the badly faulted Pierre shales at costs comparable with those accrued using conventional methods.

The Poatina Tunnel[1]

The story of the Robbins 161-108 machine (Figure 214) for the Poatina tunnel is significant. It broke existing world tunnelling records at the time but, more important than that, it was a milestone both in the developmental history of the Robbins Company's machines and as a trend-setter for other tunnelling machine manufacturers throughout the world. It also served to convince a very sceptical tunnelling industry that tunnelling through hard rock was a viable proposition — given the right machine and cutters.

Designs for the physical layout of the Great Lake power development scheme at Poatina were first mooted by the Hydro-Electric Commission (H.E.C.) of Tasmania during the mid-1940s and early 1950s. In 1955 P. T. A. Griffiths, who was leading designer in hydraulics, went to America for a period of about 12 months on a Dominion Civil Service Fellowship. It was during this time that he saw the Robbins tunnelling machines at work on the Oahe dam project. The obvious advantages of a machine-bored tunnel were apparent. On his return Griffiths persuaded the H.E.C. to consider the possibility of using a tunnel borer for the proposed Poatina project.

In 1959 the H.E.C. instructed their Mr. Tulloch (Engineer for Civil Construction) to proceed to America to investigate the subject, as well as other matters of interest. Under the able auspices of A. H. Ayers who was hired as a consultant, Tulloch saw several machines of various types in America. These included some coal-mining units, soft-ground machines (i.e. mechanized shields, some medium-hard ground units and the Robbins unit which was being used by Morrison-Knudsen, Kiewit, and Johnson on the Oahe dam project. Unfortunately, the Robbins machine was not operating at the time Tulloch inspected it as it was jammed in a shattered fault zone. Large boulders had intruded at the face and prevented rotation of the cutter head. Moreover, the 'Mole', as it was christened, could not be withdrawn as it was acting as a breast wall. However, Tulloch was able to view the areas which had been cut by the machine prior to its entry into the fault zone and he also spoke to various personnel on the job. Their opinion at the time was that when approaching a faulted area or anything suspiciously like one, it was advisable to keep the Mole going as fast as possible, in the hope that it would break through before becoming stuck. Apart from the fault areas which had delayed progress by about two weeks, the Mole had recorded penetration rates of approximately 1.52 m (5 ft) per hour in good ground.

Tulloch also made an extended journey to Europe in order to inspect what was available on the Continent. There he saw a Bade (German) machine. The Bade machine was operating in a hard red clay in the salt-mines. This consisted of crystals of salts in a clay matrix. The machine cutter head carried fixed cutters only and was built in three concentric sections to reduce torque and control the cutter speed variation. The middle ring rotated in the opposite direction from the centre and outer rings. Like the Robbins unit, grippers and hydraulic thrust jacks were used. According to Tulloch, the machine was also designed for steering round small radius curves.

The tour served to convince Tulloch of the advantages to be gained from a machine-bored tunnel and the H.E.C. was advised accordingly.

The H.E.C. was then faced with the problem of which machine to purchase. It appeared from information available that there was, on the whole, little to choose between one manufacturer's hard-ground machine and another's. The decision, then, hinged upon the cutters, i.e. either the Robbins Company's patented disc cutters (which came with the machine) or a machine from one of the other manufacturers, using drag teeth bits. Milled tooth or carbide button insert roller cutters were also considered, but these were rejected because, apart from their high cost, they were not considered to be suitable for the type of material anticipated in the Poatina tunnel.

Initially David B. Sugden (H.E.C. Chief Plant Engineer) had been called upon to test the capabilities of the disc cutters. To do this he had designed a miniature machine.

The cutter head of the 51 cm (20 in) diameter borer was made in the H.E.C.'s plant workshop at Moonah, Hobart, Tasmania, and it was then assembled and tested at the portal of the tailrace tunnel at Poatina. It achieved advance rates of 8.9 cm (3½ in) per minute with baby disc cutters, and 21.6 cm (8½ in) per minute with drag bits (fixed cutters).

As the performance of the discs was satisfactory the H.E.C. decided to purchase the Robbins unit, confident that provided their engineers were able to influence the design work of the machine during the initial stage they could handle any problems arising from the operating side of the tunnelling machine.

When the unit, which was built in the Pacific Car and Foundry Company's workshop in Seattle was ready, H.E.C. personnel travelled to America to see the machine being put through its paces in the factory, before accepting delivery.

It was during these preliminary trials that an amusing incident occurred. Sugden, his intimate knowledge of hydraulics* perhaps giving forewarning of impending disaster, stepped well back and out of the way of the hydraulic cylinders which were being tested. His intuitive feelings were obviously well founded as the cylinder suddenly sprang a leak and the entire party — with the exception of Sugden — were liberally sprayed with oil. The faulty cylinders were immediately returned to the manufacturers for the ends to be strengthened.

Basically the 4.9 m (16 ft 1 in) diameter machine (Model 161-108) was the same as that used on the Toronto project, but the steering system had been modified so that it was more positive and was not dependent on flexure of the main frame as had been the case with earlier machines. When it arrived ex factory it was equipped with 34 rolling disc cutters and a small central panel of 32 fixed chisel cutters which were designed to deal with the central section of rock too small in radius to permit the use of rolling cutters. The final 5.1 cm (2 in) rib of rock which remained would then be broken off by an angle plate welded to the centre of the cutter head. Access was provided by a 91 × 46 cm (3 ft × 18 in) rectangular door positioned just off the centre of the cutter head. A diaphragm or dust shield made from light steel segments was mounted around the cutter-head supports. It was sealed around the periphery with rubber sealing strips.

The machine was 12.8 m (41 ft) long and, in the event of its jamming, the single head could be reversed. Six 75 kW (100 hp) motors supplied power to the head which revolved at approximately 5.3 rpm.

Four buckets on the periphery were designed to lift the cut material on to the 16 cm (30 in)

*The heavy-duty hydraulic drive system now used by the Robbins Company was designed by D. B. Sugden. It received the Prince Philip Prize Merit Award in 1973.

236

wide conveyor belt which travelled at 64 m (210 ft) per minute and deposit it on to an auxiliary conveyor behind the machine.

Steering was effected by hydraulic jacks in a gripper shoe assembly which allowed for both horizontal and vertical movement of the machine frame relative to the grippers. Additional correction of the steering was available through the front support and side steering rams.

The gripper assembly was mounted in guides which ran between the two main beams. These guides allowed the grippers to slide backwards and forwards relative to the main frame during the propulsion of the machine and the retraction of the gripper. The main propulsion jacks were linked directly with the gripper shoes, thus preventing the gripper assembly itself from becoming damaged by heavy thrust loads. The grippers consisted of opposing pairs of box members telescopically mounted on a heavy box beam and supporting heavy spring-centred universally mounted gripper pads or shoes at their outer ends. The gripper assembly was mounted on a central trunnion in a saddle frame which ran along the guideways on the main jumbo box beams. The pads were thrust against the tunnel walls by means of four 20 cm (8 in) diameter hydraulic cylinders which expanded the two telescopic members of the assembly.

In addition a pair of hydraulic cylinders was connected to the saddle and to one shoe assembly to enable the saddle to move sideways on the gripper assembly, while the trunnion itself was also free to slide a limited distance vertically in the saddle. Vertical and axial adjustments were handled by four hydraulic rams arranged in pairs between the saddle and the gripper assembly.

Supporting the cutter-head system at the front and the power-pack system at the rear were two main box beams. Two diaphragms, one at the front and one at the rear were fabricated with and therefore formed an integral part of the box-beam structure.

Although a Bud-type ring beam jig (capable of assembling the four segments of a ring beam from the top of the machine where it was carried on brackets bolted to the cutter-head support) was supplied with the machine, it was never used at Poatina.

Work on the Headrace tunnel was already under way using conventional methods, but an unexpected rise in the base of the dolerite penetrated brought incompetent sedimentary rocks into the tunnel line. Conventional drilling and blasting methods caused extensive overbreak in this type of material necessitating heavy support. Moreover, drilling became a dangerous procedure due to failure of the face. Work was therefore stopped pending the arrival of the Mole.

The machine was delivered to Tasmania on 18 March 1961 and it was assembled at the Headrace tunnel portal. The Poatina power development scheme in northern Tasmania was divided into three separate lengths of machine-bored tunnel namely:

First length of Headrace tunnel — 180 m (600 ft).
Tailrace tunnel — 4450 m (14,600 ft) and
Second length of Headrace tunnel — 2500 m (8300 ft).

(These were excavated from 17 April 1961 to 2 July 1961; 31 August 1961 to 25 September 1962, and 12 December 1962 to 28 May 1963, respectively.)

The T.B.M. operated as follows. The machine was advanced until its cutter head was within a few centimetres of the face. Then, with its weight resting on the rear support shoes and the front and side steering shoes, the forward gripper assemblies were positioned. Alignment of the machine in the tunnel bore was then effected by the rear support shoes, and the gripper pads were extended against the walls. The rear support shoes were then retracted. With its weight supported by the front side and vertical steer shoes and the gripper assembly, the conveyors and the main head drive were started. The cutter head was moved forward by the main hydraulic propulsion rams until it was resting against the face. As the cutter head turned the rams slowly increased their forward pressure until the cutters began biting into the face. The cut material dropped to the invert where it was picked up by the four buckets fitted to the periphery of the cutter head. It was then lifted until the bucket passed over the hole at the top of the machine where it was dropped on to the Mole conveyor belt and carried to the Dixon conveyor at the rear.

The rams continued to extend themselves until they reached the limit of their stroke, which was about 4 ft (120 cm). At this stage the cutter head and the conveyors were stopped. The rear support shoes were lowered to the ground, thus relieving the gripper assemblies of the machine's weight so that they could be retracted and moved forward into their new position on the tunnel wall, ready for the next boring cycle.

Of interest was the fact that the amount of spoil produced by each cutting stroke could be carried away in a single train load. Repositioning the grippers (after a cutting cycle) was completed in approximately 10 minutes, and the same amount of time was required to switch the loaded train for the empty one.

The Headrace tunnel

Almost immediately after the Mole commenced work in the Headrace tunnel on 17 April 1961 problems were encountered. Briefly these included damaged seals caused by excessive water. This later led to main bearing failure.

The dust shield was damaged and the buckets were blocked by rocks which fell from the crown and face of the tunnel. Extensive cavities formed in the roof and walls and these holes needed large amounts of packing. In addition, the provision of adequate support in these regions proved difficult and time-consuming.

Although silica dust, a serious problem which reared its ugly head in the very dry rocks, was reduced somewhat by water sprays, tests indicated that smaller particles still remained suspended in the air and masks were necessary in bad areas. To further reduce this nuisance the operator's cabin was supplied with filtered air at the rate of 85 cu m/min (3000 cu ft/min) and air was drawn from the face at 110 cu m/min (4000 cu ft/min). The main tunnel exhaust vent was modified to allow it to vent right at the face through the conveyors which were enclosed, and an additional fan and a telescopic section of duct were added to the Dixon conveyor.

The Mole conveyor was made retractable to facilitate belt cleaning by mounting it on rollers. Finally, a dust separator was installed at the rear end of the Dixon conveyor where the trucks were loaded.

The Tailrace tunnel

Work on the Tailrace tunnel through Permian mudstones was started on 31 August 1961. The material in the Tailrace tunnel proved harder than that encountered in the Headrace, causing rapid damage to the central panel of fixed cutters. These were replaced by a new cutter block containing 11 large cutters made by Titan in Newcastle to H.E.C. specifications.

The fan which had been installed on the Dixon conveyor tended to recirculate the dust, so it was removed in order to keep the telescopic duct below atmospheric pressure. It was found to be impracticable to seal this duct section against a positive pressure. Later, as the tunnel progressed, axial-flow fans which were positioned to keep the entire air duct below atmospheric pressure were installed.

To facilitate floor clean-up which had not proved satisfactory, the steel plate segments between the buckets were removed and a 'bulldozer' blade (shaped to conform with the tunnel profile) was fitted to the head and front shoe of the T.B.M.

Bucket blockage in wet ground was a continual problem despite the installation of the retractable conveyor rollers, as the rollers themselves also blocked.

During January 1962 overheating in the area around the main bearing pointed to extensive play in the bearing caused, apparently, by grit which had been flushed past the head seals by excessive water at the face. This in turn prevented free lubrication of the rollers by the bearing grease which had hardened.

During the overhaul improved seals which proved satisfactory were fitted, and the conveyor clearance was modified so that retraction of the rollers was no longer a problem.

A major source of concern arose when the front steering shoes fell into cavities formed by pieces of rock which had dropped from the tunnel sides. As the Mole continued to advance, the entire thrust of the propulsion rams would then be accepted by these shoes which could not be adequately strengthened to carry this tremendous load. To overcome the problem a special cut-out device was fitted to the hydraulic system which automatically reduced the pressure to the propulsion jacks to a safe level, if the current being absorbed by the main drive motors fell below a predetermined value.

After finishing the Tailrace tunnel, several significant modifications were made to the machine prior to its third run in the remaining section of the tunnel.

The front support shoe, steering shoes, and bulldozer blade were replaced by a large articulated front shoe extending across the invert.

Another important modification was the replacement of the 4 original buckets supplied by the manufacturers with 16 smaller buckets. Each bucket carried a disc cutter, and when the buckets were increased in number, the number of disc cutters on the periphery of the head was increased from 12 to 16.

The six motors driving the head were rewound so that the speed of the cutter head was reduced from 5.3 rpm to 3.6 rpm to reduce gearbox troubles.

Cutters

Various experiments were carried out to determine the best design of disc cutter, and as many factors influencing cutter life as was possible, were taken into account during this work. These included the type of material used, the relative cutting angles of the discs, the position and angle of the cutter in relation to both the cutter head and the tunnel face, etc. The edges of some disc cutters were fitted with tungsten carbide inserts, but the time available proved to be too short to evaluate their true worth when comparing cutter life with cost.

Although cutter life on the whole varied considerably, depending upon the type of material being penetrated, the average life of a roller cutter in the Tailrace tunnel was about 181 m (595 ft) of tunnel. But it should be appreciated that this figure was considerably reduced when hard limestone concretions of 90 cm (3 ft) diameter were encountered. These damaged all cutters that came in contact with them. Strangely enough, although several pebble bands of hard quartz were met with in

the latter stages of the Tailrace tunnel, these did not affect cutter wear as much as had been anticipated.

Because the gauge and outer cutters travelled further and faster than the inner cutters they naturally wore more rapidly than the latter. By and large it was found that cutter life was extended if the angle of the disc cutter at the edge was flattened more as the rock increased in hardness. However, when the angle became too flat and the cutter thus presented too wide a cutting edge to the face it caused bearing failures. The originally supplied sharp serrated-edged roller was found to be effective in weak sandstone.

After the various modifications had been incorporated, the machine completed the tunnel without further problems.

During the boring of the second section of the Headrace tunnel a world tunnelling record of 229 m (751 ft) in a six-day week was achieved by the crews, and from 2 January 1963, 1 mile (1.6 km) of tunnel was driven in less than 11 weeks — also considered to be a world record. In fact the tunnelling rate of 229 m/week (751 ft/week) was claimed by the H.E.C. as being a new world record for any size or shape of tunnel, in any rock conditions.

The maximum boring rate was 9.3 cm/min (4 in/min) and the best shift 18.2 m (67 ft).

The Rhyndaston Tunnel[2]

After completing the Poatina tunnels the Mole was later used by the Tasmanian Transport Commission to enlarge what was reputed to be one of the smallest rail tunnels in the world (Figures 213 and 214). As such the tunnel was too small to handle the carriage of container traffic and other large loads over the main north-south route of the Tasmanian Government Railways.

The Mole increased the size of the tunnel from 3.20 m (10 ft 6 in) wide and 4 m (13 ft) high to 4.93 m (16 ft 2 in) wide and 5.18 m (17 ft) high. To do the job the Mole was only permitted to work during the weekends from 05.00 on Saturday to 08.00 on Monday when normal rail traffic was at a standstill.

A special skid plate (Figure 215) to catch the cut material was designed by the H.E.C. staff.

Figure 213. Rhyndaston tunnel — reputed to be the smallest railway tunnel in the world before being bored to a larger diameter with the Robbins 161-108 Mole. (Courtesy *The Mercury,* Tasmania.)

Figure 214. Robbins Mole Model 161-108. Used at Poatina tunnel. Toppled on to its side when railway siding collapsed under its weight just prior to its being used to bore the Rhyndaston tunnel, Tasmania. (Courtesy *The Mercury* Tasmania, Australia.)

240

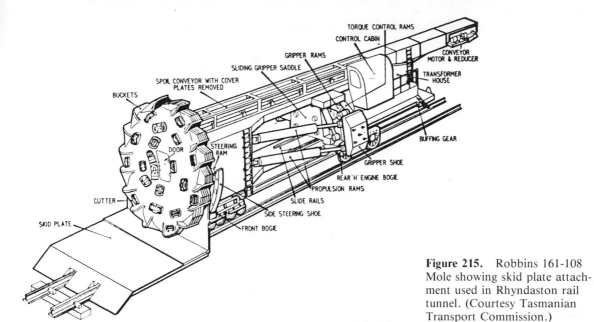

Figure 215. Robbins 161-108 Mole showing skid plate attachment used in Rhyndaston rail tunnel. (Courtesy Tasmanian Transport Commission.)

It was mounted below the cutter head and extended forward covering the rails ahead of the machine for a distance of 4.6 m (15 ft). So far as is known this was a unique arrangement which had not been used previously on any other T.B.M.

During the working week the Mole was stored on a siding at the northern end of the tunnel, the skid plate and other support equipment was stored on a short siding which had been constructed for the purpose at the southern end of the tunnel.

It took the Mole 11 weekends to hole through the original clay-brick-lined tunnel. Behind the bricks was sandstone, deposited there during the early Triassic period. Initially, during the boring process, this was left unsupported, but rock bolts were installed after an enterprising 3 t sandstone boulder from the tunnel crown hitched a ride in the truck of a goods train while it was passing through the tunnel. The 'stowaway' was discovered when the train reached the next station.

The Tasmanian T.B.M. was the first to incorporate a full floating gripper and propulsion unit which was articulated, thus permitting the operator to steer the machine during the boring phase, while the grippers remained expanded against the tunnel walls. In addition it was the first time discs fitted with

permanently sealed oil-lubricated bearings were used.

It is of interest to note that after the 161 machine, the Robbins Company built a 7 ft (2.1 m) diameter T.B.M. featuring an automatic walking gripper mechanism which, if it had been successful, would have allowed for truly continuous operation of the machine. However, the system was too complex to function efficiently under the difficult conditions encountered in a tunnel. It was subsequently replaced by an articulated Tasmanian-type gripper system described in the preceding pages. After the change the unit operated satisfactorily and completed three more tunnels.

The Mangla Dam Mole Robbins Machine No. 371-110[3]

The Mangla dam machine was basically similar to previous Robbins machines (Figures 216 and 217).

When specifications for the design of the machine had originally been laid down they had been based on all the information which was available at the time.

The Siwalik materials encountered at Mangla were generally of low compressive strength, and it was known that layers of hard rock with

Figure 216. Model 371-110. This 11.18 m (36 ft 8 in) diameter unit designed by The Robbins Company was used by Guy F. Atkinson and Partners for the Mangla dam project. (Courtesy The Robbins Company.)

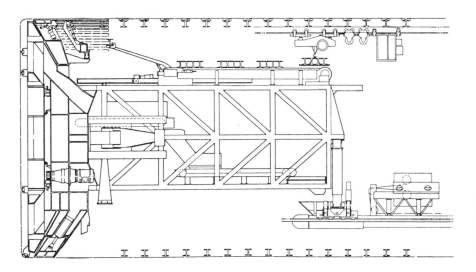

Figure 217. Model 371-110 — drawing — Mangla dam machine. (Courtesy The Robbins Company.)

compressive strengths up to 40 MPa (6000 lb/in²) existed within the softer strata. Because of these streaks, disc cutters were selected for the head of the machine.

However, when the giant 11.2 m (36 ft 8 in) diameter Mangla dam unit (Figure 216) first went to work for the contractors, Guy F. Atkinson and Partners, it was found that the majority of the material being penetrated fell into the weaker compressive strength range. As the ground was not entirely self-supporting, a considerable quantity of it dropped into the 30 cm (12 in) space which existed between the leading edge of the disc cutters and the solid diaphragm of the cutter head. Some of the material from the fallouts remained in large pieces which blocked the buckets and stalled the head. Most jams could, however, be cleared by the operator because the rotation of the head could be reversed.

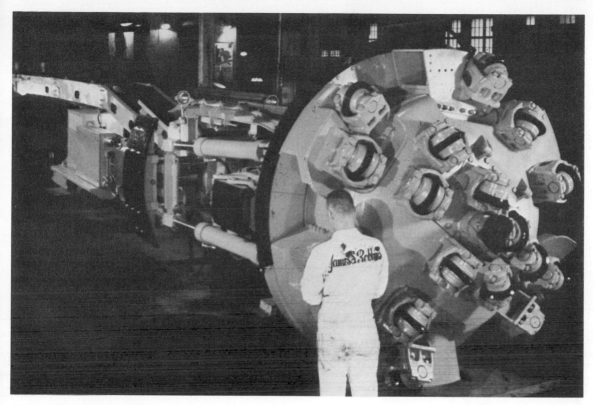

Figure 218. Robbins Model 81-113. This 2.59 m (8 ft 6 in) diameter machine was used in the Sooke lake-Goldstream water tunnel, Victoria. B.C. (Courtesy The Robbins Company.)

Modifications were made to the bucket system which enabled it to work more efficiently but, nevertheless, progress was so slow that a decision to abandon the machine was very seriously considered by the management. Finally it was decided to continue boring the first tunnel with the unit as it was, and then carry out extensive modifications after the entire range of material to be encountered could be observed. This also allowed ample time for engineering and the procurement of whatever additional equipment, etc. was necessary for the overhaul job.

Work on the modifications was begun as soon as the first tunnel was holed through. *Inter alia,* these changes consisted primarily of replacing all the disc cutters with fixed cutters which could be changed or adjusted from the rear of the head. An additional set of forward thrust jacks was added so that it was not necessary to use extreme hydraulic pressures for propulsion purposes. Finally, the sliding front

shoes were enlarged to enable the machine to cope more efficiently with the material at the bottom of the tunnel, which turned out to be softer than had been expected. As a result of this work and also because of improved mechanical supervision of the unit, the fifth tunnel was holed through in one month, compared with over three months needed to complete the first tunnel.

The Mersey Tunnel[4]

During the driving of the first Mersey tunnel (designed for rail traffic) a 2.24 m (7 ft 4 in) diameter Beaumont/English T.B.M. (which commenced work there in March 1883) was used for the construction of the drainage tunnels and the ventilation tunnel. Work on the second tunnel (13.41 m (44 ft) in diameter, and known as the Queensway road tunnel) was commenced, using conventional methods of drilling and blasting, etc. in 1925. The

243

Figure 219. Robbins T.B.M. Model 111-117. Used initially in Baden sewer tunnel in Switzerland and then later to bore the pilot tunnel for the Wirth 'Enlarging machine' at Sonnenberg, Lucerne, Switzerland. (Courtesy The Robbins Company.)

Figure 220. The 'Oso' T.B.M. Model 104-121A. Both the 'Oso' T.B.M. and its mate the 'Blanco' T.B.M. broke existing T.B.M. records in 1966-67, their rate of advance being largely governed by logistical problems. (Courtesy The Robbins Company.)

244

contractor was Edmund Nuttall Ltd. Work on the second (Kingsway) road tunnel project was started in 1965 with the driving of a pilot tunnel by drilling and blasting (contractor: Marples Ridgeway). However, it was not until November 1967 that the Robbins Mole was ready to go to work there. In fact, it was basically the same machine originally used by Guy F. Atkinson and Partners on the Mangla dam project. The main tunnel contracts for tunnels 2A and 2B were let to a consortium of Nuttall/Atkinson & Co.* and Sir Alfred McAlpine & Co. Ltd. — McAlpine attending to the main part of the Wallasey approach, while Nuttall/Atkinson were responsible for the tunnels and ancillary works and a small section of the approach. Mott, Hay, and Anderson were appointed as consulting engineers.

Apart from a short length of hand-driven tunnel at the Liverpool end, it was proposed to drive the remainder of the tunnel by machine.

The pilot tunnel had confirmed that the machine would be working in middle and upper Bunter sandstone in wet conditions under the river, and allowed advanced grouting to be done, while the proposed hand-section would rise above the rock and penetrate boulder clay. Cast-iron segmental rings were to be used to support the hand-driven section and precast concrete segmental rings the main machine-driven section.

During the driving of the 2A tunnel the highly fissured rock was supported in the crown in advance of the Mole by grouting bamboo rods into holes drilled radially from the pilot tunnel, but this had limited success. Bamboo rods were used because they could be cut through afterwards without damage to the T.B.M.

As the Mersey tunnel was to have a 9.62 m (31 ft 7 in) diameter segment-lined bore, the diameter of the Mangla cutter head was reduced to 10.33 m (33 ft 11 in). In addition a 1.68 m (5 ft 6 in) diameter centrally positioned snout which extended about 1.5 m (5 ft) in front of the machine, providing access to the pilot tunnel, was fitted to the cutter head. The propulsion system was modified and the

T.B.M. was fitted with segment handling and erection equipment.*

During the boring of the first section of the 2A tunnel, fallouts from the crown began to present a problem. From Ring No. 53 to Ring No. 83 the condition worsened and the overbreak problem was aggravated by the ingress of water which increased as the under-river section was approached. After Ring No. 83 the condition improved somewhat, although occasional fallouts from the crown, together with the slurry arising from the sandstone and the huge ingress of water, were to present a continuing problem in both tunnels, but more so in the 2A drive. It turned out that the extensive ground treatment for 2A which had been carried out from the pilot tunnel was largely ineffective, and none was done for the 2B tunnel. Sometimes rockfalls occurred in the pilot tunnel in front of the Mole and, while most of the largest pieces could be broken by workmen, smaller pieces got through, blocking and damaging the muck chute and conveyors. This in turn led to stoppages when the Mole was halted to clear the blockage and effect repairs.

When the pilot tunnel was driven, steel arch ribs and timber lagging supports were erected and, of course, these had to be removed before the Mole could penetrate the area. But sometimes these supports were trapped by falling material and then extraction ahead of the Mole became a lengthy and dangerous undertaking.

There was considerable concern that the main bearing of the Mole might become damaged by the downward force of broken rock** which fell on to its canopy or roof shield and also on the cutter head, causing torque resistance, particularly at the commencement of an excavating cycle.

The 4.88 m (16 ft) diameter main bearing

---

*The redesign of the Mangla machine for work in the Mersey tunnel was handled by Guy F. Atkinson Company's engineering staff and the modifications were carried out by Atkinson's partner, Edmund Nuttall Ltd.

**The machine at this point was passing through faulted ground, which was the main cause of the fears concerning the broken rock. In fact all the work carried out in connection with the construction of the twin Mersey tunnels had the problem of dealing with a mixture of rock blocks and slurry.

---

*Edmund Nuttall Ltd. and Guy F. Atkinson Company.

Figure 221. The Robbins Company Model 181-122 articulated 'White Pine copper mine' T.B.M. used in Michigan during 1968-72. Diameter 5.49 m (18 ft). (Courtesy The Robbins Company.)

showed signs of serious wear early in February 1969, shortly after work had been resumed following a long delay while precautionary work was executed in a fault zone. Boring operations with the damaged bearing continued, however, until 20 February 1969 when, at Ring No. 776, the bearing failed completely and it became necessary to carry out an immediate emergency replacement job. It was discovered that failure was due to the continuous deluge of abrasive slurry wearing away the paths of the labyrinth of seals, which then allowed sand to penetrate the bearing cavity. At this stage the Mole was under the river approximately half-way between Liverpool and Wallasey, having advanced some 954 m (3130 ft).

Normally a job such as a main bearing change would be carried out by extracting the machine from the tunnel. However, long before the failure occurred, plans were under way to make the replacement underground, it being considered impracticable to remove the entire machine. So far as the author* is aware, this constituted the second time* a main bearing was changed in an actual tunnel and the various elaborate steps needed to effect the change are fully described in Messrs. McKenzie and Dodds' Paper No. 7481.[4] (A main bearing change was also made during the driving of the 2B tunnel.)

*A main bearing was changed on a Robbins 121-116 machine in the Azotea water tunnel during its construction in 1965.

246

Periodically the cutter head became jammed and at these times it was found to be exceedingly difficult to gain access to the cutters for maintenance purposes.

In one section where the invert consisted mainly of broken rock interspersed with silt and sand combined with water, a soft bed formed. The tremendous weight of broken material above the Mole forced the machine to sink about 7.6 cm (3 in) into the soft bed and all efforts to lift it by jacking failed. Eventually excavation was proceeded with, despite the difference in levels, the operator attempting to steer a gradual upward path as he broke out of the bad ground. As may be imagined, the erection of rings within the specified tolerance became somewhat of a nightmare for the contractor. The rock mass at the crown further aggravated the situation when distortion of those rings already in position but not yet grouted, occurred.

At one point the tunnel passed within 8.2 m (27 ft) of the surface bedrock below the river. Salt water which seeped into each of the pilot tunnels from the river through joints and faults and was collected in a sump at the lowest point, was pumped out at the rate of approximately 150,000 gal/hour (682,500 litres/hour).

White Pine Copper Company

The Robbins machine model 181-122 (Figures 221 and 222) was built for the White Pine Copper Company for two reasons. Firstly the company needed a 2438 m (8000 ft) ventilation and access tunnel between two of their mines and, secondly, they wished to determine the application of a boring machine as a developmental and production tool.

The White Pine Copper machine was radically different from previous models in that it was articulated to allow it to negotiate curves of 30 m (100 ft) radius, both in the horizontal and the vertical planes. This was important if the machine was going to be used as a future mining tool. The 5.49 m (18 ft) diameter machine carried a domed head fitted with 47 discs and one tri-disc.

The forward portion of the machine consisted mainly of the head and gripper system, while the rear section carried the power pack and the conveyors, etc.

The head power was 895 kW (1200 hp) and the propulsion rams were capable of exerting 7031 kN (1,580,000 lb) of thrust. It was 5.4 m (18 ft) long, weighed 227 t, and was fitted with two hydraulically operated high-power rotary/percussive roof pinner drills.

The company tried the machine in two types of strata — highly abrasive sandstone (55 per cent quartz) having a compressive strength of 120 MPa (17,500 lb/in^2) and a laminated shale formation comprising different structural zones ranging in compressive strength from 138 to 195 MPa (20,000 to 28,000 lb/in^2). (The shale had an average quartz content of 25 per cent and the country was highly stressed and intersected by numerous fault zones.)

Although it completed boring the requisite 2438 m (8000 ft) of tunnel, the company decided that for economic reasons the machine was not suitable as a permanent mining tool.

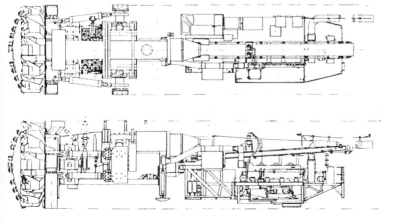

Figure 222. Plan view (top) and side elevation of Robbins T.B.M. Model 181-122 White Pine copper mine machine. (Courtesy The Robbins Company.)

Figure 223. The Robbins Company T.B.M. Model 112-124 used in the Reussport highway tunnel, Lucerne, Switzerland during 1968-69. Diameter 3.30 m (10 ft 10 in). (Courtesy The Robbins Company.)

Figure 224. Robbins Mole — Model 162-130/131. Used at Tajo-Segura water project, S.E. of Madrid, Spain. (Courtesy The Robbins Company.)

Conclusion

Since its humble beginnings in 1953 the Robbins Company has gone from strength to strength, pioneering many innovations (including that of the single-edge disc cutter — now the world standard) and chalking up an impressive number of machines to its credit on the way.* More than this they have proved to the world and to the tunnelling industry in particular that the full-face tunnel borer is a viable tool in medium and hard-rock ground conditions, depending upon ground quality, tunnel length, and diameter.

Table 7 gives a list of world records attained by Robbins tunnel borers.

The Hughes Tool Company

After the introduction of rotary flush drilling, the next major innovation in vertical drilling came in 1909 when Howard R. Hughes, founder of the Hughes Tool Company, invented the rolling-cone type of rock bit. Its

---

*During the period 1953-80 the Robbins Company produced 114 tunnelling machines which were used on 233 projects in various parts of the world, boring a total of 558 miles (898 km) of tunnel.

first successful use was in an oil well at Goose Creek, Texas. Shortly afterwards Hughes acquired a small property in downtown Houston where plant was set up in a 12.2 × 18.3 m (40 × 60 ft) machine shop. From this humble beginning the Hughes Tool Company has grown and expanded, until today the main plant occupies approximately 34 hectares (84 acres) containing 60 buildings with 12 hectares (30 acres) of floor space, situated on the east side of Houston.

For over 64 years Hughes have continued to lead the field with major advances in the design and manufacture of cone-type rock bits. (Figure 581 shows the evolution of Hughes rock bits and rolling rock cutters from 1909 to 1965.)

In 1946 the Hughes Tool Company in conjunction with the mining industry discussed the feasibility of extending Hughes's idea to other activities, most particularly the excavation of large-diameter shafts. Hitherto drilling for this purpose had been confined to blast-holes.

Theory was put into practice and in 1953 the first tools and bits specially designed to bore large-diameter shafts were marketed by the Hughes Tool Company.

Table 7 World records set and broken by successive Robbins tunnel borers: advances in rock

Model	Size m	(ft in)	Year	Tunnel	Best shift m	(ft)	Best day m	(ft)	Best week (6 day) m	(ft)	Best month m	(ft)
930-102	8	(26 3)	1954	Oahe dam, Pierre, South Dakota	18.6	(61)	42.6	(140)	193.5	(635)		
261-107-1	7.82	(25 8)	1961	Saskatchewan river dam, Canada			62.3	(204.5)	203	(666)		
161-108	4.92	(16 2)	1963	Poatina, Tasmania, Australia	18.2	(60)			229	(751)	767	(2517)
121-116	4	(13 3)	1965	Azotea, Chama, New Mexico	29.9	(98)	73.6	(241.5)			1243	(4077)
104-120	2.99	(9 11)	1966	Blanco, Colorado	41	(134.8)	114.3	(375)	533	(1748)	2046	(6713)
104-121A	3.09	(10 2)	1966	Oso, Colorado	47.5	(156)	127.7	(419)	581	(1905)	2088	(6849)

Total *in situ* volume of rock excavated and removed in one day

Model	Size m	(ft in)	Year	Tunnel	Amount cu. m (cu. yd)
353-197	10.8	(35 4)	1979	Chicago Underflow, 59th to Central Chicago, Illinois	3261 (4265)

Figure 225. Robbins Mole — Model 125-135. Used at Nchanga water diversion and mining, Zambia. (Courtesy The Robbins Company.)

Figure 226. Robbins Mole — Model 133-146-1. Used for water tunnel and pipe head at Potts Hill, Sydney, Australia. (Courtesy The Robbins Company.)

Figure 227. Robbins Mole — Model 133-146-1. Cutterhead (Courtesy The Robbins Company.)

That same year the Robbins Company of Seattle made history by constructing their first rock tunnel borer carrying both discs and conventional fixed drag bit cutters (see The Robbins story).

Hughes Tool Company followed the progress made by the Robbins Company during the period 1953 to 1957 with interest and in 1958 decided to build a small 1.07 m (40 in) diameter tunnel boring machine for ·experimental purposes.[5] The idea being to develop further the basic information already gained in this field. (It is of interest to note, however, that this was not the first attempt by the Hughes Tool Company in the area of tunnel boring machines. H. R. Hughes senior built a horizontal boring machine for military purposes during the latter period of the First World War. Unfortunately the war's end halted further production of this machine and no efforts were made to sell it commercially. The one and only unit built is now on display in the Hughes Company Museum) (Figure 382).

The 1.07 m (40 in) trial machine which was manufactured by the Hugh B. Williams Company (a Hughes Tool Company subsidiary), carried milled tooth and carbide rollers and had a thrust of 41,000 kg (90,000 lb). Total power was 78 kW (105 hp). The machine's average penetration rate when put to work boring through granite, limestone, sandstone, and chalk with compressive strengths ranging from 34 to 145 MPa (5000 to 21,000 lb/in^2) was 0.52, 1.07, 3.66, 3.96 and 6.10 m (1.7, 3.5, 12, 13 and 20 ft)/hour respectively. This 1.07 m (40 in) unit is reputed to have been the first tunnelling machine in the world to cut granite successfully. It was used initially in a number of locations in the United States, including the Texas granite quarry. Later the machine was shipped to England where it was tested by the National Coal Board at a limestone quarry. The machine was subsequently returned to the United States.

Gilsonite[6-9]

Gilsonite or uintaite, named after S. H. Gilson, is an asphalt — a bituminous material which is defined as a mixture of hydrocarbons of natural or pyrogenous origin. It is completely soluble in carbon disulphide. Asphalt (bitumen)

251

Figure 228. Robbins T.B.M. Model 114-163 used in the West Driefontein gold-mines in South Africa — fitted with rotating grippers to enable it to negotiate cross-tunnels in the mine. For a description of rotating gripper system see section on Subterranean Equipment Company. (Courtesy The Robbins Company.)

or tar may be produced by refining petroleum, but asphalt may also be obtained by mining the material in its natural state. Though minor deposits of a gilsonite-like material have been found in Wheeler and Crook counties of Oregon, the only commercially important deposits of gilsonite in the world are located in the Uintah basin. The gilsonite veins in Utah and Colorado (Rio Blanco County) vary from a few inches in width to as much as 5.5 m (18 ft). Running in a north-west-south-east direction the veins are nearly vertical and many extend downward as far as 240-270 m (800-900 ft). Though faults and inclusions of rock do occur now and then, most of the rock walls of the veins are quite smooth and regular. Gilsonite is friable and breaks into a fine penetrating dust in mining or crushing operations. It is black in colour and has a high lustrous surface.

Its origin, particularly its virtually unique occurrence in the Uintah basin has been the subject of a considerable amount of discussion, the general opinion being that large oil shale deposits in that area were converted into free-flowing masses of gilsonite by tremendous quantities of heat and/or pressure. This forced the gilsonite into small cracks and fissures or pockets in the ground. It has a specific gravity of 1.03 to 1.09, a softening point between 121°C and 260°C (250°F and 500°F) and a high dielectric strength.

Initially gilsonite was used in varnishes and paints and applied to horse-drawn carriages and vehicles, it was also added in a 50/50 mixture to liquid asphalt and used for roads as well as being combined with rubber to form buggy and coach tyres. Later, however, more than 100 patents were registered covering its use in such commodities as paint, varnish, electrical insulators, moulding compounds, paving material, vulcanizing, printing ink, etc.

Individual veins of gilsonite have been given a wide variety of names such as 'Little Emma', 'Black Dragon', 'Pride of the West', 'Bonanza', 'Rainbow', and 'Cowboy'.

Originally gilsonite ore was mined by pick and shovel, a method of recovery similar to that of early coal-mining, but with an important

Figure 229. Komatsu/Robbins rock tunnelling machine — Model TM 350G. (Courtesy Komatsu, Ltd.)

difference. The gilsonite could be extracted from the near-vertical slopes on a rising plane which allowed the dislodged pieces to roll to the bottom of the slope. These pieces were then collected and deposited in large burlap bags which had a holding capacity of 91 kg (200 lb). An average of 2 ton of gilsonite per day was, nevertheless, recovered by the miner, despite the primitive methods of collection.

When compressed-air drills came into general use, these new machines were enthusiastically tried by the gilsonite miners with unfortunate and frequently fatal results. This was due to the explosive character of the dust. As a consequence alternative mining methods were sought to speed recovery of the ore. The answer came with the introduction of hydraulic mining.

This method of mining, which is particularly suited to gilsonite, had reached a high state of development by 1961 when vertical reaming was introduced at the American Gilsonite Company's Cowboy vein in Utah. Blocks of ore approximately 1500-3000 m (5000-10,000

ft) long were laid out and then pre-drilled with a number of vertical holes 15 cm (6 in) in diameter, spaced approximately 6.10 m (20 ft) apart. The holes were drilled along the centre-line of the vein and intersected a collection drift at the bottom of the mine. A rotary rig and mast supported by a skid-mounted platform would then be moved over a hole. When the platform was in position with its bearing surface supported by firm ground on either side of the vein, 1.7 cm (4⅝ in) drill pipe would be lowered into the hole. The bit end of the drill pipe had two jet nozzles which allowed the pre-drilled 15.2 cm (6 in) diameter holes to be hydraulically reamed from the collection drift back towards the surface. The resulting slurry and cuttings dropped to the sloping collection drift and were then channelled to the shaft.

Disadvantages of the System

The system was effective theoretically except that nature frequently contrives to 'throw a spanner into the works'. At the 'Cowboy vein',

Figure 230. Komatsu/Robbins tunnelling machine — Model TM480. Daily advance rates as high as 47.9 m (157 ft) per day were attained with this machine. (Courtesy Komatsu, Ltd.)

this 'spanner' took the form of a rock jumble or matrix of boulders and gilsonite, often as much as 60 m (200 ft) thick, located near the bottom of the 210-270 m (700-900 ft) deep vein.

The presence of these boulders, which could not be blasted on account of safety factors, prevented the complete recovery of all ore, hindered the extension of the collection drift at the desired rate, and hampered maintenance of the hydraulic gradient.

When rock inclusions of this sort were encountered it became necessary to raise and then drive subdrifts over these obstacles. Naturally ore extraction near the bottom of the vein in such cases would be difficult and a low yield resulted.

Mechanization[9]

In 1960 the American Gilsonite Company in an effort to solve some of the above problems, bought the 107 cm (40 in) Hughes Tool Company machine, recently returned from its successful trials in the United Kingdom. The following year (1961) the 107 cm cutter head was modified to 137 cm (54 in) to enable the machine to bore tunnel diameters of this size. It was used by the American Gilsonite Company until 1964.

In 1963 the American Gilsonite Company also negotiated with the Hugh B. Williams Company for the design and manufacture of a special large-diameter shaft drill. Basic design parameters called for drilling 152 cm (60 in) diameter shafts to depths of 300 m (1000 ft), and a 163 cm (64 in) diameter machine was built. This machine, which was still in use in 1975, was fitted with a flexible drill stem which enabled the heavily weighted drill bit (carrying 22 convex mounted shaft cutters) to follow the ore vein. Though holes often deviated, the bores were kept moderately straight by skilled operators.

The large-hole shaft drills enabled American Gilsonite to bore 157 cm (62 in) diameter holes to a depth of 270 m (900 ft) in two to three weeks, whereas several months were required to complete a three-compartment production shaft using conventional methods.

These new methods of collection drift advancement and hole drilling have proved economically viable and have solved many of the mine's extraction and ore recovery problems.

Arizona Tunnel

In 1963 Hughes Tool Company built two 203 cm (80 in) diameter tunnelling machines. These were designed specifically for use in a water diversion tunnel in Arizona. The production rates and basic economics of this project were

Figure 231. Komatsu/Robbins tunnelling machine — Model TM450G. Advance rates averaging 42.0 m (138 ft) per day were attained with this machine. (Courtesy Komatsu, Ltd.)

carefully annotated and results proved encouraging.

White Pine Copper Company[10]

Shortly after the completion of the Arizona tunnel the White Pine Copper Company, which is situated in the upper peninsula of Michigan near the south shore of Lake Superior, decided to investigate the potential of driving mine openings by boring. At that time development was normally carried out by conventional drilling and blasting methods. The White Pine Copper Company acquired one of the used 203 cm (80 in) diameter Hughes machines and had it modified to 213 cm (84 in). Compressive strengths of from 120 to 210 MPa (17,000 to 30,000 lb/in²) were encountered in the area of the tests which were conducted in the shale ore body and the underlying sandstone. These tests emphasized one of the major problems facing hard rock tunnel men of today, namely that of dust. It was obvious that counter-measures beyond those needed for conventional methods of drilling and blasting would be required. Though water was initially sprayed on to the face, this proved to be insufficient to cope with the nuisance and a Krebs-Elbair scrubber was installed in the ducting system of the machine. Approximately 99 per cent of the dust was extracted and conditions at the face immediately improved.

Mine development

As many of the mine-development openings in the White Pine Copper Company mine reached gradients of 1 in 5, and as conventional rail-haulage methods for horizontal tunnels could not therefore be used, muck removal facilities in the form of specially designed belt conveyors were also installed at the mine to handle this aspect of the project.

Betti I[5]

The various problems encountered in Arizona and at the White Pine Copper Company mine were carefully collated. It became clear that if the use of hard-rock tunnel borers was to be more widely accepted then these difficulties would have to be overcome. Extensive research on dust control, ventilation, direction control, and other related problems was undertaken in an effort to find the most logical solution. The results of this work culminated in the manufacture of a new 6.1 m (20 ft) diameter machine (Figure 232) which was christened the Betti I. This was the machine used by Fenix and Scission, of Tulsa, Oklahoma on the Navajo project, and a few of the many improvements incorporated into the basic design of the machine are listed below:

1. *Direction control:* The Betti I was capable of being positively steered while boring to

255

Figure 232. 'Betti I' — used by Fenix & Scission on the Navajo project. (Courtesy Hughes Tool Company.)

the very close tolerances required, i.e. within 16 mm (⅝ in) of line and grade. This was accomplished by the introduction of a direction control system which incorporated a patented laser guidance system. By this means the operator was visually able to determine his position in relation to the projected line and grade at all times.

2. *Maintenance:* To minimize maintenance problems caused by the notoriously poor environment experienced in most hardrock tunnels, central automatic lubrication to all bearings was incorporated in the new machine. These precautionary measures were backed by the use of positive air pressure in critical bearing areas.

3. *Dust control:* The operator's cabin was fully air-conditioned as were the electrical and hydraulic cabins, and a negative pressure ventilation system was installed at the head.

In order to reduce power requirements when starting, the Betti I had five 150 kW (200 hp) motors, which could be started separately and then combined to produce power at the face as required. The machine carried milled-tooth rollers and multiple discs, had a forward thrust of 5124 kN (1,400,000 lb) and achieved rates of from 2.4 to 6.1 m/hour (8 to 20 ft/hour) through sandstone ranging in compressive strength from 38 to 48 MPa (5500 to 7000 lb/in²). Total power was 860 kW (1150 hp). In addition the cutter head and pick-up buckets could be hydraulically expanded or retracted thus enabling the operator to bore tunnel diameters of from 5.79 to 6.45 m (19 to 21 ft 2 in).*

Some statistics relating conventional tunnel boring methods to those obtained by the Betti I on the Navajo project are listed in Tables 8 and 9.[5] These figures were obtained by the Hughes Tool Company when a second tunnel of almost identical size and in near-identical ground conditions was driven by conventional means about 610 m (2000 ft) downstream of the machine-bored tunnel.[5]

*This feature was not a success.

256

Table 8. Tunnel performance and related cost information — Navajo project

	Tunnel No. 1, boring machine	Tunnel No. 2, conventional
Specifications		
Length, (ft)	10,000	25,720
Section	Circular	Horseshoe
Size B line (ft-in)	20-7	20-1
Average advance		
Per working day (ft)	47	39
Best day's production (ft)	171	94
Elapsed time to drive 10,000 ft (days)	269	318
Average overall advance rate (ft)	37.2	33.3
Lost-time accidents per 10,000 ft (man-days)	28	525

Table 9. Cost-related information — Navajo project

	Tunnel No. 1, boring machine	Tunnel No. 2, conventional
Average crew size	85	65
Heading crew (men-shift)	11	10
Tunnel supported (%)	44	32
Steel support (lb/linear ft of tunnel)	44	59
Percentage of overbreak		
Non-supported sections (%)	−19	±15
Supported section (%)	2.4	15-20
Clean up time (ft/day)	220	147
Concrete requirements, lining		
Non-supported (cu yd/ft)	1.92	4.8
Supported (cu yd/ft)	3.1	5.7
Capital Investment ($ million)	1.5	0.8

Though at first glance it would appear that the use of the tunnel borer necessitated higher costs for manpower and capital outlay, it will be seen that these expenses were more than compensated for by the smaller quantity of concrete and steel required and the fact that boring and clean-up rates increased.

Summary

Hughes's expertise has also been extended towards two Mitsubishi machines which were built with Hughes's technical assistance and utilized cutters manufactured by Hughes. In addition, a machine built by Alfred Wirth and Company (TB11-300H) and used by Peter Kiewit on the Colorado Rocky Mountain Project, was fitted with a cutter head, material handling, and shield system designed and built by the Hughes Tool Company.

Hughes also produced a 259 cm (102 in) machine (Figure 233) incorporating the latest improvements. This machine was sold to the Reynolds Electrical and Engineering Company, Las Vegas, in 1974, and was used to bore drifts through sandstone-type formations. Apart from this the company has, during the late nineteen-seventies, been devoting its main efforts towards the development of cutters and consideration of factors influencing their efficiency, rather than actively pursuing the manufacture and sale of tunnel-boring machines. Nevertheless, the fact that a great number of machines now being built utilize fittings which were originally introduced by the Hughes Tool Company on the Betti I indicates the extent to which this company has contributed towards the development of the modern tunnel borer.

Lawrence Manufacturing Company (subsidiary of Ingersoll-Rand Co.)

Alkirk Hardrock Tunneller[11, 12]

In the early 1950s M. B. Kirkpatrick and A. C. Swalling of Anchorage, Alaska, invented and patented an engineering technique which became known as the 'pilot-pull' principle.

In order to prove the feasibility of this principle a coal-miner machine was built under Kirkpatrick's supervision. The various components for the unit were made by a number of job shops and were then ultimately assembled by the Clough Equipment Company of Seattle.

As initial trials in the Suntrana mine at Healy, Alaska, (which was owned by Kirkpatrick and Swalling) proved encouraging, Kirkpatrick decided to negotiate a contract with the Lawrence Machine and Manufacturing Company for the design and manufacture of a full-scale model.

Figure 233. 'Hughes 102' — hard-rock tunnel-boring machine — used by the Reynolds Electrical and Engineering Co., Las Vegas, 1974. (Courtesy Hughes Tool Company.)

During the building of this full-scale machine Kirkpatrick's and Swalling's friendship with Lawrence deepened and resulted in their purchasing a number of Lawrence Company shares, so that forces could be joined in order to exploit the Alkirk pilot-pull principle more fully. Patent rights were at that time owned jointly by the Alkirk Corporation in which Lawrence was an equal partner with Kirkpatrick and Swalling.

Though not particularly successful, the second machine when tried in the Suntrana mine again showed its potentials, provided certain major problems encountered could be overcome with modification.

Accordingly, a third machine was designed. This was built by the Lawrence Machine and Manufacturing Company Inc. in their Seattle workshop. Christened the 'Alkirk cycle miner', the unit which was a 2.13 m × 3.66 m (7 × 12 ft) twin-borer, was used as a production machine in the Suntrana mine, where it performed satisfactorily for one and a half years. Penetration rates in excess of 4.57 m (15 ft)/hour were achieved and at normal operating speeds it mined coal at the rate of 15 t a minute — its rate of extraction being limited by the handling capacity of the ore removal system of the mine at that time.

From this stage onwards the Lawrence Manufacturing Company carried on their own development of the patented Alkirk pilot-pull principle. Two additional 2.13 × 3.66 m (7 × 12 ft) twin-bore machines were built. One was used by the U.S. Army Corps of Engineers for tunnelling through permafrost (frozen rocky silt) and the other was used by Morrison-Knudsen Co., Boise, for slot mining on −20° slope in the Kemmerer coal mine, Kemmerer, Wyoming. Penetration rates exceeded 3.0 and 4.6 m (10 and 15 ft)/hour respectively for these projects.

After building a fourth unit — the Alkirk rectangular borer, 1.2-1.8 m (4-6 ft) high × 1.8-2.4 m (6-8 ft) wide (which was used by the Thompson-Creek Coal Co. as a production miner in soft coal), the first Alkirk hardrock tunneller was produced in 1964. This was Model HRT-12. It was put to work in the Richmond tunnel, New York City, where it bored through pegmatite with quartz intrusions ranging in compressive strength from 140 to 260 MPa (20,000 to 38,000 lb/in^2). (It was later removed as it proved uneconomical due to problems associated with the cutters and the mucking system.)

In operation the Alkirk tunneller is positioned in the tunnel with the pilot cutter bit against the face. The rib jacks are then extended and they hold the machine steady while the pilot bit and cutter head advance, boring the initial pilot hole. When this is completed, the auxiliary jacks are used to hold the cutter head while the rib jacks and anchor are moved forward. The anchor and rib jacks are then expanded against the rock. This forms a stable arbour about which the cutter head rotates. The cutter head and the pilot bit then advance again, working simultaneously as they cut both the main tunnel and the new section of pilot hole ahead of the machine. (Cuttings from the pilot cutter are flushed by either air or water through the pilot drive shaft on to a conveyor.) Thrust for the cutter head is provided by cylinders which apply force to the cutter head. As both anchor and rib jacks are directly coupled together by a common shaft, thrust reaction from the cylinders may be taken by both or either of these components. Maximum thrust capability is thus provided at all times, even when the rib jacks or anchor are in sections of bad ground. The boring cycle is recommended by the extension of the auxiliary jacks and the repositioning of both rib jacks and pilot anchor in the forward position.

Advantages claimed by the manufacturers for this method of boring are that continuous samples of the ground ahead of the machine are provided, thus giving warning of gas, high-pressure water zones, faulty or running ground, or unexpected voids. In addition, the manufacturers claim that the pilot-pull system offers improved steering control which is especially applicable under mixed-face conditions. An example of this being when a new face is attacked and initial contact with only one side of the cutter head is possible.

Naturally when the HRT-12 was first put through its paces the usual teething problems arose. Nevertheless the potential of the machine was evident.

In 1966 the Lawrence Manufacturing Company became a wholly owned subsidiary of the Ingersoll-Rand Company. The injection of additional finance at that time towards the development of the Alkirk tunneller led to the production of the HRT-13 (Figures 234 and 235) in 1968. A 1970 m long sewage tunnel 4.2 m (13 ft 8 in) in diameter was bored by this machine through dolomitic limestone ranging in compressive strength from 110 to 255 MPa (16,000 to 37,000 lb/in^2). Average penetration rates being approximately 1.4 m (4 ft 7in) per hour.

In December 1969 a new 4.1 m (13 ft 5 in) diameter unit incorporating several design improvements was built. It carried an Ingersoll-Rand tunnel machine serial number — 007. This machine was installed in a 'continuation section' of the same tunnel which had previously been bored by the HRT-13. The new section of tunnel, 4.9 km (3 miles) long, was finally completed in September 1971, and the machine clocked penetration rates of 1.9 m (6ft 3 in) per hour.

The successful debut of these two machines was followed in rapid succession by the manufacture of five more units. Penetration rates achieved by these machines are shown in Table 10.

Early in 1972 the 006 machine was modified and relabelled 006R. This unit was originally used in the Port Huron Project in 1969. After modification it was put to work in Rochester, New York.

Summary

The numerous advantages of a machine-bored tunnel as opposed to one cut by conventional methods (i.e. drilling, blasting, etc.) are well known. However, equally well known are the factors governing the cost of machine tunnelling in really hard rock, namely the harder the rock, the greater the cost of cutters, etc. As a consequence, during the early nineteen-seventies, few contractors were willing to risk the high capital expenditure involved

Figure 234. Alkirk hard-rock tunneller — Model HRT-13. Used on Lawrence Avenue sewer system tunnel, Chicago, Ill. Built by Lawrence Mnf. Co. (Sub. of Ingersoll-Rand Ltd). (Courtesy Ingersoll-Rand Ltd.)

Table 10

Model No.	Project	Average penetration rate per hour		Best weekly advance	
		m	(ft in)	m	(ft)
006	Port Huron tunnel	3.81	(12 6)	278	(912)
008[a]	Dorchester	1.52	(5 0)	32	(105)
009	Magma Copper	1.91	(6 3)	156	(512)
010	Cookhouse tunnel	1.83	(6 0)	154	(504)

[a]This machine was installed in the Dorchester tunnel in July 1966. However, the unit did not commence work until March of the following year (1970) due to various problems such as strikes, etc. The entire project thereafter was beset with difficulties — bad ground conditions, flooding, and other extraneous problems unassociated with the machine. Reluctantly the contractors and Lawrence were eventually forced to withdraw the unit, as it was considered that the efficiency of the borer was being adversely affected by these conditions. In all the machine cut through 1165 m (3823 ft) of rock ranging in compressive strengths from 166 to 350 MPa (24,000 to 51,000 lb/in^2)

with a machine-bored tunnel while the economic viability of such an undertaking remained doubtful. Thus, in company with several other manufacturers of hard-rock machines at that time, Ingersoll-Rand found this market to be unprofitable.

The question of entering the field in competition with manufacturers producing machines capable of cutting through the softer rock formations was also considered, but as this market was already well served, Ingersoll-Rand decided to suspend the manufacture of

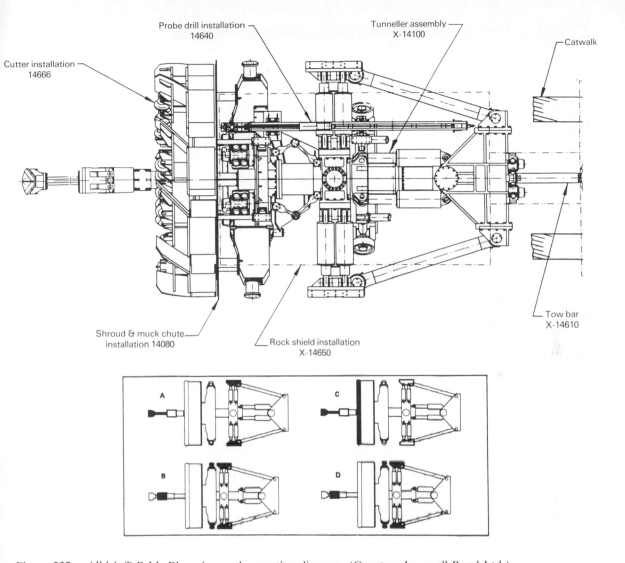

Cutter installation
14666

Probe drill installation
14640

Tunneller assembly
X-14100

Catwalk

Shroud & muck chute
installation 14080

Rock shield installation
X-14650

Tow bar
X-14610

Figure 235. Alkirk T.B.M. Plan view and operating diagram. (Courtesy Ingersoll-Rand Ltd.)

tunnelling machines and concentrate instead on the production of raise borers, etc. (which machines, incidentally, evolved as a direct result of the experience gained by the Lawrence Manufacturing Company in the development of their miners and tunnellers). Later (1979) the Raise Boring Division of Ingersoll-Rand was taken over by the Robbins Company.

Jarva Inc.

Early History

When the Italian firm S & M Constructors Inc. (originally founded by Joseph Scaraville) began

bidding for tunnelling projects, the urgent need to mechanize the excavation work became apparent and, as a result, developmental work on the first soft-ground machine was begun during the early 1950s by Joseph Scaraville and his son Victor Scaraville.

The first machine — Model SM-1, a 3.5 m (11 ft 6 in) diameter soft-ground wheel-type unit, enclosed in a shield, was produced in 1961 for a Cincinnati, Ohio, sewer project (see also Chapter 11, Mechanized shields — Jarva Inc.). However, all tunnels are not driven in soft ground and the need for a machine capable of penetrating hard ground or rock also became obvious.

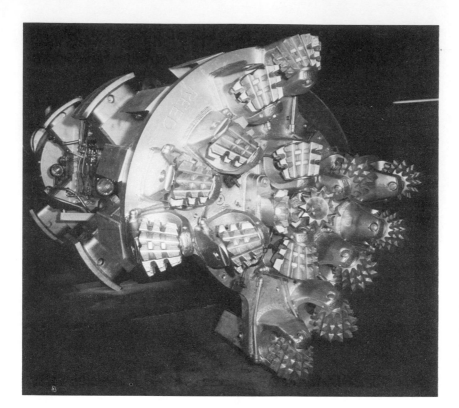

Figure 236. First hard-rock tunnel borer Model SM-2 designed by the Scaravilli combination of father and son. Both the soft-ground and the hard-rock machines proved unsuccessful. (Courtesy Jarva, Inc.)

Figure 237. Model SM-2. (Courtesy Jarva Inc.)

While the company's first attempt at a rock T.B.M. — a 2 m (6 ft 6 in) diameter unit (Model SM-2) (Figures 236 and 237) built in 1964 for a Philadelphia, Pennsylvania project — was, in the words of Jarva's General Sales Manager, E. W. Brickle, doomed from the start because of its size — too much iron for too small a hole — it nevertheless paved the way for the later development of Jarva's first successful T.B.M., the Mark 14 (Figure 238).

The Jarva Company was formed from S & M Constructors Inc. in April 1966 by Joseph V. Scaraville who, incidentally, was still with the company in the late nineteen-seventies.

The Mark 14, which was put to work in a 497 m (1630 ft) long sewer tunnel in Philadelphia, carried carbide insert kerf cutters. It weighed 73 t and its hydraulic propulsion cylinders were capable of producing 3854 kN (866,000 lb) of thrust. Total cutter-head power was 373 kW (500 hp) and the material penetrated in the tunnel consisted of mica schist and hornblende with compressive strengths in the 41-172 MPa (6000-25,000 lb/in²) range.

During its initial developmental phase the machine's design, which later was to serve as a basic pattern for future Jarva hard-rock machines, began taking form. Modifications which were effected at this stage included increasing the size of the thrust bearing from a small roller-bearing about the size of the 41 cm (16 in) shaft to a long 1.5 m (5 ft) diameter double-tapered roller-bearing.

The propulsion rams were increased in number from two to four to provide even thrust on each quadrant of the cutting wheel. This was

Figure 238. Jarva Mark 14 — first successful Jarva rock machine. Used initially in Philadelphia in 1964 on project not listed on Jarva production list. (Courtesy Jarva, Inc.)

Figure 239. Jarva Mark 22-2100 — used for Kaimai Railroad deviation tunnel by New Zealand Ministry of Works in 1972. (Courtesy Jarva Inc.)

designed to correct the imbalance in the forward thrust system which had produced 'turning moments', thus overloading the main thrust bearing.

The hydraulic system was also modified and work was carried out on the gear reducers.

When the Mark 8 machine was constructed for a sewer tunnel project in St. Louis the hard ground there pinpointed several weaknesses which had not been apparent in the softer Philadelphia ground.

Premature cutter failure became quite a severe problem, which was corrected by mounting scraper blades on the wheel to clean the invert of the bore more effectively.

Although many other important changes and modifications have taken place over the years, the basic design pattern of the Jarva developed during the construction of the Philadelphia and St. Louis tunnels has remained practically the same.

In 1971 the Jarva Mark 21 was used on the Bay Area rapid transit system project in San Francisco, California. It carried carbide insert kerf cutters on its head and weighed 195 t. The head power was 560 kW (760 hp) and its rams were capable of producing 7561 kN (1,700,000 lb) of thrust for the 6.10 m (20 ft) diameter head. The material penetrated ranged in compressive strength from 7 to 276 MPa (1000 to 40,000 lb/in^2).

This machine was followed by the Mark 22 (Figure 239) which was used early in 1972 by the New Zealand Ministry of Works for a rail tunnel. The 6.4 m (21 ft) diameter Mark 22 cutter head carried carbide insert and rotary disc cutters. Its 'clamp legs' or grippers consisted basically of a telescoping box with a hydraulic cylinder mounted on each side. Each leg was individually controlled to facilitate steerage, etc.

Eight hydraulic cylinders mounted between the clamp legs provided the head with forward thrust totalling 9789 kN (2,200,000 lb). The four cylinders at the front thrust against the rear of the bearing housing, while the four at the rear pulled on the gear housing. The pull force was transmitted forward by a tube which passed through between the clamp leg section connecting the gear housing with the bearing housing.

Within the tube was the main drive shaft which transmitted the power from the motors at the rear to the cutter head at the forward end of the machine.

A cutting cycle was commenced with the thrust cylinders closed and the pull cylinders extended.

Angled paddles on the periphery of the cutter head deflected the cut material into scoops mounted behind the wheel. The scoops lifted the material to the top where it was dropped on to a conveyor running along the top of the machine. At the rear a second conveyor system fed the muck into mine trucks or dump cars.

At the end of the push-and-pull stroke the cutter head was stopped and hydraulic jacks under the gear case and bearing housing were extended to the floor to take the weight of the machine, the wall grippers were then retracted and the flow of oil into the thrust cylinders was reversed to reposition the gripper section in readiness for the next boring cycle.

Melbourne Underground Rail Loop Project[13]

Jarva Inc. were commissioned to design a tunnelling machine for the Costruzione del Favero (Codelfa) Construction Company, main contractors for the largest section of the Melbourne Underground rail loop project in Australia.

The machine, a Mark-24-2303, was built in Melbourne and commenced work in 1974. In the specifications allowance had to be made for the installation of steel ring beams with 1 m (39 in) centres, immediately behind the face and also shotcreting of the tunnel walls 1 m (39 in) from the face, within an hour of exposure.

The unit's basic design was different from the standard type of hard-rock machine normally produced by Jarva. The head of the 7 m (23 ft 4 in) diameter machine consisted of a square-faced steel-plate structure on which four radiating spokes or arms were mounted.

Each arm or spoke carried triple-disc cutters and a gauge cutter at the periphery. Muck from the face was moved to the rear via a screw conveyor which passed through the main torque tube. Five hydraulic cylinders provided forward thrust for the unit.

During the first few months until March 1975 when the machine broke through into Parliament Station, progress was slow.

After Parliament Station much harder rock was encountered. One of the early problems was the failure of the screw conveyor to handle the materials. It was replaced by a conventional belt conveyor but, when this jammed, as it tended to do periodically, it proved to be almost impossible to free it, as there was not sufficient room for a workman to enter the torque tube and effect repairs. Gauge cutters were consumed sometimes within 2 m (6 ft 6 in) and the triple-disc cutters failed rapidly. Problems also arose with the face-plate which had to be replaced.

Because of inadequate support, the front of the unit tended to rest on one of its spokes when it was being repositioned in readiness for the next cutting cycle. This damaged the gauge cutter and forced the spoke to sink as much as 300 mm (12 in) into the ground on occasions. The only remedy for this was laboriously to jack up the head and use sleepers to pack the area beneath.

Another problem which, to some extent, was similar to that experienced by Grandori with the Robbins Company's tandem mechanized shield in southern Italy (see Chapter 26, Full-face hybrid machines) was that voids of up to 14 m³ (18 yd³) were created when large blocks of rock were dislodged from the face or crown by cutters on the spokes on the cutter head. These voids sometimes had to be packed with sandbags and concreted. The same remedy, namely the insertion of grill bars between the spokes, was tried and this was partially effective.

Muck from the face and rebound from the

shotcreting, which collected in the invert and could not be properly cleared by the buckets, had perforce to be removed by hand.

Other problems which beset the project included the frequent bursting of hydraulic hoses, when they were subjected to hydraulic pressures said to be up to 35 MPa (5100 lb/in^2) and the breaking free of one of the spokes from the square-face structure.

Codelfa supplemented the unit's conveyor system with a chain of 30 m (98 ft) long conveyors suspended from the tunnel crown. But this was not entirely satisfactory because a hold-up on one of them meant the entire chain was stopped.

Finally, on 6 June 1976 after some 250 m (820 ft) of tunnel had been driven in about 14 months, the main bearing failed.

In the meantime M.U.R.L.A.'s Director of Engineering and General Manager, Frank Watson, growing concerned with the delays, invited Tasmanian consulting engineer, David Sugden, to appraise the situation and give his advice on what options he considered to be available from a technical point of view.

It was estimated that the modifications and redesign programme suggested by Sugden would cost at least £350,000. Sugden's report and recommendation was made available to the Codelfa Company. In view of the urgency of the matter and with the machine in the state it was, Codelfa decided to order this work to proceed.

New drawings were then produced by Sugden and the Robbins Company's Kingston branch in Tasmania, and the modification work was put in hand. This included, *inter alia,* the following:

Figure 240. Jarva Mark 30 — used by Kenny/Paschen/S & M (joint venture) in 9.17 m (30 ft) diameter sewer project for Wilmette Metropolitan Sanitary District of Greater Chicago. (Courtesy Jarva Inc.)

266

Figure 241. Jarva Mark 11-1102 — used for Milwaukee sewer, Wisconsin, 1969-72. (Courtesy Jarva Inc.)

1. A complete new cutter head in the form of a flat disc with 12 scoops which discharge to the rear into '2.' below.
2. A novel side-mounted conveyor.
3. A heavy sliding front shoe to support and stabilize the cutter head and to keep the invert clear.
4. Single-edge disc cutters in place of the triple-disc cutters.
5. New simpler handling systems for supports and segments.
6. Isolation for shotcreting applications area.
7. Completely redesigned main bearing support and sealing installations.
8. Elimination of front vertical shoe and one thrust jack, leaving four forward thrust jacks.
9. Redesigned hydraulic drive system which included the replacement of the original gearboxes and eight variable-speed hydraulic motors with high-torque low-speed hydraulic motors.
10. Replacement of long conveyor spoil-handling system with conventional rail system.

Other changes involved sloping the arch ribs forward to facilitate shotcreting operations and provide better face support.

The use of a sliding front shoe prevented the erection of support within 1 m (3 ft 3 in) of the face as originally specified. However, the use of forward sloping ribs supported from the concrete invert segments restored this facility in the practical sense.

The modifications were completed by February 1977. Soon after a settling-down period, under the guidance of a new superintendent and a new management team, penetration rates of up to 73 m/week (240 ft/week) were achieved — 33 m/week (108 ft/week) better than those originally attained. By the end of May 1978, when the author visited the site, record penetration rates of 100.3m/week (330 ft/week) (11.3 m (37 ft) per eight-hour shift) had been achieved by the operating crew.

The current Jarva T.B.M., a Mark 30 (Figure 240) was in 1978 at work in Illinois, driving an 8.6 km (5.4 miles) long, 9.1 m (30 ft) diameter tunnel for the metropolitan sanitary district of Greater Chicago.

A consortium composed of Kerry Construction Co., Wheeling, Illinois; Paschen Contractors Inc., Chicago, Illinois; and S & M Constructors Inc. of Solon, Ohio, successfully bid for the U.S. $63 million contract.

The contract included a slightly smaller (6.7 m (22 ft)) diameter tunnel, 7.4 km (4.6 miles) long, which was also driven with a Jarva unit. Solid dolomite limestone ranging in compressive strength from 97 MPa (14,000 lb/in²) to 269 MPa (39,000 lb/in²) was penetrated by the Mark 30 machine.

The Mark 30 unit was basically similar to the Mark 22. Its cutting head carried 68 disc cutters and the propulsion jacks provided 13,350 kN (3,000,000 lb) of forward thrust.

Figure 242. Calweld T.B.M. Model TM-66. Used on E.S. Railway project, Sydney, by Codelfa Construction Company. Bored 4572 m (1500 ft) of tunnel. Undergoing reconditioning in preparation for new job for the Coal Cliff Collieries Pty Ltd. (Courtesy Codelfa Construction Pty Ltd.)

Figure 243. Calweld 3.96 m (13 ft) diameter hard rock tunnel-boring machine fitted with Smith Tool Company cutters, muck pick-up buckets, and stationary dust bulkhead (Model TM-40). Used in Climax mine tunnel, Climax. (Courtesy Smith International, Inc.)

268

Figure 244. Dresser tunnel borer. Model MSED 205. (Courtesy Dresser Mining Services and Equipment Division.)

Dresser Industries Inc.

In May 1971 Dresser's first tunnel borer (Model 205) designed by Doug Winberg (an ex-Robbins chief engineer) was built by their Mining Service and Equipment Division and put to work in New Mexico on the Navajo irrigation project.

The machine (Figure 244) was basically similar in concept to the Robbins Poatina machine (Model 161) (Figures 214 and 215). It was originally planned for use in 40,000 lb/in^2 (280 MPa) rock with a 4.88 m (16 ft) diameter head. Work on the manufacture of the unit had already begun when Fluor Utah Inc. (who were awarded the U.S. $8.7 million contract by the Bureau of Reclamation in 1970) indicated their interest in the unit.

The machine's cutter head was then enlarged to 6.25 m (20 ft 6 in). The cutter head carried 36 double-disc cutters and 32 fixed conical picks (radially arranged in a central block). Buckets fitted to the periphery of the head lifted the cut material and deposited it into a conveyor at the top of the machine.

The unit successfully cut through the 1043 m (3420 ft) No. 3A tunnel and also through the 5 km (3 mile) long No. 3 tunnel.

Despite problems associated with fallouts, etc. the unit averaged 59.2 ft per operating day. The best eight-hour shift was 33 m (107 ft) on 5 July 1972 and during the five-day week ending 30 June 1972, 325 m (1066 ft) of tunnel was driven by the machine.

To date no further T.B.M.s have been produced by this company.

Subterranean Tools Inc.

In 1974 the author was advised by Subterranean Tools Inc. that work was under way on the development of a hard-rock tunnelling machine. Since then Subterranean Tools Inc. have ceased to manufacture mechanical excavation machines and a new company, Subterranean Equipment Company, was formed in 1978 to cater for the manufacture, sale, and service of rock-cutting machines.

The hard-rock tunnelling machine (Model TB-11) had by that time been completed and

Figure 245. Subterranean Equipment Company rock tunnelling machine, Model TB-11. (Courtesy Subterranean Equipment Company.)

was sold to the Anglo-American Corporation of South Africa Limited. It was installed in the Vaal Reefs gold mine on level 66, where it commenced operation in February 1978.

The machine was scheduled to bore approximately 2 km (1.2 miles) of tunnel, traversing various rock strata, the hardest being Denny's quartzite which has a compressive strength of up to 414 MPa (60,000 lb/in²).

The machine (Figure 245) is fitted with a cutter head which in profile is semi-elliptically shaped. Except for the centre and gauge cutters which are button type only, either button or disc-type cutters may be used on the head. The six gauge cutters are equipped with scrapers to minimize regrinding of cuttings in the tunnel invert.

A sealed two-row tapered roller-bearing (with an oil-circulating lubrication system) supports the cutter head. The cutter-head support carrying the main roller-bearing is mounted on a bottom shoe positioned immediately behind the cutter head. The shoe rests in the tunnel invert.

The main frame or 'tubular backbone' of the machine is formed in two sections bolted

together and consisting of a forward beam and an aft beam. The forward beam is bolted to the cutter-head support while the aft beam is bolted to a jack leg unit. The jack leg has retractable legs and acts to support the rear of the machine during the repositioning cycle. The forward and aft beams are provided with sliding way surfaces for the front and rear grippers respectively.

The cutter-head drive gearbox is attached to the rear of the jack-leg assembly. A tubular drive shaft (powered by eight hydraulic motors through double planetary reduction gears and eight bull pinions meshed with a single bull gear) extends forward through the hollow beams or backbone of the machine. It is supported in the centre at the joint between the forward and aft beams by a bearing.

Side shoes which extend laterally against the tunnel walls for horizontal steering purposes are positioned directly behind the cutter head.

The grippers, consisting of two independent systems, each with two gripper pads or shoes are unusual in that the pads are extended vertically towards the floor and roof of the tunnel instead of laterally against the sidewalls, as do those on most other rock T.B.M.s.*

To compensate for precession due to driving torque, the rear gripper is also capable of rotating the machine on its longitudinal axis.

The rear gripper shoes which react against the thrust of the main propel cylinders are slightly larger than those of the front gripper shoes, which provide thrust reaction for a smaller set of auxiliary thrust cylinders used in certain applications.

While steering is normally carried out with the side steering shoes the grippers may, of course, also be used for this purpose if necessary.

An advantage gained by the use of vertically extending grippers, especially in the case of mining operations, is that where the side wall is absent, i.e. when crossing another tunnel, the forward grippers may be used for steering in lieu of the side shoes.

Another feature of the machine is that the front gripper system may be dismantled and removed if boring conditions are favourable, thus increasing the workspace immediately behind the cutter head.

Total installed power for the machine is 670 kW (900 hp) of which 450 kW (600 hp) is delivered to the 3.5 m (11 ft 6 in) diameter cutter head.

The machine which weighs approximately 90 t (excluding the back-up system) is capable of developing a torque of 544 kN m (400,000 lb ft) and thrust totalling 4890 kN (1,100,000 lb).

In 1979 R. N. Taylor (Manager, Engineering Department of the Anglo-American Technical Development Services, Welkom, South Africa) advised the author that the unit was boring satisfactorily. Only minor teething problems had been encountered during commissioning and, apart from the fitting of a tunnelling shield as a protective measure, no other modifications had so far been found to be necessary.

Note: As Robbins T.B.M. Model 114-163 (built in 1974) (Figure 228) was destined for the South African gold mines where periodically it would need to negotiate cross-tunnels in the mines, it was also necessary to provide the machine with vertically extending grippers. To do this the Robbins Company produced a machine with rotating grippers which, during normal boring operations extended sideways, but which could be rotated 90° to extend vertically during the negotiation of a cross-tunnel. This was achieved by the substitution of the main box beam with a cylindrical beam having a pair of hardened steel ways in which the way bars on the saddle carrying the rear gripper assembly travelled fore and aft. Torque from the cutter head through the main beam was transferred via the saddle to the carrier and thus the grippers through a set of teeth. During normal operation these teeth were locked in position by draw bolts, thus preventing any fore/aft movement between the saddle and the carrier. When the gripper assembly needed to be rotated so as to change the gripper operating position from say the lateral to the vertical position, the draw bolts locking the carrier to the saddle were withdrawn. The two propel cylinders mounted on large spherical balls at the cutter-head support end were disconnected and temporarily supported by special brackets attached to the gripper assembly. Once the draw bolts were withdrawn the gripper assembly could be slid forward relative to the saddle then rotated 90°. The propel cylinders were then relocated in their new positions at the cutter-head support end and the saddle and carrier teeth re-engaged and locked by the replacement of the draw bolts. Finally the temporary supporting brackets were withdrawn and secured out of the way on the gripper assembly. The grippers were then ready to operate in their new position.

During the years that this machine operated in South Africa the gripper mechanism was rotated only once from horizontal to vertical gripping mode, but not to traverse a lateral drive. The reason it was done was to obtain a better grip on the hanging and foot wall because the side walls were scaling badly and the grippers could not get a purchase. The T.B.M. was then modified for cutter-head

---

*Standard Robbins machines usually feature a box-type main frame on which the saddle carrying the gripper assembly is mounted. Guide rails on either side of the box frame allow the saddle fore and aft movement on the beam during the gripper-repositioning phase. The grippers on the saddle extend laterally towards the side walls of the tunnel for anchoring purposes.

grizzly bar, finger shield, etc. The gripper was returned to its original position before the machine was put back in operation. The grippers remained in this mode until the machine was taken out of service.

Sweden

Atlas Copco

Early History — Rock Drills and Compressors

The Atlas Company was founded in Stockholm in 1873, originally to manufacture railway rolling-stock. This was to meet the needs of Sweden's expanding railway network which was being built at that time.

During the 1890s the local demand for railway stock declined sharply and Atlas were forced to change to another manufacturing line, simply to survive. They began producing machine tools. At the same time an Atlas engineer travelled to America to study the newly developing markets there. He was particularly impressed with some pneumatic riveting and chipping hammers he saw demonstrated and returned to Stockholm with two of these tools. A few years earlier, Atlas had bought a pneumatic riveting press and a compressor for their boiler shop, and the new tools were put to good use in the boiler shop where they operated very well, contributing towards a substantial reduction in boiler manufacturing costs and improving the overall quality of the product.

A new line of business developed when the American tools broke down and Atlas were forced to manufacture parts for them, as these were not available locally. To ensure against further costly stoppages, several spares were made at the time. Rumours of this spread to other machine shops in Stockholm and resulted in a request by an old-established engineering firm for the production of the entire tool. Atlas readily agreed and during the ensuing years their own designs of drilling and grinding machines and rock drills, etc. were added to the riveters and chipping hammers.

In 1901 a special department for rock drills was established and three years later the first Atlas air compressor was built at the factory.

In 1917 Atlas merged with AB Diesels Motorer, which had been producing diesel engines since 1898, based on one of the original licences granted by Rudolf Diesel. After the merger Atlas Diesel produced diesel engines as well as the original line of pneumatic tools and compressors, etc.

The Second World War brought serious economic problems because the export business — the mainstay of the company — suddenly collapsed.

One of the major products of the company at this time was the rock drill, which was sold extensively in Sweden because the construction industry had come to appreciate its sturdy performance in the hard granite and gneiss rock outcrops which predominated throughout Scandinavia. However, while the drill itself withstood the harsh treatment meted out to it, the drill steels did not. Most steels needed reforging after only 0.5 m (20 in) or, at most, 1 m (39 in) of hole drilled. In addition, great numbers of heavy steel rods were constantly being transported from the face to the surface for reforging, which also added to the overall running costs of drilling.

To overcome these problems Atlas Diesel and Sandvik (a well-known Swedish manufacturer of drill steels and tungsten carbides) began a long-term joint developmental programme aimed at producing a rod with a brazed tungsten carbide drill bit.

Although a similar type of bit had been used in post-war coal-mining applications, it was considered impracticable at the time for use in hard-rock drilling work.

Eventually, after numerous costly trials, a carbide bit capable of withstanding the heavy shock loads of hard-rock drilling was produced. By the end of the war Atlas Diesel were producing light-weight, self-rotating, pusher-feed rock drills in combination with drill steels which had cemented tungsten carbide bits brazed on to their tips. The service life of the bits was extended enormously and news of the new 'Swedish method' of drilling as it came to be called, spread throughout the world with simultaneous economic benefits for the company.

Such was the demand for these tools that in 1948 Atlas Diesel sold out its interests in the

Figure 246. Atlas Copco full-facer — circular machine. (Courtesy Atlas Copco AB.)

diesel engine business in order to finance the expansion of its fast-growing rock drill section. In 1956 the company's name was changed to Atlas Copco.

Switzerland

Habegger Limited

In 1962 an Austrian engineer named Wohlmeyer began developing a boring system which utilized the principle employed in the milling of metals. This was, of course, different from the method used by American manufacturers of rock T.B.M.s in which the disc cutters or roller bits 'crushed' the rock.

In 1964 the Habegger engineering firm of Thun, in Switzerland, became interested in Wohlmeyer's work. At the time Habegger were producing aerial ropeways, ski-lifts and some tool machines.

The Wohlmeyer patents were assigned to Habegger, and Habegger Ltd. continued with his development of a rock-cutting tool, aiming at producing a more robust machine for use in harder rock formations.

The first tunnelling machine incorporating these concepts was completed early in 1966 and was tested in practice in sandstone and limestone formations in Thun. These ranged in compressive strength from 117 to 172 MPa (17,000 to 25,000 lb/in²) for the sandstone and limestone respectively.

The unit was then sent to Japan where it was used to bore the pilot tunnel for the Honshu-Hokkaido railway connection beneath the sea.

The head consisted of four circular cutter plates each fitted with fixed picks around their peripheries. The cutter plates which rotated, were mounted on a large disc or cutter head which also rotated. The machine weighed 86.3 t and was capable of delivering 2295 kN (515,000 lb) thrust and 679 kN m (500,000 lb ft) torque.

The 3.51 m (11 ft 6 in) diameter head had a power of 485 kW (650 hp). It successfully bored through 20,100 m (66,000 ft) of tuff and andesite with compressive strengths ranging from 35 to 234 MPa (5000 to 34,000 lb/in²).

273

Subsequently, four more machines (incorporating improvements based on the experience gained from preceding models) were built and used in Churn, Switzerland (Model 836 — in 1967), Stuttgart, Germany (Model 829 — in 1967) and Japan (Models 836 and 840 — in 1968).

In 1968 Atlas Copco acquired the Habegger patents and a new company, Atlas Copco Maschinen AG, located in Thun, Switzerland, was established early in 1969, to deal with this side of the business.

Atlas Copco Machines[14]

Approximately five more circular full-facer machines (Figure 246) on the Habegger principle with head diameters ranging from 3.40 m (11 ft 2 in) to 4.50 m (14 ft 9 in) were built by Atlas Copco Maschinen AG and used in Switzerland, Greece, and Japan, including one rectangular full-facer machine Model 4826 which cut a 4.80 × 2.60 m (15 ft 9 in × 8 ft 6 in) arch-shaped tunnel.

While the gripper system of the Atlas Copco full-facer machine is somewhat similar in concept to that used by American manufacturers, the head is radically different. Following the same principle of milling, initiated by Dr. Wohlmeyer and subsequently developed by Habegger, the Atlas Copco circular fullfacer is provided with a revolving drum on which are mounted a number of cutter heads, usually four. Each cutter head is fitted with a series of radially projecting cemented carbide bits. The axes of the cutters are inclined to the axis of the drum. During the boring operation the cutter heads rotate in the opposite direction to the drum while the machine is advanced at a speed synchronized with the rotation speed of the drum.

As the machine advances the cutters bite into the rock following concentric helical paths. The rock is thus attacked radially so that about a quarter of the rock is removed from the face. the rest, i.e. the ridge of rock remaining between the grooves, is broken off by the resulting tensional force generated. The manufacturers claim that because of this, only 20-40 per cent of the thrust needed for the same amount of work by other types of machines

using a crushing action, is required to do the job.

The main assembly is mounted on a sliding shoe and lateral hydraulic jacks reacting against the tunnel walls provide forward thrust. The hydraulic equipment and control cabin are situated at the rear.

A pair of single-flight chains, running in a conveyor trough under the machine, carry the muck from the invert at the face to hauling units at the rear. The conveyor system is similar to those used on some coal-mining machines such as roadheaders or continuous miners.

After the development of the full-face rectangular machine a *mini-full-facer* was manufactured capable of cutting a 1.30 × 2.10 m (4 ft 3 in × 6 ft 11 in) size tunnel (see section on mini tunnelling machines).

So far as the full-face rectangular machine was concerned the cutter-head drum was designed to swing about an axis rather than to rotate.

Germany

Bade & Co. GmbH[15, 16]

In 1956 the Bade Co. GmbH of Germany produced their first full-face hard-ground T.B.M., Model SVM-33-RM, which was tested in a potash mine in Hangsen, Hanover. The machine cutter-head power was 210 kW (282 hp) and carried fixed cutters and breakers (drag teeth). The head was built in three concentric sections to reduce torque and control cutter-speed variation. The middle ring rotated in the opposite direction to the centre and outer rings. To enable the machine to cut a horseshoe-shaped tunnel two auxiliary cutter heads which rotated on transverse axes were positioned immediately behind and below the main cutter head on both sides of the machine. These auxiliary cutter heads cut a groove in the circular tunnel to form a partial horseshoe shape. Two hydraulic thrust jacks which reacted against the grippers or 'dogging plates' provided forward thrust totalling 796 kN (179,000 lb). The machine was steered through an articulated main beam steering system which enabled it to round small-radius curves. The 28

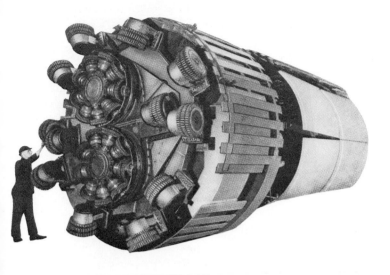

Figure 247. Bade hard-rock tunnelling machine. (Courtesy Bade & Theelen GmbH.)

t, 3.30 m (10 ft 10 in) diameter machine also featured an enclosed operator's cab.

In 1961 a larger unit (SVM-40-RM) with a 4 m (13 ft 1½ in) diameter head having a power of 550 kW (737 hp) was built for use in a coal-mine in Essen. It weighed 105 t and its propulsion jacks were capable of 4487 kN (1,008,800 lb) of forward thrust. The head was fitted with toothed rock bit roller cutters for use in harder ground.

The rotating motion of the cutting zones on the head was effected by a hydraulic drive from two explosion-proof electric motors, one of which was located at the front of the machine and the other at the rear. The front motor rotated the inner cutting zone of the head mechanically through a gear, while the other two cutting zones were driven through two hydraulic pumps and four hydraulic motors. All the rotating motors were interconnected through a planetary drive. The hydraulic motors had variable speeds regulated by the output of the hydraulic pumps.

Debris broken from the face by the cutters fell to the invert and on to a blade or shovel which rested on the floor immediately beneath the cutter head. The spoil collected in the shovel was directed by two scrapers inside the shovel towards the front end of a fixed conveyor which formed an integral part of the machine.

The fixed conveyor then transferred the spoil to an adjustable conveyor beneath the machine which carried the material to the rear.

Fried. Krupp GmbH[15, 17]

The first Krupp Machine, Model KTF 280,[18] was put to work in a water-supply tunnel during May 1967 in Talheim, Swabian Alb, South-west Germany.

Its basic design was similar to the Habegger machine in that it, too, utilized the milling method of breaking away the rock face.

The entire head of the Krupp machine revolved about a central hollow axle. On this head were mounted two inner and two outer cutters and a small central pilot cutter, which all rotated around their own axles. Each cutter consisted of a blade-disc, fitted with radially arranged replaceable teeth on its periphery. The cutters were positioned in a stepped arrangement on the head, with the pilot cutter leading. The bore was then widened by the two inner cutters followed by the two outer cutters, which cut gauge.

The hydraulic drives were designed to provide three superimposed movements, namely the axial advance of the entire machine, the rotation of the head, and the revolution of the individual cutters on the head. Circular

Figure 248. Krupp Tunnelling Machine, Model KTF 340. (Courtesy Fried. Krupp GmbH.)

helical paths were cut by these cutters as the machine was advanced into the face.

The machine was crawler-mounted and two crawlers were also mounted above the canopy or roof shield so that they bore against the tunnel roof. Forward thrust was provided by the crawlers running on the floor. The side jacks were also crawler-mounted to permit continuous boring and advance of the machine.

The muck was moved to the rear by a single-flight drag-chain conveyor system.

During the period May 1967 to November 1968 the Krupp machine bored a total of 9350 m (30,700 ft) through brown Jurassic clay. Maximum advance rates were: 5.8 m/hour (19 ft/hour), 64 m/day (210 ft/day), 845 m/month (2770 ft/month).

The second Krupp machine, Model KTF 340, (Figures 248 and 249) was put to work underground in a mine in Germany in 1967. Despite frequent problems associated with cave-ins, etc. the machine managed to bore a total of 1350 m (4400 ft) by December 1968.

In a test run in hard limestone with compressive strengths up to 177 MPa (25,700 lb/in²) the KTF 340 achieved advance rates of 3 m/hour (10 ft/hour).

In all, some 13,000 m (43,000 ft) of tunnel were driven by Krupp machines before production was finally halted in 1969 due, the author understands, to an altercation over patent rights. The matter was apparently settled and thereafter Krupp ceased to manufacture tunnelling machines.

Demag Aktiengesellschaft[19, 20]
(now Mannesman Demag Bergwerktechnik)

Demag's first full-face hard-rock tunnelling machine (Model TVM 21 H), was completed in 1966 and put to work driving a sewer tunnel in Dortmund, through Green Sandstone with compressive strengths ranging from 29 to 78 MPa (4300 to 11,000 lb/in²). The 2.10 m (6 ft 11 in) diameter head had a power of 110 kW (147 hp) and the machine was capable of

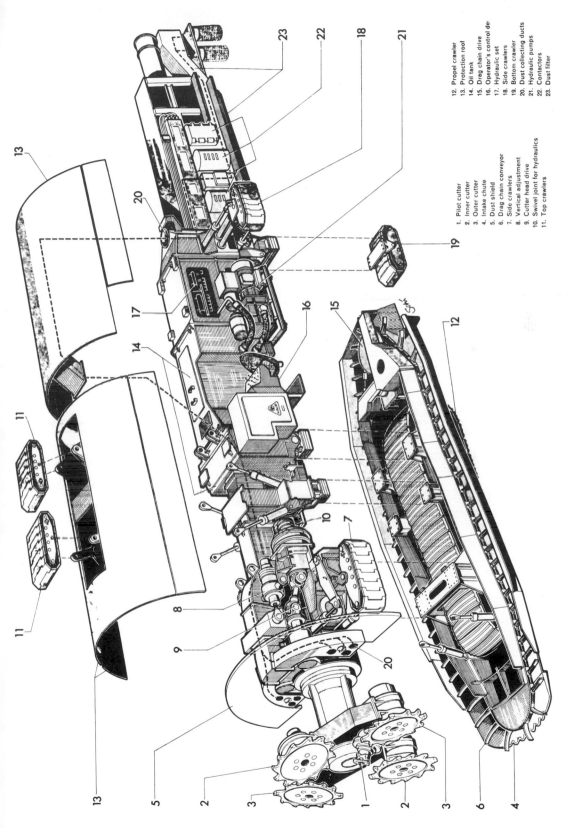

1. Pilot cutter
2. Inner cutter
3. Outer cutter
4. Intake chute
5. Dust shield
6. Drag chain conveyor
7. Side crawlers
8. Vertical adjustment
9. Cutter head drive
10. Swivel joint for hydraulics
11. Top crawlers

12. Propel crawler
13. Protection roof
14. Oil tank
15. Drag chain drive
16. Operator's control desk
17. Hydraulic set
18. Side crawlers
19. Bottom crawler
20. Dust collecting ducts
21. Hydraulic pumps
22. Contactors
23. Dust filter

Figure 249. Exploded drawing of Krupp Tunnelling Machine (Courtesy Fried. Krupp GmbH.)

277

developing 981 kN (220,000 lb) of thrust; 2800 m (9190 ft) of tunnel was driven by this unit.

In May 1968 the second Demag T.B.M. (Model 28-31 H), with a 3.15 m (10 ft 4 in) diameter head began work on the Oker-Grane tunnel in Harz.

The monthly boring rate up to the 1870 m (6135 ft) mark was approximately 267.16 m (877 ft), but this figure dropped to 74 m (243 ft) and 154 m (505 ft) during November and December because of problems caused by a large influx of water and various other factors including the necessity for a complete overhaul of the unit and the repair of some components. The abrasive properties of the Kahleberg sandstone and the hard quartzite, encountered at this stage for the first time, caused excessive wear of the roller cutters. Because of this the original 'Verdur' multi-disc-type roller cutters were replaced by newly developed tungsten carbide button cutters.

During 1969 boring rates were again affected by water. Although the water problem continued during 1970, improved average heading rates of 230 m (750 ft) per month were attained by the operating crews.

Regardless of head diameter the Demag full-face rock T.B.M. follows a basic design. Its cutting head is cone-shaped* and muck is transferred by a series of 'spoons' and deflector plates mounted on the periphery of the head to a chute which directs the spoil to a scraper-chain conveyor located under the machine. A central flushing bore capable of delivering approximately 40 litres (9 gal) of water per minute for cooling the cutter tools and for dust suppression, is available for use if required, during the boring stroke. Behind the head a dust shield with a manhole separates the cutting assembly from the remainder of the machine. A blind flange is provided for the manhole which permits the connection of a dust-filtering system there if desired. The cutting head is powered by four electric motors.

A front and rear gripper assembly, comprising two clamping units which extend laterally against the tunnel walls, hold the machine steady in the bore while four hydraulic

---

*The latest Demag machine Model TVM 55H has been fitted with a flat-faced cutter head (Figure 252).

Figure 250. Demag T.B.M., Model 54-58/61H, being assembled in factory (serial no. 37). (Courtesy Mannesmann Demag Bergwerktechnik.)

cylinders reacting against the rear gripper assembly provide forward thrust for the boring stroke.

Four additional supports which also extend laterally against the tunnel wall hold the machine while the grippers or clamping units are being repositioned in readiness for the next boring cycle.

The rear section of the unit is pivotally mounted behind the cutter-head support structure to enable the unit to be steered through the rear gripper system.

Since the introduction of the 21H and the 28-31H units Demag have produced a number of machines of varying diameters which have been used in Germany, Scotland, Sweden, Czechoslovakia, Yugoslavia, South Africa, Norway, and Switzerland (Figures 250 and 251).

The latest generation of Demag full-face machines come in a standard range of Models TVM 25H, 30H, 35H, 45H, 55H, and 65H with head diameters ranging from 2.4 m to 7.4 m (7 ft 10 in to 24 ft 3 in).

A new type of cutting tool has also been developed to cater for varying ground conditions. The tools are based on the 'unit construction principle' whereby one or two discs, with or without buttons, may be fitted to

Figure 251. Demag T.B.M. — breakthrough after heading a 35° inclined section. (Courtesy Mannesmann Demag Bergwerktechnik.)

the tool holder, depending upon the hardness of the rock.

During the early 1980s, two Model TVM54-58/60H machines with boring diameters of 6 and 6.10 m (19 ft 8 in and 20 ft) were operated in West German coal-mines driving pilot headings. The electrical and hydraulic systems of these machines were built to comply with the stringent requirements of German mining regulations. Both machines utilized 'lagging erectors' which permitted segmented annular 'lagging'* to be mechanically installed close behind the cutting head.**

---

*Quoted from Demag letter to author (apparently 'lining').

**According to Martin Hunt in an article entitled 'Rapid road construction in a German mine',[21] tunnel support rings were used for the first 800 m (2624 ft) of drivage. Each ring consisted of five sections of rigid steel arches. After the 800 m (2624 ft) mark, the rigid arches were superseded by 'yielding arches'. Both the rigid and the yielding arches were supplied by Bochumer Eisenhütte Heintzman GmbH & Co. Steel mesh in a continuous layer was inserted behind the rings as a protective measure to prevent rockfalls and this was sprayed with a 130 mm (5 in) thick lining of concrete.

A peak daily progress of about 35 m (115 ft) was frequently achieved, while monthly rates as high as 450 m (1480 ft) were attained on several occasions. One of these units bored over 10,000 m (33,000 ft) of mechanically trouble-free tunnel during the years 1973 (when it was installed) to 1978.

Alfred Wirth & Company, KG

Alfred Wirth & Co. KG of Erkelenz, Germany were originally a Hughes Tool Company licensee for oil-well tri-cone bits and other rotary drilling equipment. They became associated with the tunnelling industry in 1966 when they bought a Hughes 2.14 m (7 ft) diameter tunnel borer and modified it for use on an Austrian project.[16]

The first rock T.B.M. built by Alfred Wirth & Company, KG (Model TB 1-2114E) was used to bore a water tunnel during the period February 1967 to June 1967 in Zimmkraftwerke, Ginzling, Austria. The machine cutter head was fitted with tungsten carbide insert rollers to

Figure 252. This Demag 6.0 m (19 ft 8 in) machine, Model TVM 55H, which commenced work in the Göttelborn mine in Sept. 1979, was used to drive approximately 17 km (11 miles) of roads (ranging between 1.6 and 4.7 km (1 and 3 miles) on No. 5 level (400 m (1300 ft) below mean sea level). Note the split grippers which allows for the erection of annular steel rib sections immediately behind the dust shield (approx. 1.60 m (5 ft 3 in) behind the gauge cutters). (Courtesy Mannesmann Demag Bergwerktechnik.)

enable it to penetrate the hard granite formations in the 263 m (863 ft) long tunnel.

In 1969 the Wirth Company pioneered the 'reaming concept' (see section on reaming machines) and several Wirth rock T.B.Ms. have been successfully used for boring inclines (see section on incline boring machines).

Basically the Wirth rock T.B.M. features a cone-shaped cutter head. Eight 'clamping shields' hold the unit steady in the tunnel during the boring phase. The cutter head is mounted on an inner kelly and slides forward within guides on the outer kelly which is anchored against the walls. After completion of the boring stroke, which ranges between 700 and 1500 mm (27 and 59 in) in length, the front and rear support shoes are extended to take the weight of the machine and the clamp shields, or grippers, holding the outer kelly are

retracted. The thrust cylinders are then actuated in reverse, pulling the machine and outer kelly towards the front.

Still in an unclamped condition, the machine is aligned in the tunnel by the rear support which is designed as a hydraulic parallelogram, alignment being made in accordance with a laser beam. The trailer containing the control stand, electrical and hydraulic drive systems, transformer, and cable drum are dragged along by means of pull rods.

During the boring operation the cuttings are lifted by scoops mounted on the head and deposited on to a conveyor belt which runs above the machine and carries the spoil to a second conveyor belt installed above the trailer. This in turn conveys the muck to removal trucks, etc. at the rear.

Since the introduction of their first unit,

Figure 253. Wirth
T.B.M., Model TB 11 H.
(Courtesy Wirth GmbH.)

Wirth rock T.B.Ms. have been used on about 35 projects in Austria, Switzerland, Sweden, Germany, America, France, Italy and South Africa (Figures 253 to 259).

France

Union Industrielle Blanzy-Ouest[22]

In 1969, Union Industrielle Blanzy-Ouest (UNI.BO) of Paris, France, undertook developmental work on an unusual tunnelling machine. The initial idea was to produce a unit which, although it cut the entire face, nevertheless used fewer cutters than did conventional full-face rotary machines. The machine which finally emerged was certainly a revolutionary concept and very different in design from those which had so far been produced by other manufacturers (Figure 260).

The main frame had a spider-like arrangement held in place in the tunnel by four radially arranged jacks which extended against

281

Figure 254. Wirth T.B.M., Model TB 11 H. (Courtesy Wirth GmbH.)

the tunnel walls. Four oscillating arms, carrying one or two cutting discs each were mounted on a rotating head plate. The arms were moved towards one another in one cutting cycle and then out towards the periphery of the tunnel face on the next cutting cycle.* In other words the discs cut a spiral with a decreasing or increasing constant pitch, depending upon whether the arms were being drawn together or moved apart. The scissor-like movement of the four oscillating arms could be synchronized either by a single oscillating or reciprocating jack positioned in the axis of the machine, or by externally positioned hydraulic jacks. Thus, while the entire head rotated the four arms were being moved together (or moved apart as the case may be) so covering the entire face.

The cutting face was therefore divided into four toroidal coaxial zones for machines having one tool per arm and into eight zones for machines having two tools per arm.

Two thrust cylinders reacting against the rear grippers and bearing against the non-rotating section of the head serve to advance the machine and hold it up against the face. Unlike conventional rotary T.B.Ms., however, the

head is not continually advanced towards the face during the boring cycle.

The weight of the machine at the front is borne by two shoes on which it slides. These are arranged at approximately 60° on either side of the vertical axis of the drift. The back of the machine consists of two tubes which slide within a guideboss connected by jacks to a three- or four-point star-frame resting on the ground (the grippers).

Steering is effected by the rear jacks. The guidance system comprises a gas laser, and an optical system incorporating two plane normal mirrors and a ground glass sight. The block of two mirrors is mounted on the machine at the height of the front shoes. The sight is placed near the rear. A beam of light from the laser, situated in the tunnel at the rear, is directed towards the mirrors on the machine along a determined axis. The light beam is reflected by the mirrors back to the sight. By observing the position of the light spot in relation to the centre of the sight, the operator may judge the machine's axial position in relation to the axis of the tunnel.

The excavated material falls to the invert where it is directed towards the lower side of a chain belt conveyor (with lateral flights) by deflectors on the boring head. The conveyor

---

*The reader is invited to compare the cutting action of the UNI.BO machine with that built by E. Talbot in 1853 (see Chapter 3, Early rock tunnelling machines).

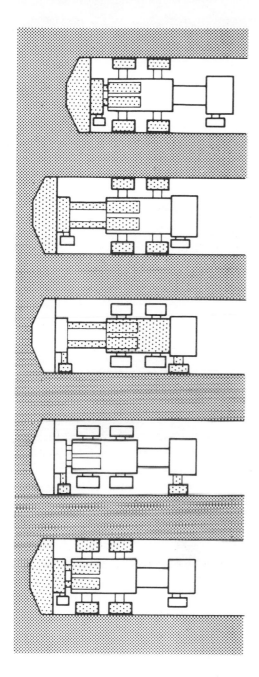

Figure 255. Drawing showing advance and repositioning cycle of Wirth tunnel boring machines. (Courtesy Wirth GmbH.)

which is suspended beneath the machine carries the material towards the rear.

An arc-shaped roof-shield may be extended by jacks against the crown if ground support is needed. As the jacks are on ball joints the machine's position in the tunnel may be adjusted without moving the shield. Holes are provided in the shield for grouting.

Jets of water for cooling and dust suppression are sprayed on to the head and on to each boring disc by a rotating nozzle.

The first UNI.BO machine (Model 3M) (Figure 260) was put to work in a 500 m (1600 ft) long R.A.T.P. (Paris Underground) R.E.P. (Regional Express System) exploration drift by the French contracting firm of Bouygues. The lime formations ranged in compressive strength from 10 to 100 MPa (1450 to 14,500 lb/in²). The machine had a cutting diameter of about 4 m (13 ft) and was fitted with four arms each bearing a single disc cutter.

The second unit with a cutting diameter of about 5.5 m (18 ft) was tried by Bouygues in an 800 m (2600 ft) long drift in Borie, Vianden (Grand Duchy of Luxemburg). The unit which had four arms was tested with a single disc cutter on each arm and also with two disc cutters on each arm.

Since then UNI.BO have produced an improved model featuring several modifications including more robust arms. Two of these new units were put to work boring a water-supply tunnel in Syria for the city of Damascus.

The project which included a 15 km (9 mile) long tunnel having two adits, one from Bassime and one from Al Ayoun, was divided into three sections. The contract for the first section was awarded to a joint venture of French and Greek contractors — Bouygues and Edok-Eter.

The UNI.BO units were operated in the Bassime and Al Ayoun adits while a Demag unit was used at Wali, where the control room, etc. was situated.

The new 30 t machines were fitted with three arms on each of which a single cutter disc was mounted. The head which revolved at 12 rpm had a power of 157 kW (210 hp). Penetration rates of 35 m (115 ft) per day have been attained with the UNI.BO units through homogeneous limestone with compressive strengths ranging from 40 to 50 MPa (5800 to 7250 lb/in²), while 15 to 20 m/day (49 to 66 ft/day) were achieved through some hard limestone formations interspersed with flint.

Two additional UNI.BO units were being built for use in Norway.[23]

Figure 256. Wirth tunnelling machine, Model TB 11, being assembled in factory before delivery to South Africa for use in the gold mines. (Courtesy Wirth GmbH.)

Figure 257. Wirth full-face boring machine, Model TB V-530E. (Courtesy Wirth GmbH.)

It is claimed by the manufacturers that because there are fewer boring discs (4-8 compared with about 20-30 on conventional rotary T.B.M.s) lower stresses are imposed upon the unit as a whole, permitting the use of a much lighter machine structure than would otherwise be necessary. Moreover, because each tool is mounted at the end of its arm, a small-diameter supporting cutter head may be used, thus giving better access to the face and front of the T.B.M.

Of course it should also be realized that while only a few cutters are used these would naturally wear at a much faster rate than those on a conventional unit of comparable size, because they would travel a much greater distance for a given length of tunnel.

French Profile Cutting Machine[24]

Stemming from the 'Berlin Wall' methods, whereby holes are drilled or excavated in the periphery of a tunnel or trench, followed by the insertion of steel beams which are then sealed with a lean-mix concrete, the French have

Figure 260. UNI.BO. tunnelling machine. (Courtesy Union Industrielle Blanzy-Ouest.)

Figure 261. UNI.BO. T.B.M. — improved model, front view. (Courtesy Bouygues, France.)

286

Figure 263. UNI.BO. T.B.M. — at work underground (Courtesy Bouygues, France.)

Figure 264. UNI.BO. T.B.Ms. in the workshop. (Courtesy Bouygues, France.)

recently developed a machine which takes this method of tunnelling from soft ground to hard rock.

Basically the purpose is to pre-cut, support, and thus isolate the area around a tunnel from the material within which is to be excavated.

The profile cutting machine (Figure 266) consists of a heavy metal arch supported by a pair of self-propelling crawler tracks. A robust cutting unit is mounted on twin rack gears fitted to the concave section of the arch. The cutting head itself is somewhat similar to the chain cutter units which were used extensively in the coal-mines for undercutting the coal seam, before the advent of the continuous miner. It consists of a jib around which a chain, carrying picks, is rotated. The rack gears on the arch enable the cutting unit to traverse the tunnel profile from the floor on one side to the floor on the other.

According to J. F. Bougard, Chief of the Underground Works Department in Paris, France, after extensive preliminary trials the machine was used for excavating the second regional subway line under the right bank of the river Seine through limestone formations. The project included two tunnels with internal spans of 5.7 m (18 ft 9 in). During the construction of the upper portion of these tunnels an 8 cm wide × 2 m deep (3 in × 6 ft 6 in) slot was first cut by the machine. A hydraulic drill jumbo was then moved in and the remaining core was drilled and blasted in the conventional way.

Later other methods were tried, including that of working loose uncohesive ground by first pre-cutting a slot. The slot was then filled with shot quick-setting concrete to form a protective shell beneath which the earth could subsequently be removed in safety.

In a letter to the author, Bougard reports that the 'mechanical pre-cutting' method had, by 1980 been in use on R.A.T.P.'s underground work sites in Paris for approximately five years, during which time it had proved satisfactory. Improvements were continuously being made. Because this is a relatively cheap method of tunnelling it offers a viable alternative to shields and full-face boring machines, particularly where short tunnel lengths are involved.

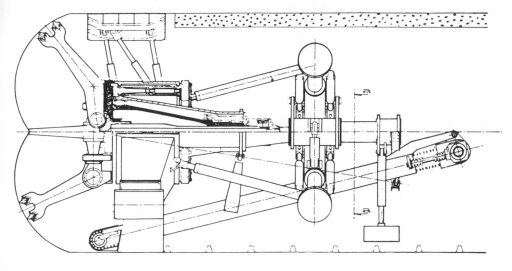

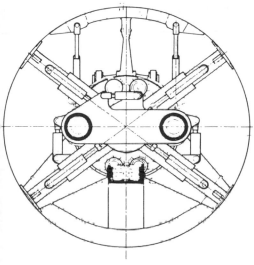

Figure 266. Side elevation and end view of UNI.BO. machine. (Courtesy Union Industrielle Blanzy-Ouest.)

Britain

Greenside/McAlpine T.B.M.[25, 26]

Following on the use by Sir Robert McAlpine of a Greenside header machine in the Birmingham road tunnel in 1968, the Greenside machine was subsequently modified by Sir Robert McAlpine & Sons Ltd. for civil engineering use to cut a circular tunnel.

Basically similar in concept to the original Greenside, the 1.22 m (4 ft) diameter rotating cutter head of the new Greenside/McAlpine machine (Figure 267) was mounted on a robust arm fitted transversely to the end of a sliding shaft. The rotating head which could be traversed

from one end of the arm to the other was fitted with 49 tungsten carbide tipped picks with two-way cutting edges (patented design).

During operation the unit was worked as follows. With the arm in a horizontal position the head was located at the centre for the commencement of a cutting cycle. The shaft was then extended to enable the head to sump 46 cm (18 in) into the face. The head then cut outwards as it moved along the bar to the left. This was repeated to the right so that a horizontal slot (1.22 m (48 in) wide by 46 cm (18 in) deep) was formed across the face. With the head at the right-hand end the material above the slot was removed by swinging the arm anti-clockwise through 180°. Lastly the head was

Figure 266. French profile cutting machine. (Courtesy Jean-François Bougard, Chief of Underground Works Dept., Régie Autonome des Transports Paris, France.)

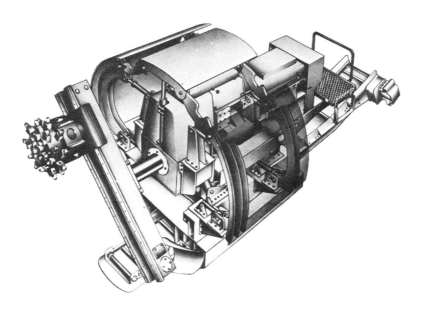

Figure 267. Green-side/McAlpine machine. (Courtesy Underground Mining Machinery Ltd., U.K.)

moved to the other end of the arm so that the remaining material below the horizontal slot could be removed by swinging the arm back through 180°.

The advantage of this type of unit is that apart from the initial sumping action where the face is attacked by the front picks on the head, the cutting head cuts sideways. Thus, except for

290

Figure 268. Double-headed Greenside/McAlpine machine. (Courtesy Underground Mining Machinery Ltd., U.K.)

the initial sumping action, the forces applied to the head are radial and torque forces only, and there is no axial stress exerted on the machine.

Material cut from the face is directed towards the conveyor at the bottom by the hydraulically operated loader blades. The conveyor moves the material to the rear under the machine, between the sides of the self-advancing base assembly.

The conveyor may be advanced or retracted so as to prevent it fouling with the cutter head and arm when these are operating across the invert of the tunnel.

The machine is supported on hydraulic jacks housed in a pair of floats or skids mounted on guideways. When the machine needs to be advanced or retracted, the skids are pushed forward (or drawn back) alternately by a pair of hydraulic cylinders mounted between the rear end of the skid and the rear end of the machine body.

The loader blades mentioned above also clear debris from the floor ahead of the walking base.

A protective canopy is fitted to the machine which serves both to support the crown and to anchor the unit during the cutting operation. Hydraulic rams raise and lower the head shield which also houses the dust extraction system.

A Greenside/McAlpine machine with one cutting head was used in 1968 at the Hinkley Nuclear Power Station to cut cooling water tunnels for the Central Electricity Generating Board (U.K.). In 1969 and 1970 respectively a Greenside/McAlpine was used at the Hunterston Nuclear Power Station and at St. Maximin, Provence.

A double head Greenside/McAlpine machine (Figure 268) was used in 1970 in the Severn-Wye cable tunnel and also at St. Maximin in conjunction with the single-head machine.

References

1. I. G. Tulloch and D. B. Sugden, *Mechanical Boring of Rock Tunnels with reference to the Robbins Tunnel Boring Machine used by the Hydro-Electric Commission at Poatina,* Parts I and II, The Hydro-Electric Commission, Hobart, Australia.
2. Enlarging a tunnel by rotary machinery, *The Tasmanian Railway Gazette,* 1 Oct. 1965.
3. Mangla, *Proc. Inst. Civ. Eng.,* No. 38 (Nov.), 1967, pp. 337-576 and No. 41 (Sept.), 1969, pp. 119-203.
4. J. C. McKenzie and G. S. Dodds, Mersey Kingsway Tunnel: construction, Paper 7481, *Proc. Inst. Civ. Eng.,* March 1972, **51**, 503-33.
5. J. M. Glass and C. D. Sholtess, The 'Hughes Tool' Mole Development, *Rapid Excavation —*

Problems and Progress, *Proc. Tun. and Shaft Con.,* May 1968, American Inst. of Min. Met. and Pet. Engs. New York, 1970.

6. *Kirk-Othmer Encyclopedia of Chemical Technology,* Vol. 10, Wiley-Interscience, New York, 1966.

7. *McGraw-Hill Encyclopedia of Science and Technology.* Vol. 6, 1960. McGraw-Hill, New York.

8. Tunnel and shaft boring at Gilsonite, *Engineering and Mining Journal,* July 1964, New York.

9. Unique tunnel borer used at ACG's hydraulic mining operation, *Engineering and Mining Journal,* July 1964 (Tunnel and shaft boring at Gilsonite) New York.

10. L. A. Garfield, Drift boring at White Pine Copper, *Rapid Excavation* — problems and progress, *Proc. Tun. and Shaft Con.,* May 1968, American Inst. of Min. Met. and Pet. Engs. New York, 1970.

11. W. E. Bruce and R. J. Morrell, *Principles of Rock Cutting Applied to Mechanical Boring Machines,* Twin Cities Min. Res. Center & Bureau of Mines, U.S. Dept of Interior, Minneapolis, Minnesota.

12. *Report on the Alkirk Pilot-Pull Principle. Application to various Types of Machines,* Ingersoll-Rand Co., U.S.A.

13. Hugh Ferguson, Tunnelling — Melbourne's troubled tunneller grinds to a halt. *New Civil Engineer.* Aug. 18, 1977, 16-17, London.

14. *Atlas Copco Manual,* Atlas Copco AB, Stockholm, Sweden, 1975.

15. *Bade Tunnelling Machines Technical Data,* Bade & Theelen GmbH, Lehrte, W. Germany.

16. Richard J. Robbins (Lecture 1), *History of Rock Boring,* S.A. Institute of Mining and Metallurgy, Feb. 1980, South Africa.

17. *Krupp — Tunnelfraser KTF 340,* Krupp Ind. -Und Stahlbau. Rheinhausen, W. Germany.

18. *Technical and Operational Data Krupp Tunnelling Machine KTF 280, Location — Talheim,* Krupp Industrie und Stahlbau, Reinhausen, W. Germany.

19. *Demag Tunnelling Machines — Full-Section Tunnelling Machine for Hard Rock,* Mannesman Demag, Duisberg, W. Germany.

20. *Construction of the Oker-Grane Tunnel with a Tunnel Heading Machine,* Mannesman Demag, Duisburg, W. Germany.

21. Martin Hunt (Ed.), Rapid road construction in a German mine, *Tunnels and Tunnelling,* **10,** No. 7, Sept. 1978, London.

22. UNI.BO *Tunnel Boring Machines Technical Data,* UNI.BO, Paris.

23. Martin Hunt, Fresh water at Damascus, *Tunnels and Tunnelling,* **9,** No. 5, 35, Sept. 1977, London.

24. Jean-François Bougard, *Urban Area Soft Ground Tunnelling,* Underground Works Dpt. Regie Autonome des Transports, Parisiens, Paris, France.

25. Bill Morse, Greenside-McAlpine tunneller, *Contractors' Plant Review,* June 1969, London.

26. N. D. Pirrie, Hinkley tunnels prove economics of machine for short distance tunnels. (Sir Robert McAlpine and Sons Ltd). *Tunnels and Tunnelling,* May-June, 1969, London.

Bibliography

Hugh Fraser, Records tumble at Navajo tunnel No. 3, *Western Construction,* Aug. 1972, Eugene, OR, USA.

Mechanical 'mole' breaks records for fast tunnelling, *Australian Civil Engineering and Construction,* 5 April 1963, 49-51, Sydney/Melbourne, Australia (publ. ceased).

Richard J. Robbins, A status report with an eye to the future, Tunnel and Shaft Excavation Conf., May 1968, Minneapolis Minn. The Robbins Company, Seattle, USA.

Richard J. Robbins, Vehicular tunnels in rock — a direction for development, Joint ASCE-ASME National Transportation Engineering Conf., 25-30 July 1971, Seattle, Washington.

Robbins tunnel boring machine, *Canadian Mining Journal,* 1955. Don Mills, Ont., U.S.A.

The Second Mersey Road Tunnel — Background Information, Mott, Hay and Anderson, Croydon, U.K.

Tunnel Boring — Development and Application of Hughes Hard Rock Tunnel Boring Machines. Hughes Tool Co, 1966, Houston, U.S.A.

L. B. Underwood, Soft rock tunnelling at Missouri river dams, *Civil Eng.,* July 1962, 40-43, London.

UNI.BO Machines à Forer Les Tunnels, UNI.BO, Paris.

Correspondence

The Robbins Company, Seattle, U.S.A.

Hydro-Electric Commission of Tasmania, Australia.

White Pine Copper Company, Michigan, U.S.A.

I. G. Tulloch, Hobart, Tasmania, Australia.

D. B. Sugden, Kingston, Tasmania, Australia.

Hughes Tool Company — J. M. Glass, Houston, U.S.A.

Ingersoll-Rand (Aus.) Ltd — J. H. Whitehead, Melbourne, Australia.

Jarva Inc. — E. W. Brickle, Solon, U.S.A.

Subterranean Equipment Co. — T. McCann, Denver, U.S.A.

Dresser Industries Inc., Texas, U.S.A.

Atlas Copco AB, Switzerland.

Bade & Co. GmbH (now Bade & Theelen GmbH), Lehrte, West Germany.

Fried. Krupp GmbH Machinen-Und Stahlbau Rheinhausen, West Germany.

Mannesmann Demag Bergwerktechnik, Duisberg, West Germany.

Alfred Wirth & Co. KG (now Wirth Maschinen und Bohrgerate Fabrik GmbH), Erkelenz, West Germany.

Anglo-American Corporation of South Africa Ltd. — R. N. Taylor.

Department of Transport, Paris — J. F. Bougard, France.

Sir Robert McAlpine & Sons Ltd. — J. Buchanan, London, U.K.

Bouygues, Paris, France.

SIX

Incline Boring Machines

While no specific records exist, it seems reasonable to assume that the first inclined slopes bored with a *full-face* T.B.M. would have been in the Stanley colliery at Nuneaton, England, where Reginald Stanley spent some 20 years (1885-1905) building and developing his tunnelling machines for use in the Stanley colliery and elsewhere.

After the Stanley came the McKinlay entry driver, two of which were used for 25 years to drive pairs of entries in the New Orient mine (U.S.A.) during the period 1926-1951.

So far as the new breed of full-face T.B.M.s is concerned (i.e. those that evolved after the introduction of the Robbins Company's 1953 unit), the Jarva Company claim to be the first to drive an inclined gallery in a coal-mine with this type of machine. It was used by the Republic Steel Corp. in their Adirondac mine, Mineville, N.Y., in 1967 to drive a slope with an inclination of 27°. The machine in question was a 3.05 m (10 ft) diameter Jarva Inc. Mark 11 unit, using carbide insert kerf cutters.

Work was commenced in April and completed in November of that year. The unit which was capable of developing a maximum thrust of 2452 kN (550,000 lb) penetrated 234 m (768 ft) of varied strata containing magnetite, hornblende biotite gneiss, and grey granite gneiss. The cutter head was powered by 298 kW (400 hp). (Later the Hanna Coal Company became the second coal company to use such a machine for an inclined gallery.)

On 1 October 1968, a 3.00 m (9 ft 10 in) diameter Alfred Wirth & Co. KG rock-tunnelling machine, Model TB11-300E, was used to bore a 33°, 1150 m (3770 ft) long inclined gallery ('Corbes'), for the Gran Emosson project at Châtelard. The material penetrated was granite and the machine was dressed with carbide insert cutters. The project was successfully concluded on 1 September 1969.

The following year, i.e. on 17 March 1969, a 2.25 m (7 ft 5 in) Wirth rock-tunnelling machine (Model TB1-214) with carbide insert cutters was used for a 42° inclined gallery ('Barberine') on the Gran Emosson project at Châtelard, Switzerland. The gallery which was some 1000 m (3300 ft) in length was completed on 16 December 1969.

Other projects involving inclined galleries or shafts bored by Wirth units include a pressure gallery at Wehr, Black Forest (inclination 28°), a ventilation shaft at St. Gotthard (inclination 42°), a glacier rail shaft in Kaprun (Figure 269) (inclination 28°) and a gallery at Tschingelmad, Switzerland, (inclination 27°).

Of interest are the Black Forest pressure gallery at Wehr and the inclined rail shaft at Kaprun, Austria.

When the Black Forest gallery was worked, a 1400 m (4600 ft) long pilot shaft was first driven up towards the Hornberg stage by a 3.0 m (9 ft 10 in) diameter Wirth TB11-300E unit. The shaft was then back-reamed from the top down by means of a Wirth 6.30 m (20 ft 8 in) diameter TBE 300/600 reaming machine. Tungsten carbide insert cutters were used on both machines — 26 on the pilot shaft machine and 33 on the reaming unit.

The Kaprun inclined railway shaft was built to cater for increased transportation demands which called for expansion to the existing glacier cableway system. Before boring work could be commenced the Wirth unit had to be lifted up a height of some 600 m (2000 ft) to the face of the proposed shaft, which was then bored upwards.

Figure 269. Wirth T.B.M. being launched for operation on the 23° inclination rail shaft in Kaprun. (Courtesy Wirth GmbH.)

In 1971 a Calweld T.B.M. Model 64 was used by McGuire Shaft and Tunnelling Corp. to bore a 5.18 m (17 ft) (o.d.) inclined gallery for the Amax Coal Co. The material penetrated in the 900 m (3000 ft) long gallery included limestone, sandstone, and shale, and the machine was dressed with disc cutters.

A Demag machine Model TVM 34-38/42HS was used to bore a 35°, 4.20 m (13 ft 9 in) diameter inclined shaft at Mapragg (Figure 251) through limestone and schist with compressive strengths in the range from 78 to 127 MPa (11,300 to 18,400 lb/in²).

Advance rates averaged 2.4 m/hour (94 in/hour) (31 m/day (102 ft/day) in a 2 × 10 hour day shift).

The project was commenced on 2 March 1973 and was completed on 19 September 1973.

So far as the author is aware the *steepest slope* (45°) driven to date (1980) is the 812 m (2660 ft) long Hydro Penstock gallery at Gunsel Oberaar, Switzerland in 1975. The machine used was a Robbins Company 4.3 m (14 ft 1½ in) diameter unit (Model 145-168) carrying 38 cm (1 ft 3 in) disc cutters. The cutter head, which could develop a torque of approximately 960 kN m (708,000 lb ft) was powered by a 634 kW (850 hp) motor, and the machine's advance jacks were capable of exerting a total maximum thrust of 862 MPa (125,000 lb/in²). The shaft was worked upwards.

Both the Robbins unit and the Wirth machines were fitted with special 'non-return' or 'anti-back slip' protective systems.

Wirth 'Non-return' Protective System[1]

The Wirth non-return protective system was patented in Germany on 23 June 1969 (No. 19 31 775.0) and in the United States on 15 February 1972 (No. 3,642,326). The patent was granted to Willi Steufmehl of Erkelenz Rhineland, Germany, and was assigned to Wirth & Co. KG in 1969.

Basically the system consists of two independent gripping or clamping systems, each capable of holding the T.B.M. in the tunnel and preventing it from slipping back down the slope.

The T.B.M. is divided into three interconnected sections comprising: (1) the cutter head and drive system; (2) the auxiliary bracing mechanism framework made of longitudinal beams on which are mounted two pairs of diametrically opposed gripper pads; (3) a trailer housing the operator's cabin and the hydraulic supply source for the entire machine.

On the front section, i.e. that carrying the rotary cutting head and drive system, are mounted two sets of four hydraulically powered gripper pads which thrust against the tunnel walls and so hold the machine in position during the boring operation. At the end of a boring stroke the front grippers are retracted and the machine is held by the auxiliary grippers on the bracing section, while the front grippers are advanced and repositioned ready for the next boring cycle. These auxiliary grippers automatically come into effect when the front grippers are retracted.

As a safety measure the auxiliary bracing system operates as follows: powerful springs housed in telescopic tubes are held in a retracted position by hydraulically actuated cylinders. Thus if for some reason, say because of a burst hose, there should be a failure in the hydraulic system, the springs would automatically be released so that they would thrust the auxiliary grippers against the tunnel walls and hold the machine in position preventing it from accidentally back-slipping.

The Robbins Anti-Back Slip System

The Robbins anti-back slip (ABS) system (Figures 270 and 271) consists basically of a structural frame and muck chute, a reaction ring, two hydraulic cylinders, and a control and interlock system.

The two ABS hydraulic cylinders are positioned on either side of the machine muck chute which runs along the floor of the unit. It accepts the muck that enters via the opening under the cutter-head support and directs it into the tunnel muck chute. The aft end of the machine muck chute is telescoped into the tunnel muck chute which is installed behind the machine as it advances up the incline. The forward end of the structural frame on which the ABS cylinders are mounted bear against the cutter-head support. When the ABS cylinders

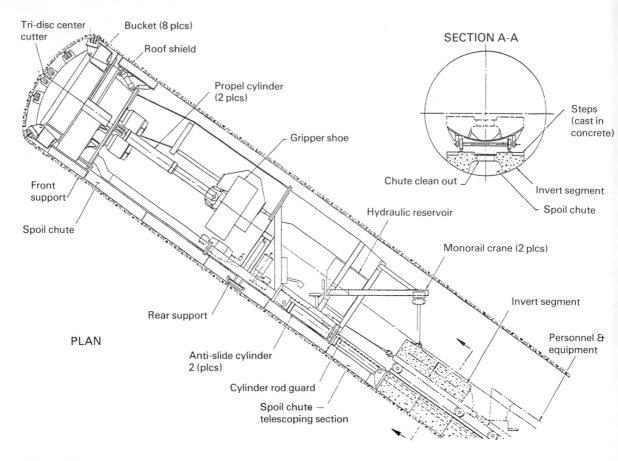

Figure 270. Side elevation of Robbins incline boring machine. (Courtesy The Robbins Company)

are extended, thrust reaction at the rear is taken by the reaction ring which in turn transmits it to the tunnel muck chute. The ABS cylinders are capable of providing a thrust force of 202 t and they have a stroke of 274 cm (108 in), which is two strokes of the T.B.M.'s propel cylinders.

A special control and interlock system ensures that the gripper and propel cylinders are prevented from retracting unless the ABS cylinders are pressurized sufficiently to hold the machine on the incline. This system also prevents the ABS cylinders from retracting unless the gripper and propel cylinders are pressurized sufficiently to hold the machine on the incline. Limit switches are provided to prevent forcing the machine past the limit of travel of the ABS cylinders. Thus when the machine is operating the following procedure is followed:

At the beginning of the cycle and with both the ABS cylinders and the propel and gripper cylinders pressurized, the T.B.M. is advanced into the face. When the T.B.M.'s thrust cylinders are fully extended the T.B.M. is held in position by means of the ABS cylinders which bear against the tunnel muck chute via the reaction ring. The grippers are then retracted and repositioned and the machine is ready for another forward thrust. At the end of the second forward thrust of the propel cylinders, the ABS cylinders are fully extended. While the gripper and propel cylinders are still pressurized and holding the machine, the ABS cylinders are retracted and a new section of tunnel muck chute is installed and bolted in place. The ABS cylinders can then be extended so that the reaction ring is bearing against the recently installed tunnel muck chute section. With the ABS cylinders still pressurized and holding the machine, the grippers are retracted and repositioned and the machine is ready to commence another boring cycle.

298

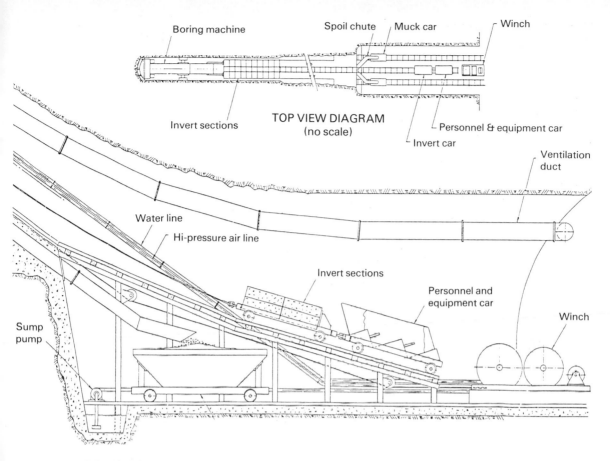

Boring machine · Spoil chute · Muck car · Winch

Invert sections

TOP VIEW DIAGRAM
(no scale)

Personnel & equipment car

Invert car

Ventilation duct

Water line

Hi-pressure air line

Invert sections

Personnel and equipment car

Winch

Sump pump

Figure 271. Side elevation showing muck and equipment car for Robbins incline boring machine. (Courtesy The Robbins Company)

Thus the ABS cylinders remain pressurized at all times except during the installation of a new section of tunnel muck chute, which occurs after every other propel stroke.

The holding valves are mounted in manifolds which bolt directly on to the cylinders, and the rod end of the cylinders needs to be pressurized before the valve opens and allows oil to exit from the piston end of the cylinder. This arrangement ensures that the holding capabilities of the system will not be jeopardized by a sudden failure of the hydraulic system due to a broken line, etc. In addition, any interruption to the electrical current will automatically cause the valves to close, producing in other words, electrically a 'fail-safe' system.

During long periods of inoperation it was recommended by the manufacturers that the machine should be supported by the tunnel muck chute with the ABS cylinders in their fully retracted position. This would mean that at such times the machine was not relying on hydraulic pressure to hold it in the incline.

The Robbins ABS system was first used in 1972 by Theiler and Kalbermater to drive the 1400 m (4600 ft) Gries gas pipeline tunnel in Ulrichen, Switzerland. A Robbins T.B.M. (123-133-1) was used to drive the incline which penetrated schist with compressive strengths ranging from 118 to 173 MPa (17,100 to 25,100 lb/in²). (This machine was originally used in the Thompson-Yarra water diversion tunnel in Melbourne, Australia.) Special precast invert segments were used for the job. These segments accepted the thrust reaction from the machine's ABS cylinders as the T.B.M. was advanced and also supported the tunnel muck chute and carriage system for personnel and equipment.

Robbins Mini Hard-rock Incline T.B.M. (Reef Raiser)

In 1977 two Robbins mini hard-rock tunnelling machines (Models Nos. 61-176 and 61-177) (Figure 272) were delivered to South Africa for use in the gold mines there.

The machines are designed to bore at a nominal angle of 30° to the horizontal, to enable them to follow the gold reef. This boring angle may be increased to a maximum of 45°, if required.

The machine can be disassembled into several components for transportation through the mine on special cars.

The cutter head is powered by a 150 kW (200 hp) motor and carries nineteen 31.75 cm (12½ in) diameter disc cutters. Thrust is applied to the cutter head by hydraulic cylinders which react against the grippers.

The grippers used on this machine differ from the standard Robbins 'floating grippers' normally used. They are similar in design to those frequently used by other manufacturers, consisting basically of rigid telescoping box members which extend laterally against the tunnel walls. Normally, when this type of gripper system is used, steering can only be carried out during the repositioning cycle and not during the boring phase, as is possible with the floating gripper system (see also section on Poatina machine). However, in the Robbins reef raisers steering is carried out during the boring phase by using side steer shoes controlled by an automatic 'micro feed system'. In addition the micro feed system controls the degree of steer to ensure that the machine is not oversteered in any direction.

The operator, controls, and hydraulic power supply are pulled behind the machine on a skid. All machine functions are controlled from the operator's console.

Because the machine is intended to work on an incline, no spoil-conveying system is used. Instead muck from in front of the cutter head is washed down into the invert of the tunnel and allowed to flow naturally away from the working area.

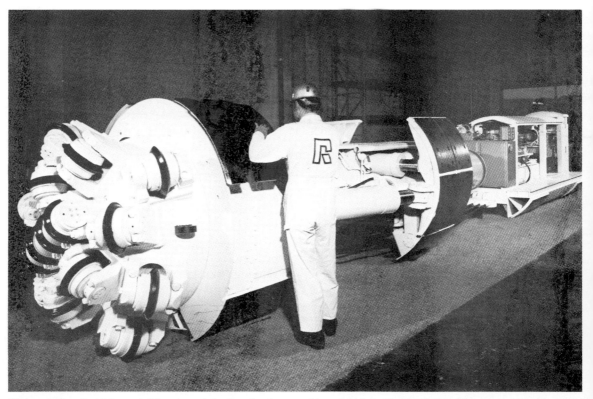

Figure 272. Robbins mini hard-rock incline boring machine, Models 61-176/7 'Reef Raiser'. (Courtesy The Robbins Company.)

The machine has a 61 cm (24 in) boring stroke and the manufacturers expect penetration rates to be approximately 1.5-3.0 m (5-10 ft)/hour (see also section entitled 'Circular full-face mini rock T.B.M.s').

Conclusion

It should be noted that there is an important difference between the tunnels driven with the Jarva machines and those subsequently bored with the Wirth, Robbins, and Calweld T.B.M.s. While the tunnels bored by the latter machines were all at a 'positive angle', those driven by the Jarva were at a 'negative angle'. In other words the Jarva machines bored *downwards* while the Wirth and Robbins machines, etc. bored *upwards*. It can be readily appreciated that quite different problems, in particular those concerning muck removal, forward thrust, and backsliding, would arise in connection with these two boring angles.

In June 1975 a 4.26 m (14 ft) diameter Mark 12 Jarva machine was delivered to the Rochester & Pittsburgh Coal Company's Urling mines at Shelocta, Pennsylvania, for use in 'slope conveyor' tunnels.[2]

Because the machine was boring downwards water naturally tended to run to the face. According to Walter E. Bullers (Division engineer for the Rochester & Pittsburgh Coal Co.) this caused outages or stoppages for the duration of a complete shift on 3 separate occasions out of 265 shifts during the boring of the first slope. These problems were apparently initiated by either a pump failure or a power cut-out. During the driving of the next slope a diesel generator was used as a spare power source and, in addition, a spare pump was kept on hand. These greatly minimized this problem. (During construction the slopes kept dry while an average of 114 litres (25 gal) of water per minute was pumped from the sumps.)

Muck removal presented no problems except when the machine was penetrating 'fire clay'.

During this period the clay, which was wet and adhesive, stuck in the muck buckets, discharge hopper, and on the conveyor belt. In a letter to the author Bullers commented that the only solution was to stop frequently and clean out the clay before the blockage became too severe. Sometimes the clay became so hard it could only be removed by means of an air chipping hammer. When the second slope was bored, the angle of the discharge belt which had apparently been operating at too steep an angle, was altered.

Initially the machine tended to nose-dive when penetrating softer areas such as coal seams, etc. However, it was found that with better control by a more experienced operator, this problem was largely overcome.

References

1. Wirth 'Non-Return' Patent No. U.S. 3,643,326/1972, Alfred Wirth & Co. KG, Erkelenze, W. Germany.
2. Walter E. Bullers, *Boring a coal mine slope*, *Mining Congress Journal*, **63**, No. 5, 40-43, May 1977, Washington, D.C.

Bibliography

Floyd H. Nickeson, *Oak Park's output goes to industry*, *Coal Mining & Processing*, Aug. 1972, 42-46, Chicago, U.S.A.
Tunnels and Tunnelling, **5**, No. 6, 549-556, Nov./Dec. 1973, London.

Correspondence

Jarva Inc., Solon, U.S.A.
Calweld (Division of Smith Industries Int. Inc.), Santa Fe Springs, U.S.A.
The Robbins Company, Seattle, U.S.A.
Walter E. Bullers — Rochester & Pittsburgh Coal Company, U.S.A.
Alfred Wirth & Co. KG (now Wirth Maschinen und Bohrgerate Fabrik GmbH), Erkelenze, West Germany.

SEVEN

Reaming Machines

T. English

So far as the author is aware English was the first to moot the idea of a reaming machine. The suggestion was put forward in the specifications which accompanied Thomas English's Patent No. 5317 (Figure 273) dated 5 December 1881, covering 'Improvements in tunnelling machinery'.

In the 1880 English patent the machine was designed to bore a tunnel of only moderate size, in other words one just large enough to admit the machinery and the workmen. English suggested that this tunnel could, if desired, be enlarged by inverting the machine, i.e. turning it end for end and then fixing a larger boring head on the shaft at what would then become the 'rear end' of the machine. By this means the smaller tunnel could be reamed out. The reader is invited to examine the illustration which accompanied English's 1881 specifications and compare it with pictures of the modern Wirth reaming machines described in this section.

English suggested further that if two machines were employed in the same tunnel, with the first at work some distance ahead of the second, a reaming machine, the debris from the first machine could be removed from the narrower

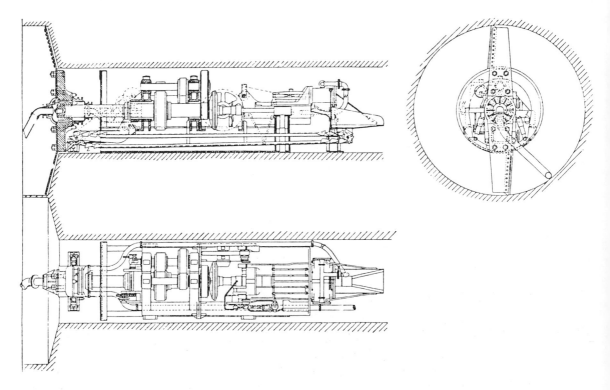

Figure 273. Reaming machine proposed by Thomas English in his Patent No. 5,317, dated 5 Dec. 1881

bore by a continuously travelling chain of buckets which received the debris made by the first machine and delivered it into the enlarged tunnel at the larger boring head, from which point it could be removed by trucks, etc.

In order that the pipes conveying compressed air, etc. to the front machine should not be interfered with, English suggested that the pipe be connected to a swivel union at the centre of the large boring head, and from this union the pipe could be bent so that it would pass the boss of the boring head to a hollow stationary sleeve fitted with stuffing boxes on the revolving shaft. From this sleeve the air could be conducted by pipes to the engines that operated either or both machines. As an alternative to

bending the pipe and carrying it past the boss of the boring head, the shaft could be bored to allow the pipe through.

Another suggestion was that the one machine could carry both the front head and the rear enlarging head, but in this case it would be necessary to counter-rotate the heads in order to prevent the machine from turning in the bore.

Alfred Wirth & Company kg (now Wirth Maschinen GmbH)

The idea was not, however, tested in practice until 1969 when the first reaming machines

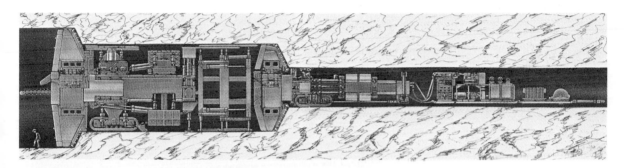

Figure 274. Wirth reaming machine TBE 350/770 and TBE 770/1046. (Courtesy Wirth GmbH.)

Figure 275. Rear view of Wirth reaming machine at the south entrance of the Sonnenberg road tunnel near Lucerne, about to commence enlarging the bore from 7.70 to 10.46 m (25 ft 3 in to 34 ft 4 in) diameter. (Courtesy Wirth GmbH.)

were built by the Alfred Wirth Company which pioneered this unusual concept in large-diameter machine-bored tunnels.

The prototype units were used for the twin Swiss highway N2 Sonnenberg road tunnels in Grienz, Lucerne, which ran from GroBhof to Gütsch. The two 1400 m (4600 ft) long tunnels were 10.5 m (34 ft 4 in) in diameter and penetrated difficult strata comprising rock which consisted partially of calcareous sandstone and quartz intercalated with marl.

The method of operation was as follows. Initially a 3.50 m (11 ft 6 in) pilot tunnel was driven through the centre of the proposed bore with a Robbins T.B.M. Model 111-117. (This machine was originally used to drive a sewer tunnel in Baden, Switzerland, in 1965 and for other subsequent projects). When the pilot tunnel had been completed, the hole was enlarged by two Wirth enlarging machines (Model TBE 350/770 and TBE 770/104) (Figures 274 and 275) with diameters of 7.70 and 10.46 m (25 ft 3 in and 34 ft 4 in) respectively. These two latter machines were coupled together and operated alternately; that is, while one machine was in the process of boring, the other was being reset and aligned in readiness for the next cycle. A walking roof support was provided between the two machines in the 7.70 m (25 ft 3 in) diameter bore as a protective measure.

All the machines, including the Robbins unit used for the pilot tunnel, were dressed with disc cutters.

The two larger Wirth machines were interlocked in the already bored tunnel and the cutter head in each case was positioned *behind* the machine (exactly as suggested by English in 1881 - Figure 275). Thus tunnel lining and other associated tunnel support work could be carried out in the relatively free area behind the last cutter head (see Figure 274).

Work on the project commenced in 1969 and was completed on 28 November 1973. (The pilot tunnel was completed in 1970.)

Subsequently, two more projects were undertaken with Wirth reaming machines TB 11-300E and TBE 300/600. These were the 28° inclined pressure gallery at Wehr, Black Forest, which was commenced in May 1970 and completed on 10 October 1972, and the 42° inclined ventilation shaft (TB 11-300E and TBE 300/600H) for St. Gotthard, which was commenced on 7 May 1972 and completed on 13 September 1973.

Bibliography

Günter Girnau, Research and Development in West German Tunnelling *Tunnels and Tunnelling,* Nov.-Dec. 1973, **5**, No. 6, 549-556.

Correspondence

Alfred Wirth & Co. K.G. (now Wirth Maschinen und Bohrgerate Fabrik GmbH), Erkelenze, West Germany.
The Robbins Company, Seattle, U.S.A.

Patents

Patent No. 5317/1881. U.K.

EIGHT

Mini Rock-Tunnelling Machines

Rectangular Mini Rock-Tunnelling Machines

Atlas Copco Mini Full-facers

The urgent need for a full-face tunnel borer for small-sized tunnels had been recognized by the industry for some time. These were required for water mains, gas ducts, electric and communication cables, and sewer lines, etc. especially where these services passed through rock formations and beneath built-up areas where cut and cover methods were impracticable and blasting was prohibited.

In 1971 Atlas Copco produced their first mini full-facer (Figures 276 to 280). It featured a single cutter head mounted on a horizontal axis at right angles to the axis of the tunnel. Using the same undercutting method described earlier (page 274), it cut an arch-shaped tunnel 1.30 × 2.10 m (4 ft 3 in × 6 ft 11 in) in size.

Thrust, drive, and muck removal were provided by a compactly designed machine with an adequate back-up system which included rock support, materials supply, and extension of service lines to the tunnel.

The cutter head disc was fitted with radially arranged cemented carbide-tipped cutter tools set on its periphery. In operation the head is swung from a starting downward position upwards through an angle of 155°, producing a tunnel cross-section with vertical walls, a flat floor, and an arched roof. On its downward movement the cutter head idles. Before a new cycle commences the machine is advanced a predetermined distance by a hydraulic propulsion cylinder.

Since their introduction, Atlas Copco machines have been used in increasing numbers in Switzerland, Italy, Norway, France, England, Canada, and Australia, in which latter country the author was able to inspect the unit driving the 2400 m (7900 ft) Engadine Water Board sewer tunnel in Sydney — to date (1980) reputed to be the longest small-bore tunnel in the world.

In this project the muck from the face was carried to the rear of the unit by the T.B.M.'s conveyor system. It was then deposited directly into a muck train which transported the spoil to the surface where it was dumped. While the train was carrying a load out of the tunnel the mini continued operating until the muck on its conveyor was detected by an electric eye set in the side of the conveyor at its rear. This automatically switched off the conveyor system until the muck train was again hitched to the conveyor, thus preventing the spoil from being deposited on to the tracks. To ensure that this device functioned properly, the maintenance engineer occasionally cleaned the eye with his hand on his way to and from the main section of the unit. According to a representative of Atlas Copco a blower was later installed obviating the necessity to clean the eye by hand.

The Robbins Mini Borer

In 1973 the Robbins Company produced a rock-tunnelling machine (Figure 281) which differed radically from its predecessors and, indeed, from those built by any other manufacturer.

Apart from the fact that it was designed to bore a 2.1 m high × 1.5 m wide (7 ft × 5 ft) rectangular tunnel, the head carried two arms, each fitted with a pair of disc cutters. The arms moved in towards the face at the top of the

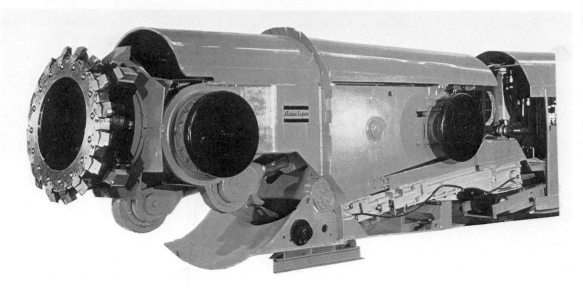

Figure 276. Atlas Copco — mini full-facer (rectangular). (Courtesy Atlas Copco.)

Figure 277. Atlas Copco — mini full-facer (rectangular). 1. Cutter head; 2. Front gripper unit; 3. Chain conveyor; 4. Protecting roof; 5. Machine body; 6. Rear gripper unit (also lateral steering); 7. Operator's cabin; 8. Trailer; 9. Hydraulic power pack; 10. Drive motor for chain conveyor (Courtesy Atlas Copco.)

boring stroke and then continued down to the invert where they moved back towards the machine and a fixed shoe blade resting on the floor beneath the head. The arms then moved up to the top of the bore and in towards the face for the commencement of the next cutting cycle.

The hydraulically powered cutter head was driven by a 104 kW (140 hp) motor. The unit was capable of exerting 445 kN (100,000 lb) of thrust and 90 kN m (66,665 lb ft) of torque. It weighed 18 t.

Four grippers, two on either side of the

machine served to hold the unit steady in the tunnel during the boring operation.

Muck from the face was swept towards the front shoe or blade by the lower cutting discs on their backward stroke and then directed on to a conveyor belt running beneath the machine to the rear. A hydraulically actuated roof canopy protected the operator in his fully air-conditioned cabin and the machine itself near the tunnel face.

The machine which was built by a five-man Robbins crew at the company's Tasmanian factory in Australia, took about nine months to

306

Figure 278. Atlas Copco mini full-facer undergoing routine maintenance operations. This machine was used to drive the 2400 m (7870 ft) Engadine Water Board sewer tunnel in Sydney, Australia. (Courtesy Atlas Copco.)

Figure 279. The Atlas Copco mini full-facer at work in the Engadine sewer tunnel — reputed to be the longest 'small-bore' tunnel in the world at that time (1977). (Courtesy Atlas Copco.)

construct. It was tested in a sewer tunnel at West Lane Cove, Sydney, Australia, by the Sydney Water Board.

However, as may be expected of a prototype unit, many teething problems developed, necessitating the withdrawal and return of the machine to Tasmania for modification. This took some time to effect and was still under way in 1974 when the Sydney Water Board, whose need for a small machine remained, were offered an Atlas Copco mini full-facer. A unit was tested at Barden Road, Sydney, in sandstone with compressive strengths ranging from 40 to 60 MPa (5800 to 8700 lb/in²).

Figure 280. Atlas Copco mini full-facer muck train dropping its load at the discharge platform. (Courtesy Atlas Copco.)

Figure 281. Robbins mini borer — Model 75-150. This machine which cuts a rectangular tunnel 1.52 × 2.13 m (5 ft × 7 ft) was designed to bore small-aperture tunnels for use as sewers, etc. (Courtesy The Robbins Company.)

This was the sixth mini full-facer produced by Atlas Copco, and understandably most teething problems had by this time been overcome. The machine sucessfully bored the 425 m (1400 ft)-long tunnel and was then promptly put to work at Como, Sydney, in a 598 m (2000 ft) sewer tunnel with comparable sandstone formations.

The Robbins mini borer was in the meantime modified. It was tested in a sandstone formation in Tasmania where it performed satisfactorily. However, due to difficulties associated with the contract the unit was never reinstalled in Sydney.

Circular Full-Face Mini Rock T.B.M.s

Jarva Mini Hard-rock T.B.M.[1]

During May 1977 Jarva's new small-diameter tunnel-boring machine, the Mark 6 (Figures 282 and 283), went into action boring a 1350 m (4430 ft) long, 1.98 m (6 ft 6 in) diameter tunnel for the city of Davenport, Iowa, U.S.A.

The unit was used by the J. E. Sieben Construction Company who were installing 1.37 m (54 in) i.d. concrete pipes in the bore.

Four hydraulic clamp legs (grippers) are fitted to the main frame of the unit while two hydraulic alignment jacks are attached to the rear section housing, the gear reducer, and electric drive motor. These legs and jacks extend laterally from the unit towards the tunnel walls.

Two lift legs (one front and one rear) support the machine, while the alignment jacks hold it in position in the bore, during the period when the unit is being repositioned for a new boring stroke and its four horizontal clamp legs have been retracted.

Thrust cylinders on larger Jarva machines are mounted between the rear of the main frame and the front of the gear housing. These cylinders pull the gear housing forward, transmitting thrust through the torque tube to the cutter wheel. The torque tube is supported and slides on bronze pads located in the main frame. When the machine is being repositioned,

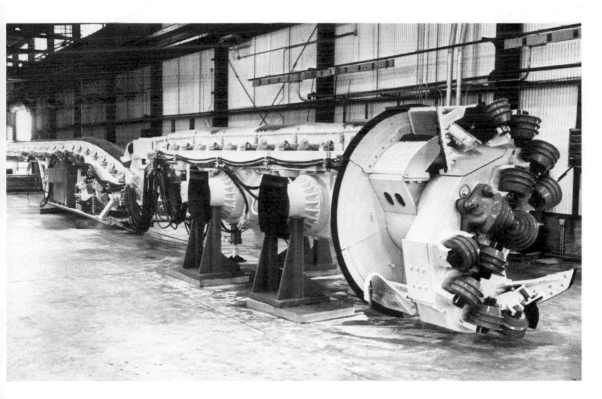

Figure 282. Jarva Mark 6 — mini Hard-rock tunnelling machine — used by J. E. Sieben Construction Company in Davenport, Iowa 2 m (6 ft 6 in) diameter sewer tunnel. (Courtesy Jarva Inc.)

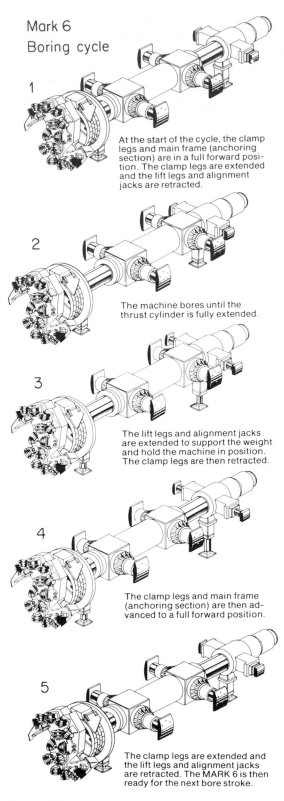

Mark 6
Boring cycle

1

At the start of the cycle, the clamp legs and main frame (anchoring section) are in a full forward position. The clamp legs are extended and the lift legs and alignment jacks are retracted.

2

The machine bores until the thrust cylinder is fully extended.

3

The lift legs and alignment jacks are extended to support the weight and hold the machine in position. The clamp legs are then retracted.

4

The clamp legs and main frame (anchoring section) are then advanced to a full forward position.

5

The clamp legs are extended and the lift legs and alignment jacks are retracted. The MARK 6 is then ready for the next bore stroke.

Figure 283. Jarva Mark 6 mini hard-rock tunnelling machine boring cycle. (Courtesy Jarva Inc.)

these cylinders operate (with reverse oil flow) to move the main frame up to its new position.

In the Mark 6, however, the hydraulic thrust cylinder is designed as an integral part of the main frame. The main frame in essence is one large hydraulic cylinder designed to handle the thrust and torque reactions. The cylinder rod (ram) is connected to the back of the cutter wheel. The rod supports the cutter wheel weight while providing thrust to the wheel.

Muck buckets mounted on the back of the cutter wheel scoop up cuttings and carry them to the top of the machine. There they are dumped on a belt conveyor which carries them back over the top of the machine to a trailing conveyor. The trailing conveyor dumps the material into muck cars for haulage out of the tunnel.

After a few initial teething problems had been dealt with, the Mark 6 performed satisfactorily, cutting its way through dolomite, limestone, and other formations in the Davenport hills. Penetration rates of between 10 and 12 m (35 and 40 ft) of advance per eight hour shift were consistently attained with the unit.

The Mark 6 is 7.3 m (24 ft) long and has a cutting stroke of 0.6 m (2 ft). Weighing 27 t, the machine can be transported fully assembled and lowered in one piece into the tunnel. After hydraulic and electrical connections are made between the power trailer sled and the machine, the contractor can begin mining.

Robbins Mini Hard-rock T.B.M. (Reef Raiser)

The same year (i.e. 1977) that the Jarva Company produced their Mark 6 unit, the Robbins Company delivered two small hard-rock machines to the South African gold mines for use as reef raisers. These units (Models 61-176 and 61-177) (Figures 284 and 285) with head diameters of 1.98 m (6 ft 6 in) have been used successfully through quartzite with compressive strengths as high as 117 MPa (26,000 lb/in^2).

They were specifically designed to follow the angle of the gold reef and are more fully described in Chapter 6 'Incline boring machines — Robbins mini hard-rock incline boring machine'.

Figure 284. Robbins mini hard-rock circular full-face T.B.M. — 'Reef Raisers' (Model 61-176/7). Used in South African gold-mines. (Courtesy The Robbins Company.)

References

1. Small rock tunneler proves out, *Highway & Heavy Construction,* Feb. 1978, 50-51, New York.

Correspondence

Atlas Copco Company, Sweden, Thun, and Australia
The Robbins Company, Seattle, U.S.A.
Jarva Inc., Solon, U.S.A.

Figure 285. Cutter head of Robbins mini hard-rock T.B.M. (Courtesy The Robbins Company.)

NINE

Water-Jet Assisted Rock Machines

Water-Jet Assisted Drag-Bit Rock Cutters[1,2]

During the mid-1970s after the development of the drag-bit rock cutter (see Chapter 21 on South Africa — gold mining), work was undertaken by the Chamber of Mines of South Africa Research Organization to investigate the effect of combining the drag-bit with strong jets of water.

This work was initiated as a result of prior investigations which were conducted to find a method of improving the life of drag-bit cutters. Initial experiments showed that in the majority of cases the bit failed at the brazed joint between the tungsten carbide insert and the steel bit body.

Using thermocouples, the temperature of the brazed joint was measured in the laboratory during the cutting action. The very high temperatures reached seemed to indicate that failure of the bit at this point was due to excessive heat. The bit was then cooled with a 7 MPa (1015 lb/in^2) 'flat-fan' water jet directed across the wide width of the tungsten carbide bit insert. This in turn led to the discovery that the penetrating force of the bit could be reduced considerably with the assistance of a water jet. Experiments to investigate this aspect were then commenced.

Nozzles were mounted along the leading edge of the tungsten carbide bit inserts (Figure 286) and various configurations of jets, i.e. single and double, directed at different points in relation to the cutting tool, were tried. The flow was maintained at a steady 30 litres min (8 gal/min), while the water jet pressure was varied between 8 and 55 MPa (1160 and 8000 lb/in^2).

Numerous tests were also conducted so that comparisons could be made between cuts taken at various fixed depths, with and without water jets.

Although highly abrasive quartzite (ranging in compressive strength from 200 to 300 MPa (29,000 to 43,500 lb/in^2) and with a quartz content of between 35 and 95 per cent) constitutes the country rock adjacent to the gold reef, a block of quartzite of the size needed for the laboratory tests was unobtainable. Therefore norite, which is similar in compressive strength (about 300 MPa (43,500 lb/in^2)) but far less abrasive, was used instead.

According to M. Hood[1] in one particular experiment where norite was being cut with 35 mm (1.4 in) wide bits, the rock was penetrated to a maximum depth of 4.5 mm (0.18 in) without water jets. With water jets the depth of cut increased to 9 mm (0.36 in). At the same time water pressures varying in range from 8 to 10 MPa (1200 to 1500 lb/in^2) and 50 to 55 MPa (7300 to 8000 lb/in^2) were tried. The *cutting force* (Figure 287) (force in the direction of cutting) was *reduced* substantially even when water jets at pressures as low as 10 MPa were used (as may be seen from Hood's graphs).

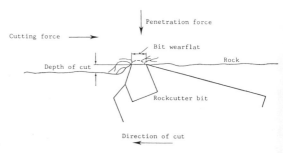

Figure 286. Diagram of a bit cutting the rock, illustrating the components of the bit force. (Courtesy M. Hood, *Journal of the South African Institute of Mining and Metallurgy,* Nov. 1976.)

However, there was no significant improvement in the reduction of cutting force when the pressure of the jets was increased to 50 MPa (7300 lb/in^2). On the other hand Hood's graphs (Figure 287) show that the *penetrating force* (force normal to the cutting direction) needed to achieve the same depth of cut was significantly *reduced* when the water-jet pressure was increased from 10 to 50 MPa.

Overall it was found that by combining fairly moderate water pressures with drag-bits, an increase in depth by a factor of two could be achieved when cutting norite in the laboratory, and by a factor of five when cutting quartzite in underground tests. The results were substantially influenced by the direction of the jets relative to the tungsten carbide inserts; two, one directed towards each corner of the leading edge of the tungsten carbide bit, apparently giving the best results.

Apart from indicating that high-pressure water-jet assisted drag-bit cutters performed more efficiently than unassisted drag-bit cutters, the tests showed that bit life with water jets was also significantly better. Unassisted bits failed rapidly either at the brazed joint or because the tungsten carbide insert fractured. However, no such premature failures were experienced with the water-jet assisted cutters, the bits remaining in use until they were eventually discarded because of wear.

Hood comments that the true reasons for the failures of the brazed joints and the carbide inserts are not yet fully understood. He suggests that the cause could be due either to the 'high rate of heat transfer to the bit causing thermal deterioration of the brazed joint and tungsten carbide, or by excess load causing shattering of the carbide during the cut'.[2] Thus the reduction in cutting force combined with the cooling effect of the water probably affected the bit life. No doubt when harder grades of tungsten carbide combined with high-pressure water jets are used, bit life will be extended even further.

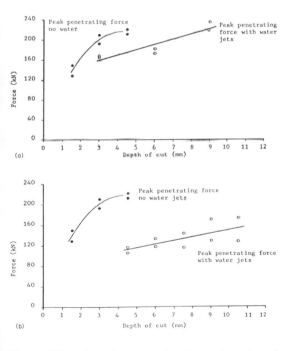

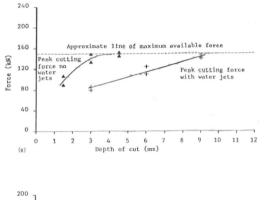

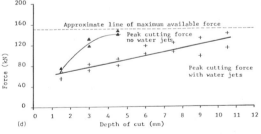

Figure 287   (a) Peak penetrating force plotted against depth of cut when 10 MPa (1450 lb/in^2) water jets were directed outside the corners of the tungsten-carbide inserts in the bit. (b) Peak penetrating force plotted against depth of cut when 50 MPa (7250 lb/in^2) water jets were directed outside the corners of the tungsten-carbide inserts in the bit; (c) Peak cutting force plotted against depth of cut when 10 MPa (1450 lb/in^2) water jets were directed outside the corners of the tungsten-carbide inserts in the bit; (d) Peak cutting force plotted against depth of cut when 50 MPa (7250 lb/in^2) water jets were directed outside the corners of the tungsten-carbide inserts in the bit. (Courtesy M. Hood, *Journal of the South African Institute of Mining and Metallurgy,* Nov. 1976.)

313

Figure 288. Robbins water jet assisted T.B.M. showing water jets in operation. (Courtesy The Robbins Company.)

Robbins Water-Jet Assisted T.B.M.[3,4]

In 1975 experiments to determine the influence of water jets used in conjunction with disc cutters on a full-face rock-tunnelling machine were carried out at the Colorado School of Mines by Dr. Fun-Den Wang and Russell Miller.

The tunnel borer (Model 74-115) (Figures 288 to 290) was provided by the Robbins Company and the water-jet equipment by Flow Industries, Inc.

This work, costing U.S. $600,000, was sponsored by the Research Applied to National Need Program of the National Science Foundation, the U.S. Bureau of Mines and the Colorado School of Mines.

The 32 t tunnel borer was a specially modified conventional rock-tunnelling machine with a 2.13 m (7 ft) diameter head having a power of 112 kW (150 hp). It was capable of exerting a total thrust of 1750 kN (322,700 lb) and 115 kN m (85,000 lb ft) of torque. The head was fitted with 16 disc cutters and 31 water-jet nozzles.

Figure 289. Cutter head of Robbins water jet assisted T.B.M. (Courtesy The Robbins Company.)

314

Various comparison tests with and without water jets were made in a granite quarry in Skykomish, Washington.

Initial laboratory tests were made to determine the influence of varying water-jet pressures, nozzle size, stand-off distance, speed and direction of the jet, and the positions of the water jets on the cutter head relative to the positions of the disc cutters, etc.

Because a conventional machine was used, the manifold and nozzles were necessarily positioned on the front face of the cutter head, though Dr. Fun-Den Wang and R. Miller commented that, ideally, the manifold should have been behind the cutter head and connections between the nozzles and the manifold made by plumbing through the cutter head.[4]

The Miller river quarry near Skykomish where the machine was tested was leased by the Morrison-Knudsen Co., who also assisted fieldwork by providing a test site and driving a pilot tunnel for the starting bore.

Water was fed to the T.B.M. via high-pressure water pipes from two 'intensifiers'* capable of delivering water at 386 MPa (56,000 lb/in²) pressure. The water was transferred from the pipes to the rotating cutter head by a swivel fitted with high-pressure seals. The swivel functioned very satisfactorily during the field tests.

Water from the Miller river was transferred by centrifugal pump to a 1900 litres (500 gal) tank which supplied the intensifiers via a two-stage filtering process.

The material at the test site consisted of extremely hard granite and it was found that jet pressures in excess of 200 MPa (30,000 lb/in²) were necessary to produce effective results, the best results being obtained in the 280-340 MPa (40,000-50,000 lb/in²) range.

Of interest were results obtained in some early laboratory tests. Apparently when the nozzles were positioned beneath the cutters instead of between them, better penetration rates were obtained. These rates were achieved

with approximately 50 per cent lower water-jet pressures than had been necessary when the nozzles were placed between the cutters. However, when a similar configuration was tested in the field, penetration rates actually dropped. Messrs. Wang and Miller comment that the reason for this may have been the different edge geometry of the particular cutters used in the laboratory from those used in the field. In addition 76 mm (3 in) spacings between cuts were used in the laboratory, whereas spacings in the field were 100 mm (4 in). Because of the difficulties involved in varying the geometry and/or cutter types in the field at the time, further tests with the nozzle positioned beneath the cutters were shelved, and work was concentrated instead on the already successful 'jets between cutters' configuration.

In their summary, Messrs. Wang and Miller noted that improved advance rates of from 50 to 60 per cent had been obtained in the field when using water jets in conjunction with conventional disc cutters. Indeed, in some tests the penetration rates were more than doubled. They expressed the opinion, however, that a great deal more research was needed to determine the influence of various factors such as cutter spacings, cutter types, jet and cutter configurations, etc. on water-jet and mechanical cutting.

Wirth Water-Jet Rock Boring Machine[5, 6]

Sponsored by the Nordhrein-Westfalen government, the Bergbau-Forschung GmbH (a mining research institute based at Essen) commenced full-scale tests in 1978 in an endeavour to improve penetration rates in hard rock, using a 2.60 m (8 ft 6 in) diameter Wirth tunnelling machine (Model TB-1-260) (Figures 291 to 293).

The water-jet equipment, of an improved design, was supplied by Flow Industries, Inc., Seattle, U.S.A.

A specially designed cutter head, fitted with 14 disc roller cutters and 94 high-pressure water nozzles, was built for the unit. Different cutter spacings (50 and 100 mm (2 and 4 in)), nozzle-stand-off distances (25-50 mm (1-2 in)), number

---

*A plunger-pump featuring a large oil-driven central piston that reciprocates and drives two smaller pistons which pump water on each end. Because of this arrangement the pump was capable of delivering water on both the intake and exhaust stroke.

Figure 290. Tunnel face showing cutting pattern made by Robbins water jet assisted T.B.M. (Courtesy The Robbins Company.)

316

Figure 291. Wirth water jet rock boring machine used by the Bergbau-Forschung GmbH in full-scale underground tests in Essen, West Germany. (Courtesy Bergbau-Forschung GmbH.)

and array of nozzles (0-60), number of nozzles under and/or between disc cutters, and water pressures (0-400 MPa (58,000 lb/in^2)), etc. were permissible with the newly designed head.

In effect the high-pressure jets were designed to act much as the fixed tools did on earlier Robbins machines, cutting concentric kerfs or grooves in the rockface and allowing the following disc cutters to deal with the remaining ridges or rock ribs.

It was expected that, as the thrusting forces would be converted to shearing forces, less load would be imposed on the disc cutters, thus leading the way towards the development of lighter and less expensive full-face rock units. Other benefits which may be derived from this change in basic design could be easier transport and assembly, and the possible use of this type of machine for short drivages of 1 km (0.62 miles) instead of about 3 km (1.9 miles) which is accepted as the shortest economical driving distance at present.

As a result of an existing agreement, data on the subject collected by the Colorado School of Mines for the U.S. Bureau of Mines during the tests carried out with the Robbins machine have been made available to the Bergbau-Forschung GmbH.

According to an article by Dr. Ing. J. Henneke,[5] extensive laboratory investigations by the Bergbau-Forschung were followed by the first full-scale underground tests which were carried out in an Upper Carboniferous sandstone quarry during the period October 1977-May 1978. The underground tests were later resumed in October 1978.

The compressive strength of the sandstone was approximately 10MPa (1500 lb/in^2) and the tensile strength was reported to be about 0.83 MPa (120 lb/in^2).

While the U.S. Bureau of Mines sought to improve the cutting speed of rock-tunnelling machines, the primary aim of the Bergbau-Forschung was to produce a machine which was

317

Figure 292. Wirth water jet rock boring machine showing jets in operation. (Courtesy Bergbau-Forschung GmbH.)

Figure 293. Roller cutter traces and waterjet kerfs made in sandstone tunnel face by Wirth water jet boring machine. (Courtesy Bergbau-Forschung GmbH.)

smaller and lighter than a conventional rock-tunnelling unit, yet which could efficiently handle the coal strata.

Four double-acting intensifiers were used to deliver high-pressure water (360 MPa (52,000 lb/in²) at 120 litres/min (32 gal/min).

During the initial tests water-jet assistance was mainly directed towards the gauge area of the cutter head because, generally speaking, the greater percentage of cutter wear and cost is concentrated in that region. However, results with this configuration were disappointing, and in the second series of tests 70 small nozzles (0.25 mm (0.01 in) diameter) were distributed over the entire cutter head. A 30 per cent reduction in forward thrust was achieved with this configuration. During the third series of tests, 12 large nozzles (0.35-0.63 mm (0.014 to 0.025 in) diameter) were evenly placed over the radius of the cutter head. This latter configuration produced a maximum decrease in forward thrust of between 55 and 60 per cent, and a corresponding reduction in the energy consumption of the cutter-head drive of between 40 and 45 per cent.

In addition to the above, simultaneous tests were carried out using water jets in conjunction with drills for rock-bolting work and in coring operations as a profiling unit for making roadways. Other research work utilizing drag-bits in conjunction with water jets was also being carried out by the Colorado School of Mines and the Bergbau-Forschung GmbH.

References

1. M. Hood, Water-jet assisted drag-bit cutting of hard rock, *Chamber of Mines Research Review,* 1975/76, 92-103, Johannesburg.
2. M. Hood, Cutting strong rock with a drag bit assisted by high-pressure water jets, *Journal of the South African Institute of Mining and Metallurgy,* Nov. 1976, 79-90, Johannesburg.
3. Robbins water jet machine, *Mining Newsmonth, American Mining Congress Journal,* Nov. 1975, Washington, D.C.
4. Dr. Fun-Den Wang and Russell Miller, High pressure water jet assisted tunnelling, *Proc. R.E.T.C.,* Las Vegas, 1976, 649-675.
5. Dr. Ing. J. Henneke, Water jets cut shear forces at the face, supplement to *New Civil Engineer,* 8 March 1979, x-xi, (Facing up to the future), London.
6. Bergbau-Forschung, Research News, *Tunnels and Tunnelling,* Jan. 1978, **10,** No. 1, 7-48, London.

Correspondence

Dr. N.G.W. Cook, Chamber of Mines, S. Africa and University of California, U.S.A.

Dr. Fun-Den Wang, Colorado School of Mines, Golden, U.S.A.

Dr. Ing. J. Henneke, Bergbau-Forschung GmbH & Mannesman Demag, West Germany.

M. Hood (Asst. Prof. Mining Eng., University of California), U.S.A.

Excavation-Soft Ground

TEN

Shields

As its name suggests, the shield simply provides within a tunnel a working area which is protected against the collapse either of the walls or roof of that section of the tunnel which has been recently excavated, and in which no tunnel lining or other means of support has yet been erected. Obviously when the tunnel is being driven through stable ground no shield is necessary. However, as its value as a tunnelling tool extends beyond that of merely supporting the ground, it is not uncommon for contractors to use a simple shield when driving tunnels in good ground, in order to provide housing for segment erectors and muck disposal units, etc. The shield also provides a bridge between the erected lining and the face so that work on extension of the lining and excavation of the face may be carried out simultaneously.

When large-diameter tunnels are excavated the shield is used as a type of travelling support for the erection of tunnel-face scaffolding, enabling workmen to gain access to all points of the face.

After the completion of each excavation cycle, thrust jacks positioned at the rear of the shield react against the newly erected tunnel lining or against a thrust ring and move the shield forward, together with all its face and trailing ancillary equipment such as sledges, gantries, and conveyors.

When a high water-table level is encountered or the soil is particularly unstable, it has also been found necessary to provide some form of protection at the working face. This may range from a simple protective diaphragm to the more complicated installation of an airlock.

Generally speaking, apart from the simple basic open-ended type, non-mechanized shields may be divided into two main groups: those with closed heads and those fitted with rigid horizontal shelves with/without vertical partitions, or cutting grids (horizontal and vertical bars with/without a sharpened forward edge).

In the closed-head type a bulkhead/diaphragm is positioned immediately behind the forward or cutting edge of the shield. This type of shield has been frequently used to tunnel through uncohesive ground such as soft-flowing mud or silt, the bulkhead acting to support the face. Apertures in the form of holes or slits in the bulkhead allow a controlled amount of material to enter the shield as it advances. Sometimes these apertures are fitted with doors or gates which reduce the size of the aperture according to the nature of the ground ahead. The material in the shield is then removed in various ways by means of conveyors, etc.

Closed-head shields may also feature conical, domed, or other shaped heads which precede the actual shield. Their basic function, however, remains the same.

Shields incorporating rigid horizontal shelves with/without vertical partitions are generally used to tunnel through uncohesive sandy or granular soils.

The soil from the face is allowed to fall onto the platform and settle at a natural slope, the slumped soil lying on the platform acting to support the face. This obviates the necessity to use timber supports in the form of breasting boards, etc.

Alred Ely Beach is reputed to have built the first shield incorporating 'slump shelves' (see section on A. E. Beach).

After Beach, several small-diameter shields with slump shelves were built and used. Though a proposal was made by the Russian engineer, V. A. Varganov, in 1938 to build a large-diameter shield with horizontal slump shelves, it was not until the theory of slump-shelf shields was supported with authentic scientific research data (provided by the Foundation and Underground Structures Research Institution in 1953 and the Ministry of Transport Construction Research Institute, U.S.S.R., in 1958), that large-diameter slump-shelf shields were built and used for tunnelling work in the Moscow Underground.

In some cases thick unsharpened steel plate is used, in others the forward edge of the shelf is sharpened to facilitate its penetration into the face.

To increase the use of this type of shield to other soil types, extensible shelves were also sometimes used, or the shelf was fitted with guides which enabled it to be thrust forward or retracted. In others the entire supporting ring or framework carrying the shelves and partitions could be slid forward independently of the main shield body.

Shields with horizontal and vertical cutting grids (bars with/without sharpened forward edges) have been used for tunnelling through clayey ground. The clay was extruded through the grid apertures into the shield where it was cut and removed.

P. W. Barlow is said to have designed the first grid cutting shield. Charles Bonnet, who patented a similar type of shield, provided a rotary blade mounted on a shaft behind the grid. As the clay extruded into the shield, it was cut into blocks and allowed to drop onto a belt conveyor via a chute.

The following chapters trace the historical development of the shield as it is today.

(*Note*: When a description only of a particular patent has been given and there is no mention of whether or not the unit was ever constructed or used, this indicates that such data were not available to the author at the time of writing, although such a device might very well have been constructed and/or tested.)

Marc Isambard Brunel — 1818[1,2]

For the invention of the first soft-ground tunnelling shield the world is indebted to Brunel. Born on 25 April 1769, Marc Isambard Brunel, a Frenchman, showed an early interest in things mechanical. Though he excelled at mathematics, his interest in Latin and Greek was almost non-existent and it soon became obvious, both to his tutors and to his exasperated parents, that their choice for him of a religious career would be highly unsuitable. Eventually a family friend suggested the French Navy. Here his exceptional ability for the design and construction of mechanical devices first became apparent.

While being tutored in trigonometry, a necessary qualification before being admitted into the Navy, Brunel designed an instrument with which he was able to measure the height of the Rouen Cathedral spire.

In 1786 after completing his studies, Brunel (aged 17) went aboard his first ship, a frigate, prior to its departure for the West Indies. While being presented to the captain, Brunel noticed a quadrant lying in the cabin. Intrigued, he longed to pick it up and examine it more closely, but dared not. Nevertheless, during the very brief period of the interview, Brunel absorbed enough to enable him to return home and make a similar instrument. But, not satisfied with this initial attempt, Brunel made another shortly afterwards which served him throughout his naval career.

In 1792 (after six years' service in the Navy) Brunel returned to France. By this time the French Revolution was well under way and Brunel, who had strong Royalist leanings, found himself compelled to flee for his life. He reached New York in September 1793, where his unusual talents were quickly appreciated and he was appointed Chief Engineer of that city. He also practised as an architect. In this capacity Brunel, amongst his other notable achievements, designed and superintended the building of New York's famous Bowery Theatre (which later (1821) burned down). He was also commissioned to build an arsenal and cannon factory for the city of New York in which he installed unusual and ingenious machinery.

Though there was undoubtedly great scope in a young and expanding country for a man with Brunel's talents, he decided to return to England in 1799. Two factors influenced

Brunel's decision. The first was that when in Rouen he had met and fallen in love with an attractive English lass named Sophia Kingdom. The second reason was that while in America Brunel was invited one evening to dine with Alexander Hamilton. There he met a Frenchman who had just recently come from England. The Frenchman told Brunel of the difficulties experienced by the British Naval Yard, which he had visited while in England, in producing pulley blocks for their ships. Brunel immediately designed what he considered would be a simpler method of production. When he left Hamilton's house he continued to work at the problem until he felt he had perfected it. Then he decided to take his ideas and present them personally in England. His decision in this respect was no doubt also strongly influenced by his desire to renew the acquaintance of Sophia Kingdom.

Brunel was successful in both areas. He married Sophia Kingdom at Holborn and the newly wed couple set up house in Portsea. Backed by the strong recommendation of Sir Samuel Bentham, Brunel also persuaded the naval authorities to try his revolutionary pulley-block making scheme. By 1808 after a preliminary period of about five years during which time the machinery for manufacturing the pulley blocks was made and assembled, Brunel had it operating successfully, producing pulley blocks at a rate which was far faster than they had ever been produced before.

During the succeeding years Brunel's fertile mind devised many different types of machines. These included, amongst others, improved machines used for sawing and bending timber, machines for winding cotton thread into balls, and machines for manufacturing nails.

At about that time Brunel also became interested in the various attempts which had been made by prominent engineers of his day to drive a tunnel under the Thames. The original idea was the brainchild of Ralph Dodd. But Dodd's project (a 4.8 m (16 ft) wide tunnel under the river which was to connect Gravesend with Tilbury) failed soon after commencement, due to financial difficulties.

Robert Vazie, a Cornishman, then took up the cudgels. Vazie was backed by the Thames Archway Company. He proposed driving a small pilot tunnel or driftway under the Thames between Rotherhithe and Limehouse.

A 3.34 m (11 ft) diameter shaft (later reduced at a depth of 12.76 m (42 ft) to 2.43 m (8 ft) diameter) was started on the Rotherhithe side, some 96.1 m (315 ft) from the river. Unfortunately, owing to various difficulties such as the repeated infiltration of land water, etc. work was stopped at a depth of 23 m (76 ft) and engineers, John Rennie and William Chapman were consulted. The ideas of these two men conflicted, so Vazie suggested that Richard Trevithick should be approached. Trevithick worked initially as Vazie's resident engineer but later the Thames Archway Company proprietors sacked Vazie and appointed Trevithick in his place. Vazie was naturally furious but, as the company held the purse strings he could do nothing about their peremptory decision. However, from the company's point of view the decision was a good one. Overcoming the current problem of water and quicksand, Trevithick began work on the pilot tunnel.

Labouring in the confined space of the 5 ft (1.52 m) high by 3 ft (0.92 m) wide tunnel, Trevithick, despite numerous further problems with both water and quicksand, drove the driftway 305 m (1000 ft) under the river. At this point, on 26 January 1808, an exceptionally high tide caused an irruption so great that the pumps were unable to cope with the vast volume of water.

The miners raced back along the tunnel, followed closely by Trevithick himself. But before they reached the entry shaft, the water had risen as high as their necks. Indeed Trevithick, who was in the rear, was very nearly drowned.

Trevithick stemmed most of the inflow with clay bags which he deposited over the hole on the bed of the river (a system later used by Brunel for the same purpose). He made various suggestions both for effecting a more permanent cure for their current problem and also for the construction of the full-size tunnel.

For the latter he proposed building a series of coffer dams within which he would dig a trench and lay a tunnel constructed of lengths of cast-iron piping. But the company rejected his proposals, preferring instead to accept the

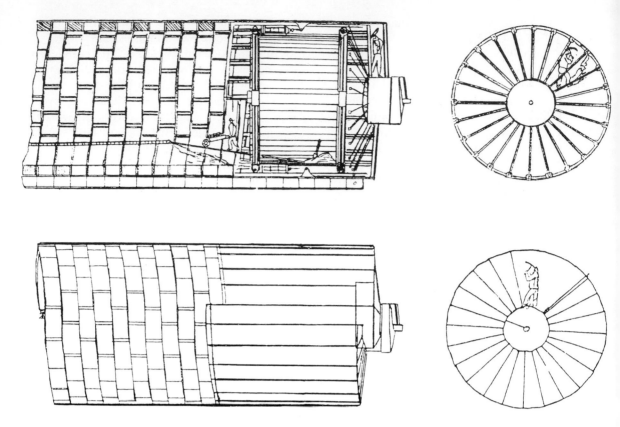

Figure 294. Brunel shields — U.K. Pat. No. 4,204, dated 20 Jan. 1818. Circular type

advice of William Jessop who announced that a tunnel under the river, though very desirable from the public's point of view, was an impracticable undetaking. (Later Trevithick's novel ideas were successfully tested by engineers in America.)

In the meanwhile Marc Brunel was considering ways and means by which a tunnel might be built beneath the river Neva at St. Petersburg (Leningrad). Here, due to the presence of block ice from Lake Lagoda after winter each year, a bridge constructed with piers would have been seriously endangered.

While working at Chatham Dockyard, Brunel frequently saw evidence of the destructive work carried out by the *Teredo navalis* or ship-worm. It was this diminutive but nevertheless powerful creature which inspired Brunel's greatest invention — the soft ground tunnelling shield.

So far as the Neva river was concerned, Brunel finally submitted plans for a suspension bridge but, on 20 July 1818, he patented his

ideas for a *circular* tunnelling sheild. (Figures 294 and 295). The 1818 patent covered Brunel's proposal for two methods of tunnelling.

Brunel pointed out in his patent that:

the smaller the opening of a drift, the easier and the more secure the operation of making the excavation must be. A drift on dimensions not exceeding three feet in breadth by six feet in height, forms an opening of 18 feet area, whereas the body of a tunnel on dimensions sufficiently capacious to admit of a free passage for two carriages abreast cannot be less than 22 feet in diameter, consequently about 20 times as large as the opening of a small drift.

Brunel then goes on to explain how by using a 'casing or a cell, intended to be forced forward before the timbering — which is generally applied to secure the work' he would form an excavation 'suitable to tunnels of large dimensions' and by dividing his casing into 'small cells ... lying alongside of and parallel with each other' he would overcome the problem of too large an opening. Each cell

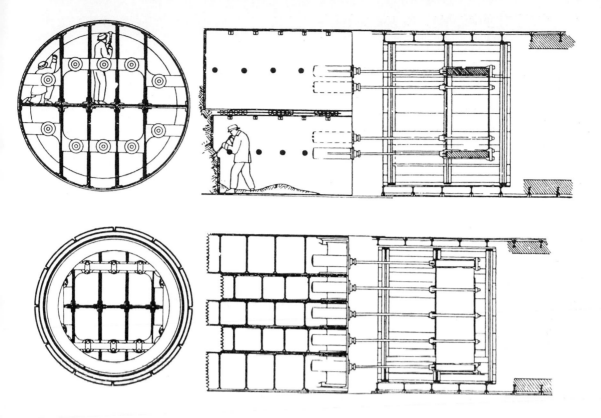

Figure 295. Brunel shield — U.K. Pat. No. 4,204, dated 20 Jan. 1818

could be 'forced forward independently of the contiguous one, so that each workman is supposed to operate in a small drift independently of the adjacent one'. The front of the work was to be 'protected by small boards ... which the workman applies as he finds most convenient'. Several men could thus work at the same time... 'with perfect security and without being liable to any obstruction from each other'.

Brunel suggested that each cell could be moved forward by any mechanical aid, but stated his preference for hydraulic presses. These presses would apply thrust against 'strong framing ... fixed within the body or shell of the tunnel'.

After all the cells had been advanced Brunel suggested that the space thus exposed between the already constructed tunnel and the rear of the cells could be protected against an irruption of a large body of water or the breaking down of the ceiling by the insertion of strong iron plates which could extend beyond the cell and overlap the shell of the previously constructed tunnel.

He proposed that the body or shell of the tunnel should be made of brick or masonry, but stated that he preferred to use pieces of cast iron which could afterwards be lined with brickwork or masonry.

In his second method* Brunel again divides his shield into small compartments, but instead of a relatively simple device consisting of a large cylinder or casing with an open cutting edge which is forced forward by jacks while workmen in each separate cell inside the shield

---

*On the subject of Brunel, W. C. Copperthwaite[3] comments: 'The second type of shield described and figured in the specification of 1818 is the one suggested, as described by Brunel himself, by the screwlike action of the *Teredo navalis,* a marine worm which can pierce the hardest woods. The shield was never practically tested, and it is one of fame's little ironies that Brunel is popularly supposed to have derived his great invention from his observations of a natural excavating machine, whereas in fact the actual shield used by him borrowed nothing from any previously known natural or other mechanism.'

remove the ground ahead of the shield, a slightly more complex machine is described.

In addition to dividing the space within the shield into cells as previously described, Brunel also divides his shield into three parts which he calls 'the cylindrical body of the larger teredo' and the 'body of the lesser teredo with its shaft' and 'the internal drum'. In effect the whole device resembled an auger bit.

Staves were attached to the circumference of the larger teredo and these moved forward in the longitudinal direction of the tunnel ahead of the shield. A separate hydraulic press behind each individual stave was designed to overcome both the friction of the earth and that caused by the staves sliding against each other.

Though the larger 'teredo' or shield did not actually rotate, a simulated rotary action was achieved when each cell was advanced in turn around the lesser teredo, giving the whole a helical forward movement. Thus, unlike the first shield proposed by Brunel, the workmen did not remove the soil which lay ahead of the shield. Instead they excavated the ground which lay radially to the side of the leading cell and the worker in fact faced the wall of the tunnel as he dug, much in the way an auger bit would remove cuttings from a hole.

So far as the lesser teredo and internal drum were concerned, Brunel obviously intended these to rotate, but did not specify how this movement was to be achieved. He says: 'The lesser teredo, so much resembling the common auger or a boring bit requires no further description, as its application is too obvious. When the lesser teredo is in action the workman stands in the body of the teredo and applies through the orifice the tools or instruments necessary to displace the earth.'

Brunel suggested that the tunnel behind his second shield be 'formed by a systematic combination of hollow pieces of cast iron which, when united together by bolts or screws, presents externally somewhat the appearance of a riband wound edge to edge round a cylinder throughout its whole length as seen in Figure 9, when it is evident the riband would take a direction similar to that of the thread of a screw'.

Note: This idea was tested in practice by E. K. Bridge and Aubrey Watson Ltd. early in 1958

(see section on mechanized shields — Bridge auger mechanized shield).

The Thames Tunnel Shields

According to L. T. C. Rolt,[1] Brunel was approached in 1823 by I. W. Tate, a director of the original Archway Company. Tate apparently persuaded Brunel to show his scheme to some prominent citizens. The result was the formation of the Thames Tunnel Company, after a meeting held on 18 February 1824 in the City of London tavern. This was followed by official approval of the project and the appointment of Brunel as Chief Engineer at £1000 per annum. A further sum of £5000 was paid to Brunel for the use of his invention, and his contract also promised a final sum of £5000 which was to be paid on completion of the project.

Brunel selected a site between Wapping and Rotherhithe some 1200 m (3900 ft) from that chosen by Richard Trevithick.

Brunel's first Thames tunnel shield was found to be unsuitable for the work and, though it was used from 1825 to 1828, it was removed when the works were temporarily halted (due to lack of finance). In 1835 a new improved rectangular shield was installed. (Figures 296 and 297) This was used throughout the remainder of the work until its completion in 1843. This second shield is the one most widely described in text or engineering books on the subject.

Like his circular shield, the cells in the rectangular shields were also capable of independent movement. The entire shield was 11.43 m (37 ft 6 in) wide × 6.78 m (22 ft 3 in) high × 2.74 m (9 ft) long (excluding the tail-plate).

Made of cast iron the shield consisted of 12 frames each about 1 m (39 in) wide divided into three sections (an upper, middle and lower section). Each frame stood upon two swinging legs attached by ball joints to two massive flat shoes (base-plates).

Chisel-shaped staves or sliders attached to the top of each section slid forward to cut and support the ground immediately ahead of the shield. (The staves or sliders on Brunel's shield were in effect an early example of Memco's

Figure 296. Brunel shield — rectangular type as used by Brunel in Thames tunnel (1825-43)

poling plate system and Westfalia Lunen's blade shield.) Those staves attached to the middle and lower sections also served as foot-plates for the workmen above. A foot- or base-plate resting on planks supported the bottom of the lowest section. These planks remained in position after the shield had been advanced along them and were subsequently used as a platform for the tunnel brickwork.

Though Brunel showed a marked preference for hydraulic jacks in his 1818 patent, the

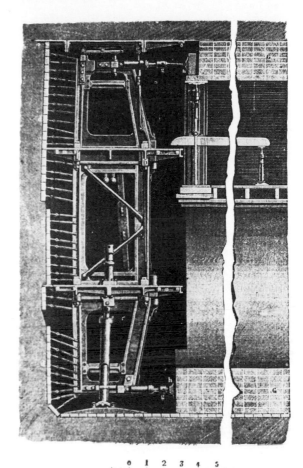

Figure 297. Brunel shield — rectangular type

Thames tunnel shield was advanced by means of screw jacks which thrust against the finished masonry of the tunnel. (Cast-iron tunnel lining also advocated by Brunel in the 1818 patent was not utilized either — probably because as Copperthwaite comments 'a rectangular tunnel was decided upon and almost as a necessary consequence a masonry tunnel was preferred to a cast-iron one'.)

In the circular shield patent Brunel pointed out that: 'In order to facilitate the progressive movement of the cells I introduce friction rollers between the opposite sides of all the cells....' This idea was tested in practice in the Thames tunnel shield by the installation of rollers between each frame and the adjacent one.

Each section supported its part of the tunnel face by means of 14 or 15 horizontal breasting

329

boards. Each board was held in position, or advanced, by means of a pair of screw jacks; one at each end.

The reaction of these jacks was normally taken by the vertical sides of the section frame, but when all boards of a section had been fully advanced and it was desired to move the frame up to the face, it was possible to transfer temporarily the support of the boards to adjacent frames and so keep the boards in place while freeing the particular section frame for forward movement. In the same way the roof plates of each section could be supported by vertical jacks while that section was being advanced.

Again, as suggested in Brunel's 1818 patent, strong iron plates were fitted between the top of the tail ends of the roof staves and the top of the already completed section of tunnel lining. These protective plates kept debris from falling onto workmen in the exposed area which appeared after the shield had been advanced and before the masonry or brickwork could be built.

The 12 frames were advanced alternately one at a time. Each advance was not more than 15 cm (6 in). Thus the shield progressed at a relatively slow pace — 4.3 m (14 ft) being the maximum distance attained in one week.

The tunnel was finally completed after years of hardship, toil, mishap, delay, etc. Yet so well was it constructed, so excellent was the workmanship that even today some 150 years after the project was commenced, the tunnel does not leak nor has it been found necessary to carry out any major repairs.

The financial and physical cost of completing Brunel's tunnel had been great. Men's lives had been lost, others had had theirs shortened as a result of the hardships they had suffered during the construction work. Moreover, the ultimate financial reward of this costly project was negligible. Although other inventors submitted patents which they believed were improvements on Brunel's original designs, no further tunnels were constructed by means of a shield until the year 1869.

Brunel and the Subject of Compressed Air

It is interesting to note that on 24 January 1828, approximately two years after Brunel's shield first commenced its hazardous journey under the Thames, Colladon wrote to Brunel suggesting to him the possibility of using compressed air for preventing an irruption of water. Brunel is accused of ignoring this advice and his reason for rejecting this proposal has been a matter for wide speculation. However, in 1976 R. Glossop, consultant of John Mowlem & Co. Ltd., submitted a paper[4] which has shed some light on the subject.

It is common knowledge to those who have researched Brunel's historical background that, from time to time during the progress of the shield, and especially after periods of stress, Brunel was inundated with letters, etc. from self-styled experts giving meticulous directions on how he should proceed with the undertaking. Colladon's suggestion came at just such a time (i.e. after the irruption which occurred on 12 January 1828) and it seems clear that Brunel, impatient and exasperated by the flood of advice, threw the baby out with the bath water.

However, Glossop, in his paper,[4] considers that acording to several diary entries made by Brunel and from various letters he wrote to the directors of the Thames tunnel,* the question of using compressed air in the tunnel was again brought to Brunel's attention when (according to a diary entry dated 7 March 1831) he 'found' while on a visit to the Patent Office 'that Lord Cochrane had taken out a Patent'. This patent described Cochrane's proposals for using compressed air when tunnelling below the groundwater table, and also detailed his schemes for building an airlock. Brunel promptly wrote the following letter to the directors of the Thames tunnel about this 'Plan of Lord Cochrane':

Gentlemen:
I avail myself of this opportunity to inform you that, Receiving occasionally additional propositions either for completing the Tunnel upon some new Plan or for rendering our own Plan quite infallible, I invariably refer the parties to the Court, although I have found nothing, as yet, that can in any way be beneficial to the concern.

One however has just come to my knowledge, which, from the novelty of its application to the

*These letters and the tunnel diaries (consisting of 17 volumes) are now stored in the library of the Institution of Civil Engineers, London.

Art of Mining, coupled with the character and respectability of the projector, cannot fail of exciting some interest and is likely to call your attention when known. The plan in question, applicable as it is presumed by the inventor for making Tunnels is, I understand for effecting the excavation under a pressure of air, capable of supporting the surrounding pressure, corresponding in some respect with the use of a Diving Bell for working under water.

This plan, as an auxiliary means of protection, under some circumstances in particular, has not, as well as some others, escaped our attention; but as the Shield is as indispensable as before, even with the plan in question, I must leave the subject to the consideration of the Court when it comes before you. The inventor is Lord Cochrane who has taken out a Patent for this new application to the Art of Mining.

I have the honour to be, Gentlemen,
Your Most Obedient humble Servant,
M. I. Brunel
To the Directors of the Thames Tunnel.

Later, according to Brunel's diary entries (quoted by Glossop in his paper),[4] a Mr. Dyer approached Brunel and sought to persuade him to use compressed air in the tunnel. Brunel, however, felt that Dyer was merely duplicating Cochrane's proposals and that Cochrane should receive what credit was due for suggesting the use of compressed air in tunnelling as he had made the prior claim. In any case Brunel made it clear that he had already given the matter serious consideration, and had decided that for his purposes the shield was best. He outlined these views in his letter to the Thames tunnel directors dated 19 September 1831, which letter is quoted below:

Gentlemen
On Wednesday last, I met by appointment, Mr. Dyer, at the office, in relation to the proposition he has made to you, lately, of a Plan, by which he intends to effect the excavation of the Thames Tunnel, by means of condensed air for supporting the ground. I have shown to Mr. D. some samples of the Strata that are in the line of the excavation, such as I have described them in a letter lately addressed to the Deputy Chairman, to which letter I beg to refer you. Now, altho condensed air might be useful for supporting substantial and solid ground, it is obvious that it cannot answer against a Shingly stratum. Condensed air cannot prevent its rolling down wherever it is as lose [sic] as the sample we have, which sample was laid before the Court in March 1827, as being met daily in large portions out of a surface of about

225 feet, independently of the Sides. The cavities that would be the consequence of such displacements would inevitably be fatal to the undertaking. It must be equally obvious that condensed air could not protect the work against the *influx of fluid ground* such as is met at the Top of the excavation.

Some persons have pretended that the fluid state of the ground had resulted from the inefficacy of the Shield, and from mismanagement and carelessness in working it, and therefore that the ground is not naturally in that state; in refutation of such assertions, it is enough to refer to the Report of Mr. Trevithick who stated, that, in the line of the Driftway he had probed the ground, by making several borings above head, for the purpose of ascertaining what the nature of the ground was, in the way in which he would have to operate, for the proposed Archway. These borings were secured with cast-iron pipes. These pipes were carefully plugged up after the water and some fluid sand had ran through. It is conclusive, from these facts, that certain portions of ground over the head of the Driftway were quite fluid. Warned as this Engineer was by his own recent probings, however, when he reached by the activity of his excavations, the level of these fluid portions, 'notwithstanding' (as he reported it), 'that every side and even the face of his excavation was closely timbered except 6 inches, and every precaution taken in every other respect, the ground burst upon him, in a fluid state, and in a few minutes the River made its way through upwards of 25 feet of intervening ground, the greatest part of which consisted of clay'.

If the best miners that could be engaged for the enterprise in question have failed in an attempt that was only 30 inches in breadth, and that too in consequence of these lodgements of water and fluid sand, can any one impute to the Engineer of the Thames Tunnel who has effected an excavation 38 feet wide through similar strata, that the accidents that have occurred at the Tunnel have occurred from the grossest neglect and from the most unpardonable carelessness; and also from the total inefficacy of the Shield?

A plan whatever it may be must be made for the bad ground, it must be calculated to meet all exigencies all disasters and to overcome them after they have occurred.

It is evident that, on these points, Mr. Dyer has not been as provident as Mr. Colladon of Geneva whose plan No. 2 dated 24th January 1828 may be referred to. Mr. C. would not dispense with the Shield, on the contrary, for greater security, he would have every poling board made like the Clappet of a foot valve, that would shut of itself in case of accident.

Several other propositions with condensed air have been communicated to us, but they were not satisfactorily explained except that of Lord

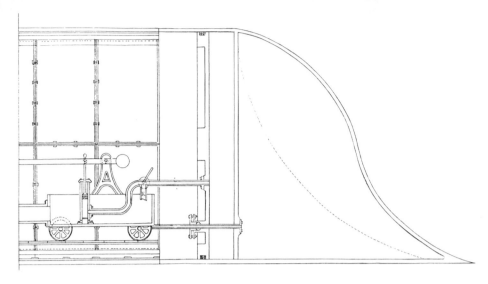

Figure 298. Samuel Dunn tunnelling shield

Cochrane who has lately taken out a patent for a plan which so far corresponds with that of Mr. Dyer that Mr. D. could not possibly maintain his patent against his Lordship's prior right. I have perused Lord Cochrane's Specification which is accompanied with a Drawing showing the Section of the bed of the Thames, as his Lordship has conceived it to be, viz. consisting of *a Stratum of blue clay to a great depth and extending quite across the bed of the River,* and until his Lordship has better information, his plan cannot but, most probably fail, in his own estimation. With reference to Mr. Dyer's plan, altho I have to express my thanks to that Gentleman and most particularly to Mr. Brunton for the offer of placing the plan in question at my disposal, yet Mr. Dyer must be aware that the Tunnel Company could not possibly treat him in the face of a patent which precedes his, besides the *prior* claims of others, besides our own for the same object.

> I have the honour to be, Gentlemen,
> Your most obedient humble Servant,
> Mc. I. Brunel.

Apart from his obvious anger at Dyer's criticisms of his own project, Brunel makes it plain in his letter that he does not believe that condensed air will hold back 'the influx of fluid ground such as is met at the Top of the excavation'.

Looking back at the Thames tunnel project with hindsight it must be realized that, despite his many problems, Brunel had already ample proof that his shield system of tunnelling was working. His failure to use compressed air may well have been the result of an instinctive appreciation of the problems which could have arisen had he used compressed air. This is corroborated by the experiences of De Witt Haskin who in 1879 used compressed air without a shield for the first time in a large tunnel,* the famous Hudson river tunnel which ran from Hoboken, N.J., to Manton St., New York. In June 1880 the 4.88 m (16 ft) wide by 5.49 m (18 ft) high northerly tube of the proposed twin-tube tunnel had reached 110 m (360 ft) from the Hoboken shaft when the compressed air blew a hole through the soft silt of the tunnel crown at this point and the subsequent irruption of water drowned the 20 men within the faulty airlock in which they were sheltering.

Note: In 1904 during the construction of another Hudson river tunnel located south of the original tunnel, silt irrupted into the shield through the open door, during the jacking process. A miner was buried but the remainder of the crew managed to reach the safety of the airlock. A spectacular column some 12 m (40 ft) high, comprising compressed air and silt shot upwards. Work proceeded after the hole was closed with clay from the surface.

Samuel Dunn — 1849

On 26 November 1849 Samuel Dunn took out a patent (No. 12,632) (Figures 298 and 299) for a tunnelling machine for working in soft sand

---

*Hersent used compressed air for the first time the same year in a much smaller tunnel, the Antwerp dock tunnel which was successfully constructed with cast-iron lining.[5]

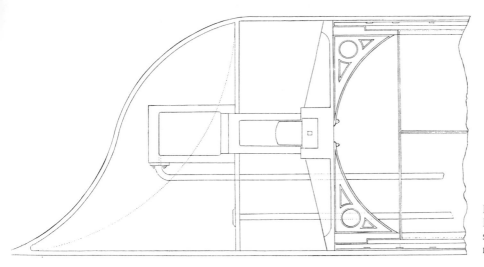

Figure 299. Samuel Dunn tunnelling shield showing suggested modification

and mud. Dunn's design was significant in that he proposed using a cylindrical or elliptical shield which was advanced in one piece.

An unusual feature of the Dunn shield was its ploughshare-shaped front and the fact that he proposed using a piston which had a head the full size of the shield for forcing the machine forward. Dunn suggested that the ram be either hydraulically or atmospherically powered. Copperthwaite commented that a piston of such dimensions would be difficult to make properly watertight and that Dunn, probably anticipating such problems, offered as an alternative, a design for a smaller hydraulic ram which could be constructed in the centre of the plough, in the axis of the tunnel. Copperthwaite[3] considered the design crude and remarked that the plough front, the main feature of the shield, would be 'quite unworkable'.

So far as muck removal was concerned, Dunn anticipated that the sand or mud through which the shield was travelling would be pushed aside by the plough as the machine advanced.

P. W. Barlow — 1864[2]

Peter William Barlow took out two patents for tunnelling shields. The first granted under No. 2,207 (Figure 300) was dated 9 September 1864. Distinctive features covered by the 1864 patent were that the circular shield was to be advanced in one piece and cast-iron lining was to be used in the tunnel. The shield was moved forward by screw jacks. In addition Barlow made the important suggestion that 'the space as it is left between the earth and the exterior of the tunnel may be filled by injecting or running in fluid cement'. So far as the author is aware, this appears to be the first time the proposal to use grouting behind the tunnel lining was made. Barlow's 1868 patent (no number) was provisional only and covered proposals for a transverse partition or diaphragm which had a centrally situated door or opening, which trapped a pocket of air in the upper part of the tunnel in the event of an inrush of water.

R. Morton — 1866

Robert Morton's Patent No. 770 dated 15 March 1866 also describes an elliptical or circular shield, the front of which was 'sharp or pointed like a wedge'. In their book,[6] B. H. M. Hewett and S. Johannesson comment that Morton's 1866 patent was distinctive in that it was the first to use the word 'shield' in describing the appliance. However a close examination of Dunn's 1849 Patent No. 12,632 will show that Dunn used the word 'shield' in conjunction with the words 'plough' and 'cylinder' many times in his specifications.

A. E. Beach — 1869[7, 8]

Alfred Ely Beach was born on 1 September 1826. In 1846 Beach and Orson D. Munn formed a partnership and purchased the paper

333

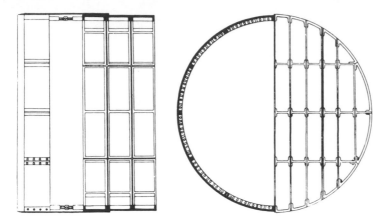

Figure 300. Peter William Barlow shield — patented in London on 9 Sept. 1864 under No. 2,207

the *Scientific American* which they ran for approximately 50 years. Beach, who had acquired a prior training in journalism while working under the guidance of his father who owned the New York paper *The Sun*, soon became Editor. In this capacity he was able to assist new and would-be inventors by a unique service which he offered through his paper. This service covered legal advice and the securing of patents through the Scientific American Patent Agency. He is reputed to have personally handled thousands of patents which he deposited in the Patent Office in Washington every fortnight. In addition, the paper featured articles of interest concerning new inventions, as well as any notable scientific or engineering achievements either within America or abroad. Nor did Beach's work end here, he was also a noted inventor in his own right. In 1847 he patented a design for a typewriter which was followed in 1857 by a patent for a typewriter for the blind. (This machine had been exhibited at New York's first World Fair in 1853.)

In 1864 Beach was granted a patent for a system of city transportation based on rail-cars drawn by stationary engines using cables. However, certain unsatisfactory aspects connected with the cable system led Beach to look in other directions, and the following year (1865) he took out a patent covering pneumatic tubes for mail and passengers. It was in proving the practicability of this latter system that Beach came to design his tunnelling shield.

In order first to gain the approval of the city administrators and the support of the public for this unique underground railway system, Beach

decided to construct a small trial section between Warren and Murray Streets in Lower Broadway, New York. He proposed building a 92m (300 ft) long tunnel which would have a diameter of 2.4 m (7 ft 10 in). However, Beach was aware that if he openly applied for a charter for a pneumatic tube for passengers, his plans would be thwarted by William Marcy Tweed, popularly known as 'Boss Tweed'. Tweed was an American politician and leader of the notorious 'Tweed Ring' — a criminal organization which flourished in New York at that time. The Tweed Ring through Tweed, Mayor A. Oakey Hall, and Messrs. B. Sweeny and R. B. Connolly, controlled city administration, public works, and education. This syndicate obtained a large amount of money in the form of graft from the various street-car companies operating at that time. Naturally any scheme which might threaten this lucrative flow of money had to be squashed promptly and Beach's proposed plan for a pneumatic underground railway system posed just such a threat. Beach therefore asked the legislature for a charter ostensibly to build a small 1.2 m (4 ft) diameter pneumatic tube to show that delivery of mail by this method was feasible. The charter was for purposes of carrying mail by tube from Liberty Street, New York, to the Harlem river, and it contained the important proviso that the street above must not be disturbed. This proviso, preventing Beach from using the conventional cut and cover method then in vogue, forced him to seek alternative methods of tunnelling through the dry sandy soil. He found the solution in the design of his 8 ft (2.4 m) diameter cylindrical shield. (Figures

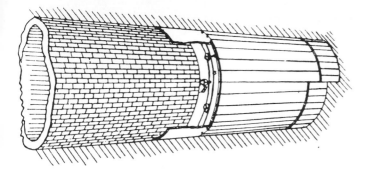

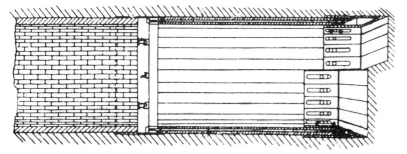

301 and 302) The machine was advanced in one piece and thrust was provided by hydraulic jacks* which pushed against the recently constructed masonry of the tunnel. The main body of the shield was built of timber fronted with a heavy chilled iron ring, brought to a cutting edge. The interior of the cutting edge was crossed by iron shelves which were also sharpened. At the rear of the timber section a wrought-iron forged ring took and distributed the thrust of the rams.

The shield and the tunnel itself (designed and constructed according to Gies[9] in 1868) were built secretly, most of the work, such as muck disposal, etc. being attended to during the night, so that as little attention as possible would be drawn to the project. When the tunnel was complete, Beach spent the following two years lavishly outfitting the reception rooms and platforms with all manner of accessories including, amongst other embellishments, a grand piano, a fountain, and a fish-tank. This was done so that the public and, most important, the newspaper reporters would be duly impressed. Only thus with full public support on his side, could Beach hope to outwit

Figure 302. Beach shield — used in Broadway tunnel, New York City. Tunnel was for a pneumatic railway which was opened for public use on 26 Feb. 1870. Shield constructed in 1869 by A.E. Beach

Boss Tweed and his gang. The pneumatic tube was proclaimed an instant success and thousands of visitors paid 25 cents each to the ladies of the Union Home for Orphans of Soldiers and Sailors who manned the entrances.

*This was the first time hydraulic jacks were used to advance a tunnelling shield.

Unfortunately, however, Boss Tweed was not amused. Through the city administration which he controlled, he attempted to force Beach to close the tunnel. Beach went to Albany and fronted Tweed in the legislature. Tweed countered by cunningly suggesting that the city's transport problems be solved his way, i.e. by building an overhead railway system which would be powered by steam. (Naturally Tweed stood to gain from this proposal as a good income flowed into his coffers from building material graft.) Tweed's proposals and Beach's daring schemes appealed to the legislature, who promptly approved *both* Bills. However, Tweed had the last word — he persuaded Governor John T. Hoffman (reputed to be a Tweed man) to veto the pneumatic tube Bill.

Beach was beaten, but did not give up the fight. In the meantime Tweed's Ring was finally exposed in 1871 by the publication in the *New York Times* of evidence handed to it by a bookkeeper. Tweed was tried and found guilty. He was sentenced to 12 years in the penitentiary. Beach continued to fight for the acceptance of his pneumatic tube transportation system, but fate was against him. The 1873 depression hit New York and finance for such a scheme became unavailable. Beach found he was unable to obtain a renewal for the charter which had in the meantime expired. Later the development of electricity rendered his mode of propulsion obsolete. He died on 1 January 1896.

J. H. Greathead — New Thames Tunnel 1869-70[8, 10, 15]

James Henry Greathead was born in the Cape Colony in South Africa in 1844. In 1859 he travelled to England in order to complete his education. Soon after this Greathead became apprenticed to Peter W. Barlow, a well-known English engineer who was connected with the London Metropolitan Railway development scheme.

In 1868, after Greathead had completed his term of apprenticeship, Barlow set forth his proposals for a second Thames tunnel. The project was approved by Parliament, but the many difficulties encountered by Brunel still lingered in the minds of building contractors, and none could be found who was willing to tender a bid. When it became obvious that there were no other takers Greathead, then only 26 years old, daringly suggested that he undertake the contract for Barlow. Barlow agreed and Greathead designed and built his own tunnelling shield for the work. This shield was very similar to Barlow's 1868 patent which had a transverse partition or diaphragm.

Greathead's shield (Figure 303) was designed to advance in one piece. The front or forward end of the shield consisted of a cast-iron ring which was bolted to the main body of the shield. In contrast to later shield designs, however, where the cutting edges were made acute, Greathead's shield had a rounded cutting edge. Behind the cast-iron ring Greathead constructed a bulkhead or diaphragm of 19 mm (¾ in) plates, having in it a doorway or opening reaching nearly to the top of the shield. The doorway could be closed if necessary by dropping across it 8 cm planks, the ends of which would be held by the vertical channel iron framing the jambs. (In Barlow's 1868 provisional patent the door or opening was situated in the centre of the transverse partition.)*

Barlow suggested in his 1864 patent that the space left between the earth and the exterior of the tunnel could be filled by injecting or running in fluid cement. In constructing the Tower subway Greathead used a hand syringe. With this clumsy device he forced grout (lime and water previously mixed in a tub) into the space left between the tunnel walls and the cast-iron lining. Later (1886) Greathead designed and patented a grouting pan which proved to be a much more efficient means of carrying out this work. It was first used in the construction of the City and South London tunnels.

---

*Copperthwaite[3] remarks that the substitution of an opening reaching nearly to the top of the shield in place of the central opening proposed by Barlow in his provisional patent of 1868 can hardly be regarded as an improvement. No doubt access to the face was easier, but on the other hand the raising of the top of the doorway did away with the safety diaphragm above. With this shield, an inrush of water would at once have filled the tunnel to the soffit of the roof. Fortunately for Greathead, protection from an inrush of water proved to be unnecessary so far as the Tower subway was concerned for, in Drinker's words, 'the water encountered might at almost any time have been gathered in a stable-pail'.

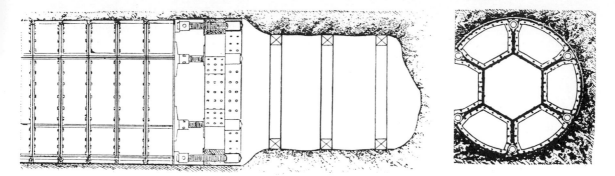

Figure 303. Greathead shield — used in second Thames tunnel. Building of the second Thames tunnel was promoted by P. W. Barlow. The shield was designed and built by J. H. Greathead and is similar in design to Barlow's 1868 patent

Of interest are the main differences between Brunel's tunnel and shield and Greathead's first tunnel and shield. Brunel's old tunnel brickwork is 11.2 m (36 ft 9 in) wide by 6.8 m (23 ft 4 in) high, the new Tower subway consists of an iron tube about 2.0 m (6 ft 7 in) in diameter. The old tunnel was constructed with a shield weighing 120 ton (109 t) and accommodating 36 workmen, while the Tower subway built by Greathead was driven by a shield weighing 2½ ton (2.3 t) and accommodating at most three workmen at a time. Brunel faced five irruptions from the river, Greathead encountered none. Between the commencement and the completion of Brunel's tunnel 18 years elapsed, while the Tower subway tunnel was completed in less than a year. Brunel's tunnel cost over £600,000 while the cost of the Tower subway was under £20,000.

For the remainder of his working career Greathead was associated with the design and construction of tunnels and shields. He devised many innovations for tunnelling in water-bearing strata. Of particular significance was his shield (built in 1874 for the Woolwich tunnel) which incorporated a water-seal and which was specially designed to work with compressed air. In this shield Greathead had an extended cutting edge, behind which he fixed a diaphragm which covered the upper section of the shield only. A small distance behind this front diaphragm he placed an airlock which had an airtight door opening forwards. The door was situated in the upper half of the shield. The outer door of the airlock which

opened into the airlock was also situated in the upper half of the shield. Behind the airlock the skin of the shield extended a sufficient distance so that the hydraulic rams could be accommodated, and also so that the skin of the shield could overlap the section of tunnel already built. For reasons unconnected with the engineering features of the tunnel, Greathead's shield was not used in the Woolwich tunnel.[2] Attempts to drive the tunnel without compressed air resulted in failure and ultimately the entire project was abandoned. Later a shield of this type was tested in practice at Antwerp and in the Hudson river tunnel.

Another innovation patented by Greathead in 1873 was a hydraulic segment-lifting apparatus which was made and successfully tested in the Woolwich subway tunnel. Greathead died in London in 1896.

Summary

Over the years the charge of conscious plagiarism, whether true or false, has been levelled at many inventors. This is particularly so in the case of Barlow, Beach, and Greathead. Barlow and Greathead in England and Beach in America were no doubt working independently at the shield problem.

Brunel was the first to suggest using a circular shield, Dunn the first to propose moving it in one piece and Barlow the first to suggest that the annular space left behind the cast-iron lining, after the advance of the shield, be filled with grout — though he did not specify how this was to be accomplished. In 1868 Barlow provisionally patented a shield having near the cutting edge a transverse partition or diaphragm. But it was Beach in New York and Greathead in England who actually built and used shields incorporating many of the aforementioned features — yet differing from each other in detail.

In the discussion which followed the presentation of his paper in 1895[10] Greathead states that he was unaware of the existence of Barlow's provisional patent of 1868, which Greathead's own shield most resembles.

Beach says he made the first designs for his shield as early as 1865, and that in 1868 he actually built and tried a small experimental model 92 cm (3 ft) in diameter.[9] Copperthwaite comments that Beach did not appear to have been aware of the existence of Barlow's shield, though he knew about Greathead's tower subway shield.[3]

Yet it also seems likely that though Greathead was unaware of Barlow's provisional patent describing the transverse diaphragm, Greathead, working as he did, first as Barlow's apprentice and then later as Barlow's contractor, would have at some stage or other discussed the subject with Barlow. It seems conceivable that such discussions bore seed which later germinated.

Beach also may very well have been unaware of Barlow's 1864 patent which Beach's shield most closely resembles, yet again one must consider the fact that Beach came daily into contact with inventors of all types through the Scientific American Patent Agency which he personally ran and the paper *Scientific American* of which he was Editor. Furthermore, he was obviously kept well informed of the latest outstanding achievements of engineers both at home and abroad. He may not have actually seen Barlow's patent, but it is conceivable that the exciting subject of shields, initiated by Brunel in 1818, would also be freely discussed by prominent contractors and engineers of his day — after all, many of the difficulties and problems associated with subaqueous tunnelling or tunnelling in loose sandy soil still remained, despite Brunel's momentous accomplishment.

Nevertheless, whether the charge of conscious plagiarism be true or false and from whatever source the ideas were originally gleaned, *credit must be given* to these two men, Greathead and Beach, for having the courage and ingenuity to put into effective practical form these theoretical ideas.

Greathead's transverse diaphragm and airlock, his grouting pan, and segment erector are features still used today. Beach was the first to use hydraulic jacks for advancing his shield and, with Greathead, shares the distinction of being the first to use a shield which advanced in one piece as do most modern units today.

M. J. Jennings and J. J. Nobbs — 1889-90

Two patents for shields Nos. 19,550 (Figure 304) and 7,374 dated 5 December 1889 and 12 May 1890, respectively, filed by Mathew Jones Jennings of Surrey and James John Nobbs of London are of interest because of their great similarity to the MEMCO hydraulic shield (see John Tabor — Mining Equipment Manufacturing company).

The Jennings patent described a shield built of iron or steel bars or 'needles'. The needles were provided with longitudinal grooves which permitted each needle to be linked to its neighbour but still allowed for their separate longitudinal advance. Jennings proposed that the needles be arranged side by side within the excavation so as to form a complete temporary lining and support to the roof and sides of the tunnel, until such time as the permanent lining or support had been constructed immediately under the needle shield. The needles could then be driven forward one at a time by hydraulic power or screw jacks.

The Nobbs patent related to a design for a tunnelling shield which consisted of a series of steel bars or piles which were fitted to overlap each other at the edges so as to form a kind of lagging and which were supported internally by a sectional rib or ribs. The bars or piles were capable of being driven forward independently of each other by means of hydraulic rams or screw jacks. According to Copperthwaite, this type of shield was actually constructed in 1901 for use in a 2.1 m (6ft 11 in) diameter sewer tunnel which was driven in the Isle of Dogs.

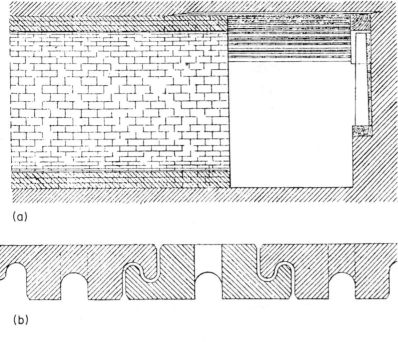

(a)

(b)

(c)

Figure 304. M. J. Jennings shield. U.K. Pat. No. 19,550, dated 5 Dec. 1889. (a) Tunnel in process of construction; (b) transverse section of three needles; (c) one needle

However, apparently results were unsatisfactory.

H. H. Dalrymple-Hay — 1896[11]

Like the name of Greathead, that of Harley Hugh Dalrymple-Hay became well known in tunnelling circles as a consulting engineer during the period of initial construction of London's underground tube railway system.

In 1894 Dalrymple-Hay was working as the Resident Engineer in charge of the Waterloo and City Railway line which was being driven through waterbearing ground. At first Dalrymple-Hay used a conventional Greathead-type shield in conjunction with what was known as the 'assisted shield' method devised by Greathead,* but after some months, during which the Greathead method was used, Dalrymple-Hay modified the cutting edge of the shield by extending the top so that it formed a 'hood' or 'vizor'. After experiments with various other methods which proved unsuccessful, Hay found that good sealing could be obtained with the modified shield if an annular space a few inches in front of the cutting edge was excavated and filled with well-tempered clay. This was of course excavated bit by bit, each section being filled with clay immediately to prevent the collapse of the material. When the annulus had been completely excavated and filled, the entire cutting edge of the machine was able to move forward within this self-sealing clay matrix. Dalrymple-Hay filed a patent (No. 622) for this technique (known as 'pocket holing') on 9 January 1896.

*When water-bearing strata were encountered 'a small heading was driven at the top in advance of the shield, stout poling-boards being used to support the top, resting at one end upon the forward end of the shield; the heading was then widened out and the polings continued until about three-fourths of the circumference and the whole of the face had been poled'.[10]

J. Tabor — Mining Equipment Manufacturing Company — MEMCO[12, 13]

When John R. Tabor (born 21 May 1932) attended the Cass Technical School during the period 1946-50 he showed a natural aptitude for electrical and mechanical engineering. As a result he enrolled as a student at the Michigan School of Mining and Technology where he subsequently obtained his M.M.E. Mining Engineering and his B.Sc. Civil Engineering qualifications in 1955. From that year until the beginning of 1961, Tabor gradually widened his experience in these particular fields while working as Chief Field Engineer and Geologist for Bert L. Smokler and Co., and then later (1956-61) as Project Manager for the Fattore Construction Company in Warren, Michigan.

In 1961 Tabor founded the Mining Equipment and Manufacturing Company, serving as its President and Chairman of the Board. The company specialized in the manufacture of tunnel supports which were produced on a machine Tabor had had installed in a nearby fabricator's shop. So successful was this initial venture that demand began to exceed the company's capacity to supply. Lacking the necessary capital at that time for the purchase of adequate accommodation to house his rapidly expanding business, Tabor obtained a large circus tent and MEMCO transferred its belongings, lock, stock, and barrel. But the inadequacies of this type of accommodation were rather sharply brought home to the company when the first and the second tent (obtained to replace the loss of the first) were damaged by gale-force winds. The third tent was destroyed by fire. Tabor was by this time ready to occupy almost anything of a more permanent nature — if he could find it. Then fate took a hand. While skiing one evening, Tabor and his wife chanced upon an old abandoned building — a one-time dance hall of the gay 1920s. The following day, after Tabor had obtained the key, he and his partner Walter Weiss returned to the site for a closer look. Tabor unlocked the doors and Weiss entered. Wiess's second step sent him crashing through the rotten floorboards.

Undaunted, the MEMCO firm bought the dilapidated old building, repaired the flooring, and moved in.

While the MEMCO firm was busy manufacturing and selling tunnelling supports and settling its accommodation problems, work was begun in April 1961 under contract on a 2440 m (8000 ft) long sewer tunnel estimated to cost U.S.$2 million. The project was a joint-undertaking by Hurley Construction (St. Paul), Winston Bros. Co. (Minneapolis), and Foley Bros. Inc. (St. Paul).

After only 15 m (50 ft) of tunnel had been excavated with a conventional excavator, Hurley struck a section of running ground which irrupted, burying the excavator and the heading. The excavator was fairly easily retrieved and, under the impression that the bad ground encountered was merely a short piece, Hurley transferred the excavating machine to the other end of the tunnel.

Approximately 1800 m (6000 ft) of tunnel had been satisfactorily excavated and lined when another irruption occurred, this time burying the machine so completely that it could not be salvaged. An attempt was made to win through the unstable ground by hand-mining and timber bracing, but this method was finally abandoned after it was found that the timber bracing would not hold back the ground. The use of compressed air was also considered, then discarded, when it was realized that the ground ahead was strewn with twisted steel beams and timbering from two lost headings including the collapsed tunnel lining.

The final section of sewer remaining between the two cave-ins was deleted from the original contract and a new contract was drawn up and re-bid. At an estimated cost of U.S.$1.5 million the new contract was awarded to the same contractors.

The answer to Hurley's problem came in the form of a new hydraulic forepoling shield designed and built specifically for the job by MEMCO. Apparently over the years Tabor had been attempting to find a solution to the difficult problem created by a mixed face consisting of materials such as running sand and limestone. Until the appearance of the new hydraulic shield, such ground was usually attacked by using hand-installed spiles which were jacked ahead of the face, thus forming a temporary roof support. But this method frequently proved inefficient and on occasions even dangerous.

Figure 305. Calweld shield. Used by Perini Corporation and Brown & Root for the construction of the twin shield-driven tunnels connecting Market Street subway and the Trans-Bay tube. View of shield being lowered into north shaft. Cutting-edge breast jacks and sliding tables can be seen in this view. In top centre pocket, control valve panel can be seen. Tunnel muck was removed through bottom centre pocket. (Courtesy Perini Corporation.)

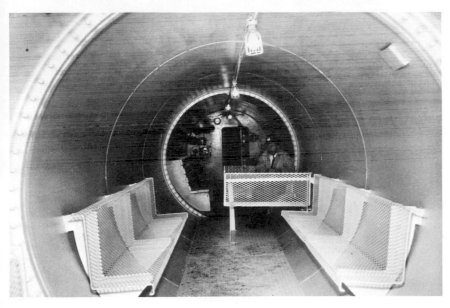

Figure 306. View in luxury manlock — looking towards the door to the transition manlock. Used with Calweld shield during construction of the twin tunnels connecting the Market Street subway and the Trans-Bay tube. (Bay Area Rapid Transit System, San Francisco, 29 Jan. 1969). (Courtesy Perini Corporation.)

MEMCO's new 5.5 m (18 ft) long shield featured seven poling plates constructed of T-1 steel with 91 cm (3 ft) cutting edges, each plate being individually powered by a 54 t hydraulic jack. The plates, which varied in total length from 3 to 4 m (10 to 13 ft) (depending upon their position on the shield's inclined front),

could be extended for a distance of 1.2 m (4 ft) beyond the shield, thus providing cover for the area above the springline. As forepoling plates tend to dive in sand and rise in clay, the 54 t jacks were mounted at an adjustable angle within the skin to compensate for this.

Hurley commenced work on the new contract

Figure 307. Markham shield being constructed for use in the second Dartford tunnel under the river Thames. (Courtesy Markham & Co. Ltd.)

at the beginning of the year (1962); by midyear the longer cave-in section had been successfully negotiated and a mere 200 m (600 ft) or so of unconsolidated sandstone lay ahead.

'The forepoling plates worked beautifully in that 129-ft stretch', Hurley commented enthusiastically. 'The material was so poor that the plates were out of sight the whole time, but we went through with no big trouble.'[12]

For a time after the debut of the original unit used by Hurley, MEMCO continued to manufacture this unique type of forepoling shield and, of course, as was inevitable, other manufacturers soon began producing their own versions of it. Though these differed from the MEMCO unit in many respects, they nevertheless incorporated the same basic engineering principles employed by Tabor in his shield.

Westfalia Blade Shield[14]

In the late 1970s Hochtief AG of Essen, West Germany, constructed single-track rapid-transit system tunnels in Frankfurt/Main using the 'blade shield' construction method.

The shield, which was manufactured by Westfalia Lünen, is basically similar in concept to the needle and bar shields designed by Jennings and Nobbs in 1889 and 1890, respectively, described earlier in this section. It is also similar to the Ennis digger shield (Patent No. 690,137 dated 1901 described in the section entitled Hybrid machines — digger shields).

Basically the Westfalia blade shield consists of a leading and a trailing section. Around the perimeter of the leading section of the shield are

342

32 heading blades, each independently advanced by a hydraulic jack, while two hydraulic cylinders advance each of the 8 base blades. The leading section is supported by a base frame featuring three working platforms divided vertically by a central column. One platform is positioned about half-way up the shield face from the base section and the other two form intermediate platforms above the centrally positioned platform.

Face rams are mounted on the central column and also on the side support frames. The side frames are, in turn, attached to the central platform and to the base section. Operating equipment for the heading blades is housed in the forward section.

The respective ends of the various cylinders are hinged to the heading blades and to the supporting frame. The 32 blades in the trailing section are also hinged to the leading blades. The trailing section acts as a protective cover beneath which the concrete lining segments may be erected.

All breasting and heading operations can be controlled from a central control point. Each heading blade may be advanced a maximum distance of 600 mm (2 ft) — the length of the cylinder stroke.

Thrust reaction is taken by the entire machine, which is held stationary in the bore by the friction between the ground and the static blades. As each blade is advanced separately comparatively little force is needed, thus both the cylinders and the pumping equipment are of small dimensions.

According to Horst Gruner,[14] when the blade shield was used in the Essen Underground municipal railway tunnels the two 'Westfalia 'Dachs' boom cutter loader machines proved inadequate in the 'extremely hard green sand marl' encountered there and were replaced by a pair of back-acters with 170-litre capacity buckets.

Horseshoe and circular-type Westfalia blade shields of various sizes have been used in conjunction with boom cutter loaders or excavators in Wiesbaden, Frankfurt, Gunzenhausen, (West Germany), Paris (France), and Tokyo (Japan) (see Figures 561-564) for water sewage, rail and highway tunnels.

References

1. L. T. C. Rolt, *Isambard Kingdom Brunel*, Longmans, Green, London, 1957.
2. H. S. Drinker, *Tunnelling, Explosive Compounds and Rockdrills*. Wiley New York, 1893.
3. William Charles Copperthwaite, *Tunnel Shields and the use of Compressed Air in Subaqueous Works,* Archibald Constable, London, 1906.
4. R. Glossop, The invention and early use of compressed air to exclude water from shafts and tunnels during construction. *Geotechnique,* **26,** No. 2, 253-280, 1976, London.
5. *Encyclopaedia Britannica,* Vol. 22, 1927, pp. 560-572.
6. B. H. M. Hewett and S. Johannesson, *Shield and Compressed Air Tunnelling,* McGraw-Hill, New York, 1922.
7. Tunnelling under city streets — the atmospheric railway, *Scientific American,* **xxvi,** No. 12, 1872, New York.
8. The new Thames tunnel — how the work is carried on, *Scientific American,* Nov., 1869, 314-315.
9. Joseph Gies, *Adventure Underground,* Robert Hale, London, 1962.
10. J. H. Greathead, The City & South London Railway, Paper No. 2873, *Proc. Inst. Civil Engs.,* Nov. 1895, 39-72, London.
11. Gosta E. Sandström, *The History of Tunnelling,* Barrie & Rockliff. London, 1963.
12. J. O. Monoghan, Hydraulic forepoling shield drives caved-in tunnel drift, *Construction Methods,* June 1964.
13. *Engineering News-record,* MEMCO: Young Turks that tackle tunnels, Aug. 1967.
14. Horst Gruner, Blade shield tunnelling in Essen, *Tunnels and Tunnelling,* June 1978, **10,** No. 5, 24-28, London.
15. H. W. Richardson and R. S. Mayo, *Practical Tunnel Driving,* 2nd edn., McGraw-Hill, 1975, p. 233.

Correspondence

Mining Equipment Mnf. Co. (MEMCO), Racine, U.S.A.
Westfalia Lünen, Lünen, West Germany.

Patents

M. I. Brunel — 4,202/1818 U.K.
S. Dunn — 12,632/1849 U.K.
R. Morton — 770/1866 U.K.
P. W. Barlow — 2,207/1864 U.K.
A. E. Beach — 91,071/1869 U.S.A.
G. T. Bousfield — 2,221/1873 U.K.
J. H. Greathead — 13,215/1887 U.K.
J. J. Jennings — 19,550/1889 U.K.
J. J. Nobbs — 7,374/1890 U.K.
H. H. Dalrymple-Hay — 622/1896 U.K.

ELEVEN

Mechanized Shields

Under the title 'Tunnelling machines' on the evolutionary chart (printed on the inside covers of this book) are two subtitles namely 'Rock Tunnelling Machines' and 'Mechanized Shields'. As may be imagined, the dividing line between these two types of machine, particularly in the case of rotary-wheel-type units, has in some cases been rather difficult to define.

However, for the purposes of this book a 'Mechanized Shield' is defined as a machine comprising a shield and a cutter head where the outside or peripheral cutters *do not* cut gauge. A hard-ground or 'Rock-Tunnelling Machine', on the other hand, is a machine where the gauge of the tunnel profile *is cut by the outer or peripheral cutters* and not by the cutting edge of a shield. Thus the T.B.M. built by R. L. Priestley Ltd for the Mangapapa hydro-electric project which featured a fully shielded unit with a cutter head which cut gauge is designated a 'Shielded Tunnelling Machine' to distinguish it from a mechanized shield and a rock-tunnelling machine.

The development of the shield did not obviate the necessity for the removal of muck and debris from the face. This still needed to be excavated and, for many years, the work was performed by men using picks and shovels.

Because this procedure was both slow and arduous, it was not long before engineers were attempting to find a mechanical solution to the problem. Again, as it was with the shield, designs for a mechanized shield were patented some time before a unit was actually built and tested in practice.

Mechanized shields may be divided into about six main groups; those with:
(a) rotating-wheel-type cutting assemblies (having either radial arms or a disc fitted with cutting tools);
(b) planetary cutting assemblies. In this case a number (one or more) of independently rotating cutter heads with tools is fitted either to the ends of a rotating beam, the arms of a rotating cross, or on a rotating disc. This action (i.e. the combined rotation of the disc, cross, etc. and the cutter heads) causes individual tools on each cutter head to describe epicycloidal paths in the plane of the face);
(c) oscillator-type cutting assemblies;
(d) auger-head cutting assemblies;
(e) active horizontal shelves or cutting grids;
(f) water jets instead of conventional cutters or tools to excavate the ground.

Only a few planetary action shields have been tested in practice and, with the exception of a Demag unit, most of these were built and used in the Soviet Union.

The Demag machine featured nine cutter heads mounted (three each) on three radial arms. The main cutter head carrying the nine heads could be advanced or retracted from the face independently of the main shield body. Cuttings were lifted from the invert by scoops on the main cutting head and fed via a chute to a central point where they were removed pneumatically. Arrangements could also be made for the hydraulic removal of the spoil, if required.

Two types of planetary machine were built in the U.S.S.R. Four of these carried six discs, and two, pairs of discs. The six-disc machine featured an extensible rotary cutter ring on which was mounted four radial arms in the shape of a cross. The arms carried six disc cutters, each fitted with a number of cutting teeth. Buckets on the main cutting ring lifted the cuttings and deposited them through a hopper onto a conveyor belt which carried the cuttings to the rear. The six-disc planetary shields were used during the construction of the Leningrad Underground Railway tunnels. Of the two-disc units, one was used during the construction of the Moscow Underground, where it tunnelled through dense clays, clayey shales, marls, and limestone, etc.

According to Marsaak and Samoilov,[1] the first active horizontal-shelf shield built was the Hant machine, used in London. The rigid shelf of the Hant shield had short protective diaphragms fitted to their fronts. These diaphragms were capable of being raised or lowered relative to the shelf, thereby allowing the operator to control the ingress of soil from the face.

Apart from this it would appear that a considerable amount of developmental work on active horizontal slump-shelf shields and those fitted with cutting grids, has been carried out by Soviet engineers. Front and rear sections which rotated in the vertical plane (i.e. flapped up and down) were fitted to rigid slump platforms.

Numerous experiments were also carried out with vibrating sections, either front, rear, or side members. However, generally speaking, it appears that these experiments were not particularly successful, as it was found that in most cases the vibrations tended to cause the soil slope on individual platforms to slip, ultimately leading to the collapse of the entire face.

A shield with an active cutting grid was also built by the Foundations and Underground Structures Research Institute (N.I.I.O.S.P.), (U.S.S.R.). The shield cutter head consisted of two wheels, one mounted behind the other within the front end of the shield body. Each wheel was fitted with a number of parallel bars, evenly spaced, across the diameter of the wheel. Each wheel could be rotated independently of the other, thus allowing the bars of each wheel to form various positions relative to those in the adjacent wheel. In this manner when the bars were directly behind each other, either horizontal or vertical apertures could be formed, or the bars could be positioned so that diamond or square-shaped apertures controlled the entry of spoil from the face. The wheels could be advanced or retracted from the face indepen-

Figure 308. J. D. Brunton and G. Brunton mechanized shield. U.K. Pat. No. 1,424, dated 3 April 1876

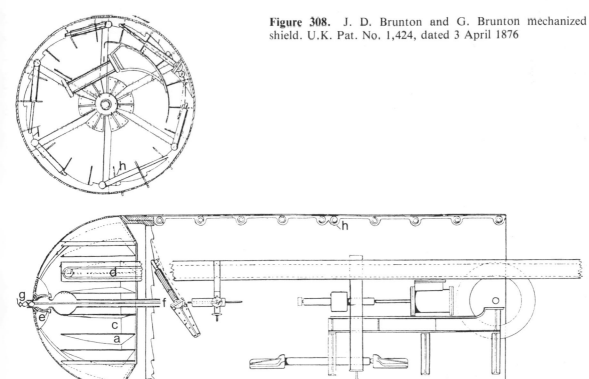

dently of the main shield body, and by using the different configurations mentioned above, a number of soil conditions from cohesion-less sand to clay could be handled by the machine. (Figures 324 to 329 show recent Japanese (Robbins Company licencee), Russian and U.S. mechanized shields being used in various parts of the world.)

J. D. and G. Brunton — 1876

So far as the author is aware John Dickinson Brunton and George Brunton of England were the first to patent a design for a mechanized shield. This was filed under Patent No. 1,424 dated 3 April 1876 (Figure 308).

The mechanized shield proposed by J. D. and G. Brunton employed a hemispherically shaped revolving cutting head built of plates. These plates [a] were so bent, spirally arranged, and inclined that their protruding edges formed cutters which were designed to pare or cut away the chalk or other material as the head revolved. The material so cut was to fall between the plates and into buckets [c] which were radially arranged on the interior of the cutting head. As the head rotated the buckets rose on the delivery side and allowed the cuttings to be discharged on to a chute situated near the top of the machine. The chute directed the debris on to a conveyor belt [d] (an endless band supported on rollers which carried the material to the rear). Messrs. Brunton proposed that the machine be guided by means of a centrally placed steering pipe [f] which passed through a spherical bearing [e] in the head of the machine and protruded slightly from its exterior. A passage for the steering pipe was provided by means of a screw auger [g] which drilled a hole ahead of the cutters. The auger was attached to the central driving shaft of the machine. (Single and multiple-auger shields similar to that proposed by Messrs. Brunton have recently been produced by Machinoexport of Russia and also by some Japanese manufacturers.)

The cutter head was rotated by six hydraulic jacks [h] attached to brackets fitted to the interior of the shield. The rams of the jacks operated against the teeth of a ratchet ring [i] attached to the cutter head.

Messrs. Brunton also suggested that the main body of the shield, which was split or divided longitudinally, be constructed of boiler plate. During operation these plates could be expanded against the tunnel walls by means of hydraulic jacks, thus resisting both the torque and thrust of the rotating cutter head. The machine could be advanced by means of hydraulic jacks placed at the rear of the shield.

Frank O. Brown — 1886

In 1886 under No. 340,759 (Figure 309), Frank O. Brown of New York filed a patent for a mechanized shield. Brown proposed using a cylindrical shield [a] made of plate-iron bolted together. The forward cutting end [c] of the shield was inclined or bevelled and was arranged with its point uppermost. Behind the bevelled end of the shield Brown placed an airtight transverse partition [d] in which were a number of manholes [e]. These holes could be closed by plates and made airtight if desired.

A conveyor tube [f] passed through one of the manholes. Within this tube was an auger-shaped conveyor shaft [g] which projected a little beyond the forward end of the tube where it acted as a drilling tool. Brown suggested that if the ground at the face was too hard for the auger to loosen and remove, the upper part of the conveyor tube could be opened. This would enable workmen to excavate the ground ahead with picks, and shovel the debris into the tube. The spiral auger blades would then carry the muck through the tube to the rear of the shield. A steam-engine would drive the auger blade. Brown proposed to move his shield forward by atmospheric pressure. When excavation work ahead of the shield was completed, the auger and tube would be removed and the manholes closed. Steam would then be pumped into the newly formed cavity through a pipe. This steam would in turn force the air within the cavity to escape by means of an air-discharge valve. As soon as steam started to emerge through the air-discharge valve the steam pipe would be closed and the steam within the cavity allowed to condense, thus forming a partial vacuum. In case this was not enough to advance the shield Brown suggested further that the forward movement of the machine could be assisted by

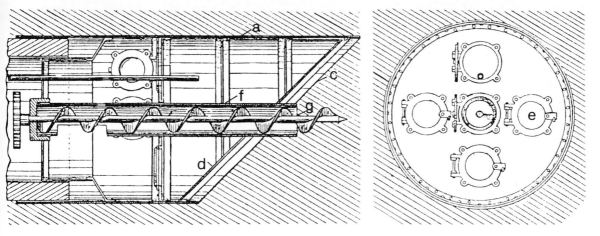

Figure 309. F. O. Brown mechanized shield U.K. Pat. No. 340,759, dated 27 April 1886

'compressed air confined in the rear of and allowed to press against' the transverse partition.

J. H. Greathead — 1887-89

Apart from his other patents Greathead submitted two designs for a mechanized shield. The first of these was filed under Provisional Patent No. 13,215 and was dated 29 September 1887. It featured a central rotary cutter which Greathead envisaged would drive a small heading or core in front of the shield, in order to provide space for the spoil which was apparently to be broken down by tapered or wedge-shaped spikes set around and slightly to the rear of the cutter head. Greathead proposed that the spikes or piles be worked independently by means of hydraulic jacks placed within the shield. That is the spikes could be forced forward either separately or in groups as was necessary. However, no mention is made of how the debris thus loosened was to be conveyed away from the face, nor does Greathead indicate the method whereby rotary power is to be supplied to the cutter head.

The second Provisional Patent No. 195 dated 4 January 1889, covers a shield carrying a central revolving shaft to which cutter blades are attached. Again apart from describing the particular angle at which these blades were to be placed, i.e. 'inclined helically to the plane of rotation' of the shaft, Greathead makes no reference in the specifications concerning motive power for the cutter head.

However, of interest is his proposal that an airlock or plenum chamber be placed immediately behind the cutter head. He suggests that water, under pressure, could then be forced into the plenum chamber through an inlet pipe and thereafter removed, together with any excavated material, by means of an outlet pipe.

This particular method of soil removal was described more fully by Greathead in Patent No. 1,738 dated 16 May 1874 (see also Chapter 13, Slurry machines).

J. J. Robins — 1893

The James Jennings Robins mechanized shield was patented under No. 1,445 (Figure 310) in London on 25 November 1893. Robins designed a roughly conical cutter head [h] which rotated about a hollow axle [i], the axis of which coincided with the centre of the tunnel. The cutter head consisted of a flat disc [j] through which the hollow axle protruded. A number of radial arms [l] were attached to the circumference of the forward end of the axle [i] and were raked rearward and fastened at their outer end at the circumference of the disc. Attached to the arms were a series of picks or chisels [m] for loosening the soil. In addition each of the arms carried a metal plate [n] which filled the triangular opening bounded by the arm, the disc, and the axle. The purpose of

347

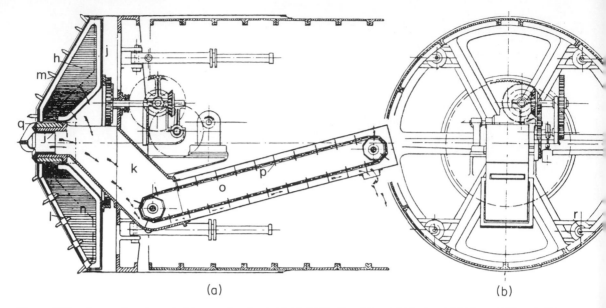

Figure 310. James Jennings Robins mechanized shield. U.K. Pat. No. 1,445, dated 25 Nov. 1893. (a) Side view; (b) rear view

these plates or scoops was to raise the material loosened by the cutters and deposit it into a central trough [k] from which it was guided by a chute on to a conveyor [o]. The conveyor (an endless chain belt with transverse blades [p] set at intervals along it) carried the muck to the rear of the shield.

Robins proposed that the cutter head and conveyor mechanism be driven by either steam, compressed air, oil, gas, hydraulic, or electric power, but stated a preference for the latter. His shield was to be advanced by means of hydraulic jacks [r] arranged around the circumference of the shield and thrusting against the completed portion of the tunnel lining. In order to enable the jacks to be swung out for readjustment, if desired, Robins suggested that they could be attached to the shield by hinge pins.

The forward end of the hollow axle was fitted with projection curved blades [q], their function being to enter the face first and scrape away the soil at the axis of the tunnel. The soil thus loosened then passed through the hollow axle into the chute and thence to the rear of the conveyor.

J. Price — 1896

The renowned Price mechanized shield was first patented by John Price under No. 13,907 in London on 23 June 1896. It made its debut in 1897 on the Shepherd's Bush to Marble Arch section of the Central London Railway line.

Though the first Price machine shared several common features with the Robins mechanized shield, there were significant differences, the most important of these being the absence of any diaphragm or transverse partition.

The cutter head [a] consisted of four radiating arms [b] bearing cutters or scrapers [c]. The arms also carried trough-shaped scoops [d] which collected the loosened material, lifted it and deposited it into a chute [e] from which it was guided directly into waiting skips.

The long driving shaft was centrally situated, its axle coinciding with the axis of the tunnel. It was driven by electrical power.

The first Price unit (Figure 311) was not particularly successful, mainly because the driving power was applied in the wrong place, that is at the axle instead of at the periphery of the machine. Owing to the presence of the long shaft [f] the machine tended to become unwieldy when working around a curve as it was unable to excavate more on one side than on another. This problem was also accentuated by the fact that the shield and excavator were independent units.

Price later eliminated these faults. He did this

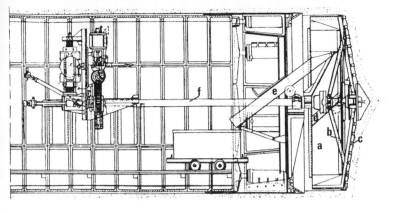

Figure 311. J. Price mechanized shield. Patented in London on 23 June 1896 under No. 13,907

Figure 312. A 3.86 m (12 ft 8 in) dia. 'Price Patent' mechanized shield with periphery drive. First built at Broad Oaks in 1901. (Courtesy Markham & Co. Ltd.)

by converting the axial drive to a peripheral one (Figure 312) and, at the same time, joining the shield and excavator together so that they combined to form a single unit. Thus instead of the cutter head being rotated by the motor through the central shaft as had been done previously, the cutter head was turned by means of a series of gears, the last of which engaged with a rigidly attached circular rack situated at the circumference of the excavating wheel. The gears were powered by a 45 kW (60 hp) electric motor (500 rpm) which turned the cutter head at approximately 1½ rpm.

The excavator wheel had six arms held together at their extremities by circumferential plates, to which the arms were fastened by bolts. (These plates also served as base-plates for the circular rack.) Attached to the spokes of the wheel were cutters or chisels and buckets. The buckets collected the loosened soil, then lifted and deposited it into the chute from whence it was dropped into a waiting skip. The hub of the excavator wheel was mounted on a short shaft which was concentric with the shield. The forward end of the central hub was fitted with a toothed scraper which, like that of the Robin's unit, was designed to enter the face ahead of the excavator and scrape away the soil at the axis of the tunnel. It also served to steady the machine during operation.

Markham & Co. Ltd. (see also Chapter 13, Slurry machines, Markham and Company Ltd) began producing the unit commercially in 1901. One of the earliest modifications made by the company to the Price machine was the attachment of an additional cutter to the outside edge of two of the arms. These extra cutters could, by means of cams or an eccentric arrangement, be projected beyond the end of the arm and the skin of the shield for a prescribed arc of the circumference of the face, thus providing, for that distance, a wider excavation into which the shield could be turned. Another improvement introduced by Markham was the addition of a belt conveyor in 1905.

Price mechanized shields were used regularly on the Charing Cross and Hampstead railway lines and attained an average penetration rate of 55 m/week (180 ft/week) (108 rings of tunnel lining). They were, moreover, found to be exceptionally reliable, giving long periods of mechanically trouble-free operation. Ten workmen, that is one ganger, two miners, six labourers, and one boy were required to operate the unit, whereas thirteen men and a boy were normally required to work an ordinary Greathead shield.

Copperthwaite,[2] whose book was published in 1906 commented that: 'The principal practical difficulty at present in the working of this machine is the tendency of the shield to get out of line, unless very carefully watched when in movement.'

G. Burt — 1897

George Burt's London Patent No. 9,549 (Figure 313) dated 14 April 1897 for a rotary mechanized shield is virtually the same as that filed by Price the previous year. That is it

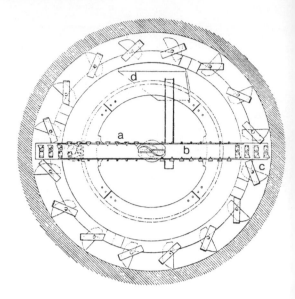

Figure 313. Cutter head of G. Burt mechanized shield. U.K. Pat. No. 9,549, dated 14 April 1897

employed removeable adjustable cutters [a] fixed to a revolving arm [b] which was mounted on a central driving shaft. A similar arrangement of buckets, [c] set behind the rotating cutter head, picked up the excavated material and dropped it into a chute [d] which guided it on to a conveyor belt for transport to the rear of the machine.

A. W. Farnsworth — 1901

On 15 May 1901, in London Alfred William Farnsworth filed a patent (Figure 314) for a mechanized shield. Farnsworth proposed that the excavator component would cut out the core of the tunnel. While his cutter head followed conventional lines in that it consisted of radially arranged arms [a] attached to a boss [d] in the centre and a rim [c] on the outside, the cutters affixed thereto were unusual. These consisted of 'plough cutters' [b] which were made to travel back and forth along the arms while the arms were rotating.

In order to achieve this, Farnsworth designed the arms so that they were, in effect, screw-threaded shafts [e] mounted on brackets and enclosed by a guard to prevent clogging. By means of gearing connected to the motor, the cutters were made to travel along the arm from

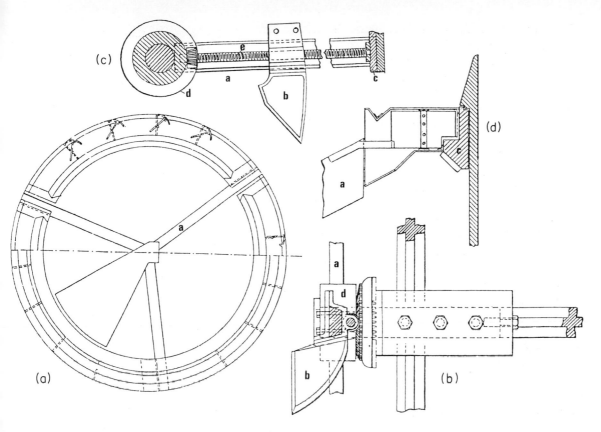

Figure 314. A. W. Farnsworth mechanized shield. U.K. Pat. No. 10,045, dated 15 May 1901. (a) Rear elevation of muck removal apparatus; (b) part plan view of thrust pedestal slideway or rotating arm and plough cutter; (c) rear elevation of plough cutter slideway and feedscrew; (d) Detailed view of the earth or muck carrier

one end to the other during the rotation of the excavator wheel. The wheel itself was placed inside and just behind the cutting edge of the shield.

In operation the cutting edge of the shield and the excavator wheel (with the plough cutters all positioned at one end of the arms) was brought up to the face. The excavator wheel was then set in motion while thrust was applied by the rams until the cutters had entered the face the required depth of 15 cm (6 in). At this stage the advance of the thrusting rams was halted and the feed mechanism for the plough cutters was engaged. (Apparently the feed mechanism for the plough cutters was not to be worked during the time that the rams were advancing the shield.) As soon as the ploughs had travelled the full length of the arms, the feed was halted and the cycle recommenced.

C. H. Bonnett — 1914: Carpenter's Excavator

Charles H. Bonnett's excavator, which was patented in the U.S. on 24 April 1914 under No. 1,093,603 (Figure 315) featured a single cutter arm which was mounted on a central rotating shaft. The machine was powered by an internal combustion engine.

The bit holder on Bonnett's unit could travel back and forth along a rack which ran the length of the cutter arm. The design pattern of this component was, to some extent, similar to that suggested by Farnsworth in his patent dated 1901. However, so far as the Bonnett unit was concerned this operation does not appear to have been automatic as, after each complete revolution of the cutter blade, it was apparently necessary to readjust the position of the bit by means of a crank-operated pinion and the rack

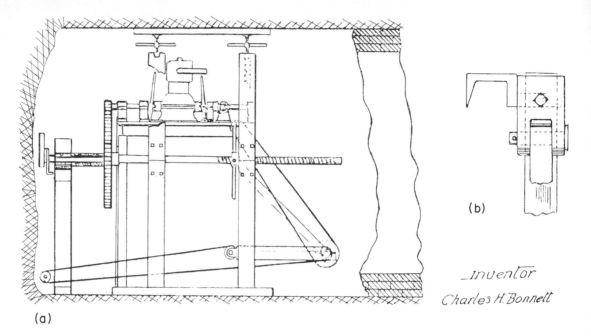

(a)

(b)

_inventor
Charles H. Bonnett

Figure 315. Carpenter's mechanized shield — designed by C. H. Bonnett. Pat. No. 1,093,603, dated 21 April 1914 (U.S.A.) (a) Crank mounting, the cutter holder and cutter are not shown; (b) bit and bit-holder

bar. In this manner the bit holder was moved outward along the cross-bar into each new cutting position. The manoeuvre would be continued until the bit reached the outer end of the cross-bar and a complete layer had been removed from the face.

Advancement of the cutter head was achieved by means of a hand-operated feed screw. Muck was carried away by a belt-type conveyor and, after each complete cutting cycle, the machine was moved forward bodily and repositioned.

In the same patent Bonnett makes the following alternative suggestions:

1. That the machine be carriage-mounted on railway tracks.
2. That power be supplied by means of an electric motor.
3. That a number of bit holders be mounted upon the cross-arm at fixed distances. The bits so positioned would then enable the machine to cut concentric annular kerfs or grooves in the face.

A Bonnett machine (known as the 'carpenter excavator') was built and worked in Detroit in 1914 during the construction there of a sewer system. The Detroit unit was carriage-mounted, but the cross arm carried only one adjustable cutting tool on it. No shield was used with the machine, which was powered by an electric motor.

D. Whitaker — 1917

Douglas Whitaker patented his first invention — a stone-breaking machine, in 1903. This particular patent was abandoned but, over the following quarter of a century, Whitaker patented over 20 inventions of various types. Of these, 9 covered significant improvements to tunnelling machines, excavators, or mechanized shields. The remainder concerned such items as conveyors, power-navvies, endless track vehicles, concrete-mixing machines, joints for drain-pipes, production of concrete slabs and drainage systems, etc.

According to an article published in *Engineering* on 26 January 1923, a few Whitaker mechanized shields were used by the War Department for excavation work during the First World War, and K. W. Adams in his report, 'Folkestone Warren: Channel tunnel trial boring machine'[3] also mentioned the use of Whitaker tunnelling excavators by the War Department. Adams comments that these machines were probably used for driving

352

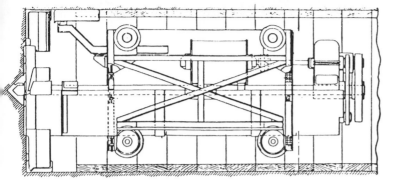

Figure 316. Douglas Whitaker mechanized shield — patented in London on 28 July 1917 under No. 130,005

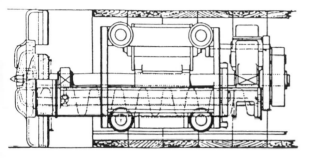

Figure 317. Douglas Whitaker mechanized shield — patented in London on 6 Oct. 1917 under No. 119,112

headings in some tunnels at Dover, such as the Priory gallery, etc.

So far as the Whitaker mechanized shield improvements are concerned, the particular innovation proposed by Whitaker (engineer) and H. C. Baxter (draughtsman) in Patent No. 130,005* is interesting. Referring to the difficulty which was experienced in changing the direction of advance of the mechanized shield by forcibly turning it bodily in the bore (thereby creating considerable pressure between various sections of the outer case of the shield and the tunnel wall), Whitaker suggested the problem could be overcome by displacing the cutter head, relative to the case or body of the shield.**

In other words, if the operator wished to, say, turn right, then the cutter head could be eccentrically positioned towards the right, thus allowing the machine to cut a greater quantity of earth or material on that side than on the other. This allowed the shield casing sufficient freedom to turn in that direction in the tunnel. To effect this it was suggested by Whitaker that

the cutter head, motor, gearing, and associated driving parts, including the thrusting jacks, be mounted on a framework which was supported within the outer case of the shield. At the rear of the shield, the framework should be gimbal-mounted which would allow for universal movement. At the opposite end (or front) the framework should be held in position by some form of adjustable device. Whitaker suggested that this could be achieved by, say, screws or jacks placed radially about the framework and carried by a ring, or some such support, which was attached to the casing of the shield.

McAlpine Mechanized Shield — 1956 (Sir Robert McAlpine & Sons Ltd.)

In 1869 a young man aged 22 years named Robert McAlpine carried out his first civil engineering job, the repair of a factory chimney, for which he was paid 49 shillings.

This was merely the start. McAlpine very rapidly showed his capabilities in this field as only five years later, at the age of 27, 1000 men were being employed by him.

In 1878 when other lesser mortals were swept into oblivion by the sudden collapse of the

*Dated 28 July 1919 (applied for on 28 July 1917)
**Also used on the Buckskin water tunnel Robbins T.B.M. 233-172 (see Chapter 26, Full-face hybrid machines).

Glasgow Bank, McAlpine courageously 'began again at his beginnings', this time devoting his attention to railway construction work and other civil engineering projects rather than to the building industry alone, as he had done previously.

During the ensuing century the company undertook a great variety of civil engineering and building projects. Their entry into the machine-tunnelling field occurred in 1956 when the company introduced the first McAlpine mechanized shield, a 3.34 m (11 ft) diameter unit, which was designed and manufactured at the company's Hayes workshop in Middlesex. The project in question was for two cooling-water tunnels for the Bradwell Nuclear Power Station. Each tunnel was some 460 m (1500 ft) in length.

At that time two important innovations were introduced by the McAlpine firm. The first of these was the utilization of cams to enable the peripheral cutters to excavate a greater radius over the crown and sides of the shield, thus relieving surface pressure and so facilitating the movement of the shield through the ground. (Also used by Markham in 1901 on a Price mechanized shield — see J. Price — 1896.)

The second was a means of preventing the thin concrete tunnel lining from being damaged due to the pressure exerted on it by the powerful hydraulic thrust jacks. To achieve this the McAlpine Company inserted an intermediate reaction thrust ring between the rear end of the jacks and the newly erected tunnel-lining segments. During operation of the cutter head the reaction ring was expanded against the tunnel walls enabling it to take the axial thrust of the main shoving jacks. The reaction unit consisted of a number of fabricated rings bolted together to form a cylinder. The rings were split on the horizontal centre line, and the upper section could be raised or lowered in relation to the lower section by means of hydraulic cylinders. After the thrust jacks had been extended to their fullest length, the ring was contracted then moved forward by the rams into a new position, where it was once again expanded against the tunnel walls in readiness for the next cutting cycle.

The cutter head of the McAlpine mechanized shield used at Bradwell was powered by a 56 kW (75 hp) electric motor and was rotated at a fixed speed by the central drive shaft on which it was mounted.

Because taking the drive through the central shaft had limitations in torque capacity for the future, the cutter head of the 1961 McAlpine machine (Figure 318) used for the Minworth sewer tunnel was mounted on a central shaft which was now static and not used for transmitting torque. The drive to the cutter head was by peripherally mounted radial-piston hydraulic motors. These motors were coupled to drive pinions which meshed with a pin wheel mounted to the rear of the cutter head. Variable delivery pumps powered by an electric motor supplied pressure for the hydraulic motors.

Basically the McAlpine mechanized shield follows conventional lines in that it has a cylindrical body and front cutting edge. The centrally mounted, peripherally driven cutter head housed within the leading edge of the shield is designed to carry either picks, disc cutters, or roller cutters, depending upon the requirements of the ground being excavated.

Material cut from the face falls to the invert where it is picked up by radial plates attached to the rim of the cutter wheel and deposited in a hopper near the top of the machine. The hopper directs the spoil into a conveyor which carries it to the rear.

Steering is effected by the use of replaceable beads on the cutting edge and selection of appropriate thrust jacks. Subsequent projects driven by McAlpine machines were:

1963 Toronto subway (5.3 m (17 ft 4 in) dia. cutter head)
1963 Victoria Line (4.1 m (13 ft 5 in) dia. cutter head)
1972 Prittlebrook diversion tunnel (2.82 m (9 ft 3 in) dia. cutter head)
1973 Chinnor mechanized tunnelling trials (5.00 m (16 ft 5 in) dia. cutter head)
1977 Cardiff cable tunnel (2.82 m (9 ft 3 in) dia. cutter head)
1979 Kakopetria chrome mines (2.92 m (9 ft 7 in) dia. cutter head)

Chinnor Tunnel Trials

During the empirical investigations carried out in the Lower Chalk at Chinnor in the Chilterns,

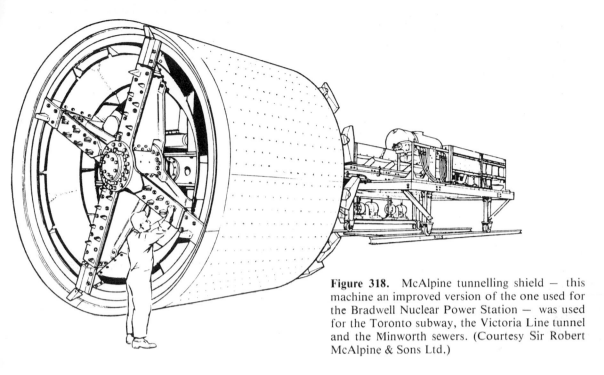

Figure 318. McAlpine tunnelling shield — this machine an improved version of the one used for the Bradwell Nuclear Power Station — was used for the Toronto subway, the Victoria Line tunnel and the Minworth sewers. (Courtesy Sir Robert McAlpine & Sons Ltd.)

England, during the late nineteen-seventies, the main contract for a full-face tunnelling machine was awarded to Sir Robert McAlpine & Sons, Ltd.

A team of engineers from the Tunnels Division of the Transport and Road Research Laboratory in association with Mott, Hay and Anderson (consulting engineers), carried out the trials, the purpose of which was to obtain meaningful data on the performance of tunnelling machines under differing conditions. During these tests comparisons were made between the relative efficiency of single and multiple-cutting tools when used in the laboratory, at the pilot stage and when fitted to a machine operating in a tunnel. Amongst other numerous experiments, tests were also carried out on the effect of various cutting-tool arrangements.

An important feature of the new McAlpine comprehensive all-purpose tunnel boring machine, used in the Chinnor tunnel trials, was the provision of trailing bars* which were

attached to the top section of the shield tail. These bars formed a protective roof support over the gap which arose behind the reaction ring and the rear of the shield, during the operation of the thrust jacks. In addition, in order to provide a safe area for the erection of steel tunnel ribs McAlpine's machine had a removable canopy which was fitted to the rear of the reaction ring.

Steering was effected by hydraulic cylinders which operated radially at 45° between the tail of the machine and the horizontal steering beam. The front end of the machine was mounted in a universal joint while the rear rested on the sledge behind the machine.

Bridge Auger Mechanized Shield[5]

An unusual mechanized shield was designed by E. K. Bridge and developed by Aubrey Watson Ltd. early in 1958. (Figures 319-321).

Instead of using motors to rotate the cutter head, as had been done previously by other designers, Bridge rotated the entire shield by

---

*In 1955 during the driving of the Thames-Lee tunnel water main 'steel poling-boards' were fitted to the rear of the shield above the skin to allow the 'Donseg' tunnel-lining segments to be erected and expanded against the clay exposed between the boards.[4]

Later (1966-67) flexible steel fingers were fitted to the roof support of the Melbourne hybrid machine during the initial modification period (see also Chapter 26, Full-face hybrid machines).

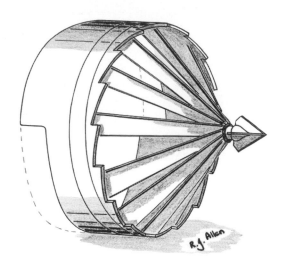

Figure 319. Bridge auger mechanized shield — drawn by R. J. Allen, Department of Design and Technology, Tasmanian College of Advanced Education. (From article by Rolt Hammond entitled Rotating shield Method of tunnelling, *The Engineer* 4 April 1958, pp. 493-495.)

means of two diametrically opposed, tangentially operating hydraulic jacks which thrust against the radial surfaces of the last placed of a series of precast concrete segments arranged in a double spiral around the interior surface of the tunnel.*

The mechanized shield was advanced in the usual manner by thrust jacks which bore against the rear of the shield and the sides of the recently erected concrete-lining segments.

Spearheading the advance of the shield was the tip of the steering mechanism which consisted of four tapered 12.7 mm (½ in) thick chamfered blades of steel. This tip formed the end of a steering rod which passed down the centre of the 152 mm (6 in) bore tube which constituted the central shaft of the shield. Eccentrics at the rear end of this hollow shaft were used to adjust the position of the rear end of the steering rod within the shaft. This could produce a change of direction of the steering point of 16 mm (⁶/₁₀ in) in 39 cm (1 ft 3 in).

Attached to the main shaft enclosing the steering rod were the radially arranged excavating blades carrying tines which were also fastened to the front end of the cylindrical body of the shield, thus forming a conical cutting head.

As the entire shield was longitudinally rotated about its axis by the diametrically positioned jacks in the rear, the ridge of material left by the tines on the preceding cutter blade was pared off by the tines of the following blade.

The machine, which was built early in 1958, was tried in a 1.8 m diameter tunnel and at that time achieved advance rates of approximately 1.8 m/hour (5 ft 11 in/hour). Muck was removed by means of a shovel-type conveyor positioned within the shield.

In his article on the subject,[5] Rolt Hammond commented that in a unit of this type careful manufacture of both the machine and the precast lining segments was of the utmost significance if high advance rates were to be attained.

For various unspecified reasons the Aubrey Watson Ltd. subsidiary company dealing with the development of this unit was wound up some 20 years ago and developmental work on the auger machine thereafter ceased.

Jarva Inc. — (S. & M. Construction)

The gradual development of the Jarva mechanized shield began in the early 1950s when the obvious advantages of a machine-bored tunnel became apparent to the S. & M.*Construction Company. However, it was not until 1961 that the first unit (Model SM-1) (Figure 322) a 3.51 m (11 ft 6 in) diameter soft-ground wheel-type machine enclosed in a shield was built for a Cincinnati (Ohio) sewer project. Inexperienced in dealing with forces of the magnitude to be encountered, they constructed the first cutter head of standard-strength pipe and when it was put to work the inevitable happened — the wheel simply collapsed. It was rebuilt with double-strength pipe which was then able to withstand the strains imposed on it. However, other problems arose. These were connected with the hydraulic system and the thrust bearings.

---

*As suggested by Marc Isambard Brunel in his circular shield patent dated 1818.

*The initials 'S. & M.' were derived from the surnames of the brothers in the Scaravilli and Mara families.

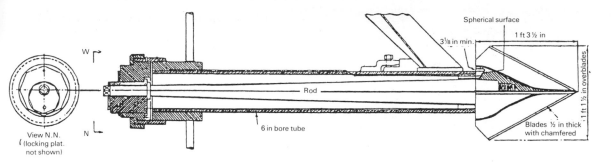

Figure 320. Section through steering point mechanism of Bridge's auger shield. (From article by Rolt Hammond entitled Rotating shield method of tunnelling, *The Engineer* 4 April 1958, pp. 493-495.)

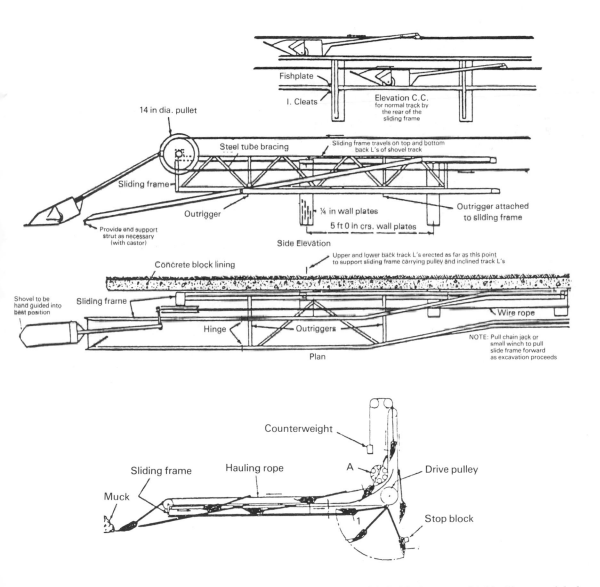

Figure 321. Diagram showing principle of shovel conveyor used with Bridge's auger shield. (From article by Rolt Hammond entitled Rotating shield method of tunnelling, *The Engineer,* 4 April, 1958, pp. 493-495.)

Figure 322. First soft-ground tunnel-boring machine Model SM-1 designed by Victor Scaravilli Jr. and Joseph Scaravilli for S & M Construction. (Courtesy Jarva Inc.)

The next machine — Model SM-2 (a 2 m rock machine) was built in 1964 for a Philadelphia, Pennsylvania project, but was not successful.

In April 1966 Jarva Inc. was formed from S. & M. Construction by Joseph V. Scaravilli, its current Acting President at that time. Between 1964 and 1971 developmental work was mostly devoted to the improvement of their rock-tunnelling machines (see Chapter 5, Modern Rock Tunnelling Machines — Jarva Inc.). However, in 1971 the Jarva Company manufactured two shields equipped with rotating cutter wheels. These were used in Chicago and at Fort Wayne. The following year a mechanized shield carrying rotary discs (Model M/F S1907 W) was used by S. & M. Construction in the Moss Point Drainage System, Euclid, Ohio (Figure 323).

Calweld, Inc. — (Division of Smith International Pty Ltd)

George F. Casey's occupation was that of a cesspool digger, which naturally included the installation of cesspools and septic tanks. At that time the digging operations were mostly carried out by hand. In the late 1920s and early 1930s Casey designed and built, in his backyard, a rig which was capable of digging cesspools.

Sometime during the Second World War or shortly thereafter (1944 or 1945), Casey was joined by George Reber who owned the California Welding & Blacksmith Company. This firm was managed by Reber as President while Casey Snr. (a silent partner of C.W. & B.) managed his own company — Casey Case Company.

Figure 323. Jarva mechanized shield — Model M/F series, used on Moss Point drainage system, Euclid, Ohio. (Courtesy Jarva Inc.)

Figure 324. Komatsu Mechanized Shield Model TM52S. (Courtesy Komatsu, Ltd.)

Figure 325. Komatsu cross-section mechanical boring or auger-type shield Model TM306S. (Courtesy Komatsu, Ltd.)

Figure 326. Mechanized soft ground tunnelling shield made in Russia. (Courtesy Machinoexport, U.S.S.R.)

Figure 327. Mechanized soft ground tunnelling shield made in Russia. Diameter 3.6 m (11 ft 10 in). (Courtesy Machinoexport U.S.S.R.)

Figure 328. Memco T.B.M. being lowered for use in section 2 of the Dandenong Valley sewer tunnel, Melbourne, Australia. (Courtesy Melbourne & Metropolitan Board of Works.)

In 1950 the company name of California Welding and Blacksmith was changed to Calweld Inc. The main reason for the change of name was that, though blacksmith work was no longer undertaken by the company, the firm was nevertheless continually being contacted by people who wanted their horses shod.

The Calweld organisation gradually expanded as crane attachments, horizontal boring machines, truck-mounted bucket rigs and auger rigs, etc. were added to their growing list of available products.

In 1961 Calweld purchased the rights and assets of the Badger Tunnel Company, a producer of horizontal boring and tunnelling machines. The tunnelling machine was known as the 'Badger' and one such unit was used to

Figure 329. Lovat M-130. The forward shield of this 3.3 m (10 ft 10 in) diameter Lovat mechanized shield is hydraulically articulated facilitating steering and machine alignment in the bore. Hydraulic cylinders are also used to operate the 'flood control doors' or face plates and the stabilizer fins. An interesting feature included to protect personnel from rollovers is a horizontal sensing device which cuts off the power supply if the machine rotates beyond 10° should the cutter head meet excessive resistance. (Courtesy Lovat Tunnel Equipment Inc.)

drive passageways between missile sites at Vanderberg Air Force Base in 1960.

Calweld's own first tunnel-boring machine (a mechanized shield carrying spade cutters) was delivered to the Kenny Construction Company of Chicago, Illinois in 1962. It was a 2.3 m (7 ft 6 in) diameter machine which bored 1680 m (5510 ft) of tunnel through clay, gravel, boulders, and sands.

Shields and mechanized shields followed in quick succession and Calweld's reputation as a manufacturer of soft (Figures 305 and 306) to medium-hard ground tunnelling machines gradually spread.

The first hard-ground Calweld unit dressed with disc cutters was Model TBM-15, which was used by Streeters Ltd. in a Coventry sewer tunnel. The material excavated was red sandstone. (Figures 242 and 243 are examples of Calweld hardground machines.)

Calweld continued manufacturing tunnelling machines of various types until the early part of the 1970s.

Oscillator-Type Mechanized Shields

On the 30 June 1965 Smith Industries International of Whittier, California, took over Calweld. The following year (1966) Calweld introduced the first of their oscillator-type units (Model TBM-18 S-Type) which was used to drive a 2.95 m (9 ft 8 in) diameter storm-water drain through silty clay, water, and sand at Anaheim, California. Since then over a dozen of this type of machine, varying in size from the baby 2.44 m (8 ft) diameter unit used to drive a Montreal storm-water drain, to the large 7.87 m (25 ft 10 in) diameter machine which was used for the Newhall water tunnel (California State water project) have been built by Calweld (Figures 330 and 331).

Markham & Co. Ltd. of Chesterfield, England, installed oscillator-type cutter heads in their three slurry machines which were delivered to Mexico City for use in the central interceptor drainage tunnel. These machines are described more fully in the section on slurry machines — universal soft-ground T.B.M.

Two oscillator-type mechanized shields built by Tunnelling Equipment (London) Ltd. were used at the Isle of Grain Power Station cooling-water tunnels with qualified success. Apparently two parallel drives were carried out concurrently with identical machines. One was successful, but the other stalled in the squeezing clay and became buried.

Finally, the first Bade mechanized shield used in Hamburg was fitted with an oscillator-type head.

361

Figure 330. View of 5.47 m (18 ft) diameter Calweld oscillating shield showing the four individual cutter arms with mounted cutting teeth. Note poling plate on centre line and bulkhead arrangement behind arms for cave-in protection. (Courtesy Smith International, Inc.)

Summary

Although this type of intermittent movement is not generally considered to be as efficient as the rotary action for tunnelling through competent ground, oscillating cutters may be advantageous in certain applications, i.e. soft and running ground conditions where the potentially unstable tunnel face should be disturbed as little as possible. Moreover, as the cutting arms operate in 'windshield wiper' fashion, the upper and lower sets being controlled independently, the shield's forward or cutting edge may be designed with a pronounced forward slope to provide greater support for the crown of the tunnel immediately behind the face.

Drum Diggers

A. W. Manton — 1900

On 11 May 1900 Arthur Woodroffe Manton filed a patent (No. 8,748 U.K.) (Figures 332 and 333) for an unusual mechanized shield — a drum digger.

So far as the author is aware, this patent covers the first proposal for a drum-digger type of machine. Although Manton's unit had not been tested in practice at the time (1900), it is interesting to compare his design with that of the Kinnear Moodie/Arthur Foster drum digger unit which was designed and built some 55 years later.

Figure 331. Calweld oscillator diamond type shield (TBM-35) used in Munich subway. The 7.90 m (26 ft) diameter machine breaks through. (Courtesy Smith International, Inc.)

Like its modern counterpart, Manton also proposed the use of a horizontally arranged cylinder [g] which rotated on a hollow or barrel bearing mounted on an abutment ring [f]. The abutment ring was capable of axial movement along the completed tunnel or it could be permanently fitted to a tunnelling shield.

Attached to the rotating cylinder was a faceplate [a] which carried either cutting knives or rotating discs [b] on radial arms [c]. The faceplate and cylinder were rotated by a circumferential worm wheel [d] which was secured to the cylinder and which was driven through a worm gear by a chain, rope, belt, or connecting rods from a motor. (Manton suggested that the motor could be either electric, hydraulic, or compressed air.) Thrust was provided by screw jacks [e] which were attached to the abutment ring and which were positioned parallel to the rotative axis. The screw jacks advanced or withdrew the bearing cylinder and the rotating face cylinder which was mounted on it.

The cutters could be arranged on the faceplate so that they either cut away the entire area or the central portion only of the tunnel face. Alternatively, they could be arranged so as to leave a central uncut core of material which

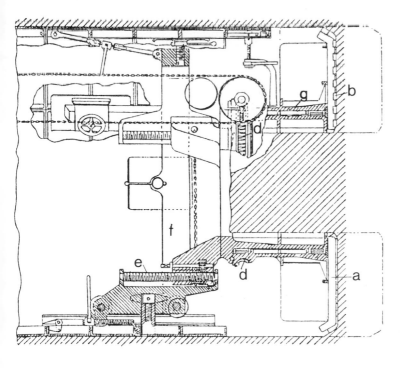

Figure 332. A. W. Manton's mechanized shield patented in London on 11 May 1900 under No. 8,747

363

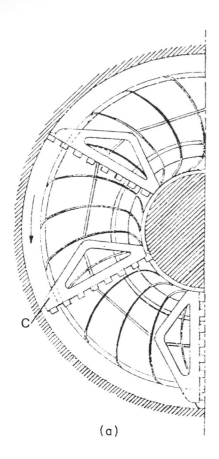

(a)

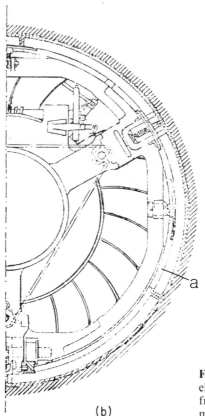

(b)

Figure 333. (a) Half front elevation of cutter head from A. W. Manton's mechanized shield; (b) Rear view of (a)

passed through the central aperture in the face-plate. This could be broken off and disposed of in any suitable manner. Rotating or fixed vanes were provided for collecting the material and delivering it to the rear of the cylinder for removal by conveyor, etc.

Kinnear Moodie/Arthur Foster Drum Digger – (1955)[4, 6-9]

In 1955 the Metropolitan Water Board was faced with the problem of driving the small diameter Thames-Lee water-main tunnel for a distance of approximately 32 km (20 miles). Use of a conventional Price-type mechanized shield, i.e. one fitted with a central shaft and with the peripheral drive mechanism occupying much of the central area of the shield was excluded. This was because the small diameter of the tunnel and the smaller cutter head meant that there was less room available for muck-removal arrangements after the insertion of the

centre shaft and drive machinery.

At about this time development work on small hydraulic motors, destined for use on coal-mining machinery, was under way. The use of these small units for driving the cutter head was conceived, designed, and developed by Kinnear Moodie & Company and Arthur Foster Constructional Engineers Ltd. (Kinnear Moodie were at that time about to undertake the construction work for the proposed Thames-Lee water-main tunnel).[4]

The result of the combined efforts of these two firms was the production of the first drum digger, the joint patent for which was filed in London on 16 September 1955 under No. 762,416, by the co-patentees and manufacturers of the drum-digger machines, Kinnear Moodie and Company Limited and Arthur Foster Constructional Engineers Limited.

Penetration rates (including the erection of tunnel lining) of approximately 110 m (360 ft)/week were achieved with the prototype machine, which cut a 2.69 m (8 ft 10 in)

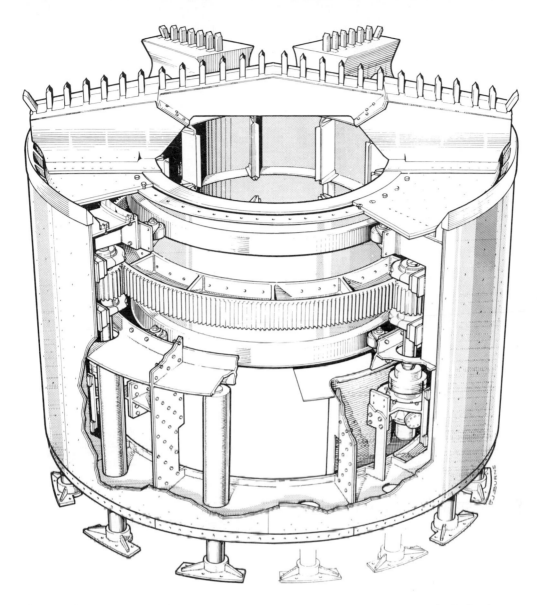

Figure 334. Cut-away drawing of Kinnear Moodie 'drum digger'. Patented in the United Kingdom on 16 Sept. 1955 under No. 762,416. Improvements filed under Pat. No. 970,016 dated 14 Feb. 1963. (Courtesy Kinnear Moodie & Co. Ltd.)

diameter tunnel. The tunnel linings — Donsegs — for this project were designed and patented by H. J. Donovan. Kinnear Moodie manufactured and developed them during the construction of the Thames-Lee tunnel. The record-breaking penetration rate achieved by the drum-digger was in part due to the successful use of the Donsegs wedge-shaped precast concrete linings, the various segments of which were interchangeable.

The Victoria Line

When serious thought was being given to proceeding with plans for the extension of the London Underground tube system by the building of the Victoria Line, the latest tunnelling methods and machines were examined. In particular, use of the Kinnear Moodie drum-digger shield was considered.

365

Figure 335. Cutter head and rear view of Kinnear Moodie 'drum digger' used on Victoria Line extensions. (Courtesy Kinnear Moodie & Co. Ltd.)

To test the efficiency of this machine and the latest type of tunnel-lining supports, London Transport decided to build two experimental lengths of twin tunnel. The line of these tunnels which commenced at Finsbury Park and Netherton Road (Tottenham) respectively, and finally met at Manor House, lay along the route of the proposed Victoria Line. The one starting at Finsbury Park was built by Kinnear Moodie & Co. Ltd., and had concrete lining while the other was built by Edmund Nuttall Limited, and was cast-iron lined.

Two drum-diggers were ordered. These machines were used in the construction of the concrete-lined tunnels (external diameter 4.27 m (14 ft)) and the cast-iron lined tunnels (external diameter 4.00 m (13 ft 2 in)). The drum-diggers in question were both Model Mark Is. Though of a larger size than the prototype drum digger used by the Metropolitan Water Board, basically the design of the units remained the same. However, the new machines incorporated certain minor modifications to the cutter head. In addition, improvements to the drive mechanism (outlined in Patent No. 970,016 dated 14 February 1963) were also built into the new diggers.

366

The Drum Digger

In effect the drum digger (Figures 334 and 335) used for the Victoria Line was a cylindrical shield with a bevelled cutting edge at its forward end. Within this shield and carried on roller races was another smaller cylinder or drum which rotated by means of an annular rack and pinion located outside the inner drum.

Placed diametrically across the central drum was a cutting arm fitted with picks. Since these tools did not remove the material in front of the outer shield nor in front of the gap between the two shields, six radially arranged arms also carrying picks, were fitted around the circumference of the inner drum to extend the cutting radius for this purpose. The loosened spoil was picked up by the buckets or chutes mounted on the cutting arms and dropped into the mouth of the inner drum. From this point it was directed inwards towards the conveyor by projecting blades or fins attached to the inner wall of the central rotating drum.

Four 40 kW (50 hp) hydraulic motors situated around the circumference of the shield provided driving power for the inner revolving drum. Thrust was provided in the usual manner by peripherally placed hydraulic thrust jacks, situated at the rear of the shield and bearing against the recently erected lining segments of the tunnel. This unique arrangement left the centre of the machine free for the important work of soil removal.

Tunnel Linings

Two types of lining were tried in the experimental tunnels. These were flexible-jointed cast-iron linings (designed by Mott, Hay, and Anderson) and concrete-lining segments (designed by Sir William Halcrow and Partners). Both these companies were acting as consulting engineers for London Transport.

An unusual feature of both the cast-iron linings and the precast concrete segments was that no grouting was used. Furthermore no bolts were used on the flexible-jointed cast-iron rings.

As soon as a complete ring had been erected, it was expanded firmly against the tunnel walls before the clay could swell and become unstable.

The cast-iron rings consisted of six separate pieces or segments, each segment being 2 ft wide and 1 in thick (61 cm wide and 25 mm thick). They weighed about 0.25 t each.

To achieve flexibility the cast-iron segments were hinged. That is each segment had a convex and a concave end which, when fitted together, formed a flexible hinge-type joint. In operation the invert segments were laid immediately the thrust rams of the shield were retracted. Then the side wall and crown pieces were inserted in that order. Retractable bars which extended behind the shield acted as a temporary support for the crown pieces while workmen manoeuvred them manually into position. The final stage, that is the expansion of the ring against the tunnel walls was accomplished by hydraulic jacks using a force of 14 t. The resultant gaps (which formed between the side and invert sections) were filled with specially designed cast-iron pieces. These consisted of a pair of tapered wedges and two knuckle-shaped pieces which fitted the concave and convex ends of the tunnel ribs respectively.

So far as the precast cement segments were concerned, these were somewhat similar in design to the cast-iron rings in that they, too, had concave and convex ends which fitted together to form a type of flexible hinge joint. When the concrete segments were expanded against the tunnel walls, the gap (which formed at the top in this case) was closed with purpose-made reinforced concrete 'folding wedges'.

During the construction of their section of experimental twin tunnel at Seven Sisters, the contractors (Edmund Nuttall Limited) using a Kinnear Moodie drum-digger shield and the flexible cast-iron rings created a new world speed record for soft ground machine-bored tunnels, which was then 142 m (467 ft) in one week.

After the successful construction and completion of the experimental tunnels, and in anticipation of government approval for the construction of the Victoria Line itself, London Transport Executive (L.T.E.) carried out numerous preliminary investigations and made arrangements so that when official sanction was finally given they would be ready to proceed with the immediate issue of tenders, etc. This authorization was granted on 20 August 1962.

The diameters of the original two Mark I drum-digger machines which were used on the experimental section were modified slightly and, in addition, a further two machines (Models Mark II) were supplied to the L.T.E. Apart from these four machines built by Kinnear Moodie & Co. Ltd., Sir Robert McAlpine & Sons Ltd. were asked to build four centrally mounted, peripherally driven mechanized shields, for use on the Victoria Line — thus making a total of eight machines for the project in question.

An opportunity arose, during construction of this line, for a direct comparison to be made between the centrally mounted mechanized shields and the drum-digger units, while these machines were operating in similar ground conditions. Relevant comments on this aspect were set out in a paper by J. A. M. Clark.[9] In the paper Clark examines the pros and cons of both machines in some detail. Generally speaking, while Clark considered the centre-shaft machine to be superior in some specific areas, he felt on the other hand that there was '. . . a threshold diameter below which the drum digger provides the only practical answer'.

Figure 336. Marcon powervane. (Courtesy Marcon International Ltd.)

Marcon Powervane – 1977[10-12]

In 1977, the designer and developer of the original mini tunnel system, M. A. Richardson, witnessed the launching of his new machine, the Powervane. The prototype was assembled by his company, Marcon International Limited, in 1977.

So far as the author is aware, the Powervane is the first true 'drum-digger'-type machine to be produced since the advent of the Kinnear Moodie unit. While the Powervane follows convention in that it is equipped with an inner drum and an outer cylindrical shell, there are significant and interesting differences between the two types of machine.

The Powervane (Figure 336) is pneumatically driven by compressed air. The air enters a closed chamber immediately behind the cutter head through ports located in a diametrically opposite pair of stator vanes. Rotor vanes, attached to the inner drum, co-operate with the stator vanes fixed to the outer shell to produce rotary motion.

By means of changeover valves[10] the air forces the inner drum to oscillate through 180°.

The rotor and stator vane seals are accessible from the interior of the machine.

As the inner drum rotates through 180°, so does a three-armed cutter head which forms an integral part of the inner drum. The cutter head is attached to the front of the inner drum.

Although the unit is equipped with four pneumatically powered 10 t steering rams, there are no conventional thrust jacks. The cutter head is kept up to the face by the air pressure between the casings which forces the inner section of the machine forward with a thrust of some 100 t, while it also oscillates the inner drum and thus the cutter arms. After driving the head, the exhausted compressed air serves to ventilate the tunnel work area.

Material cut from the face is rolled progressively rearwards by the oscillating motion of the inner drum and fed on to a conveyor, the front roller of which is located immediately behind and below the level of the bottom of the inner drum.

A rotary screw assembly may be fitted between the conveyor and the cutter head in the event of the operator encountering sticky or cohesive material. If desired, a diaphragm may be installed behind the cutters which will enable the machine to use the Japanese earth pressure balanced shield method (see Chapter 14, Earth pressure balanced machines). Screw-on type rock picks are fitted on to the cutter teeth mountings. Finally, where conditions warrant it, the cutting arms may be removed and the unit used as a hand shield.

The prototype machine was tested in Cheadle in a waste quarry tip through sandy clays and gravels. At Pontefract a tunnel was driven with the Powervane beneath a railway line through wet shale with bands of ironstone (up to 100 m (330 ft) thick) at advance rates of approximately 25 mm/min (1 in/min). Problems were encountered with the triangular web stiffeners and these were replaced on production machines with a continuous fillet stiffener.[10] Subsequently the machine was tested at Stockton-on-Tees where it bored a 1.37 m (4 ft 6 in) diameter tunnel through clay interspersed with pockets of wet sand and boulders up to 300 mm (12 in) diameter.

In medium-hard soils, production machines (according to the manufacturer) excavate at about 300 mm/min (12 in/min), the machine consuming 10 m³/min (13 yd³/min) of air at a supply pressure of 0.58 MPa (84 lb/in²). The Powervane is intended for tunnels up to 1.5 m (5 ft) internal diameter of either segmental or pipejacked form.

Robert L. Priestley Limited

In 1901 a small general engineering shop was founded by Robert L. Priestley at the Canal Basin, Gravesend. In 1921 the firm became known as Robert L. Priestley Ltd. As the years passed the works of the company were gradually expanded. The main turning point came in 1946 when Robert L. Priestley decided to try to meet the needs of underground civil engineering contractors. This led to their debut into the field of underground construction works and the manufacture of airlocks, tunnel shields, medical locks, tunnel shutters, etc. (Figure 337)

The influx of new business from this area soon forced the company to seek larger premises and in 1952 they moved to Mark Lane.

The development of Priestley shielded tunnelling machines commenced in 1967 with the introduction of the Ely-Ouse unit. The basic machine was cylindrically shaped and was advanced by 11 hydraulic rams capable of a maximum forward thrust of 762 t. Thrust reaction was taken by the recently completed tunnel-lining supports. Another ram, positioned at the crown of the shield, just slightly forward of the 11 main thrust rams, was provided for inserting the key block of the tunnel-lining segments. The insertion of this key block acted to expand the lining outward against the tunnel wall.

There was a significant difference between the centre-shaft and drum-digger machines and the new Priestley Ely-Ouse unit. This concerned the cutter head which, in the case of the Priestley machine, was supported on a cross roller-bearing specially sealed against ingress of dirt or liquid. The advantages of this particular

Figure 337. First Priestley mechanized shield — used at Ely-Ouse for Essex River Authority 2.82 m (9 ft 3 in) diameter tunnel. (Courtesy Robert L. Priestley Ltd.)

369

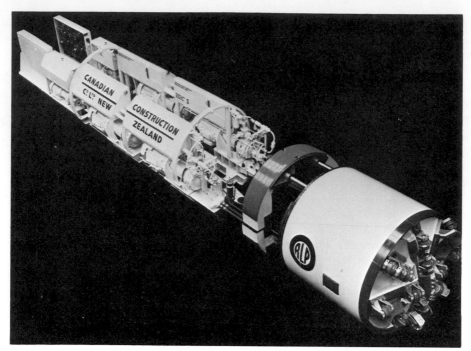

Figure 338. Priestley shielded tunnelling machine used at Mangapapa Power Station, New Zealand. Note anchor thrust ring. (Courtesy Robert L. Priestley Ltd.)

method of cutter head support were that maximum stability and support was provided for the cutter head and, in addition, there was a clear central area with access to the machine.

The cutter head consisted of four radially arranged arms carrying drag picks. The cutter head was powered by four 37 kW (50 hp) motors. Muck cut by the drag picks was picked up on the arms which formed an integral part of the cutting-head arrangement and was deposited into a chute. The chute directed it to a twin conveyor system, which transported it to the rear of the shield train for disposal into trucks.

This machine, which successfully bored a 2.81 m (9 ft 8 in) diameter, 9.7 km (6 mile) long aqueduct tunnel for the Essex River Authority, broke the existing world record for soft-ground tunnelling by excavating and lining 434 m (1400 ft) of tunnel in one week.

Other notable machines built by Priestley include units suitable for variable ground conditions such as the shielded tunnelling machine built for use on the Mangapapa Hydro-Electric project for the Tauranga Joint Generation Committee, where both soft ground and medium-hard rock were encountered.

Distinctive features of this machine were that the cutting head was dressed with rotary disc cutters which cut gauge and the shield was equipped with an expanding anchor thrust reaction ring (Figure 338). The use of this latter device meant that instead of the machine advancing by pushing against the newly erected tunnel lining, as it usually does when tunnelling in soft ground, pressure was instead exerted against a circumferentially placed reaction ring. (The ring was expanded against the tunnel walls (see Chapter 11, Mechanized Shields — McAlpine Mechanized Shield).) This method of shield advance allows the contractor to erect relatively thin concrete lining not robust enough to withstand the powerful pressures exerted by the machine's thrust jacks, but capable of supporting the tunnel walls.

The California State Water Project — a strenuous proving ground for machines and techniques[13]

At an estimated cost of approximately U.S. $1½ billion one of the largest, and certainly one of the most interesting, water distribution schemes was undertaken by the Metropolitan Water District of Southern California (M.W.D.). It is designed to tap vast quantities of surplus northern Californian water and channel it south to the southern Californian coastal plains. This area 1,200,000 ha (4800 sq. miles) which has been supplied by the M.W.D. since 1941 with 4.5 billion litres (1 billion gallons) of water per day via the Colorado river aqueduct, has within its boundaries 122 cities including that of Los Angeles.

Conservative experts anticipate that the number of people in this area will nearly double by the year 1990, and it was to meet this population explosion that the existing water supplies were to be increased to 9.1 billion litres (2 billion gallons) per day.

Over 480 km (300 miles) of tunnels, pipelines, and other ancillary civil engineering projects connected with this undertaking would be constructed before the entire distribution scheme was eventually completed.

During the construction of this vast new feeder system, new machines and unusual tunnelling techniques have courageously been tried; to be either accepted or discarded as the case may be. Already the benefits of these hard-learned lessons are gradually spreading throughout the tunnelling industry and in many instances these new techniques are becoming accepted as standard practice where similar conditions are believed to exist.

The Newhall Tunnel

Calweld Machine

The first machine-bored tunnel of significance in the California State Water Project (C.S.W.P.) scheme was the 5500 m (17,950 ft) long Newhall tunnel which ran through varied sedimentary rock strata. For this the contractor used a 7.80 m (25 ft 7 in) diameter Calweld mechanized shield (Model TBM-19) dressed with spades and disc cutters and capable of exerting a total thrust of 33,540 kN (7,500,000 lb) and a torque of 1800 kN m (800,000 lb ft). The shield was advanced and steered by peripherally placed hydraulic jacks with the reaction being taken by the tunnel support system.

The spoke-shaped cutting wheel, which could be rotated in either direction (in case of

blocking or jamming), was mounted in the conventional manner on a central shaft and was driven at variable speed by hydraulic motors.

The machine commenced work early in 1967. After successfully negotiating approximately 3000 m (10,000 ft) of difficult ground comprising sandstone, siltstone, and mudstone, badly caving ground was encountered. This effectively halted the forward progress of the machine so that the unit had, perforce, to be abandoned.

While moderately stable ground conditions had prevailed, the machine had progressed satisfactorily, but as soon as the unit entered non-cohesive material it tended to stall. One reason for this was that the wet unstable ground in front of the shield was disturbed by the excavator wheel which protruded beyond the roof-supporting cutting edge of the shield. As a result the crown of the tunnel collapsed and the face sagged and dropped to the invert, jamming the cutter head.

In his article, 'Construction of a two-billion gallon per day water distribution system,'[13] Maynard M. Anderson, Chief Engineer of the Metropolitan Water District of Southern California, suggests that the unit might have been more successful in such conditions had it been fitted with a bulkhead instead of an open spoke-type excavator wheel.

Thus in the first attempt at machine-boring a tunnel for the project, one of the major difficulties connected with such work was highlighted, i.e. that of providing a versatile unit capable of dealing with both cohesive and non-cohesive ground in the same tunnel.

The Sepulveda Sewer Tunnel

Before work on the second section of the Newhall tunnel was commenced, Drummond and Bronneck used one of Calweld's latest innovations, an oscillator-type mechanized shield to drive the 3.66 m (12 ft) diameter, 2230 m (7300 ft) long Sepulveda sewer tunnel (also part of the C.S.W.P. scheme). The machine in question (TBM-41) was a 77 t Oscillator Diamond, capable of delivering a total thrust of 1825 kN (410,000 lb) and a torque of 711 kN m (520,000 lb ft). Since its debut in 1966 in the Anaheim sewer drain tunnel, five models of this type of machine had been used with varying success in different projects. Amongst these was the 5.5 m (18 ft) diameter San Francisco subway tunnel, the 7.80 m (25 ft 6 in) diameter Munich subway tunnel and the 2.74 m (9 ft) diameter Calumet sewer tunnel (Oscillator S-Type unit). Ground conditions in these tunnels had varied from non-cohesive silty clay and flowing watery sand to moderately consolidated marl, limestone, flint, and alluvium.

The Calweld unit used in the Sepulveda tunnel consisted of a conventional-type shield, the forward or cutting end of which was inclined or sloped. As the protruding end of the shield face was positioned uppermost, this provided a temporary support for the tunnel roof at the working face.

Instead of a rotary wheel the cutter head consisted of two sets of oscillating arms, an upper and a lower unit each having three arms which operated in windshield wiper fashion. Drag teeth, constructed of T-1 steel, studded the cutting arms. However, in practice it was found that this type of tooth wore rapidly in the abrasive slate and sandstone formations. To overcome this problem the original steel teeth were replaced with new ones carrying tungsten carbide inserts, but even these succumbed to the difficult ground conditions. Excessive wear became evident in the vicinity of the inserts while the teeth themselves tended to twist or snap off.

(During the late 1960s a Calweld 3.34 m (11 ft) diameter oscillator machine came to grief in Melbourne, Australia, because of similar problems. So far as the Melbourne unit was concerned, a major problem was that the cusps of rock, which tended to form where the arcs of the two cutting arms intersected, were of such hard material that they did not readily break away as the shield was advanced and this considerably slowed the progress of the unit.)

Unfortunately, the further the Sepulveda machine advanced along the tunnel line and into the depths of the mountain, the more solid the layers of slate encountered there became, with little or no evidence of fracture or cleavage lines to aid the progress of the drag-bit cutters. Consequently, advance rates decreased until the machine was finally halted, leading to the dis-

mantling and removal of the excavator head and drive mechanism. For a short period attempts were made to continue tunnelling with the shield alone, but due to the hardness of the slate which made control of the shield difficult, it was eventually abandoned in the tunnel. The contractor then altered the tunnel profile from a circular to a horseshoe shape and work progressed by conventional methods.

In the first section of the Newhall tunnel, the machine, designed as it was for driving through dry moderately consolidated material such as sandstone, etc. failed to function efficiently when confronted with wet sloughing ground. On the other hand the machine used in the Sepulveda tunnel was primarily built to handle softer formations of clay, silt, and sand, including non-cohesive materials needing immediate support of the crown at the working face. This machine proved to be inefficient when put to work in the unexpectedly hard ground conditions experienced there. Thus, once again the important problem of providing a versatile machine which could cope with varying ground conditions within one tunnel, was emphasized.

The San Fernando Tunnel[14]

Ground consisting of moderately consolidated sedimentary rock and alluvium lay along the route of the proposed San Fernando tunnel line. In addition the strata in some sections of the route were expected to contain a fairly high percentage of water.

The machine selected in 1969 by the contractor for this work was the recently designed Robbins Company 'ripper scraper', a digger shield. This machine evolved as a result of a joint venture which was originally undertaken by Scheumann Johnson and Traylor Brothers, contractors. During the early 1960s some rudimentary digger-type machines were developed by these contractors for sewer tunnelling in Seattle. This developmental work was later continued under licence by the Robbins Company which produced its first ripper scraper (Model 221S-132) in 1970.

Basically similar in concept to the MEMCO 'Big John' excavator, the Robbins unit (Figures 339 and 340) also consisted of a shield within

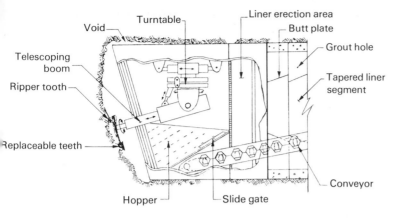

Figure 339. Side view of Robbins digger shield showing turntable, dewatering hopper, and conveyor arrangement as depicted in M. Anderson's paper presented at the ASCE National Meeting in Memphis, Tenn., in January 1970. (Courtesy M. Anderson and Metropolitan Water District of Southern California.)

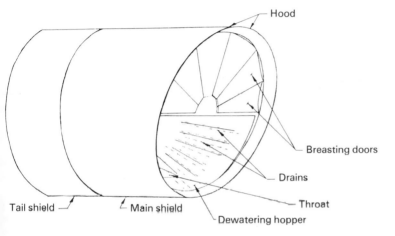

Figure 340. Front view of Robbins digger shield showing dewatering hopper and breasting doors as depicted in M. Anderson's paper presented at the ASCE National Meeting in Memphis, Tenn., in January 1970. (Courtesy M. Anderson and Metropolitan Water District of Southern California.)

which a digger arm or boom operated. However, there were significant differences between the two machines. The MEMCO excavator had a knuckling bucket which cut the material from the face like a back-hoe excavator then picked up a bucketful and carried it rearward to a point where it was dropped directly into a mine car, or later on to a belt. The Robbins unit, on the other hand, was fitted with an apron or hopper. The material was initially drawn by the ripper arm up on to the apron and then pushed directly from the apron on to the conveyor belt — a somewhat faster excavating technique than the knuckling bucket method.

Another significant difference between the two types of machine was the method of supporting the face material. Whereas the MEMCO machine relied on its forepoling plates to protect the face in soft and caving ground, the Robbins unit was equipped with powerful breasting doors which supported the face from the springline to the crown of the tunnel. These 'flower petal'-like doors were set back a short distance from the shield's cutting edge, which edge formed part of a protective hood support immediately behind the face. Within the shield of the Robbins unit was a telescopic boom mounted on a turntable attached to the roof of the shield. The telescopic boom was armed with a single ripping pick fixed to the top of a mouldboard.

Behind and below the vertically arranged breasting doors was a trough-like after-bulkhead placed at an oblique angle, so that it formed a perforated dewatering hopper. The hopper's open end was presented to the material in front of the shield. The dewatering hopper was specifically designed for use in the wet and running ground conditions anticipated in the San Fernando tunnel. In well-consolidated material the spoil was raked

373

directly on to the conveyor belt by the digger arm. (In later models the sloping hopper or apron was not perforated.)

(Note the similarity between the breasting doors and dewatering hopper arrangement on the Robbins machine (Figures 339 and 340) and the forward bulkhead and trough or after-bulkhead arrangement proposed by Theodore Cooper in his 1902 patent (see Figure 341) — described in Chapter 12, Pressurized plenum chamber machines — Theodore Cooper — 1902.)

Forward thrust totalling some 31,186 kN (7 million lb) was available from the shield's jacks which reacted against a thrust ring which, in turn, bore against the precast tunnel-lining segments.

The tunnel-lining segments were lifted and positioned by an erector mechanism which operated within the protective tail section of the shield.

Work with the Robbins unit commenced in 1970. Initially all went well. The machine averaged advance rates of approximately 66 m/day (220 ft/day) (during five consecutive working days) and attained a maximum of 84 m (276 ft) in a three-shift day. A performance which at that time constituted a world record in soft ground shield tunnelling. This record was broken by the Priestley Channel tunnel machine which excavated at the rate of 10 m/hour (33 ft/hour) in the course of the 100 m (300 ft) that were bored before the project was shelved.

By 23 June 1971 the unit had excavated some 8230 m (27,000 ft) so that only 610 m (2000 ft) of unexcavated tunnel remained. Then suddenly during the third shift of that day an explosion occurred which injured an M.W.D. inspector and three workmen. But this was only the preliminary to an even worse disaster, a second explosion of far greater intensity, which rocked the tunnel at 55 minutes past midnight on the following day, 24 June 1971. As a result 17 men lost their lives and the excavator machine was extensively damaged.

According to the findings of a court of enquiry, the explosion was deemed to have been due to the presence of heptane gas, which was either present in the tunnel at the time, or was released by the excavator from a pocket in the rock, just prior to the accident. It was suggested by Dr. Gordon B. Oakeshott, Deputy Chief of the California Division of Mines and Geology, that as the epicentre of the February earthquake lay near the route of the tunnel which passed beneath the Sylmar foothills, the earthquake may have been indirectly responsible for causing the accident. According to Dr. Oakeshott, any fluid (i.e. gas, oil, or water) trapped in fissures or pockets in the rocks could have been disturbed by the quake and its 300 or so after-shocks, so that it seeped through newly created pathways in the fractured rock strata.

A lengthy enquiry, followed by an even lengthier trial, the proceedings of which dragged on for over a year, culminated in the conviction by the jury of the Lockheed Ship-building and Construction Co., a Seattle-based subsidiary of the Lockheed Aircraft Corp., and two of the men, on charges of multiple mis-demeanour violations which stemmed from the accident.

After the accident state legislation ruled that the area in which the remainder of the tunnel route lay be proclaimed as 'gassy' ground. As a result certain stringent safety precautions had to be followed when mining or tunnelling within this area. For instance, the Robbins ripper scraper which had since been repaired, as the contractor had found the damage was not as extensive as had previously been thought, now needed to be powered hydraulically from a pump situated some distance from the machine. This involved the installation of telescopic feed lines, hydraulic motors, and supply pipes, etc., and the purchase of a 23 t Plymouth locomotive which could be operated in a 'gassy' area.

Altercation then arose as to who was to foot the bill for these additional costs, the contractor or the M.W.D.? After many months of legal wrangling which commenced towards the end of 1972 and ended in April 1973, Lockheed and the M.W.D. agreed to share the additional costs involved in completing the tunnel, without either party admitting liability or accepting responsibility for the accident. Work was to proceed in September or October of that year (1973). However, the rate of progress thereafter was to be restricted within certain limits, so as to discourage any tendency towards carelessness in the observation of the safety regulations.

The Tonner Tunnels[15]

As a result of the San Fernando accident and also those which occurred in the Sunshine mine, Kellogg, Idaho, and the Port Huron tunnel, Michigan, stringent safety regulations covering excavation or mining work in known gassy areas were enforced. These new regulations came fully into effect when excavation work commenced in the Tonner tunnels.

Wide interest was shown in the application of these new regulations by a diverse group of people consisting of agency representatives, engineers, state and federal government officials, manufacturers, contractors, and safety experts, connected with the mining and civil engineering industries.

Amongst other requirements stipulated by the new laws was the proviso that adequate fresh air should be available at the working face to disperse any gas: 2.1 cu.m/min (75 cu.ft/min) per brake horse power was needed, and in addition 5.66 cu.m/min (200 cu.ft/min) per man was to be provided. So far as the contractor J. F. Shea Co. Inc. was concerned, this amounted to a total of some 514 cu.m/min (18,150 cu.ft/min), i.e. for the 12 men required per shift and to cater for the Plymouth locomotive used by J. F. Shea which was rated at 156 kW (210 hp). J. F. Shea was able to provide 849 cu.m/min (30,000 cu.ft/min) of air through a 91 cm diameter pipeline which was fed by two 75 kW (100 hp) Joy fans situated at the tunnel entrance. The pipeline was divided into two 56 cm (1 ft 10 in) pipes just behind the T.B.M.'s trailing gear. One of these pipes led to the face area immediately in front of the T.B.M.'s dust shield, while the other supplied fresh air to the area behind the machine. In addition, two methane gas monitoring instruments manufactured by Bacharach Instrument Co. were installed. These enabled safety inspectors, etc. to take readings from eight different points situated between the face and a point located some 91 m (300 ft) from the face. (When the concentration of methane gas reaches 5 per cent it is liable to explode if exposed to incendiary sparks. This point is referred to as the lower explosive limit (L.E.L.). To ensure that the gas does not reach this danger level, the monitoring equipment is designed to sound a warning hooter when the gas reaches only 10 per cent of the L.E.L. If this warning is not heeded and the level should go as high as 20 per cent of the L.E.L., the monitor automatically cuts off the power and all personnel are directed to leave the tunnel immediately.)

The haulage locomotive was not permitted closer to the working face than 91 m, (300 ft) i.e. within the area under constant surveillance by the methane gas detection instruments. This meant that a conveyor was needed to transport the muck from the T.B.M.'s conveyor to the muck trains.

A 61.3 t Calweld T.B.M. No. 70 was used by the contractor J. F. Shea Co. Inc., for the excavation work. The machine was a 3.43 m (11 ft 3 in) diameter mechanized shield-type unit dressed with kerf cutters. The cutter head was powered by a 300 kW (400 hp) motor and could develop a torque of approximately 400 kN m (300,000 lb ft) while the machine's jacks were capable of exerting a total thrust of 64,040 kN (14 million lb). The strata through which the machine tunnelled consisted of a mixture of shale, limestone, and sandstone with some boulders.

The Castaic Tunnels

Work on the first of the four 8 m (26 ft 3 in) (o.d. rough bore) Castaic tunnels which run through varied strata comprising mudstone, sandstone, and conglomerates was commenced in January 1967. Construction work for the four tunnels which totalled approximately 8 km (5 miles) in length, was carried out by the Italian firm of Vianini-S.P.A., who used the newly designed MEMCO (Mining Equipment and Manufacturing Company) mining machine, known as the 'Big John' excavator for the projects. The machine, which falls into the hybrid class, consists of a combination of a boom-type machine and a shield. However, in the case of the MEMCO unit the shield section of the machine was further refined by having 10 poling-plates mounted above the springline at the forward end of the shield. Each plate was individually powered by a 318 t hydraulic jack. The upper or crown section of the shield's cutting edge projected approximately 2 m (6 ft 6 in) ahead of the invert section and the poling-plates added another 1.5 m (4 ft 11 in) of support to the crown of the tunnel when they were fully extended. (See also the description of MEMCO's hydraulic forepoling shield given in Chapter 10, Shields — J. Tabor — Mining Equipment Manufacturing Company.)

As a similar machine had been tried and abandoned in the Carley Porter tunnel, some criticism was apparently raised in tunnelling circles when MEMCO's unusual unit was chosen for the Castaic tunnels. According to the manufacturers, the poling-plates were designed to act as a support in soft and running ground. They were not intended to be used as a mining tool. When so misused, as they apparently were in the Carley Porter tunnel, they became buckled and distorted to such an extent that eventually the entire unit was put out of action.

Apart from its poling-plates, the shield followed conventional lines in that gauge was cut by the shield's 3 cm (1.2 in) thick cutting edge. It was advanced by 30 jacks which reacted against the precast primary lining of the tunnel. The jacks were capable of exerting a total forward thrust of 4080 t.

The unit's hydraulic excavating arm was equipped with a type of hoe or digging bucket similar to those used on conventional 'back hoe' excavator machines. Material which was sliced from the periphery of the tunnel by the shield's cutting edge was directed towards the conveyor belt by the raking action of the digger arm. The arm also operated within the tunnel profile to dislodge or break material from the face or, if necessary, dig out and remove any harder inclusions such as floaters or boulders. The bucket, which was equipped with two powerful ripping teeth at its bottom edge, could be used for either digging or ripping.

Special tunnel-lining segments (designed by MEMCO's founder and President, J. R. Tabor) which were capable of withstanding the high pressures exerted upon them by the shield's thrust jacks, were used for the tunnel's primary and final lining and support system. These consisted of precast concrete segments which had been mounted around a steel rib arc. The segments were so constructed that the ends of the steel ribs extended a short distance beyond the ends of the precast concrete form. These steel end pieces were temporarily bolted together during erection in the tunnel to facilitate alignment, after which they were welded together. The space left between the precast segments and steel ribs was filled with pneumatically applied concrete to a depth of 46 cm (1 ft 6 in). This effectively enclosed the exposed steel rib ends with concrete so that they were no longer subject to corrosion. Longitudinally positioned guide rails allowed the boom unit axial movement within the shield.

In practice, it was generally found preferable to dig a pilot hole in the face of the tunnel, which was enlarged as far as practicable using the ripping teeth for the purpose. When as much material as possible had been removed by this means, the shield was advanced so that its cutting edge could shear off the remainder of the material left behind around the edges by the digging arm.

During the driving of the first three tunnels the MEMCO unit averaged approximately 30.5 m (100 ft) per working day, its rate of advance being limited on occasions by the capacity of the muck-removal system. However, some difficult strata encountered in the last tunnel proved troublesome. During the negotiation of an area consisting of saturated unstable sandstone, two bad runs caused the shield to sink and divert from its true course. In another area hard shale and sandstone, beyond the capabilities of the digger arm, slowed the unit's progress to such an extent that the contractor was eventually forced to use small quantities of explosives to lightly blast the area ahead of the shield until the difficult section had been passed.

References

1. S. A. Marsaak and V. P. Samoilov, *Shield Tunnelling,* Nedra, Moscow, 1967.
2. William Charles Copperthwaite, *Tunnel Shields and the use of Compressed Air in Subaqueous Works,* Archibald Constable, London, 1906.
3. K. W. Adams, Folkestone Warren: Channel tunnel trial boring machine, unpublished.
4. E. W. Cuthbert and F. Wood, The Thames-Lee tunnel water main, Paper No. 6578, joint meeting of the Institution of Civil Engineers and the Institution of Water Engineers, London, 6 March 1962.

5. Rolt Hammond, Rotating shield method of tunnelling, *The Engineer,* April 1958, 493-495, London.
6. John R. Day, *The Story of the Victoria Line,* London Transport, London, 1969.
7. H. G. Follenfant *et al.,* The Victoria Line, *Proc. Inst. Civ. Eng.,* Paper No. 7270S, suppl. vol. 1969, 337-356, London.
8. C. E. Dunton, J. Kell and H. D. Morgan, Victoria Line — experimentation design, programming and early progress, *Proc. Inst. Civ. Eng.* **31,** 1-24, 1965, London.
9. J. A. M. Clark, Some modern developments in tunnelling construction — aspects of mechanical shield tunnelling and comparison with hand shield tunnelling, *Proc. Inst. Civ. Eng.,* Paper No. 7270S, Suppl. Vol. 1969, 397-451, London.
10. Ian Atkinson, All-air mole makes its mark, *Contract Journal,* 17 Jan. 1980, 32-33, Surrey.
11. *A Status Report on the Development of the Marcon Powervane Tunnel Boring Machine,* Marcon International Ltd, Guildford, Surrey, U.K.
12. Piers, G. Harding, Powervane — a unique TBM, *Tunnels and Tunnelling,* Sept. 1979, London.
13. Maynard M. Anderson, Construction of a two-billion gallon per day water distribution system, Paper, A.S.C.E. National Meeting on Water Resources, Memphis, Tenn., Jan. 1970.
14. *Los Angeles Times,* 25 June 1971, 26 June, 1971, 9 Aug. 1973, 17 May 1974.
15. The Tonner tunnels: new method for invert placing: disaster and aftermath, *Western Construction,* Aug, 1973, 36-60, Eugene, OR, U.S.A.

Bibliography

G. C. Archer, Rapid tunnelling in soft ground, *The Contract Journal,* May 1958, Surrey, U.K.

P. E. Garbult, *How the Underground Works,* London Transport, London 1968.

Japan Tunnelling Association Secretariat, Tunnelling in Japan, *Tunnels and Tunnelling,* **10,** No. 5, 19-22, June 1978, London.

R. J. Robbins, *Vehicular tunnels in rock – A direction for development,* Joint A.S.C.E.-A.S.M.E. National Transportation Engineering Conference, July 1971, Seattle.

R. J. Robbins, Soft ground tunnelling machines, University of Wisconsin, Ext. Nov. 1974, U.S.A.

D. B. Sugden, Tunnel boring machines and systems — a survey, *Journal of the Inst. of Engs.,* Australia, Nov./Dec. 1975.

D. B. Sugden, Tunnel boring machines — their advantages and disadvantages, *Contracting Construction Eng.,* Aug. 1973, Sydney, Australia.

D. B. Sugden, Ground support in mechanically bored tunnels, *Journal of the Inst. of Engs.* Australia, Aug. 1973.

Tunnelling machine holes through four months early, *Engineering News-Record, McGraw-Hill's Construction Weekly,* March 1969.

Correspondence

Sir Robert McAlpine & Sons Ltd. London, U.K.
Jarva Inc. Solon, U.S.A.
Calweld Inc. Santa Fe Springs, U.S.A.
Kinnear Moodie (1973) Ltd., Peterborough, U.K.
Robert L. Priestley Ltd., Gravesend, U.K.
The Robbins Company, Seattle, U.S.A.
Maynard M. Anderson (Met. Water District of S. California), U.S.A.
London Transport Executive, London.
Mitsubishi Heavy Industries, Tokyo, Japan.
Markham & Co. Ltd., Chesterfield, U.K.
Marcon International Ltd., Surrey, U.K.
Lovat Tunnel Equipment Inc., Rexdale, U.S.A.
Zokor International (UK) Ltd., London, U.K.

Patents

J. D. and G. Brunton — 1424/1876 U.K.
F. O. Brown — 340,759/1886 U.S.A.
J. H. Greathead — 195/1889 U.K.
J. J. Robins — 1,445/1893 U.K.
J. Price — 13,907/1896 U.K.
G. Burt — 9,549/1897 U.K.
A. W. Manton — 8,748/1900 U.K.
A. W. Farnsworth — 10,045/1901 U.K.
J. E. Ennis — 690,137/1901 U.S.A.
C. H. Bonnett — 1,093,603/1914 U.S.A.
D. Whitaker — 130,005/1918 U.K., 119,112/1918 U.K.
Kinnear Moodie/Arthur Foster Con. — 762,416/ 1955 U.K., — 970,016/1964 U.K.

TWELVE

Pressurized Plenum Chamber Machines

The use of compressed air when tunnelling through water-bearing strata is no longer considered to be a novelty, and numerous improvements have been introduced since the first subaqueous tunnels were driven in 1879 with compressed air. These were the Antwerp Docks tunnel (1.5 m × 1.2 m (5 ft × 4 ft)) with cast-iron lining constructed by M. Hersent and the famed Hudson river tunnel (5.5 m × 4.9 m (18 ft × 16 ft)), New York, started by De Witt Haskins.

The initial attempt to drive the Hudson river tunnel, sans shield, with 0.24 MPa (35 lb/in²) air pressure failed when a hole was blown through the soft silt in the crown of the tunnel.

The second attempt was made with a shield and compressed air and Sir Benjamin Baker and Henry Greathead were appointed consulting engineers for the British company which undertook the work. By mid-1891 when the northern tube had been driven a distance of some 1130 m (3700 ft) from the Jersey shaft the venture was halted for economic reasons. It was finally completed by William McAdoo's newly formed company under the guidance of consulting engineer Charles M. Jacobs.

During the period when the second attempt was being made in 1889 many of the miners were succumbing to the 'bends' (diver's palsy or caisson disease) because they were moving too rapidly from the compressed-air section through the airlock to normal air pressure. As a result the nitrogen which had been absorbed in their body fluids or blood during their exposure to pressure was allowed to escape too rapidly. This problem was largely eliminated after the construction of a 'hospital lock' (in effect a surface airlock). The lock was designed by E. E. Muir, a site engineer working on the project.

As soon as a miner showed symptoms of the bends, he was quickly placed in the hospital lock where the air pressure was increased until it equalled that being used in the tunnel. Then it was slowly decreased at approximately half the rate used in the tunnel airlock.

Despite the various improvements introduced over the years, several important problems associated with compressed-air tunnelling remained unsolved until fairly recently. Paramount amongst these was the difficulty experienced in choosing a satisfactory air pressure when driving large-diameter tunnels which lay below the groundwater table, because of the difference in the hydrostatic head at the top and bottom of the tunnel. Although compressed air kept out the groundwater, it did not always hold up the tunnel face. Frequently the use of unwieldy types of support such as breasting plates, etc. were necessary in large tunnels or where running ground was encountered.

Stringent safety regulations were enforced governing decompression time spent in the airlocks. However, apart from the workmen's susceptibility to the bends they were also prone to the long-term effects of bone necrosis. This painful and crippling disease was the more insidious because its effects only became apparent many years after exposure.

As a result, special penalty rates were awarded to those men who were obliged to work in such areas. Because of this and the other factors mentioned above, the cost of tunnelling under pressure rose dramatically, and both manufacturers and contractors were encouraged to find an alternative to tunnelling with the entire tunnel under pressure. In particular, manufacturers turned their

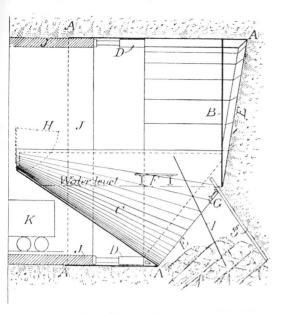

Figure 341. Theodore Cooper shield — U.K. Pat. No. 23,417 dated 27 Oct. 1902. (a) Circular shell of shield; (b) forward bulkhead; (c) sloping and tapering after bulkhead; (d) thrust rams; (e) cutting edge of shield; (f) working platform; (g) removable beam for supporting poling bars, when necessary; (h) emergency door for closing mouth of after bulkhead: (i) drill or other tool; (j) completed tunnel; (k) material car

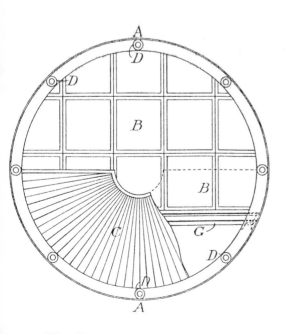

Figure 342. Front view of Cooper shield

attention once more to pressurizing only the front compartment or face of the tunnel.

Early History

Henry Greathead's Plenum Process Shield

According to Gösta E. Sandström,[1] the 'plenum process' was first used by Henry Greathead about 1875. Only the face area was pressurized, the compressed air at the face being separated from the rest of the tunnel by an airtight bulkhead built into the front of the shield. But because of the difficulties experienced in confining the air pressure to a front compartment, this method was not further developed at that time and contractors reverted to pressurizing the whole tunnel.

Theodore Cooper (1902)

On 27 October 1902 (under U.K. No. 23,417) (Figures 341 and 342) Theodore Cooper filed a patent covering 'Improvements in pneumatic tunnelling shields'.

Cooper suggested that the need for using poling-bars, grouting, etc. at the face of a tunnel, when working under compressed air could, to some extent, be eliminated by the following method. He proposed that a trough-like after-bulkhead or diaphragm be placed at an oblique angle below a vertical forward-bulkhead (which was cut away at its bottom for this purpose). The trough's forward open end was to be presented to the material in front of the shield (i.e. the face) while the rear end of the trough or after-bulkhead was to rise upwards to a point above the water-level. The water-level was to be maintained by the tunnel's internal air pressure which was to act as a counter-balance against the external hydrostatic pressure of the groundwater. The forward lower end of the trough or after-bulkhead was to be attached to the shield near its cutting edge. The shield's cutting edge was to be cut back to a point immediately behind the front bulkhead. This was to bring the forward end of the trough as close as possible to the tunnel face. Water was then allowed to enter the trough up to the required level, (which level was

379

maintained by the cushion of air in the tunnel). Cooper suggested that workmen could stand on a platform situated in the trough behind the forward-bulkhead and, from this position, reach through the water to the material in front of and below the shield. He further suggested that the excavated material could then be brought backward and upward (by conveyor mechanism, if desired) through the trough to a point where it could be conveniently removed — preferably a point on the extension of the trough back in the completed tunnel.

While the above method of tunnelling lacks many refinements, it is nevertheless interesting as a similar idea was used by Wayss and Freytag to maintain the slurry pressure in the forward compartment of their Hydroshield some 72 years later.

The need for men to work under pressure was partially eliminated by the introduction of two machines. The first of these was built in 1961 by Entreprises Campenon Bernard, and although excavation work in this unit was carried out hydraulically, it was, basically, a pressurized plenum chamber machine. The second machine was the Robbins Company's Etoile unit, Model No. 341-111, (Figure 343) a pressurized plenum chamber machine with a rotary excavating head.

The introduction of these two machines meant that, generally speaking, tunnel crews could operate the units under normal atmospheric air pressure most of the time. The exception occurred when it occasionally became necessary to replace rubber seals, change the cutters, or carry out maintenance work on the conveyor system, etc. At such times the work was done by placing the entire tunnel under pressure and for this purpose a bulkhead and airlock were installed in the tunnel, usually at the entrance.

The Robbins Etoile Machine (1964)

The Etoile unit was reputed to be one of the largest and heaviest mechanized shields of its time. A contract for the production of this machine was signed in November 1962, but it was not completed until March 1964. It was delivered on site in July of that year and work was commenced towards the end of 1964. The 10.3 m (33 ft 10 in) diameter mechanized shield (Model 341-111) which weighed 450 t was used by the French contractor, Etablissements

Figure 343. The Robbins Company 'Etoile' mechanized shield Model 341-111 — first rotary shield machine with pressurized face. (Courtesy The Robbins Company.)

Billard, to drive an express subway tunnel under the city of Paris.

So far as is known, this was the first time the use of compressed air was confined to the front compartment of a *rotary* mechanized shield.

The excavator mechanism and loading buckets were positioned within the shield a short distance back from the forward cutting edge of the shield hood. Behind the cutter-head mechanism was an air-tight steel plate or bulkhead erected transversely across the forward section of the machine so that in effect an air-tight compartment was formed at the tunnel working face. This compartment contained the excavator head and loading buckets. The cutter head was dressed with disc cutters and 180 fixed tools. Thrust of 71,000 kN (16 million lb) provided by 37 rams reacting against the tunnel-lining segments and torque totalling 7120 kN m (5,200,000 lb ft) was available.

A double-hopper airlock system mounted on a dolly carriage unit incorporated in the rear end of the conveyor tube was used to transfer the muck from the pressurized section to the free-air section beyond the bulkhead with minimum loss of compressed air. Later the Komatsu Company of Tokyo (a Robbins Company licensee) designed a rotary muck discharge system for their unit (a pressurized bulkhead machine) which was required for the construction of a pedestrian underpass tunnel. The rotary system (a rotary drum feedwheel — similar in concept to the helical feed wheel used by Priestley/Nuttall in their slurry machine), occupied less space than did the hopper system.

The Shake-Down Period

Very few innovations have been introduced into the tunnelling field without at least one or two initial teething problems being encountered. The Etoile unit was no exception.

During the first six months the hydraulic system controlling the precast concrete tunnel-liner segments erector proved to be faulty, but after extensive modification it operated satis-factorily and gave no further trouble.

Another problem concerned the seals. Initially the unit was designed to withstand 2.5 atm (37 lb/in^2) pressure, but in practice it was found that only 1.5 atm (22 lb/in^2) was necessary. As a result the original steel reinforced seals (built to withstand 2.5 atm pressure) which were too stiff were replaced by simple rubber ones which were found to be quite satisfactory.

When penetrating certain areas of mixed beds, containing lignite and pyrites in sand, the water was held back by compressed air and, as a result of exposure to air under pressure, the pyrites oxidized and the sulphuric acid produced, contaminated the water fairly heavily. In addition, high concentrations of CO_2 and CO were produced when the lignites ignited as they did from time to time. The sulphuric acid attacked the bolts holding the lining segments in place and these bolts had to be replaced. To overcome this problem, the face was heavily sprayed with water and in addition grout was injected into the face ahead of the machine over the remainder of that area.

Despite these and other similar problems which arose during the initial shake-down period, the machine completed its drive by tunnelling through 2870 m (9400 ft) of varied and difficult ground, comprising areas of hard crystalline limestone, mixed limestone, and clay and silty sand below the groundwater table. After most of the initial problems had been met and dealt with, either by modification of the unit itself, or by other means such as the alteration of the tunnel route to a rising grade in order to avoid certain of the more difficult ground areas, average advance rates of approximately 8 m/day (26 ft/day) were achieved.

While these two machines (i.e. the Campenon Bernard shield and the Robbins Company's Etoile unit) were significant in that they spearheaded the initial development of pressurized bulkhead machines, they also served to appraise the tunnelling industry of the need for a machine which could operate efficiently with liquid at the face.

Bade Mechanised Shield — 1968

Bade & Co. GmbH, Germany[2]
(now Bade & Theelen GmbH)

When underground construction work was planned for the Hamburg subway tunnels and

Figure 344. Bade fully mechanized shield Model MDS-61. Used in Vienna Underground Railway. Similar models used in Hamburg subway and in São Paulo. (Courtesy Bade & Theelen GmbH.)

Vienna's underground railway system, and later also for work under São Paulo, the problems of tunnelling successfully and safely through the difficult ground conditions which existed there seemed formidable.

In Hamburg it was found necessary to tunnel through complicated unstable geological formations subject to high water pressures. While in Vienna the problem was further compounded by the fact that many of the tunnels lay at comparatively shallow depths beneath Vienna's historically priceless old buildings. Obviously adequate face support at all times was essential for the safeguard of both men and equipment, and also to prevent any surface settlement which might endanger the stability of the precious buildings.

The Hamburg unit (Model MDS 61) (Figure 347) commenced work in 1965 while the Vienna (Figure 344) and São Paulo units (Model MDS 62) (Figure 348) commenced work in 1973. Two units with cutting diameters of 6.2 m (20 ft 4 in) were put to work in São Paulo at the beginning of 1973. By the summer of that year both machines were achieving advance rates of approximately 15 m (49 ft)/day (two shifts).

The Hamburg unit differed in some respects from the Vienna and São Paulo machines in that it had an oscillating cutter head instead of a rotating one. While this type of cutter head was found to be satisfactory if the ground was soft and homogeneous, difficulty was experienced when harder ground was encountered. Therefore when the Vienna and São Paulo machines were built they were fitted with rotating cutter heads to enable the machines to cope with a wider variety of ground conditions. In addition the face support plate layout was altered to that of a star-type pattern. In most other respects, however, the design features of these three machines was the same.

Basically the Bade units consisted of a reversible rotating cutter head which was equipped with a sequence of movable steel support plates and adjustable scraper blades. (Figures 345 and 346). The steel support plates were held against the face by jacks which operated independently.

When operating in non-cohesive soils, the pressure exerted by the main thrust jacks could be so regulated that it corresponded to the pressure exerted by the soil at the face. Thus during the rotation of the cutter head, pressure was applied to the face by the support plates which both supported and consolidated the soil. At the same time the scraper blades were inclined inwards and a gap was created which relieved the pressure on the face at that point, thus encouraging ingress of the soil there. However, when cohesive ground was encountered the cutting or scraper blades could be repositioned so as to project in the direction

382

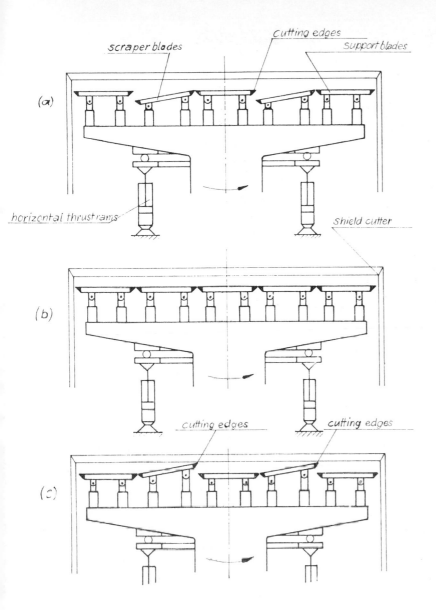

scraper blades

Cutting edges

support blades

(a)

horizontal thrust rams

shield cutter

(b)

cutting edges

cutting edges

(c)

Figure 345. Bade system: (a) in non-cohesive strata; (b) completely enclosed support base; (c) in cohesive strata. (Courtesy Bade & Theelen GmbH.)

of advance of the rotating head. This repositioning set the blades in a helical cutting arrangement which allowed them to pare off the material at the face.

The steel cylinder of the shield was a double-skinned welded construction which was provided with a cutting edge at the front. An important feature in common with all double-skin type machines was that it was possible to detach the outer skin at the end of a drive and leave it *in situ* as a segment of tunnel lining, while the inner skin, carrying all the equipment, could be dismantled and removed for re-use with a new outer skin elsewhere at a later date.

If inclusions of a harder type of material were encountered such as boulders, etc. it was possible to remove the entire cutting head. In addition, should it be found necessary to tunnel through material which was below the groundwater table, then the Bade shield could be worked with compressed air which was confined to the front compartment, housing the mechanical borer and muck removal installation. A safety wall or transverse diaphragm, situated behind the cutter-head operating motors, effectively sealed this front area from the rest of the shield.

In hard self-supporting ground the

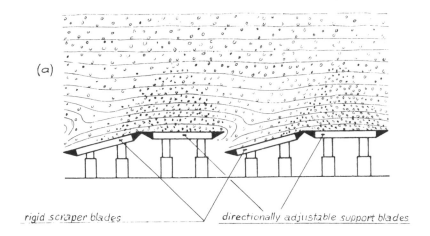

(a)

rigid scraper blades   directionally adjustable support blades

directionally adjustable support blades   rigid scraper blades

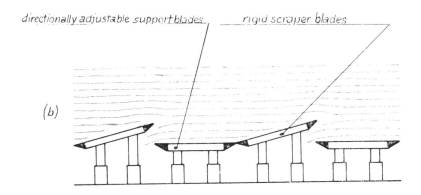

(b)

Figure 346. Bade system: (a) in non-cohesive strata; (b) in cohesive strata (Courtesy Bade & Theelen GmbH.)

mechanical boring head could be advanced a maximum distance of 400 mm (16 in) ahead of the shield. Alternatively in non-cohesive soils, the cutting head could be withdrawn into the shield a maximum distance of 400 mm (16 in). The advance and retraction of the cutting head was effected by means of four hydraulic jacks. Thrust reaction for these jacks was taken by the rear diaphragm.

In the original unit used in Hamburg the muck entered into the scoops or pockets situated immediately behind the support plates, through the gaps which were adjustable according to the needs of the soil conditions encountered. These scoops then lifted the spoil and deposited it on to the conveyor belt by means of a slide or chute. A tube-like opening in the central area of the borer head also allowed for the entry of muck from the face. This opening could be closed from the control panel if necessary.

A large-capacity chain scraper conveyor situated about a third of the way up the shield then conveyed all the muck through a double-sluice arrangement from the pressurized bulkhead compartment to the free-air area at the rear. The conveyor system was completely enclosed.

However, in the unit used in Vienna the inlet of the conveyor was moved to the invert and the muck was raked in from either side of the invert by gathering arms which deposited it on to a slide where it was guided to the main conveyor.

The cutting wheel was rotated by eight rams which utilized the rack and pinion principle, while the entire shield was advanced by 25 thrust jacks which reacted against the newly erected tunnel-lining segments. In the São Paulo subway tunnels conventional cast-iron tunnel-lining rings were used which consisted of eight segments and a key segment. Erector arms in the tail of the shield lifted and positioned the

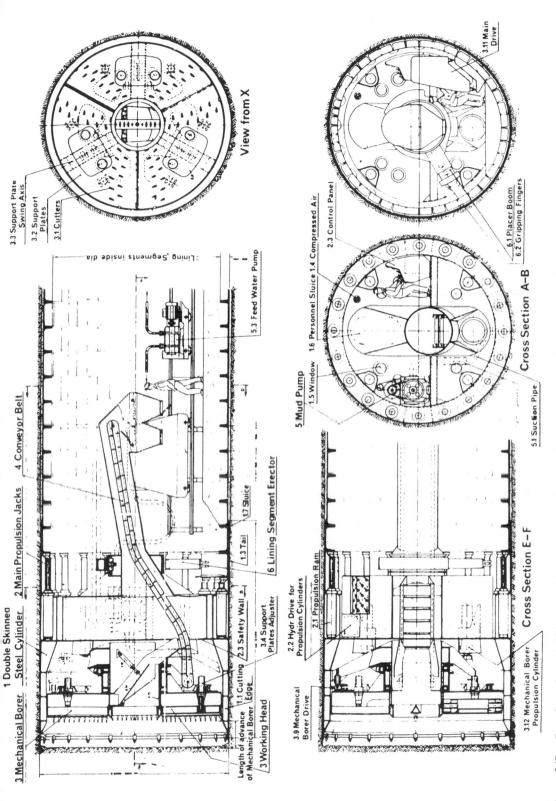

Figure 347. Drawing of Bade Shield used in Hamburg. Note position of conveyor belt and cutterhead plate support system layout. (Courtesy Bade & Theelen GmbH.)

385

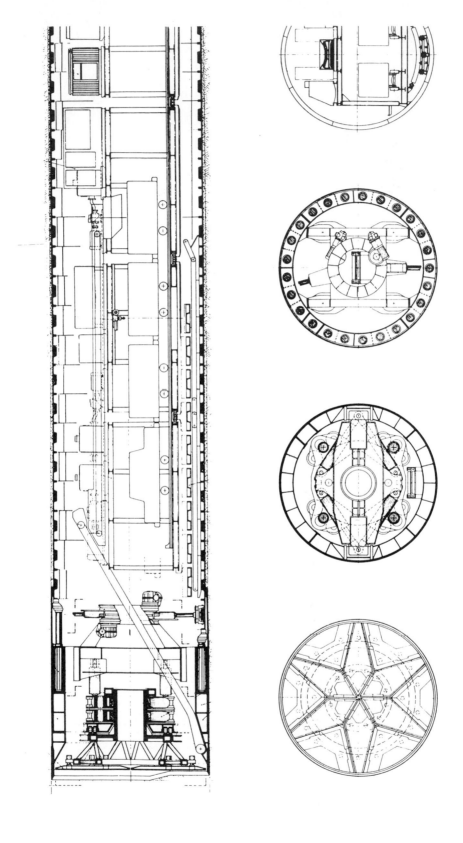

Figure 348. Drawing of Bade shield used in Vienna and São Paulo. Note cutter-head plate support system layout and position of conveyor belt. (Courtesy Bade & Theelen GmbH.)

Table 11

Year	Type[a]	Project
1965	MDS-550-GS	Subway, Section St. George, Hamburg
1967	MDS-550-GS	Subway, Section Karl-Muck-Platz, Hamburg
1967	MDS-560-GS	Subway, Section Karolinestrasse, Hamburg
1968	MDS-300-GS/C	Main sewage collector, Solothurn
1968	MDS-550-GS	Subway, Section Rosenstrasse, Hamburg
1969	MDS-560-GS	Subway, Section Grosse Bleichen, Hamburg
1970	MDS-610-GS	Subway, Section Karlsplatz, Vienna
1970	MDS-650-GS	Express railway, Section Davidstrasse, Hamburg
1972	MDS-610-GS	Subway, Section Südtiroler Platz, Vienna
1972	MDS-620-GS	Metro, North-South Line, São Paulo
1973	MDS-620-GS	Metro, North-South Line, São Paulo
1974	MDS-610-GS	Subway, Section Schwedenplatz, Vienna
1975	MDS-650-GS	Subway, Section B, Hannover
1975	MDS-620-GS	Metro, East-West Line, São Paulo

[a]The diameter of the shield in centimetres is shown between the abbreviations of the type of shield and its configuration, e.g. MDS-610-GS refers to a fully mechanized tunnelling shield, diameter 6.1 m (20 ft), with full-face support.

tunnel-lining segments and lead caulking was inserted between the segments to seal the joints.

The projects listed in Table 11 were carried out using a Bade full-face mechanized tunnelling shield.

At the beginning of 1975 a new company was formed which became known as Bade & Theelen GmbH. The main aim of this company was the development of tunnelling machines and tunnelling techniques. This firm operates in close association with the production company of Bade & Company.

References

1. Gösta E. Sandström, *The History of Tunnelling*, Barrie & Rockcliff, London, 1963.
2. Dipl. Ing. W. M. Braum, The art of tunnelling in Vienna, *Tunnels and Tunnelling*, May 1964, London.

Correspondence

The Robbins Company, Seattle, U.S.A.
Bade & Co. GmbH (now Bade & Theelen GmbH) Lehrte, West Germany.
Mitsubishi Heavy Industries, Tokyo, Japan.

Patents

T. Cooper — 23,417/1902 U.K.

THIRTEEN

Slurry Machines

Of all tunnelling problems surely the label 'most difficult' must be attached to that of dealing with wet running ground which lies well beneath the water table. Both Brunels encountered such conditions during the excavation of the Thames tunnel and, as a result, barely escaped with their lives. Other tunnellers working on similar projects have not been so fortunate.

Over the years many attempts have been made to solve the various problems associated with tunnelling through this type of ground. Included amongst these attempts was the use of compressed air, which proved very effective. The recent advent of the pressurized bulkhead was another important step forward in the march of progress but, nevertheless, many problems still remained.

If compressed air is used only at the face area, this generally overcomes the necessity to expose miners or workmen to a high air pressure during their work period. On the other hand, confining the air to the face area alone is not always entirely successful. The air has the capacity to escape through small channels made by it between the outer skin of the shield and the tunnel wall. It forces its way past the tail seal at the rear of the shield and enters the tunnel, resulting in a failure of the plenum process. Such failures have on occasion necessitated the pressurizing of the entire tunnel before the project could be successfully completed.

Assume for example, that a conventional shield or a shielded T.B.M. with compressed air was contemplated for use in a large-diameter tunnel which was to lie close beneath the bed of a lake or a river. Assume in addition that the strata through which the route of the proposed tunnel was to run, consisted of non-viscous mud beneath a high groundwater table. Because the route of such a tunnel would lie in close proximity to say the bed of a lake or river, and because there was a marked difference between the hydrostatic head at the top and bottom of a large-diameter tunnel, it was generally found difficult in such circumstances to keep the air pressure in the tunnel or at the working face balanced against the different pressures exerted by the groundwater at the top and bottom of the face. This meant that during excavation the workers faced the very serious risk of a 'blowout' if the air pressure was too high, or a run at the bottom of the face if the air pressure was too low.

Apart from the use of compressed air, under certain favourable conditions other methods could be used to overcome some of the above-mentioned problems. These included grouting the ground ahead of the shield or T.B.M. or, alternatively, the ground could be excavated by the cut and cover method. However, the disadvantages of using these latter methods included such factors as time, cost, disruption to traffic or shipping, etc. Engineers were thus prompted to think of a better solution and the answer came with the introduction of the 'slurry shield' or, as some Japanese companies have called it, the 'mud circulation shield'.

The modern slurry machine which has now evolved is capable of safely supporting and excavating non-viscous mud strata lying beneath a high groundwater table. In most cases operators may work under normal atmospheric pressure, and good direction control of the machine itself can be maintained. Most important, de-watering of the ground with consequent ground subsidence can be kept to a minimum. Units incorporating these important features, which were pioneered and developed by Japanese engineers, are presently operating in many projects in Japan today, while engineers in Britain and Germany on the other hand have developed and produced their own particular versions of the slurry machine.

Although the concept of a slurry machine is certainly an innovation, the use of clay, mud, bentonite clay, or slurry, etc. in drilling and tunnelling is not new, as will be seen from the following chapters.

Early Use of Slurry

In 1845 a French engineer (Fauvelle) reported that water, forced down a drill stem had been used to flush drill cuttings to the surface via the annular space existing between the exterior of the pipe and the bore-hole wall. This soon became an accepted method of flushing a hole which had previously been drilled by percussion (i.e. by drop drill and rod, etc.)

Though rotary drilling had been attempted in the latter half of the nineteenth century (Sweeney filed a U.S. patent for a rotary drill in 1866), it was not generally accepted as an efficient alternative to percussion drilling until after 1901, when the Lucas oil-well at Spindletop, Beaumont, Texas, was successfully drilled with a rotary drill.

As mentioned earlier (see Chapter 1, Drills — Vertical drills — The hydraulic rotary drill or rotary flush drilling), water alone was at first used to flush the cuttings to the surface. However, it soon became apparent that mud, formed from the pulverized drill cuttings and flushing water, actually improved the condition of the bore-hole walls and, in addition, carried the debris more readily to the surface.

Within five years of that time, operators were deliberately adding measured quantities of surface clay to their drilling water in order to improve its viscosity. From that point onwards various experiments were carried out in an effort to increase the density of the mud and, amongst other ingredients, iron oxide and barite were tried. By the early 1920s ground barite was used almost exclusively in place of surface clays and other substances. The drilling mud industry proper did not, however, begin any serious development or expansion until

Harth demonstrated that the addition of bentonite clays to the ground barite and water mixture was of significance in that it helped to suspend the barite.

Bentonite[1,2]

Although the term 'bentonite' is now often widely used to describe clays of the montmorillonite group, no matter where they are found, strictly speaking, the name bentonite applies to the type of clay which was found by Knight in Wyoming, near Fort Benton (from which it derived its name).

There are more than six clay minerals which form the montmorillonite group but, by far the most popular, because of their excellent capacity for increasing the viscosity of drilling mud, are the calcium montmorillonites and the sodium montmorillonites. Sodium montmorillonites in particular (of which Wyoming bentonites are chiefly composed) are very effective in this respect. While calcium bentonites have a swelling capacity (after the addition of water) of from 1 to 1½ times their original size, Wyoming bentonites may increase to as much as 14 to 16 times their original size when mixed with water.

In addition to this property, bentonite is also, like many of the clays, highly thixotropic. That is, it may, when sufficient water has been absorbed, change rapidly from a solid to a liquid if subjected to a sudden shock or vibration. Indeed this change may even occur spontaneously. Several major disasters in the form of landslides have been caused by the thixotropic properties of some clays — Surte, Sweden (1950), Knockshinnoch, Ireland (September 1950) and Turnagain Heights, Alaska, (March 1964) are but a few examples. Conversely, this process can be readily reversed. When agitation ceases the clay particles gel and what one moment is a creamy liquid can rapidly firm up and become consolidated. Use has been made of this property for many years in the stabilization of the walls of trenches and ditches.

In his Patent No. 1,738 dated 16th May, 1874, Greathead was, so far as is known, the first to suggest that *muck may be removed from the working face of a tunnelling machine by*

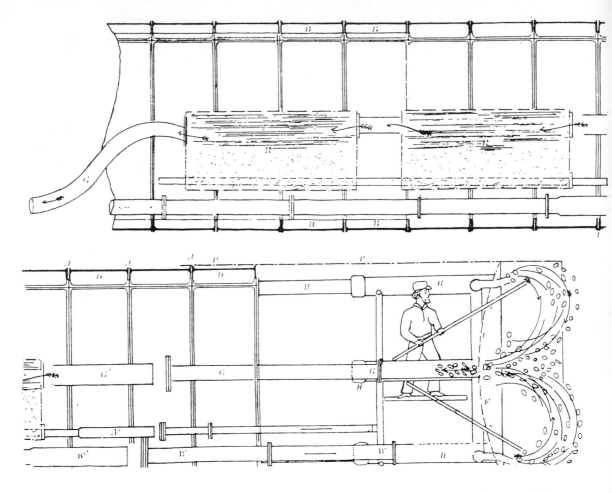

Figure 349. Greathead's proposed slurry shield. Patented in U.K. on 16 May 1874 under No. 1,738

pumping it out as slurry. Later Prof. H. Lorenz (Berlin) proposed in a patent (now expired) that *a tunnel face could be stabilized by the application of a mixture of bentonite clay and water under pressure.* Neither of these ideas was, however, tested in practice until the late 1950s when Elmer C. Gardner of Houston put his teredo T.B.M. to work in a 3.35 m (10 ft 11 in) diameter storm-water drain tunnel.

Greathead's Patent

Greathead's patent (Figure 349) described a double-skinned cylindrical shield, the outer skin of which overlapped and enclosed the inner tubular skin so that it was both air- and watertight. At the front end of the shield Greathead placed a diaphragm or bulkhead through which a number of holes were cut and

fitted with stuffing boxes, devices which allow freedom of movement of a shaft or piston rod through openings in a plate, while yet keeping the openings air- and watertight. This sealing is effected by means of tight wadding or packing held around the rod, etc.

Greathead also provided two inlet holes and one large central outlet hole. It was proposed that a number of manual digging tools (i.e. crowbars, etc.) be inserted through the spherical sockets in the stuffing boxes and used to excavate or loosen the material from the tunnel face. The material thus loosened was to fall into the water which had been pumped into the forward section through the two inlet tubes. The resulting suspended matter (slurry) was to be removed through the central outlet hole and carried to settling tanks situated at the rear of the shield. Greathead further suggested that if,

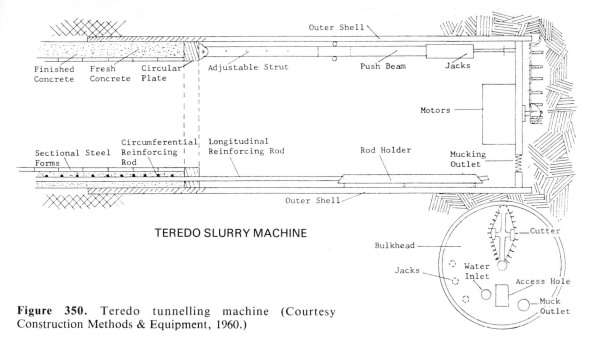

Finished Concrete Fresh Concrete Circular Plate Adjustable Strut Outer Shell Push Beam Jacks

Motors

Sectional Steel Forms Circumferential Reinforcing Rod Longitudinal Reinforcing Rod Rod Holder Mucking Outlet

Outer Shell

TEREDO SLURRY MACHINE

Bulkhead Cutter

Jacks Water Inlet Access Hole Muck Outlet

Figure 350. Teredo tunnelling machine (Courtesy Construction Methods & Equipment, 1960.)

instead of running ground, loose dry sand was encountered, air could be used to force the sand out through the central outlet hole.

He proposed that a manhole be built into the bulkhead to provide access to the front section. Should it be necessary for men to work at the face under compressed air for prolonged periods, due to the particular ground conditions encountered, a second bulkhead could be constructed behind the first to form a type of airlock.

Greathead suggested that the shield be advanced by means of either screw or hydraulic jacks bearing against the recently constructed tunnel-lining supports, which could be iron rings, iron frames filled with cement, concrete, or other material made in 'voussoir segments bolted together'.

Gardner Engineering Corporation

Teredo T.B.M. — 1959/60[3]

Elmer C. Gardner attributes the successful debut of his slurry machine to the combined efforts of his vice-president and design engineer, E. A. Horstketter, his hydraulics specialist, F. Fuller, and his field and shop

superintendents M. Boehning and W. H. Duval respectively.

Basically the Teredo* (Figure 350) consisted of a full shield which was cylindrical in shape. It measured approximately 9 m (30 ft) in length

*The Teredo T.B.M. was so named after the dreaded *Teredo navalis* (ship-worm) which used the same method of muck removal. The *Teredo navalis* has a long worm-like body which may vary from 30 cm (1 ft) in length when found in temperate seas to as much as 2 m (6.5 ft) in length in tropical waters.

At its inner end (i.e. the face of the tunnel) is a pair of small shell-type plates (valves) which are almost hemispherical in shape and which form a deep notch at the front end. In this gap is a round sucker-like type of foot with which the worm grips various parts of its tunnel. On the sides of the valves which form the edges of the notch are two sets of ridges which, when magnified, are seen to be minute rows of teeth, somewhat similar to those of a rasp. In operation, while the sucker foot is maintaining a firm grip on the mother-of-pearl lined wooden walls of its tunnel, the worm imparts a type of rocking motion to its teeth by means of anterior and posterior adductor muscles. The teeth then cut or rasp away the wood from the head of the tunnel.

The tail of the worm is always in contact with the sea water outside its tunnel so that the pair of siphons at this end may operate and either draw in or expel water. When the wood has been cut from the face it is swept into the mouth of the worm and then passed on into the stomach where it is digested and the waste eventually expelled.

The amazing similarity of the modern slurry machine's muck removal system to that of the ship-worm need not be further emphasized.[4]

and was divided into three compartments. At the nose or front end was the combined cutter and bulkhead. The forward or cutting edge of the shield itself extended a short distance ahead of the steel bulkhead and cutting mechanism. Employing somewhat similar engineering principles to the early type breast-chain machine (used in coal-mining), the cutting mechanism of the Teredo was simply a pick-carrying cutter chain. The cutter chain revolved about a cross-shaped radial arm which extended from the central shaft in the bulkhead to the outer edge of the shield. While the cutter chain revolved about the arm, the arm itself rotated about the central shaft. (Initially the cutter chain mechanism was open, but later, due to problems which arose as a result of excessive wear on the teeth and the consequential 'downtime' involved when these needed changing, the teeth were replaced by rotary cutters and the drive chain was covered by protective casing and lubricated with oil. Thereafter downtime due to cutter wear was noticeably reduced.)

In operation the spoil cut from the face dropped to the invert of the tunnel and into water which was kept at a constant level of 1.52 m (5 ft). Fresh water from the city mains was constantly being pumped into the nose section beyond the bulkhead, through a pair of 10 cm (4 in) diameter holes in the bulkhead, and the slurry or muck which formed in the water was sucked out through a 15 cm (6 in) diameter hole cut in the bulkhead a few centimetres from its base. Access to the face was provided by means of an opening situated at the bottom of the bulkhead between the twin water inlets and the muck outlet hole. The slurry chamber in front of the bulkhead was not pressurized.

Thrust was provided by eighteen 25 ton hydraulic rams which abutted on steel push beams and 2.5 m (8 ft) adjustable struts. These push beams and struts were positioned around the circumference of the second compartment and the tail-end of each strut terminated in a heavy 23 cm (9 in) steel-plate ring which encircled the shield just inside its outer skin. Thus the thrust from the jacks was transmitted through the push beams, the struts, and also through the steel ring which, in turn, bore evenly against the recently poured concrete lining of the tunnel. In order to enable the tunnelling crews to extend the length of the concrete lining to as much as 2 to 2.5 m (6.5 to 8 ft) before it became necessary to reposition the entire shield, the following process was carried out: when the jacks were retracted, after completing their full 60 cm (2 ft) stroke, the push beams were repositioned by the simple expedient of removing the hinge-pins connecting the push beams to the struts and then reconnecting the beam on to the strut at a new point 60 cm (2 ft) ahead of its original position.

Beyond the 4.9 m (16 ft) long middle compartment of the shield lay the 3 m (10 ft) long tail section. Within this tail section an inner lining was erected from short prefabricated sections, leaving an annular space 23 cm (9 in) wide between it and the tail section of the shield itself. It was into this space that the concrete was poured and then moulded into a continuous lining for the tunnel.

Upper and lower movable fins positioned outside the shield near the front end prevented it from rolling. They could also be used for steering the machine. In addition, the shield's direction could be changed by realigning the central cutter shaft in an eccentric position and, finally, the machine's direction could also be controlled by the thrust rams in the normal way.

Initially pneumatic pumps supplied fresh concrete to the T.B.M. from the street level through hoses which entered the tunnel at its portal end. However, as the tunnel grew in length and the distance from the portal became excessive, the concrete was supplied to the T.B.M. through manholes which lay at 152 m intervals along the route of the storm-water drain tunnel.

Despite Gardner's promise not to disrupt city traffic and the fact that his bid was the lowest by about 30 per cent of those received, the City Council of Houston, Texas, was extremely reluctant to award him the contract and eventually only did so on condition that he would undertake to complete the job by cut and cover methods at no additional cost to the City Council, if his machine failed. Gardner unhesitatingly agreed to this proviso because he had the courage to back his own judgement and his new invention financially.

Occasionally during the execution of the contract, the tunnelling crews encountered pockets of quicksand. When this occurred it was possible to draw out the entire pocket without any appreciable advancement being made by the T.B.M.! Maintaining the correct grade in such areas was a further attendant problem. However, despite these difficulties, Gardner's machine successfully completed the drive.

Markham & Co. Ltd[5]

The firm of Markham & Co. Ltd. was founded in 1845. It moved to its present site at Broad Oaks works in 1872. The works occupies a site of some 9.7 hectares (24 acres) with a covered area of 23,000 sq. m (250,000 sq. ft).

In 1887 the firm of Markham & Co. Ltd. received an order from a certain J. W. Williams for the construction of 8 tunnelling shields, and, during the succeeding 10 years 60 more shields were built for various customers.

When the company was first formed it specialized in mine winders and sundry colliery equipment (i.e. fans, fan engines, headgears, head sheaves, etc.). However, the demand for shields encouraged the firm to enter into the tunnelling equipment field and such items as airlocks, bulkheads, caissons, hoists, etc. were added to their list of products.

By the outbreak of the First World War, Markhams had chalked up over 200 shields on their production list. (This included mechanized shields made to the famous Price patent design which were first built at Broad Oaks in 1901.)

The vast extension programme undertaken between the First and Second World Wars to London's Underground tube system boosted this number to 400 shields (Figure 307 depicts one of Markham's later shields), including the manufacture of about 40 mechanized shields. These shields varied in diameter from a mere 1.65 m (5 ft 6 in) to over 9 m (30 ft).

In 1967 Markhams received an unusual order from the Mitchell Construction Kinnear Moodie Group Ltd, who at that time were awarded an export contract to supply tunnelling equipment and construction methods to Mexico City, where the city fathers were experiencing considerable difficulty in constructing their deep-level drainage system.

Mexico City — Early History[4,5]

Mexico City which was about 2240 m (7349 ft) above sea-level in 1974 has a population of some 6 million inhabitants. It stands on a small plain which occupies the south-western part of a depression known as the Valley of Mexico (El Valle de México), and lies on the western shore of Lake Texcoco. At one time the waters of Lake Texcoco covered most of the area on which the city now stands.

From its earliest recorded history Mexico City has had severe drainage problems and was rated as one of the worst-drained cities in the world. The city is only 2 m (6 ft) above the level of the lake. During the year 1629 the city was inundated by floodwaters which rose to a height of 1 m (3 ft) and did not recede for five years thereafter.

Some effort was made to drain the city by the Spaniard, Maartens. As Lake Zumpango was the nearest natural drainage basin in the area, Maartens proposed that a drainage channel be cut through the northern hill of Nochistonga so that the overflow from Lake Zumpango could be directed into the river Tula which was a tributary of the Pánuco. The cutting (21 km (13 miles) in length) was commenced in the year 1607 and was finally completed in 1789. It was named the Tajo de Nochistongo.

As these steps only partly alleviated the problem, tenders for further drainage works were called for by President Ignacio Comonfort. The proposal was for a 69 km (43 miles) long drain which was to commence as an open canal at a depth of 1.2 m (4 ft) below Mexico City on its eastern side and then run north towards Zumpango, at which point it would enter a tunnel 94 m (308 ft) deep under the mountain rim and veer eastwards for 9.5 km (6 miles) until it eventually emptied into a tributary of the river Pánuco.

In 1900 the work was completed and water began flowing down the canal. However, the drainage system as such was not entirely satisfactory as it created other more serious problems which had not been foreseen. In effect, the city was built on a vast lake of mud which had collected in and filled the crater of a volcano. Though this mud had dried on the surface, the subsoils were still permanently

Figure 351. Universal soft ground tunnelling machine — designed by K.M. Tunnelling Machines Ltd and constructed by Markham & Co. Ltd. (Courtesy Markham & Co. Ltd.)

Figure 352. Universal soft ground tunnelling machine — constructed by Markham & Co. Ltd. for the Mitchell Construction Kinnear Moodie Group Ltd. Used under Mexico City for the construction of sewer tunnels. (Courtesy Markham & Co. Ltd.)

saturated and contained as much as 80 per cent water.

In the year 1929 there were 11 public and some 1375 private artesian wells operating in the city. As the population expanded so did the number of wells increase. The result was that greater and greater quantities of water were daily being drawn up from the subsoil for use in households, or extracted and channelled away via the canals and tunnel for irrigation purposes, etc. As the water was removed, the subsoils began drying out and contracting and the city commenced sinking at the alarming rate of approximately 30 cm (1 ft) per annum.

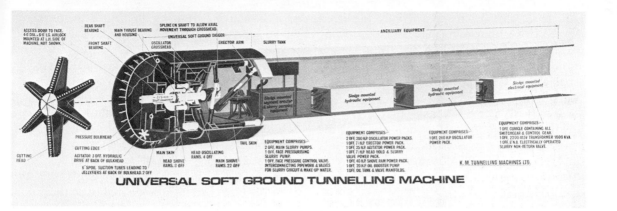

Figure 353. Cut-away view of universal soft ground tunnelling machine (Courtesy Markham & Co. Ltd.)

According to the *Encyclopaedia Britannica* of 1929 the city was some 2260 m (7415 ft) above sea-level in 1911. By 1974 the level of the city was about 2240 m (7349 ft) above sea level, which shows a drop of approximately 20 m (65 ft) in 63 years. But the problem was not generally noticed nor taken very seriously until buildings began leaning over. By 1935 the Palace of Fine Arts had actually sunk a complete storey. Over the succeeding years citizens and officials watched helplessly as various parts of their city sank lower and lower. By 1950 it was 6 m (20 ft) and more below the original level of the canal tunnel. Eventually it became necessary to pump all the sewage uphill, and dire warnings were issued by eminent engineers that unless something was done soon, a severe rainstorm could flood their sewers and overflow into the city to the height of about 1m (39 in) carrying with it the vile-smelling effluent. Nevertheless it was only after the experience of the 1951 floods that this advice was heeded. Many of the artesian wells were permanently closed and the city's water was drawn from the hills nearby. In addition plans were laid for the construction of a 6 m (9 ft 8 in) diameter central interceptor drainage tunnel 30 m (98 ft) below the city surface, its purpose being to draw off the majority of the storm waters in the event of a bad flood.

A part (48 km (30 miles)) of the proposed tunnel route lay through the mud under the city and the remainder (a distance of some 64 km (40 miles)) travelled through the rock strata which lay beneath the mountain ridge.

It was for the mud section of the tunnel that engineers from Markham were asked to submit designs for three T.B.M.s based on the combined design concepts of engineers from Kinnear Moodie Tunnelling Machines Ltd, (a member of the Mitchell Construction Kinnear Moodie Group) and the Federal District Department of Mexico City.

Work on the first machine was started in July 1967. By 14 February 1968 the prototype had successfully completed its works tests.

The Slurry Machine — Universal Soft-ground T.B.M.[5]

The machine, (Figures 351-353) built by Markhams to tunnel through Mexico City's unstable subsoil under soil-water pressures nearing 3 atms (44 lb/in²), was basically a mechanized shield, but of an unusual type. The head, or excavating section of the machine, was separated from the main body by a heavy steel bulkhead which was designed to withstand the full hydrostatic head of the ground water.

The excavator cutting head, consisting of six radially arranged cutting arms was mounted on a central drive shaft, which transmitted an oscillatory movement to the arms. This particular type of movement was chosen in order that the unstable ground ahead of the shield should not be unduly disturbed and also in order to counteract any tendency for the machine to rotate within the bore.

Two pairs of hydraulically powered rams attached to radius arms at the rear end of the

395

main drive shaft and reacting against the invert and crown sections of the shield, respectively, provided the necessary oscillatory action to the cutting head. When in operation the arms described an arc of some 60-70° across the tunnel face.

Twenty-two powerful jacks placed circumferentially at the rear of the unit and reacting against the recently constructed tunnel lining, provided a total forward thrust of 1300 t to the shield. (In addition, a spline on the main drive shaft permitted a 40 cm (1 ft 4 in) axial movement of the head within the shield. This movement was effected by eight rams capable of a total thrust of 500 t.)

The space between the bulkhead and the tunnel face (which, of course, contained the cutter-head mechanism) was filled with water under pressure. In operation the cuttings dislodged from the face were turned into a slurry by three hydraulically driven agitators situated behind the cutter head at the bottom of the bulkhead. The drive for the agitators was placed behind the bulkhead.

The resulting slurry formed by the combined action of the cutting arms, the water and the agitators was then sucked from the excavation compartment through two 15 cm (6 in) diameter outlet pipes placed near the bottom of the bulkhead. These pipes conveyed the slurry to a slurry tank and slurrifying unit mounted on a sledge which formed part of the ancillary equipment behind the shield itself. From this point the spoil was pumped away from the working area and out of the tunnel. However, some of the slurry was redirected back into the digging compartment through a pressure-actuated bypass valve which maintained the hydraulic pressure at the face.

Concrete tunnel-lining segments (designed by Kinnear Moodie (Concrete) Ltd.) were automatically positioned by a hydraulic erector arm situated in the tail section of the shield.

While the remainder of the shield and, of course the tunnel, remained in free air, access to the face for maintenance purposes, etc. was provided through a small airlock built into the bulkhead. The airlock was 1.2 m (4 ft) in diameter and 1.8 m (6 ft) long.

Soon after the universal soft ground T.B.M.s reached Mexico City, Markham lost contact with their progress, as the contractor to whom the units had been supplied ceased to be involved in the actual tunnelling process. During the intervening years the author has made various attempts to discover the fate of these machines but to no avail, as all letters sent to Mexico were ignored. Finally some news filtered through via a group of Japanese delegates (who had visited Mexico and who were attending a shaft and tunnelling conference with the author). They advised that two of the units had been converted to conventional shields and the third had been abandoned. Then in October 1976 the author received a letter from D. H. Scotney of Kinnear Moodie which is quoted below:

The fate of the Mexico machines is far from clear but from the information available, is as follows:

The first machine started work in July, 1969, and broke out of the shaft into a full face of pumice. Hydraulic face support was not necessary but water and slurry were pumped into the cutting chamber to act as a carrier for the excavated ground. Work continued throughout 1969 on a single shift basis and was eventually stopped in March, 1970, after 100 m of tunnel had been built. After standing for two months with no face pressure, a 'chimney' formed to the surface interrupting an existing high level sewer. The resultant land slip flowed into the machine through the open bulkhead doors creating a pretty nasty mess but no damage to the machine. The area was cleared and the machine recovered but for some reason, work was not restarted on this section.

The second machine started work early in 1970 in soft ground with face pressures of 40 to 45 lb/in². Average progress was 3 m/day and 40 m of tunnel were constructed before a major disaster occurred. Due to the use of excessive and unequal shield thrust, the primary lining was badly damaged and ultimately failed, flooding the tunnel and shaft, fortunately without loss of life. This machine was then abandoned.

The third machine was assembled at the bottom of a shaft, but never started work.

As a result of their problems, the Mexican Engineers abandoned the original scheme and the tunnel was re-aligned at a higher level. Dewatering was introduced and tunnelling carried out with hand shields in compressed air at up to 1 atmosphere. Six hand shields were eventually used of which two were probably produced from our original machines.

On reading through the case histories of new machines which have been introduced over the years one cannot but be struck by the fact that most of these units succeeded — not because of the perfection of the prototype model, though naturally it was essential that all engineering principles involved be sound — but because of the ingenuity, courage, perseverance, faith, and financial backing of one man or a group of men who had a personal and dedicated interest in the success of the unit. The story of the old McKinlay entry driver is a classic example of this and there are, of course, many others. Under the circumstances one is left with the thought that the ultimate fate of these three machines might very well have been different had the project been undertaken in England where both the designers and manufacturers of the unit would have been available.

Edmund Nuttall Limited

The Edmund Nuttall firm of contractors was founded by James Nuttall in 1865. In 1902 Nuttall's two sons joined the business as partners and the firm was renamed Edmund Nuttall & Company.

Many significant civil engineering projects were carried out by the company during the ensuing years. Amongst these was the Manchester main drainage scheme project where an Arrol-Whitaker T.B.M. was used in 1924 (see also Rock-tunnelling machines — Douglas Whitaker). The Manchester project was started in 1913 and completed in 1927.

In 1968 the firm of Robert L. Priestley Ltd. became a wholly owned subsidiary company of Edmund Nuttall Ltd.

The Bentonite Tunnelling Machine[6,7]

The bentonite tunnelling machine owes its development to two significant factors. Firstly the need to extend their underground railway network into strata consisting of large proportions of non-cohesive gravelly soil and silt, which lay to the south of the Thames river, prompted London Transport Executive to seek a more economical method of tunnelling through such strata than had hitherto been available. Secondly, at about this time, J. V. Bartlett (a senior partner in the consulting engineering firm of Mott, Hay, and Anderson)

patented a process whereby a bentonite slurry mixture could be maintained at a selected pressure in a forward or plenum compartment of a clay-type shield excavator machine in order to stabilize the tunnel face (Figure 356).

A patent for this process was taken out by Bartlett in 1964 under No. 1,083,322.

Plans for the new Fleet Line subway were already well advanced. In view of this it was not surprising to find that London Transport were willing to provide at least half the cost of building a trial machine and driving an experimental tunnel with it at New Cross. The New Cross site was selected because the experimental section could ultimately be incorporated into the Fleet Line route when that project was completed.

Negotiations were entered into with the National Research Development Corporation (N.R.D.C.) who, for their part, agreed to provide the remainder of the capital necessary for this venture. (Mott, Hay, and Anderson were naturally appointed as consulting engineers.)

Tenders were sought from six contractors and in January 1971 the firm of Edmund Nuttall was asked to build (through their subsidiary company, Robert L. Priestley Limited) a machine based on the unusual design concepts submitted by the consulting engineers, Mott, Hay, and Anderson. (It was originally proposed by London Transport Executive and the N.R.D.C. that an existing tunnelling machine should be used for the experiment. However, this suggestion was strongly opposed by Edmund Nuttall and their subsidiary company, Robert L. Priestley Limited, who felt that in order to give the experiment the maximum chance of success a new and different machine was required.)

The prototype unit (Figures 354 and 355) was delivered to the New Cross site on 12 December 1971. Basically it consisted of a 4.1 m (13 ft 5½ in) diameter cylindrical mechanized shield weighing 70 t. Near the forward end of the shield a steel diaphragm or bulkhead separated the excavating mechanism from the remainder of the shield. Provision was made for the forward compartment to be pressurized up to 0.2 MPa (28 lb/in²) with bentonite slurry, but during the experiment the actual pressure used

Figure 354. Bentonite shield — process patented by J. V. Bartlett in 1964. Constructed by Robert L. Priestley Ltd., subsidiary of Edmund Nuttall Ltd. — used at New Cross. (Courtesy Robert L. Priestley Ltd.)

excavated spoil. Slurry, carrying suspended particles of sand and silt, left the chamber via a pipe which led from the top of the bulkhead through a 7.5 cm (3 in) fullway pressure-control valve and into a sump situated behind the bulkhead.

The cutter-head design was of the standard type used by Priestleys for their normal tunnelling machines and was driven by four hydraulic motors capable of variable speeds. It was supported on a cross roller-bearing designed to withstand both the radial and axial stresses on the cutter head. Large material cut from the face was lifted by the rotating cutting arms and deposited onto the centrally positioned chute, whilst smaller pieces which remained in suspension passed through the upper bentonite discharge pipe. The chute conducted the spoil to the feed wheel, a device for removing solids without loss of bentonite or pressure, which led it through the bulkhead and delivered it into the sump through a 10 cm (4 in) mesh screen. From the sump all of the excavated material was pumped to the surface treatment plant. Settlement of material at the bottom of the sump tank was prevented by an agitator which was fitted to the sump outlet pipe.

At the surface the excavated material was passed through vibrating screens and hydrocyclones (centrifugal separators) which

Figure 355. Bentonite shield — used at Warrington New Town. (Courtesy Robert L. Priestley Ltd.)

for most of the time was nearer 0.1 MPa (14 lb/in²). Bentonite slurry was pumped into this plenum chamber or forward compartment through two 10 cm (4 in) pipes situated at the bottom of the bulkhead. The bentonite slurry served the dual purpose of supporting the face and acting as a vehicle for the removal of

398

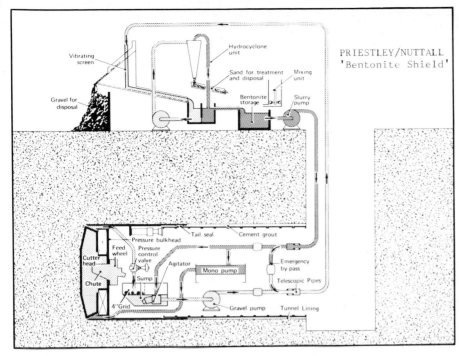

Figure 356. Bartlett bentonite process. (Courtesy National Research Development Corporation.)

separated the bentonite slurry from the excavated spoil. The slurry, plus whatever additives were required (i.e. additional bentonite, chemicals, etc.) was then redirected back into the pressurized chamber.

As this was an experimental drive and detailed inspections of both the face and the ground behind parts of the lining were envisaged, no provision was made for an access airlock in the bulkhead. When carrying out maintenance or inspections ahead of the bulkhead, it was therefore necessary to bring the entire tunnel up to the required pressure by means of an airlock situated at the tunnel entrance.

The machine was advanced by 16 rams capable of providing a total maximum thrust of 2023 t. However, during the experiment only approximately 600 t of thrust were required at any stage. Thrust reaction was taken by the newly erected cast-iron tunnel-lining rings which were lifted into position by a conventional erector-gear mechanism positioned at the tail end of the unit.

If desired, the machine could be converted for work as a normal mechanized shield for operation in cohesive ground by the removal of the feed wheel and the sump tank and the substitution of a mechanical conveyor system.

Teething Problems

The Tail Seal

As the machine was advanced it was found that slurry began to ooze into the tunnel around the tail of the shield. This slurry was forced into the annular space between the outside of the shield and the excavated tunnel wall by the pressure in the plenum chamber. It then worked its way around the end of the shield and back along the inside of the shield through the annular space created between the inside of the tail of the shield and the newly erected tunnel lining to emerge where the next lining section was due to be erected. This obviously proved most annoying and attempts were made to stem this undesirable flow of slurry. Several thick grouts were pumped into the offending spaces at the tail of the shield without great success and, even when lengthy research yielded a quick-setting grout which proved effective, two frustrating side effects soon became apparent. Firstly, the time consumed by the grouting process doubled the cycle time of the machine and this, of course, halved the penetration rate! Secondly, some of the injected grout followed the original slurry route in reverse and entered the plenum chamber where it promptly solidified, agglomerated, and gave the whole machine a rather nasty bout of constipation!

The final solution to the problem proved to be the design and installation of a nylon skirt or brush (similar to those used on the Japanese slurry machines) around the inside of the tail of the shield, to seal the cavity between the shield and the newly laid segments of tunnel lining. This skirt was attached to the inside of the shield and held firmly in contact with the outside of the tunnel lining by a pneumatically inflated continuous rubber tube which operated rather like a garter around a stocking. The whole seal was thus dragged along the lining as the tail of the machine advanced and remained in continuous contact keeping the offending slurry at bay.

The feedwheel

Another problem encountered during the early shake-down period was that, although material was effectively extracted from the pressure chamber by the feed wheel, this component had some inherent design faults which eventually led to the complete removal of the feed wheel and the temporary substitution of an hydraulically operated extractor mechanism. It was in effect a miniature airlock having two doors or 'plates' which opened alternately, thus allowing the spoil to enter and leave the chamber without unduly affecting the pressurized compartment. So far as the original feed wheel was concerned, the main problems were:

(a) that the seals at the blade ends and at the periphery were ineffective; this allowed material to work its way into the annular space surrounding the rotating feed wheel and the wheel tended to jam;

(b) that in addition to jamming, which occurred fairly frequently, the feed wheel components wore rapidly as a result of the abrasion to which they were exposed.

After modification the helical feed wheel was replaced and thereafter performed satisfactorily.

Another problem which became evident during the trials concerned the difficulty of maintaining accurate pressure control in the plenum chamber. This problem was mainly due to the faulty helical feed wheel allowing a feedback of air which was not satisfactorily controlled by the air-bleed valve. The modified feed wheel reduced the amount of air feedback and pressure control in the chamber improved considerably.

With the bentonite tunnelling machine, 144 m (473 ft) of tunnel were driven, and of this, 132.5 m (435 ft) required pressurized slurry. Periodically the entire tunnel was put under compressed air (0.05 MPa (7.3 lb/in^2)) in order that the face could be inspected. It was found that in most instances the bentonite slurry had penetrated and consolidated the tunnel face to a depth of about 500 mm (20 in).

A 2.82 m (7 ft 6 in) diameter bentonite tunnelling machine was used to drive a 1400 m (4600 ft) tunnel through fine sand in Warrington. This was the first commercial application of the bentonite process in the United Kingdom. During the initial drive the machine bored through 250 m (820 ft) of sandstone by using disc cutters in place of the drag picks and by the installation of a conveyor system in place of the feed wheel and pipes. When it became necessary to reconvert to the bentonite process, the conveyor was removed, the feed wheel refitted, and the disc cutters were again replaced by drag bits.

Wayss & Freytag Aktiengesellschaft[8]

The contracting firm of Wayss & Freytag was founded in Neustadt an der Weinstrasse in June 1875. Later the company moved to Frankfurt am Main.

During its formative years the company specialized in the construction of reinforced concrete buildings in Germany. In 1901 Emil Morsch (later appointed Professor at the Technical University of Stuttgart) was taken on by the firm as a director of their technical department. During this time he carried out numerous scientific studies on the theories of reinforced concrete. (These theories were later published, and are still considered to be valid today.)

As a result of Prof. Morsch's influence and his close association with Wayss & Freytag, the company became well known as contractors in the specialist field of prestressed concrete. The firm was awarded major contracts which included prestressed concrete buildings, cantilever bridges, climbing and sliding framework for industrial chimneys and towers, and, most significant, underground construction work, including compressed-air work and tunnelling.

In the mid-1970s the Wayss & Freytag piloted group of contracting firms with Dyckerhoff & Widmann and Hochtief was awarded the contract work for a 4.6 km (3 miles) long sewage collector tunnel for the new Hamburg collector system in Hamburg-Wilhelmsburg. Its proposed route ran under the harbour area through sand and gravel deposits lying beneath a layer of clay, peat, and back-fill material. The ground-water table in this area was some 16 m (52 ft) above the proposed tunnel floor. (This measurement excluded the influence of the tide.)

As the tunnel was to be driven close to the surface of the bed of some waterways the use of a conventional shield and compressed air was out of the question, in view of the risk of a blowout.

The Hydroshield

The machine (Figure 357) eventually chosen for the project was the Hydroshield bentonite machine which was designed by Wayss & Freytag. General feasibility studies for the Hydroshield were commenced in 1971 and by 1972 design specifications were ready in time for the bidding for the Hamburg-Wilhelmsburg sewer contract. The contract was won in 1973. A year later, i.e. February 1974, the first Hydroshield was assembled on site and ready to start work. Basically the unit was a 4.50 m (14 ft 9 in) cylindrical shield with an open spoke-type excavator wheel mounted on an inclined shaft. The excavator wheel was fitted with six radially disposed arms carrying fixed cutting tools.

The excavating compartment was separated from the remainder of the shield by a pressure bulkhead. This forward compartment was further divided into two sections by a transverse wall or diaphragm which extended from the crown of the shield to near the invert.

During operation a bentonite slurry filled the forward working compartment completely, while the rear chamber was allowed to fill to approximately half-way up.

The active pressure of the slurry was kept constant by a cushion of pressure-controlled air which acted on the free horizontal fluid level in the rear compartment. The slurry

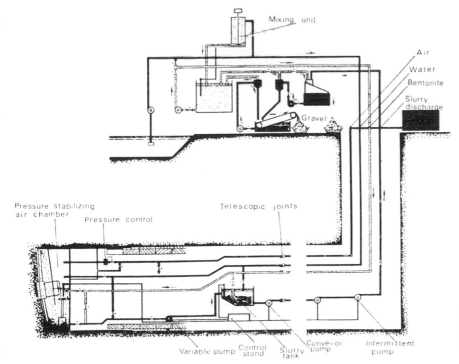

Figure 357. Wayss & Freytag 'Hydroshield' slurry system. (Courtesy Wayss & Freytag AG.)

401

Figure 358. Wayss & Freytag mechanized shield — used in Munich Underground rail system — installation of the cutting wheel. (Courtesy Wayss & Freytag AG.)

volume was regulated by a level-measuring device which acted on the speed of the feed pump. The feed or spoil removal pumps automatically cut out if either the upper or lower extreme positions of the level were reached. Despite the fact that the liquid level behind the diaphragm wall could alter, air bubbles were nevertheless prevented from passing into the front compartment by the diaphragm which extended well below the minimum liquid level.

After the material had been cut from the face it was pumped out of the pressurized compartment through a discharge pipe located near the bottom of the bulkhead. Stones or cobbles measuring more than 10 cm (4 in) diameter

were prevented from entering the spoil-removal pipes by an oblique grid placed over the outlet pipe. The excavated material was then pumped into a stone collecting box where an 8 cm (3 in) sieve acted as an additional screen, before the spoil was transported to the separation plant at the surface. This was accomplished in the normal manner by means of gravel pumps. The speed of the front gravel pump could be regulated and automatically cut out if the liquid level between the diaphragm wall and the bulkhead dropped below a certain height. Supply pipes were specially fitted with nozzle-shaped heads so that the spoil was prevented from settling on the floor of the chamber.

Behind the upper section of the main bulk-

Figure 359. Partly completed Munich Underground railway tunnel showing special lining designed by Wayss & Freytag. (Courtesy Wayss & Freytag AG.) *Note*: Construction of the Munich railway tunnels was commenced in 1965 and initially the lining consisted of reinforced concrete castings with a site-mixed inner skin. It was used for 11 km (7 miles) of tunnel, but in 1974 the double-shell lining system was discontinued and watertight single-shell reinforced concrete segments, developed by Wayss & Freytag, were used in the construction of two single-track tunnels. These tunnels, which have an outside diameter of 6.9 m (22 ft 8 in) lie beneath the river Isar at a depth of approximately 15 m (49 ft) and pass through mixed strata of marl-, silt-, and water-bearing sands. A complete tunnel lining ring consisted of eight segments and a keystone. The ring was 1 m (3 ft 3 in) in width. Joints between the segments were sealed with solid rubber sections which were bonded into a peripheral groove on the mating faces. In addition the contact faces of the concrete were given a plastic coating to equalize pressure during assembly and thus prevent cracking

head was a small airlock which gave access into the plenum chamber. The airlock could accommodate two workmen if necessary. The process of pumping the slurry out of the working chamber to allow workmen to enter it soon became a routine operation. At the surface the excavated material was separated from the bentonite slurry and disposed of. The recovered bentonite, together with additives, was then redirected into the pressurized excavating chamber.

The shield was advanced by means of 20 rams capable of exerting a maximum thrust of 1800 t. Another unusual feature of the Hydroshield was that the entire cutter head mechanism was capable of being extended or retracted from the face for a distance of 30 cm (1 ft).

At the time the Hydroshield was built, Wayss & Freytag felt that a reliable tail seal had not yet been developed. However, the shield was so designed that a tail seal could be fitted if and when a suitable one became available. According to the terms of the contract, therefore, compressed air was used in the tunnel.

The machine commenced work in 1974 and during the initial stages of its journey encountered a great many difficulties. The route penetrated a reinforced concrete pile foundation and a bomb crater full of refuse in the Ross canal, which had to be cut through. This

Figure 360. Tunnel lining being erected in the Munich Underground railway tunnels behind a Wayss & Freytag fully mechanized shield (Courtesy Wayss & Freytag AG.)

procedure was, of course, time-consuming. In addition, during the period when the machine had to negotiate the bomb crater, a total loss of bentonite was experienced and soil irrupted into the front compartment of the shield. These problems rendered adaptation of the control and feeding process difficult, and the ingenuity

of the tunnelling crew was severely taxed in its efforts to cope with these adverse conditions.

During the first eight months the shield was advanced at average rates of approximately 150 m/month (500 ft/month) (i.e. a total of 1200 m (4000 ft)). However when the difficult area had been safely traversed, average penetration rates rose to some 200 m/month (700 ft/month).

By June 1976 a 3560 m (12,000 ft) long, 4.34 m (14 ft 3 in) (o.d.) tunnel had been excavated. In the meantime notable progress had been made in the development of tail seals which had been installed. This allowed the contractors gradually to lower the pressure in the tunnel. It was then anticipated that the final 1000 m (3300 ft) of tunnel would be driven under normal atmospheric pressure.

Mitsubishi Heavy Industries

Mitsubishi Heavy Industries started business in 1857 as the Nagasaki Forge of Tokugawa Shogunate. Over the succeeding 77 years the name of the firm changed several times, until in 1934 it became Mitsubishi Heavy Industries Ltd. From 1934 to 1964 several other companies merged with Mitsubishi to produce a large variety of products. These companies traded as subsidiaries of Mitsubishi under various names but, in 1964, the company reformed as Mitsubishi Heavy Industries and on 1 June 1970, the parent company was reorganized so that the former Motor Vehicles Headquarters could become an independent company which now trades under the name of Mitsubishi Motors Corporation.

According to the company's shield machine production list, they built their first shield in 1939 for use in the Kammon tunnel by the Japanese National Railways. Their first confined plenum process shield, a partially pressurized shield, was built for research purposes in 1965. Later that year a similar shield was made for Hasama Gumi. It was used in a subway tunnel under Tokyo.

Their first slurry shield or mud-circulation machine as it is sometimes called by Japanese companies, was tested by the contracting company Kajima Kensetsu in February 1967. Subsequently, two more such units were built, the first for use by the contracting firm of

Figure 361. Front view of the 'Hydroshield' with cutting wheel. A thixotropic fluid is used to support the working face. (Courtesy Wayss & Freytag AG.)

Figure 362. Regeneration plant. (Courtesy Wayss & Freytag AG.)

Nishimatsu Kensetsu in a sewer tunnel and the second for use by the contracting firm of Kajima Kensetsu in a subway tunnel in Tokyo.

In September 1970 two 6.55 m (21 ft) long slurry machines were built by Mitsubishi Heavy Industries for Nishimatsu Construction Company. These machines were put to work in twin railway tunnels running beneath Tokyo bay. The tunnels were to be 853 m (3000 ft) long and the slurry machines had a cutting diameter of 7.29 m (23 ft 11 in).

Before the selection of the slurry machines the use of conventional shields and compressed air was considered, but speedily rejected in view of the close proximity of the tunnels to the bed of the bay. Under such adverse conditions the risk of a blowout was extremely high. Nor was it possible to grout ahead of the shields and thus prevent such an occurrence without jeopardizing the passage of shipping with the grouting equipment, which would need to be positioned on barges above the area being

405

treated. As another alternative, sunken tubes were also considered. However, as the laying of these tubes would necessitate the use of dredges and crane barges, this idea was also discarded because they would be hazardous to both shipping and aircraft. (Part of the tunnel route lay close to the northern end of Tokyo International Airport's main runway.)

Mitsubishi's slurry units consisted of a cylindrical shield with a forward compartment which was separated from the rest of the shield by a watertight bulkhead. The compartment ahead of the bulkhead was kept pressurized with water and the crew worked under normal air pressure behind it.

Within the pressurized compartment and lying immediately to the rear of the cutting wheel were a pair of 4 ft (1.2 m) diameter agitators equipped with rotary mixing blades. The cutting wheel itself consisted of a flat disc in which a series of 24 in (61 cm) wide radially positioned slits had been made.

The unit was forced forward by 27 powerful rams exerting a total thrust of 3920 t (4320 ton). These rams reacted against the newly erected tunnel lining.

As the machine advanced, the silt entered the pressurized compartment through the slits in the cutting wheel and was there turned into a pumpable slurry by the stirring action of the agitators. The slurry so formed was then pumped out through an 8 in (20 cm) diameter pipe to a treatment plant. In the treatment plant the spoil was separated from the water which was returned to the pressurized compartment while the spoil was disposed of. Fresh sea water was also added at this point to maintain the correct pressure in the chamber. The rate of flow of silt into the chamber through the slits in the cutting wheel was controlled by the water pressure in the chamber.

Urethane-foam packing and a rubber plate in the shape of an L, placed in the space between the tail-plate of the shield and the tunnel lining, prevented the water under pressure from leaving the pressurized chamber between the tunnel wall and the outside of the shield and so entering the tunnel at the tail of the shield. Average penetration rates of approximately 6.4 m (21 ft)/day in a 24-hour shift were attained by the tunnelling crews.

Tekken Kensetu Automatic Slurry Shield[9]

Driving a tunnel through strata containing gravel, clay, silt, and quicksand, lying well beneath the groundwater table, has long been considered one of the most difficult and hazardous operations in soft-ground tunnelling work. If, in addition, the surface area is cluttered with buildings, etc. the problem becomes even more aggravated as land settlement must, perforce, be kept to a minimum.

Such was the problem confronting the Tekken Construction Company of Tokyo when it undertook to drive the following five tunnels:

1. Horikiri sewer main No. 4. for the Sewerage Bureau of Tokyo Metropolitan Government.
2. Ayasegawa sewer main No. 2. for the Sewerage Bureau of Tokyo Metropolitan Government.
3. Water main in the Suginami Ward, Tokyo, for the Waterworks Bureau of Tokyo Metropolitan Government.
4. Continuous-flow-type sewage pipeline (Route No. 49, auxiliary highway) for Jiro Iwakami, Governor of Ibraki Prefecture.
5. Yotugi sewer branch work for the Sewerage Bureau of Tokyo Metropolitan Government.

The routes of the first two tunnels, i.e. the Horikiri sewer main tunnel and the Ayasegawa sewer main tunnel lay along the main south-north trunk road, east of the river Arakawa in the district of Katsushika-ku, Tokyo. All of the above conditions existed here, and the problem was compounded by features associated with the highway, namely, dense traffic, complicated underground installations, and even the crossing of an overhead railway. Lying below sea-level, the area consists of alluvial beds of sand some 30-40 m (100 to 130 ft) deep which, along the Horikiri route, is particularly fine-grained. Groundwater in this latter section, moreover, is only 1 m (39 in) below the surface and the area is so unstable that the slightest disturbance may result in the manifestation of quicksand.

The third project (i.e. a water main in the

Figure 363. Tekken Kensetu Company Limited automatic slurry tunnelling machine. Constructed by Mitsubishi Heavy Industries Limited, Akashi Works, in 1973. (Courtesy Tekken Kensetu Company Ltd.)

Figure 364. Tekken shield — support system — hydraulic transportation of muck. (Courtesy Tekken Kensetu Company Limited.)

Suginami Ward) was to be driven through the Upper Tokyo gravel formation which lies in the plains of Musashino. Although normally stable, this ground, which lies beneath an 8 m (26 ft) layer of Kanto loam, consists mainly of gravel intercalated with clay. It has a high groundwater level. Because of this, the excavation of a channel through the gravel/clay strata is likely to cause a flash flood of underground water leading to surface subsidence. The last two projects run through similarly difficult terrain.

As was usual, most conventional methods of soft-ground tunnelling, i.e. shield and compressed air, grouting, cut and cover, etc. were considered and rejected for various reasons which included risk of a blowout, cost, time factors, the disruption of traffic, and the possibility of land settlement. Eventually it was decided that a slurry machine would be used. The prototype unit was designed by the Tekken Construction Company Ltd. and was built by Mitsubishi Heavy Industries Ltd. in 1973 at their Akashi works factory.

Figure 365. Tekken shield — surface control panel. (Courtesy Tekken Kensetu Company Limited.)

Figure 366. Tekken shield — slury disposal plant. (Courtesy Tekken Kensetu Company Limited.)

While slurry machines were no novelty, design engineers of the Tekken Kensetu Company incorporated into their prototype machine several significant features which were then unique and which brought the dream of remote-control automatic tunnelling nearer to reality.

The Tekken unit (Figure 363) was similar in basic design to the slurry machines used by the Nishimatsu Construction Company under Tokyo bay. It consisted of a cylindrical shield with an outside diameter of 5.05 m (16 ft 7 in). A full-face cutter head or wheel with a conical centre was dressed with twin-action drag picks disposed along a pair of diameters set at right angles across its face. Slits in the cutter-head

wheel beside the cutting tools allowed muck to enter the pressurized slurry chamber. The cutter head was capable of being rotated in either direction. Four access openings were available if necessary. The access openings which were situated in the spaces between the rows of cutting tools were normally kept closed by means of a plate which fitted across the opening and which was bolted to the cutting wheel. The plenum chamber was sealed fom the main area of the shield by a watertight steel bulkhead positioned about 56 cm (1 ft 10 in) behind the cutting wheel.

An incoming 20 cm (8 in) diameter pipeline situated near the top of the bulkhead injected a slurry mixture of silt, clay, and water into the plenum chamber. This served as a vehicle to carry the cuttings which were stirred into the slurry by two agitators placed near the bottom of the plenum chamber. The muck-laden slurry was then discharged through a pipe at the bottom of the bulkhead and transported to the surface by gravel pumps. The machine was kept true to course by means of a gyrocompass and laser guidance equipment.

The heading was supported by slurry which was maintained at a pressure slightly greater than that exerted at the face by the groundwater. Supplementary support was also provided by the rotating cutting wheel which during operation was forced against the face. Slurry pressure was controlled by an automatic pressure-control valve installed on the inlet pipe.

Where possible, the naturally occurring clay in the excavated mud was used to form the slurry but, occasionally, when sandy ground devoid of this clay was encountered, additives such as CMC,* polymer, etc. were introduced.

On some projects, due to the limited amount of space available at the treatment site, the size of the treatment plant was somewhat smaller than desired. As a result, the muck-laden slurry took longer than usual to be processed and the machine had to be stopped periodically.

Under normal working conditions the slurry machine advanced 1 m (39 in) during each cycle and approximately 15 to 20 minutes were required to carry out this amount of

---

*Na-carboxyl methyl cellulose.

408

excavation. Segment erection took 20 to 30 minutes and grouting, extension to rail for transporting slurry feed and discharge pipes and segments, another 10 to 20 minutes. All in all it took approximately one hour to complete a cycle of work. However, two hours were required for the treatment, separation, and disposal of the muck by the plant. Thus the ultimate driving speed was necessarily determined by the treatment plant's ability to handle the muck.

During this enforced period of idleness when the wheel was no longer providing supplementary support for the face, the face was maintained only by the pressure of the slurry in the excavating chamber. This meant that a slurry leakage, say through the tail section, could cause a drop in pressure of the slurry in the front chamber, thus inviting a collapse of the face. To avoid such an occurrence, Tekken installed a valve which automatically adjusted the slurry pressure in the chamber during this critical period.

Leakage into the tunnel through the tail end or through joints between tunnel-lining segments was effectively controlled, in the first instance by a special tail-seal packing consisting of an L-shaped wire brush made from piano wire and in the second, by a polyurethane-type sealing tape which was applied to one side of each segment, while a butyl-type sealing tape was applied to the other side. When these two types of tape were brought into contact with each other a good watertight seal was formed.

After leaving the pressure chamber, the slurry muck was first passed through a trommel or gravel eliminator which effectively eliminated all the larger pieces of gravel or stone, allowing only particles measuring some 40 to 50 mm (1½ to 2 in) or less to pass through into the pumping system. These smaller pieces, together with the slurry mixture, were transported to the surface by gravel pumps while the larger pieces of gravel and stone were collected in the trommel until it was filled. They were then removed and deposited beneath the sleepers where they acted as a temporary ballast support for the sleepers. The installation of the trommel thus made possible the excavation of strata containing gravel up to a diameter of 200 mm (8 in).

A major problem connected with soft-ground tunnelling was the risk of over-excavation of the face area (i.e. when a run of ground occurred and more spoil was removed from the face than was warranted by the machine's rate of advance).

As the operator was virtually travelling 'blind' since the face was out of view in front of the bulkhead and cutting wheel, there was an even greater risk of such an occurrence going unnoticed until considerable damage had been done by ground subsidence when using this type of machine. To prevent this, Tekken installed measuring devices on either side of the slurry-feed-and-discharge pipelines. These devices calculated the net volume of slurry remaining, after the volume of slurry which was fed into the pressurized chamber was subtracted from the volume of excavated spoil and slurry which was being discharged per metre of advance. This information which was indicated digitally was monitored in the central control room. In order to obviate human error, the central control system was designed to operate automatically. Only one operator was needed in the shield to advance the unit. All instructions were phoned to him from a central control room at the surface. In this room the slurry pressure at the face and the mucking and discharge systems were effectively monitored and controlled.

At the surface the muck was separated from the slurry and disposed of after a dehydration and compression process had converted it into cakes. The slurry, together with whatever additives were required, was directed back to the pressurized face.

When work commenced it was necessary for the slurry mixture to bypass the pressure compartment until the controller was satisfied that all pumps and pipes were functioning satisfactorily. Then the bypass valve which operated to cut the pressure chamber out of the system was closed and the bulkhead valve was opened. At the same time the shield operator was instructed by control to activate the cutting wheel.

Back-fill grouting, particularly in non-viscous mud strata, is extremely important. If grouting around the tunnel lining should be neglected, or carelessly executed, after the passage of the shield, there is a serious risk that the ground may subside, or indeed, the tunnel itself may subside and become badly distorted.

During mid-1977 a new Tekken-designed slurry shield was due to commence operations. In an attempt to minimize this risk, Tekken made arrangements whereby the back-fill grouting in the new unit would be controlled from the central control room. If this method of grouting control proved effective, ground subsidence should always, according to Tekken, be reduced to a mere 10 mm (0.4 in).

In addition, it was planned that an automatic surveying system which would also be monitored in the central control room would be installed on the new 7 m diameter machine. Another modification proposed by the Tekken Company at that time was a circumferentially supported cutter wheel. In previous models the cutting wheel had been mounted on a central shaft.

Summary

On looking back at the various descriptions of machines from the first pressurized bulkhead units used in France and Germany to the latest slurry machines now operating in Japan, the developmental trends are fairly clear to see, as are the different methods used by various companies to maintain a constant pressure at the face.

Pressure in the digging compartment of the Markham universal soft-ground tunnelling machine was maintained by a pressure-actuated bypass valve through which some of the separated slurry was redirected back to the face.

Both the Mitsubishi machines used under Tokyo bay and the Tekken Kensetu automatic units maintained the pressure in the plenum chamber through control of the inlet valves, although in the case of the Tekken Kensetu machine this process was further refined by the provision of a valve which automatically adjusted the slurry pressure in the chamber and which could be monitored in the control room.

Pressure in the forward compartment of the British bentonite machine was controlled by the pressure-control valve situated immediately behind the bulkhead. This device regulated the amount of slurry *leaving the plenum chamber*.

The German Hydroshield, on the other hand, controlled the slurry pressure in the digging compartment by means of the cushion of air which was pumped into the chamber behind the transverse diaphragm.

One difference between the Markham T.B.M.s and the Priestley/Nuttal bentonite machine was that whereas the universal soft-ground T.B.M. used in Mexico utilized the naturally occurring thixotropic clays found in the tunnel to form a stabilizing slurry mixture, the bentonite machine was of particular value in sandy or gravelly ground devoid of such clays. While no provision was made in the universal units for the introduction of bentonite, the author was advised by Markham that, if necessary, bentonite could have been introduced at the face.

The Japanese, for the most part, use natural clays with additives and utilize very sophisticated surface treatment plants.

O & K (Orenstein & Koppel)/ Holzmann Thixshield[10] — Boom Shield Slurry-type Machine

The slurry machines so far described in this section have been either rotary-wheel or oscillator-head-type machines.

In 1978 a boom shield slurry-type machine was built by O & K (Orenstein & Koppel AG) of West Germany. It was developed on a patented process owned by the West German contractor, Philipp Holzmann AG.

The machine was tested in practice in 1978 during the driving of a 4.2 m (13 ft 9 in) diameter tunnel beneath the canal in the port of Hamburg.

As may be seen from the drawing and photographs (Figures 367-369) of this machine, the unit consists basically of an extensible cutting boom centrally mounted in a circular face-plate or bulkhead.

During operation the space ahead of the bulkhead is filled with a bentonite slurry under pressure, which counteracts the pevailing ground pressure, thus supporting the working face. The area behind the bulkhead is maintained at normal air pressure. The slurry control system is a combination of centrifugal pumps and air cushion mufflers which operate

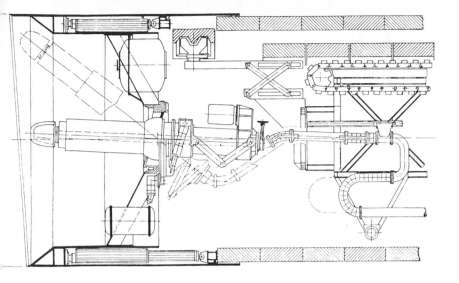

Figure 367. Drawing of O & K/Holzmann Thixshield. (Courtesy O & K Orenstein & Koppel AG.)

Figure 368. Front view of O & K/Holzmann Thixshield. (Courtesy O & K Orenstein & Koppel AG.)

Figure 369. O & K/Holzmann Thixshield on trials in the Lübeck Works. (Courtesy O & K Orenstein & Koppel AG.)

before the slurry reaches the face. As the ground ahead of the unit is cut and loosened by the rotary cutter head mounted on the end of the telescopic boom, it is drawn through the hollow drive shaft of the boom into a suction line connected to the base of the drive shaft. The suction line connection with the drive shaft is articulated to allow the telescopic cutter boom full freedom of movement.

The electro-hydraulically driven drive shaft is mounted in the bulkhead in a ball and socket joint which allows it to cover the working face in level concentric passes. Automatically controlled computer monitoring may be carried out via a central operator console with indicator lamps. The slurry is pumped to the tunnel outlet where the bentonite mixture is separated from the cuttings and recirculated. With a tunnel diameter of 4.2 m (13 ft 9 in) the machine discharges approximately 250 m³/hour (8800 ft³/hour) of slurry and during the driving of the Hamburg tunnel a maximum driving speed of 3 m/hour (10 ft/hour) was achieved.

References

1. *The Encyclopedia America,* International edn., Vol. 3, p. 564; Vol. 19, p. 416.
2. *Kirk-Othmer Encyclopedia of Chemical Technology,* Wiley-Interscience, New York, Vol. 6, 1979, p. 190.
3. Unusual tunnelling machine cuts and lines storm sewer, *Construction Methods and Equipment,* March, 1960, 163-167, New York.
4. *Encyclopaedia Britannica,* Vol. 15, 1929, p. 399.
5. A. Armstrong, Markham tunnelling equipment from 1887-1968, *Broad Oaks Mag.,* Spring 1968.
6. J. V. Bartlett, A. R. Biggart and R. L. Triggs, The bentonite tunnelling machine, Paper 7670, *Proc. British Tunnelling Society,* February 1974. London.
7. J. V. Bartlett. Soft-ground tunnelling system proved practicable, *S.A. Mining & Eng. Journal,* Aug. 1973, 15, Johannesburg.
8. Dipl. Ing. Erich Jacob, *The Bentonite Shield,* Wayss & Freytag AG, Main, West Germany.
9. *Tekken Slurry Mole Automatic Control System,* Tekken Kensetu Co. Ltd., Tokyo, Japan.
10. *O & K News,* **3,** No. 2, 1979, Orenstein & Koppel, Dortmund, West Germany.

Correspondence

Robert L. Priestley Ltd., Gravesend, U.K.
Mitsubishi Heavy Industries, Tokyo, Japan.
Markham & Co. Ltd., Chesterfield, U.K.
Edmund Nuttall Ltd., London, U.K.
Wayss & Freytag AG, Main, West Germany.
Tekken Kensetu Co. Ltd., Tokyo, Japan.
Gardner Eng. Corp. (Gardner/B/H/Constructors), Houston, U.S.A.
O & K (Orenstein & Koppel), Dortmund, West Germany.

Patents

J. H. Greathead — 1,738/1874 U.K.
Edmund Nuttall Ltd — 934,966/1962 (Priestley T.B.M.) U.K.
J. V. Bartlett — 1,083,322/1964 (Bentonite Tunnelling Process) U.K.

FOURTEEN

Earth Pressure Balanced Machines

Another type of machine which has recently been introduced in Japan is the earth balancing machine. These units, of which there are two types, are a further development on the lines of the Japanese slurry machine.

The original concept was pioneered by the Sato Kogyo Company Limited, a Japanese construction firm, who were seeking a method of tunnelling through soft and running ground below the water table.[1]

Although numerous tunnelling projects with shields and compressed air or with slurry machines had already been successfully accomplished in Japan, there were disadvantages and limitations associated with these methods which the Sato Kogyo Company sought to eliminate. In particular, they were anxious to find a machine which would tunnel efficiently yet comply with the environmental regulations and laws in force in many of the major cities of Japan. These included air and water pollution control laws, industrial water, waste disposal and public cleaning laws, and prevention of oxygen deficiency and prevention of compressed air hazard ordinances, etc.

Developmental work on the earth pressure balanced shield was begun by the Sato Kogyo Company in 1963 and, after considerable research both in the laboratory and in the field, a unit was finally built by the Ishikawajima-Harima Heavy Industries Company Ltd (I.H.I.) in 1966.

Earth Pressure Balanced Shield and Earth Pressure Balanced Shield — Water Pressure Type[1, 2]

By 1977 the systematic development of the earth pressure balanced shield — water pressure type had also been completed and the machine was ready for practical use.

Depending upon the expected ground conditions, either of the above units, (Figure 370) i.e. the earth pressure type or the water pressure type may be selected.

The machines themselves are basically similar in design. Externally the shields resemble the slurry unit in that the cutter head consists of a rotating disc fitted with drag teeth. The drag teeth are positioned along both edges of five radially arranged arms. Slits on either side of each cutter arm allow the material excavated from the face to enter the cutter frame, a drum-like chamber behind the cutter head. The material excavated from the face is collected and compressed in this drum chamber. This material in turn forms a plug which acts to support the face and prevent the ingress of groundwater.

By means of an auger or screw conveyor the material in the drum is then moved upwards and passed through the bulkhead and through a hydraulically operated sliding gate at the rear end of the screw conveyor. When the gate is closed, the face and drum compartment are completely sealed from the rear of the shield and the tunnel.

In order to maintain a constant pressure at the face the rotating cutter frame and screw conveyor are kept constantly filled with earth. The earth and water pressure at the cuttng face is constantly monitored and cutter-head torque, rotation speed, screw conveyor torque, and driving speed, as well as the opening of the gate jack are regulated to maintain cutter operation stability, despite any changes which might occur in the ground conditions.

The earth pressure type shield is

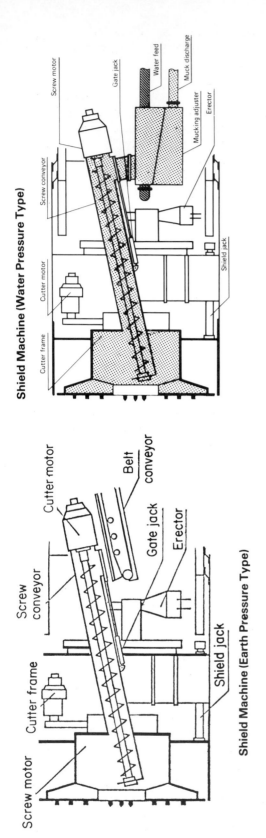

Shield Machine (Water Pressure Type)

Screw motor

Gate jack

Water feed

Muck discharge

Mucking adjuster

Erector

Screw conveyor

Cutter motor

Cutter frame

Shield jack

Cutter motor

Belt conveyor

Screw conveyor

Gate jack

Erector

Cutter frame

Shield jack

Screw motor

Shield Machine (Earth Pressure Type)

Earth Pressure Balance Shield Method (Schematic)

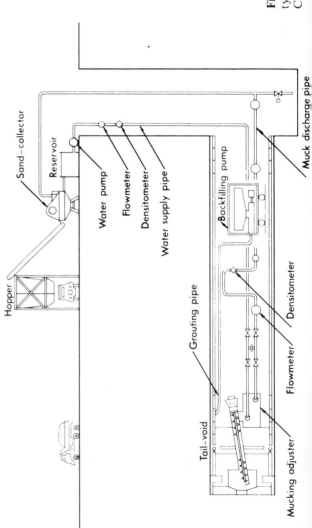

Sand-collector

Reservoir

Hopper

Water pump

Flowmeter

Densitometer

Water supply pipe

Backfilling pump

Muck discharge pipe

Grouting pipe

Tail-void

Mucking adjuster

Flowmeter

Densitometer

Figure 370. Sato Kogyo earth and water pressure type shield machines. (Courtesy Sato Kogyo Company Limited.)

414

recommended for silt and clay soils which, when compressed in the drum and screw conveyor, will form an impermeable barrier, capable of supporting the face and holding back the groundwater. When highly permeable ground consisting mainly of sand or gravel strata is encountered the Sato Kogyo Company suggest that the water pressure type shield be used. This unit, although similar in other respects to the earth pressure type shield, has in addition, a special 'mucking adjuster'. It permits the supply of pressurized bentonite-free water which acts to support the face and prevent a run of water through the cutter frame and screw conveyor. The mucking adjuster is attached to the gate-end of the screw conveyor. A device in the mucking adjuster separates the cobbles from the fine material.

From the mucking adjuster the cuttings are pumped as a slurry to the surface where the water is separated from the spoil and recycled. (Muck from the earth pressure shield is carried by belt conveyor from the screw conveyor gate and is then deposited into muck cars or skips which may be lifted to the surface through an access shaft.)

Because no bentonite slurry is used for the water pressure type unit, the necessity for an expensive slurry treatment plant is obviated. In addition, as there is no need to treat the spoil chemically, this material is relatively environmentally clean.

An important contribution towards the success of these units has been the development of an improved tail seal, a sandwich-type packing consisting of a rubber seal reinforced by steel cords. Steel plates on either side of the packing act as additional strengtheners for the seal. Another contribution was a new type of grouting pump. This latter device allows pressured back-filling material to be forced into the void between the tunnel lining and the surrounding ground. This void is created by the forward movement of the shield as it advances. The amount of grouting inserted at the tail-end of the shield is carefully monitored and regulated to ensure that this void is kept to a minimum. Another avantage associated with this type of unit is that the compressed material at the cutter face prevents the ingress of grouting into the front compartment, as was

frequently the case with conventional-type slurry machines (see section on slurry machines).

Development of both the earth pressure and the water pressure type units by the Sato Kogyo Company has been in conjunction with I.H.I. who have been responsible for the manufacture of these machines. According to Sato Kogyo, who have used this system at several job sites in Japan, the method is particularly suited for work in sandy, water-bearing strata or permeable earth with a high groundwater table.

Confined Soil Shield

Similar machines are also presently being manufactured by Mitsubishi Heavy Industries and several other Japanese companies. The Mitsubishi unit (Figure 371) is known as a 'confined soil shield'. This unit and most of those made by the other manufacturers operate in much the same manner as does the earth pressure shield developed by the Sato Kogyo Company and I.H.I. However, according to the Sato Kogyo Company, the majority of these units do not have facilities for the installation of a mucking adjuster or slurry transportation equipment such as are provided with the water pressure type shield.

Earth and Mud Pressure Balanced Shields[3]

A further development of the earth balancing concept was initiated by the Daiho Construction Company Limited.

Generally, earth balancing (Figure 371a) and pressure slurry shields are fitted with a cutter-head plate which acts as a direct support for the face. Slits in the plate allow for the passage of the excavated material. The Daiho earth and mud pressure balanced shield (Figures 372 and 372a), however, does not have a cutter-head face plate. The cutter head features four 'cutter wings' or arms which extend radially from the axially mounted hub. Apart from the cutter teeth, paddles or mixing blades extend rearwards from the cutter wings towards a bulkhead. The bulkhead forms the rear wall of

415

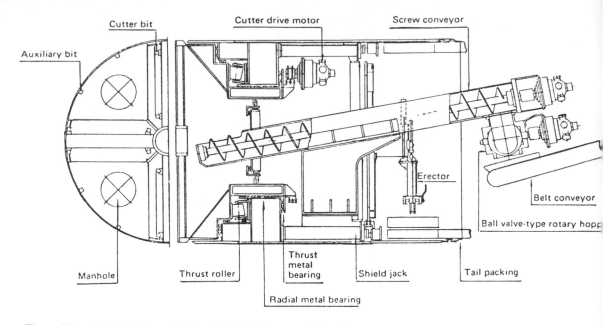

Figure 371. Mitsubishi confined soil shield. (Courtesy Mitsubishi Heavy Industries Limited.)

Figure 371a. Daiho earth balancing shield. (Courtesy Daiho Construction Company Limited.)

a 'kneading chamber' which confines the excavated soil at the face. Into this chamber mud is introduced. The mud is mixed with the sand and gravel excavated from the face by means of the paddle blades attached to the rear of the cutter head. This muddy mixture, which is pressurized, acts as a non-permeable support to hold back the face and prevent the ingress of ground water.

The material is then removed through a sealed screw conveyor in much the same manner as that utilized by the earth balancing machine.

This type of unit is considered to be particulary effective in sandy or gravel soil devoid of natural mud and thus very permeable. As conventional trucks or skips are used to remove the spoil the provision of expensive pumping and treatment plant equipment is unnecessary.

An earth and mud pressure balanced shield was used by the Daiho Construction Company in 1976 to drive a 2.44 m (8 ft)-diameter, 165 m (540 ft)-long tunnel in Aoto, Katsushika-ku, Tokyo, Japan, and the successful accomplishment of this project has encouraged further orders for this type of unit. Applications for patent rights have recently been filed in the major countries of the world.*

---

*In an article entitled 'The Japanese experience',[4] I. Kitamura (Secretary of the Japanese Tunnelling Association), reported that up to July 1978 Japanese manufacturers had among them produced a total of 160 slurry machines and some 79 earth balanced units. Of the latter type machine I.H.I. had produced 48, Hitachi 27, Kawasaki 3, and Komatsu 1.

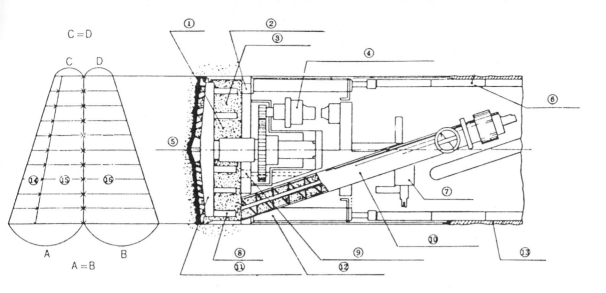

Figure 372. Daiho earth and mud pressure balanced shield — diagram. (1) Muddy earth; (2) ground pressure gauge for measurement of muddy earth pressure; (3) ground face mud chamber; (4) hydraulic motor for rotating cutter; (5) natural ground; (6) segment; (7) erector; (8) mixing blade; (9) valve for pouring hole of muddy earth making agent; (10) screw conveyor; (11) cutter wing and tooth; (12) shield jack; (13) back filling; (14) water pressure; (15) stationary ground pressure; (16) muddy ground pressure. (Courtesy Daiho Construction Company Limited.)

Figure 372a. View of cutterhead and 'kneading chamber' of Daiho earth and mud pressure balanced shield. (Courtesy Daiho Construction Company, Limited.)

References

1. K. Ishihara, *Earth Pressure Balanced Shield Tunnelling Method – Water Pressure Type,* Sato Kogyo Company Ltd., Tokyo, Japan.
2. T. Abe, Y. Sugimoto, K. Ishihara, *Development and Application of Environmentally Acceptable New Soft Ground Method,* Tunnel Symposium, 1978, Japan Tunnelling Association, Japan Society of Civil Engineers and International Tunnelling Association, Tokyo, Japan.
3. *The D.K. (Drücken-Kneten Böden) Shield driving method. (Mud Pressurized Shield),* Japan Tunnelling Association, Tokyo, Japan.
4. I. Kitamura (Secretary, Japanese Tunnelling Association), The Japanese experience, *New Civil Engineer Tunnelling Supplement – Facing up to the future,* March 1979, London.

Correspondence

Sato Kogyo Construction Company Ltd., Tokyo, Japan.
Japan Tunnelling Association, Tokyo, Japan.
Daiho Construction Co., Ltd., Tokyo, Japan.
Mitsubishi Heavy Industries Ltd., Tokyo, Japan.

FIFTEEN

Mini-Shields and Mini Mechanized Shields

Mini Tunnels International Limited[1]

As a result of design and development work carried out by M. A. Richardson of the Rees Group, Mini Tunnels International Limited produced an integrated mini tunnel system for soft ground in 1973 (Figures 373 and 374). It was designed specifically for sewer and pipeline construction in the 1000 mm to 1300 mm (3 ft 3 in to 4 ft 3 in) diameter range.

The mini tunnel is constructed of three identical unreinforced concrete segments per tunnel ring, each completed ring being 600 mm (1 ft 11½ in) long. Each segment ring is built within the rear section of a mini tunnelling shield. The cylindrical steel shield is advanced by six 12 t hydraulic rams mounted within the shield in a mobile rig, consisting of front and rear thrust rings. The rear ring bears against and pushes off the last-erected concrete segment ring. The front thrust ring incorporates spring-loaded catches which engage with corresponding pawls within the inside face of the shield, transmitting the thrust loads from the segments into the shield.

The mobile rig may be moved to the front of the shield during the erection stage by releasing the spring-loaded catches and to the rear when the shield is to be advanced. This particular arrangement was necessary because of the very limited working space available for the miner. Plough slots are provided near the cutting edge to assist in controlling shield roll and attitude, and steering is effected by a hydraulic differential-ram pressure system. A shield hood, face tray, and full-face diaphragm which may be bolted on to meet differing ground conditions are also supplied by the manufacturers.

A special rail track with sleepers shaped to conform to the tunnel invert is pulled by the shield as it progresses. Muck is moved out of the tunnel along this track. The track also acts to protect the tunnel invert during construction. Additional track lengths are latched on at the drive shaft as required. The track also incorporates a steel air main and water-drainage pipes for the shield.

After the erection of the segment rings an overbreak cut by the shield amounting to approximately 6 per cent of the tunnel bore is pneumatically filled with small-grade gravel through preformed blind grout holes in the concrete segments. The gravel is fed from a pressurized hopper and injector at the drive shaft and transported to the shield in another steel pipe accommodated in the rail track.

In 1975 the Queen's Award to Industry was conferred upon Mini Tunnels International in recognition of their outstanding achievement in technological innovation in a small-diameter shield-driven tunnelling system. In addition the company have been awarded the Design Council Award (1975), the Export Award for Small Manufacturers (1974-75) and the *Engineering News-Record* Award, U.S.A. (1973) for design, export performance, and service to the construction industry, respectively.

Compressed air was recently used for a 130 m (426 ft) long, 1.0 m (3 ft 3 in) i.d. mini tunnel which was driven beneath a housing development at Burnham-on-Sea[1]. To achieve this, a full-face diaphragm was specially designed for the work. The diaphragm was fitted to the forward end of the mini shield.

The front end of the pyramid-shaped diaphragm was pushed forward ahead of the

418

Figure 373. Mini tunnelling shield — used for driving small-diameter tunnels ranging in size from 1 to 1.3 m (39 to 52 in) in soft ground. (Courtesy Mini Tunnels International Limited.)

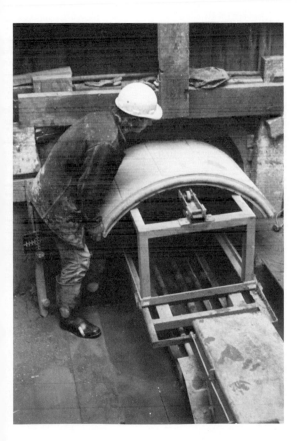

Figure 374. Mini Mule — a self-contained electric locomotive used in conjunction with the mini tunnelling shield for the transport of lining segments to the tunnel face and for muck removal. (Courtesy Mini Tunnels International Limited.)

shield by hydraulic thrust jacks which reacted against the recently installed tunnel-lining segments.

As the apex of the pyramid was thrust forward through the ground, the soil was forced down along the sides of the pyramid and into openings in the diaphragm at the base of the pyramid. These openings (four in number) were protected by hydraulically controlled sliding doors which allowed the material to squeeze through into the mini tunnel where it was hand-spaded into a waiting skip.

This system of tunnelling is restricted to use in soil conditions where the soil resistance does not exceed the thrust capacity of the jacks. In addition, of course, the thrust capacity of the jacks cannot be increased beyond the point where the precast lining segments already erected can provide adequate reaction to the thrust forces.

The author has been advised by T.V. Manlow, Director of Mini Tunnels International Limited that a 'fully mechanized' mini tunnel system has also been developed.

Machinoexport, U.S.S.R.

At about the same time as the mini tunnelling system was being introduced, Russian engineers designed and developed a mini mechanized shield with a rotating cutter head for use in soft

419

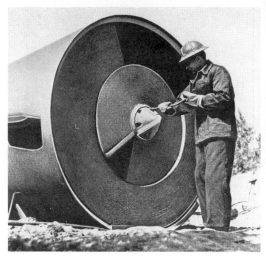

Figure 375. Mechanized small-diameter soft ground tunnelling shield made in Russia. Tunnel size on completion 1.8 m (5 ft 11 in) complete with reinforced concrete lining. Outer diameter of shield 2.1 m (6 ft 11 in). May be used at minimum tunnel depths of 6 m (19 ft 8 in) or less where special circumstances warrant it. (Courtesy Machinoexport U.S.S.R.)

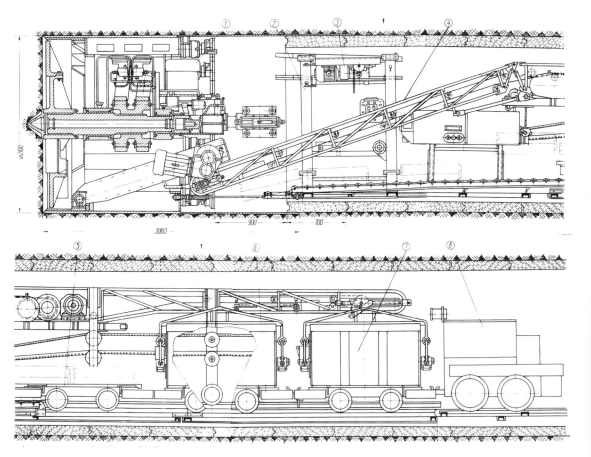

Figure 376. Small-diameter mechanized tunnelling shield schematic diagram. (1) Shield; (2) process platform; (3) winch; (4) belt conveyor; (5) base plate for ring sections; (6) wheeled truck; (7) bucket; (8) electric locomotive. (Courtesy Machinoexport U.S.S.R.)

420

ground. The mechanized shield (Figure 375 and 376) was similar in concept to conventional mechanized shields except that the cutter head of the unit was only 2.1 m (6 ft 10 in) in diameter.

The cutting plates rotated about a central shaft which terminated in a cone at its forward end. The cone was fitted with cutting blades. The cone extended towards the face, slightly in advance of the cutter plates. The head consisted of an inner cutting plate and an outer annular plate, both apparently revolving in the same direction. These cutter plates formed an auger-shaped head which fully supported the face. As the head revolved, a cutting knife on each of the plates sliced off a section of the face and the material was then forced through an aperture behind the cutting edge where it dropped on to a conveyor belt which travelled beneath the operational section towards a second conveyor, on to which the spoil was dropped. The second conveyor carried the muck upwards to the rear of the machine and deposited it into rail trucks.

References

1. Charles McCaul and Graham West (Transport and Road Research Lab. Berkshire) and Terance V. Manlow (Mini Tunnels International Ltd.),Driving with a full-face diaphragm, *Tunnels and Tunnelling,* **6,** 23-25, July 1978, London.

Correspondence

Mini Tunnels International Limited, Surrey, U.K.
Machinoexport, Moscow, U.S.S.R.

SIXTEEN

Pipe Jacking

Early History[1-3]

According to M. W. Loving[1] and Howard F. Peckworth,[2] pipe jacking with cast-iron pipes was pioneered by the Northern Pacific Railroad Company between 1896 and 1900. Cast-iron pipes were jacked under the Great Western Railroad tracks at Ingalton, Illinois, in 1911 and also under the Southern Pacific Railway tracks in California in 1915. By the 1930s Northern Pacific had standardized the use of 107 to 183 cm (42 to 72 in) diameter reinforced concrete culvert pipes for jacking purposes. Pipes loaded with ammonal were also jacked through into the enemy's tunnels by the Allies during the First World War (see Chapter 17 — 'Military use of the tunnelling machine').

From its humble beginnings in America, pipe jacking has developed into a highly skilled art, now used extensively in the U.K., Germany, Japan, and other parts of the world — its success largely depending upon a combination of good equipment and the skill and experience of the contractor and operating crew.

Simply, the system (Figures 377 to 379) involves the pushing or thrusting of a specially fabricated cylindrical steel shield through the ground ahead of pipes. Concrete pipes are normally used unless an access tube or 'carrier pipe' is required to carry a number of smaller diameter pipes for gas, water, electricity, etc. when steel sleeves are utilized instead of the concrete pipes.

Thrusting is accomplished by special high-pressure hydraulic jacks which react against a thrust wall of concrete or timber built at the rear of a thrust pit or shaft excavated at the commencement of a thrust section. As the pipes move forward through the soil behind the shield, new pipes are added to the end of the string. The new pipe is inserted between the jacks and the end pipe when the jacks have been retracted in readiness for their next forward stroke. The forward end of the shield is sometimes hooded for added protection at the face.

The first concrete pipe immediately behind the shield is called the lead pipe. It is specially constructed with a rebated front to enable it to slip into the trailing end of the shield. The outside diameters of the shield and the pipes are the same. This minimizes any tendency towards overbreaking.

A steel thrust ring placed against the end of the concrete pipe between the rear pipe and the jacks, distributes the jacking loads evenly so that the concrete pipes are not damaged. In addition to the thrust ring, a steel pressure plate or plates may sometimes be placed between the rear end of the jacks and the thrust wall to assist in spreading the reaction loads. The number of jacks used depends upon the diameter of the pipes being installed.

Matched sets of 'spacer blocks' are also sometimes used between the thrust jacks and the thrust ring in order to increase the reach of the jacks and so enable the operator to install a full pipeline length. Additional jacks are incorporated within the shield to steer and thus control line and grade.

As the art of pipe jacking has developed, so have manufacturers of concrete pipes met this growing need by producing specially made pipes which are centrifugally spun and incorporate concentric reinforcement cages. The pipes are made with flexible joints and both the joints and the pipes themselves are designed to resist the large forces imposed on them during installation.

Before a job begins, the contractor assesses

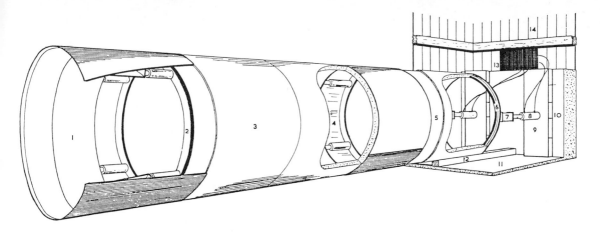

Figure 377. General arrangement of pipe-jacking equipment: (1) jacking shield; (2) steel thrust ring; (3) concrete lead pipe; (4) interjack station; (5) interjack pipe; (6) steel thrust ring; (7) spacer blocks; (8) hydraulic jacks; (9) steel pressure plate; (10) concrete thrust wall; (11) concrete floor; (12) guide rails; (13) power pack; (14) thrust pit. (From *Bulletin No. 1,* Courtesy Pipe Jacking Association, London.)

Figure 378. Westfalia Lünen pipe-jacking equipment operating in main jacking station. (Courtesy Westfalia Lünen.)

the number of intermediate jacking stations which will be needed to complete the work. The intermediate jacking station consists of a rolled steel shield the same overall diameter as the

Figure 379. Westfalia Lünen intermediate jacking unit in operation during the construction of a 2 m (6 ft 6 in) i.d. sewer at Bad Godesberg. (Courtesy Westfalia Lünen.)

pipes. Hydraulic jacks inside the intermediate station thrust against distribution pads or a thrust ring which is inserted between the jacks and the pipes. The station moves forward with the pipes in the normal way until jacking loads at the rear indicate that it is time to use the intermediate station. Forward thrust is then effected from the intermediate jacks. At the end of the thrust stroke of the intermediate

423

jacks, the pressure in the jack cylinders is relieved and the main jacks at the thrust wall push the rear pipes forward so that additional pipes may be added to the line.

'Inter jack pipes' are inserted on either side of the intermediate jacking station (Figure 379). These pipes are specially designed with a rebate on one end to allow them to fit inside the ends of the intermediate jack station shield.

Excavated material from the face is carried to the rear through the pipes by skips, etc. Mining at the face is effected with hand tools or pneumatic clay spades.

Modern suppliers of pipe-jacking equipment now provide shields which incorporate mechanized shields, bucket excavators, and boom-cutter loaders for loosening and loading the cut soil. But this type of equipment can only be used in conjunction with the larger-diameter pipes. The Westfalia Lünen boom-cutter loader (Model 'Wühlmaus') (see figure 526) may be used with pipe diameters as small as 1.40 m (5 ft). One operator is required to control the boom-cutter loader, which may be dismantled into parts small enough to allow it to be carried through the pipes to the shield.

Assessment of ground conditions prior to pipe jacking is, of course, essential. Problems may occur with water, both in the shield (i.e. flooding or running ground) or above or around the jacked pipes, (when the movement of fines and gravels disturbed by the progress of the pipes through the ground may start a water flow — this in turn would cause surface ground settlement).

Sticky cohesive clays may cause frictional forces to build up, necessitating the placement of additional intermediate jacking stations. Face or breasting boards may be needed if the ground ahead is uncohesive. Rocks may be encountered which need breaking by explosives or by splitting devices, etc.

In some cases frictional resistance on the pipes may be lowered by pumping bentonite and other chemical preparations around the outside of the pipes. The bentonite mixtures are thixotropic and act as lubricants which ease the frictional forces on the pipes.

At the end of the jacking line is the receiving pit where the jacking shield is removed. After its jacks have been dismantled, the intermediate

station remains in the bore where it forms part of the permanent pipeline.

Standard nominal jacking-pipe bores range from 900 mm (36 in) to 2550 mm (102 in).

Where soil conditions are favourable pipe jacking may be used for installing sewer, gas and water piping under roads, railways, canals, and buildings etc.

In addition to circular sections, rectangular* concrete sections and jacking shields may also be used if required.

An Unusual Pipe-jacking Project[4,5]

In 1960 A Californian contractor, Johnson Western Constructors of San Pedro carried out an unusual jacking project. A concrete storm sewer pipe 3050 mm (120 in) diameter, was jacked through 360 m (1200 ft) from a single jacking pit.** The U.S. $375,000 contract was for a storm drain for the National City Municipality of California.

---

*According to *Tunnels and Tunnelling,* dated July 1978[6] a major pipe-jacking project, involving the construction of two rectangular access tunnels beneath the A41 highway at the Brent Cross Interchange, North London, was carried out during the late nineteen-seventies.

Concrete boxes 9.75 m (32 ft) wide by 6.7 m (22 ft) high by about 12 m (40 ft) long were used for the south tunnel at the east and west ends. Three concrete units measuring 9.4 m (30 ft) wide by 7.0 m (23 ft) high were used for the north tunnel under the A41. These three sections which were jacked in a line varied in length and were 11.0 m, 16.8 m, and 18.3 m (36 ft, 55 ft, and 60 ft) long, respectively. This latter tunnel was reputed to be the largest jacked tunnel of its type in the world at that time (1978).

**According to Martin Hunt[7] a 1200 m (3940 ft) long, 2.6 m (8 ft 6 in) diameter pipe-jacking project for an interception sewer scheme involving 15 interjacking stations was undertaken in 1978 in Harburg, a suburb of Hamburg, Germany. This is reputed to be a record length jacked from a single thrust pit.

Six 300 t jacks at the thrust pit assisted by twenty-four 70 t jacks at each of the fifteen interjacking stations were used to move the pipes through the ground. Because the water table was at ground level, compressed air was also used. A shield fitted with face-plates and divided into three compartments headed the pipe-jacking string. The muck was either hand-excavated or flushed out on to a conveyor which deposited it into a water tank positioned a few metres from the face. From the tank the muck, in the form of a slurry, was pumped to the surface where it was dumped into a holding pond and later removed.

During the weekend when the compressed air was shut off, the face-plates of the shield were closed to support the face. However, in order to prevent the groundwater pressure from causing the pipes to backslide, thrust on the pipes was sustained during this period.

A gyrocompass was used for guidance purposes.

Although the specifications called for open-trench installation, the jacking bid was the lowest economically. Various other factors influenced the choice towards pipe jacking. These included the fact that the soil formation which was almost a conglomerate of dry clay, cemented sand, gravel, and cobble, was favourable and, so far as was known, was devoid of running or squeezing ground or rock. Furthermore, because the overburden remained undisturbed, less reinforcing steel was necessary than would otherwise have been the case in an open-trench installation system.

A mechanized shield led the pipe string, including four intermediate jacking stations, which boosted the total jacking capacity to (U.S.) 3800 ton (3400 t). The thrust reaction from the four, 4.11 m (13½ ft) stroke, 250 ton (230 t) jacks in the main jack pit was taken by 69 cm (27 in) beams backed by vertical timbers and grouted soil. Old conveyor belting was inserted between the 19 mm (¾ in) welded steel plate thrust ring and the rear pipe to protect the pipe from damage.

The mechanized shield had a cutter head consisting of five radially disposed arms which rotated about a central drive shaft. Each arm was fitted with a number of cutting picks or chisels. The gauge picks were turned outwards to enable the excavator to cut a hole slightly larger than the outside pipe diameter. In addition the steel cutting edge on the forward end of the mechanized shield was also slightly larger in diameter than the pipes so that an annular space approximately 9.5 mm (⅜ in) was left around the pipe. This helped ease the frictional forces around the pipes and also provided space for external lubrication to be inserted. This consisted of a mixture of bentonite, water, crude oil, and a patented jelly compound. The mixture, including the jelly compound, was prepared and supplied by Ken Corp. It was inserted at three points, namely at the cutting edge as already mentioned, at the jacking pit around each pipe as it entered the tunnel, and about 19.8 m (65 ft) from the main jacking pit, where a 91 cm (36 in) diameter hole drilled directly over the pipe was used as a lubricant reservoir.

As the head rotated, material cut from the face fell on to vanes or fins fitted to the inside of the mechanized shield at its forward end. The vanes lifted the material and deposited it on to a conveyor belt which carried it to a hopper situated 3.6 m (12 ft) behind the tunnel face. Hydraulic gates on the hopper controlled the dumping of the material into a removable 1.5 cu. m (2 cu. yd.) capacity skip on a battery-driven rubber-tyred muck car. When the car reached the jacking pit the skip was hoisted out.

The muck-removal system proved adequate for the first 270 m (900 ft), but during the driving of the final 90 m (300 ft) periodic delays were experienced as the car could not get rid of the muck and return fast enough to keep up with the boring rate of the mechanized shield.

One operator who was in telephonic communication with the mechanized shield operator, controlled all four intermediate jacking stations at a central post. The jacking station operator was given a signal to operate the first jacking station when the excavator's thrust jacks had completed their boring stroke. The pipes between the front jacking station and the boring machine were then moved forward until the boring machine operator called a halt. After which each following string of pipes was moved forward successively, the entire operation being controlled by a system of valves and signal lights.

Average advance rates of about 1.5 m (5 ft) per eight hour shift were attained. The work was of necessity carried out on a continuous three shifts a day, seven days a week, basis, because it was found that frictional forces tended to build up and bind the pipes if they were left stationary for too long a period.

Nishimatsu Construction Co. Ltd[8]

In 1972 a patent (U.K. 1,403,033) (Figure 380) was granted to the Nishimatsu Construction Company Limited of Tokyo, Japan, for an unusual pipe-jacking system invented by Enakichi Suzuki and Hiroshi Yoshida.

Basically the method consisted of driving through the ground a number of parallel steel pipes which were linked or joined to each other by couplings running along the entire length of the outside surface of each pipe. Each pipe was coupled to those already in position then driven horizontally by hydraulic jacks along the line of

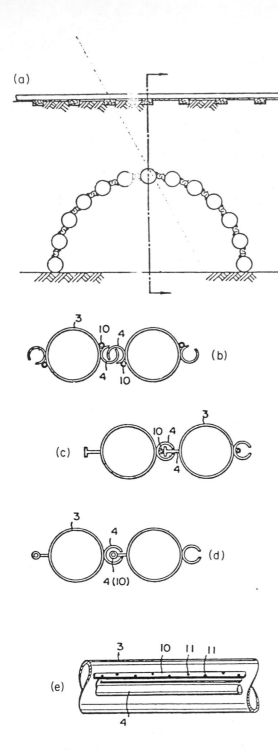

the tunnel. On each longitudinal coupling (sometimes on the female section and sometimes on the male section of the coupling, depending upon the particular type of coupling design used) was a water-jet tube. The water jet was designed to soften the ground ahead of the pipe as it was being pushed through.

When all the pipes were in position, they formed a protective arch which supported the ground so that the material within the protected area beneath the arch could be safely excavated without disturbing the ground outside.

In some cases the coupling design could include a water-jet tube which has a number of holes drilled in its wall. After the pipes are in position the area around the couplings could be rendered watertight by pumping grout through the water-jet tube and out the specially drilled grout holes.

This method of pipe pushing has been successfully used by the Nishimatsu Construction Company in Japan for a number of projects under roads and railway lines, etc. where very little clearance existed between the tunnel and the surface because of the proximity of other service tunnels, and where it was essential that the extremely busy traffic flow should not be interrupted, or endangered by ground subsidence.

Alternative Pipe-Laying Methods[9]

One of the most significant disadvantages associated with horizontal boring is that the cuttings are frequently left in the hole after it has been bored, causing difficulties when the pipes are being installed. To overcome this disadvantage, Gilbert Turner, President of the Boring and Tunnelling Company, sought other methods of pipe installation.

Several methods of pipe installation by horizontal boring are now used by the company and these are described below.

Slurry Boring

In this method a pilot hole is first drilled through from the starting pit to the reception pit. This hole is then either 'back-reamed' by attaching a 'slusher/reamer head' to the end of

Figure 380. Nishimatsu parallel steel pipe-jacking system. (a) Sectional view of pipes in position in tunnel; (b)-(d) sectional view of couplings; (e) side view of steel pipes shown in (b); (3) steel pipe; (4) coupling; (10) water jet tube; (11) grout holes. (Courtesy Nishimatsu Construction Company, Tokyo, Japan.)

the drill stem, or the drill string is withdrawn and the hole is reamed from the starting pit.

The spoil from the reamer, which is passed back through the reamer head, helps to support the hole walls until the hole has been cleaned and casing inserted. After completion of the reaming a swab, attached to a cable, is drawn through the hole, followed immediately by the casing. Occasionally, depending upon soil conditions, no swab is used. Instead open-ended pipes are drawn into the hole. The pipes are later cleaned by means of a rubber swab or, alternatively, the muck is washed out by strong water jets.

A refinement of the above method used by Gilbert Turner's Boring and Tunneling Company, is the introduction of a bentonite clay and water mixture into the hole during the reaming operation. The bentonite and water form a slurry which carries the cuttings and also supports the walls until the pipes are thrust or pulled into the hole. It also acts to lubricate the hole, thus facilitating the entry of the pipes.

Horner Method

According to an article in *World Construction* (February 1978)* the company also use the 'Horner method (basic patents)' when boring through colloidal clay. Less water is needed for this method than for the method described above.

After completion of the initial pilot hole the reamer is attached and about a 1 m (39 in) of hole is bored. Still keeping the reamer rotating, the head is drawn back towards the power plant carrying the cuttings with it and compressing them into a plug. The water remains behind in the newly created chamber. The borer is then advanced another metre and the process is repeated so that a second plug is formed, thus providing an airtight seal for the inner water compartment. Air is pumped into the inner pocket until sufficient pressure is provided to force the cuttings out of the hole into the starting or reception pit, as the case may be.

While this method (which was developed during the 1950s in order to conserve water) is considered by Turner to be particularly effective in short bores, it has also been found useful in long bores where it starts the movement of the cuttings towards the rear opening, thus facilitating their expulsion with a swab.

When working in soft ground the reamer is fitted with pick-type teeth. A heavier reamer with a correspondingly larger-diameter drill stem (up to 12 cm (4.7 in)) is used for harder ground. When drilling through rock, conventional hard-rock tools (i.e. disc cutters or milled tooth roller cutters, etc.) may be used.

Saddle- or Cradle-Type Boring Machine

This method utilizes a power unit supported by cables suspended from a side boom on a crawler tractor. The 'saddle' method has been found useful on some cross-country projects where ample working space is available for the tractor and sideboom.

An auger reamer is used within the leading casing section to clear the ground. Forward thrust for the casing is provided by a cable on a winch, the end of the cable being secured by an anchor or 'dead man' buried in the ground. Material cut by the reamer head is mixed with drilling fluid and then passed back to the rear by the auger conveyor within the pipe. It emerges as a slurry in the starting pit. An advantage associated with this method of boring is that no base or bed is necessary in the starting pit, as is required for the other methods described.

References

1. M. W. Loving, *Concrete Pipe in American Sewerage Practice,* Bulletin 17, American Concrete Pipe Association, 1938, Chicago, Ill. U.S.A.
2. Howard F. Peckworth, *Concrete Pipe Handbook,* American Concrete Pipe Association, Chicago, Ill, 1958.
3. *Jacking Concrete Pipes,* Bulletin No. 1, Pipe Jacking Association, Feb. 1975 (Concrete Pipe Association of Great Britain), London.
4. *Unusual jacking setup pushes big pipe 1,200 ft, Construction Methods and Equipment,* March 1960.
5. C. M. Hough, PJA Bulletin Number Two. Pipe Jacking Case Histories, *Tunnels and Tunnelling,* Jan. 1978, **10,** No. 1, 91, London.

*Based on a paper presented by G. Turner at the Small Size Tunnel Conference held at the University of Wisconsin.[9]

6. Paul Whitehouse (Marketing Manager, Tube Headings), (a) A41 interchange. PJA Bulletin number two. Pipejacking case histories, *Tunnels and Tunnelling,* July, 1978, **10,** No. 6, 65, London.
7. Martin Hunt (Editor, *Tunnels and Tunnelling,* Pipe jacking the Harburg sewer, *Tunnels and Tunnelling,* July 1978, **10,** No. 6, 19, London.
8. Patent 1,403,033 (Nishimatsu Construction Company, Limited, Tokyo, Japan) dated 11 Oct. 1972.
9. Pipelines installed without trenching, *World Construction,* Feb. 1978. (Based on paper presented by G. Turner at the Small Size Tunnel Conference, University of Wisconsin-Extension, Milwaukee, U.S.A.), pp. 46-48.

Correspondence

G. Turner (Boring and Tunnelling Co.).
J. M. Skaggs, Houston, U.S.A.
Nishimatsu Construction Co. Ltd., Tokyo, Japan.

SEVENTEEN

Military Use of the Tunnelling Machine

From the time when man built the first substantial walls and towers to defend his home, counter-measures were undertaken by his enemies to gain entrance and destroy his defences. One of these measures took the form of mines and tunnels which he dug to undermine specific defence positions or to gain entrance to the fortifications themselves.

When a particular section needed to be destroyed, a chamber was dug beneath the building in question which was supported by timber as the soil was removed. When the excavation was completed the timber was ignited and in due course the tower or wall collapsed into the hole. To hasten the procedure oil was very often poured over the wood before ignition.

In retaliation the besieged would dig tunnels out towards the enemy and so intercept their attackers before they reached a vital defence position, or they would tunnel towards an approaching wooden 'belfry' and collapse it in the manner already described.

Occasionally mediaeval sappers used a 'cat-castle'*[1, 2] or *Chat-châteil* (Figure 381) to protect themselves from falling debris and rock during mining operations.

Later gunpowder was used to blow up the fortifications and the art of mining and counter-mining was more fully developed. Camouflets, small explosive charges of such a size that they are capable of damaging the enemy's tunnel without breaking through to the surface ground, are frequently used for this latter type of underground warfare.

The First World War

Messines — The Miner's War[3-6]

During the early part of 1916, Sir Douglas Haig was given the mammoth task of planning the proposed British campaign on the Somme front. However, so formidable was this task, which involved the capture of a heavily guarded and well-positioned ridge, that Haig, though he expected victory, nevertheless laid alternative

Figure 381. '... Chat-châteil, or cat castle, which we have seen was a gallery under which miners worked, tearing up the ground like the claws of a cat. ... These engines were thus made in the manner of cars covered with great frames of thick timber, and planking upon them; within were the miners.' (From Sir Samaul Rush Meyrick, *A Critical inquiry into ancient armour*, Plate XXVI, London, 1824

*This was a movable building or penthouse which ran on wheels. It was used above ground as a protective cover for miners to enable them to approach close to the fortification walls under which they wished to tunnel. They were then able to carry out digging operations unmolested. Very often the 'cat' was constructed with a steeply sloping roof, reinforced with stout iron straps. The purpose of the slope being to deflect the rocks, etc. thrown down from above. Some 'cats' were constructed with crenelles or loopholes so that archers stationed within could fire through them. 'Cats' were also occasionally used to protect those operating the battering rams.

plans in the event of a defeat. The plan was to move his troops to Flanders if the ridge assault failed.

At Ypres in Flanders, General Sir Herbert Plumer, who was in command of the Second Army, was busily holding a position which was highly vulnerable to the German guns. Despite this, Plumer had as early as 1915 managed to so fortify and defend the salient, that eventually the enemy abandoned their fruitless attempts to dislodge him from his holes and concentrated mainly upon defending their own lines.

The result of the Somme assault was inconclusive and Haig, whose hopes remained high, despite various setbacks, was reluctant to abandon the operation. Instead of moving his troops to Flanders as had originally been planned, he decided to continue his offensive and tackle the Belgian problem later.

Naturally Plumer was intensely disappointed that the long-expected reinforcements did not materialize and that the proposed offensive operations for which he had been preparing, had been temporarily postponed. Nevertheless he continued to carry out the necessary work involved. Most especially he concentrated on measures concerned with the taking of the Messines-Wytschaete ridge (popularly referred to as 'White sheet' by the Allied troops).

At that time the front consisted of a long line stretching some 400 miles. It ran from the Belgian coast starting at a point approximately half-way between Dunkirk and Ostend and continuing on through Belgium and France until it reached the Swiss border.

Some parts, such as those in the south-east, which ran through rugged mountainous country were but sparsely held and guarded. In others troops and equipment were heavily concentrated. One of these latter areas was the Messines-Wytschaete ridge.

The ridge itself was no more than a spur of ground running for some 24 km (15 miles). It partly encircled the village of Ypres. Not particularly high so far as altitude was concerned, nevertheless it was practically the only raised ground in that part of Flanders which is well known for its flatness. It was therefore considered by both sides to be an extremely important military acquisition.

Between Ypres and the neighbouring village

of Cormines a single-track railway line had been laid some time before the war. The line for the track had been cut through the Messines ridge and during construction work gangs had laboriously moved the soil from the cutting and deposited it in three separate dumps beside the track. Some enthusiastic person had later measured the height of the largest of these dumps of soil and had painted the measurement — 60 m (197 ft) — on a signboard which had been placed alongside the biggest dump. It became known simply as 'Hill 60'. The two smaller piles of earth had also been named. They were situated on the opposite side of the track from Hill 60 and were known as the 'Caterpillar' and the 'Dump'.

Fierce battles were fought and many sacrifices were made by both sides, involving the loss of countless thousands of lives in the effort to gain possession of these three coveted man-made peaks, the most important military view in that part of Flanders. Such was the advantage to the particular side which held these dumps, and of course the remainder of Messines ridge, that they were able to fire into almost every point of the positions held by the opposing army down on the lower ground.

During the long and weary years when the Allies had been compelled to hold the salient, General Plumer and his Chief of Staff, Major-General Sir Charles Harrington, had examined every angle, almost every depression of the ridge. They knew it better perhaps than most men would know their own homeland. More than any others, these two men were determined that the ridge would ultimately be captured — whatever the cost.

Because of the peculiar circumstances surrounding events at Messines, it was fought as few wars have been fought before or since. No words describe the entire operation more aptly than those penned by Captain W. Grant Grieve and Bernard Newman when they said: 'Messines is to the tunneller what Waterloo was to Wellington. Never in the history of warfare has the miner played such a great and vital part in a battle.'[4] For though there certainly were surface skirmishes and engagements, *the war at Messines was fought mainly underground.*

Initiated by the Germans the art of mining and countermining soon became an accepted method of attack and defence. At first both the Allied troops and the Germans laid mines and counter-mines in shallow trenches and tunnels barely 5 to 6 m (15 to 20 ft) below the surface. But as the war developed these tunnels and galleries tended to become deeper and the mines themselves larger and more powerful. Nevertheless, during the early stages of this fantastic underground war, though specific points were attacked and blown, no co-ordinated mining plan was carried out by either side.

Due mainly to the ceaseless efforts of Major John Norton Griffiths who recruited many thousands of miners from all parts of England, Britain's famous Tunnelling Companies were finally established officially. They developed from a small nucleus into an efficient and integral part of the main army. Later they were joined by the 1st, 2nd, and 3rd Tunnelling Companies of the Australian and Canadian armies and the New Zealand Tunnelling Company who also soon proved their capabilities in this specialized field. It was due, too, to Major Griffiths that the idea to 'blow' or 'earthquake' the entire Messines ridge was originally initiated.

In the early stages Griffiths suggested[3] that a series of about half a dozen large mines (carrying approximately 9000 kg (20,000 lb) of ammonal each) be laid in strategic positions along the front. To confuse the enemy he further suggested that these mines be officially known as 'deep wells' which had ostensibly been dug to provide water for the army.

As time went by the plan was further developed until the idea of blowing up the entire ridge was envisaged. This plan was summarily rejected by a high-level conference of army commanders held on 6 January 1916. Griffiths, Fowke, and Harvey who were all intimately concerned with the scheme were bitterly disappointed. Then, miraculously, a few hours later, General Fowke hurried to inform Griffiths that the generals had decided to use the plan after all. They would blow the ridge.

Thus, finally it was planned that at precisely 3.10 a.m. on 7 June 1917, 22 enormous mines, each containing approximately 40,000 kg (90,000 lb) of ammonal and placed in specific positions along Messines ridge would be detonated.

However, before this ambitious plan could be carried out many problems had to be solved. An important one concerned the saturated, waterlogged subsoil through which the tunnels leading to the mines would have to pass. The difficulties at that stage appeared to be insurmountable. This problem was finally solved when Captain Cecil Cropper, through a process of trial and error, discovered that it was easier to tunnel through a particular band of impervious blue clay than through the running sand. Cropper was working on a scheme of his own. He planned to place four mines which he hoped would blow up five positions marked on the British maps as Hollandscheschuur, Petit Bois, Peckham, Spanbroekmolen, and Kruisstraat. In his various efforts to carry out this plan Cropper had discovered the difficulties of tunnelling through the upper subsoil of Flanders. Quite accidentally he had come across a stratum of heavy blue clay which was impervious to water. The advantages of tunnelling through this type of ground soon became apparent and Cropper thereafter deliberately sought it. After a thorough examination of the area he found that in order to gain access to it he had to retreat well to the rear of the front line to a position where a stratum of the firm blue clay rose nearly to the surface. Mostly the clay surfaced in the beds of streams. However, this meant that Cropper's men would have unprecedented distances of from 600 to 900 m (2000 to 3000 ft) to tunnel in order to reach their final objectives. Quite a feat under those difficult conditions. Undaunted he told his superior officer of his ambitious plan. Later, when Griffiths came to learn of it, he was delighted, as it fitted very neatly into his own overall plan to blow the ridge. Cropper was told to go ahead but Griffiths stressed that on no account were any of the mines to be blown before the order was given.

Partly as a result of Cropper's successful tunnelling work in the blue clay, and partly as a result of the studies carried out by the geologist Major Edgeworth David and Brigadier-General G. H. Fowke, a plan was laid whereby deep shafts (varying in depth from 20 to 40 m (80 to

120 ft) would be dug in order to gain access to the firm blue clay discovered by Cropper. Once this stratum was reached, galleries and tunnels would be sent out towards the specific objective points along the ridge.

Success of the entire operation hinged upon secrecy and so, also at Griffiths's suggestion, as each new shaft was dug, it was isolated by protective rails and conspicuously labelled with signs which warned 'Deep well. Keep out'. This ruse apparently worked, for though raiding parties occasionally penetrated the Allied trenches, they failed to discover that the important shafts led to the main tunnels and galleries further down below.

In the meantime while this feverish tunnelling activity was going on, the normal mining and counter-mining continued higher up. In order to further distract the attention of the enemy from the work being carried out far below, Plumer's men intensified their efforts. Many shallow trenches and galleries were dug and camouflets detonated in enemy trenches and tunnels. The enemy, naturally, returned the compliment.

Amongst other ingenious devices tried at that time was the art of 'pipe pushing', or as it was also known 'torpedoing'. When the men stationed at the face heard the enemy working nearby a small hole was bored with a hand auger. Into this a 2.4 m (8 ft) long pipe ranging in size between 150 and 200 mm (6 and 8 in) in diameter (and fitted at its tip with a charge of approximately 45 kg (100 lb) of ammonal) was pushed into the hole and forced through to where the enemy was suspected to be tunnelling. Then, at the right moment, the charge would be detonated. Sometimes the Germans acted first and blew their mine before the British could detonate their own charge. In May 1916 a special device called a 'sentinel jack' was used to push the pipes through the soil. Excerpts from the daily diary of the 251st Tunnelling Company, Royal Engineers, who used this device are quoted below.[6] The excerpts cover the period May 1916 to August 1917.

4th May 1916... with Lieut. Martin who gave the opinion that pipes could not be pushed in our chalk* with Sentinel Jack, but this could be done successfully with clay.

20th May 1916. Visited pipe pushing at Cambrian Chateau. Pushed 40' with Hughes and Lancaster machine which stuck at 13 tons pressure. Pipes pushed in to 56' with Sentinel Hydraulic Jack. A second hole pushed in 56' with Sentinel Jack. Depth 16' strata rather sandy clay.

23rd June 1916. Pipe pushed 50' by Sentinel Jack. Pressure then reached 25 tons and push had to be abandoned. Pushed in clay at about 14'.

8th July 1916. Two charges each 180 lbs Ammonal in 3" pipes pushed 91' by hydraulic Jack.

28th August 1916. ...enemy heard working by water electric listening apparatus.

8th September 1916. Pipe exploded 1.40 a.m.; complete 18' wide cut through enemy wire enabling raiding party to enter enemy trenches. Pipe pushed by Hughes and Lancaster machine.

14th September 1916. ... also pipe pushed by Sentinel Jack 112' and trench blown to edge of crater.

11th November 1916. 2/Lieut. Barratt giving demonstration of listening apparatus.

15th June 1917. La Bassée Road Hole closed in and electric listening apparatus installed.

10th August 1917. Leads from seismo-microphone, which was left in charge, were brought back to firing point. Enemy heard working at estimated distance of 15' away at moment of firing.

That the enemy continued to remain unaware of the plan to blow the ridge during the long period of preparation, seemed incredible. Yet though the Germans feared an attack of this

---

*The 251st Tunnelling Company were at that time stationed at La Bassée. Attached to the 251st Tunnelling Company was the 3rd Australian Tunnelling Company commanded by Major Leslie Coulter. Thick beds of chalk lying beneath narrow bands of surface clay were prevalent in several areas along the front. Mining in these chalk areas called for vastly different techniques from those required in the sticky clay. Though there were several advantages in mining chalk there were, too, many disadvantages not usually associated with civil mining. It was for instance more noisy to work than clay, which meant that the enemy could hear your approach. Of course the enemy could also be heard, which was perhaps some small consolation. But a particular hazard was its brilliant whiteness which made debris from the tunnels very difficult to conceal. When this was spotted by the pilots of enemy aircraft the Germans knew at once where to direct their next counter-mine attack.

nature more than any other, such was indeed the case. Later it became known that in May of that year (1917), barely one month before 'zero hour', Lieutenant-General von Kuhl had strongly advocated the withdrawal of their troops from the ridge because he feared it might be mined and blown. Fortunately for the Allies, however, Crown Prince Rupprecht had received unreliable reports from his intelligence agents that the Allies were apparently no longer carrying out mining activities on a large scale. In fact, so his agents confidently informed the Crown Prince, the only mining work of any significance was that which was currently being undertaken in the vicinity of Hill 60. Rupprecht's fears concerning this particular area were soon allayed by his officers in the field who reported that all mining activities by the Second Army had been very effectively nullified by the efficient counter-mining carried out by the Germans.

That preparations were undoubtedly in hand for a full-scale frontal assault, the Allies made no attempt to conceal. So far as they were concerned, the more the Germans believed this mode of attack to be the one and only one intended, the less inclined they would be to suspect the true intentions of Plumer and his men. The Crown Prince for his part was, of course, confident from past experience that a surface assault would not succeed. Moreover, he was determined that, come what may, the area would be held, even if they were cut off all round. Thus his order was given during those last few critical months prior to 7 June, to 'hold the ridge at all costs'.

Though reassured by their own agents that all was well, nevertheless Haig and Plumer worried incessantly, knowing full well that once the Prince was appraised of the true situation, he would most certainly abandon the area and all their efforts would be lost.

Several times the Germans sent special raiding parties out — not to capture prisoners, but to check on the type of soil being mined by the Allies. However, Plumer and his men had not overlooked this very important point. Cunningly they hauled every particle of blue clay raised from their deep tunnels and galleries well to the rear of their lines where it could not be seen by enemy raiding parties. And this,

considering the miles of tunnel dug, was no mean feat. The work of removing and concealing the clay muck was carried out mainly at night.

Occasionally in the war of mining and counter-mining which was taking place some 4.6 to 6.1 m (15 to 20 ft) below the surface, the Germans descended as low as 18 m (60 ft). This depth brought them dangerously close to some of the galleries dug by the Tunnelling Companies. These were anxious moments for the Allies who, perforce, had to cease work and listen to the advance of the German galleries through their microphones. On one memorable occasion when the Germans seemed closer than ever and in imminent danger of breaking through, Plumer was faced with the terrible decision of whether or not to blow prematurely the mines already laid. He decided against it and after approaching as close as 460 mm (18 in) in the northern corner of the ridge, the Germans unaccountably changed direction and the tunnellers relaxed.

While fighting in the form of mining and counter-mining went on unabated in the upper sublevels, the 250th and 251st Tunnelling Companies continued working feverishly lower down. Now and then German miners would break through into their tunnels and a free-for-all would ensue. Lacking guns or rifles the miners used hands, picks, shovels, or whatever implement was handy to put down his foe. And many a man died — not from a bullet or shrapnel wound but from the business end of a heavy pickaxe.

Morale remained high, though conditions in the galleries below were deplorable. So bad were they that it was acknowledged tacitly that no miner could maintain the pace for more than a few months at a time without succumbing to the strain both mentally and physically.

The Tunnelling Machine

Every effort was therefore made to alleviate the situation and if possible speed the work. Griffiths, his brain ever active, had another bright idea, *Why not use a mechanical borer?* No sooner said than done. He approached Brigadier General Robert Napier Harvey and outlined his plan. He requested that an order

433

for several machines (at least four or better still about six) be put through immediately, as he felt it senseless to get only one. Most especially he felt that Cropper needed assistance, as his company had by far the greatest distance to tunnel.

Harvey, while listening sympathetically to Griffiths's eloquent plea would not allow himself to be pressured into ordering more than one tunnelling machine. It was at this stage that a colossal error of judgement was made, for it seems that unknowingly Griffiths had confused rock-type machines with mechanized shields. The first intimation of this is given in Alexander Barrie's book *War Underground* where he says:

> ...Norton Griffiths had another major inspiration: in certain collieries, mechanical borers were used to drive mine headings and roads through coal seams. Could these not be adapted to bore through clay? It seemed very probable; larger models of broadly similar design had already been used on the London underground railway tunnels in clay... Plans to obtain it, the Stanley Heading Machine (Figure 382) manufactured in Nuneaton, went forward with a rush. A special cutting head designed for the hard Ypresian clay had to be made, but was promised by the makers for delivery in six weeks[3]

And so the obvious choice, a mechanised shield machine such as was used so successfully in clay in the London Underground tunnels, was ignored.

Harvey and Griffiths made a joint visit to Cropper to advise him of their decision. Still unaware of the overall plan to blow the ridge at that stage, Cropper was somewhat piqued at Griffiths's and Harvey's attitude towards what he considered was 'his' plan. However, as the idea of using a tunnelling machine appealed to him, he swallowed his annoyance and said he would carry out whatever preparations were necessary to receive the machine.

When Griffiths again visited Cropper on 10 February 1916 he found that the promised digging of a chamber at the foot of the Petit Bois shaft (S.P. 13) to receive the machine had not yet been attended to. Nor was the site above ground prepared to receive the ancillary equipment. Griffiths stressed the urgency for this work, and then left for London hoping Cropper would carry out his promise to rectify the matter.

Griffiths arrived in England on 14 February 1916 and found that the machine (packed in 24 separate packages and weighing approximately 7½ ton) was lying at Marylebone Station. After innumerable delays and frustrating mishaps, Griffiths finally got them transported across England in time to catch an ammunition boat leaving Newhaven for France on the evening of 15 February. Griffiths left for France after further arrangements had been made for a suitable engineer to accompany him. The engineer, a man named Carter, was to assist with the assembly and installation of the machine and ancillary plant and also instruct Cropper's men how to operate the borer.

When the machine at long last arrived at its destination, further difficulties were encountered as Cropper and his men struggled to install the machine in the specially prepared chamber 24 m (80 ft) below the surface. A make-shift track had been laid, along which the various components were hauled in hand trolleys. Unfortunately the Germans had an excellent view of the new railway line and

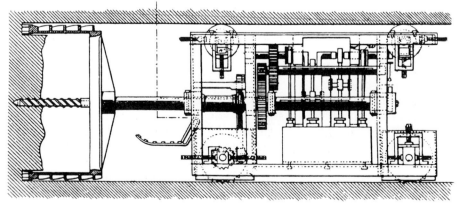

Figure 382. Type of Stanley tunnelling machine used in the Stanley Colliery at Nuneaton, England. This machine was patented in America on the 7 Aug. 1894 under No. 524,149

immediately began delivering a heavy and sustained barrage of fire in that direction. The parts then had to be moved at night, an exhausting and nerve-racking procedure, as the flimsy rails often gave under the weight of the heavy components causing frequent derailments of the trolleys. This would entail a laborious reloading process which was done by hand in the pitch-dark. The immense difficulty experienced in reloading some of the heavy pieces of equipment after a derailment had occurred often meant that day would dawn before it could reach the shaft. Then frantic action to camouflage the component had to be taken.

Somehow, as a result no doubt of all this feverish activity, word got out to the Germans that a mysterious new underground weapon was being installed at Petit Bois. Instead of attempting to quash these rumours, Cropper and his men wisely encouraged them, thus diverting the Germans from the true purpose of the new machine.

Assembled at long last, the machine was started on the evening of 4 March and for some time it ran exceptionally well. Progressing at the rate of approximately 610 mm/hour (2 ft/hour) it cut steadily through the clay, leaving a clean 1.8 m (6 ft) diameter tunnel in its wake. After working satisfactorily for several hours, the machine was halted in order that a routine maintenance check could be made. As everything appeared to be in order the machine's motors were restarted. However, during the interval when the maintenance work was being undertaken, the clay in the tunnel walls had time to absorb some of the heavy moisture present in the air. It swelled to such an extent that the machine became wedged tight and was unable to progress forward, its cutters jammed fast in the clay.

From that stage onwards the entire project was beset with difficulties, such as a too low power output and failure of the electric motor driving the compressor, the fuses of which blew so frequently that their entire store of fuse wire was soon depleted. An attempt to overcome this new problem by using barbed wire as a substitute ended disastrously with yet another stoppage of the machine. And each time the machine's operation was stopped because of these various problems, the clay, like some innate alien monster, quickly took over and gripped it tenaciously so that it was unable to proceed until once more freed by the laborious process of digging it out. Eventually after nearly five frustrating months of battle the machine was abandoned and Carter returned home.

In their book, *Tunnellers,* Grieve and Newman refer very briefly to 'A miniature "great shield" such as was used to cut the tunnels for the London tubes, was with difficulty, man-handled to No. 2 shaft.'[4] And, as will be seen from the following excerpts taken from the daily diary of the 250th Tunnelling Company, Royal Engineers, during the period 31 January 1916 to 18 June 1916, very little description is given of the actual type of tunnelling machine used at S.P. 13 in their notes either. In other words was it a conventional mechanized shield which failed because the ground conditions were too difficult or because the operators were unskilled in that type of work, or did the fault lie rather with the type of machine selected?

31st January. A special Tunnelling Shield mechanically driven has been ordered for one of the deep shafts at S.P. 13. This will be the first of its kind to be installed in France on Military work. It is hoped that it will be able to excavate quickly so as to allow of long tunnels being driven in a short time.

7th February. Work at No. 2 Shaft to be laid out for Mechanical Excavator, Electric power point and compressor air hoist.

18th February. 24 cases of machinery comprising 'shield excavator' for No. 2 shaft S.P. 13 arrived at Coy. Hd. Qrs. from England.

21st February. Electric lighting set and engine and generator for driving 'shield excavator' being put in position at No. 2 shaft. No. 2 shaft completed at 88 feet. Chamber cut for excavator.

28th February. Generator sets for light and mechanical shield excavator completed. Excavator parts lowered down shaft and erected in position for work. Awaiting delivery of sufficient lining for gallery before starting machine and also delivery of air hoists and compressor oil.

6th March. No. 2 shaft — carry on with completion of surface arrangements for 'excavator Plant' (Well Boring Apparatus). Machine commenced to cut clay.

13th March. No. 2 Shaft Well Boring Plant machine stopped pending mechanical adjustment. Cutting tools not satisfactory — Engine giving trouble.

27th March. Well Boring Plant running under the most favourable conditions has cut 18″ in 20 minutes with 13 minutes to put 1 ring of lining in position. Machine's tendency to dive necessitating working on the two bottom rams with 2,500 lbs. pressure has retarded progress. Hydraulic pressure pipe bursting frequently. Intermittent engine troubles.

3rd April. 'Wellboring Plant' machine being levelled up and chamber being cut to allow of dismantling for alterations.

3rd April. Well boring plant working unsatisfactorily — machine tilted to obviate tendency to dive on a down grade, hand drive continued.

10th April. Cleaning and clearing cutter, also fixing on Hood to well-boring machine.

16th April. Back shield for Well Boring Plant fitted on machine.

18th June. Compressor and Lighting plant running well (i.e. after enemy damage to tunnels resulting in 11 deaths through entombment.)*

---

*This refers to an incident which occurred just prior to the Petit Douve affair. Apparently severe damage was done by enemy mining to an area above and in very close proximity to gallery S.P. 13, as a result of which 12 men were entombed.

Frantic efforts were made by their comrades to dig them out, though they knew the position was well-nigh hopeless. Nevertheless, as miners all over the world will do under similar circumstances, they continued to toil feverishly and ceaselessly for a period of seven days until at last they broke through. But it was too late, for eleven of the men were found almost at once, all dead. There was no sign of the twelfth miner and they naturally assumed he had been crushed and hidden by the tons of fallen debris scattered about the area. As the air in the damaged tunnel was heavily polluted, the rescue team retreated temporarily to allow fresh air to enter before continuing with their work. When they returned they were startled to find the groggy figure of the twelth man — Sapper Bedson — crawling painfully towards them. He was alive though somewhat weak from his ghastly ordeal. Bedson attributed his amazing survival to the fact that he had had considerable experience as a miner before the war. He remembered how once some coal-miners had actually survived a 13-day

As has already been noted Griffiths apparently could see little difference between a mechanized shield designed to cope with swelling sticky clay conditions such as may be found both in Flanders and in the tunnels dug beneath London for the underground tubes, and a conventional rock machine made to bore through the hard abrasive Nuneaton coal seams. The error is understandable, as no doubt to the layman one type of tunnelling machine is much like another. It is also perhaps typical of the mistakes made by the Army during this period that they should have placed an order for such a machine with the Stanley Bros., makers of rock-tunnelling machines, rather than with acknowledged experts in this field, such as the firm of Markham and Company, England, who even at that time were well known for their reliable mechanized shields (Price design) which were currently tunnelling through the London Clay. Stanley Bros., though possessing considerable experience in the manufacture of hard-rock machines, had had no previous experience at building an efficient mechanized shield. It is also inconceivable that they could produce such a unit in six weeks from a workshop which had up to that time been producing Stanley heading machines. It seems likely therefore that when Stanley Bros. received the order from the Army for a machine which had to be completed within six weeks, they procured a cutting head from some other manufacturer. This was then fitted hastily to one of the Stanley machines and the whole affair shipped forthwith to France.

Griffiths's description of the machine* given

---

*'The machine consisted of a chassis on wheels carrying a pair of rotating cutters at the front. These cutters were powered by a two-cylinder compressed-air engine, which was in turn driven by a miniature generating station installed on the surface. As the compressed air blasted into the cutters they screwed themselves forward tearing out the face ahead and feeding the spoil back. While cutting went on, the machine was stabilized by top and bottom jacks. It was a massive and complex-looking affair.'[3]

---

entombment period. He knew too that his comrades would not spare themselves until all that could be done was done. With only two army biscuits and a bottle of water for sustenance, Bedson lay quietly near the working face where the ground was at its highest. During the first three days his comrades died one by one until eventually only Bedson was left to greet his rescuers on that memorable day of the breakthrough with the words, 'It's been a long shift, for God's sake give me a drink.'[3]

by Alexander Barrie in *War Underground* appears to fit very well with that of a Stanley header (see the section on the Stanley heading machine in Chapter 18). And though one gains the impression both from the diary and from Grieve and Newman's book[4] that a mechanized shield was used, Griffiths's subsequent complaint* that the machine's unaccountable tendency to dive was due to 'wings extending sideways into the clay...'. seems to indicate that this was indeed a rock machine fitted with grippers designed to thrust against the hard unyielding walls of the tunnel in order to counteract torque and allow the machine to cut into the face. These grippers had apparently been modified into fins or 'wings' to attempt to perform their function in clay. Mechanized shields which can cut through clay are, of course, constructed quite differently. The first type of shield — Greathead design— (see Chapter 10, Shields) was basically a cylindrical shell with a sharp cutting edge which was driven through the ground usually by the use of hydraulic jacks pushing against some form of rigid lining assembled immediately behind it. Later (Price — 1896) mechanized shields were invented. Price mechanized shields were first manufactured by Markham & Company and consisted of a Greathead-type shield with a robust cutting head installed at the forward or face end. The method of propulsion remained the same. That is, thrust was applied to the rear end of the shield by hydraulic jacks pushing

against rigid sections of tunnel lining already in place.

The two entries made on 10 and 16 April in the daily diaries relating to the 'fixing on of a Hood' and the 'fitting on of a Back Shield' seems to indicate a rather belated effort to turn the rock machine into a mechanized shield. No wonder the Petit Bois unit failed!

Incidentally, one other Stanley unit was ordered at that time. It was sent to the new 3rd Canadian Tunnelling Company stationed at the Bluff. But this machine also apparently failed and, like the former Petit Bois unit, was eventually abandoned. Of interest is the fact that Markham & Company did receive an order for the manufacture of eight 'attack shields' (Figure 383) from the War Ministry, some time during the year 1916. This order was executed. These units, they understood, were for delivery to Flanders for use under Hill 60. But no record of their ultimate fate appears to be available. Certainly there seems to be no evidence that they ever arrived in France.

Though in the original plan it was proposed to detonate 22 mines on 7 June, in fact only 19 were blown on the day. The reason for this was that one mine, at Petit Douve, near Messines, was accidentally put out of commission by the Germans. They detonated a camouflet which so damaged the tunnel leading to the land mine as to render it inaccessible. The mine, perforce, was written off by Plumer. The two (30,000 lb (14,000 kg) of ammonal each) southern mines situated to the extreme right, though prepared, were not fired as Plumer considered they were too far to the south to be of any value. One of these subsequently blew up on the 17 July 1955, supposedly detonated by a bolt of lightning. Fortunately no one was injured at the time. The other, its exact position unknown, is thought to be still there lying dormant.

Gradually each mine was laid and electric detonators set. As zero hour approached, activities above ground increased as the British, Australian, and New Zealand divisions marshalled their forces into battle position. They would be ready to move the moment the word was given.

Desultory artillery fire broke out just prior to the crucial minute. Then at 3.09 all, unaccountably, fell silent. The minute seemed

---

*'But by far the worst difficulty was the machine's baffling tendency to dive. "We never discovered why", wrote an eye-witness, anonymously, in an Old Comrades Association Bulletin, 'but this machine showed a complete disinclination to proceed towards Germany, but preferred to head to Australia by the most direct route." Prolific advice was offered to the sweating Carter and Talbot, and prominent among the givers of it was Norton Griffiths. During a visit on 12 March, he decided that almost everything was wrong with the equipment and reported in the gloomiest of terms to Harvey. He blamed the 'scratch motor engine and dynamo' which, he said, was a hastily gathered together outfit never intended to work with the cutter; also, the adjustment of the cutter tools in the face, he thought, 'was not quite right'; he felt there was something wrong, too, with the position of the arm of the cutting tool; lastly, he gave a complex, lengthy, and improbable explanation for the machine's tendency to dive, blaming wings extending sideways into the clay and designed to check a tendency for the whole contraption to rotate. Oddly, he had still not lost faith, and two days later was urging Harvey to set up a special six or eight-man team to install future cutters.'[3]

Figure 383. A 1.82 m (6 ft 0 in) dia. attack shield built for the War Office during 1915-16. Eight of these shields (special excavator type) were manufactured by Markham & Co. Ltd. Chesterfield, England. (Courtesy Markham & Co. Ltd.)

to stretch into eternity for the Allied troops who lay concealed behind their various shelters. Some were nearer, some were further away, but none was permitted closer to the first mine than 180 m (200 yd.)

Almost as a signal a few of the larger German guns again began firing a few seconds before 3.10 a.m. Then the vast charges buried deep beneath the ridge were detonated and the earth exploded and erupted like some giant volcano. Messines and Wytschaete were flung high amid smoke, flames, and tons of debris, and were no more. The sound of that vast explosion continued travelling outwards like the ripples on a lake after a stone has been thrown into the centre. Echoes of it even reached as far as London and were heard, so it is said, by the Prime Minister, Lloyd George himself, as he sat busily working late that same night in his office at 10 Downing Street.

When the air had cleared, the Allies moved in to find the Germans devastated and demoralized. Those that still remained alive were in such a state of deep shock that they offered no resistance whatever.

As soon as the British had gained the ridge they consolidated their position and prepared to clear the many pockets of Germans who still held outlying posts in the neighbourhood. It was in this most unfortunate phase of the exercise that many of Plumer's men were

mistakenly killed by their own comrades.

After the first attack confusion reigned, and in disbelief that the conquest could have been so complete, the troops held themselves alert and suspicious in readiness for expected counter-attacks. It was this extreme suspicion which led to the tragic slaughter of their own men as one party took another to be an enemy patrol and opened fire.

Of interest is the fact that though the Allies gained territorially and psychologically, they actually suffered slightly more casualties than did the enemy during that period. Thus ended the assault on Messines, surely one of the most remarkable campaigns of the First World War.

Whitaker Tunnelling Machines

Several Whitaker tunnelling machines were also purchased by the War Office during the First World War. These were used to drive some headings at Dover (presumably for the fortifications), one of which is known as the Priory gallery.

Hughes Tool Company Machine

During the First World War a tunnelling machine was also built in the United States for military purposes but the war ended before it could be used (Figure 384).

Figure 384. Horizontal boring machine built by H. R. Hughes (Snr.) for military purposes during the latter part of the First World War. Hughes is seen standing just inside the protective tent. (Courtesy Hughes Tool Company.)

The Austrian Machine[7,8]

The idea of mechanical assistance in the form of a tunnelling machine occurred to the Austrians as well as it did to the Allies during the First World War, and a machine was apparently constructed and tested on the Galician front by the Germans.

According to photostat copies of photographs of the machine (Figure 385) received by the author from the editor of the magazine *Weapons and Warfare,* the head consisted of a plate on which four arms were mounted in the shape of a cross. Each arm carried a row of radially arranged fixed cutter blades. Muck from the face was lifted by a scoop positioned at the periphery of each arm and transferred to a belt conveyor running along the top of the machine, from the forward end to the rear.

From the photostats available it appears that no gripper system was provided, torque and thrust no doubt being reacted by the dead weight of the unit. The tunnelling machine was mounted on wheels which ran on rails and the head diameter was approximately 1.8 m (6 ft).

The Second World War

Churchill's Excavator[7,9]

During the early phases of the First World War the Germans retreated and entrenched themselves on the Aisne, in a position heavily protected by barbed-wire entanglements and machine-guns.

Repeated assaults by the French and British troops proved this type of defence to be almost impregnable — at least with the equipment available at that time. Even attempts to outflank the Germans merely resulted in the defence line being extended until it reached the sea. Thus a type of stalemate developed.

There were, in the opinion of E. D. Swinton, a lieutenant-colonel in the Royal Engineers, only two ways for the stalemate to be broken so that the Allies could advance once more. Either they had to increase their artillery support to the point where they could literally blast their way through, or they could do it with some sort of protected mechanical contrivance which was capable of crossing the barbed wire-entanglements and the trenches.

Figure 385. Austrian T.B.M. used in First World War. (Courtesy Kriegsarchiv, Vienna.)

After several negative responses from various army officials in the hierarchy, Swinton's scheme eventually reached the desk of Winston Churchill. Churchill was, at that time, already in the process of experimenting with armoured cars for the naval detachment on the Belgian coast and he immediately appreciated the significance of Swinton's proposal. At Churchill's instigation, a committee was formed and, as a result, the first tank fitted with caterpillar tracks and christened 'Little Willie'

was produced in 1915. It was designed by Lieutenant-Colonel W. G. Wilson and was built by Ruston and Proctor of Lincoln (later to become Ruston-Bucyrus).

During the battle of the Somme some 420,000 British soldiers lost their lives. At that time, Churchill had suggested in a memorandum, dated November 1916, that there was a need for machines which could move ahead of the tanks, troops, and guns at a speed of say 5 km/hour (3 miles/hour), digging protective trenches within which the army could advance until it was comparatively safe, *beneath* the heavy artillery guns of the enemy.

Thus, as the Second World War approached, the German western defence along the Siegfried Line was the cause of grave concern. Churchill, as First Lord of the Admiralty, swore he would avert if possible the carnage suffered by the troops during the Somme battle.

In 1936* Churchill approached Sir Stanley Goodall (Director of Naval Construction at that time) with his proposition for a trenching machine.

What was needed, Churchill told him, was a machine which could cut a trench wide enough to admit a tank and also deep enough to offer comparative safety within which the army could advance until it reached the enemy lines.

Six drawings were ultimately submitted, all depicting machines which were basically similar in concept to the tank. After a 1 m (39 in) long model constructed by Basset-Lowke (a firm which specialized in the manufacture of model trains) had proved the feasibility of the design, Ruston-Bucyrus of Lincoln were ordered to produce 240 units. Of the 240, 200 would be 2 m (6 ft) wide (infantry type) and 40 would be 2.30 m (7 ft 6 in) wide (officer type). The prototype machine was originally code-named 'White Rabbit No. 6', but later Churchill suggested that it be changed to 'Cultivator No. 6' as 'White Rabbit' was more conspicuous as a code-name. However, the machine's construction came under the jurisdiction of the Admiralty's Naval Land Equipment Section, and so it was known as the 'Naval Land Equipment trenching machine, Mark I' or

*In 1916 Sir Tennyson d'Eyncourt (Director of Naval Construction) chaired the committee which investigated Swinton's scheme.

440

Figure 386. Front view of Churchill's Excavator showing Nellie I's plough blade. (Courtesy Imperial War Museum.)

Figure 387. Plan view of Nellie looking towards the front. (Courtesy Imperial War Museum.)

Figure 388. Churchill inspecting Nellie I during trials. (Courtesy Imperial War Museum.)

441

'N.L.E. Mark I'. Employees at the Ruston-Bucyrus plant where it was made immediately dubbed it 'Nellie' and Nellie it remained.

Although it had been the intention to power the machine with a Rolls-Royce Merlin engine (marine type), these powerful units were in short supply and all were earmarked for the R.A.F. planes. An alternative was found in the 600 hp (448 kW) Davey-Paxman diesel engine, but two diesels were needed to produce approximately the same power output as a single Rolls-Royce Merlin could have provided. Thus the weight of Nellie was increased until, in its final form it weighed 132 t and was 23.6 m (77 ft 6 in) long by 3.10 m high (10 ft 2 in) by 2.6 m (8 ft 6 in) (overall) wide. It cut a 1.52 m (5 ft) deep by 2.29 m (7 ft 6 in) wide trench at the rate of 800 m/hour (2600 ft/hour), removing some 8000 t of soil during this period.

Nellie (Figures 386-388) ran on caterpillar tracks powered by one of the diesel engines and her overall design, especially her rear, resembled the slope-backed tanks of the First World War. Nellie's front, however, was radically different. It boasted a robustly constructed plough-shaped leading cutter blade surmounting a transversely positioned rotary cutter powered by the second diesel engine.

As the machine was advanced on its crawler tracks, the plough sliced through the ground which was directed sideways by the vee-shaped blade. At the same time the horizontal cross blades of the rotary excavator positioned behind and below the ploughshare, cut the ground on their downward stroke and then lifted the material on to transversely positioned conveyors which deposited it at the sides. By this means a 2.44 m (8 ft) deep × 2.29 m (7 ft 6 in) wide protective trench was made by the excavator.

Though Churchill's Nellie was built and tested in practice, it was never used during the war. By the time the prototype machine had been put through its paces and modified where necessary, the Siegfried Line was reached and passed. Nellie thus became obsolete before she could be put on the production line.

References

1. Sir Samuel Rush Meyrick, *A Critical Inquiry into Ancient Armour,* London, 1824.
2. John Quick, *Dictionary of Weapons – Military Terms,* McGraw-Hill, 1973.
3. Alexander Barrie, *War Underground.* Frederick Muller, London, 1962.
4. Captain W. Grant Grieve and Bernard Newman, *Tunnellers,* Herbert Jenkins. (Anthony Shield & Associates), London, 1936.
5. Leon Wolff, *In Flanders Fields,* Longmans, Green, London, 1960.
6. Daily Diaries — 250th and 251st Tunnelling Companies, Royal Engineers. 20 Oct. to 31 Dec. 1915; 1 Jan. to 30 June 1916; Jan. 1916 to Jan. 1919, Public Records Office, W.O. 95/551.
7. *Weapons & Warfare,* Issue No. 13, Phoebus Publishing Co., 1976, London.
8. *The Illustrated War News,* Part 21, New Series, 1 Nov. 1916, p. 9, London.
9. *100 years of Good Company,* Ruston and Hornsby Ltd., Lincoln, U.K.

Correspondence

Bucyrus-Erie Company, South Milwaukee, U.S.A.
Ruston-Bucyrus Company, Lincoln, U.K.

Excavation – Coal-mining

Introduction

Just as the drill, both rotary and percussion, has played a vital part in the development of the hard-rock tunnel borer, so too has mechanization at the coal-face influenced the design of tunnelling machines. Early ancestors of both the Robbins and the Habegger (Wohlmeyer) 'Moles' cut their teeth in coal-mines.

For hundreds of years the common pick and shovel joined later by explosives, were the only tools wielded by the miner. But with the dawn of the industrial era, a few enterprising men turned their attention to the development of a machine which, it was hoped, would lighten the work of the miner. One of the first was Michael Menzles of Newcastle who, in 1761, filed a patent for a type of coal-cutting machine. The Menzies patent proposed the use of cutting tools fixed to a series of heavy chains which would be powered by man and horse. His invention encouraged other men to try their hand, and a number of variations to Menzies's original design were proposed. Though the majority of these early model coal cutters were designed to be worked manually, it is believed that at least one inventor also mooted the use of a horse-operated treadmill as a power source.

At the time when Menzies[1] and other such enterprising inventors were making their first efforts towards mechanization, coal was being won by men, women, and children who worked long hours in appalling conditions. It was not uncommon for girls and boys, some not much more than seven years old, to be employed hauling coal.

Menzies and his fellow inventors were beaten by the lack of a suitable power source. Yet with cheap labour so readily available, it is doubtful whether even the introduction of compressed air at that time would have availed, as few changes in those days were made for purely humanitarian reasons.

However, with the turn of the century and the simultaneous growth of the Industrial Revolution, the voices of men such as Lord Shaftesbury began to influence contemporary thinking. A Bill prohibiting coal operators from employing women and girls underground was passed. So, too, was a Bill which stated that no boy under the age of 10 years could be worked underground.

Despite this, conditions underground showed no real change for the better. Long hard hours were the common lot of the miner, who automatically regarded his employer as his enemy. Mine-owners for their part looked upon the miner as little better than a piece of equipment, this attitude being fostered by the general public's attitude towards these people. It was one of aloof indifference to their sufferings. Not surprising, therefore, that the seeds of unionism fell on fertile ground here. Not only did this movement draw an ostracized community together, but it provided the means, indeed the weapon, with which these oppressed people could fight and regain human dignity. All over the world similar conditions existed, the United Kingdom, America, and Australia being no exceptions.

Reference

1. F.S. Anderson and R.H. Thorpe, A century of coal-face mechanisation, Inst. of Min. Engs., *The Mining Engineer,* **126,** No. 83, 775-785, Aug. 1967, London.

EIGHTEEN

Great Britain

The Development of the Longwall Method of Mining[1-7]

When coal-mining in Britain was in its infancy and the majority of seams worked were thick or near the surface, the 'bord and pillar' or 'room and pillar' method of mining was mostly used. As mines grew deeper or the seams worked became generally thinner, the longwall system was gradually adopted until, eventually, nearly all mining carried out in British collieries was of the longwall type.

A contributory factor in the development of longwall mining was that it was found easier to direct a continuous stream of air over a long face than over a series of short openings or entries, in order to clear the pockets of high gas concentrations that tended to collect at the working face.

To begin with, the handworked walls were stepped, that is a series of short sections of wall were worked and then joined to form a more or less continuous line. The advent of the coal-cutter, which was hauled along the face, encouraged miners to straighten the line. Incidentally, this operation also facilitated better roof-supporting methods.

There are two basic methods of working longwall mining, i.e. by advancing or by retreating.

In longwall advance work, the coal beds in an ore body are mapped out in panels and then headings or stable holes are driven out from the main entries into the ore body and the panels left lying between the stables are excavated. These headings or stable holes extend a short distance beyond the actual working face and are used initially to house the face-cutting and loading machinery during the period when the conveyor and roof supports, etc. are being advanced towards the face in preparation for the next cut. As the face advances the stable holes or headings become access roads or gates which provide entries to either side of the longwall working face for men, supplies, and conveyors. Cross gates or roads are also made in the 'goaf' area for the conveyance of coal, etc. or for ventilation purposes. (The 'goaf' or 'gob' is the waste area left behind after the coal has been extracted and the roof has been allowed to collapse. The collapsing of the roof occurs after the roof supports have been advanced into their new positions along the working face of the longwall.)

In longwall retreat mining two headings or roadways are driven just beyond the end of the ore body and a further roadway is then driven along the back of the ore body to link the ends of the two headings. This exposes the rear face of the seam and this face is worked back towards the original points of entry.

The advantages derived from using the retreat method are that the exact size and condition of the ore body is known in advance and that production work is not held up or hampered by the need to drive headings or stable holes in advance of the face for housing the face equipment, etc.; also access roadways or gates driven through the solid bed of coal are easier to maintain than those which have been made through the disturbed goaf or mined-out area behind the face.

However, the deeper the mine gets the more susceptible it becomes to ground pressures, and as the face retreats, open areas close to the face such as roadways and gates are particularly affected by these pressures which cause the roof to collapse and the floor to heave. Indeed in some cases when a mine has been left unworked for a period of time, such as over a holiday weekend or because of an industrial dispute, miners have returned to find loaders and other equipment jammed tightly between the floor and roof.

As this tendency is more prevalent close to the working face it is understandable that roadways which have been standing for long periods, such as those giving access to panels which are being worked on the retreat system, will be more susceptible to the face pressure wave than the newly excavated stable holes or headings used for advance work.

In any event, regardless of which system of longwall work is selected, be it advance or retreat, the actual method of working the face remains the same and the same equipment is used.

Originally this method consisted of hand-picking a horizontal groove or kerf along the bottom of the face and also, when necessary, at an intermediate height or along the top of the coal seam. Holes were then drilled and explosives inserted. When the explosives had been detonated the coal was shovelled into tubs and removed. As mechanization increased, the hand picks were replaced with pneumatic picks or other types of coal-cutters, and conveyors were installed immediately behind the cutting machinery. Compressed air or carbon dioxide was also sometimes used instead of explosives. It was forced into holes drilled in the face and caused the coal to break or shatter.

In longwall work the roof is well supported immediately behind the face, and these supports together with the conveyor are advanced as the face is worked. Waste material is thrown into the goaf or mined-out area behind the face. This area is allowed to collapse as the roof supports are advanced into their new positions.

Coal-Cutters

Despite the work of many brilliant inventors, mechanization in Great Britain was notably slow. Efforts were mainly directly towards the development of the coal-cutter during the period 1850-1900.

During the initial stages of this era, however, very few of these inventions were successful. But there is no doubt that, like the early drills of Couch, Fowle, and Cave, etc. these clumsy

and ineffective machines showed the way towards new and better models.

Naturally this trend towards longwall mining influenced the work of designers in Britain to a considerable extent, so that mining machinery generally was designed with this method in view.

William Peace, William Firth, and William Baird[1, 8]

William Peace of Wigan filed a patent in 1853 for a machine carrying picks on a chain and, in 1861, after first marketing a pick machine, William Firth of Leeds produced a machine which attempted to imitate the manual operations of hand-holing. This was Firth's renowned 'Iron man'. As late as the year 1893 it was considered by George Blake Walker to be among the most successful compressed-air machines invented at that time. One of these machines was put to work at the West Ardsley colliery near Leeds. So satisfactorily did the Firth pick function that 30 years later the West Ardsley colliery was still using such a machine in its mines. Though several of these particular machines were tried in various other mines, they did not find favour with the men. Nevertheless this invention inspired others and several models employing similar principles of operation were built soon afterwards. One of these was the Howit percussion machine. Invented in 1867, it was operated by compressed air and inflicted hammer blows upon the coal.

In 1864 William Baird and Company produced a machine called the 'Gartsherrie' (Figure 392). This was a chain machine which had a jib or arm around which ran an endless chain carrying teeth or cutters. In operation this arm would project beneath the coal for distances varying from 840 mm (2 ft 9 in) up to approximately 1.5 m (5 ft). The machine was capable of cutting a slot 60 mm (2½ in) wide of say 840 mm (2 ft 9 in) in depth by 91 m (260 ft) long in about 8 to 10 hours, using two men and a boy as operators. It was installed in Lockwood colliery where it successfully managed to cut the coal. Ironically the British did not at first exploit this machine, they left this to the Americans. It was entered in the

Figure 389. Early Anderson Boyes electric disc-type coal cutter (Courtesy Anderson Strathclyde Limited.)

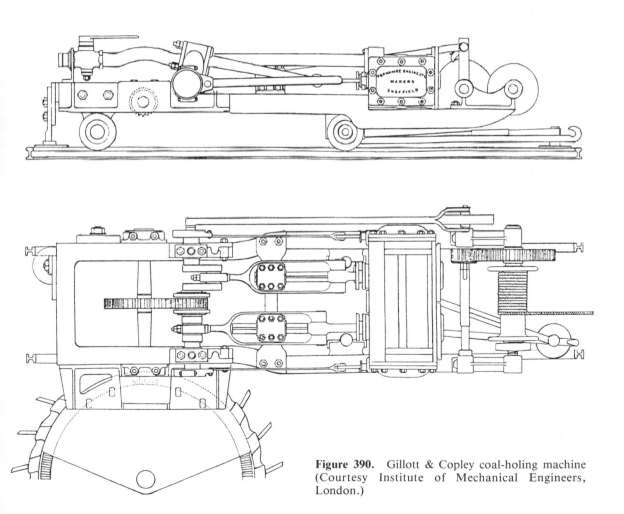

Figure 390. Gillott & Copley coal-holing machine (Courtesy Institute of Mechanical Engineers, London.)

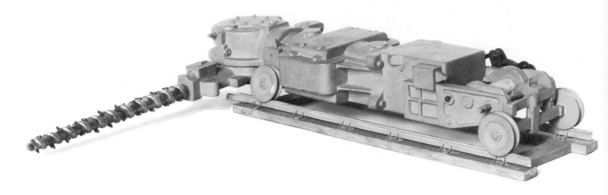

Figure 391. Early electric bar-type coalcutter (Courtesy Anderson Strathclyde Ltd.)

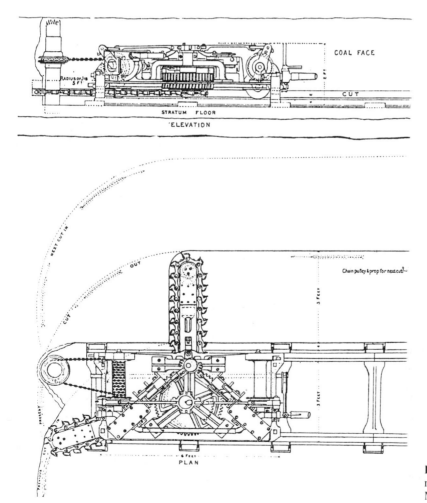

Figure 392. Baird's coal-holing machine (Courtesy Institute Mechanical Engineers, London.)

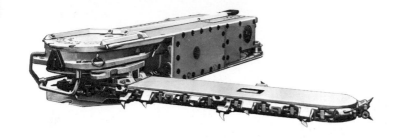

Figure 393. Anderson Boyes fifteen longwall coal cutter. (Courtesy Anderson Strathclyde Ltd.)

Philadelphia Exhibition in 1876 and America was introduced to the principles of the chain cutter for the first time.

Development of the Coal-Cutter

Disc-type Machines

The disc-type or rotary machine (Figure 389) made its initial appearance in 1868 when J. S. Walker of Wigan, using Michael Menzies's original saw design as a guide, produced a machine which cut a 100 mm (4 in) kerf to a depth of 610 mm (2 ft). It was powered by compressed air delivered at 0.13 MPa (18 lb/in^2), and undercut the coal by means of teeth or chisels set on the periphery of a wheel.

Walker's machine was followed by the Winstanley, the Gillott and Copley (1872), the Rigg and Meiklejohn, and many other successful disc-type machines. The disc machine was especially popular in longwall work and, during the 1880s, names such as Gillott and Copley, Rigg and Meiklejohn, Clarke and Steavenson and the Yorkshire Engine Company were closely associated with these particular machines.

During operation, the Rigg and Meiklejohn machine cut from the front to the back, while the Gillott and Copley machine cut in the opposite direction. Thus loose debris from the Rigg and Meiklejohn machine was carried into the cut, whereas (provided the cut was above floor level) debris from the Gillott and Copley machine was carried out. Some material would be carried into the cut if the Gillott and Copley machine was worked below floor level. To prevent this a guard was sometimes attached to the rear of the wheel.

As the depth of cut produced by the disc machine was limited, the bar machine was preferred for overcutting work or whenever a very deep cut was required. So far as the Winstanley disc machine was concerned, though it possessed a few commendable features such as the ability to cut into the face, it was not particularly successful, mainly because the machine lacked power.

Bar Machines

Bar machines (Figure 391) were capable of making cuts of up to 2.74 m (9 ft) in depth. The first patent for a bar machine was taken out in 1856, but it was not until 1888 that this type of machine was successfully tested. The model in question was the 'Goolden' bar machine which was put to work at T. R. and W. Bower's Allerton main collieries, near Leeds.

Not only was the Goolden the first coal-cutter to be driven by electricity, but its gearing as well as its 7.46 kW (10 hp) motor were completely enclosed. This refinement was not extending to most early models of coal-cutters where the cast-iron gearing was generally left exposed to the dust and dirt of the coal-mine, or occasionally afforded the doubtful protection of a sheet of galvanized iron across the top. The motor of the Goolden machine was capable of 600 rpm and operated a serrated cutting bar.

Percussion Machines

While Britain was developing the chain cutter and other similar types of machine, the Continent as a whole preferred percussion machines (see Figure 461) such as pneumatic hammer picks. These were especially popular in

451

Figure 394. Anderson Boyes fifteen longwall coal cutter fitted with gummer. (Courtesy Anderson Strathclyde Ltd.)

Figure 395. Anderson Boyes fifteen longwall coal cutter fitted with gummer at work underground. (Courtesy Anderson Strathclyde Ltd.)

the Ruhr district, Westphalia, and Belgium. Though these machines were not favoured in Britain, a few collieries did use them for a limited period, and one British manufacturer at least, produced a pneumatic hammer pick called the 'Eloy'.

These machines were not used to cut a kerf or groove, but were used for breaking down the coal from the face in both longwall and room and pillar work, etc. Moreover, where this type of pick was used it was often the practice to break down the face without either undercutting the seam or using explosives.

Basically there were two kinds or groups of percussive machine. One lot was hand-held and weighed about 12 to 13 kg (26 to 29 lb), while the other was a much heavier machine which was mounted on a frame. The Siskol and the Little Hardy (both manufactured in the U.K.) fell into the latter group, while the well-known American machine, the Ingersoll, was included in the first group.

The Chain Coal-Cutter[1, 3]

Gradually the chain cutter superseded other types of cutter such as the disc and bar machines, and became the major cutting tool in British collieries.

The simple cutter (Figure 393) consisted of a unit measuring approximately 2440 mm (8 ft) long by 760 mm (2 ft 6 in) wide and about 300 mm (1 ft) to 460 mm (1 ft 6 in) high. It was powered by a 30 kW (40 hp) or 45 kW (60 hp) motor having at one end a rope haulage unit and at the other a gear head. The haulage unit had two rope drums, situated one on either side of the machine. The cutting unit, which could be horizontally rotated thorugh an arc of about 180° to allow the unit to work either way across the face, consisted of a jib bracket which was mounted on the gear head. The jib bracket in turn supported the projecting jib, around the perimeter of which a toothed chain ran. The jib varied in length from 1.07 m to 2.74 m (3 ft 6 in to 9 ft) and the chain could be driven at speeds which ranged from 120 to 150 m/min (390 to

452

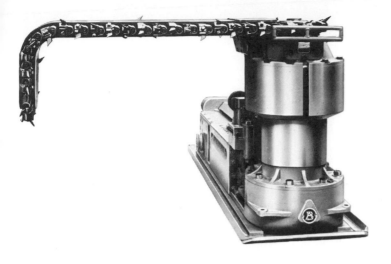

Figure 396. Anderson Boyes fifteen longwall overcutter fitted with hydraulic elevating turret and curved jib. (Courtesy Anderson Strathclyde Ltd.)

Figure 397. Anderson Boyes fifteen longwall coal cutter fitted with spade jib. (Courtesy Anderson Strathclyde Ltd.)

Figure 398. Anderson Boyes fifteen longwall overcutting coal cutter fitted with twin jibs and overcutting gum stower. (Courtesy Anderson Strathclyde Ltd.)

490 ft/min). Under normal working conditions the projecting jib was perpendicular to the face of the machine.

As the individual picks or tools in the cutting chain could be removed and replaced or their position altered so that different 'lacing patterns' were made, advantage could be taken of this facility to arrange the picks to suit the conditions of particular coal seams. For instance when soft material was encountered, the picks were set further apart on the chain than they would have been if harder material was being dealt with.

During operation the cutter was placed directly on the floor of the mine and the towing ropes were attached firmly along the working face. The cutting jib was then driven into the seam at floor level and moved across the face, cutting a horizontal groove or kerf at the bottom of the seam while the machine was being pulled along by means of the haulage unit. As the teeth on the cutting chain rotated about the jib, the coal was both cut and cleaned from the groove. (Later models were fitted with a 'gummer'.) (Figures 394 and 395). This was a screw or paddle-type device which collected the coal cuttings thrown out by the cutting chain and directed them towards the back of the machine, thus keeping the undercut clean.)

Depending upon local conditions the coal-cutter could be hauled through the coal at speeds varying from 0.30 to 1.52 m/min (1 to 5 ft) and was capable of 'flitting' speeds of up to 12 m/min (40 ft).

As the use of these machines grew, manufacturers began developing useful variations which could be applied to specific jobs.

Curved jib

In order to enable cuts to be made at intermediate or roof levels in the seam, coal-cutters with hydraulically operated turrets were built, the cutting jib extending in the usual way at right angles to the face of the machine from the top of the turret. Sometimes a down-curved jib arm would be fitted to the turret (Figure 396) to enable the operator to round the corners between the hanging wall (roof) and the vertical wall, thus increasing the roof's holding ability. It was also used to make a shear cut at the back

of the web while the horizonal cut was being made at the same time.

Twin-jib units

The introduction of the two-armed or twin-jib (Figures 398 to 399) coal-cutter allowed the operator to cut two grooves simultaneously, one at ground level and the other at an intermediate level in the seam. The vertical distance between the two arms could be adjusted by means of rising crossbars which regulated the height of the upper arm.

Mushroom arm cutter

Another special type of arm was the mushroom arm (Figure 400). This consisted of a standard jib with a vertical tree or turrett mounted near its end. The turret was fitted with teeth and rotated about its axis. It was driven by the cutting chain on the main jib. The mushroom device was designed to fit on to any standard-type chain coal-cutter and, like the curved jib, it allowed the operator to make a shear cut at the back of the web at the same time as the horizontal cut was being made.

Other variations on this theme included units with twin mushroom jibs and units with both a curved arm and a mushroom arm on the same machine.

Multi-disc cutter

The multi-disc cutter was also an interesting development and good examples of these units are to be seen in the Anderson-Boyes Anderton shearer loader (some models of which were fitted with a hydraulic turret overcutter or curved arm which was mounted on top of the unit) and the Anderson-Boyes multi-disc cutter-loader (which was fitted with a projecting cutting jib for precutting).

Some Anderson-Boyes machines were fitted with a special gummer that diverted, to the back of the machine, the material which was cut from the soft rock or clay bank which often lies immediately beneath a coal seam.

The disc cutter head consisted basically of a number of discs mounted side by side, either vertically or horizontally on the machine. The

454

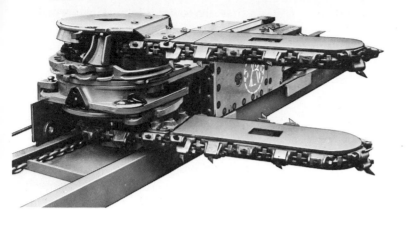

Figure 399. Anderson Boyes fifteen longwall coal cutter mounted on armoured conveyor and fitted with two jibs. (Courtesy Anderson Strathclyde Ltd.)

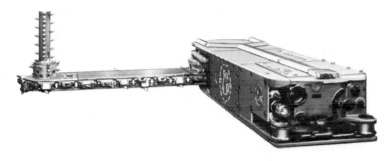

Figure 400. Anderson Boyes fifteen longwall coal cutter fitted with mushroom jib. (Courtesy Anderson Strathclyde Ltd.)

discs were fitted with cutting picks, and as the machine traversed the face the rotary action of the discs discharged the coal onto the conveyor.

In operation the discs rotated about their axes and so sheared coal from the face as the unit was advanced along it. When the multi-disc cutter had completed its journey, a coal plough was fitted and the machine was moved back along the face loading the coal lying on the ground onto the conveyor by means of the rotary action of the discs and aided by the coal plough. These multi-disc machines were the forerunners of the modern drum shearers which are used extensively throughout the coal industry in Britain today.

Dranyams

During the early 1960s a longwall power loader fitted with a cutting drum which rotated about a vertical axis was designed by a N.C.B. mining engineer named Maynard Davies — the title 'Dranyam' is the reverse of M. Davies's first name. It was designed to produce larger coal

than the Anderton shearer and several units were built in the Bretby workshops and then tested underground. Unfortunately, however, according to H. M. Hughes of the N.C.B., the action tended to force the machine out of the coal, and because stone bands are normally horizontal a few of the picks had to contend with this continuously. These picks wore more rapidly than the rest and after various trials these problems ultimately led to the abandonment of the unit.

Growth of the Industry

Before the First World War, British inventors established for themselves the enviable reputation of being in the forefront so far as engineering techniques were concerned, and many basic engineering principles which had originated in Britain were copied elsewhere. This was especially so in the area of coal-mining machinery. However, their efforts in this direction were extensively curtailed, largely as a result of the First World War.

In 1910 some 83 million t of coal were being

exported from the United Kingdom. This figure climbed to about 85 million in 1911 and 1912, and rose again to about 95 million t in 1913. From this point onwards there was a steady decline which reached its lowest point of some 43 million t in 1918. After the war there was a brief period of recovery when figures climbed to 48 million t. The advent of the depression years again affected the export market and there was another steeper dip to 35 million t in 1921.

Because of the conflict, supplies of coal from Britain almost stopped during the war years, but the world's demand for coal increased. As a result, new coalfields were brought into production in other areas of the world.

Though Britain experienced a brief period of recovery during the post-war years of 1922 and 1923, the gradual development of overseas coalfields encroached on Britain's export market, which continued to deteriorate for several years thereafter. With the heavy drop in output and consequent lack of ready capital for expansion programmes, colliery owners were unable to engage in extensive modernization plans. This in turn affected the work of inventors and engineers in this field.

During the succeeding years Britain gradually recovered, only to be plunged once more into the depths of another world war in 1939. Her export trade reached 53 million t of coal in 1937, was 48 million t in 1939 but plummeted down to 27 million t in 1940. But the worst was yet to come for in 1941 it was 9 million, 1942 — 7 million and by 1944 had slumped to 6 million t.

The vital importance of a continued and adequate supply of coal, especially for the war effort, was recognized soon after the outbreak of war. In 1939 emergency powers designed to control industry were conferred upon the ministers and government departments specified in the Defence (General) Regulations 1939, No. 927. The coal industry, amongst others, came under this order.

As a result of these regulations, and despite opposition from several quarters, the government assumed operational direction of all British coal-mines during the early 1940s. (The Coal Nationalisation Bill was a post-war Labour measure and was not passed until 1946. It came into effect officially on 1 January 1947.)

As a first step towards improving production the Mines Department appointed a joint committee which became known collectively as the Mechanization Advisory Committee. Its members consisted of a selection of machinery manufacturers, mining engineers, and Ministry officials.

Coal was needed for the war effort and it was needed urgently. Obviously Britain's conventional methods of winning coal were proving totally inadequate for meeting the current high demands, and the Mechanization Committee recommended a vast programme of modernization and mechanization of the coal-mines. However, because coal production generally had deteriorated over the preceding 20 years, the design and development of new mining machinery had also stagnated. To rectify this situation, manufacturing companies were promised government assistance in producing new machines, provided they could show reasonable proof that these devices would increase the output of coal. In addition, new trends in the United States were meticulously examined. Some American equipment was brought to Britain under a Lease-Lend arrangement with the United States. This was not altogether satisfactory, as British coal-mining methods differed considerably from those practised in America. There a great proportion of the coal is mined by the room and pillar system of mining, which did not suit conditions in Britain where coal generally is extracted from great depths and the longwall method of mining is adopted.* In addition, if the American machines were to perform at their best it was essential to select the thicker seams of coal. However, because of the American Agreement Britain found herself committed to giving the new machinery a thorough trial. Use of the new equipment also required radical and expensive alterations to the general layout of the mines. These were strongly opposed by many mining officials. It was at this critical juncture that the new A. B. Meco-Moore cutter-loader was introduced.

---

* Ironically the longwall system of mining is only now being introduced into the United States. This 'new' system of mining is slowly gaining popularity there (see Chapter 19 — America).

Figure 401. Meco Moor coal cutter loader — view from gob side showing machine in undercutting position. (Courtesy Dowty Meco Ltd.)

Figure 402. Early A.B. Meco Moore cutter loader. (Courtesy Anderson Strathclyde Ltd.)

A. B. Meco-Moore Cutter-Loader[1,2]

In 1930 Matthew S. Moore of Mining Engineering Company Limited designed a cutter-loader (Figure 401) which was first tested at Chiswall Hall colliery in 1933. This machine undercut the coal seam in the usual way. The coal was then broken down by blasting, after which the machine was run back along the face and the loading mechanism was put into operation, collecting the coal and loading it on to the face conveyor. Later the mining Engineering Company Limited and Anderson Boyes and Company Limited of Motherwell pooled ideas and in 1943 the A.B. Meco-Moore machine, (Figure 402) which was a combined coal-cutter and loader of superior design, was introduced to the market.

When layers of coal are lying at great depth below the surface, considerable pressures are exerted upon these seams from the material above. It has been found that when the internal stresses produced by these pressures are relieved on the underside only, by the cutting of a horizontal slot at the bottom of the seam with, perhaps, another approximately half-way up, the remaining stresses will force the coal to slab off the face on its own to nearly the depth of the slots. This will, of course, only take place freely if natural, vertical, or oblique cleavage planes exist in the area being worked. However, if no such natural planes are in evidence, the same effect can be produced by cutting vertical slots at intervals across the face of the seam. The A.B. Meco-Moore cutter-loader was specially designed to cut such artificial shear lines in order to reap maximum benefit from the existence of the very high stresses present in the coal lying at these deep levels. Of interest is the fact that the basic principles employed in this machine were to some extent used subsequently by the Mining Research Laboratory of the Chamber of Mines of South Africa in their 'drag-bit rock cutters'. (These machines are referred to more fully later.)

In the case of the A. B. Meco-Moore cutter-loader, two cutting jibs (one located at floor level the other at a predetermined position

457

higher up the seam) cut two slots along the working face in a horizontal direction. In addition, a shearing jib mounted at the back of the loader made another cut in the seam, this time vertically. Then, as the machine travelled along the working face the fallen coal was automatically gathered by the loader and dropped on to the face conveyer.

At about the same time that the new A. B. Meco-Moore cutter-loader was introduced, the Huwood loader and Shelton loader also appeared on the market. But these machines were not generally favoured, whereas the A. B. Meco-Moore machine found wide acceptance.

The A. B. Meco-Moore machine constituted the first real step forward from the coal-cutter along the road towards the evolution of the continuous miner proper.

Comminuting Machine

During the post-Second World War period (i.e. the late 1940s), after the A.B. Meco-Moore had become established as an efficient coal-getting unit, Anderson Boyes were requested by the Domainal potash mines of Alsace to build a machine (Figures 403 and 404) which was capable of cutting, in a single operation, the full thickness of a potash bed and loading it on to a conveyor. Such a machine would, they felt, alleviate a great deal of the drudgery involved in winning potash underground and, in addition, could eliminate some of the surface work concerned with crushing and grading.

The machine which evolved was called a 'comminuting machine' and was the first approach to the application of the shearer for modern longwall work. (A variant of the comminuting machine called the 'Dechisteuse' was also developed by Anderson Boyes. Its purpose was to cut out the schist or shale band in the seam of potash and stow this schist in the gob.)

The Samson Stripper

This machine (Figure 405) was developed in 1948 by Mavor and Coulson and the prototype was tried in the Parkgate seam of a South Yorkshire colliery. The machine, which was

Figure 403. Anderson Boyes comminuting machine 'Dechisteuse'. Developed to cut out 'schist' or shale band in the seam of potash and stow this schist in the gob. (Courtesy Anderson Strathclyde Ltd.)

Figure 404. Anderson Boyes comminuting machine with ranging drum (Courtesy Anderson Strathclyde Ltd.)

Figure 405. Mavor & Coulson Samson stripper. (Courtesy Anderson Strathclyde Ltd.)

self-propelled, ran along the face cutting out a 610 mm (2 ft) web of coal which it loaded on to the face conveyor. The machine was equipped with a wedge head at each end.

Basically the unit consisted of three sections, namely the two wedge heads between which was a hydraulic jack, and a hydraulic propulsion unit. Running laterally between the two wedge heads and through the centre of the hydraulic jack and the propulsion unit were two robust rods. In operation the jack was thrust firmly against the floor and the roof then the wedge heads were moved horizontally across the face by means of the rods which were supported in the middle of the unit by a bearing. When the rods had completed their lateral movement, the jack was retracted and the wedge heads dropped to the floor. The central unit, comprising the jack and the propulsion unit, was then able to move along the rods to a new position while being supported by the wedge heads. After the jack had again been extended against the floor and roof the machine was ready to commence the next cut. It was necessary to provide a stable at each end of the face, although the machine could work in either direction and therefore did not need to be turned at the end of a cut. When a cut was completed, the face conveyor was advanced 610 mm and the roof supports were repositioned.

Each wedge head was fitted with a number of sharp detachable, adjustable blades or chisels which, if necessary, could be positioned one in advance of the other, thus providing a bursting effect.

The Dosco Miner — Dominion Steel & Coal Corp. Ltd.

This Canadian-based company owned steelworks and coal-mines on Cape Breton Island in Nova Scotia. In the early 1950s they developed a machine for the mining of coal by the longwall method called the Dosco miner. Initially this machine was installed in the Dosco mines in Nova Scotia where eventually about 20 were put to work.

In order to market the Canadian-built machine in the U.K., Dosco Overseas Engineering Ltd. was formed there in 1953, and a Dosco miner was introduced that year at the Rawdon colliery. It was somewhat similar in concept to the continuous miners which were developed in America in the early 1950s and its modern counterpart may be found in the Dosco/N.C.B. dintheader which made its debut in 1967.

The original Dosco miner was mounted on crawler tracks. The 1.45 m (4 ft 9 in) wide cutting head was supported by a pivoted sliding

459

frame which enabled it to move backwards and forwards in a horizontal plane for a distance of about 460 mm (1 ft 6 in). It could also pivot up and down vertically, but could not move from side to side. Seven cutter chains running side by side and fitted with picks ran around the sliding frame and over the cutter-head drum.

Before the machine could be installed at Rawdon, several modifications, particularly in regard to the electrical apparatus, had to be carried out so that the machine would conform to British Standards. In addition, previous operational experience at the Dosco mines suggested that other minor modifications to the machine's crawler tracks, its cooling system and cross belt, would improve its performance. The original seven chains on the jib were replaced by two pick mats as a result of development work at C.E.E. (now M.R.D.E.).[9]

In operation the Dosco miner was positioned against the longwall face where it cut along the face in much the same manner as the continuous miner attacks the face from the front. That is, the head was lowered to the floor and allowed to sump into the coal. (The sumping distance was usually about 460 mm (18 in).) Then the cutting head was worked upwards to the top of the seam, after which the crawler tracks repositioned the miner and the cycle was repeated. The broken coal was loaded on to the conveyor for removal.

By 1957 Dosco miners were each producing approximately 254,000 t (250,000 ton)/year of saleable coal, and one unit at Donisthorpe colliery actually produced as much as 345,000 t (339,217 ton) of saleable coal in a year. The units were also sold fairly extensively in Germany.

However, during the early 1960s the attention of H. M. Inspectorate was drawn to various aspects of mine safety, in particular that of adequate roof support over the working area. Because of this, criticism was levelled at the large amount of roof exposed by the Dosco miner's wide-buttock system of coal-getting, which was necessary for the operation of a machine of this type. As a result the unit fell out of favour and ceased to be used in the U.K. Ironically, at about the same time developmental work was begun on powered supports which today provide sufficient cantilever support to allow a 3.04 m (10 ft) wide continuous miner to work on a longwall installation. The suggestion is put forward by Reid of Dosco that if these supports had been available when the Dosco miners were introduced, the development of longwall machines might well have taken a different turn, and the Dosco miner principle might have been widely accepted for longwall mining in medium-height seams.

Supports

Originally timber props and waste material or stone were used to support the roof. Later hydraulic jacks were used, and this greatly speeded up the work as they could be extended or retracted at will. But the greatest impetus to modern longwall mining came with the introduction of the self-propelled powered support.

According to C. T. Jones,[10] developmental work on mechanized support systems commenced as early as 1912 and, in 1928, Hamel devised a prop unit which utilized two endless belts run on rollers. The upper one was designed to support the roof and the other provided traction on the floor. The top belt assembly also extended a short distance beyond the lower belt assembly and acted as a sort of protective canopy over the face conveyor. In 1943 the first 'walking support' was invented by Winkhaus. It consisted of two parallel bars for supporting the roof. These bars were attached to each other by a sliding joint which enabled each bar to be moved forward individually when the unit needed to be repositioned. The horizontal roof bars were supported by vertical metal props which were hinged at their connection point with the roof bar. The prototype unit was tested at Hansa colliery in Germany, but apparently did not prove to be satisfactory.

These were followed by a variety of supports which included the *Kuzbass* system (Soviet Union) — also a walking support, and by a caterpillar track unit designed and patented by the Joy Manufacturing Company. This latter unit was basically similar in concept to the Hamel roller belt device. However neither the Hamel device nor the Joy unit was ever tested in practice underground.

460

Figure 406. Twenty-five ton yield load *single* hydraulic props in use with roof bars and conveyor. (Courtesy Gullick Dobson Ltd.)

Figure 407. Old-type roof support. (Courtesy Gullick Dobson Ltd.)

During the 1950s a great deal of developmental work on mechanized supports was carried out in Britain, Germany, the Soviet Union, and in various other countries, and several designs of mechanized support units were tested. Attention was directed at that time to the problem of floor and roof irregularities and undulations and also lateral pressures. While most supports in Britain up until the mid-1950s used both articulated (hinged) and rigid roof and floor bars, Klockner-Ferromatic of Germany were testing spring steel bars for both floor and roof components. The prototype units, consisting of two hydraulic props operating between spring steel roof and floor bars, were installed in closely positioned pairs along the face. They were tested at the Victor-Ickern colliery in Germany.

In addition to developmental work on the roof and floor components, the props themselves have over the years undergone changes. Originally the hydraulic system was an open-circuit one, which meant that fluid was discharged on to the floor, often causing the feet of the props to sink. Later closed-circuit systems were introduced which eliminated this problem. In addition, in an attempt to provide added flexbility and stability to the roof and floor members, the hydraulic props were attached by ball-and-socket joints within a rubber bush. The bush ensured that the leg would return automatically to its natural vertical position after each move. Some props were also attached with a dome-and-spigot joint. This consisted of a prop which was dome-shaped at each end. In the centre of each dome was a hole which accepted the end of a spigot attached to either the roof or floor member, as

461

Figure 408.   Installation of old-type roof supports. (Courtesy Gullick Dobson Ltd.)

the case may be. This swivel mounting arrangement was also surrounded by a rubber bush which ensured that the prop returned to the vertical position after each resetting. Later both the ball-and-socket and the dome-and-spigot joints were eliminated and replaced by rigid connections, as it was found that they only tended to increase the susceptibility of the prop to collapse rather than to prevent it.

When wooden props were used, these were wedged into place until they were firm and this was known as the 'setting load'. As the face advanced, the roof began settling and the timber gradually yielded to the pressure until it became stabilized or, if the downward pressure continued and no further chocks were inserted to arrest this movement, the props would collapse. When hydraulic jacks were introduced the jacks were designed to accept a certain setting load, if this load were exceeded the hydraulic fluid was slowly bled away to enable

the prop to 'yield' to the pressures. Two methods were used. Either the fluid drained completely away until the jack was no longer able to yield in which case it became solid, or the fluid was bled from one chamber into another until a state of equilibrium was reached when the jack also became solid. This state of affairs must naturally be avoided in a mine, and usually long before a jack becomes solid other jacks are inserted to provide additional support. When a jack does become solid it is usually removed by dismantling it from the roof and floor members and it is then taken to the workshop before the hydraulic system is opened and repaired. This is to prevent the hydraulic system from becoming contaminated with dust, etc.

With the introduction of the hydraulic jack came the necessity to understand more fully the various factors influencing pressures in a mine. The setting of supports in various mine

462

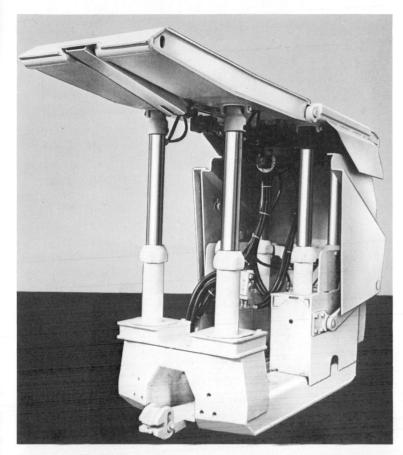

Figure 409. Six-leg, 240-ton hydraulic roof support. (Courtesy Gullick Dobson Ltd.)

Figure 410. Six-leg, 240-ton hydraulic roof supports, underground in a north-eastern colliery. (Courtesy Gullick Dobson Ltd.)

463

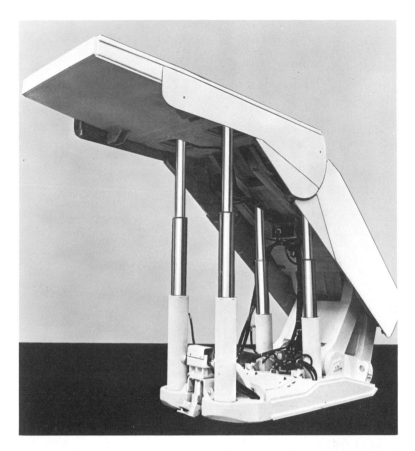

Figure 411. Four-leg, 500 (US) ton shield-type roof support. (Courtesy Gullick Dobson Ltd.)

Figure 412. Five-leg, 200-ton supports, underground in a northern colliery. (Courtesy Gullick Dobson Ltd.)

Figure 413. Five-leg, 200-ton supports, underground in a North Yorkshire colliery. (Courtesy Gullick Dobson Ltd.)

conditions is critical and may vary from mine to mine. Supports which are set at too high a load setting may cause premature failure of the roof because they may be thrust into the floor. On the other hand, if the load setting is too low this may produce stresses at the face which could cause the face to break away prematurely. The degree of yield is also important. Supports are designed with specific yield loads for specific purposes. Thus the support must be carefully chosen for the particular conditions existing in a mine. Its function is to adequately support the roof, act as an anchor for the conveyor, to hold it in position during the mining phase, provide thrust reaction for the jacks as they advance the conveyor against the face after a cut has been made, and, finally, move itself into place beside the advanced conveyor in readiness for the next cut.[7]

A typical modern self-advancing powered support (Figures 406 to 414) for longwall work consists of a cantilever prop unit or roof beam section (to allow for a prop-free front) which is supported by hydraulically activated legs (varying from two to six in number). These legs are similar to hydraulic jacks. The unit has, in addition, another jack, lying horizontally at floor level, which is used both to advance the conveyor and to pull the powered support unit into its new position behind the conveyor.

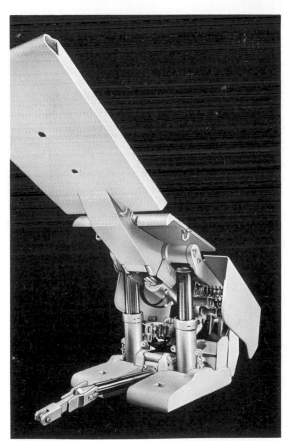

Figure 414. Voest-Alpine two-leg hydraulic roof support. (Courtesy Voest-Alpine AG.)

Figure 415. Westfalia Reisshaken plough. (Courtesy Westfalia Lünen.)

Conveyors

At first, coal from the face was loaded into tubs and carried out of the mine by animals or men then, for many years, it was carried in hand trucks on rails. In the early 1900s the first belt conveyors were used. They were placed along the face and coal was loaded on to them and then carried to the gates or roadways where it was delivered into skips. As belt conveyors improved they were used more extensively. However, like the powered support the greatest impetus to longwall mining came with the introduction in the early 1950s of the flexible armoured conveyor which made practicable the concept of a prop-free front. These conveyors moved the broken coal from the face area and also carried power loaders and coal-cutters running on their side rails. Basically they consist of 1.52 m (5 ft) long sections of steel plate flexibly joined to form one continuous length along the face. Supporting side plates run down each side of the conveyor. Coal is moved along the conveyor by means of scraper chains and/or flights. Its flexible joints allow it to be snaked into a new position along the face by the power supports immediately after the cutting and loading equipment has passed

across. When a coal plough is used, the conveyor holds it against the face during the cutting operation.

Germany was the first to use this type of conveyor in conjunction with its 'plough'-type machines (Figures 415 and 416). These units pared off layers of coal with a rigid plough or blade as they were hauled along the face. As the plough moved across the face the blades also deflected the coal on to the conveyor. They were tried in Britain in 1947 but were not popular as they were found to be more suited to the softer coals. However, later models have been developed which will deal with the harder coals.

Huwood Slicer Loader

There are two types of coal plough, low-speed ploughs and high-speed ploughs. Low-speed ploughs cut approximately 380 to 510 mm (1 ft 3 in to 1 ft 8 in) from the face as they move across, whereas high-speed ploughs will only cut off about 50 to 100 mm (2 in to 4 in) of coal during one pass.

In later models the blades were activated with either percussive picks or oscillating blades. A good example of this latter type of machine is

466

Figure 416. Westfalia Gleithobel plough. (Courtesy Westfalia Lünen.)

to be found in the Huwood slicer loader which is mounted on an armoured conveyor. Vertically arranged picks fitted to a bearing bar on each end of the machine oscillate and cut the coal as the unit traverses the face. As the coal drops from the face it is deflected on to the armoured conveyor by the wedge head or blade which is positioned immediately behind the cutting tools.

The A. B. Anderton Shearer Loader
(Anderson-Boyes & Co. Ltd), and
The B. J. D. Anderton Shearer Loader
(British Jeffrey-Diamond Ltd)

These machines made their debut in the early 1950s and they were mounted on armoured conveyors. The first machines were developed by Anderson-Boyes & Co. Ltd. (Figures 417 to 422), but later British Jeffrey-Diamond Ltd. also made significant contributions towards the development of the unit. When the first

Anderton shearer loader was made, the cutting head was fitted with a series of six discs each equipped with picks. Later the disc cutter was superseded by a revolving drum, around the periphery of which were set the cutting picks. In both types of machine the picks were so arranged that they cut upwards into the coal. This action tossed the broken coal back across the top of the drum and on to a specially shaped plate or inverted hopper which deflected the coal on to the conveyor. The machine was hauled along the armoured face conveyor by a wire rope.

Ranging Drum Shearers

Gradually over the years the longwall shearer has been developed further. Manufacturers began producing first single-ended and then double-ended shearers on which the cutting drum was carried at the end of a ranging arm which could be raised or lowered by a hydraulic

467

Figure 417. Anderson Boyes Sixteen Anderton shearer loader. (Courtesy Anderson Strathclyde Ltd.)

Figure 418. Anderson Boyes bi-directional shearer loader. (Courtesy Anderson Strathclyde Ltd.)

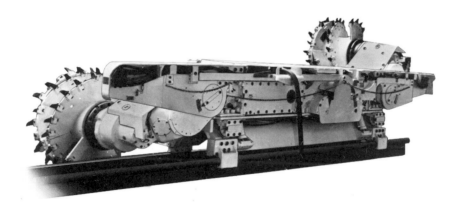

Figure 419. Anderson Boyes double-ended ranging drum shearer with roll steering under-frame. (Courtesy Anderson Strathclyde Ltd.)

cylinder, thus allowing thicker seams to be worked. In some cases longer arms have been provided to enable the drum to cut past the conveyor drive head and so eliminate the stable. This also made it possible in suitable conditions, to cut out the roof material and form the roadway, thereby avoiding the need to use a separate machine for an advance heading or for ripping behind the face.

Chainless Haulage Systems

Another important innovation is the chainless haulage system now being produced by various mining machine manufacturers and used in conjunction with all types of drum shearers. Although employing different design techniques, these systems are basically similar in concept. The 'Rollrack' system produced by Anderson Strathclyde Ltd. (U.K.) is described below.

468

Figure 420. Anderson Boyes double-ended ranging drum shearer at work in a potash mine. (Courtesy Anderson Strathclyde Ltd.)

Figure 421. Anderson Mavor Sixteen buttock shearer at work underground. (Courtesy Anderson Strathclyde Ltd.)

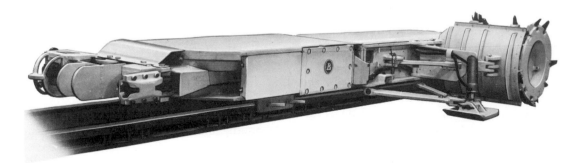

Figure 422. Anderson Boyes 10/12 thin seam shearer. (Courtesy Anderson Strathclyde Ltd.)

Figure 423. AM16 single-ended ranging drum shearer with roll-rack traction unit. (Courtesy Anderson Strathclyde Ltd.)

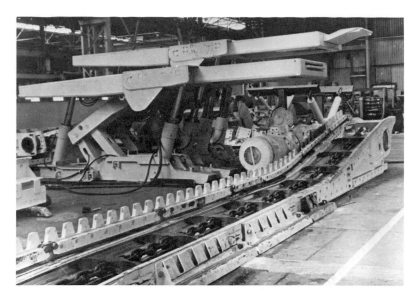

Figure 424. Centre-strand conveyor with racks for roll-rack haulage system. Note the double strap links which connect the rack bars. (Courtesy Anderson Strathclyde Ltd.)

This system consists of a rollrack traction unit and a rack bar (Figure 423). The rack bar extends the length of the conveyor and is bolted to the goaf side of the armoured face conveyor (A.F.C.) pan. The rack bars are interconnected by means of simple double strap links to maintain the pitch of the rack teeth within the permitted tolerances. The links also allow the A.F.C. pan articulation in both the horizontal and vertical planes, sufficient for all normal operations (Figure 424).

In one arrangement, for a single-ended machine, the traction unit is mounted on a vertical plate attached to the haulage end of the machine underframe. A wheel with five equally spaced, hardened steel rollers is mounted on the output shaft of the traction unit. The rollers on this wheel engage with the teeth on the rack bar

470

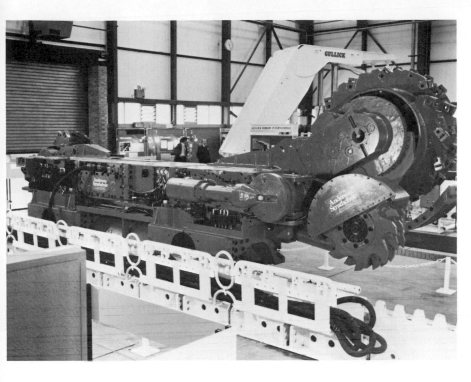

Figure 425. AM500 fitted with lump breaker and double-drive roll-back traction underframe. (Courtesy Anderson Strathclyde Ltd.)

and will drive the machine in either direction. The rollers are maintained in contact with the rack by a trapping shoe bolted to the traction unit, usually vertically below the centre-line of the output shaft.

In the case of double-ended ranging drum shearers, a traction unit may be fitted to each end of the machine underframe. Alternatively two, three, or four driving units may be incorporated in the underframe to provide an independent traction underframe. Attached to the unit and travelling just ahead of the roller wheel is a rack cleaner which prevents dirt from packing into the teeth of the rack bar.

Lump Breakers

Another innovation is the provision of a lump breaker which may be mounted on the machine at one end (Figure 425) when required to break down the material on the conveyor to a size which can pass through the tunnel beneath the machine.

Anderson-Boyes Longwall Trepanner

This machine (Figure 426) was also introduced in the early 1950s at about the same time that the Anderton shearer loaders came on the market. Although the trepanner was floor-mounted, it nevertheless made use of the armoured face conveyor to guide it along by means of a rail bolted to the bottom of the trepanner, which ran along and bore against the face side of the conveyor.

In operation the trepanner-type cutting head which consisted of two trepanning arms equipped with picks, rotated about its axis parallel to the face. As the machine moved across the coal seam it cut a 690 mm (2 ft 3 in) deep curved groove from the face. The cuttings were directed onto the conveyor by the trepanner wheel. In the meantime the shear jib and roof cutting disc cut through the remaining coal at the roof, which caused it to fall onto the sloping top of the trepanner machine from where it slipped down onto the conveyor. At the same time the seam was undercut by the floor jib.

As soon as the trepanner had passed and any coal which might have been spilt during its passage cleared from the area immediately behind the face, the power supports snaked the conveyor into its new position behind the face and also repositioned themselves against it.

471

Figure 426. Anderson Boyes HD Type longwall trepanner (floor mounted). (Courtesy Anderson Strathclyde Ltd.)

Remotely Operated Longwall Face (R.O.L.F.)

In January 1963, two remotely operated longwall-face schemes were launched. One was in the High Main Seam at Newstead Colliery No. 4 Area N.C.B. East Midlands Divison, and the other was in the East Midlands Division Pipe Seam at Ormonde Colliery, No. 5 Area.

Since then a considerable amount of developmental work has taken place in this field and radio control systems which enable the operator to control the machine from up to 15 m (50 ft) distance are now available. As a precautionary measure in the event of the operator being outdistanced by the machine, manufacturers such as Anderson Strathclyde Ltd have designed the units to fail to safety.

Another device offered by Anderson Strathclyde Ltd is an 'automatic horizon system'[11] which uses an isotope sensing device to maintain the shearer in a predetermined horizon throughout the length of the coal face.

Heading Machines

During the late nineteenth and early twentieth centuries British collieries used predominantly longwall machines, but heading machines increased in number as their use in developmental work was greatly extended. They were also extensively used during this period, and up to the advent of the second generation of chain coal-cutters, for the purpose of driving advance headings or stables at each end of a longwall face to allow the longwall machine to be housed and set up for the next cut.

Later chain coal-cutters eliminated the need for stables at the face ends.*

Brunton's Heading Machine[12]

The idea of using a heading machine was mooted as early as October 1875 by William Johnson of Bretby collieries in a paper presented by him at a meeting of the Chesterfield and Derbyshire engineers. Johnson suggested that the time had come for a mechanical header to be introduced. Johnson commented that while holing machines** were making steady progress (and he hoped they would succeed) they were of no use until some considerable extent of headings had been driven and stalls at the ends of working faces had been opened out. Johnson suggested the use of Brunton's tunnelling machine for this purpose. This machine he said, had been recommended

---

* Apart from the British heading machines described in the following chapter, Figures 452 to 454 depict examples of some 'continuous miners' which are currently being used for both extraction and heading work in Russia.

** The mining term for all coal-cutters, such as those described at the beginning of this section, which 'hole' or undercut the coal face by means of a deep kerf.

472

by the engineers for the proposed Channel tunnel and had been well tested. He was certain that if properly applied such a machine would be well suited to such work. Johnson claimed to have witnessed a 2.13 m (7 ft) diameter model of Brunton's machine at work on the grey chalk of Snodland in Kent. He stated that it had driven 52 cm (20½ in) in 25 minutes. Johnson felt that the chalk was in every respect as difficult to work as the coal would be.

Despite Johnson's eloquent plea, so far as is known, Brunton's machine was not tried in any of the British collieries.

Stanley Heading Machine[13, 14]

On 10 January 1885 under No. 335 'An improved boring machine', Reginald Stanley patented his first tunnelling machine. During the succeeding five years, that is from 1885 to 1890, Stanley designed several different types of cutter head and made various improvements to the machine in general (Figures 427 to 429). Prototypes of these machines were built and most were put to work in the Stanley colliery at Nuneaton.

The first Stanley machine (Figure 427) was basically a trepanner which cut an annular

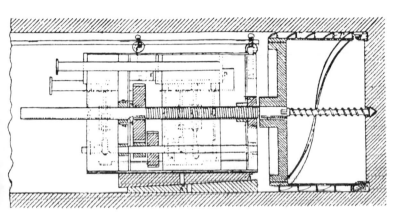

Figure 427. Reginald Stanley's first machine. U.K. Pat. No. 335, dated 10 Jan. 1885

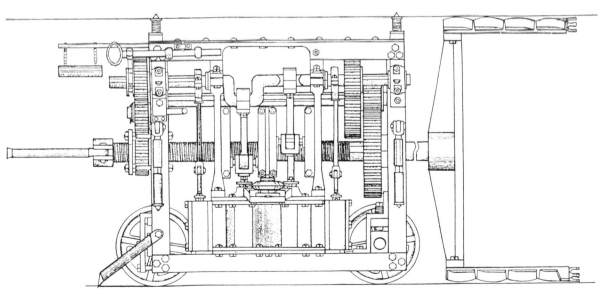

Figure 428. Stanley's 1888 machine — U.K. Pat. No. 1,763 dated 21 Dec. 1888 (application filed 6 Feb. 1888)

473

groove 1.52 m (5 ft) in diameter in the face of the heading. The cutter head consisted of a cross-head which was fastened to the end of the main driving shaft. Attached to the cross-head and set at right angles thereto were two projecting arms which extended forward towards the face and carried cutting tools on their ends. The length of the projecting arms (i.e. 0.91 m (3 ft)) controlled the maximum depth of the cut.

A drill fitted with a spiral worm projected from the centre of the end of the main shaft. Its purpose was to bore a central hole for blasting out the core and also to assist in steadying the

end of the central shaft while the machine was in operation. The cross-head was advanced to drive the cutters into the face by means of a screw thread cut along the main shaft for most of its length.

To maintain its position and to provide reaction to the thrust applied against the face, Stanley's machine was anchored to the sides and floor of the tunnel. Stanley also provided a wide curving blade of metal shaped to form a spiral sweeper, of the diameter of the tunnel, to break away loose coal and move it back from the face. This constituted part of a cycle of a worm conveyor and ran from the forward end of one cutting arm in a spiral sweep back to the cross-head at the rear end of the other cutting arm.

In Patent No. 1,449 dated 1 February 1886 Stanley replaced the screw-threaded central shaft with a plain one which extended beyond the main frame of the machine, thus allowing space behind the arms for a workman if this became necessary. Or, alternatively, the arms could be retracted and worked close to the frame if desired.

Stanley's 1887 machine (Patent No. 2,312) was mounted on central wheels or rollers which were either spiked or roughened and run on the floor of the heading instead of on rails as had hitherto been the case. This was designed to provide a more rapid means of advance between the periods of cutting and repositioning. In addition the central drill had been eliminated. The cutter head of the new machine consisted simply of a massive bar lying parallel to the face and carrying arms set at

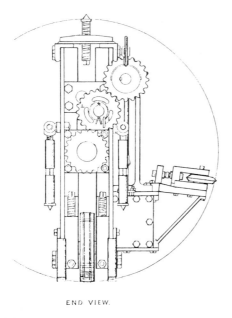

END VIEW.

Figure 429. Rear-end view of Stanley's 1888 machine

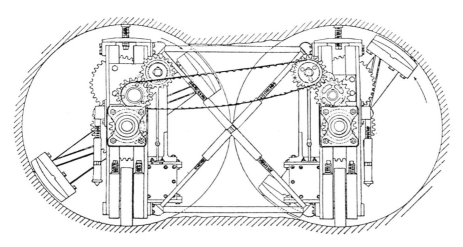

Figure 430. Rear view of Stanley's double header. U.K. Pat. No. 3,595 dated 6 Jan. 1891 (application filed 6 Mar. 1890)

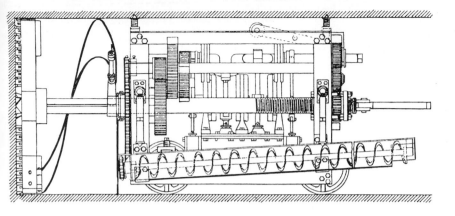

Figure 431. Stanley's T.B.M. showing Archimedean screw in a pipe conveyor. U.S. Pat. No. 577,331 dated 16 Feb. 1897

right angles thereto, which extended towards the face for a distance of 610 mm (2 ft). On these arms were attached the cutting chisels or knives. Later that same year Stanley filed a patent for a full-face tunnelling machine. In this model the extending arms carrying the cutters were dispensed with and the head consisted of a long heavy bar lying parallel to the face and attached to the main central driving shaft. Along the length of this bar Stanley inserted a series of chisels. (At Dixon's suggestion a machine of this design was tried for a short period at the Hamilton Palace colliery, but Dixon found it almost impossible to keep the machine up to the face and it was therefore returned to the manufacturer.) Stanley's 1887 full-face machine carried no spiral sweeper but 10 years later this was reintroduced when Stanley filed a patent covering this design in America under No. 577,331.

Possibly as a result of the pioneering work carried out by Dixon at the Hamilton Palace colliery, Reginald Stanley also designed a machine which became known as the 'Stanley double-header' (Figure 430), which he patented in London on 6 March 1890 under No. 3,595 and in America in 1893 under No. 504,179. This consisted of two machines — a right and left hand which were coupled together by stays and braces at the top and bottom. With such a machine a heading 3.4 m (11 ft) wide could be driven through the coal, leaving 1.2 m (4 ft) between the coupled machines for the passage of material in tubs, etc. (The suggested use of such a machine appeared in the discussion dated 11 October 1893 which followed J. S.

Dixon's initial report on work done by the Stanley heading machines at Hamilton Palace colliery.)

In December 1888 it was found necessary to carry out developmental work at Hamilton Palace colliery at a much faster pace than was currently possible with hand labour. As favourable reports had been received of the work performed by the new Stanley heading machines, it was decided to install one of these units as a trial. The model chosen was a 152 cm (5 ft) trepanner without the central drill and spiral sweeper. It was driven by compressed air.

Soon after installation it was found that the exhaust air from the machine was not sufficient to keep the face clear of fire-damp and a continuous blower of compressed air was installed. Apart from its expense, this, too, proved inadequate.* As haulage roads were normally 3.35 m (77 ft) wide, James S. Dixon (in charge of the operation) decided to use two Stanley headers, one working immediately in advance of the other, to cut parallel tunnels side by side leaving a drift of coal approximately 300 mm (1 ft) thick between them. The idea was to utilize this curtain of coal as a form of bratticing for guiding the air. When the drift was afterwards removed it left the roof flat with curved sides which stood well.

This method of driving, claimed Dixon, gave additional room for ventilation and loading, etc. as well as space for the trailing pipes and

---

*It is of interest that out of a normal 8 hours 14 minutes of work, the machine itself was only at work for 1 hour and 18 minutes so that a complete stagnation of the ventilation system lasted for 7 hours. This accounts for the fact that the machine exhaust was inadequate for ventilating the face.

other accessories connected with compressed-air machines.

The machines were used for a total period of approximately five years at Hamilton Palace colliery. During this time an average distance of 3.6 m (11.96 ft) per shift was maintained and five workmen were employed to operate the machines at a cost of 5s. (1s. per man) per shift. This meant that the coal cost the colliery 2s. 6d. per ton to produce. Dixon's conclusion was that, while the Stanley heading machines drove a place 3.35 m (11 ft) wide at about four times the rate of hand labour, the cost to the colliery was double. However, he felt that in special circumstances such as existed at the Hamilton colliery, it was worth the extra cost in order to achieve the desired rapid rate of advancement.[13]

(In the discussion[14] which followed Dixon's report Stanley Bros., who were present, suggested that some of Dixon's cost problems at the Hamilton Palace colliery might have been averted had he employed competent English workmen instead of Scotsmen. A. Blyth (Hamilton) thereupon retorted that an English expert had been sent to Hamilton Palace colliery to instruct the workmen there, yet when the Scots crew commenced work they actually cut about 0.5 m/day (1.5 ft/day) more than had the English expert.)

Further improvements to Stanley's machine

For a period of approximately 20 years commencing from the year 1885, Reginald Stanley continued to work at improving his machines, most of which were used in his own colliery at Nuneaton. The central shaft and cutting arms were strengthened, and later he also provided for a slow rotation of the cutting arms for use when the machine was working in extremely dense material, and a faster rotation of the arms for work in soft material. This latter improvement applied particularly to his full-face machines.

In 1905 he filed his final patent covering improvements to his duplex coal heading machine (double-header). These improvements were concerned with the mechanical removal of the coal from the face.

By turning the duplex machine upside down (i.e. the engines and gear were positioned at the top instead of at the bottom as previously), a clear space was provided underneath, from the coal face through to the back of the machine. In this space a plate iron trough was suspended. This trough was pivoted and could be raised or lowered at either end to provide the desired inclination. Along one side and over its entire length almost within the trough, Stanley ran an endless chain carrying a series of blades or scrapers which were fastened to its upper edge. These blades, which projected traversely across the trough and almost scraped the bottom, pushed the coal or debris which had been cut at the face to the rear of the machine. (This was a complete change from Stanley's 1897 model patented in America under No. 577,331 in which he used an Archimedean screw (Figure 431) in a pipe to convey debris to the rear, rather as a wood-boring bit conveys chippings out of a hole. Beaumont proposed this method for his machines, but does not appear to have incorporated it in any models actually built. English describes the screw method more fully in the Provisional Specification covering his improvements to Beaumont's second machine. However, between the period when this was filed and the date the specification was finally granted (25 October 1880 − 23 April 1881) English appears to have changed his mind, as the specification mentions a chain of buckets which is driven by pulleys on a shaft mounted at the back end of the upper frame. The drawings accompanying the patent also show what appears to be a bucket chain).

Although Stanley did not file any further patents covering tunnelling or heading machines after 1905, he continued with his inventive work for some years thereafter and several additional patents were lodged by him dealing with moulding bricks and tiles and other similar subjects. These inventions related to machines in use at the Stanley Brickworks at Nuneaton.

Stanley died on 21 July 1914 at Bexhill-on-Sea following a prolonged illness. In the obituary written after his death Stanley was remembered as the founder of the Nuneaton firm of colliery proprietors and brickmakers and mention was made of the fact that he had built various churches and donated 1000 guineas to the Wesleyan Twentieth Century

Fund. It was also mentioned that he had once declined an invitation to enter politics as a candidate for the Nuneaton Division of the Liberal movement. Strangely enough no reference was made in the obituary to the extensive work on tunnelling or heading machines which had been carried out by Reginald Stanley for two decades.

The following list of patents taken out by R. Stanley from 10 January 1885 to 28 December 1905, shows the great contribution made by this man towards the development of the modern tunnelling machine.

335	10 Jan 1885	An improved boring machine.
12,715	23 Oct 1885	An improved boring and air coursing machine.
1,449	1 Feb 1886	Improvements in boring and tunnelling machines.
2,312	14 Feb 1887	An improved boring or tunelling machine.
8,375	10 June 1887	Improvements in tunnelling machines.
1,763	6 Feb 1888	Improvements in boring or tunelling machines.
15,466	27 Oct 1888	Boring or tunnelling machines (application abandoned).
14,348	11 Sept 1889	Improvements in boring or tunnelling machines.
3,595	6 March 1890	Improvements in boring or tunnelling machines.
20,638	17 Dec 1890	Improvements in boring or tunnelling machines.
27,068	28 Dec 1905	Improvements in or connected with coal heading machines.

Summary

The Stanley heading machine was doomed for three important reasons. Firstly, at that time British colliers in the main concentrated on longwall mining and so heading machines were not particularly exploited. Secondly, coal cutters were cheaper and replacement parts considerably less expensive than for the larger heading machines. Thirdly, labour was plentiful and still extremely cheap — an important factor. As Dixon had pointed out, *though the heading machine drove a place at four times the rate of hand labour, the cost to the colliery was double that when performed with hand labour.* So long as the labour market

remained at that level, mechanization would perforce be held at bay.

Later with the gradual growth of the trade unions, a change occurred and the use of heading machines was extended slowly. Today many new models, designed and built by British manufacturers, have made their appearance on the market. Included amongst these are the Dosco-N.C.B. dintheader, the NCB-Dosco in-seam miner, and the Anderson Mavor RH1 roadheader,* etc.

Though designed primarily with the coal industry in mind, these machines are essentially *tunnelling machines* within the soft to medium hard-rock range, dealing effectively with such materials as clay, gypsum, coal, salt, potash, etc.

The Dosco/NCB Dintheader

The Dosco dintheader (Figure 432) was developed by the N.C.B. (National Coal Board) who hold the patent for it and Dosco (Dominion Steel and Coal Corporation Ltd.) to deal with floor heave and other similar problems and also as a roadheader for low-height rectangular openings. The basic design was carried out by C.E.E. (Central Engineering Establishment of the N.C.B.) and it is manufactured by Dosco under licence.

Two prototypes were built during the period May-November 1967 under a development contract and one unit was tested initially at the N.C.B. Swadlincote test site. Later the machine was moved to the Calverton colliery, South Nottinghamshire Area, to dint 343 m (1120 ft) of roadway, and subsequently it was put to work at Chislet colliery where it drove 366 m (1200 ft) of a 4.27 × 1.98 m (14 ft × 6 ft 6 in) rectangular heading. The other unit was tried at New Lount colliery where it drove a 54.9 m (180 ft) heading through coal and stone to produce a 3.96 × 1.98 m (13 ft × 6 ft 6 in) rectangular opening.

Like its predecessor, the Dosco miner, the new dintheader was mounted on crawler tracks. It was 1240 mm (4 ft) high and weighed 14 t.

---

* These three machine titles were coined at M.R.D.E. to cover N.C.B. patented developments. M.R.D.E. received a Queen's Award for Technological Achievement for the in-seam miner in 1977.

Figure 432. Dosco/NCB dintheader. One of these machines was exhibited at the Poznan Exhibition in Poland. (Courtesy Dosco Overseas Eng. Ltd.)

The cutterhead consisted of a continuous pick mat which ran up the jib, over the cutter drum, down the underside of the jib, around the driving drum and back. The jib was pivoted in a vertical plane by three-stage telescopic hydraulic cylinders (in later models the jib was pivoted by two two-stage hydraulic cylinders). Bollards which extended on each side of the jib headshaft gave an effective cutting width of 1710 mm (5 ft 7 in) while the jib could be raised to cut a height of 2130 mm (7 ft). The prototype units contained about 340 picks on the mat and bollards, but this number was increased to 540 on the latest units now operating.

The unit operated as follows. As the material was cut from the face by the cutting jib it dropped to the floor and was pushed forward by a plough or plate in front of the machine. The material was then fed via a hopper behind the pick mat on to a scraper conveyor which ran through the body of the machine and discharged on to a bridge belt conveyor towed behind the machine.

Its mode of operation was somewhat similar to that of the original Dosco miner except that the unit did not work along the face. The traction system was left unchanged, but the sumping action of the jib was eliminated. The machine cut in the same way from the floor to the ceiling.

Since its introduction at Calverton colliery some 350 units have been produced. The traction on the latest models has been strengthened and, as a result of improved chain designs, the multiple-chain configuration for the jib has been reintroduced.

Ripping and Stablehole Machines[15, 16]

The first serious attempt to mechanize the ripping lip operation in Britain was put in hand in 1958 and, by 1959, a unit with two arm-

478

mounted shearer-type cutting drums was built to a design submitted by C. V. Peake, Area Production Manager, No. 5 Area, East Midlands Division. The unit could be used to cut an arc around the profile of a roadway.

In 1960 a shearer drum machine (Mark II) (Figure 435) was developed by Joy-Sullivan Ltd., in collaboration with the C.E.E. This unit, which was built by January 1961, consisted of a skid base section on which was mounted the main frame or carriage of the unit. This latter section could be hydraulically advanced or retracted a distance of about 600 mm (2 ft) for sumping purposes. Beyond this

distance the base needed to be repositioned. This operation was accomplished by attaching chains to the front of the machine and to props under the ripping lip. The unit was then advanced by the retraction of the sumping cylinders.

Three shearer-type cutting drums fitted with tungsten carbide insert picks were mounted on the cutter arm, which was capable of turning 180° in the vertical plane in order to cut the whole face. The peripheral picks were fitted with water sprays for dust control.

Also during the early 1960s the C.E.E. designed a vertical drum-type machine (Mark

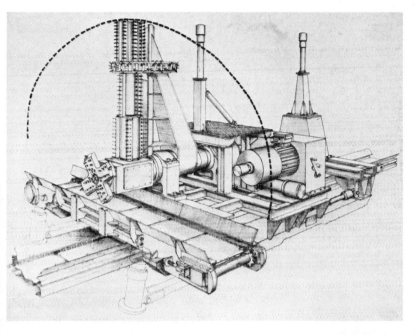

Figure 433. Drawing depicting the Mark III vertical drum ripping machine built by the C.E.E. in 1962

Figure 434. The Mark III undergoing trials at the ripping lip in the brooch seam of the Coppice colliery, Cannock Area, West Midlands Division. The lip height is approximately 1 m (39 in) and consists of shale and ironstone bands

479

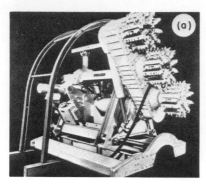

Figure 435. (a) Shearer drum-type ripping machine. (Joy Sullivan Ltd.); (b) Miller-type ripping machine (Bretby/Richard Sutcliffe Ltd); (c) vertical drum type (Bretby/Meco.) These machines were basically similar in concept. That is all three types utilized a hydraulically operated arm-mounted cutting mechanism which swung back and forth across the ripping face. However, while both the shearer and the vertical drum machines made fairly deep cuts of approximately 30 cm (12 in) and 46 cm (18 in) respectively, the milling machine made shallow cuts of about 5 cm (2 in)

Figure 436. Dinnington colliery showing the condition of the roadway before the installation of a ripping machine, after the installation of the ripping machine, and its condition after the machine had been withdrawn and some hand work had been carried out

III). (Figures 433 and 434). It was built by the C.E.E. in 1962. This also consisted of a main frame or carriage, mounted on two base skids. The rotating cutting drum mounted in front of the unit was capable of swinging through 180° in the plane of the cutting face and was fitted with picks.

A hydraulically operated rack-and-pinion arrangement allowed the cutting drum to swing through its arc. When it was necessary for the unit to be advanced, four hydraulically operated jacks raised the machine and the sumping cylinders were retracted. The prototype machine was first tried in 1963 in a 45 m (147 ft) drivage at Coppice colliery, West Midlands Division, and, on completing the

drivage was installed in a main gate at Cannock Wood colliery in the same division.

Another machine produced at that time was the miller drum-type machine (Mark IV) (Figure 435) which was built by Richard Sutcliffe Limited. It was basically similar in concept to the ripping machines already described and was also mounted on a sliding platform, except that the cutting arm which, by means of a rack and pinion arrangement could arc through 196°, was fitted with four milling discs. The hydraulically operated rack and pinion was coupled to the drive shaft of the main arm. It was known as the 'continuous ripper'. An interesting feature associated with the Bretby/Sutcliffe machine was that in order

Figure 437. The Bretby/Meco vertical drum machine (without cross-conveyor). Five of these machines were designed and built by the Mining Engineering Company Limited (Meco) in conjunction with the C.E.E. of the National Coal Board after the successful debut of the Mark III which was produced by the Central Engineering Establishment. The first of the Bretby/Meco units was installed towards the end of 1964 in the Langton colliery, East Midlands Division

to provide access to the face the conveyor was designed so that it could slide a distance of approximately 46 cm (1 ft 6 in) across the front of the machine and then stand vertically out of the way.

The prototype unit was tried in a limestone quarry at Bulwell, Nottinghamshire, in 1963 and was later tested by the C.E.E. at their experimental face at Cadley Hill colliery.

Thus at that time (1963) there were about five different ripping machines being tested, namely the original Peake ripping machine; the shearer drum type manufactured by Joy Sullivan Limited, the miller type manufactured by Richard Sutcliffe Limited to a C.E.E. design; the vertical drum type or Bretby/Meco (Figures 435, 437, and 438) and the Greenside, of which apparently only about two were made. These machines were all designed to cut an arched roadway with a flat floor.

The Greenside machines (Figure 440) tip forward so as to be able to head a full tunnel. The drums were similar to the Joy arrangement, but traversed the arm so that only one or two were necessary and the size of heading could also be varied. In 1968 Contractors Sir Robert McAlpine & Sons Ltd. used a Greenside in the Birmingham road tunnel and subsequently modified the unit for civil engineering use (see section on rock tunnelling machines — Britain — Greenside McAlpine T.B.M.).

The Dawson Miller Stablehole Machine[16]

This machine (Figure 439) was developed by G. B. Dawson and L. J. Mills of the N.C.B. during the early 1960s. The prototype machine (Mark I) consisted of a frame on which was mounted a conveyor, a chain haulage system, and a carriage. To this carriage was fitted a rotary cutting head carrying six radially arranged cutting arms. Each arm was fitted with a pick.

In operation the carriage traversed the base frame and sheared off a narrow web of coal from the face. The machine's advance was automatically controlled by two toe-plates which resisted the thrust of the hydraulic push rams. As the cutting head moved past these two plates and a further narrow web of coal was sheared off the face, the resistance in front of the toe-plate was removed, causing the push rams to automatically advance the unit until the plates were once again bearing against the face.

481

Figure 438. The Bretby/Meco vertical drum ripping machine with cross-conveyor

Figure 439. The Dawson Miller stable-hole machine. Developed by G. B. Dawson and L. J. Mills in conjunction with the N.C.B. during the early 1960s

N.C.B./Dosco In-seam Miner[17]

Another unit which was developed for coal headings, face drivage or stablehole application (particularly where shallow seams were being worked) was the N.C.B./Dosco in-seam miner (Figures 442 to 445). The prototype unit was tried at Donisthorp mine in 1972. Basically the unit consisted of a two-part main frame construction, the right and left-hand sections of which were bolted together to form the base from which the cutting jib was pivoted. This unusual arrangement enabled the contractor to increase the machine's cutting width from 5.18 m up to 9.14 m (17 ft up to 30 ft) by adding spacer sections.

The cutting jib frame carried a hydraulic motor at each end and a continuous sliding track which ran along the top of the face,

482

Figure 440. Greenside machine. (Courtesy Underground Mining Machinery Ltd, U.K.)

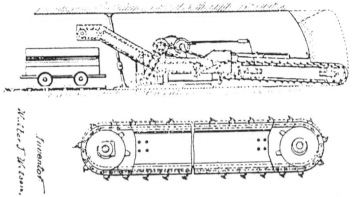

Figure 441. W. J. Wilson Pat. No. 1,549,699. Note the similarity of the cutter head on Wilson's 1925 patent to that used on the Dosco in-seam miner. (Courtesy E. M. Warner, Joy Manufacturing Company.)

traced out a semicircle at each end and joined up back along the bottom. This track carried a series of muck-collecting buckets interspersed with pairs of picks so arranged that each pick traced out a parallel but different track from every other pick. Thus the whole of the face was served by the set of 14-30 picks, depending upon the width of the heading and the nature of the face. The picks and buckets were towed around the sliding track by a 150 mm (6 in) pitch chain powered by hydraulic motors.

Rock Tunnelling Machines

The Bretby T.B.M. [18, 19]

During the early part of the 1950s British collieries were driving their tunnels forward at the rate of about 48 to 64 km (30 to 40 miles) per year. However, the N.C.B.'s 'reconstruction programme' advocated an advance of something closer to 160 km (100 miles) per year.

It was obvious that if these requirements were to be met, drastic changes in the advance methods used would need to be effected.

Recent technical developments on the Continent, and in the United States, indicated that mechanization could be substituted for the conventional advance methods currently being used. Though a mechanized system would naturally necessitate heavy capital outlays, this initial expense was considered to be justified, particularly in the case of new collieries where production could not proceed until the tunnels had been driven. In such cases it was estimated that a saving of as much as 20 per cent could be effected on a £5 million project.

Because of the stringent regulations governing working conditions in the British coal mines, specifications for the proposed new tunnelling machine were unusual. Amongst other requirements the machine would need to comply with the Mines and Quarries Act of 1954 by incorporating such features as flameproof enclosures,[*7] advance methane gas detection, dust and ventilation control, and the elimination or control of high heat conditions at the working face to prevent ignition of methane or coal dust. The machine would also be required to advance through rocks ranging in compressive strength from about 28 MPa (4,000 lb/in²) to about 170 or 200 MPa (25,000 or 30,000 lb/in²).

Before the initial designs were drawn up, Bretby engineers made an extensive and detailed study of some 15 tunnelling machines. These included Wilson's first machine (Patent No. 14,483 dated 18 March 1856), the Whitaker, the Schmidt, a Russian machine, and the original Robbins 'Mole'.

A limestone quarry at Breedon-on-the-Hill, Leicestershire, was selected as the trial site because of its easy access and convenient location for the C.E.E. In addition the compressive strength of the rock ranged from 60 to 250 MPa (9,000 to 36,000 lb/in²) but averaged about 140 MPa (20,000 lb/in²). This fell within the range required in the machine specifications.

Roller cutters (introduced by Howard R. Hughes in 1909 and used extensively for oil well and bore hole drilling — see 'Hughes Tool Company') were fitted to the cutting head.

The 300 t, 5.4 m (18 ft) diameter Bretby machine (Figures 446, 448 and 449) carried 88 roller cutters on a conical head divided into inner and outer sections which rotated in opposite directions. The 2.74 m (9 ft) diameter inner head had 22 cutters in 11 paths and the outer head carried 66 in the same number of paths. The machine was 18 m (59 ft) long and was advanced by means of hydraulic thrust cylinders which reacted between the fixed head and the anchor unit. The hydraulic supply pressure of 14 MPa (2000 lb/in²) applied to the machine's thrust cylinders produced a total forward thrust of 440 t. The inner head was driven by one 120 kW (160 hp) electric motor while the outer 'annular' cutter head with an outside diameter of 18 feet was driven by three 120 kW (160 hp) electric motors. The angular velocity of the inner head was three times that of the outer: speeds could be varied but 9 and 3 rpm respectively were used.

During the 14-month trial period, which commenced in 1962, several teething problems made their appearance.

---

* According to the General Regulation issued under the Coal Mines Act all equipment must be approved before it can be used underground. Regarding electrical apparatus British Specification No. 229 states: 'A flameproof enclosure for electrical apparatus is one that will withstand without injury any explosion of the prescribed inflammable gas that may occur within it under practical conditions of operation within the rating of the apparatus (and recognized overloads, if any, associated therewith), and will prevent the transmission of flame, such as will ignite the prescribed inflammable gas (firedamp) which may be present in the surrounding atmosphere.'

These provisions for safe apparatus include the requirement that gaps between any joints, flanges, etc. on flameproof equipment must not under any circumstances *exceed* 0.5 mm (0.002 in) in addition the breadth of the gap as measured across the parallel flanged edges must not be *less than* 19 to 25 mm (¾ to 1 in). Nor is it permissible for any cables or wires of external circuits to pass through the casing of the flameproof enclosure. These must all be fitted with approved socket and plug or terminal box connections. Screws, nuts, bolts, studs, etc., which are used to fasten or attach doors or cover plates to any flameproof enclosures must be shrouded.

To determine the efficiency and strength of such equipment it is submitted to various tests which are carried out at the Health and Safety Executive Testing Station, Buxton, England. (Although originally carried out at the Ministry of Power Testing Station, Buxton, all government safety and inspection agencies in Great Britain have now been brought under one Health and Safety Commission which is run by the Health and Safety Executive.)

The area inside the 'flameproof enclosure' is allowed to fill with an explosive mixture of methane gas (firedamp) and air. The 'enclosure' is placed inside another chamber filled with a similar combustible mixture which is kept circulating around the 'flameproof enclosure' by means of a pump or a fan. The gas inside the apparatus is then ignited by a spark or perhaps the blowing of a specially arranged fuse.

Provided the external gas which is being circulated around the apparatus does not ignite, the apparatus itself is undamaged by the shock of the internal explosion and, provided also the design specifications of the apparatus concerned are approved, then a flameproof certificate is issued to the manufacturers. In addition a licence plate is attached to the equipment. This plate bears the number of the certificate and depicts a crown within which the letters F.L.P. appear.

Statutory certificates for 'intrinsically safe electrical apparatus' are also issued following similar testing. These are for apparatus in which the power of any spark which could be caused is insufficient to ignite methane. Circuits of this type have low voltage and inductance, but enable control and signalling circuits to be used without massive flameproof enclosures.

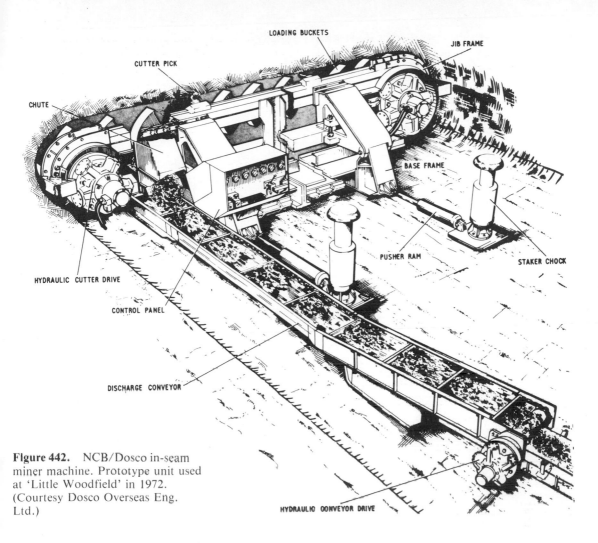

Figure 442. NCB/Dosco in-seam miner machine. Prototype unit used at 'Little Woodfield' in 1972. (Courtesy Dosco Overseas Eng. Ltd.)

LOADING BUCKETS

CUTTER PICK

JIB FRAME

CHUTE

BASE FRAME

PUSHER RAM

STAKER CHOCK

HYDRAULIC CUTTER DRIVE

CONTROL PANEL

DISCHARGE CONVEYOR

HYDRAULIC CONVEYOR DRIVE

Clearly the steering of a tunnel borer of this length involved the cutter head applying a radial cutting action to part of the circumference of the tunnel. As the roller cutters at the periphery of this machine faced forward like the others, they applied their sides and not their cutting edges in this attempt to cut radially. The result was that these cutters were badly worn, their mounting brackets distorted, and even their bearings exposed. One was torn from its studs. This led to a reduction in the effective cutting diameter of the head and the ultimate seizure of the machine in the reduced bore. The problem was effectively overcome by installing three reaming rollers which were slightly conical cutters placed just behind the face with their axes in the direction of the progress of the machine, their front ends just

inside the designed cutting diameter, and their rear ends just outside.

Unfortunately, developmental trials by the N.C.B. on its new machine were abruptly halted by a change in the condition of the local coal market. The current expansion programme was of necessity abandoned. Instead, existing collieries* were to be worked to their fullest extent. As this meant an immediate reduction in the demand for new tunnels, the N.C.B. decided that further expenditure on the Bretby machine at that critical stage was unwarranted.

As the United Steel Cos. Ltd. wished to test the strength and mechanical efficiency of the Bretby on a proposed 3.2 km (2-mile) tunnel drive for winning iron ore, negotiations were

*The number was also halved between 1960 and 1970 and of course there were no new ones created.

485

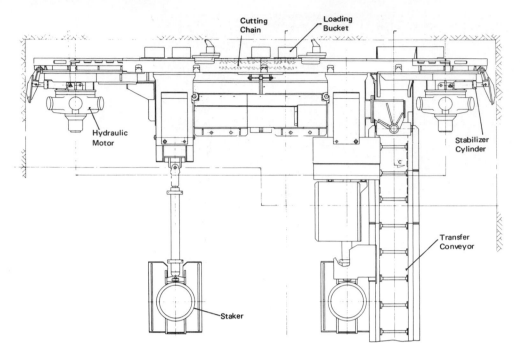

PLAN VIEW

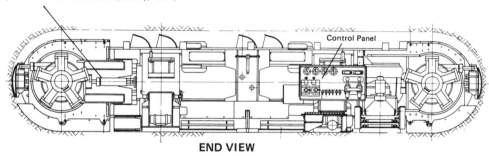

END VIEW

Figure 443. Plan and end view drawings of NCB/Dosco in-seam miner machine. (Courtesy Dosco Overseas Eng. Ltd.)

entered into with the N.C.B. for the transfer of the machine to their Drangonby iron ore mine. Work on this section was scheduled to commence in 1965.

It was anticipated that the machine would be required to work ore which averaged approximately 28 MPa (4000 lb/in²) unconfined compressive strength.

Experience gained on the number and type of tools or cutters required to cut a given face led to the replacement at Dragonby of 77 of the roller cutters by twenty 216 mm (8½ in) diameter discs. These discs decreased the number of

main cutting tools from 88 to 31. In addition there were 6 reaming roller cutters around the periphery. This modification resulted in an increase of approximately 10 per cent in the machine's rate of advance with the same total thrust, at the same time reducing the power consumption by about 30 per cent.

Debris cut from the face by the Bretby machine was lifted in buckets and deposited, via a hopper at one side of the unit, onto an S-bend plate conveyor which carried the material to the rear under the main structure of the unit. This system operated very well at

486

Figure 444. NCB/Dosco in-seam miner being assembled in factory. (Courtesy Dosco Overseas Eng. Ltd.)

Figure 445. New type of NCB/Dosco in-seam miner dirt stower being developed by the M.R.D.E. of the N.C.B. The purpose of this machine is to cut a wide enough entry so as to provide access after the stone from the lip has been deposited at the sides. This obviates the necessity for carting the stone ripped from the lip to the surface

Breedon where conditions were dry. However, when the machine was put to work at Dragonby conditions were found to be very wet. Material on the S-bend conveyor solidified on the plates, thus causing a great deal of trouble. Progress was brought to a halt. To rectify the situation it was decided to replace the S-bend plate conveyor with a straight belt conveyor. The belt could not be taken beneath the machine — if it was to remain straight — and could only pass over the top. This replacement operation necessitated extensive modifications which included removing the bucket and hopper system, and replacing it with a series of 12 curved scraper blades which were inclined at 45° to the axis of the machine. These blades rotated with the outer head and collected the material cut from the face. The debris slid

Figure 446. The Bretby T.B.M. on the launching pad at Breedon in 1961 with the sledge behind

Figure 447. Tunnel face at Dragonby cut by disc cutters which were first used there on the 5.5 m (18 ft) Bretby in 1966

towards the lower or rear portion of the scraper blades on to a fixed cylindrical head and was then moved by the blades to the top of the machine, where it was dropped onto the belt conveyor through a rectangular opening in the cylindrical head.

The belt carried the material to the rear straight across the top of the T.B.M. and had to be loaded at top centre within 50 cms (20 in) of the roof.

After these modifications the Bretby performed very well and during the remainder of its service at Dragonby and until the iron ore mine ceased operating in 1969, required very little mechanical maintenance. Moreover, so improved were the machine's steering and general handling capabilities that it was maintained, operated, and steered by one mine electrician who only occasionally required a survey check on direction.

488

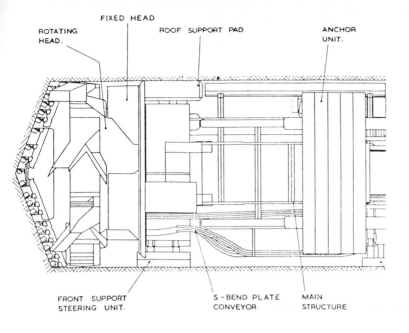

FIXED HEAD

ROTATING HEAD.

ROOF SUPPORT PAD

ANCHOR UNIT.

FRONT SUPPORT STEERING UNIT.

S - BEND PLATE CONVEYOR.

MAIN STRUCTURE.

Figure 448. Bretby tunnelling machine — side elevation

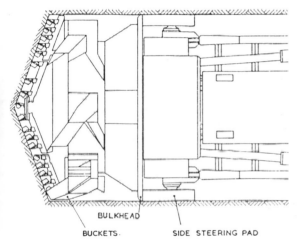

BULKHEAD

BUCKETS.

SIDE STEERING PAD

Figure 449. Bretby tunnelling machine — plan view.

The Thyssen T.B.M. [20]
(Thyssen (Great Britain) Limited)

In 1970, on their own initiative and quite independently, the Thyssen* Company of Great Britain conducted an extensive research into the future needs of the coal-mining industry in the U.K. From this study the management of Thyssen concluded, after

---

* Thyssen were shown the Bretby tunnelling machine operating at Breedon in 1961 and were aware of N.C.B. developments with tunnelling machines.

taking into consideration the economic climate, that present driveage rates (using conventional methods of development) were totally inadequate. What seemed to be needed was a faster, more efficient, means of progress.

Hard-rock tunnelling machines all over the world were breaking new records. Within the Thyssen Group itself, considerable experience had already been gained by the use of a Demag machine (Model 20-23/24 H— 1968) for the driving of water tunnels in Loch Lomond, Scotland, and several further tunnels had also been driven in Europe using Robbins (U.S.A.), Wirth (German), and Krupp (German) tunnelling machines.

In 1972 Thyssen approached the N.C.B. with their proposal. It was discovered that Thyssen and the N.C.B. had covered virtually the same ground in their initial studies. A common interest bond was soon established and the N.C.B. put at the disposal of Thyssen the valuable experience they had gained during the manufacture and trial of the Bretby machine.

Naturally, the problem was thoroughly examined and the various options considered, including that of conversion to use in coal. (The viability of this had already been proved by the successful conversion of some Robbins machines for specific use in Germany).

Many of the parts, such as gearboxes, motors, etc. which would be needed for the new tunnel borer, were already in use in other coal-mining machines and therefore readily available. A full investigation showed that the cost of adapting a 'ready-made' machine to the particular requirements of the N.C.B. would be no higher than that of building an entirely new model from scratch. Nevertheless it was felt that the stringent requirements of the U.K. Coal Mines Act could be more readily met with a brand new machine than with a converted one.

Amongst other specifications the machine was required to drive circular roadways 3.6 m (12 ft) in diameter through varying ground conditions and, like the Bretby, would need to comply with the exacting regulations of the Mines and Quarries Act of 1954. This Act demanded high safety standards and flameproof operating conditions. Thyssen agreed to build such a machine.

Two years were needed to finalize the design and build the first operational machine. During this time a close technical liaison developed between N.C.B. engineers (i.e. mining, mechanical, hydraulic, electrical, etc.) and those from the Thyssen Company. In addition both parties followed the guidance of H.M. Inspectorate in those areas concerned with operational safety.

Technical specifications for the Thyssen F.L.P. (1975) (Figures 450 and 451) are as follows:

Diameter	3.65 m (12 ft)
Length	
Overall	6.90 m (22 ft 8 in)
First point of permanent ground support	5.0 m (16 ft 5 in)
Weight	90 t
Total main drive	240 kW (320 hp)
Hydraulic system	Aquacent emulsion
Thrust	406,000 kg (895,000 lb)
Torque	42,000 kg m (304,000 lb ft)
Cutters	20/12 twin disc cutters and six tooth cutters for centre and reaming

Amongst other components installed on the Thyssen machine was a laser guidance instrument. This instrument (a F.L.P. Mark II made by Vickers and the International Research and Development Company for specific use by the mining industry) was built to meet the stringent regulations laid down by the N.C.B.

The Thyssen machine was completed and ready for its initial trials at Dawdon colliery, County Durham, in January 1975, and by June 1975 had completed over 1000 m (3300 ft) of tunnel through rock ranging from 24 MPa to 165 MPa (3500 to 24,000 lb/in²) crushing strength.

During the 1960s the presence of thick beds of coal was discovered by the N.C.B.'s offshore drillers in an area which is situated approximately 4.8 km (3 miles) off the coast of Durham. This area lies beyond two extensive faults and two parallel roadways have been driven through these faults. It was the N.C.B.'s intention to use the Thyssen to drive exploratory roadways through the area which lies beyond the faults. It was hoped that these drivages would accomplish two important things, namely provide access to the field and produce vital data on the extent and quality of these new submarine coking coal reserves, which, incidentally, experts believe hold some 50 million t of coal. This new deposit forms part of the southern section of the 550 million t reserve area already mapped by the N.C.B. during their offshore drilling programmes which were undertaken in the last decade.

The Thyssen FLP/35 machine is claimed to be capable of handling varying ground strata of up to 240 MPa (35,000 lb/in²) crushing strength, and support of the tunnel is possible up to 5.0 m (16 ft) from the cut face. This means, in effect, that the Thyssen is not limited to use within the coal-mines only and it has attracted the interest of civil tunnelling engineers.

At the 1976 Tunnelling Symposium held in London, a paper was submitted by P. B. Rees, H. M. Hughes and J. D. Hay on 'Full-face tunnelling machines in British coal mines'. The paper described, *inter alia,* the progress which had been made by the Thyssen machine.[21]

It was anticipated that each of the exploratory roadways would cover a distance of approximately 3000 m (9900 ft) and that the

Figure 450. Thyssen tunnelling machine — Model FLP/12/35. (Courtesy Thyssen (G.B.) Ltd.)

Figure 451. Thyssen tunnelling machine — cutter head. (Courtesy Thyssen (G.B.) Ltd.)

machine would need to traverse about 1000 m (3300 ft) before reaching the fault zone. In fact at 1040 m (3400 ft) the first major fault was encountered. The machine continued working through this area and during the following 10 weeks bored through 40 m (130 ft) of badly faulted ground which was being subjected to high lateral pressures. During this period the crew had to contend with four major roof falls (consisting in total of over 2000 t of debris). After the fourth roof fall the tunnel was driven conventionally for about 27 m (90 ft). This brought it through the fault zone and into the coal seam.

The machine was then taken back out through the 3.43 m (11 ft 3 in) diameter completed tunnel. To accomplish this two special bogies were built to carry the unit and, in addition, the diameter of the machine itself was reduced by the removal of the roof and

491

Figure 452. Russian continuous miner. Drives either single-track or double-track headings of arch shape. Used mainly for coal but can handle mixed faces with rock inclusion strengths of up to 40 MPa (5700 lb/in²). (Courtesy Machinoexport, U.S.S.R.)

dust shields, the grippers and steering jacks and the outer section of the cutting head. By March-April 1976 the machine had been reassembled and was ready to start work on the second exploratory tunnel.

Later a joint marketing and manufacturing agreement was entered into between Thyssen (G.B.) Ltd., and Markham & Co. Ltd. for the production and sale of the Thyssen rock machine under the name 'Thymark'.

Romotely Controlled Continuous Miners

The Durham Continuous Miner[22, 23]

Many coal seams in the United Kingdom, particularly in Scotland, are thin and the operational problems and costs associated with extraction of these seams are well known.

Until 1955 most attempts to win this coal depended upon the miniaturization of conventional equipment. But still the operator, who could not be reduced in size, was needed at or near the face. In an attempt to overcome this barrier J. Gibbon and H. E. Collins designed a remotely controlled machine which could enter into these narrow areas and cut and extract the coal sans the presence of an operator. This was the Durham continuous miner.

The Durham miner was pneumatically powered, the air, after use, also serving to ventilate the stall. It was propelled by hydraulic rams which reacted against a launching platform and conveyed forward thrust to the miner via a series of push rods.

The cutting head of the unit consisted of a number of picks fitted to a chain which in operation described a sort of elliptical path somewhat like that cut by the head of the N.C.B./Dosco in-seam miner. That is, the chain ran along the top of the face, traced out a semicircle at each end, and joined up back along the bottom.

The machine was launched into the seam

Figure 453. Russian continuous miner (PK.8). This machine bores tunnels ranging in size from 3.0 to 3.1 m (9 ft 10 in to 10 ft 2 in) in height to widths of from 3.0 to 3.2 m (9ft 10 in to 10 ft 6 in). It's total power is rated at 325 kW (436 hp) and it weighs 55 t. (Courtesy Machinoexport U.S.S.R.)

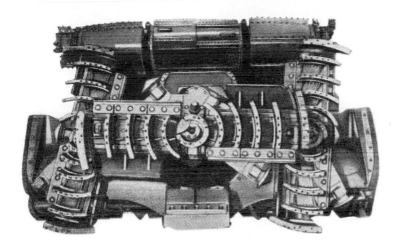

Figure 454. Russian miner (PK.10). Used for winning potash ores and for developmental work. Cutting height 2.5-3 m (8 ft 3 in-9 ft 10 in). Cross-section area 12-14 m² (129-151 ft²). Cutter head unit has three drill bits, upper breaking drum and lower cutters for floor levelling. (Courtesy Machinoexport, U.S.S.R.)

from a roadway running at right angles to the series of parallel self-supporting stalls driven by the 460 mm high × 2 m wide (18 in × 6 ft 6 in) miner. Coal cut by the miner was automatically loaded on to a conveyor belt which ran along above the push rods and was then turned through 90° into the winning roadway. The length of the stall driven was limited because the machine was not steerable.

The Collins Miner[22, 23]

Though this prototype machine (Figure 456) was by no means perfect, its potential as a remotely controlled underground mining machine was evident. As a result developmental work was begin in 1959 on a continuous miner based on the Durham machine. It was designed specifically to extract ultra-thin seams of about 900 mm (3 ft) or less.

A special branch or section headed by G. B. Dawson was created by Bretby to handle the developmental work of the Collins miner. Design was specified by this branch. Detailing and manufacture by Crawley Industrial Products Limited was under the control of this branch of Bretby.

Although design for the instrumentation and control equipment was handled initially by the N.C.B. (M.R.D.E.) this latter part of the work

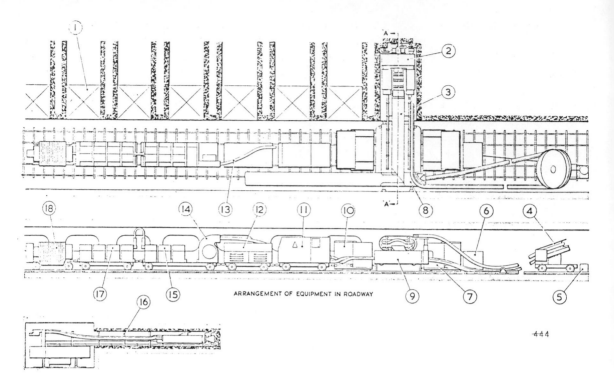

Figure 455. Collins thin seam miner layout: (1) Sand-bag stopping; (2) cutter unit; (3) conveyor belt; (4) control cable, water hose, and ventilation duct return pulley; (5) return pulley anchor station; (6) winch and mobile bench unit; (7) pushing rod storage bogie; (8) conveyor angle stalton; (9) launching platform; (10) pushing rod storage bogie; (11) control cab; (12) hydraulic power pack; (13) ventilation duct; (14) ventilation fan; (15) gate-end boxes; (16) vertical rollers; (17) gate-end box trolley; (18) transformer

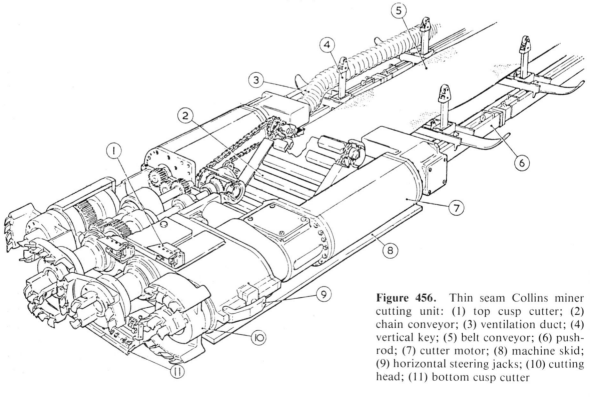

Figure 456. Thin seam Collins miner cutting unit: (1) top cusp cutter; (2) chain conveyor; (3) ventilation duct; (4) vertical key; (5) belt conveyor; (6) push-rod; (7) cutter motor; (8) machine skid; (9) horizontal steering jacks; (10) cutting head; (11) bottom cusp cutter

494

was subsequently handed over to Dowty Mining Equipment Limited, while the N.C.B. assumed responsibility for ventilation and cable-handling equipment.

The prototype equipment was tested initially at Swadlincote test site and then later installed underground at New Lount colliery.

Although the pictorial print of the Collins thin seam miner shows the machine with four heads, in fact, the actual unit built carried only three rotating heads fitted with picks. (These resembled the cutting mechanisms used on the American borer-type machines such as the Goodman borer or the Marietta miner.) The two outer heads were contra-rotating and were set slightly in advance of the central head.

Thrust and propulsion were provided by a system of hydraulically operated push rods which fitted together with male and female joints linked by a key.

Launching was from a platform positioned on a dinted roadway and, in general, a similar mining pattern to that undertaken by the Union Carbide and Joy machines was followed, except that the Collins miner was launched underground, whereas the American units were launched into a highwall and their designers, therefore, were not hampered by the need to fit all the equipment into the confined space of an underground roadway. Nevertheless the Collins miner bored a series of 90 m (long) × 76 cm (high) × 190 cm (wide) (100 yd × 2 ft 6 in × 6 ft 3 in) stalls or tunnels in the coal seam, leaving a supporting rib of approximately 2 m (6 ft) between the parallel tunnels.

The Sensing Device

A nucleonic sensing device,* which was developed at Bretby for keeping a machine within the coal seam, was fitted to the miner. The sensing device measured the thickness of the coal, and the operator kept it to a value of say 25 to 50 mm (1 to 2 in). It consisted of radioactive material (i.e. a thulium 170 disc) which discharged gamma-rays into the coal and surrounding rock. These rays were partially deflected and partially absorbed by the coal and

---

* Nucleonic devices were originally used to sense the thickness of the coal left by the machine at the floor, then at both roof and floor, and finally at the roof.

rock, respectively, in varying degrees. The strength of the reflected rays (back scatter) was measured by Gieger tubes. After amplification, the signal from the Geiger tubes was relayed to a meter in the control cab which was calibrated in inches. By this means the operator was able to judge the machine's position in relation to the material above and below the coal seam. If necessary the operator could then correct the machine's position by transmitting a signal to the electro-hydraulic valves operating the lifting jacks which controlled the position of the cutting heads in the seam.

Equipment

The equipment was controlled from a cabin located on one of nine rail-mounted bogies positioned in the roadway lying at right angles to the stalls. The other eight bogies carried the remainder of the equipment which included the launching platform with the thrust equipment (i.e. push rods) and conveyor angle station, etc. conveyor idlers, the hydraulic power pack, the gate end boxes, and a transformer.

Coal cut by the heads was directed by the blades on to a central chain conveyor running beneath the machine. The chain conveyor emerged behind the motors where it lifted the coal up on to the main belt conveyor. This belt conveyor straddled the push rods and carried the material over the rods and out of the stall to a 90° angle station which turned the belt into the main roadway. As the machine advanced into the stall it drew the belt conveyor after it. (Spare belting was stored in a long 'loop-take-up' at the angle station.) At the same time push rods were taken from the storage bogies and a pair was manually placed on the launching platform by a man on either side. The rods were then thrust forward again a distance of 1.37 m (2 ft 3 in) by two hydraulic rams.

The push rods were tied together by cross-members which fitted over the push rods and which were spaced approximately 2 m (6 ft 8 in) apart. A roller was mounted on each cross-member. This roller performed a dual purpose. It carried the top belt and also a pair of 'ears' (one on each side) which acted as a lock to connect both the roller and the cross-member to the push rod. Furthermore, in order to prevent horizontal buckling of the rod and belt structure, the 'ear' rested 'against the side of the bore'. Buckling in the vertical plane was prevented by vertical keys positioned at every alternate push rod joint. Each of these keys was fitted with a small wheel which ran along the roof of the stall.[23]

Two 'arresters' held the machine steady in the bore while the hydraulic rams were being retracted in preparation for the next forward thrust stroke. When the machine needed to be withdrawn the 'arresters' were 'locked out' to enable the rams to retract the machine and at the same time the push rods were removed and restored to the bogie. With the machine once more back on its launching platform, the operator was able to remove the stelling rams holding the platform in place during the mining cycle. The entire train was then moved by means of two hydraulic jacks to its new launching position. Then the stelling jacks were again positioned and the platform was aligned both in the horizontal and vertical planes in readiness for the next launch.

Developmental Period

During the period of development, i.e. between 1959 and 1967, three machines capable of cutting from 76 cm to 99 cm (2 ft 6 in to 3 ft 3 in) were built. These were Prototype A (which was tried in the Yard seam at New Lount colliery), Prototype B (installed at Fishburn colliery, Durham Division, in November 1963, and Prototype C (which was installed at Rothwell colliery, Yorkshire Division, about June/July 1964). Also in 1964 an ultra-thin seam Collins miner was constructed which was capable of cutting from 46 cm (1 ft 6 in) to 56 cm (1 ft 10 in) in height.

In October 1967 the Collins miner was tested in a longwall application at Manor colliery. The coal-getting method used, which is known as the 'Buttock system' was similar to that employed by the Dosco miner machines of the mid and late 1950s. Working as a single or unidirectional machine, the Collins miner cut along the length of the face.

Various problems were experienced during the developmental period. These mainly concerned the thrust assembly. It was found that approximately 71 t of thrust was needed to enable the fixed blades to break the cusps formed by the rotating cutter blades. This in turn placed high stresses on the multiple-thrust assembly of push rods, and both strength and clearance defects were made evident during the trials. This problem of inadequate thrust was amplified when the cutter head itself jammed occasionally.

Another problem associated with the Collins miner concerned methane gas. The machine was fitted with methane-detection equipment and air was fed to the face from the roadway through louvres on the machine. However, because of the confined nature of the heading, gas freed from the coal seam during the boring operation did not readily disperse, despite the provision of the above-mentioned ventilation facilities.

While horizontal steering effected by lateral jacks positioned behind the cutting heads proved no problem, vertical steering, on the other hand, caused some considerable embarrassment. This was because it was necessary to keep the machine in the coal seam which often undulated in its path. The difficulty was further compounded by the fact that the machine itself was practically the same height as the seam, and very close tolerances therefore needed to be followed if the roof was not to be disturbed. (If a stable roof was to be maintained, it was essential that the stone band above the coal seam was left intact.)

Control of the machine in the vertical plane was effected by changing the angle or range of the cutting heads with respect to the skid base on which the unit was mounted. Other factors which indirectly affected the steering included torque and thrust and the shape and angle of the upper and lower cusp cutters. Generally speaking, these factors contributed towards the machine's tendency to rise.

The sensing equipment also caused difficulties, because when this equipment was installed it was tested on a sample of coal which did not contain pyritic nodules. These nodules, which made their appearance later, gave false readings indicating that the machine was entering the stone band when, in fact, it was actually some inches away. As a result of these problems the coal-sensing system was abandoned and skilled operators learned to judge the machine's position in the seam by observing the outgoing coal on the conveyor. The machine was steered by the amount of stone cut by the static floor blades. (*Note:* Similar problems were experienced by operators of the American remotely controlled machines and this led to the ultimate development of 'in-sight-of-the-machine' cable and radio remote control equipment: see Chapter 19 — Remotely controlled continuous miners.)

Developmental work on the Collins miner ceased in 1967.

References

1. F. S. Anderson and R. H. Thorpe, A Century of Coal-face Mechanisation, Inst. of Min. Engs, *The Mining Engineer,* **126,** No. 83, 775-785, Aug. 1967, London.
2. F. S. Anderson, *The mechanisation of collieries, Engineering,* 381-384, 17 Oct. 1947, London.
3. Le havage moderne et son influence sur le chargement mecanique, M.M. Anderson, Boyes & Co. Ltd., *Mines,* No. 2, 1953, Paris, France.
4. F. S. Anderson, Progress in longwall face mechanisation, Inst. of Min. Engs., *Mining Engineer,* **122,** No. 35, 797-803, 1963.
5. Sir Richard Redmayne, K.C.B., The mechanisation of British collieries, *The Engineer,* 20 Oct. 1944, London.
6. R. Shepherd and A. G. Withers, *Mechanized Cutting and Loading of Coal,* Adlard and Son, London and Dorking, 1960.
7. H. N. Kalia, Support considerations for the longwall mining of coal, *Mining Congress Journal,* 126-128, Sept. 1974, Washington, D.C., USA.
8. George Blake Walker, Coalgetting by machinery, *Transactions Federated Institution of Mining Engineers,* **1,** 1889-90, Scotland.
9. R. W. Wrathall, The work of the NCB Central Engineering Establishment, *Min. Elec. Mech. Eng.,* **39,** No. 460, 199-204, Jan. 1959, London.
10. C. Treharne Jones, *Mechanized Roof Support at the Coal Face.* Con. National Assoc of Colliery Managers, London, 30 May 1956.
11. Anderson Mavor Ltd., Leaflet AM2/76, Anderson Strathclyde Ltd, Motherwell, Scotland.
12. William Johnson, Brunton' heading machine, Bretby Collieries, *Chesterfield and Derbyshire Engineers Proc.* 2 Oct. 1875.
13. James S. Dixon, *Stanley heading-machines,* Inst. Mining Engs., *Trans. Mining Inst. of Scotland,* **6,** 12 Aug. 1893/4.
14. Discussion, Stanley heading-machines, Mining Inst. of Scotland, Inst. of Min. Engs., Glasgow, 11 Oct. 1893.
15. J. D. Hay and P. Shuttleworth, *The development of ripping machines, Trans. Min. Eng.,* **1241,** 1964/65, London.
16. G. B. Dawson and L. J. Mills, The development of the Dawson Miller stable hole machine, Inst. of Eng., 1961, *The Min. Eng. Trans.,* **21,** London.
17. G. E. Boast, *In-seam Miner,* National Coal Board, Tech. Mem. MD 75(6), N.C.B. Burton-on-Trent, U.K.
18. J. H. Hay, H. M. Hughes and R. W. Wrathall, The Bretby tunnelling machine, Paper No. 6830, Institutions of Civil, Mechanical and Electrical Engineers, Joint Meeting 25 May, 1965.
19. H. M. Hughes, Mechanized stone work, *The Mining Engineer,* 689-698, Sept, 1969, London.
20. *Thyssen Tunnelling Machine,* Thyssen (G.B.) Ltd., Llanelli, U.K.
21. R. B. Rees, H. M. Hughes and J. D. Hay, Full-face tunnelling in British coal mines, Tunnelling 1976 Symposium, London, March 1976, Inst. of Min. and Met., London.
22. R. F. Lansdown and G. B. Dawson, The development of the Collins miner, *Trans. Inst. Min. Eng.,* **122,** No. 36, 841-860, Sept. 1963, U.K.
23. W. J. Charlton, M. Riddell and J. Nixon, Experiences with the Collins miner at Rothwell colllery, *Min. Eng.,* **126,** No. 84, 793-808, Sept. 1967, London.

Bibliography

G. M. Gullick. Coal mines mechanisation, *Engineering,* 15 Feb. 1946, 162.
Seth D. Woodruff, *Methods of Working Coal and Metal Mines,* Vol. 3, Pergamon Press, London, 1966.
R. T. Deshmuki (1920), B. M. Vorobjev (1928) and K. Keshmukh (1966) *Advanced Coal Mining,* Asia Publishing House, Bombay, India.

Correspondence

Anderson Mavor Ltd (now Anderson/Strathclyde Ltd) — F. W. Anderson, Motherwell, Scotland, U.K.
Dosco Overseas Engineering Ltd., Tuxford, U.K.
Gullick Dobson Ltd. — A. Purdy, Inc. U.K.
National Coal Board — H. M. Hughes, Burton-on-Trent, U.K.
Thyssen (G.B.) Ltd., London.
Sir Robert McAlpine & Sons, London.

NINETEEN

America

Longwall Mining[1-5]

The first longwall machine to be made in America was designed by Jonas P. Mitchell who applied for a patent in 1891 to cover his invention. Before the patent was granted it was purchased by John C. Osgood (a coal operator) and F. K. Copeland. These gentlemen then employed Mitchell to assist them build and develop his machine. As Copeland was at that time President of the Diamond Prospecting Company of Chicago, Mitchell was in effect working for the Prospecting Company. During the period 1885 to 1892 the Diamond Prospecting Company acted as agent in the territory of New England for the Sullivan Company. In 1892 the Sullivan Machine Company and the Diamond Prospecting Company merged. With the Sullivan Company at that time was *Albert Ball,* an engineer of outstanding ability.

The first machine produced by Mitchell (No. 100) was able to cut a clean square kerf along the length of the face. In other words it did not leave a wedge of uncut coal at the back of the groove. The cutting mechanism consisted of a bar fitted with picks. The bar projected from the machine at right angles. This machine was put to work in the White Breast Fuel Company mine in Forbush. The mine was owned by Osgood who employed the longwall method of mining.

Great difficulty was experienced in keeping the cutter bar in the coal and, as a result, several modification were made to the original unit. Eventually this difficulty was overcome by the provision of jacks and cables to hold it in place.

By 1893 Mitchell had built his fourth machine. This unit incorporated various modifications which had been tried on the earlier models. These included a new cutter bar, a cleaner chain, a floating sheeve, and a friction clutch (the clutch was designed by Albert Ball for Machine No. 101). Machine No. 103 had been sent to the Dominion Coal Company, Sydney, Cape Breton, on 29 March 1893, while Machine No. 104 was shipped on the 21 July of that year to the Elsworth Coal Company, Scott Haven, Pennsylvania, where it was used to widen a heading previously cut by a Stanley machine (see also section on Stanley heading machines).

At Cape Breton trouble was experienced with the rotary cutter bar which continually broke. Mitchell had for some time prior to this advocated the use of a cutter chain instead of the bar, but because a number of other companies had made unsuccessful attempts to build breast machines with cutter chains, Copeland had dissuaded Mitchell from using one and insisted he stick to the rotary cutter bar with which they were all familiar. Nevertheless, the rotary bar was also unsatisfactory as, amongst other problems associated therewith, bits would sometimes break off and these, in turn, caused the set-screw heads to fail. This led to an uneven strain being placed on the cutter bar itself so that it flexed during operation and eventually broke.

That year the Jeffrey Manufacturing Company abandoned the rotary cutter bar and replaced it with a breast-cutting chain machine which was successfully used by the Elsworth Coal Company. Mitchell was influenced by this, and in 1894 he fitted a cutter bar and chain to the Cape Breton machine (No. 103) (Figure 457). Initially a cutting jib measuring only 107 cm (42 in) in length was tried, but this did not

Figure 457. J. P. Mitchell's compressed-air longwall rotary bar machine No. 103, Class 'C' built at Claremont. This machine was converted in 1894 to a chain cutter by Mitchell. The original rotary-bar unit was displayed at the 1893 Chicago World Fair. (Courtesy Joy Manufacturing Company.)

prove satisfactory and so Mitchell replaced it with a 183 cm (72 in) long cutter bar and chain which improved the performance of the machine.

However, it soon became obvious to Mitchell that his longwall units were not popular because the main trend at that time was towards room and pillar work. (This fact was also recognized by the Goodman Manufacturing Company who built their first longwall coal-cutter in 1900, but discontinued the line as only a few of these units were sold each year.)

In 1895 Model No. 104 was completely redesigned and became Model CB, Mitchell's first room and pillar machine. During the succeeding years Mitchell continued developmental work on this machine. He was joined in 1897 by Albert Ball who again redesigned the unit, converting it to an electrically driven chain cutter which was known as the 'Sullivan CE' machine. It was put on the market by the Sullivan Company on 27 March 1899.* This unit was later recognized as

*The Sullivan Machinery Company was taken over by the Joy Manufacturing Company in 1945.

being the first of the American 'shortwall' machines. (However, it is of interest to note that this term, i.e. 'Shortwall' was initially used by the Jeffrey Manufacturing Company as a trade mark for a similar type of machine made by them at a *later* date.)

Mitchell's machine had serious disadvantages. It was moved around the mine on a carriage which ran on rails or tracks. When it was needed it was taken off the carriage, pulled across the mine floor, and positioned at the face. After the face had been cut the machine was returned to the carriage and reloaded. This was a laborious and time-consuming operation and in 1909 the Jeffrey Manufacturing Company produced their arcwall cutter, a track-mounted unit, which could operate from the track. This was followed in 1914 by the Goodman Manufacturing Company's square face cutting machine, Model 11B, which was also a track-mounted machine. The next year the Goodman Company brought out their No. 24 slabbing machine which could cut in an arc or make slabbing cuts. (The Goodman Company produced their first shortwall machine in 1910

499

Figure 458. Joy's 15RU. (Courtesy Joy Manufacturing Company.)

Figure 459. Joy's 16RB. (Courtesy Joy Manufacturing Company.)

and, like the Goodman longwall and breast machines, it employed a rotating pick-filled chain on a jib which cut at floor level.)

Generally speaking, the shortwall machine is basically similar in concept to the longwall unit. It is designed to operate as a heading machine using longwall methods. In other words the machine cuts a type of 'short long wall'.

During the succeeding 40 years or so, American designers and manufacturing companies kept pace with British developments and produced a variety of chain-cutter-type machines to cater for the diverse needs of the mining industry. These included units for top, centre, and bottom cutting and also for shear or side cuts, etc. Some of these machines were hauled across the floor, some were track mounted, while others were equipped with tractor tread or rubber tyres for floor operation. Regardless of their type these

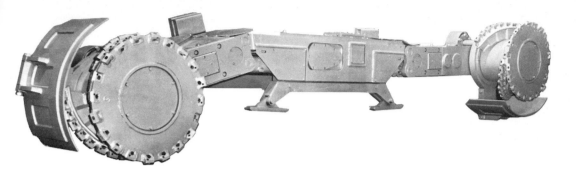

Figure 460. Joy longwall shearing machine. (Courtesy Joy Manufacturing Company.)

machines were designed mainly for room and pillar work as were the continuous miners which made their debut in 1948-50. Today two U.S. firms have emerged as producers of chain-cutting machines, the Goodman Manufacturing Company and the Joy Manufacturing Company, with Joy being the major producer (Figures 458 and 459). (The Jeffrey Manufacturing Company stopped producing chain-cutting machines in 1967.)

In 1950 the Eastern Associated Coal Corporation of West Virginia began using a coal plough for longwall extraction work, and since then the longwall method of mining has gradually expanded so that by 1970 there were approximately 40 units being used in various U.S. mines. By 1975 this number had doubled.

To meet this need the Joy Manufacturing Company began selling longwall coal cutters of various types towards the end of the 1960s. (Sales agreements were made between Joy and Gullick-Dobson of England for face supports, and with Eickhoff of Germany for shearing machines and face conveyors. In 1975 the Joy Company began marketing its own design of longwall shearer (Figure 460), followed shortly thereafter by American designed face conveyors and roof supports.)

No doubt as seams become thinner and mines go deeper the longwall method will become even more widely accepted in America, and other manufacturers there will enter this field. Increased productivity and greater safety factors derived from the longwall method of mining will also influence the market.

An indication of this growing trend towards longwall mining may be seen in John Reeves's article 'Deep mining of coal in Colorado'.

Reeves described the current method of coal extraction used by his company which, he stated, exclusively utilized continuous miners and shuttle-car haulage. Up to the date of his article most extraction work in his company's mines had been carried out at depths of 600 m (2000 ft) or less. However, it had recently become necessary to exceed this depth and already problems of floor heave, etc. had become manifest. Although so far none of their continuous miners had become so badly trapped that it was unable to extricate itself, this possibility nevertheless posed an ever-increasing threat. In addition, greater quantities of methane gas were evident in conjunction with the manifestation of floor heave and this problem was compounded by the fact that gob areas at these depths tended to compact more tightly than they did at lesser depths, thus rendering the bleeder entries less effective for ventilation purposes.

Reeves felt that the obvious answer to this problem was to change to the longwall mining system. While he recognized that such a step would, of course, mean increased coal extraction costs (i.e. capital outlay and maintenance costs of powered supports, conveyors, and equipment, etc.) he felt that these factors were outweighed by the undoubted benefits (i.e. greater safety, etc.) which would be derived if the longwall method of mining was adopted.

501

Room and Pillar Mining

In room and pillar work the coal is extracted by driving a series of roads and rooms in straight and perpendicular lines within the ore body, leaving square or rectangular pillars of coal which either remain in place or, in favourable situations, are extracted on the retreat and the roof allowed to collapse.

In some cases, instead of being driven at right angles to the heading the cross cuts or roads are made at an oblique angle to the rooms or headings which leaves behind pillars of coal in the shape of parallelograms.

The sizes of the rooms, roads, and pillars, etc. may vary according to local ground conditions. Areas are usually divided into the main haulage roads, which run at right angles to each other, and the panels of coal lying between these haulage roads. These panels are then subdivided into rooms and pillars. As soon as possible after the initial work has been completed the pillars are extracted.

Because coal seams in America are generally thick and fairly close to the surface, room and pillar work has been favoured there.

Early Coal-Cutters[1-5]

Horace F. Brown

Horace F. Brown, Indianapolis, is credited with inventing the first American coal-cutting machine. The patent for this was filed in 1873 after Brown had spent nearly four years perfecting his machine. This became known as the 'Monitor coal-cutter'. It was, in fact, a revolving wheel set on a solid frame of cast iron. Around the periphery of the wheel were the cutting teeth. The machine, which was mounted on a track, was powered by either steam or compressed air. It was tested for the first time in the Coal Brook mines of Brazil, Indiana, in 1873.

J. W. Harrison

In 1877 J. W. Harrison filed a patent for a pick machine. Later the George D. Whitcombe Company obtained the patent and after effecting several important changes produced the machine in their factory. It was fairly successful.

Joseph A. Jeffrey

The same year that Harrison filed his patent for a pick machine, that is in 1877, the Jeffrey Manufacturing Company produced the first American cutting machine or undercutter. It became known as a 'breast machine'.

Joseph A. Jeffrey of Columbus, Ohio, the inventor of the machine is reputed to have been inspired by the sight of Francis M. Lechner's (a well-known inventor of mining machinery) wooden model of a coal-cutter which had been displayed in a shop window in High Street, Columbus. It was put there ostensibly with the idea of encouraging some wealthy and also, perhaps, talented inventor to produce such a machine. The result of this chance sighting was the founding of the Lechner Mining Machine Company on 18 July 1876, in which Jeffrey played a major role. Lechner was President of the company.

Though Lechner's model was crude, Jeffrey was convinced that the engineering principles involved were sound and he at once set to work on the design of his own machine. It was built by the Franklin Machine Company of Columbus. When it was complete it was tested at the New Catfish mine in Clarion County, Pennsylvania, and also at another mine belonging to the New Straitsville Central Mining Company, Straitsville, Ohio.

The first Lechner/Jeffrey machine was fitted with a 75 cm (30 in) long square bar on which were set chisels or picks. The bar, which rotated, was mounted on a frame which slid backwards or forwards on the main chassis of the machine. In operation the cutter was positioned with the rotating bar against the face. The bar would then be forced into the coal until the sliding frame was extended to its fullest length, after which it was drawn back and the cutting machine was levered into a new position for another cut.

Unfortunately Jeffrey's machines were badly constructed so that when they were put to work in hard anthracite the weak components fractured and broke. Nevertheless they did work tolerably well in the softer coals. However, these were not the only problems, for

502

it was found to be almost impossible to keep the cuts parallel. The result was that wedges of coal were frequently left at the back of the cut. These invariably remained in position after blasting. As this coal had to be removed before another cut was commenced, this necessitated additional pick-work for the miner. Moreover, it was seldom that one cut was made on the same level as another, thus rendering shovelling work more difficult.

On 11 July 1887, Lechner sold all his intersts in the company to Jeffrey and the Jeffrey Manufacturing Company was formed. Despite the above-mentioned problems the Jeffrey Company never abandoned hope for their machine, though the units were frequently being returned to the workshop for major repairs.

In 1888 the Jeffrey Manufacturing Company brought out its first improved model. This was a breast-cutting machine powered by electricity, the first of its kind in the United States. Another important improvement made by the Jeffrey Manufacturing Company at that time was the replacement of the rotating bar with a cutter chain carrying picks.

In the meantime the Jeffrey Manufacturing Company was slowly developing and in the process was attracting men such as B. Legg, H. B. Dierdorff, A. Hoermle, and H. H. Bliss.

B. Legg

Legg's machine weighed about 510 kg (10 cwt) and was so designed that apart from normal undercutting work it could also, in a limited capacity, be used as a heading machine. Basically it consisted of a frame upon which was mounted a revolving cutter bar carrying 26 steel bits, each of which was secured by a set screw. The cutter bar was operated by an endless steel chain attached to the driving shaft which also controlled the feed mechanism. This mechanism could be engaged or disengaged as required, by a lever. A 1.5×1.1 m (5 ft $\times$ 3 ft 6 in) deep cut was made by this machine in about four to six minutes and all material was cleaned from the cut by scraper chains.

It was through the efforts of such men as these that major improvements were effected. For instance, in Legg's machine, the cutting action in the earlier models of this type of coal-cutter was performed by a revolving bar which was driven by a chain. The cutting 'teeth' being rigid tools set in the bar. Later, in 1893, Dierdorff designed a coal-cutter in which the 'cutter bar' was replaced by a travelling endless chain in which the cutting teeth were inserted.

New Breast Machine

In 1897 several companies independently produced a new type of chain breast machine. Numbering amongst these were well-known firms such as the Sullivan Machinery Company, the Jeffrey Manufacturing Company, the Morgan-Gardiner Electric Company, and the Link-Belt Company (later to become the Goodman Manufacturing Company).

This new revolutionary design in breast chain machines was the forerunner of the modern chain coal-cutter, and many of the principles involved in its construction are still used today. Simply, the machine had a long flat blade around which ran an endless chain carrying teeth. In operation the machine would be placed against the coal face and the chain set in motion. A horizontal slot would then be cut in the coal face running to the depth of the bar. When this operation was complete the machine was moved along the face and another slot was made. The work would proceed thus until the entire face had been undercut, after which the machine would be withdrawn and explosive charges placed and fired in the usual way.

Distinctive features of the 1897 chain breast machine made by Link-Belt's Electrical Mining Machine Department were an enclosed motor and a stationary frame. This was so designed as to enable cuts to be made close to the bottom of the coal face. In place of the original ratchet feed and chain, Link-Belt introduced a double rack and sprocket mechanism which enabled the movable frame to be fed forward. When a cut had been completed, rollers were placed under the rear end of the stationary frame so that it could be repositioned for the next cut.

Three types of coal-cutting machines were available on the market by 1900. These were percussive, breast, and longwall machines. Familiar names connected with breast machines were Brown-McHugh, Jeffrey, Morgan-

Gardner and Goodman. Longwall machines included, amongst others, such names as the Jeffrey electric, the Sullivan and the Goodman electric.

Percussion Machines

While conventional percussive rock-drilling machines were being developed in the United States, several American manufacturers also produced special machines designed specifically for use in the coal-mines. Amongst the first of these to be introduced to the market were the Harrison pneumatic (1880), the Sperry coal digger, the Ingersoll-Sergeant, the Morgan-Gardner, and the Sullivan. Later, quite a number of these pneumatic tools found their way on to the British market.

The Harrison puncher or Harrison pneumatic, patented by J. W. Harrison on 25 December 1877 was a pick machine which delivered approximately 180 to 200 hammer blows a minute. It was operated by compressed air. Initially the pressure was 0.3 MPa (45 lb/in²) but later Harrison increased this to 0.5 MPa (75 lb/in²) pressure. The first machine was built by the George D. Whitcomb Company of Chicago in 1880. An important modification to Harrison's machine was made in 1882. This consisted of an inclined wooden stage on which the puncher was placed. This staging helped to eliminate a great deal of the recoil which had made operation of the original unmounted machine so unpleasant. But the platform brought its own brand of problems. Splinters caused by the vibration of the puncher found a ready target in the posterior of the operator who was forced to sit with his legs held against the machine in order to steady it. Thus the operator's mate, whose official job was that of assisting in the repositioning of the machine for a new cut and shovelling the coal, also acquired proficiency in the more delicate task of removing splinters from the operator's buttocks.

Harrison punchers were used fairly extensively in the coal-mines of America at that time and by 1898, 1700 of them had been marketed.

Sperry Puncher[1]

The Sperry coal digger or puncher (Figure 461) (designed by Elmer H. Sperry, brother-in-law of Herbert E. Goodman) was installed in an Illinois mine on 14 March 1889. (A Sperry puncher was also exhibited at the Paris Exhibition). It was reputed to have been the first electric coal-cutter in the world. However, J. D. A. Morrow points out in his article entitled 'History of the development of underground machines'[3] that there is strong evidence indicating that the Jeffrey breast-cutter, mentioned earlier, was the first electrified coal-cutting machine to be produced in America. In support of this contention Morrow quotes the existence of a photograph dated 1888 which is kept by the Jeffrey Manufacturing Company in their file of new machines.

The Sperry puncher consisted basically of a pick which could deliver powerful and rapid blows at the coal face in a straight line. Operation of the pick was effected by means of a spring which was wound up by an electric motor. According to the Sperry Coal Digger Company, their machines could deliver from 200 to 250 blows per minute and, by this means, in 10 hours of operation, produce a 1.22 m (4 ft) depth of undercut 36.6 m (120 feet) long.

Modern Trends in Coal-Cutters[7]

In 1961 a new coal-cutting machine was marketed by the Mighty Miner Company, a division of the Shallway Corporation of

Figure 461. The Sperry Puncher. (Courtesy Goodman Equipment Corporation.)

Connellsville, Pennsylvania, the original prototype having been manufactured by Sherman G. Martin of Dunbar, Pennsylvania. Weighing less than 90 kg (200 lb) the machine was designed not only in the interests of one-man operated mines, but also for use in larger mines where economics was forcing proprietors to work thinner seams.

As this model was only 43 cm (16 in) in overall height, it was possible to work seams as narrow as 50 cm (20 in) with it. The high-powered electric motor which drove a carbide-tipped auger bit was mounted on a slide on top of a square tubular frame 1.5 m (5 ft) long and almost 1.2 m (4 ft) wide. This slide was capable of positive movement both axially and laterally, using racks and pinions and controlled by two hand cranks. The auger operated through a cylindrical sleeve and could thus be accurately positioned.

Making use of index marks the auger could be moved laterally after each hole was bored in such a way as to remove a slightly serrated kerf roughly equal in length to the width of the machine. A small charge placed above would readily drop the coal into the undercut.

John L. Lewis — his influence on Mechanization in America and the Birth of the Continuous Miner[8]

If the early convicts of Australia suffered in the coal-mines, no less terrible were the hardships endured by the coal-miners of Wales. They toiled for a pittance from sunrise to sunset under conditions so abominable and so hazardous that few lived beyond the age of 30 years.

During the latter years of the nineteenth century America's frontiers were forging outwards, and embroidered tales of this marvellous land reached Europe. Those that could packed their meagre belongings and sailed for the land of promise. Amongst them were a large number of Welsh miners who found, when they landed in America, that they needed a job, and that urgently. For want of something better to do they naturally turned to the only trade they knew, coal-mining.

Amongst those who made America their new home were the parents of John L. Lewis. Lewis was born on 12 February 1880 and in his lifetime he is said to have done more to alter and shape the destiny of both miners and operators than any man has done before or since. At the time of his birth the trade union movement in America was undergoing its own tumultuous birth pangs. Lewis's father, Thomas Lewis, was blacklisted when the boy was only two years old for organizing a strike against the White Breast Fuel Company of Lucas, Iowa. The miners won their fight but Lewis's father was forced to move from place to place afterwards, branded as a radical and unionist and unable to hold a job in a coal-mine. Eventually in 1897 these iniquitous blacklists were destroyed and Lewis's father was able to resume work as a coal-miner. Nevertheless Lewis never forgot that in his father's hour of need, when the miners had won their fight, they had passively stood by and allowed his father to take the brunt of the company's wrath. Thereafter Lewis's enthusiasm for the miner's cause was always tempered by the knowledge of how fickle their support could be.

When young Lewis reached adulthood, he left home and for a period of five years, roamed the continent, working in various types of mine throughout America. These included copper in Montana, silver in Utah, gold in Arizona, and coal in Colorado. After this he returned to Lucas and joined his father and fellow Welsh miners in the coal-mines. These additional experiences also made a deep impression on Lewis, for he had discovered that by and large, conditions in the mines of America at that time were not significantly better than those endured by miners in other countries. In America, as elsewhere, the miner was regarded as something less than human, a being apart, whose sufferings and tragedies were unimportant and whose rebellious stirrings were to be crushed and destroyed at all costs, even where lives were involved.

In 1909 Lewis and his wife moved to Panama, Illinois. They were followed by the rest of the family including Lewis's five brothers. This marked the beginning of Lewis's career with the mine unions. For it was here that Lewis began his long and bitter campaign for the rights and dignity of the coal-miner.

Figure 462. Hoadley-Knight machine. (Courtesy Joy Manufacturing Company.)

Figure 463. Colonel O'Toole machine. Both the Hoadley-Knight (Fig. 462) and Col. O'Toole machines were early American attempts at the production of a continuous miner. The Hoadley-Knight built in 1912 was electrically powered. It featured a water spray, but was not particularly successful because it could only cut powder-to-nut-size coal. Colonel O'Toole's machine was also built during the 1920s. This unit is interesting because the cutter head is somewhat similar to the rotary drum head units produced by the Jeffrey Manufacturing Company in 1965. Note the scroll-type gathering device. The machine mined 140 t in 9 hours and 47 minutes. According to E. M. Warner (Joy Manufacturing Company.), an attempt was made during the trials to use pneumatic conveying but although the conveyor handled the output for several hundred feet it tended to powder the coal. (Courtesy E. M. Warner — Joy Manufacturing Company.)

Assisted by the vociferous support of his brothers who applauded or jeered when necessary, Lewis was elected President of his local union branch. His dynamic personality and natural talents for oration soon won for him the approval of his colleagues. Gradually his influence spread so that by 1920 he had become President of the United Mine Workers Union of America.

This was only the beginning. During the years that followed, Lewis campaigned and fought across the length and breadth of America, using strikes, subterfuge or any other political weapon available in order to gain his ends.

In the North, coal operators were forced by Lewis to increase the pay of the coal-miners. By 1922 he had won for them a daily wage of $7.50. Unfortunately, other non-union low-wage mines in the South which paid their workers only $3.00 a day, still operated at that time, and these were able to produce coal at a much cheaper rate than their Northern competitors. The situation threatened economic ruin for the Northern operators. The result was chaos and Lewis learned the hard way that it was as important to have a *stable and efficient industry* as it was to have a properly controlled and organized mine union. Lewis's answer was *mechanization*. He reasoned that though this would probably mean less men employed, ultimately they would benefit. It was better to have fewer well-paid men with a job than more men working at substandard wages — or not working at all.

In 1949 the coal industry in Australia was dubbed as 'tottering'. In America it was decidedly 'unhappy'. Lewis had ridden on the crest of the wave from strength to strength. His epic battles first with Roosevelt and then to a somewhat lesser degree with Truman, were behind him. During the years 1935-39 the coal-miners' take-home pay of $22.16 per week was third from the bottom of a long list of some 15 basic industries which included steel, printing, automobiles, and shipbuilding. In 1949 with a take-home pay of $76.84 Lewis had lifted the miners' wages to the top of the list — an increase of approximately 250 per cent.

On the other side of the picture were the coal operators who in the year 1949 were on the brink of disaster. Production had fallen from 600 million to 435 million t. This was the lowest it had been since 1939. Gas and oil were in their ascendancy with oil for the first time surpassing bituminous coal as an energy provider. Adding fuel to the fire were the latest coal wage and welfare increases. (The Welfare and Retirement Fund was considered to be one of Lewis's greatest triumphs.) These new increases pushed the cost of coal production up another 21 cents.

The coal industry was literally sitting upon the horns of dilemma. In order to stop the flow of business now being channelled towards the new gas and oil industries, indeed if they were to survive at all, the coal industry knew they would be forced to drop their prices. And coming as it did at the height of the greatest production cost rises known to the industry, it seemed only a miracle could save them.

The miracle arrived unobtrusively and ringed with question marks — *continuous miners*.

Continuous Miners[9]

At that time most U.S. coal was won by undercutting, drilling, blasting, and loading in that order. The question asked was whether these U.S. $50,000 to $85,000 machines (a considerable sum in those days) could increase production and at the same time justify the enormous expense involved in their purchase.

Four new machines had been introduced to the market. These were the Joy (1948), the Colmol (1948) the BCR (Bituminous Coal Research Inc.) (1950), and the Marietta miner (1950).

Though the mechanical loader was already well established, the continuous miner was looked upon with extreme scepticism, as indeed had been the loader when it was introduced in the early 1920s. But then, as in 1949 (some 30 years later), Lewis's influence played a major part in persuading the operators that they must mechanize or go under.

The Joy Continuous Miner
(Joy Manufacturing Company)[3]

The idea of a mechanical loading machine was the brainchild of Joe Joy who, during the First

Figure 464. The 'Silver machine'. (Courtesy Joy Manufacturing Company.)

507

World War, thought up the brilliant notion of a loader capable of drawing coal on to a conveyor with the help of two steel grabbers (arms) that moved forward and back again. Unfortunately, at the time that Joe Joy set up his company in 1921 and built his first prototype machines, the industry was virtually penniless. So too was Joe Joy by 1926. However, though Joe Joy abandoned ship, the Joy Manufacturing Company continued to operate, albeit somewhat shakily.

Then John L. Lewis began his campaign and from that point onwards the company flourished. As one Joy official is said to have remarked, 'John L. Lewis was the best salesman Joy ever had.' Lewis with his avowed 'no backward step', swept from victory to victory. While wages and coal prices were being drastically cut in Kentucky and West Virginia, operators in Illinois and Indiana (a Lewis stronghold) on the brink of disaster, gambled

and bought the new Joy loaders. Then, as Lewis moved east and south, Joy loaders followed in his wake, becoming accepted for the most part as standard mine equipment.

The Joy continuous miner developed as a result of a search by Carson Smith, President of Consolidated Coal, for such a machine. With the assistance of a sugar-beet machinery engineer named Harold F. Silver, a continuous miner (Figure 464) was designed by the two men. It was eventually built by Silver who used his own shop for the final engineering work. By 1940 it was being tested in practice.

In the meantime dreams of producing an efficient continuous miner filtered through the minds of the Joy Manufacturing Company's executives. Their engineers spent their spare time sketching continuous miners of all types or examining the drawings of hopeful inventors.

In 1946 the company became aware that a new machine, the property of the Consolidated

Figure 465. Harold Silver and Joy's first 3JCM. (Courtesy Joy Manufacturing Company.)

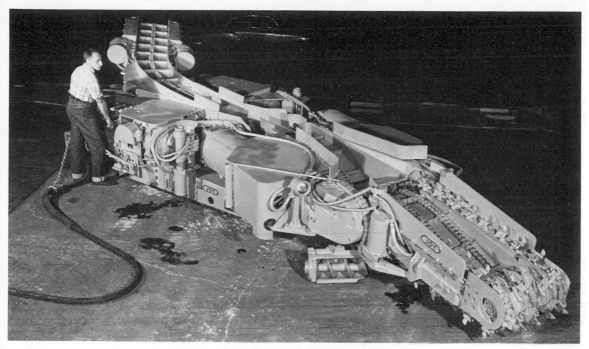

Figure 466. Joy continuous miner — Model 3JCM. As built in 1951-52. (Courtesy Joy Manufacturing Company.)

Figure 467. Latest type Joy miner (1975) — Model 12CM. (Courtesy Joy Manufacturing Company.)

Coal and Coke Company, was operating in Denver, Colorado. As soon as Joy's Vice-President, Arthur Knoizen, saw the machine, he knew it was what they were seeking and by 1947 the Joy Company had bought the patent rights. Yet one Joy engineer admitted quite candidly that he would not have looked twice at the machine if he had only seen the drawings, as the machine was based on the same principle as that of the earliest coal-cutters, namely a

Figure 468. Joy 8CM. (Courtesy Joy Manufacturing Company.)

Figure 469. Joy 15CM. (Courtesy Joy Manufacturing Company.)

mechanism operating a cutting chain. Joy found that the new machine actually worked. Moreover, it even cut almost the right size coal chunks — not the slack normally produced by such methods. (Slack is small coal sold by operators at a lower market price than the size normally considered acceptable.) In addition it was extremely versatile. At a cost of nearly a million dollars Joy built 25 prototypes of these ripper machines and installed them in a variety of mines of differing conditions at a nominal cost to the mine.

Of the 25 prototype machines made, 12 were Model 3JCMs (Figure 465) and 13 were 4JCMs. The 3JCM was an 86 cm (34 in) high machine designed for medium to low seams of coal, whereas the 4JCM was a 122 cm (48 in) machine designed for higher seams.

510

Figure 470. Joy CU43. (Courtesy Joy Manufacturing Company.)

Figure 471. Joy 2BT6. (Courtesy Joy Manufacturing Company.)

The mines were assured that any necessary modifications or replacements would be borne by the Joy Company. The investment was justified by the highly successful results which these early trials achieved. Naturally, like all new machines, there were plenty of teething problems such as excessive bearing wear and burst hydraulic hoses. Another serious difficulty was the fact that a great quantity of coal was left behind on the floor after an operation cycle. But these and other problems were gradually overcome and soon the machine was operating at a faster pace than the coal could be transported to the surface.

The successful debut of the 3JCM and 4JCM

was followed by a variety of Joy continuous miners (Figures 466 to 471) which included, amongst others, Model 10CM. This machine was produced in 1968 and constituted the first of the new Joy rotary drum cutter head types. They were basically similar in concept to the rotary drum head machines introduced in 1965 by the Jeffrey Manufacturing Company. Cutting bits were mounted on scrolls on either end of the drum and on a wide chain which ran around the centre of the drum. The machines were capable of mining 8 to 12 t of coal per minute. (The first Joy continuous miners produced 2 t of coal per minute.)

Summary

Joy presently offer both chain ripper miners and the rotary drum head-type miners introduced in 1968. However, the rotary drum head machines are proving to be the more popular and look to be superseding the chain ripper models produced by Joy.

In operation the chain ripper machine works as follows. The machine is trammed* forward into position then, while the main chassis of the machine remains stationary, the cutting head

---

*Technical term used for the driving backward and forward of such machinery.

511

(a)

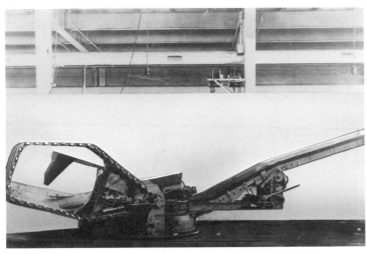

(b)

(c)

Figure 472. (a) Jeffrey entry driver; (b) Jeffrey MM39-A; (c) Konnerth miner. As early as 1911 the Jeffrey Manufacturing Company began development work on a combined cutter loader, the first model of which was designated MM32-A. This was followed in 1913 by the M34-A which was known as the 'Jeffrey entry driver'. Virtually following the same process used by the miner, the unit under- and side-cut the face with picks on an endless chain, followed by a percussive pick-type device which broke the coal at the top. In 1915 the MM39 featured a 'swinging hollow cutter bar' which cut a block of coal. The block was broken by picks on a pneumatic percussive breaker. Later in 1952 the company built 40 34F model machines, known as the 'Konnerth miner'. According to E. M. Warner (Joy Manufacturing Company), these units were not particularly successful because the vibrating hammers could not handle all types of coal seam conditions. (Courtesy E. M. Warner — Joy Manufacturing Company.)

moves forward, sumping into the coal at the *bottom of the face*. The rotation of the chains carries the broken coal back over the head and deposits it on to the chain conveyor which then moves the coal to the rear end of the machine. When the sumping operation is complete, the head travels upwards towards the top of the seam, breaking out the coal as it moves. As soon as the cycle is complete the head is withdrawn and swung to the side in order to make another cut.

The rotary drum miner, on the other hand, sumps into the coal face *at the top of the seam*. When the predetermined sumping distance is reached, the head moves downwards breaking out the coal as it progresses. Gathering arms on the loader sweep the fallen coal from the floor on to the chain conveyor where it is carried through the throat of the machine to the rear end of the miner.

The Colmol Continuous Miner (Jeffrey Manufacturing Company)

The Joy Manufacturing Company was well aware that at least 10, possibly more, ideas for a continuous miner were currently being developed at the time it was negotiating for the rights of the Smith/Silver machine. But, appreciating the complications involved in building such a machine, it felt it highly unlikely that any other company would market a competitive model before its own miner was ready. Therefore, apart from making public the information that it had acquired a continuous miner, no advance pictures or other details were published.

To the complete surprise of all concerned — well nearly all — the Colmol, a rotating breaker-arm-type of machine was put on the market in 1948 and became, for many Americans, the first of this type of machine ever seen by them.

The Colmol was conceived jointly by Vincent J. McCarthy, an inventor of drilling machines, Clifford Snyder, President of Sunnyhill Coal Company, Pittsburg, and Arnold Lamm, Vice-President of the Sunnyhill Coal Producers Association.

Snyder who, at the age of 38 was managing a successful strip-mining business which he had built up during the war years, was observing the

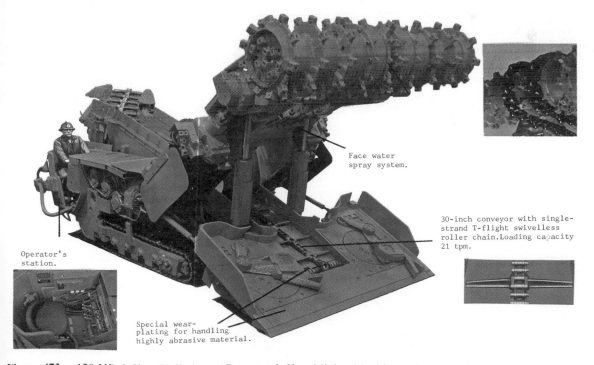

Face water spray system.

30-inch conveyor with single-strand T-flight swivelless roller chain.Loading capacity 21 tpm.

Operator's station.

Special wear-plating for handling highly abrasive material.

Figure 473. 120-HR Jeffrey Heliminer. (Courtesy Jeffrey Mining Machinery Company.)

Figure 474. Jeffrey 120HR Heliminer at work in an iron ore mine of A.R.B.E.D., Mines Françaises in Moselle, France. (Courtesy Jeffrey Mining Machinery Division of Dresser)

15 cm (6 in) diameter horizontal drills bore easily through the toughest shales. This drilling operation was a necessary part of strip-mining because, before the power shovels could be put to work laying bare the coal seams, all overlying rock and debris had to be broken up and removed. This was accomplished by horizontal drilling and blasting.

It occurred to Snyder one day that this idea of horizontal drilling might be incorporated into a type of continuous miner. When Snyder next saw his friend McCarthy, he asked him if he had ever measured the quantity of rock debris produced by one of his drills after one minute's drilling. McCarthy replied that the same ideas had recently been running through his own mind. They approached Arnold Lamm and the three men put their heads together and began exchanging ideas on how to convert a drilling machine into a continuous miner. Eventually the required bit was designed. Ten of these were mounted on to a converted army tank which had been purchased by Snyder for $4000. At this point McCarthy lost interest in the project, but Snyder and Lamm persevered with the assistance of a Pittsburg patent attorney, William D. Carothers. When they

were satisfied that the idea was feasible, a completely new model was built. This was the machine exhibited to the industry in 1948.

By this time a sum of $750,000 had been spent by Snyder and Lamm in developing the machine. Though they had originally intended to manufacture and market the machine themselves, economic circumstances forced them to accept a fifty-fifty offer from the Jeffrey Manufacturing Company.

Jeffrey had already built two continuous miners (Figures 472 and 473), but these were unsuccessful. The new agreement was signed in 1949 and within a short time the company had built six trial models of the Snyder/Lamm machine. They hoped it would establish their contention that their machine, the 'Colmol', as it became known, could bore out the coal at a faster and more economic rate — and in bigger chunks — than could the Joy. These machines were successful. However, the Joy Manufacturing Company considered that their machines enjoyed advantages not shared by their rivals at that time, namely greater manoeuvrability and flexibility. These added qualities, they contended, would make them particularly attractive to operators contem-

514

plating the use of a continuous miner within the extremely confined spaces of a coalmine.

The early Colmol (1949 model) which Snyder described as a 'giant mole' bit into a 2.89 m (9½ ft) face, 1.2 m (4 ft) high and devoured the coal at a rate of 46 to 61 cm (18 to 24 in)/min. Moving on caterpillar tractors it hungrily advanced, cleaning up the mine floor as it progressed and leaving only 1 per cent of the coal in the outer corner of the room. As much as 100 t of coal per man-day was taken by the Colmol's rotary chipping heads. These heads carried teeth which were set in a progressively receding pattern, widely spaced, so as to enable them to chip out the coal in overlapping annular concentric grooves.

Since the introduction of its first continuous miner, the Jeffrey Manufacturing Company has produced two types of miner of the Colmol design. These are classified under Model Nos. 76B and 76AM. The machines which fall into the former category, namely 76B, are designed to deal with coal seams ranging in width from 1.2 to 1.8 m (4 to 6 ft) while those models falling into category 76AM deal mainly with coal seams of from 1 m to 1.4 m (3 to 4½ ft) widths. Though there are minor variations in width, length, and weight between these models, the basic overall design principle has remained the same. Both the 76B and the 76AM incorporate two rows of rotating heads. Eight of these heads are mounted on the top row and five are positioned on the bottom row. As the miner progresses, a hole is first bored by the central leading drill bit of each head and then the remaining coal is 'levered' out by the following picks. In this way large marketable pieces of coal are produced and fines are reduced to a minimum.

The entire cutting unit, which is hydraulically operated, can be elevated or depressed or, if necessary, tilted. In addition the distance between the two tows of cutter heads can be varied slightly according to the thickness of the seam being attacked.

In 1968 the Jeffrey Manufacturing Company became known as Jeffrey Galion Inc., which corporation utilized three unincorporated divisions known as the Jeffrey Manufacturing Company, the Jeffrey Mining Machinery Company, and the Galion Division. (The unincorporated division known as Jeffrey Mining Machinery Company produced machinery equipment for coal extraction at the face.) On 31 May 1974 Dressor Industries Inc. officially acquired the assets of Jeffrey Galion Inc., including the assets of the unincorporated divisions.

The Jeffrey Mining Machinery Division of Dresser Industries Inc. is currently producing a new type of continuous miner known as the 'Jeffrey Heliminer', which made its debut in 1965-67. This machine is basically similar in design to the Joy Manufacturing Company's rotary drum cutter-head-type machine which appeared on the market the following year (See 'The Joy continuous miner').

Several types of Jeffrey Heliminer machines are currently available. These range from the 120-M Model powered by two 205 kW (275 hp) motors to the large 120-HR machine which is powered by two motors rated for steady cutting at 224kW (300 hp) each.

The Heliminer 120-HR is capable of producing 15 tons per minute from seams up to 3.5 m (12 ft) high. The helical cutting head, from which the Heliminer takes its name, is nearly 3.3 m (11 ft) wide and has a 93.3 cm (36¾ in) diameter drum. In operation the machine is trammed forward and commences its cycle by sumping 61 to 71 cm (24 to 28 in) into the face at the roof line. When the required sumping distance has been reached, hydraulic cylinders force the cutting head down to the floor line. The machine is then reverse trammed approximately 15 to 20 cm (6 to 8 in) to smooth the floor. After which the head is raised in order to commence another cutting cycle. The fallen coal or other debris is swept on to the loader by gathering arms where it is carried by the chain conveyor through the throat of the machine to the rear.

The BCR Machine

As is so frequently the case when a need arises — such as that faced by the coal industry in its battle for survival — the idea of a machine that would raise production and yet lower the costs, was uppermost in the minds of many coal operators of that time.

In 1947 when both the Colmol and the Joy

were suffering their first teething troubles, a group of such operators met and decided that it was time that coal industry knowledge be brought to bear on the development of a continuous miner. An amount of $258,000 was collected within a year from some 71 different companies, including coal, railroad, and land companies (representing some 60 per cent of the total tonnage produced by the whole nation), and a Committee, the Mining Development Committee, was appointed. This Committee operated within the industry's own general research organization, namely Bituminous Coal Research Inc. The Mining Development Committee, composed principally of leading U.S. mining engineers, drew up specifications for the type of machine desired and the performance that was required. Then Gerald Von Stroh was appointed as Director of the project.

The Mining Development Committee conducted a detailed examination of the various known methods of advancing a coal face. The method eventually selected, described as the 'bursting' principle, was completely different from operation principles employed by either the Joy miner or the Colmol. The Mining Development Committee was convinced that a machine using this principle would far surpass in tons per hour per unit of energy any machine available to the industry. *At that time the only other machine ever to make use of the 'bursting' principle was the old McKinlay 'entry driver' (see section on McKinlay) which had used two large boring heads of fixed diameter.*

When the Colmol had first appeared on the scene, many people in the coal industry had been under the impression that the 'Colmol' was merely a modification of the McKinlay. However, this was not so as the McKinlay machine used a wedge-shaped bursting wheel which followed behind the rotating bits.

The Mining Development Committee considered many alternatives before eventually agreeing to a final design. This group of mining engineers had stipulated stringent requirements concerning the type of machine they needed. They imposed various limitations such as seam thickness under which they considered the machine should be capable of operating efficiently. This in turn affected motive and cutting power in relation to the space required for motors, etc.

All these points were considered in depth before the final design was passed. Specifications for the proposed new BCR machine (figure 475) showed remarkable promise. It seemed to have combined qualities such as flexibility with the then exceptional ability of allowing the machine

Figure 475. BCR machine — being readied for testing. (Courtesy Bituminous Coal Research, Inc.)

516

to win coal from seams which were as low as 71 cm (28 in) — 10 to 25 cm (4 to 10 in) lower than the early Colmol or Joy machines were capable of mining.

The development of the prototype machine by the time it was eventually built consumed the entire sum of $258,000. collected for the project. Initial trial results were particularly encouraging, but a further amount of $500,000 at least was needed to finance the project before a thoroughly tested machine could be produced.

Bituminous Coal Research Inc. felt that development had proceeded to a point where commercialization should be in the hands of a manufacturer and in July 1952 the Le Roi Company licensed the machine from Bituminous Coal Research, Inc.

Subsequently the Le Roi Company was purchased by Westinghouse Air Brake Company. With the transfer of its other holdings went the original agreement between Bituminous Coal Research Inc. and Le Roi, covering the licensing rights to the prototype machine. This in turn led to a study which was sponsored by Westinghouse Air Brake to ascertain the market potential for continuous miners. Though this was found to be favourable, Westinghouse Air Brake decided not to exploit the BCR machine further, as such a market could not be served compatibly with other markets currently being served by Westinghouse Air Brake's sales organization.

So far as the Bituminous Coal Research Inc. organization was concerned, the committee decided not to further promote the commercialization of the BCR machine as, by that time, there were sufficient manufacturing and marketing programmes initiated by various companies such as the Joy, the Jeffrey, and the Goodman manufacturing companies, to satisfy the needs of the industry. Thus ended the story of what might, perhaps, have been a truly remarkable machine.

McKinlay Entry Driver

Edward Scofield McKinlay[10,11]

Edward Scofield McKinlay was an English inventor and engineer. His most brilliant concept, the McKinlay entry driver was designed by him when he was still in England. Of interest is the fact that *many of the principles in McKinlay's machine were initiated by Reginald Stanley*. An unusual innovation introduced by McKinlay was the set of rotating discs or 'wedging wheels' which were placed behind the cutters in such a position and at such an angle that they were able to exert a sideways pressure against the ridges of coal left by the cutters, thus breaking them away from the face. It must be noted that although the idea of using rotating discs on a tunnelling machine was not new (Wilson made use of them as cutters as early as 1856 on his machine, which he built for the Hoosac tunnel and, in 1866, Brunton also used rotating wheels or discs as cutters on his first tunnelling machine) their use as wedging wheels after the conventional cutters had done their work, was original.

Though McKinlay designed his machine in England, he neither attempted at that stage to patent his invention there nor to build a prototype. Perhaps realizing the futility of offering such a machine to conservative English operators, McKinlay took his designs to America where he patented and built his first entry driver. U.S. Patent rights were granted to McKinlay under No. 1,284,398 (Figure 479) on 12 November 1918.

During the early 1920s McKinlay built and installed about 20 of his machines in various mines. Unfortunately, the majority of these units proved to be unsuccessful, mainly because the coal industry at that time was in its infancy and lacked the essential resources needed to develop such a device.

Over the years McKinlay worked continuously at improving both his machines and the conveying and loading systems associated with it. On 15 October 1926 he filed his first U.K. Patent (No. 2,575/26) in London, covering 'Improvements in or relating to coal mining and loading machines'. This was eventually accepted on 5 January, 1928 under No. 282,921. (Figures 476 and 477).

Basically McKinlay's machine was built on a well-braced framework of substantial channel iron. Two or more cutter heads were mounted on shafts at the front of the machine. These shafts were driven by an electric motor through

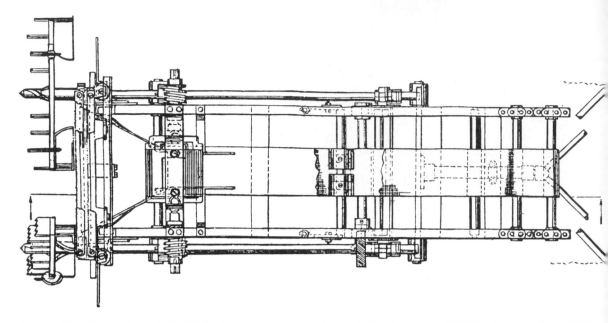

Figure 476. Edward Scofield McKinlay tunnelling machine. U.K. Pat. No. 282,921 dated 5 Jan. 1928, (application filed 15 Oct. 1926). Plan view of machine

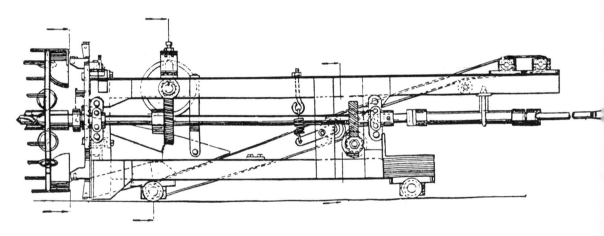

Figure 477. Side elevation of McKinlay's T.B.M.

a series of worm gears. The cutter heads were counter-rotating. Each cutter head carried a boring and centring bit which was mounted in the tip of a heavy cone. During operation the cone exerted a bursting pressure against the innermost and smallest ring of coal cut by the bit and the two cutter blades nearest the centre of the cutter head. Behind the second and outer cutter blades on each head, McKinlay attached his wedging wheels or rotating discs. These discs followed the cutting blades and broke off the concentric rings of coal left by the cutters.

In operation the machine (Figures 476 to 479) was rolled into the mine on rollers or wheels running on axles mounted at the bottom of the frame and moved up to the working face. A hydraulic cylinder was situated at the rear of the machine. Attached to the cylinder were two 'claw bars' which dug into the side walls of the tunnel. These provided anchor points against which the thrust of the piston could be applied. As the rotating heads were forced against the face by the action of the hydraulic cylinders, the cutters, assisted by the breaking wedges, cut

518

Figure 478. McKinlay entry driver. (Courtesy National Mine Service Co.)

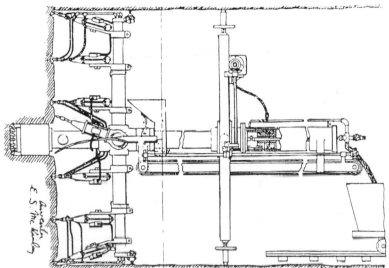

Figure 479. E. S. McKinlay tunnelling machine. First patent filed by McKinlay in America on 12 Nov. 1918 under No. 1,284,398

two circular cavities in the face. This left material above and below the cutting heads, roughly triangular in section. McKinlay dealt with these remaining sections by mounting two pairs of saws, one pair above and the other below the heads at the front of the machine behind the cutting tools and wedge breakers. The saws operated with a reciprocating action and effectively cut through the triangular sections of material left untouched by the rotating cutters and discs.

Scoops or shovels fixed to the rear edges of the cutter heads collected the coal or other material as it fell from the face and deposited it in a chute which channeled it on to the conveyor belt running in a central position through the machine. This belt in turn conveyed the cuttings to the rear of the machine where the material was dropped into trucks or cars.

In Patent No. 282,921, dated 15 October 1926, McKinlay provided the machine with hollow shafts and pipes for either extracting dust from the face or, alternatively delivering liquid to it.

The duct required to convey these substances in either direction was provided as follows. The shaft was hollow from the front main bearing to the cutter head. At the cutter head, channels built into the head branched the duct out so that an opening occurred just before the cutting end of each cutting tool. Access to the hollow tube at the main bearing was through an opening in the outer cover of the bearing, about halfway along its length. This connected to an internally cut circular groove in the bearing surface at this same position. Holes drilled radially through the wall of the hollow shaft gave final access to the central duct. Air was thus drawn out or liquid pumped in through this opening in the main bearing.

Like so many of the tunnelling machines invented before the McKinlay entry driver was introduced, McKinlay's device might well have died a natural death if chance had not directed two of his entry drivers into the New Orient mine which at that time (1926) was owned by the Chicago, Wilmington & Franklin Company (now a part of the Freeman Coal Corporation).

Unlike other operators, the company and its employees accepted the units as a challenge. They were at once firmly resolved that, come what may, the machines would be made to operate successfully. Furthermore, they had at their disposal the necessary facilities such as staff and engineering equipment, with which to accomplish this task.

The New Orient mine was at that time considered to be one of the largest in the world. It held the record of lifting some 15,000 t of coal through a vertical shaft in a single day. Naturally, developmental work in such a mine would far exceed that normally undertaken elsewhere.

Many problems arose after the initial installation of the machines. The majority of these were gradually overcome. In his paper 'The Marietta miner'[11] C. C. Conway mentions that the McKinlay entry drivers were invariably '... used in adjacent places to drive pairs of entries. By careful planning the projected paths were so arranged as to keep the machines always advancing because abandoning a pair of entries meant partially dismantling the units so that they might be moved to a new point of beginning'.

Though the two machines were gradually improved and refined by the company over the succeeding years with the result that they operated successfully for a total of 25 years, the problem of restricted manoeuvrability still remained. However, so great was the overall improvement that at the end of the 25-year

period (1926 to 1951) the machines were able to progress at twice the speed attained during their early years of installation. It was estimated, though no record was actually kept at the time, that during their period of operation the McKinlays travelled a total distance of more than 160 km (100 miles).

Despite the fact that many thousands of eminent personages from America and overseas viewed the machines during their operation period few, if any, really appreciated the potential which the McKinlay entry driver possessed.

The Marrietta Miner[11]

Harry Treadwell who was at that time Vice-President of the Chicago, Wilmington and Franklin (C. W. and F.) Coal Company, had seen the installation of the first McKinlay machines and, during the following years, had contributed a considerable amount of effort towards their successful operation. Though he was undoubtedly well aware of their limitations, he also recognized that the potential was there, in the basic design of the machine, for an outstanding continuous miner. When McKinlay died, Treadwell, who had been assigned the task of developing a completely new mine, decided to plan this new development 'for 100 per cent continuously mined extraction,[11] using for the task continuous miners which he intended to have built. Treadwell hoped to be able to overcome the remaining disadvantages associated with the McKinlay by various modifications of his own.

Perhaps one can more fully appreciate the dynamic qualities of the man when it is realized that, while he was busy organizing the sinking and development of the new mine, Treadwell was also gathering about him a group of engineers. These men were employed by Treadwell and commissioned to design and build a new machine, which was based on the McKinlay, but which nevertheless embodied all the innovations which Treadwell envisaged would overcome the problems associated with the old miner.

The machines were built by the Marietta Manufacturing Company who, for their part, contributed much towards this project in the

Figure 480. Marietta continuous miner — Model 780-AW4. (Courtesy National Mine Service Company.)

area of engineering and manufacturing skills.

Of particular interest in the light of subsequent events *was the presence there at that time of a brilliant consulting engineer, namely James S. Robbins (Senior), founder of the Robbins Company of Seattle.* Robbins was Consultant for the Marietta Boat and Barge Company (Marietta Manufacturing Company) at Point Pleasant. During the building of the new machine, Robbins also acted as consultant for Treadwell's firm the C. W. and F. Coal Company. He continued to serve the C. W. and F. Coal Company as Consultant until the old McKinlay entry drivers at the C. W. and F. Coal Company's Orient No. 2 mine were eventually taken out of service.*

The new machine, christened the 'Marietta miner' was such that it aroused the interest and admiration of many all over the country. There were several significant improvements. These included alterations to the top trim chain and lower bar extensions, permitting greater flexibility and a fully manoeuvrable hydraulic steering system. Most important the old

*Subsequently Robbins consulted for the Old Ben Coal Corporation and the Goodman Manufacturing Company. See chapters 5 and 19 on the Robbins Company and Goodman Manufacturing Company, respectively.

cumbersome rollers had been replaced by modern crawler tracks and its overall power had been increased considerably.

The first Marietta miner was manufactured in August 1950. To Treadwell's organizing ability must go the further credit for the delivery at the mine site of this first machine on practically the same day that the coal seam in the new mine was reached.

So successful was this initial unit's performance that 10 more machines were built in the following two years. There were no basic changes in the design of these additional machines.

After the 11 units had been operating successfully for several years it was decided that a further increase in power and flexibility would be advantageous. These improvements were incorporated into new machines which the Marietta Manufacturing Company were producing.

Gradually other manufacturers entered the field. At this stage, for various reasons, the Marietta Manufacturing Company decided to sell their interests in mining machinery. The Clarkson Manufacturing Company, which is now known as the Clarkson Division of National Mine Service Company, bought the mining interest as well as drawings, patents, goodwill, and inventory of both the Marietta miner and the McKinlay entry driver. Today this latter company is still producing Marietta miners (Figure 480). The modern machines have since undergone minor refinements which have given them added power, flexibility, and greater structural strength. But otherwise the new Marietta remains basically the same as the original model which, on its own crawlers, travelled from the car to the face to begin work on that momentous day in August 1950.

Other patents taken out by E. S. McKinlay in chronological order are given in Table 12.

Table 12

British

| 282,921 | 15 October 1926 | Improvements in or relating to coal-mining or loading machines |

United States

978,610	13 December 1910	Concentrating apparatus
1,153,804	14 September 1915	Screen for ores and other material
1,284,398	12 November 1918	Tunnelling machine
1,379,637	31 May 1921	Cartridge explosive

McKinlay, Edward S. assignor to McKinlay Mining and Loading Machine Company

1,603,621	19 October 1926	Coal-mining and loading machine
1,726,963	3 September 1929	Mining and loading machine
1,748,301	25 February 1930	Conveyor system for mines and other purposes

The Goodman Manufacturing Company[1]

The Goodman Manufacturing Company had its beginning in the spring of 1889 when Herbert Edward Goodman and his brother-in-law, Elmer Ambrose Sperry formed the Sperry Electric Mining Machine Company in Chicago, Illinois.

Like the Smith and Rand Company,* the

*See Chapter 1, Drills — Horizontal drills — the Rand Bros.

gradual expansion and progress of the original Sperry Electric Mining Machine Company during the succeeding years brought many changes both in name and directorship.

Of particular significance were the changes effected during the period 1890-1900. In April 1890 certain executives of the Link-Belt Machinery Company acquired shares in the enterprise. By 1895 half the interest in the business was under their control, and finally by 1898 they had assumed full financial control.

During this period of transition, Herbert Goodman (founder of the original firm of Sperry Electric Mining Corporation) had continued his management of the business but, in December 1899, he decided to establish his own company by purchasing the Mining Machinery section of the business.

As Link-Belt Machinery Company (owner of the patent rights) wished to concentrate on its regular line of products, they were agreeable to the sale. But it was not until 23 April 1900 that the transaction was eventually finalized and the Goodman Manufacturing Company was officially formed. It carried with it full rights for the manufacture and sale of the complete line of mining machinery.

The following years saw a gradual expansion of activities, the Goodman Company matching other manufacturers in meeting the ever growing demands of the mining community for more and better machines and equipment. Apart from the different types of coal-cutter already mentioned, Goodman also produced amongst other items, improved haulage locomotives, belt and chain conveyors, and several kinds of loading machines.

When the thoughts of various mining machine companies, including those of the Goodman Manufacturing Company, were turning towards the production of a continuous miner, the Goodman Company became associated with James S. Robbins. Robbins acted as Goodman's consultant during the design and development of their new machine. He brought with him the expertise and experience which he had gained while working with other engineers on the development of the Marietta miner.* The result was a machine

*See section on the McKinlay entry driver.

which was similar to the Marietta and incorporated many important features which had originated with the old McKinlay entry driver. This was Goodman's new borer.

Goodman Continuous Borers

Over the past 20 years Goodman have developed several different models of their machine (Figure 481). These machines, while retaining the essential basic features, vary according to specific seam height requirements. Thus Model 302, for instance, has a variable cutting height of from 137 to 152 cm (54 to 60 in), and the path cut in a single pass may range from 3.86 m (12 ft 8 in) to 4.37 m (14 ft 4 in), depending upon the mining height selected. That is, it is 4.37 m (14 ft 4 in) wide at 137 cm (54 in) height and 4.52 m (14 ft 10 in) wide at 152 cm (60 in) height. However, the flat roadway, which is cut by the horizontal bottom chain cutter, remains at a width of 3.86 m (12 ft 8 in) no matter what seam height is selected. (This width varies with different models and would be 3.29 m (10 ft 7½ in) in, say, Model 405).

The variation of mining heights is effected by simple external mechanical adjustment of the outer cutter on its telescoping arm, and by positive individual hydraulic control of top and bottom cutter bars.

In the case of the 405, the 429, and the 430 machines, cutting heights vary from 1.83 m to 2.29 m (6 ft to 7 ft 6 in) while the width of the tunnel may vary from 3.48 m to 4.06 m (11 ft 5 in to 13 ft 4 in), depending upon the particular model used and the actual cutting height selected.

The 302 is a four-rotor machine, each rotor of which cuts three concentric kerfs by means of an inner core barrel, a fixed intermediate cutter arm, and a telescoping outer arm. The 405 and 429 have two rotor borers which rotate in opposite directions, while the 430 can be equipped with either two three-arm rotors, or two two-arm rotors which are interchangeable.

The Goodman Manufacturing Company has also produced full-face continuous borers for use in harder rock formations such as salt and potash. These machines carry disc cutters as well as the standard bit-filled cutting arms, used on other Goodman machines.

In these cases the McKinlay 'bursting principle' is used. That is, the normal toothed cutters are followed by the bursting discs. These discs then shatter the inter-kerf ridges.

In operation, coal or other debris drops from the face to the floor of the tunnel and is swept to the centre by ploughs on the outer arms of the rotating units. The material is then moved up the throat of the machine by conveyor, the discharge end of which may be raised and lowered, or swung 40° to either side of the centre line.

The Lee-Norse Miner[12]

A chance meeting in 1939 between E. M. Arentzen, a mechanical engineer from Norway

Figure 482. The Lee-Norse CM-46 — an early model of the 'Koalmaster'. (Courtesy Lee-Norse/Ingersoll Rand.)

Figure 483. The Lee-Norse Junior, a further development of the original 'Koalmaster', sumped hydraulically as the crawlers stood still. Note the scroll-type clean-up unit which resembles the loader mechanism used on the Belgian cutter loader produced about the same time (1954-55). However the system was not satisfactory and in later models the company reverted to the gathering arm clean-up method. (Courtesy E. M. Warner— Joy Manufacturing Company.)

and Arthur Lee, an electrical engineer from Pittsburgh, led to the establishment of the Lee-Norse Company on 5 March 1940. The 'Lee' is, of course, Arthur's surname while the 'Norse' was Arentzen's way of proclaiming his pride in his Norwegian heritage.

Initially the company specialized in the production of coal loaders, personnel carriers, and mine jitneys but, like most other mining equipment manufacturers of that time the company recognized the need for a machine which would both cut and load coal. In 1951 Lee-Norse introduced their 'Koal-master' (Figure 482) to the market.

While the main chassis of the machine, comprising a flexible-type conveyor and gathering head, was basically similar in design to that of the Joy, the cutter head itself was significantly different. A heavy steel boom, operated and supported by two double-acting hydraulic jacks, could be raised and lowered between the floor and the ceiling. This boom was attached by means of a hinged joint at its rear to the main chassis of the machine above the gathering head.

Two rotary-type oscillating cutter heads were mounted on the forward end of the boom. They were each powered by a 60 hp (45 kW) motor, and were geared together so that the two cutter heads would run synchronously. Five wheels spaced 38 cm (15 in) apart were fitted to each cutter head. These wheels carried a number of

524

arms (up to a maximum of eight) which in turn carried cutting bits or picks.

The heads, while rotating, oscillated horizontally a distance of 38 cm (15 in); moving towards one another and then apart. This produced a pincer-like movement as the boom moved up and down against the face. This ripping-type action resulted in the production of coarse-grained coal, devoid of fines, as the bits traced a diagonal intersecting pattern of grooves or kerfs across the face.

In the prototype machine rib jacks were fitted to provide additional forward thrust for sumping purposes, but modifications of the cutter-head arrangement later obviated the necessity for these jacks. The machine was also fitted with two hydraulically powered rotary drills for roof-bolting purposes.

During operation the cutter heads are raised to the roof and sumped into the face to a depth of about 76 cm (30 in). The boom is then lowered slowly to the floor and the machine is retracted slightly so that the operator may smooth out the floor. The coal is swept into the throat of the machine by the action of the cutter heads and by the gathering arms on the apron, and is then carried to the rear by the flexible conveyor and discharged into shuttle cars, etc. The original machine, produced in 1951, weighed 18 t and was 122 cm (4 ft) high × 244 cm (8 ft) wide × 7,80 m (25 ft 7 in) long. It was capable of making 3.66 m (12 ft) cuts in seams which varied from 152 cm to 244 cm (5 ft to 8 ft) in height.

In 1952 a Lee-Norse oscillating head was fitted to a 14BU Joy chassis. It was christened the 'Junior miner' (Figure 483) and was designed specifically for low coal seams which varied in height from 91 cm to 122 cm (3 ft to 4 ft). The prototype unit was put to work in an eastern Kentucky coal-mine where it cut coal in 112 cm (3 ft 8 in) high seams. The Junior miner was somewhat different from the original machine in that the cutter heads carried only four 61 cm (2 ft) diameter wheels (two on each head) which cut a place approximately 244 cm (8 ft) wide. It was 81 cm (2 ft 8 in) high × 183 cm (6 ft) wide and was roughly the length of a conventional loading machine. Each cutter head was powered by a 19 kW (25 hp) motor. These were again synchronized so that both heads ran in unison.

As a result of the successful trials of the Junior miner, which cut and loaded the coal at the rate of about 1 ton/min, Lee-Norse designed and built a new and more powerful machine which was mounted on a Lee-Norse caterpillar chassis. Its dimensions were 122 cm (4 ft) high × 183 cm (6 ft) wide × 8.23 m (27 ft) long, and it was capable of making a 244 cm (8 ft) wide cut in seams ranging in height from 152 cm (5 ft) to 244 cm (8 ft).

The company, which in 1964 became a subsidiary of the Ingersoll-Rand Company,

Figure 484. Wilcox prototype. (Courtesy E. M. Warner — Joy Manufacturing Company.)

currently produces a line of continuous miners some of which are capable of cutting seams as low as 76 cm (2 ft 6 in) while others can handle seams up to 4.3 m (14 ft) in height. These include, apart from the oscillating head-type machines, a new range of fixed-head miners which were introduced by Lee-Norse in 1970.

In company with several other major mining machine manufacturers Lee-Norse also supplies a remotely operated system for use with the low chassis height machines. The unit can thus be radio controlled or alternatively electrically controlled by means of a 6.10 m (20 ft) cable connection (see also Chapter 19, America — Remotely controlled continuous miners).

Auger Machines[13]

Auger machines were manufactured in both Britain and America. The main British machine was the Cardox Hardsorg.

Developmental work was begun in America in 1955 by A. G. Wilcox of West Virginia and then later carried on by the Jeffrey Manufacturing Company. The early Jeffrey unit was known as the Jeffrey 100-L auger drill (Figure 485).

Attempts by Wilcox to build a shortwall cutting machine suitable for low seams led to the production of the Wilcox miner (Figure 484). The cutting mechanism of this machine consisted of two contra-rotating auger-type heads which could be varied in height from 655 mm (2 ft 2 in) to 1300 mm (4 ft 4 in). Cutter bits were fitted along the periphery of each auger and also at the front end of the augers. During operation the machine sumped into the face at the centre then the chassis was moved sideways mechanically, thus enabling the auger heads to cut across the face from right to left.

Wilcox produced the miner for some 20 years until the company was bought out by Fairchild Incorporated in the mid 1970s. Fairchild continued to produce the Wilcox miner, but changed the name of the unit to 'Fairchild' towards the end of 1978 after major modifications to the machine produced a new model, the Mark 21, (Figure 486) which provided for the remote setting of the jacks which serve as the anchor points for the machine to pull itself across the face.

In addition to the 'Fairchild' machine the company also market the 'Badger'. According to P. L. McWhorter (Vice-President, International Marketing and Export Sales of Fairchild Incorporated) this single auger head machine is a 'custom-designed' unit for the particular working conditions where it will be applied, namely the recovery of coal from peripheral areas and pillars which do not lend themselves to removal by any other means. A second unit of this type called the 'Kerf cutter/I' is now under development. This unit is designed as a 'cutting machine'. Instead of cutting a long rectangular slot across the face, this machine will auger one or more holes which

Figure 485. Jeffrey 100-L. (Courtesy E. M. Warner — Joy Manufacturing Company.)

Figure 486. Fairchild auger machine Mark 21. (Courtesy Fairchild Inc.)

will serve as relief for coal extraction by explosives.

Though the auger-type unit was fairly successful in America where it was used for open-cut mining and also for low seam work, its use was rapidly discontinued in Britain due to the potential danger of gas ignition.

Basically the auger machine consists of an auger-type cutter head which utilizes the Archimedean screw principle to cut into and extract the coal from the face. In some machines the auger cutter head is open and unprotected as it is with an auger wood drilling bit, whereas in others it is enclosed in a cylinder.

In operation the machine is braced between the floor and roof by hydraulic jacks and the auger is fed into the face by hydraulic rams capable of developing 4 to 5 t of thrust. The thrust jacks have a forward stroke of about 1.5 m (5 ft), and by means of sections which can be joined to the auger cutting head via a square male and female joint, the auger head can be fed into the face up to a depth of about 30 m (100 ft).

Remotely Controlled Continuous Miners

Very often when the time is ripe for the development of new machines, new systems or whatever else may be required, this need is simultaneously recognized by scientists, engineers, or designers, etc. in various parts of the world. Thus where, in England and Europe, developmental work was commenced on remotely operated longwall machinery, similar studies were undertaken at about the same time in America, first by the Union Carbide and Carbon Corporation and then shortly thereafter by the Joy Manufacturing Company on the development of a remotely controlled continuous miner.

The Union Carbide Corporation Machines

Models No. 1 and 2[14]

In 1946 the Union Carbide and Carbon Corporation (now known as the Union Carbide Corporation), began developmental work on a remotely controlled machine which was to bore parallel holes in coal seams to admit air for experimental underground gasification of coal. Developmental machines Nos. 1 and 2 were actually hole-boring machines which cut holes in the coal seam, 914 mm (36 in) diameter, the lower corners of which were squared off to produce a flat surface measuring about 863 mm (34 in) wide upon which the machine could travel. The second of these units was the 'first truly remotely controlled mining machine'.

Patents covering the remote control system were filed (No. 2,699,328, Alspaugh,

527

Figure 487. Union Carbide Corporation No. 3 Model remotely controlled mining machine— shown on special launching platform. (Courtesy Union Carbide Corporation.)

Heimaster and McNeill) in 1949 in America, and were issued by the Patents Office in 1955. Similar patents were also granted by the U.K. Patent Offfice and several other countries. These patents covered guidance devices, control devices, conveying means, etc. which were incorporated in the remotely controlled mining system.

By the time the second machine had been built it became evident to the Union Carbide Corporation that underground gasification of coal was not an economic proposition and this influenced their decision to make their third machine a 'coal recovery unit' rather than one which simply bored holes.

Model No. 3

The third machine (Figure 487) went into operation in 1949. It was built to carry out highwall mining. This type of mining was necessary when there was found to be so much overburden over the coal seam that it was uneconomical to mine this in the conventional way with strip mining machines. The new machine was designed to work along the highwall. The coal was removed by boring a series of parallel tunnels which were divided by ribs measuring between 61 cm and 121 cm (2 ft and 4 ft) in thickness. These ribs were left as a support for the tunnels.

The machine was similar in design to the Goodman borer or the Marietta miner, in that the cutting head consisted of four overlapping cutter heads which cut a hole 965 mm (3 ft 2 in) high by 2.95 m (9 ft 8 in) wide with rounded corners. If the seam thickness exceeded 965 mm (3 ft 2 in) a second pass was made at the seam (usually below the first pass). Fixed blades positioned at the top and bottom of the head dealt with the cusps between the holes cut by the rotary cutters. The coal was directed into the throat of the machine by paddles mounted on the two outer cutter heads. It was then carried to the rear on the 102 cm (3 ft 4 in) wide conveyor which was centrally positioned. So that no coal could find its way into the tunnel except via the conveyor the front of the machine was shielded. The machine moved on crawler tracks. Vertical and horizontal steering was controlled by a hydraulic cylinder which tilted or raised the entire cutting assembly, and by guide shoes on each side of the front of the machine. The horizontal direction of the machine could thus be altered by making one of the shoes bear against a side wall.

The coal was transported by a train of 9.14 m (30 ft) long belt conveyors which ran on rubber-tyred wheels. The conveyors followed the boring unit into the hole. When the maximum depth had been bored the machine was retracted and, as each conveyor emerged, it was lifted by a portable crane and moved out of the way. The crane also served to add conveyors to the train when the unit was advancing. As the machine advanced into the seam a continuous stream of coal emerged on the belt conveyors. When it left the last conveyor in the train the coal dropped through an opening in the floor of the launching platform on to a flight conveyor which carried it up to the truck loading hopper. The launching platform served as a housing for the control station, cable reels, electrical switchgear, and other ancillary equipment. It was used to position the boring unit at the correct height in relation to the seam being recovered. Four hydraulic jacks supported the platform. When it was necessary to move the platform to a new position in order to start another hole, the platform was lowered on to rails by the retraction of its jacks. The platform also lifted and repositioned the rails in readiness for the next cut when the hydraulic jacks were again extended.

Initially strain gauges were attached to specific picks which were located on the sides of the machine. These gauges were, in turn, connected to cathode-ray oscilloscopes located on the operator's control panel. By this means the operator was instantly made aware of any vertical or horizontal deviation of his machine. However, later in all of the 'Stratascope' models developed by the Union Carbide Corporation the signal from the resiliently mounted cutter tooth was detected and transmitted by means of a linear variable transducer instead of the strain gauges. This device, which was called a 'stratascope', measured the amount of pressure exerted on the bit by the material being cut. One of these devices was mounted on each outer rotary cutting head. An electrical signal was relayed from each stratascope to an oscilloscope screen in the control cabin. The signal was recorded on each screen as a moving spot of light which traced a circular path matching the path of the cutting bit. While the cutting bit remained in coal the light spot followed an even circular pattern; however, if the cutter bit moved into rock or material which was harder than coal it began tracing an erratic pattern — jumping in and out of its normal circular path. By this means the operator soon came to recognize the particular message being portrayed on his oscilloscope screens and he could judge whether the coal seam he was following was rising or dipping and so adjust the direction of his mining machine accordingly.

The rib thickness between the tunnels was periodically measured by means of an electric drill which bored through to the adjacent tunnel and sent a signal back to the operator when it holed through. Other devices measured and relayed information concerning the machine's inclination, and the angle between the machine and the first conveyor. Thus, provided the operator steered his machine so that it maintained a straight line in relation to the first conveyor, the tunnel itself was kept straight.

The holes were drilled to a maximum depth of 210 m, this depth being determined by the number of portable conveyors available at that time.

Model No. 4

In 1953 the fourth remotely controlled machine was built and put to work on the No. 5 block seam of the Blue Creek area in Kanawha County, West Virginia, where it successfully

529

produced about 900 t of coal per day. This machine and the ancillary equipment embodied a number of significant improvements which included the following:

(a) It bored the same size hole as the previous model, but was more robust and a great deal more powerful.

(b) Two 45 kW (60 hp) motors in parallel drove the cutter heads of the earlier model through a speed reducer, while the cutter heads of the later machine were driven by a single 150 kW (200 hp) oil-cooled motor.

(c) The fixed blade at the top was replaced by a long horizontal rotary drum-type cutter fitted with picks, but the fixed blade at the bottom was not changed.

(d) Two parallel conveyors, each individually powered by a 5.6 kW (7.5 hp) reversible motor, were fitted to the machine. Thus in the event of a blockage occurring, due to the entry of an oversize piece of coal, the conveyor could be reversed and the obstruction would be fed back to the cutter heads for recutting.

(e) The crane was dispensed with and a 244 m (800 ft) long continuous track was laid along the highwall to carry the complete conveyor train. The train ran along this wooden track until opposite the tunnel entry when it rounded a sharp right-angle bend, mounted the launching platform, and entered the tunnel. This arrangement caused some problems in that it was found to be difficult to convey the coal around the sharp bend. To overcome this the coal was discharged through a hole in the launching platform on to a flight conveyor before reaching the bend. The hole extended for the full length of one conveyor. Thus as the coal poured on to the empty conveyor approaching the hole in the platform, the operator flicked a switch which caused the flow of coal on that particular conveyor to discharge at its forward end into the hole on the platform. When the platform conveyor reached the end of the hole, the operator again flicked his switch which reversed the flow of coal on the conveyor so that it began running in unison with the rest of the conveyors in the tunnel. Simultaneously the succeeding empty conveyor (which by this time had reached the beginning of the discharge opening in the platform) was made to convey in the reverse direction, so that it in its turn

discharged off its forward end into the platform hole. The entire process was repeated each time an empty conveyor approached the platform discharge hole. When the conveyors were empty and running along the track approaching the platform, the conveyor mechanism did not operate.

(f) The 61 cm (24 in) wide belt conveyors were replaced by 152 cm (5 ft) wide chain and flight conveyors. This was a significant improvement as the new conveyors were easily able to handle minor roof falls up to 1.83 m (6 ft) wide by 20 to 25 cm (8 to 10 in) in depth. It obviated the need for men to enter the tunnels in order to remove obstructions or falls as had occasionally been necessary with the previous unit. To carry out this type of work meant temporarily supporting the roof with timber props until the obstruction was cleared, when the workmen would retreat, removing the timber props as they moved back out of the hole. However, if the roof fall exceeded 30 cm (1 ft) in thickness then it was still necessary to clear it manually. Fortunately however, falls of that magnitude were comparatively rare in that area.

(g) In the event of the machine jamming in the hole, the entire train, which was designed with a breaking strain of 50 t, could be hauled out of the tunnel by means of a winch. According to J. W. Heimaster, on only two occasions during the five-year trial period with these remotely controlled prototypes did the winch fail to move the machine after it had become jammed. In both cases it was found that a crawler chain had broken and it was necessary to fit a new one before the machine could be moved.

The Joy Push-button Miner[15]

In 1958 the Union Carbide Corporation signed a patent licensing agreement with the Joy Manufacturing Company, allowing Joy to produce the remotely controlled mining equipment under U.C.C. Patents. The first commercial system was manufactured by Joy and sold to the Peabody Coal Company, one of the largest coal producers in the U.S.

Before building the equipment the Joy Manufacturing Company made extensive time

Figure 488. Joy pushbutton miner advancing into the seam. (Courtesy Joy Manufacturing Company.)

Figure 489. Joy pushbutton miner. Inside the control cabin. (Courtesy Joy Manufacturing Company.)

and motion studies of the entire system (Figure 490) and the results of these studies significantly influenced the design work of the new equipment. Basically the Joy push-button miner consisted of three main parts.

1. The Mining Machine

This (Figure 488 and 489) was similar in concept to the previous U.C.C. units except that it was capable of cutting seams varying from 91 to 122 cm (3 to 4 ft) in height by 2.97 m (9 ft 9 in) in width. The four overlapping rotary cutting heads cut a fixed seam height of 91 cm (3 ft), but it was possible to cut an extra 30 cm (1 ft) of seam with the top rotary drum-type cutter, which could be positioned to cut either the minimum 91 cm (3 ft) in conjunction with the overlapping rotary cutting heads, or raised so that an overall height of 122 cm (4 ft) was cut. The floor of the tunnel was kept clean by a fixed plough blade which cut the coal beneath the four rotary cutters.

Two 93 kW (125 hp) motors drove the rotary cutters and two 11 kW (15 hp) motors drove the top rotary drum cutter. The remotely controlled cutter head was adjustable to enable the operator to follow the dips and rolls of the coal seam. In addition it was possible to position either crawler lower than the other, so that any tendency of the machine to spiral could be immediately corrected. The chains on the crawlers were independently driven by two 19 kW (25 hp) motors.

Twin individually powered conveyors carried the coal to the rear of the machine and deposited it on to the portable conveyor system.

2. The Conveyor System

Sixty 5.2 m (17 ft) long portable conveyors were linked together to form a 305 m (1000 ft) long train. Each conveyor, which was mounted on shock absorbers and supported by a rigid axle,

531

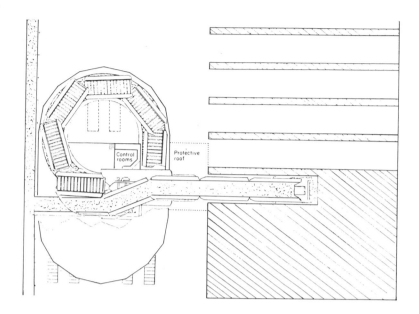

Figure 490. Joy pushbutton mining system, (Courtesy Joy Manufacturing Company.)

Figure 491. Joy pushbutton miner heli-track conveyor system. (Courtesy Joy Manufacturing Company.)

ran on a pair of solid rubber-tyred wheels. The conveyor consisted of two decks. The lower deck housed the intricate wiring which controlled the operations of the miner and the conveyor traction system. Tubular steel sections enclosed this delicate apparatus and protected it from damage. Traction was supplied to every second conveyor by a 2.3 kW (3 hp) motor.

The removable top deck of each conveyor could be swung to one side to divert the coal on to an auxiliary conveyor system.

Coal was moved by chain and flight conveyors in a continuous stream from conveyor to conveyor until it reached the launching platform, when it was discharged onto the conveyor system which carried it to the truck loading hopper.

3. The Conveyor Storage Mechanism

This portable mechanism was called the *heli-track* (Figure 491). It consisted of a helically shaped steel structure 13.7 m (45 ft high) × 14.6 m (48 ft) wide × 23.5 m (77 ft) long, weighing over 680 t (669 tons) fully loaded. It housed the entire conveyor system, the boring machine, the launching platform, and the operator's air-conditoned control cabin (which was located beside the launching platform beneath the lower frame of the heli-track). The entire heli-track, fully loaded, could be moved from location to location on its three crawler trucks.

The conveyors were housed on a six-level spiral ramp which encircled the structure in a helical pattern. When the heli-track was moved to a new location it was automatically levelled both transversely and longitudinally by means of three large levelling cylinders. Additional clearance up to a maximum of 152 cm (5 ft) beneath the machine could also be provided by extending the three cylinders an equal amount.

The centre of the heli-track structure housed the generating equipment, a completely equipped machine workshop for servicing the apparatus, and other ancillary equipment including two large reels of cable, which provided power for the traction system and conducted the electrical signals to the control circuitry of the boring machine.

Suspended from the top of the heli-track structure was a screen which could be extended over the control cabin and launching platform to deflect rocks and other falling debris from the working area.

In operation the launching platform was positioned at the correct seam height and the machine commenced boring into the highwall. As it advanced coal began emerging from the tunnel in a continuous stream over the top decks of the conveyor trucks. It continued flowing from truck to truck until it reached the conveyor truck on the launching platform. The top deck of this conveyor was angularly deflected so that the coal was discharged over the edge of the truck on to the auxiliary conveyor system. The remainder of the empty conveyors were stored on the spiral ramp of the heli-track structure and, as the machine

penetrated deeper into the highwall, the conveyor on the launching platform was pulled into the hole, its top deck automatically swinging back into line. In the meantime, the succeeding empty truck was pulled onto the launching platform and its top deck was deflected to the side so that the coal (now pouring onto it from the truck which has been on the platform immediately before it) could be discharged onto the auxiliary conveyor system.

The conveyor motors were inoperable when stored on the heli-track, but started up automatically as soon as each truck reached the launching platform.

The rib thickness between the parallel tunnels was periodically measured by the opeator with his 1.73 m (5 ft 8 in) remotely controlled probe auger drill. To do this the boring machine was stopped. Penetration rates averaged between 1.83 m (6 ft) and 0.914 m (3 ft)/min, depending upon whether the machine was mining at its minimum or maximum seam height.

The fact that the top deck was removable meant that in the event of a heavy roof fall this deck could be removed and replaced if damaged, without interfering with the lower deck or disrupting the remote-control system.

If the machine became jammed and it was found necessary to winch the entire train out of the hole, the straight tubular sections enclosing the lower decks of the conveyors were useful as an additional tension member.

A stratascope device was mounted on each of the outer rotary cutting heads. Signals from these devices indicating whether the machine was cutting in the coal seam or rock, was relayed to a screen in the control cabin, and the operator was thus able to make whatever vertical adjustments were necessary in order to keep the machine in the seam. Various other devices relayed messages concerning the machine's direction, its grade alignment, and its angle of inclination (spiralling).

Only one man was needed to operate the heli-track conveyor system and carry out the mining process.

From this point onwards the system was gradually refined and improved and work was begun on the development of remotely controlled miners for use underground. Initially the continuous miners were controlled

533

via a series of cables which linked the machine to the mobile cabin, but because the sophisticated coal-rock sensing system was not entirely reliable simpler hand-held 'in sight of the machine' controls were developed. However this hand-held control was still linked to the machine via cables and it was soon recognized by Joy that there was a serious disadvantage in this method of control. The 15.2 m (50 ft) long cable connection was highly susceptible to damage or severance during mining operations. Because of this the Joy Manufacturing Company began work on the development of a miner which could be remotely controlled by radio (Figure 492). Parallel with the development of remotely controlled miners and longwall units was the introduction of printed circuits and solid-state electronic components. The beneficial use of the miniature electronic controls was quickly recognized by the mining industry and solid-state electronic components were incorporated in both the radio and cable systems. This made possible the manufacture of lightweight control boxes for both cable and radio systems which weighed 4.3 kg (9 lb 8 oz) and 5.6 kg (12 lb 6 oz) respectively.

In his article,[16] 'Cable and wireless remote control of continuous miners', E. M. Warner gives some of the reasons why there has been initial resistance to the use of remotely controlled machines by operators. These include the fact that a great many operators

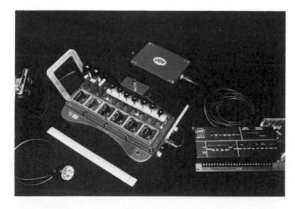

Figure 492. Joy radio control for 12CM miner. (left) Battery; (centre) operator's control/transmitter; (top) receiver at machine; (right) (at machine) interface relay panel. (Courtesy Joy Manufacturing Company.)

appear to drive their machines by 'the seat of their pants'. In other words they respond automatically to the different vibrations of the machine as it cuts into the ground. The answer to this has been the suggestion that operators be properly trained — above ground — before they are allowed to use the system underground. In addition, when the system was first introduced an attempt was made to adapt the man to the machine rather than the machine to the man. To some extent this has since been rectified, and on the latest Joy machines the position and operation of the miniature levers on the control box have been carefully designed so that they resemble as closely as possible the functions and arrangement of those in the machine cab. (In time manufacturers may also come to recognize the importance of sensation and touch, and perhaps the distinctive vibration patterns of the machine will be transmitted to the controller's box).

Owing to the possible presence of extraneous noises which might accidentally trigger the machine's radio link and cause erratic operation, the Joy Company has developed a number of safety systems which are incorporated into its radio control system and which are designed to negate this risk.

In addition, because the radio box receives its power from the miner's cap lamp battery (which according to various tests appears to give a useful life of four hours when used simultaneously — i.e. the cap lamp and the radio apparatus), the system becomes inherently safe if the operator leaves the site, since this automatically 'de-energizes' the mining machine.

Because the operator is able to choose his position and so keep within prescribed safety areas (i.e. areas which are adequately supported by timber or jacks) the extraction of coal pillars is facilitated. Indeed it has now become possible to extract coal pillars which would otherwise have had to be left standing because of the potential danger to the operator if they were extracted by conventional means. Thus the advent of the radio- or cable-controlled miner may raise the percentage of coal recovered — an important factor in view of the world's dwindling energy resources.

While many problems relating to radio-

534

controlled mining systems still exist, their wider application and acceptance by contractors and operators alike will ensure that developmental work in this direction continues. Gradually, as machines and mining systems improve, the many disadvantages which for so long have been associated with mining (i.e. discomfort, health hazard, danger, etc.) will disappear.

Jarva Hard-Rock Slot Machine: Model SM-1500[17, 18]

The declining productivity in the majority of eastern United States coal-mines has led to the expansion of coal production west of the Mississippi to meet current high demands for steam coal. As a result the price of coal (f.o.b. the mine) rose from U.S.$4.99 per ton to over U.S.$20.00 per ton during the period 1969-77. However, at the same time, labour productivity declined from 19.9 tons per man-day to under 12 tons per man-day, thus considerably reducing the profit margin. This problem of low productivity was most noticable in highly labour-intensive, thin-seam, underground Appalachian mines, which recently have also had to cope with new Federal safety and environmental impact regulations.

Various methods to offset these adverse factors were studied, including that of removing bottom rock. In effect what was needed was a machine designed specifically for removing bottom rock. Amongst other functions such a machine could, say, open up main haulage ways, thus preventing them from becoming bottlenecks during maintenance periods, or it could assist in providing additional air intake capacity by opening up existing roadways, etc. With this end in view Jarva Inc. began a developmental programme in 1975 which culminated in 1978 in the design and production of their first unit designated the 'Jarva slot machine'.

The prototype machine was installed in the Jarva test pit. During its initial trials the prototype unit cut through shales, concrete (cured to 34.5 MPa (5000 lb/in²) compressive strength), and sandstone (of about 83 MPa (12,000 lb/in²) compressive strength). The testing of the prototype unit was witnessed by the Westmoreland Coal Company which issued a purchase order for the first production machine — Model SM-1500.

The first production machine for the Westmoreland Coal Company featured improved mucking and drive systems and consisted of three main components which could be easily assembled or dismantled for transport underground through narrow entries.

After the removal of the coal seam (approximately 96 cm (38 in) thick) the company envisage using the machine to cut a 5.6 km (3½ mile) long entry for an easily maintainable belt haulage system. The entry is formed by increasing the head room from 96 cm to 203 cm (38 in to 80 in), using the slot machine to remove a 106 cm (42 in) deep × 4.57 m (15 ft) wide 'slot' along the path of the entry. This entry is to be between the company's MacAlpin mine and the coal preparation plant at the East Gulf mine in its Winding Gulf Division near Beckley, West Virginia. Anticipated completion time for this project was expected to be approximately nine months.

The machine is being developed to assist in the later creation of seven 2.1 m (7 ft) high entries, consisting of three air intake entries, one main haulage entry and three return air entries. The six air entries will be 5.5 m (18 ft) wide and the main haulage entry 7.3 m (24 ft) wide.

If the unit is successful it will obviate the necessity for the mine company to sink two 244 m (800 ft) air shafts, thus saving some U.S. $6 million, at the same time it will create a haulage system capable of handling the output of nine continuous mining sections.

Other uses for the slot machine suggested by the manufacturers are for enlarging head and tail entries in low-seam longwall systems to facilitate the installation of longwall equipment, and the flattening of tunnel bottoms after they have been cut by circular full-face tunnel-boring machines.

Description[19]

The Jarva machine cuts a 4.57 m (15 ft) wide by a nominal 107 cm (42 in) high slot in the coal-mine bottom rock (Figure 493). Its horizontally mounted cutter wheel which cuts a full face is

Figure 493. Jarva slot machine. (Courtesy Jarva Inc.)

octagonal in shape. Eighteen disc-type rock cutters are carried on eight spokes which are bolted to the eight faces of the octagon. Four of the spokes are 'short' spokes and four are 'long' spokes. The two spoke types are identical except that the long spokes have additional face area for mounting an extra disc (to cut gauge) and a gauge scraper assembly. These gauge cutters cut gauge at the bottom of the slot. Of the remaining 14 'inside' cutters, two are mounted on each of a pair of long spokes opposite each other on the octagon, two on each of the four short spokes and one on each of the remaining pair of long spokes.

The gauge scrapers are spring-loaded and are designed to clean the floor immediately ahead of each 'outside' or gauge cutter. The removal of the debris ahead of the cutter protects the cutter from unnecessary abrasive wear and also prolongs the disc cutter bearing life by preventing them from being stressed axially, which would occur if they were forced to ride over the accumulation of debris.

A steel scraper is also bolted to each of the other cutter spokes. As the wheel rotates, the scrapers move the cuttings along the floor of the slot and up a ramp on to a semi-circular muck apron at the rear of the cutter wheel. From the muck apron the material is directed into a hopper which deposits it on to a 61 cm (24 in) belt conveyor. The conveyor, which is mounted on the main frame of the machine, moves the material over the clamp carrier

assembly towards the rear where it is dumped into a truck or rail-car.

A draw-bar is also provided for towing muck removal equipment, if this is required. It is fitted with an overload device which trips out at about 9 t.

The main bearing carrying the cutter wheel is mounted in the centre of the octagon. Drive (provided by two 112 kW (150 hp) motors mounted in the left and right front compartments of the main body of the machine) is transmitted, via the drive shafts through Cincinnati gearboxes bolted to the bottom of the cutter wheel support, to the main drive ring gear bolted and dowelled to the underside of the octagon. The octagon acts as a housing for the ring gear and all open gearing mounted on the top side of the cutter wheel support structure.

A heavy welded structure supports the weight of the cutter wheel via the main bearing. The bearing is mounted on a large cylinder welded to the centre of the support structure. About 10 cm (4 in) of vertical steering at the front end of the wheel is provided by this cylinder. The rod of the cylinder bears on the wheel support shoe, spreading out the weight load at the front end of the machine to reduce ground bearing pressures.

The muck apron forms an integral part of the cutter wheel support structure. Side plough assemblies which direct broken rock to the muck ramps are positioned on each side of the muck apron. They are pushed against the walls by a pair of leaf springs and are connected to the muck ramps by heavy Armorite sheeting. This arrangement provides a flexible assembly which assists the side ploughs to adjust to the varying contours of the floor and walls. The muck ramps are hinged at the rear and scrape along the slot floor as the machine advances. Wedge-shaped feet are fitted to the muck ramps to enable them ride over any geological irregularities in the floor.

Mounted on the cutter-wheel support are the left and right front steering mechanisms, each consisting of a link, a pad, and a front steering cylinder. (The steering cylinder is attached to the pad at one end and to the main frame at the other.)

During the cutting phase (i.e. the 'hold' mode) the cylinders are extended by low-pressure hydraulic fluid and act as ancillary gripper pads. When cutting a curved slot the cylinders are operated in the 'steer' mode and push the front of the machine to one side or the other.

The rear end of each of the front steering cylinders is attached to the main body of the machine, a large box-shaped weldment at the centre. Also carried by the main body are the thrust cylinders, pump platform, conveyor, and several filters. Located within the main body are the drive motors, heat exchangers, oil lubrication pump, and several hydraulic and water lines.

Behind the main body is the clamp carrier assembly which supports the rear of the machine during the cutting cycle. The carrier assembly includes the slide tubes, clamp cylinder carrier, clamp cylinder, and clamp pads, etc.

During the cutting phase, thrust from the propel cylinders is reacted by the clamp pads or grippers. The grippers are expanded laterally against the walls by a large double-ended hydraulic ram, capable of exerting up to 500,000 lb (2224 kN) force per side. A ball-and-socket connection between the gripper pads and the hydraulic cylinder enables the pads to swivel, thus permitting them to follow variations in the slot walls as the machine is advanced. Excessive swivelling is prevented by the restraint arms and the provision of a hard nylon restraint ring at the ball-and-socket connections. In addition, to enable the unit to be clamped in a curved slot, the clamp or gripper cylinder is mounted in a central trunnion. This allows it to pivot approximately 3° to either side of the centre.

The clamp carrier rides over the two slide tubes which fit into sockets at the rear of the main body.

Dust is controlled by a 'wethead' water spray system, a double dust shield, a 5000 cu.ft/min (142 cu.m/min) exhaust fan, a 'flooded bed scrubber' and a 'Euroform' mist eliminator. The dust shield completely encloses the cutter wheel, the two sets of double rubber flaps sealing to ground around the periphery of the shield. Dust which leaks past the first seal is picked up by the exhaust air stream flowing between the two sets of flaps. The exhaust air draws the dust into the flooded bed scrubber which removes most of the dust from the airstream. Water droplets are then removed by the Euroform mist eliminator so that relatively clean, dry air is discharged at the fan outlet into the mine entry.

Operation

In operation the two gripper or clamp pads are extended laterally, thus anchoring the clamp carrier assembly between the slot walls. At this stage the machine is supported by the cutter-wheel support shoe in front and the clamp pads in the rear. The front steering pads are then extended to stabilize the front of the machine and the cutter wheel is started. It is fed into the face by the two propel cylinders which react off the clamp assembly and thus off the gripper pads. At the completion of its 60 cm (2 ft) cutting stroke, the rear lift legs are extended to support the machine and the grippers are withdrawn. The propel cylinders are then retracted, thus drawing the clamp carrier forward on the two slide tubes. With the grippers and front steering pads once more extended, and the rear lift legs retracted, the machine is ready for its next cutting cycle.

Specifications

Cutter-wheel drive 224 kW (300 hp).
Cutter-wheel speed 10.7 rpm.

Cutter-wheel torque 200 kN m (147,000 ft-lb).
Cutter-wheel thrust 667 kN (150,000 lb).
Machine weight 65 t (72 U.S. tons) (less dust shield and drawbar assemblies).
Overall length 13.26 m (43 ft 6 in) (including conveyor).
Overall height* 160 cm (63 in) (assembled, less dust shield).
'Exposed' height in operation 53 cm (21 in) (at rear of cutter wheel).

References

1. Half a century underground, *Electrical Mining,* **47,** No. 4, 23 April 1950 (Goodman Manf. Co.), Chicago, U.S.A.
2. F. S. Anderson and R. H. Thorpe, A century of coal-face mechanization, Inst. of Min. Eng., *Min. Eng.,* No. 83, Aug. 1967, London.
3. J. D. A. Morrow (Joy Mnf. Co.), History of the development of underground machines, *Proc. of the Illinois Min. Inst.,* 62-81, 16 Nov. 1962, U.S.A.
4. Joy Mnf. Co., *Joy Mining Machine History – Cutting Machines,* Claremont, N.H., 7 March 1910, U.S.A.
5. Waldemar Kaempffert (Ed.), *A Popular History of American Invention,* Charles Scribner's Sons, New York, 1924.
6. John Reeves, Deep mining of coal in Colorado, *Mining Congress Journal,* **60,** No. 10, 14-20, Oct. 1974, Washington D.C.
7. New coalcutting machine, *Colliery Eng.,* Dec. 1961, U.K.
8. Saul Alinsky, *John L. Lewis,* G. P. Putman's Sons, New York, 1949.
9. Continuous coal mining, *Technology,* (1949/53?), (author unknown), Australia.
10. J. Karlovsky, Developments in cutter heads or boring type continuous miners, Presented at West Virginia Coal Mining Institute and Central Appalachian Section A.I.M.E. 3 Nov. 1967, New York.
11. C. C. Conway, The Marietta miner, Presented at West Virginia Coal Mining Institute, 21 April 1961, U.S.A.
12. E. M. Arentzen, Coal Mine Modernization, *Continuous Mining — 1951* and *Continuous Mining — 1952.* pp. 143-146 (1951); p. 87 (1952). Lee-Norse/Ingersoll-Rand, Pittsburgh, U.S.A.
13. E. M. Warner, An historical review of American continuous miners. Paper presented at International Conference on Mining Machinery, Brisbane, Australia, 4 July 1979, Institution of Mining Engineers, Barton, A.C.T., Australia.
14. J. W. Heimaster (Union Carbide Corp.), *Remote control in highwall mining, Mining Congress Journal,* May 1959, Washington D.C.
15. K. E. McElhattan (Joy Mnf. Co.), The push-button miner, *Mining,* Sept. 1961, U.S.A.
16. E. M. Warner, Cable and wireless remote control of continuous miners, *Mining Congress Journal,* 10 Oct. 1974, 84, Washington, D.C.
17. Jarva Inc., *Jarva Incorporated/Robertson Associates Mine Productivity Improvement Study,* Salon, U.S.A.
18. Richard H. Mason, Slot machine pays out faster haulage, *Coal Mining and Processing,* 1979, Chicago, U.S.A.
19. Jarva Inc., *Jarva Slot Machine Specifications – Model SM-1500 (Serial No. X002),* Salon, U.S.A.

Bibliography

George J. Bancroft, A history of the tunnel boring machine. *Mining Science,* July 23, pages 65-68; August 6, pages 106-108; August 13, pages 125-127; August 20, pages 145-146; August 27, pages 165-166; 1908, Denver, U.S.A.

R. T. Deshmuki (1920), B. M. Vorobjev (1928), and K. Keshmukh (1966), *Advanced Coal Mining,* Asia Publishing House, Bombay, India.

Correspondence

United States Dept. of the Interior — Bureau of Mines, U.S.A.
Goodman Mnf. Co., Chicago, U.S.A.
Bituminous Coal Research, Inc., Monroeville, U.S.A.
National Mine Service Co. (Clarkson Div.) — J. Karlovsky, Nashville, U.S.A.
Joy Manufacturing Co., Franklin, U.S.A.
Jeffrey Mining Machinery Div, Dresser Industries Ind., Columbus, U.S.A.
Jarva Inc. — E. W. Brickle, Solon, U.S.A.
Union Carbide Corp. — R. L. McNeill, New York, U.S.A.
Lee-Norse Company (Ingersoll-Rand), Pittsburgh, U.S.A.
Dresser Industries, Inc., Texas, U.S.A.
Fairchild Incorp. — P. L. McWhorter, Beckley, U.S.A.

*Although the overall assembled height of the machine is listed as 160 cm, when the machine is cutting the 'exposed' height of the machine above the slot in which it operates is only 53 cm (21 in). Height may be reduced for transport underground.

TWENTY

Australia

Early History — Discovery of Coal[1,2]

While men, women, and children were dying in English coal mines from overwork, inadequate safety precautions, and other contributory factors, the story of coal-mining in Australia was about to unfold.

The first known sighting of coal in Australia by white men was made by the courageous party of William and Mary Bryant. Bryant, though he had already served his seven-year sentence (given for assisting smugglers) foresaw hard times ahead. He correctly anticipated that the young colony would shortly endure a prolonged period of misery and starvation and decided to leave while he was fit and able. With his wife and two young children, Bryant gathered a small party of seven other convicts about him then during the dark hours of the night of 28 March 1791, they sailed from Sydney.

The only food they took with them was a small quantity of pork and rice which was soon consumed. Thereafter they were forced to rely on Bryant's noted fishing skills and such berries, etc. as they could gather *en route*.

The party were constantly being threatened by either storms at sea, aborigines, or island natives, and several times nearly died from lack of food and water.

During their voyage and aided only by a rough chart, a compass and a quadrant, which Bryant had somehow managed to obtain from the Captain of a Dutch brig before they left, they sailed 17,000 km (3254 miles) along the east coast of Australia until at last on 5 June 1791 the entire party arrived safe and sound in Koepang, Timor. This, considering the voyage was undertaken in a small six-oared ketch by a party whose members were totally unskilled in the arts of navigation and seamanship, was a remarkable accomplishment. It was during this epic journey that coal was first discovered at Hunter river by Bryant and the others. But naturally, considering the circumstances of its discovery, nothing could come of the find and the settlement had to wait until it was 'rediscovered' before it could mine its first coal.

Newcastle — Hunter River

Six years later some convicts, aided by crewmen of the *Cumberland* (which was travelling from Sydney to Hawkesbury with stores for the farmers there), overpowered the *Cumberland's* Captain and those of the crew who had refused to join them. After putting these people ashore the renegades sailed north.

The *Cumberland* was a schooner and considered by the Governor to be the best boat in the colony. Infuriated at losing her, Hunter sent two cutters after them. One headed north, the other south.

Shortland was in charge of the boat which sailed north. His search was fruitless and he returned empty-handed. However, on the way back his boat was caught in a sudden squall which arose as they were passing a small island. This diminutive piece of land had been noticed by James Cook some 27 years earlier. At the time Cook had not made a detailed study of the area, but had merely recorded the island's presence in the ship's log as he sailed past.

In search of shelter from the storm, Shortland sailed around the island. He came upon a large natural harbour partially blocked by a sand bar over which the sea broke. Skilfully he manoeuvred his boat through the breakers and over the sand bar which lay between the island and the mainland (at low tide this bar was covered with less than 4 m (13 ft) of water). At one end of the harbour a strongly flowing river emptied its waters into the bay. Shortland beached his boat and then went ashore to explore the area. He discovered a good quantity of surface coal on the foreshore of the southern end of the inlet. Strangely enough, subsequent surveys revealed that Shortland's find was merely part of some 40,000 sq. km. (15,000 sq. miles) of a roughly saucer-shaped deposit of coal encompassing the entire area from Coalcliff on the Illawarra in the south, Newcastle in the north, and including Lithgow in the far west.

Shortland christened the river 'Coal river', but this was later altered and it is now known as the 'Hunter river'. On the banks of the river mouth, the city of Newcastle was born.

Newcastle

In October 1801 a settlement was established in the area. But this initial attempt was unsuccessful and after a stormy period of continued mutiny, etc. Newcastle was abandoned. Two years passed, and during this time very few people other than free-lance miners visited the area. These men were attracted to Newcastle by the coal which they mined haphazardly and then sold in Sydney. However, in March 1804 a serious uprising occurred at Castle Hill. This forced Governor King (successor to Hunter) to seek another location for the ringleaders of this dangerous Irish faction. He looked again at Newcastle and saw in the area commendable qualities. Not only was there a good supply of coal, cedar, and lime and an easily accessible sea route close to Sydney town, but the ruggedness of the surrounding terrain made land escape difficult and untempting. An excellent spot in fact to which to send a batch of cantankerous prisoners.

Rigorous steps were taken to prevent the convicts from escaping from Newcastle. Though the bush presented a natural barrier, which would have deterred most, the Governor was well aware that he was dealing with a hardy and desperate group of men and women who would attempt even the seemingly impossible.

Local tribesmen were therefore rewarded with gifts of food or clothing if they were successful in recapturing and returning an escapee. Even free settlers living nearby were not permitted to keep hunting dogs as these were coveted by convicts who stole them and then used them for hunting food. Thwarted, the convicts looked towards the sea. But here again avenues were closed as laws came into force requiring all owners of small craft or rowing boats, etc. to strip their vessels of oars, rudders, and sails so that these could not be used by escapees. In addition, in order to ensure that captains and masters also maintained strict vigilance, heavy penalties in the form of fines were imposed upon them if they allowed escaped convicts to board their ships. Even if the captain or master was ignorant of a prisoner's presence aboard his ship when he sailed, he was not absolved from blame. The responsibility for finding the stowaway therefore rested entirely with him.

As far as the convicts themselves were concerned, penalties for attempting to escape were severe indeed. The least they might expect would be a flogging of 100 lashes followed by a diet of bread and water and a week or more in leg irons. However, so terrible were conditions at Newcastle in those early days that, despite the fear of dire retribution, many prisoners tried to flee. During the year 1819 the settlement held a total of 91 prisoners and that same year no fewer than 34 faced punishment for escape attempts.

Now and again a few actually managed to win through and some even reached Sydney. Others lived as bushrangers in the Hawkesbury district or inhabited areas in the south, where they became a menace to farmers and other settlers.

During the formative years of the colony, no free settlers were permitted to reside in Newcastle without the express permission of the government. In 1824 a proclamation was made allowing people to settle there without having to obtain governmental approval first. Immediately the atmosphere of the settlement began to change, though they continued to use convicts in the coal-mine. Later these men were gradually replaced by qualified miners and conditions generally improved. Actual mining methods, however, remained basically the same.

Coalcliff — Illawarra River

During the period of six years which intervened between the fruitless discovery of coal at Newcastle by the party of fleeing convicts and its rediscovery by Shortland, there was a major find to the south which for reasons to be explained was not exploited for many years.

In 1797 the ship *Sydney Cove* sailed from Bengal to Port Jackson. During the voyage the ship sprang a leak and was eventually beached on Preservation Island (Furneaux Group). Using the long-boat, 17 of the ship's men, including the Chief Officer, Hugh Thompson and W. Clarke the supercargo, attempted to reach Port Jackson in an effort to secure help.

They were forced off course by rough weather and the long-boat was driven aground on strange and unexplored land, now known as 'Ninety-mile beach'. This lies along the Victorian coastline between Wilson's Promontory and Lakes Entrance. The 17 men were then faced with a long and arduous trek northwards along a virtually unknown coastal strip.

The rigours of the journey claimed many, others died when no longer able to keep pace with the main group. These exhausted stragglers became easy prey for marauding tribesmen who speared them to death one by one. Occasionally a few friendly tribes were encountered, but the majority were hostile so that by the time the party reached a place some 56 km (35 miles) south of Port Jackson, only a handful of the original group of 17 remained alive. At this point the wet and thoroughly weary men paused for a rest and, those who could, set about seeking material for a fire. In his search for wood, the attention of one of the men was attracted towards some intriguing black stones which were lying at the base of a nearby cliff. His curiosity aroused, the sailor examined his find more closely and to his amazement saw what he felt certain must be *coal*. Returning to his companions he placed the black rocks on the newly made fire. Within minutes they were burning brightly, thus confirming his initial identification.

In spite of the warm comfort of the fire and the rest, Thompson and at least one other castaway died there. Of the 17 men who started this perilous journey only three, including Clarke the supercargo, won through to safety.

When appraised of the situation, Governor Hunter immediately sent a ship to rescue the stranded men on Preservation Island. However, the Governor considered the story of a coal find of sufficient importance to warrant the despatch of another ship under the command of George Bass to investigate the Coalcliff area. Clarke volunteered to accompany the party as a guide.

Unfortunately Bass believed the coal, a 2 m (6 ft) wide seam which ran across the face of the cliff for a distance of nearly 11 km (7 miles) to be economically inaccessible. Sixty-four years later (1861) the rising demand for this commodity encouraged prospectors to have a second look at the find and Coalcliff on the Illawarra came into being with the construction of a jetty immediately below the coal-bearing cliff face: This meant that the coal could be loaded straight from the workings into the holds of the moored vessels.

Mechanization — Early History

The Greta Company, formed in 1886 is reputed to be the first Australian company to attempt mechanization. In 1890 they daringly ordered a Stanley header from England and installed it at Gretna. Then in 1894 a Jeffrey coal-cutter was put to work in the Hetton-Merthyr mine owned by George Clift.

Two years later (1896) sales representatives of the Jeffrey Manufacturing company approached the A.A. Company (Australian Agricultural Company) with offers of Jeffrey coal-cutters. It was not until 1904, however, that the first machines were put to work at Hebburn and also at Pelaw Main. Hebburn was owned by the A.A. Company and Pelaw Main by J. and A. Brown.

The mine-owners' decision to try mechanization was to a large extent influenced by the numerous strikes which disrupted production during that period, the worst being those which occurred in 1900, 1905, and 1906, and also by other subsequent labour problems. There was hope that these new machines would somehow alleviate these problems.

The A.A. Company and J. and A. Brown

followed by other owners of thick seams in the Maitland field such as the Caledonian Coal Company, the Abermain Colleries, etc. who had the necessary financial backing and facilities to carry out trials, installed a few of these coal-cutters in their mines. As these proved to be successful their use gradually spread to other mines. Percussion machines such as the Ingersoll-Sergeant and Morgan-Gardiner also made their appearance. Later Sullivans (one of the new breed of breast machines produced in 1897 in America — see 'American developments') were installed in the southern field. These machines were either driven by compressed air or operated by electricity.

Although there is no doubt that these first halting steps taken towards mechanization had far-reaching effects, they did not provide the panacea for all the ills of the industry. For in its wake mechanization brought with it its own brand of difficulty. Numerous disputes over new wage determinations for the operators of the cutters arose and had to be settled and, of course, the cutters themselves needed special attention such as repair and replacement of worn parts, etc. Dust, always a problem and a threat to health increased and perforce had to be controlled. And last, but by no means least, was the suspicion and antagonism of the miners themselves against mechanization that mine-owners had continually to fight. Despite these teething problems the use of the coal-cutter continued to grow, the breast-chain type apearing to be the most popular. These gradually superseded the cutter bar and most percussion machines.

Mechanisation — 1930-1950[3]

During the depression years of the early 1930s, when hundreds of men were out of work as a result of 'speed-up' policies and other contributory factors, mechanization was looked upon by union leaders as a dire threat to the unions. It would, they believed, deprive men of their jobs.

Mechanization in Australia during the following decade remained almost static. The coal-cutter reigned supreme as a major coal-winning device. Nevertheless, the machines themselves gradually improved as year after year various manufacturers strove to produce better equipment in a widely competitive field. It thus developed from what was a crude and awkward mechanism into a highly practical and efficient unit capable of cutting either horizontally or vertically as required at any position in the seam. Rubber-tyred and self-propelling, the modern coal-cutter was a far cry from the early models which needed to be 'barred' into position when a new cut was required. Indeed, even today when the latest trend seems to be more and more towards the use of continuous miners of various types, a few highly efficient Australian mines still use coal-cutters in association with the high-capacity coal loaders now available.

During this period, too, the old furnace shafts were replaced by mechanized ventilation systems, and rope and locomotive haulage methods were generally improved. Spontaneous combustion, the greatest scourge of the coal-miner, was to a large extent overcome by the introduction of significant advances in mining techniques such as that of special 'mine-layouts'. At the same time the appearance of flame-proof electrical equipment and the general improvement of permitted explosives also alleviated this serious problem.

Despite all this activity in other areas, actual mechanization at the coal face was slow to proceed. An indication of this is in the fact that though mobile loaders were introduced in the early 1930s, by 1944 only 21 per cent of coal in Australian mines was loaded mechanically. However, in 1947 with the establishment of the Joint Coal Board, a radical change occurred. In an effort to save what Justice Davidson had referred to in his report of 1946 as a 'tottering industry' the Joint Coal Board advocated large-scale mechanization programmes.

Naturally enough these ideas met with severe antagonism from many colliery proprietors who lacked the capital to implement such programmes. There were, too, other problems associated with mechanization which tended to dissuade colliery proprietors from proceeding.

When an area had been blocked out, permission from the Minister for Mines was required before the pillars (coal left behind in the course of the board and pillar method of mining) could be extracted by mechanized means. Mines which had become mechanized often blocked out the entire colliery, leaving only the pillars behind. Permission was then sought to remove the remaining pillars by mechanization. When this was refused, hand-mining was perforce resorted to. But because of its slowness, this method resulted in many roadways remaining untouched for several years following the initial period of extraction. Inevitably, neglected areas deteriorated and became unworkable. As a result a great quantity of good coal was wasted. This was particularly so in the South Maitland District where spontaneous combustion imposed a constant threat. Coal containing a high sulphur content was prevalent in the upper portion of the seam in this district. This type of coal was not readily sold and so was usually left behind. After a period of time these unwanted pieces of coal would fall from the ceiling, sometimes igniting and causing a considerable loss of good coal.

Mechanization — 1950-1975[3,4]

In 1954 the Coal Industry Tribunal issued a decision permitting the mechanical removal of pillars. This in effect nullified legislation governing the mechanical extraction of pillars, (though ironically the actual clause still existed in the Coal Mines Act and was only removed from the regulations in December 1973).

Its effect was dramatic. The introduction of the continuous coal-miner now made it possible to extract the coal at a much faster rate than had hitherto been achieved. This meant that before the remaining coal had an opportunity to deteriorate and become heated, as had frequently occurred when conventional hand-mining methods were used, the roof was allowed to drop and cover the danger area. A great quantity of coal which would otherwise have been lost was thus saved.

Continuous Miners Arrive in Australia[4,5]

In an article dated 10 January 1950 entitled 'Continuous miners for Australian coal mines', B. Nicholls[5] reported that five trackless (not requiring rail tracks) Joy continuous miners had been ordered from America. One was destined for work in Tasmania, one for Amalgamated Colliers (Western Australia Limited, Colliers), one for Australian Iron and Steel Limited (Port Kembla), and the remaining two machines for the Joint Coal Board, New South Wales. These were the first continuous miners to be put to work in Australia.

The 14 ton ripper miner machines were crawler mounted and were capable of mining seams ranging in thickness from 1 m (3 ft 6 in) to 2.5 m (8 ft 4 in). The 76 cm (30 in) wide cutting and loading head was equiped with six parallel cutting chains which simultaneously cut and loaded the coal as it progressed. Two 48 kW (65 hp) motors drove the cutting chains at 150 m/min (500 ft/min) while hydraulic controls moved the head 46 cm (18 in) forward and swung it 13° either side of the centre line.

By the end of that year (1950) three such machines were in operation in New South Wales, the Tasmanian machine having later been transferred for use on the mainland.

During the following five years, no further continuous miners of any type were ordered as the mining industry paused to take stock and note the effect of these expensive innovations. Their undoubted success can nevertheless be attested to by the spectacular increase seen in the rate of introduction of continuous miners into New South Wales coal mines alone, which occurred from 1955 onwards:[3]

1955	3
1956	15
1957	32
1958	39
1959	48
1960	71
1961	88
1966	146
1970	191

Although the Joy ripper was the first machine of its kind to be put to work in Australian coal-mines, it was followed fairly rapidly by improved Joy models and a variety of machines made by other manufacturers. These included the Jeffrey Colmol, the Goodman 500 miner and the Lee Norse miner.

The introduction of continuous miners using

a milling action was considered to be a particularly significant advance from the point of view of design. Manufacturers of these machines claimed that considerable savings in both maintenance costs and power requirements were effected by the superior efficiency of these milling-type miners, especially when hard coal was being won. This, they contended, was due to the fact that instead of working from the floor up as did most miners, the milling-type miners operated from the top down.

The oscillating action of the head of the milling type of continuous miner became unnecessary with the introduction of the Jeffrey Heliminer, Joy 10CM, and Lee Norse Hard head machines which were hailed as important contributions towards design development. These machines employ wide cutter-heads with picks on a revolving helix.

Though a few borer-type miners were tried in the Southern District of New South Wales, and one Newcastle District colliery claimed satisfactory production results, borers were not generally favoured by the majority of collieries, who preferred the other designs of continuous miners. The initial reason for this given in the early 1960s was that various difficulties still needed to be overcome before colliery-owners would accept them. These involved dust extraction, the necessity for the suspension of normal mining operations while roof support work was being carried out, and certain other economic factors. Though many of these problems have now been satisfactorily solved, manufacturers are still encountering resistance by the majority of mine operators to borer-type machines. But whatever type of machine is used, be it milling, boring or ripping the impact of mechanization in Australia has been considerable. With the advent of the continuous miner and modern methods of loading and transporting coal to the surfce, it has been found to be more economical to limit the number of coal faces being attacked at any one time. The result has been a considerable reduction in mine staff while actual production rates have risen dramatically. Though at first glance these trends may have appeared to be detrimental to the mine-worker, time has proved that in fact these 'inevitable changes' have ultimately led to better overall working conditions and an improved take-home pay-packet for the individual.

References

1. M. H. Ellis, *A Saga of Coal,* Angus and Robertson, Sydney, 1969.
2. Robin Gollan, *The Coalminers of New South Wales* (A History of the Union 1860-1960), Melbourne University Press in assoc. with the Australian National University, 1963, Melbourne, Australia.
3. T. M. Clark, The mechanisation and development of the New South Wales coal industry over the last twenty years, Paper No. 20, Annual Conference, N.Z. — Aust. Inst. of Mining and Met., 1 March 1971, Barton, A.C.T., Australia.
4. A. H. Hams and H. L. Pearce, Recent developments in continuous mining and underground transportation of coal in New South Wales, Paper 14 II. 1 2/3, Sixth World Power Conf., 20 Oct. 1962, Melbourne, Australia.
5. B. Nicholls, Continuous miners for Australian coal mines, *Chemical Engineering and Mining Review,* 10 Jan. 1950, 143, Melbourne, Australia.

Bibliography

L. J. Thomas, *An Introduction to Mining,* Hicks Smith & Sons, Sydney, Australia, 1973.

Correspondence

Joint Coal Board — T. M. Clark, N.S.W., Australia.

South Africa – Gold-Mining

Drag Bit Rock Cutters[1-4]

Although coal-cutters were destined to fall out of favour as a major coal-winning device, there was one interesting development which has maintained for this principle a lasting place in the annals of mining.

Of great interest are the new drag bit rock cutters (Figures 494 and 495) produced by Anderson Mavor Limited and Klockner Ferromatik (S.A.) Pty. Ltd. These machines, a recent development extending the principle of coal-cutters into *much harder rock,* were designed to cope with a problem facing the gold-mining industry of South Africa.

A considerable amount of gold in the new South African mines is won from narrow seams of 500 mm (1 ft 8 in) or less (sometimes as little as 100 mm (4 in)), running at great depth below ground. Owing to the nature of mining with explosives, narrow seams can only be effectively won by blowing out a much wider slice, including waste. This not only produces a dilution of two or even five to one of gold-bearing reef with waste rock but it also mixes them so thoroughly that hand pre-sorting becomes virtually impossible. The whole has then to be processed through the mill as ore. It is, however, necessary in order to provide sufficient working space within which miners and their equipment can operate, to mine the stopes at widths close to 1000 mm (40 in). What was needed was a machine which could remove the rich gold-bearing reef without removing the waste rock as well. The stope could then be widened to 1000 mm (40 in) and the waste removed separately and used as fill.

After a great deal of experiment and trial, a machine called a 'drag bit rock cutter' was designed by the Mining Research Laboratory of the Chamber of Mines of South Africa and prototypes were built. The first of these was tested in a stope at Luipardsvlei Estate and Gold Mining Company in October 1967. From these initial trials it was established that it was possible to cut quartzite in a stope underground.

By 1968, Dr. N. G. W. Cook, Director of the Mining Research Laboratory, was able to report at a tunnel and shaft conference held in Minnesota, that though minor engineering problems associated with these new machines still existed, they were confident that the actual cutting principles adopted were sound. He emphasized, however, that the efficient operation of the new machines was only a part of the overall problem. There still remained other factors associated with the consequent changes to the entire mining system. These would need to be looked at very closely.

On 16 October 1973, in reply to a query from the author, Dr. Cook wrote advising that their drag bit rock cutters had progressed considerably in the preceding five years and that the number of machines operating in two pilot production trials had been increased to 18. He described their mode of operation as being somewhat similar to that of undercutting with a coal-cutter, but said each kerf was cut with a single tungsten-carbide bit which reciprocated back and forth over a distance of 3 to 4 m (10 to 13 ft), increasing the depth of the cut at each pass by between 3 and 12 mm (0.12 and 0.47 in) depending upon the hardness of the rock, until a maximum depth of about 600 mm (2 ft) was reached. Dr. Cook mentioned, however, that rarely was a slot of this depth formed, the rock tending to slab off the face above and below the

Figure 494. Anderson Mavor rockcutter — mounted for transport underground. (Courtesy Anderson Strathclyde Ltd.)

Figure 495. Anderson Mavor rockcutter — viewed from gob-side. (Courtesy Anderson Strathclyde Ltd.)

kerf as the machine progressed. In fact the maximum depth of kerf seldom exceeded 100 mm (4 in). (He added, as a matter of interest, that the Habegger tunnelling machine, developed by Atlas Copco, used practically the same mechanism and virtually identical cutting bits to those on the new drag bit rock cutters.)

Their method of operation, which is described more fully by Dr. Cook and Dr. N. C. Joughin in an article entitled 'The role of rockcutting in strata control',[1] is given below.

The rock-cutting machine commences by making a cut at, say, the north end of the panel. Should the face be subject to parting planes or slabbing as a result of stress, only one cut made anywhere above or below the reef would be necessary, as most of the rock would fall off the face of its own volition. However, in the event of either the hanging or footwall being found to be free of such a parting, then another cut could be made to allow the rock to fall. Assuming that only one cut was necessary,

when this was completed, the machine would be moved 4 m (13 ft) down and a second cut made.

The machine would be followed by a man who would toss the gold-bearing reef together with all fines into the conveyor, while the waste rock would be thrown across the conveyor. Such rock as still remained on the face after the cutting operation had been completed would be removed by two additional men wielding pneumatic picks. This would again be sorted and any gold-bearing reef would be thrown on the conveyor while the waste was thrown over the conveyor. The process was repeated until the entire panel had been dealt with, after which the machine would again be set up at the north end and the cycle recommenced.

Swing Hammer Miner and Reciprocating Flight Conveyor[1-4]

The prototype unit of the swing hammer miner was developed by the Chamber of Mines in

546

Figure 496. Swing hammer miner — a view of the swing hammer miner mounted on a reciprocating flight conveyor and showing the hammer set inclined in the rotor. (Courtesy Chamber of Mines of South Africa.)

Association with Anglo-Transvaal Consolidated Investment Co. Ltd., and Anderson-Mavor (S.A.) (Pty.) Ltd. It commenced work in 1975. The unit is mounted on a reciprocating flight conveyor which was built in 1974 for specific use with the swing hammer miner.

The Reciprocating Flight Conveyor

The conveyor is basically similar in concept to the armoured conveyors which have been used extensively in longwall coal-mining work in Germany, Britain, and elsewhere. The main difference between the two units is that conventional armoured conveyors are usually equipped with chains and flights which move in the one direction and return on the underside of the conveyor. The new conveyor, however, is fitted with a single chain equipped with flights which move back and forth across the conveyor. The flights on the reciprocating conveyor are hinged so that they stand upright or at right angles to the chain on the forward stroke, to facilitate the moving of material, and collapse parallel to the chain on the return stroke.

The prototype reciprocating flight conveyor used in the Grootvlei mine was tested in a stope which had previously been mined by blasting. Approximately 3000 m² (32,000 ft²) were successfully mined in nine months with the conveyor and, according to the South African Chamber of Mines Research and Development Council, they found that the new conveyor could carry out all normal functions associated with stoping. They deemed it to be more compact than the conventional armoured face conveyor and, in addition, while being both lighter and less expensive, it also appeared to be wearing better than the armoured conveyor which was tested in similar mining conditions.

Swing Hammer Miner

The prototype swing hammer miner (Figure 496) was put to work the following year in a stope 3 km (2 miles) below the surface. Basically the miner consists of a vertical rotor with two rows of three hammers (total of six) eccentrically mounted on the rotor, (similar to a hammer mill on its side). The hammers swing freely and each hammer is strengthened by a tungsten carbide tip. If the rock struck by one of the hammers does not break, the hammer merely rebounds, thus allowing the unit's hammerhead to continue its normal rotary movement without jamming.

An impact energy of approximately 500 J (370 ft-lb) is delivered by the miner which operates at a speed of 120 rpm. It is specifically designed to break large pieces of stress-fractured rock away from the full face of a longwall gold reef and in operation moves across the face in much the same manner as do conventional longwall units. If the seam is thick enough, the hanging wall and top section of the reef are broken away on the first cut, then the machine is returned to the start and put to work breaking out the lower section of the reef and

547

also part of the footwall. Each cut is normally about 10 cm in depth. After the full face has been dealt with, the conveyor is advanced in readiness for the next cut.

References

1. N. G. W. Cook and N. C. Joughin, The role of rockcutting in strata control, *Chamber of Mines of South Africa Research,* Johannesburg, S. Africa.
2. *Chamber of Mines of South Africa Research & Development Annual Report, 1975,* Johannesburg, S. Africa.
3. N. C. Joughin, The use of face conveyors in gold mines, *Journal of the South African Inst. of Mining and Metallurgy,* Feb. 1975, 315-324, Johannesburg, S. Africa.
4. N. C. Joughin, Potential for the mechanization of stoping in gold mines, *Journal of the South African Inst. of Mining and Met.,* Jan. 1976, 285-306, Johannesburg, S. Africa.

Correspondence

Chamber of Mines Research Department — N. G. W. Cook, Johannesburg, S. Africa.

TWENTY-TWO

Boom-Type Machines

Boomheaders

Boomheading machines may be divided into two basic types, namely those with longitudinally rotating cutting heads and those with transversely rotating cutting heads.

In the mining industry boomheader machines with longitudinally rotating and transversely rotating cutting heads are commonly referred to as 'milling-type' and 'ripper-type' units respectively. For the purposes of this book and in the interests of simplicity the latter terminology shall be used.

The milling type utilizes a cylindrical or pineapple-shaped cutter head which rotates on a longitudinal axis, tearing away the material at the face and ejecting the cuttings sideways. The ripper-type head, on the other hand, rotates on a transverse axis and thus rips out the material in much the same manner as rotary drum-type machines or those having cutter chains fitted with picks. The material is either thrown directly downwards or it may be ejected upwards so that it passes over the top of the head before it drops to the ground on the other side. Most of these machines are crawler mounted.

Some models are fitted with 'stelling jacks' or 'spragging rams' (horizontal hydraulic jacks which extend against the side walls to provide additional stabilization for the machine when cutting through harder material), while others are fitted with vertical jacks which anchor the unit between floor and ceiling.

Modern dust-suppression methods vary and incorporate external sprays (i.e. a number of jets placed immediately behind the cutting head which direct a spray of water on to the face as the cutter rotates) or pick-tip flushing systems (whereby the water is transmitted through each or a number of pick mountings on the head, supplying a spray of water to the tip of the pick).

The cutting heads are fitted to robustly built hydraulically controlled booms or arms which are generally centrally positioned on the unit and extend cantilever fashion in front of the machine. These booms may be vertically raised and lowered or swung in an arc from side to side.

The coal is either collected by gathering arms on the apron at floor level and fed into a centrally positioned conveyor which carries it through the throat of the machine to the rear, or it is lifted by means of a single-strand flight conveyor which encircles the machine and carries the coal to a bridge belt conveyor, towed at the rear. In some of the more modern machines the gathering aprons which extend beneath the arms or flight conveyors, may be raised or lowered hydraulically to enable the machine to negotiate gradient changes.

Although, generally speaking, the boom-type header has a lower rate of production than conventional drum or rotary-pick-type continuous miners, to which it is related, it offers certain advantages over the other miners. It can mine selectively and can cut any shape or size of tunnel. In addition, because of the unit's ability to concentrate the entire power and thrust of the machine at one single point instead of spreading it across the width of the cutting head, it can cut into much harder material than can be handled by the other type of miner. As a result the boom-type header which was developed initially as a ripping machine or a road heading machine for the coal industry is now finding wide applications in civil engineering works of various types. Indeed, many have been incorporated into shields or

549

mounted on gantries or trucks to meet special requirements.

In a paper entitled 'Twenty years — production and loading machine, Type "F"[1] which was submitted for publication to the Hungarian Mining Research Institute, Dr. Z. Ajtay describes the developmental history of the boom-type cutter loader. Dr. Ajtay comments that according to N. V. Melhyikov, a Russian academic, the prototype of the modern boom-type cutter-loader machine was built in Hungary in 1949 and was based on developmental work carried out initially by Menzies of Newcastle in 1761 and then later by Walker of Wigan in 1868.

The Hungarian Machine

The cutter head of the original Hungarian machine which was mounted at the end of a boom or arm, consisted of a number of contra-rotating discs which were chain driven and fitted with picks.

On 5 October 1949, Dr. Z. Ajtay, a mining engineer, obtained the patent covering this machine and with the assistance of I. Koszorus (who carried out the main construction work) they built the first of the F-type machines. Also on the 5 October Koszorus took out a provisional patent (No. 141.911.55.b.Class 34-40-AA-39) for the hydraulic steering system of the new machine.

When F-2 was constructed J. Korbuly was Technical Director. At Korbuly's suggestion the motor was built into the base of the cutting boom and the chains were eliminated.

The machine cut the coal and then loaded it onto a conveyor which carried it to trucks at the rear. As the cutter head could be swung either horizontally to left or right or vertically upwards or downwards, it could be positioned at any point on the face. Moreover, because the coal was fed directly onto the conveyor which ran straight to the rear, there was no necessity for the machine to be moved away to recover the coal. When compared with other similar coal-getting machines of the day this constituted a significant breakthrough in design.

The cutter head of the F-2 model had hemispherical discs fitted with picks. The machine moved on crawler tracks and was extremely manoeuvrable, as it was fitted with front steering axles which allowed it to move backwards and forwards and also to turn from side to side.

The unit was driven by two independent motors, one for the boom and one for traction. Coal was cut at the rate of about ½ to 1 t/min, the cutting head rotating at the rate of 27 to 45 rpm, depending upon the type of material being cut.

After a short trial period Models F-3, F-4, and F-5 were built and then some machines carrying the serial zero were constructed, as well as about 55 machines which were known as 'Serial-1'. During this period provision was made for the machine to be remotely controlled, if desired. In addition the electrical system was safeguarded against shorting out if immersed in water, and protection against rock falls, etc. was provided for the drive systems.

Towards the end of 1950 and at the beginning of 1951 new machines were built by the Voros Csillag (Red Star) Tractor Works, and these once again were known as 'F-type' machines. The work was directed by Korbuly and his assistant, B. Liszony, the inventor Koszorus, and the Doroger Collective for Developing Machines.

The Russian PK3 and PK-3M Machines[1]

In 1953 a complete documentation of Model F-4 from the new series, covering fabrication technology and experience gained, was given to the Soviet Union. This led to the development and manufacture of the PK3 and the PK-3M (Figure 497). Of interest is the fact that during the 15 years prior to the Soviet Union obtaining the F-4 plans, etc. of the Hungarian cutter loader, coal-getting machines of various types had been developed in Russia as they had in other countries. However, according to A. V. Dokuken, after the receipt of the F-4 plans, more PK3s and PK-3Ms were built and used in Russia than any other type of coal-getting machine. For instance there were 370 PK-3 (PK-3m), 73 PK-2ms (PK-2sz), 10 PK-7s and PK-9s, 6 PKG-3s, 2 Karaganda 7/15s and about 51 sundry other machines (Figures 497 to 499).

Figure 497. Russian continuous miner (PK.3M) — a machine similar to this model was used in Kuzbass for developmental and extraction work. (Courtesy Machinoexport, U.S.S.R.)

Figure 498. Russian continuous miner (boom-heading machine, Model K.56MG). Drives headings of any shape in seams 1.9-2.5 m (6 ft 3 in-8 ft 3 in) thick. (Courtesy Machinoexport, U.S.S.R.)

According to Machinoexport, Moscow, the PK-3M unit has been exported to Rumania, Poland, Yugoslavia, and Spain since 1965.

The Continued Development of the Hungarian Version.[1]

So far as the Hungarian version was concerned, developmental work there continued and the next phase of the F-type machine was the production in 1956 of the F-5. This machine was built under the direction of Professor A. Jurek of the Technical University, in conjunction with Technical Director J. Korbuly and his special advisors. The F-5 was technically more modern, more robust, and, as a result, more efficient in its operations than its predecessor the F-4, and the experience gained in the successful use of this model, both in Russia and in foreign countries, served as a guideline for the construction of Model F-6.

In 1964 the Austrian firm Alpine Montan AK was granted a licence by NIKEX in conjunction with Országos Bányagépgyártó Vállalat to build and sell the F-6 in Western European countries. The contract stipulated, however, that the parties to the agreement, namely the Hungarians and the Austrians, must exchange all new ideas and experiences concerning the technological development of the F-6 with each other.

Developmental work on the F-6 continued and units F6H and F6A (Figure 501) were built. These latter machines were basically similar in concept to the F-6, but were made stronger. Operational experience with the two latter models indicated that the F6A was the better of the two units.

Originally the cutter head consisted of flat discs which rotated in opposite directions, but when the F-2 was built the cutterhead was changed to two hemispheres studded with picks and these also rotated in opposite directions. On Type F-6, F6H and F6A twin heads were introduced. A machine fitted with twin heads was first used in the Oroszlány mine. Alpine Montan AK used these machines for a variety of purposes which included mining, road construction work, and also for shafts. Later one was used in Salzburg during the construction of the Neutor tunnels.

Figure 499. Model PK.9. Like most of the Russian machines which are designed for work in coal-mines, the electrical equipment of this unit is explosion-proof in compliance with safety regulations adopted in the U.S.S.R. (Courtesy Machinoexport, U.S.S.R.)

Another interesting experiment was carried out with the assistance of the Germany Company of Schäffer and Urbach GmbH who did the construction work. Two machines were bolted together to form one unit and were driven by one motor. The double machine, which was mounted on rubber-tyred wheels, thus had two booms. It was used in the Hamburg and Munich underground railway systems.

According to Dr. Z Ajtay, one of the strongest boomheads was developed by the German firm, Eickhoff. It was designed for tunnel work and was mounted on their Model EV100. Dr. Ajtay also comments that a German Patent No. 339 was taken out on 6 March 1902 for a machine which had a cutting head on a boom, and in 1913 Meinhardt worked on the same principle and also F. Bago

in 1911. Bago is reputed to have actually constructed a unit. Another machine was also said to have been built by I. Szeman and to have carried parallel plate discs.

In America Chester T. Drake was granted a U.S. patent (No. 747,869) (Figure 500) on 22 December 1903 for a boom-type machine. Drake's machine had a cutter head on the end of a long shaft[e] mounted on a ball joint[a] at its rear end and rotated by a flexible drive which enabled the cutter head to be swung around in a circle while being rotated on its own axis at the same time. This planetary motion was achieved by passing the shaft through a revolvable frame[f]. The circular frame carried a pair of perforated parallel guides[b] forming a slot along a diameter of the frame. A journal on the cutter head shaft slid in the slot along the diameter of the frame and could be secured in

552

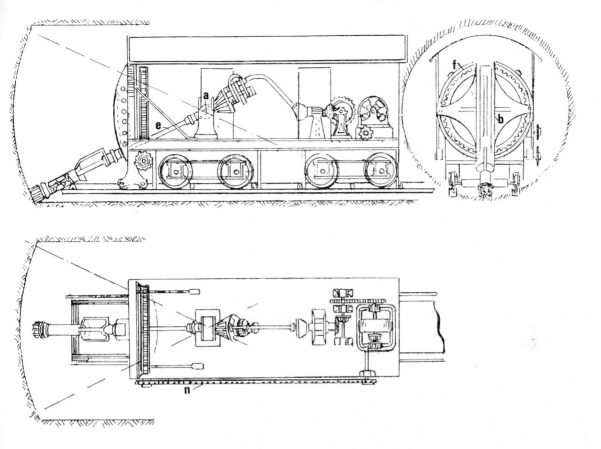

Figure 500. C. T. Drake T.B.M. U.S. Pat. No. 747,869, dated 22 Dec. 1903

position by passing a pin through corresponding holes in the guides and a hole in a lug on the shaft journal. As the frame revolved the cutter head described larger or smaller orbits depending upon the actual position in which the sliding journal was secured.

In operating the machine the journal was first fixed at the centre and a central hole the size of the cutter head was cut in the face. After this the journal was moved off centre sufficiently to move the cutter head off centre by its own diameter. The cut thus made opened out the hole already cut to three times its diameter. This process was repeated until the desired tunnel diameter was reached. The machine was then moved forward and the process repeated.

The circular frame was revolved by means of a large gear and a pinion driven by a long sprocket chain[n] which connected it to the engine at the rear.

Drake proposed that the debris cut by the head would be carried to the rear by an endless belt system running beneath the machine. A spiral conveyor within a cylinder mounted around the cutter shaft immediately behind the head directed the debris on to the belt.

The Alpine Miner AM50

In 1973 Vereinigte Osterreichische Eisen-und Stahlwerke combined with Alpine Montan Aktiengesellschaft to form the Vereinigte Osterreichische Eisen-und Stahlwerke-Alpine Montan Aktiengesellschaft company. (Later during the early 1970s the company's name was abbreviated to 'Voest-Alpine AG'.) Developmental work on the successful Model F6-A was continued and in addition in 1971 the first of the new Alpine miner machines, namely the AM50, was produced. This machine is a direct descendant of the F6-A unit but is slightly larger and heavier. The AM50 carries

553

Figure 501. Voest-Alpine miner Model F6-A. (Courtesy Voest-Alpine AG.)

twin ripper-type cutting heads which rotate at right-angles to the axis of the 3170 mm (10 ft 5 in)-long cutter boom on which they are mounted. The boom is mounted on a hydraulically controlled turret which allows vertical and horizontal swing for the arm. The cutter heads are fitted with carbide-tipped picks which 'snap-on'. A cross-section 4.00 m (13 ft 1 in) high by 4.8 m (15 ft 9 in) wide may be cut by the machine. However, by fitting an optional boom and turret extension both the cutting height and width may be extended to 4.20 m (13 ft 1 in) and 5.05 m (16 ft 7 in) respectively. Similarly the cutting width and height of the smaller F6-A machine may be extended from 4.5 m (14 ft 9 in) and 3.37 m (11 ft 1 in) to 4.90 m (16 ft 1 in) and 4.00 m (13 ft 1 in) respectively.

Traction is provided by crawlers and the material is directed onto an apron by gathering arms and is then moved into the throat of the machine and carried by single-strand chain conveyor to the rear.

In 1973 an AM50 unit was put to work in the Edgecliff to Bondi Junction Section of the Eastern Suburbs Railway Tunnel in Sydney, Australia, by the contractors Codelfa Construction Pty Ltd. It was used to cut out and square the circular 5.48 m (18 ft) diameter tunnel, which had previously been bored with a Calweld T.B.M.

The successful application of this machine in Sydney led to the further installation by Codelfa Construction of eight more of these units in the M.U.R.L. (Melbourne Underground Rail Loop) (Figure 502) tunnels

Figure 502. Alpine miner used on the Melbourne Underground loop project — Model AM50. (Courtesy M.U.R.L.A.)

in Melbourne, where an F-6A machine had already been utilized on a pilot scheme to prove feasibility. Since 1964 when Alpine Montan AG first began manufacturing this type of machine over seven hundred F6-A and AM50 Alpine miners have been delivered to various countries throughout the world.

The company's latest Alpine miner (model AM100) introduced in 1976 is basically similar in principle to the original F6-A and the medium-sized AM50, but is the heaviest and most powerful of the three miners, having a total installed capacity of 450 kW (600 hp). It can cut a 7 m (23 ft) wide by 5.4 m (17 ft 9 in) high cross-section. Like the AM50 and the F6-A both the cutting height and width of the AM100 miner may be extended slightly by fitting special optional equipment. The width of the loader apron may also be extended from 3.7 m (12 ft 2 in) to 5.5 m (18 ft) by the use of additional optional equipment.

Material cut from the face is directed towards the double chain conveyor by two gathering arms. This carries the spoil through the throat to the discharge end at the rear.

Since its introduction the AM100 has been used successfully in several European countries for both civil construction and deep coal-mining work.

By using the various Alpine miner cutting units and a combination of the main components of these machines it is possible to assemble multi-head tunnellers to suit a variety of tunnelling methods or cutting patterns. For example an Alpine miner model F6-A may be used to cut the lower part of the heading to a height of say 3 m (10 ft). The upper cross-section, to a height of say 7 m (23 ft) may be worked with the cutting unit and turret mounted on longitudinal and transverse supports which straddle the lower machine. The upper cutting unit is capable of lateral movement on the supporting framework, while the supporting frame can in turn be moved lengthwise along the tunnel floor on its wheels.

Bretby Roadheader

In March 1961 representatives of the National Coal Board (N.C.B.) journeyed to Russia to

555

inspect current developments there in heading machines of the F type. The PK3 selective heading machine was examined and a unit was subsequently bought and installed in Ellington colliery, in order to observe its operational performance under various conditions with different types of cutter picks. The strata at Ellington consisted of hard coal and strong shale.

In 1962 as a result of these preliminary studies the Bretby selective heading machine (later called the 'Bretby roadheader Mark I') was built and installed underground at Daw Mill colliery, West Midlands Division. While following the basic concept of the F-type machine the Bretby unit was more robustly constructed than the PK3 with a slower pick speed but a greater pick force and was powered by a 30 kW (40 hp) motor instead of a 22.4 kW (30 hp) motor.

In November 1963 the second Bretby roadheader Mark II, embodying several modifications was tried in Shotton colliery, Durham Division. Improved driver visibility simplification of the general structural design, and the addition of hydraulic spragging rams for anchoring purposes during operation, were among the improvements included in the new machine.

In 1964 Anderson Boys Limited and Distington Engineering Company Limited were asked to build two machines each. These were known as the Bretby Mark IIA roadheaders. In addition Mavor & Coulson Limited were asked to manufacture two Mark IIB roadheaders with Lee Norse tracks in place of Dosco tracks.

In Table 13 supplied by H. M. Hughes (N.C.B.-M.R.D.E.) a comparison may be made between the PK3 machine and the Bretby roadheader Mark I.

The power required for the same output is independent of pick speed, but if penetration rates are to be maintained while pick speed is kept below 1.5 m/sec* (4 ft 11 in/sec) it is necessary to design a stronger and heavier machine capable of proportionately higher thrust and torque; and commercial manufacture of these units in Britain has followed this trend set by the Bretby roadheader developments (Figures 503 and 504).

It will be noticed from Table 13 that during the early 1960s when the Bretby Mark I roadheader was built the weight of the machine (W) was approximately equal to the power of the cutting head motor (P) over the pick speed (v), e.g.

Weight of machine = 10 t and P/v = 9.95 kN
(2100 lb)

whereas in the modern roadheaders now being built in Britain the weight of the machine is approximately equal to three-quarters of the ratio of motor over pick speed, i.e. $W = 0.75P/v$.

This is due to the greater reserve of power now installed to deal with duties such as attacking the rock face at floor level while the cutting head rotates buried under cuttings which may be wet.

With the formation of the Anderson Mavor Limited company (now operating as Anderson Strathclyde Limited), which incorporated the firms of Anderson Boyes and Mavor & Coulson,

---

*No frictional ignition has, according to H. M. Hughes, been reported with a machine using a pick speed below 1.5 m/sec (4 ft 11 in/sec). The name 'roadheader' was introduced by M.R.D.E. in 1963 to describe machines of this nature safe for stone work in Coal Measures. Frictional ignitions in stone adjacent to coal seams are likely at speeds in excess of 1.5 m/sec. (4 ft 11 in/sec.)

Table 13

Machine	Power of cutting head motor (P) (kW)	Speed of cutting head (rpm)	Pick speed (v) (m/s)	Weight of machine (t)	Pick force (p/v) (kN)
PK3	25	73	2.52	10	9.95
Bretby Roadheader	30	36.7	1.49	20	20.1

556

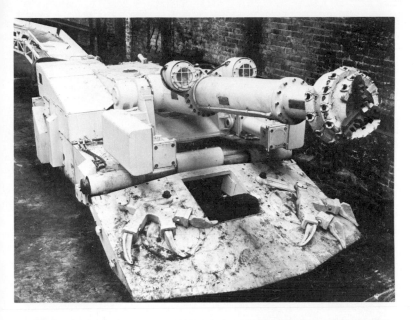

Figure 503. Bretby roadheader — front view

Figure 504. Bretby roadheader — rear view — note the hydraulically operated stelling jack on the left of the unit. This was a Bretby innovation

developmental work on the Bretby roadheader was assigned to the Mavor & Coulson Division.

The first commercial machine produced by the Mavor & Coulson Division in 1969 was designated the RH1. This unit was basically similar to the Bretby machine in that it incorporated a 48 kW (65 hp) telescopic boom, gathering arm loading system, centre conveyor, and front and rear side sprags. An added advantage was the availability of either a crawler or a walking base as the means of propulsion. Production on the RH1 ceased in 1973.

A smaller more compact unit with the same basic parameters as the RH1 (designated the RH10) was produced in limited numbers during the period 1970-71. It was specifically designed to cope with the smaller sized gate roads preferred by the N.C.B.

The RH10 was superseded in 1975 by the production of the RH20, which was even more compact in size, but again included the same basic features, i.e. the alternative of a crawler or walking base and a 48 kW (65 hp) telescopic boom. However, a new loading system which incorporated twin flight loading chains and

557

Figure 505. Anderson Strathclyde multi-purpose boom ripper with axial rotating cutting head and transversing cross conveyor for stowing the cut material. (Courtesy Anderson Strathclyde Ltd.)

Figure 506. Anderson Strathclyde boom miner with forward rotating cutting head. (Courtesy Anderson Strathclyde Ltd.)

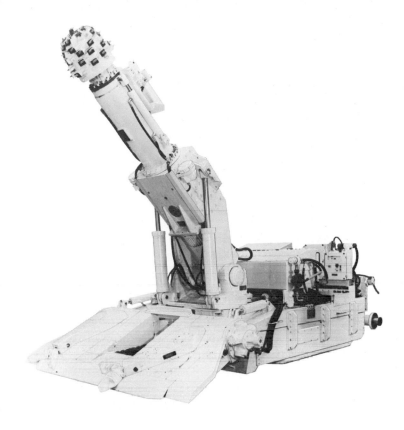

Figure 507. Anderson Mavor roadheader RH22. (Courtesy Anderson Strathclyde Ltd.)

variable apron widths was installed to enable the machine to handle different roadway conditions.

According to Forrest S. Anderson of Anderson Strathclyde Limited, developmental work on these units during the early 1970s was greatly influenced by the following factors:

1. Increase in retreat mining with requirement for rapid in-seam drivages.
2. The increasing application of Roadheaders in mixed coal and stone drivages.
3. Application of Roadheading machines in civil engineering application and major projects with the N.C.B.
4. Requirement for a Roadheading machine to operate satisfactorily on slope drivages at ¼ in favour and against.

Three new machines (designated the boom miner, the RH22 and the RH1/3) were designed

and built by Anderson Strathclyde Limited to meet these requirements.

The Boom Miner – for In-seam Drivages

The boom miner (Figure 506), which is a small compact machine weighing 22 t with total installed power of 120 kW (160 hp), has proved to be extremely efficient and manoeuvrable, allowing for rapid in-seam drivages and easy transport about the mine. It has a cutting width of 2.2 m (7 ft) and its operational height ranges from 1.4 to 3.6 m (4 ft 6 in to 12 ft).

The RH22 – for Coal/Stone Applications

The increased slew and cutting forces of this unit (Figure 507), a rugged, heavy, powerful machine weighing 35 t with installed power of 180 kW (240 hp), allows for the handling of harder strata found above and below the coal seam.

559

Figure 508. Anderson Mavor roadheader RH1/3. (Courtesy Anderson Strathclyde Ltd.)

The RH1/3 — for Civil Engineering and Major Projects

The RH1/3 (Figure 508) is a heavier machine than the RH22 weighing 51 t but, apart from the fact that it has a cutting height of 5.8 m (20 ft) (making it suitable for civil engineering and major projects) it is basically similar to the RH22 with a maximum installed power of 180 kW (240 hp). In addition the crawler units for both the RH22 and the RH1/3 were redesigned to improve their efficiency on slope drivages.

Forrest Anderson reports that future development of these machines will be aimed at 'improving their effectiveness in tackling harder strata'.

The Dosco Roadheaders

Dosco Overseas Engineering Ltd. which had formed a company in the U.K. in 1953 to market their Canadian-built Dosco miners, had by 1963 established manufacturing facilities in the U.K. That year they introduced the Dosco roadway cutter loader to the N.C.B. This machine was derived from the Russian PK-3m unit, but embodied various modifications which included the replacement of most of the electric motors with hydraulic drives,* so that

the new machine had only two electric motors, one of 37 kW (50 hp) for the cutting head and one of 48 kW (65 hp) for the hydraulic system.

The prototype machine was put to work at Snibston colliery where it performed satisfactorily with very little additional modification.

During the succeeding years the use of this type of machine grew. By 1971 over 250 units had been manufactured and sold, most for use in the U.K., but some were also exported to Spain, Germany, and Sardinia and two were installed in the Cape Breton Mines in Canada.*

Except for the strengthening of the unit by a wider use of cast steel in place of fabrications and the fitting of a single gear-type hydraulic pump in place of the more complex three-pump system driven through a gearbox, no significant changes were made to the Dosco roadway cutter loader during this period.

In 1970 a new Dosco roadheader known as the Mark IIA** (Figure 509) roadheader left the factory and was installed in the Salsgirth

---

*In 1958 Dosco was purchased by the Hawker Siddeley Group and shortly thereafter the Cape Breton mines, which had previously been owned by Dosco, were taken over by the Cape Breton Development Corporation.

**In addition to the Alpine AM50's two Dosco Mark IIAs were in use in 1977 on the M.U.R.L.A. project in Melbourne, Australia.

---

*This was taken from the Bretby roadheader.

Figure 509. Dosco Mark 2A. (Courtesy Dosco Overseas Eng. Ltd.)

Figure 510. Dosco Mark III. (Courtesy Dosco Overseas Eng. Ltd.)

colliery. While the basic concept remained the same, the new machine embodied many significant modifications which effectively improved its performance. The modifications included an increase in weight from 18 to 23 t, the electric motor for the boom was replaced by a 48 kW (65 hp) unit, and the one driving the hydraulic system by a 56 kW (75 hp) unit. The length and width remained unchanged but its height was reduced from over 1.82 m to 1.62 m (6 ft to 5 ft 4 in). Horizontal stelling cylinders were installed to provide greater stabiity when cutting on rising, dipping, and cross gradients and also for harder rock strata.

It was during this period, i.e. the early 1970s, that there dawned a greater recognition of the

561

Figure 511. Dosco universal tracked ripper. (Courtesy Dosco Overseas Eng. Ltd.)

Figure 512. Dosco LH155. A low-height machine utilizing one of the booms from the twin-boom miner. (Courtesy Dosco Overseas Eng. Ltd.)

potential value of this type of machine for use on civil engineering projects such as rail, water, sewer and irrigation tunnels and for road construction work (Figures 509 to 512). As a consequence Dosco found a ready market for their new unit, selling some 370 machines by 1976. In addition the introduction of the boom-type hybrid machine (i.e. a shield combined with a boom-type cutter (mounted either on a turntable or a sledge) within the shield) encouraged the sale of the Dosco TM 1800 unit. This was basically a shield within which was mounted a Mark IIA cutting boom. (This type of machine, i.e. shields combined with boom-type cutters, is described more fully in Chapter 25, Boom-type hybrid machines.

Dosco Twin-Boom Miner

An attempt to increase the productive capacity without, at the same time, surrendering any of the significant features which set the boom-type header apart from its cousin the American continuous miner, led to an interesting development in roadheaders.

The suggestion was mooted that the header could be fitted with two boom cutters. As a result the Dosco twin boom miner (Figure 513) was conceived and the first machine was produced in 1971. The prototype model, which was fitted with two standard Mark IIA cutting assemblies and a flight loading conveyor system, was put to work in the British Gypsum mine at Sherburn. The flights of the conveyor

Figure 513. Dosco twin-boom miner (TB600); (Courtesy Dosco Overseas Eng. Ltd.)

were unusual in that they folded for the return run after the material had been discharged.

The machine weighed a total of 40 t. The initial application proved to be a difficult one, and while results were not entirely satisfactory they did indicate that the concept of two booms was viable. However, it was obvious that more strength was needed and a greater degree of sophistication was called for. The heads were contra-rotating and the cutting pattern of one would normally be a mirror image of the other. Although this action increased the stability of the unit, it also tended to make it very difficult for the operator to watch both heads at once, and it was to overcome this problem of watching two booms at once that the profile control device or guidance system was installed. It was manufactured for Dosco by Hawker Siddeley Dynamics Ltd. and it provided the operator with an illuminated display screen showing the profile to be cut, and the relative position of each cutting head, thus virtually eliminating the need for the operator to actually see the heads. A digital readout from 20 to 0 for each head was displayed on the screen when each head reached 20 cm (8 in) from the desired profile edge.

After a considerable amount of developmental work the TB600 was produced and delivered to the British Steel Corporation's iron ore mine at Santon in Lincolnshire for preliminary trials. The total installed power of this machine was approximately 650 kW (900 hp).

Paurat GmbH

Another company favouring the milling-type cutting action is Paurat GmbH of Friedricksfeld, West Germany (Figures 517 and 518).

The Paurat Company was founded by Friedrick W. Paurat its present Director, on 1 January 1947 to produce roadway arches for German mines.

During the mid-1950s, Hollybank Engineering, a British company, entered into an agreement with Paurat, whereby Hollybank were licensed to produce Paurat arches in the United Kingdom.

Apparently Hollybank Engineering was and still is affiliated with Dosco Overseas. Both these companies have the same management and, indeed, are also producing under the same roof.

As a result of this amicable association between Paurat and Hollybank, a seven-year contract was entered into giving Paurat the right to introduce British-made Dosco machines or to produce them under a licence agreement for the European Common Market countries (consisting at that time of West Germany, France, Italy, Belgium, The Netherlands, and Luxemburg).

The first British-made Dosco machine was sold to the Rheinpreussen AG mining company for driving preparation roadways in a coal seam. In 1967 a Paurat-built Dosco boomheading machine was delivered to the Harpener Bergbau AG company for use in the Gneisenau mine.

In a great many of the West German coalmines 1 m (3 ft) thick seams, lying at considerable depths below the surface ground are worked. Often very hard adjacent floor and roof strata need to be cut. Unfortunately, according to Paurat, the light Dosco machines of the early 1960s and late 1970s were not found to be particularly suited to this type of heavy work. In all, about 20 Dosco machines of both British and West German origin were sold at that time to various Common Market countries and also to Argentina.

Because of urging by the mine industry for the production of heavier machines, Paurat began developmental work on their E134 series after the seven year licence agreement with Hollybank Engineering had expired on 22 January 1972.

The first of the Paurat E134 Series machines was built in 1975 for Ruhrkohle AG, Rheinland Group, and was used in the Pattberg-Schächte mine. Since then Paurat have made 30 of these units which have been sold in West Germany, Great Britain, France, and Australia.

Basically the E134 features a boom-mounted rotary cutting head consisting of a cylindrical high-tensile steel core on which is mounted a double spiral array of armoured pick boxes. The boxes are fitted with tungsten-carbide-tipped picks. The head is axially driven by a two-speed water-cooled 200 kW (268 hp) motor through an epicyclic gearbox.

The cutting boom is horizontally pivoted in a pedestal carried on a large slewing ring mounted on the central section of the main frame. Vertical and horizontal arcuate movements of the head are controlled by pairs of heavy-duty lifting and slewing jacks.

According to the manufacturers the machine operates from a static position and can cut a section 2.6 m (8 ft 6 in) high by 4.5 m (14 ft 9 in) wide in rock with compressive strength up to 138 MPa (20,000 lb/in²). The cutting section may be extended to 6.3 m (20 ft 8 in) high × 7.2 m (23 ft 7 in) wide by the substitution of modular components. (For smaller sections the Paurat E169 is available).

Muck is directed by the cutting head up a hydraulically adjustable flap at the front of the machine on to a shallow ramp. Two recessed single-chain armoured conveyors on the ramp then move the debris around two deflector sheaves and under the main frame of the machine to a loading jib bolted to the frame at the rear of the unit.

The E134 is crawler mounted and is capable of tramming speeds of 6 m/min (20 ft/min).

Alignment and profile control systems were recently exhibited by both Paurat (on a 'Roboter') and Eickhoff (on an EVR 160) at the Düsseldorf Mining Exhibition, but according to H. Boldt,[12] these systems are not yet available in a 'flame-proof' form and are still being developed. The Eickhoff system which is more fully described later in this chapter under the heading 'Eickhoff heading machine — profile control' was tested in practice in Geostock, but the project was prematurely abandoned due to the wet conditions encountered there.

Towards the end of 1975 a licence agreement was entered into between Paurat GmbH and Thyssen (Great Britain) Ltd., whereby Thyssen was to manufacture and market the E134 in the United Kingdom under the brand name 'Thyssen-Paurat "Titan" '. In Germany the E134 is sold under the name 'Paurat "Roboter" '.

The first of the Titan machines was built in Germany and transported to Gateshead where it was used by the Thyssen-Taywood Consortium for work on the Tyne and Wear Rapid Transit System. However, the second Titan was built by Thyssen Engineers Ltd. in Llanelli, South Wales. It was used by the N.C.B. for the construction of its Jubilee drift in the North Derbyshire area.

In accordance with the Mines and Quarries Act of 1954 the machine was flame-proofed to N.C.B. standards and 'aquasent heavy hydraulic fluid (water and emulsion)'[2] was used instead of a 'water-glycol'[2] mixture which is specified by Ruhrkohle, the German equivalent of the N.C.B. A water-glycol mixture was used on the Gateshead unit.

A Paurat Roboter (slightly larger than the Gateshead Titan machine) was used at Appen colliery, New South Wales, Australia. The machine is remotely controlled by cable connections. It weighs 62 t and can cut a 4.7 m

Figure 514. Paurat twin-boom heading machine — Model PTF70. (Courtesy Paurat GmbH.)

(15 ft 5 in) wide × 3.7 m (12 ft 2 in) high section of tunnel. It is fitted with a roof bolting accessory which has a 360° vertical and horizontal turn.

Paurat Twin-Boom Header

In 1970 an Italian construction company named Girola asked Paurat to build a heading machine capable of fast production rates. However, if the machine was to be of practical use in the construction of the row of railway tunnels situated on the Direttissima Roma-Firenze line, running between Rome and Florence, delivery had to be as soon as possible.

To cut production time to a minimum, therefore, Paurat used a normal excavator undercarriage supplied by O & K Orenstein & Koppel for supporting the cutting head which consisted of twin booms (Figure 514) (to increase productivity) pivotally mounted on a robustly constructed rectangular frame pedestal carried on a slewing ring which was positioned at the end of the main machine boom.

An independent lifting jack controlled the vertical movement of each boom, while the slewing motion was common to both cutting booms.

The excavator machine carrier was naturally designed for fast tramming speeds and arcing movements, and for instant stops. These were unsuitable for a heading machine. It was therefore necessary to convert the unit to a slow-moving machine and, as a result, a somewhat sluggish slewing ring for turning the upper structure had, perforce, to be tolerated, plus several other compromises which were contrary to accepted practice in the design of boom headers.

Despite these disadvantages and the fact that it did not meet the normal manufacturing standards set by Paurat for its machines, this

565

Figure 515. Paurat trench-cutting machine. (Courtesy Paurat GmbH.)

Figure 516. Paurat trench-cutting machine. (Courtesy Paurat GmbH.)

unit has been operating in Rome for about 5 years, successfully cutting 8 m (26 ft) high tunnels in tuff at the rate of approximately 11 m/day (36 ft/day). No teething problems were experienced with this unit.

Paurat Trench-Cutting Machine

Another interesting machine introduced by Paurat is the trench-cutting machine (Figures 515 and 516). To some extent this is similar to the excavator unit produced by Demag in 1971. Demag's excavator featured an articulated cutting boom which was designed primarily for handling larger tunnel cross-sections (see section on Demag articulated boomheader machines). The trench cutter made by Paurat evolved from the E134 series and, in fact, uses a similar cutting head on its articulated boom as was used on all the E134 units.

It is, however, a heavier unit weighing approximately 100 t, while its tramming speed is 4.7 m/min (15½ ft/min). The cutting arm is pivoted to the lower end of a short turret which, in turn, is swivel-mounted on the end of the main boom. The swivel mounting is carried in a pedestal mounted on the main frame of the machine through a heavy-duty slewing ring. All the movements of the boom assembly are controlled by heavy-duty hydraulic jacks.

During operation the head is sumped or thrust axially into the rock or other material and then moved laterally so that the side picks are brought into play. The cuttings are removed by the spiral vanes on which the pick boxes are mounted.

For trenching work the main unit moves ahead of the cutting boom which trails behind with its head in the trench. The machine is then trammed forward a short distance and the head sumps to the bottom of the trench face. With the machine stationary the head is gradually raised so that a vertical slot is cut to the surface. This process is repeated, with additional cuts being made alongside the first slot, until the full trench width has been cut. After the machine

566

Figure 517. Three of these Paurat boomheader machines fitted with impact rippers were built during 1974-75 for driving railway tunnels in abrasive rock. (Courtesy Paurat GmbH.)

moves forward for the next series of cuts a backhoe is utilized for mucking the trench.

Developmental work on the unit was begun in 1974 at the instigation of the Ballast-Palensky Consortium which needed a machine capable of cutting a trench through fairly hard limestone.

The joint-venture company of Ballast-Palensky had been awarded a major sewerage contract in Riyadh, Saudi Arabia, but the company was dissatisfied with the slow rate of progress — approximately 25 m (80 ft) of trench per week — currently being made by the 50-man labour crew, using manually operated

jack hammers. (The ground was too hard for a backhoe or a conventional trenching machine.) The trenches were being cut through country consisting of 0.5 m (1 ft 6 in) of boulder sand beneath which lay a massive limestone bed ranging in compressive strength from 58.6 MPa (8500 lb/in²) to 75.8 MPa (11,000 lb/in²). At that rate it was estimated that the 11.5 km (7 miles) of trenches would take about 12 years to complete. After the installation of the Paurat unit, progress rose to approximately 200 m (656 ft) of trench per week, thus cutting the expected completion time from 12 years to about 18 months.

Figure 518. This Paurat boomheader fitted with a platform was a special-purpose machine designed for circular cross-sections. (Courtesy Paurat GmbH.)

567

So successful was the prototype machine that five more units were built by Paurat and are being used by Middle East Countries — five in Saudi Arabia and one in Libya). According to the manufacturers, another six machines were due for delivery there by the end of 1978.

Demag Aktiengesellschaft (now Mannesmann Demag Bergwerktechnik)

Face Heading Machine VS3

Demag Aktiengesellschaft of West Germany produced the first of its boom-header units in 1962. These machines featured milling-type cutting heads and were walking prop support machines designated face heading Model VS1 'Unicorn'. Model VS1 was followed in 1967/68 by Model VS2E. Both these machines incorporated a boom-heading component which operated in conjunction with walking supports and, as such, constituted what is defined by the author as a 'hybrid machine'. They are described more fully in Chapter 25, 'Boom-type Hybrid machines'.

During the ensuing years considerable developmental work was undertaken by this company on its machines and as a result several improved models of the boomheader have been marketed.

The latest Demag face heading machine, Model VS3, is rigged with a robust cutting jib or boom capable of universal movement which is effected by two pairs of lifting and slewing jacks. The cutting head which, as mentioned, rotates about the longitudinal axis of the main jib is driven by a 160 kW (215 hp) motor. The head carries point attack cutting tools arranged in a spiral pattern. The tool spirals terminate in flat-pitched conveying spirals which move the material towards the conveyor.

The machine is crawler mounted on two independent hydraulically driven travelling gear units which are interconnected by a split-machine frame. Screw connections permit rapid dismantling and assembly of the three main components, namely the two travelling gear units and the machine frame, for easy transportation underground. Independent direction and speed control of each crawler provides machine manoevrability. For normal cutting operations the machine travels at a speed of 3.5 m/min, but this speed may be increased if required to 7.0 m/min (23 ft/min), when relocating the unit.

Two single scraper chain conveyors mounted in recesses on the front loading ramp move the material through a conveyor entrance, carrying it beneath the machine to a spoil discharge area at the rear. To obviate blockages or jamming, the conveyor entrance is fitted with a special 'retaining device' or screen which prevents the entry of oversize material into the conveyor system.

A hydraulically operated lining erector system which is designed to run on a rail similar to a monorail along the roof or sides of the tunnel, is provided. This will handle lining for arched or rectangular tunnel cross-sections. The erector and pneumatic trolley are independent of the boomheader. Pre-assembled lining elements are moved from a special assembling table situated behind the boom header and carried by the erector over the machine to the installation point at the face.

The machine is operated from a static position.

Two front supports carried on the loading chute structure are actuated by hydraulic cylinders mounted on the crawler units. These supports may either be thrust against the side walls to brace the unit, or alternatively, if support is not needed, they may be used as guides to convey the material lying at the base of the wall towards the loading ramp and scraper chain conveyors. Hydraulic stelling jacks fitted on the travelling gear section may be extended 1 m (3 ft 3 in) on each side of the machine to provide support at the rear of the unit.

Where necessray a roof support is available which utilizes hydraulic cylinders on the machine frame to thrust a cantilever arm with a supporting cap against the tunnel roof.

Articulated Boomheader or Excavator

In 1971 as a result of suggestions made by civil engineering contractors required to make tunnels with large cross-sections, Demag produced an unusual heading machine which featured an articulated boom.

Figure 519. (top) Demag excavator — articulated face heading machine with cutting boom mounted on basic main jib. (bottom) Demag excavator with cutting jib mounted on super-structure. (Courtesy Mannesmann Demag Bergwerktechnik)

569

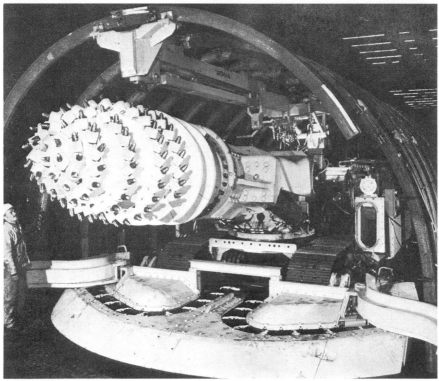

Figure 520. (above) Demag VS3 face heading machine. (below) Demag VS3 face heading machine and lining erector equipment. (Courtesy Mannesmann Demag Bergwerk-technik.)

This machine (Figures 519 and 520) was unique in that the cutting boom could be lowered to enable it to cut below ground level in much the same manner as did a normal face shovel or back-acting excavator.

Utilizing the same design as the cutting unit of the original Demag 'Unicorn' the cutting jib and head was hinged on the basic jib of the excavator and was connected to the power supply and control system of the latter. An additional hydraulic cylinder was fitted to provide more power for the slewing gear of the

Figure 521. Eickhoff selective cut heading machine, Model EVA-160 with profile and direction control system, remote control and measuring and monitoring system, on display at the International Mining Exhibition, Düsseldorf, during the period 22-29 May 1976. (Courtesy Gebr. Eickhoff Maschinenfabrik.)

Figure 522. Eickhoff EVR 200. (Courtesy Société d'Etudes Techniques et Industrielles.)

excavator and to absorb the stress forces generated during the rock-cutting operations.

Depending upon where the cutting jib was mounted, i.e. either on the basic jib or on the superstructure, the machine could work a cross-section 11 m (36 ft) in width × 10 m (33 ft) in height (subgrade cut 4.7 m (15 ft 5 in)) or 4.8 m (16 ft) in width × 3.7 m (12 ft) in height (subgrade cut 0.3 m (1 ft)) respectively.

So far as the author is aware Demag were the first company to produce a heading machine with a manoeuvrable articulated cutting boom. Later, in 1974, the Paurat Company of West Germany also produced a heading machine with an articulated cutting boom. The machine was used for trenching work in Saudi Arabia (see Chapter 22, Boom-type machines — Paurat trench-cutting machine).

Demag face heading machines, Models VS1 and VS2E, and those falling into the TSSM

571

Figure 523. Eickhoff EVR 200 used at Geostock near Paris. (Courtesy Société d'Etudes Techniques et Industrielles.)

series are hybrid machines and are described more fully in that section of the book dealing with hybrid machines (Chapter 25).

Westfalia Lünen

Westfalia produce two types of boom-heading machines, namely heading and tunnelling machines and boom cutter loaders.

Heading and Tunnelling Machines

There are two models in this series, the 'Büffel' (buffalo) and the 'Bison' (bison).

Except for one or two features these two types of machine are basically similar in design. Both are crawler mounted and utilize a pendulum-type loading arm. The loading arm directs the cuttings from the twin ripper type heads to a pair of straight 'flat-link' chain

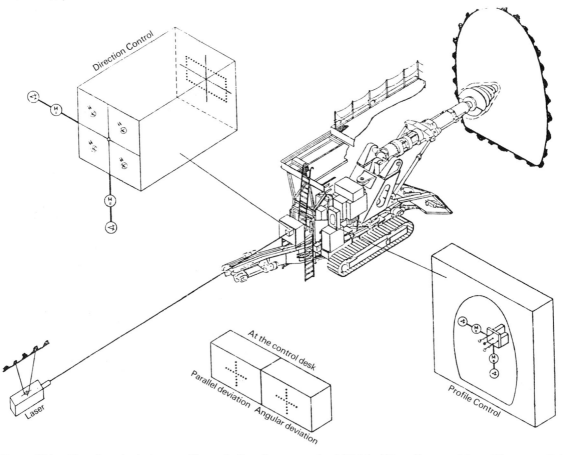

Figure 524. Drawing depicting profile and direction control of Eickhoff heading machine. (Courtesy Gebr. Eickhoff Maschinenfabrik.)

572

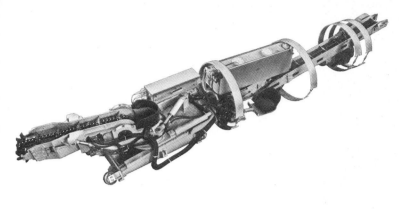

Figure 525. Westfalia 'Frettchen' (ferret) — Model FL-1-20. This mini boom cutter loader weighs 3.5 t and cuts a 1 m (3 ft 3 in) dia. minimum cross-section of roadway. (Courtesy Westfalia Lünen.)

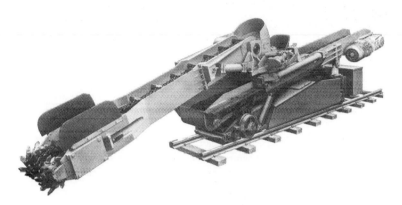

Figure 526. Westfalia 'Wühlmaus' (fieldmouse) — Model FL-2-25. The 'Wühlmaus', a slightly larger machine than the 'Frettchen', weighs 5.8 t and cuts a 1.6 m (6 ft 3 in) dia. minimum cross-section of roadway. (Courtesy Westfalia Lünen.)

Figure 527. The Westfalia 'Dachs' which weighs 12.5 t is the second largest of the Westfalia boom cutter loader machines. (Courtesy Westfalia Lünen.)

Figure 528. A Westfalia 'Dachs' (badger) shown cutting the crown of a tunnel. (Courtesy Westfalia Lünen.)

Figure 529. The Westfalia 'Luchs' (lynx) is the largest of the Westfalia Lünen boom cutter loader machines. (Courtesy Westfalia Lünen.)

conveyors which run on to a swivelling loading unit. The loading unit (which is fitted to the centre of the loading apron) facilitates the delivery of cut material from the machine's conveyors on to any subsequent conveying system, i.e. lorries, dump trucks, or belt conveyors, etc.

Both machines are fitted with hydraulically operated stabilizer legs at the rear. These legs relieve the crawlers and provide ancillary anchorage for the machine during the cutting operation.

Westfalia claim the Büffel will operate efficiently on gradients of about 14° while the

574

Figure 530. Westfalia produce two types of boomheading or selective cutting machines namely the 'Büffel' (Buffalo) and the 'Bison'. The 'Büffel' weighs some 53 t while the 'Bison' (pictured) weighs about 74 t. (Courtesy Westfalia Lünen.)

Bison can operate on gradients in the range of a dip of 9° and a rise of 14°.

The cutting head of the Büffel is powered by a 170 kW (230 hp) motor while the cutting head of the Bison is powered by a 200 kW (270 hp) motor.

A distinctive feature of the Bison is the extensible cutting boom which can be sumped forward up to 500 mm (17.7 in) or rotated through 30° for cutting the roadway profile.

Both machines are claimed to be capable of cutting minerals in the 100 MPa (14,500 lb/in²) range.

Boom Cutter Loaders (Figures 525 to 529)

Westfalia have produced five types of boom cutter loader machines namely:

'Frettchen' (ferret) — 3.5 t
'Wühlmaus' (fieldmouse) — 5.8 t
'Westfaliafuchs' (fox) — 6 t
'Dachs' (badger) — 12.5 t
'Luchs' (lynx) — 24 t

The smallest of these, namely the Frettchen and the Wühlmaus have been used in conjunction with pipejacking equipment where the pipe diameter was as small as 1.5 m (5 ft).

All of these machines incorporate a muck-removal system which forms an integral part of the cutter boom and, in this connection, the reader's attention is drawn to the Thomson

'Ladder' excavator (Figure 547) described in chapter 25 — Boom-type Hybrid machines.

Except for the Luchs the drum ripper-type cutter head of all these machines is rotated upwards by the conveyor scraper chain on the cutting boom. The two ripper-type cutting heads on the Luchs, however, are powered directly by a 90 kW (120 hp) electric motor which is mounted in the cutter boom, thus enabling this unit to cut material ranging in compressive strength up to 50 MPa (7,250 lb/in²). The material cut by the machines is transferred from the cutter boom to a swivel-mounted rear boom.

AEC Inc. (formerly Alpine Equipment Corp.)[11]

Towards the latter part of 1976 as a result of increased environmental pressures designed to curtail or even halt mining in the Death Valley National Monument, U.S.A., stringent laws were introduced prohibiting mining which could cause surface disturbance in the area. This restriction was to be in force for a period of four years.

The introduction of this law, together with other factors, affected the American Borate Company's planned mining method when they decided to work a deposit of borate ore in the Billie Mine which is located in the eastern section of Death Valley National Monument (about 225 km (140 miles) west of Las Vegas, Nevada).

575

Figure 530a. AEC Super ROC-MINER with field-interchangeable ripping and milling type cutter heads. Machine weighs 41 t and has 160 kW (215 hp) cutting head motor. (Courtesy AEC Inc.)

Figure 530b. AEC gantry miner (Jumbo) with twin booms (ripping type). This machine which has a cutting height of up to 10 m (33 ft) is used for driving tunnels with large cross sections in one single pass and for mining thick seams in coal, potash, rock salt, oil shale, etc. (Courtesy AEC Inc.)

Apart, of course, from the selected mining system being economically viable, the company had the following factors to consider in their choice of a particular mining method.

It was essential that the ore be mined selectively to reduce trucking costs. This was to comply with the regulation prohibiting any surface disturbance at the mine site, which forced the company to transport the ore a distance of some 64 km (40 miles) to the nearest mill, located outside the restricted area.

In selecting a mining method consideration also had to be given to the peculiar characteristics of the Billie ore deposit, and the adaptability of the chosen system to those characteristics.

In addition because the area to be mined was remote — a distance of some 225 km (140 miles) from the nearest town — and thus unattractive to skilled labour, it was important for the system to be highly mechanized.

Figure 530c. H-Series ROC-MINER (Model AEC-250H) with twin roofbolters. Machine weighs some 24 t when equipped with 112 kW (150 hp) cutting head motor. (Courtesy AEC Inc.)

The company decided to use boom-heading machines for the excavation work. Apart from ordering a Dosco TB600 twin-boom unit (Figure 513) fitted with milling-type heads from the United Kingdom, the company also assisted in the development and subsequent purchase of two Super Roc-Miners 330 (Figure 530a) from AEC Inc.

The unusual feature of the AEC machine which, like the Dosco TB600 was designed for work in hard rock, was that either a milling-type or ripping-type cutting head could be fitted. Moreover the heads could be changed on site to meet the particular requirements of the ground at the face.

Both units cut up to 110 MPa (16,000 lb/in²) unconfined compressive strength rock successfully.

During the initial trials various cutter-head lacing patterns and different cutting tools (i.e. conical point attack bits and flat kerf type cutters) from several manufacturers were tested. According to G.B. Sparks,[11] when the higher-powered Dosco was used, it was found that the unit gave the best productivity in rocks under 35 MPa (5000 lb/in²) unconfined compressive strength, if the head was fitted with kerf type cutters, while long conical shank point attack bits were preferred for work in hard rock up to 140 MPa (20,000 lb/in²) unconfined compressive strength. However, the AEC machine which was smaller and lighter than the Dosco, performed best with small modified plumb-bob type point attack bits in rock up to 40 MPa (7000 lb/in²) and with heavier modified plumb-bob bits in rock within the 69 to 140 MPa (10,000 to 20,000 lb/in²) strength range.

Both milling and ripping cutter heads were tested on the AEC machine, the milling head proving to be the more productive (about 25—30% higher) in the test ground selected.

Both machines met with the company's requirements for selective mining of ore and waste in the higher compressive rock strength ranges. So far as the American Borate Company was concerned, the machines were not found to be competitive as each was designed for a specific mining or tunnelling purpose. The highly mobile AEC machine was used by the company for developmental work, while the heavier Dosco was installed in the production stopes where the company required a machine for maximum productivity combined with selective mining of hard rock.

AEC Inc. produced their first Super-Rock Miner Model 330 in 1977. Since then some twenty-seven of these have been built. The following year (1978) saw the introduction of a twin boom gantry miner (Jumbo) (Figure 530b) and in 1979 the E and H Series Roc Miners (Figure 530c) fitted with twin roof bolters were put on the market. By the end of 1980, some 31 of these roof bolter type units had been manufactured.

Eickhoff Heading Machine[3]

The latest innovation in guiding systems (profile/control) emerged from the drawing boards of the German firm, Eickhoff. The machines and new guidance system were to be tested in France by the petroleum industry (Ministere du Developpement Industriel et Scientifique — Direction des Mines) who became interested in the Eickhoff machines. These machines (Eickhoff Models EVR 160, 200, etc.) (Figures 521 and 522) are basically similar in design principle to the Anderson Mavor roadheader and the Demag soft-medium ground tunnelling machines which feature a rotating conical drum studded with picks (sometimes referred to as a 'pineapple head') which is mounted on a manoeuvrable cutting boom.

So far as the EVR is concerned, the cutting picks on the head are of tungsten carbide which are secured in quick-release pick boxes. The boom is operated by three hydraulically powered and electrically controlled twin jacks which allow of universal movement.

Profile Control (Figure 524)

Just as a mechanical jig is used to restrict the movement of a tool in accurate repetitive machinery operations, so an ingenious electronic jig restricts the movement of the boom-mounted cutter to the desired tunnel profile. This is accurately drawn to $\frac{1}{20}$ scale as a transparent surface within a frosted glass screen in the cab.

As the operator swings the boom about in the tunnel, electronic sensors send signals back to the jig device and a luminous spot on the frosted screen duplicates the movements of the cutter head exactly to scale.

The operator is free to guide the cutter head in any manner he pleases. However, should he attempt to move it beyond the desired boundary in any direction, the luminous spot on the screen would, of course, strike the boundary line between the transparent surface and the frosted glass screen, whereupon a system of electro-hydraulic valves is electronically triggered and further movement of the cutter head in that direction is prevented as though controlled by a jig or template. Thus a perfect outline is preserved.

The exact position of this cross-section profile in relation to the designed axis of the tunnel is maintained by another system known as the guidance system. An He-Ne laser at the mouth of the tunnel directs a coherent beam of light parallel to the tunnel axis. This beam maintains a constant diameter of 16 mm ($\frac{5}{8}$ in) for over 400 m (1300 ft). A set of light receptors on two parallel metal plates attached to the machine frame act as a target for the beam, and electronic circuits detect angular or parallel deviations of the whole profile outline from the desired line of the tunnel. Any deviation is automatically corrected by electronic shift of the outline as a whole into the required position relative to the beam.

An EVR 200 (Figure 523) was put to work at Geostock near Paris, but unfortunately before the new guidance system could be tested the entire project was abandoned because of problems with flooding.

H.B. Zachry Company

Boom-type Trench-Cutting Machine

An unusual boom-type trench-cutting machine of which two models were produced, namely Jaws I and Jaws II, was developed about 1980 by the H. B. Zachry Company.

These units, which are more fully described in the section entitled 'Hybrid machines', feature an articulated boom mounted on the telescoping boom of an excavator superstructure. The undercarriage of a crawler-type grading machine is used as a base for the superstructure. Two transversely rotating cutting blades are fitted to the end of the articulated boom.

Mounted Impact Breakers[4-6]

The concept of mechanical impact breaking is not, of course, new. It dates back to the first percussion machines built by such men as Couch (1849), Fowle (1849), Cave (1851) and Bartlett (1854).

Over the years the method of delivering the blow energy to the tool has been refined and improved so that machines became stronger, more powerful, and more efficient. Some of these percussive drill units were fitted to carriages or jumbos and used as drifters for blast-hole work in tunnels, etc. These machines were designed to drill comparatively small-diameter holes of great depth and so the blow energy needed for the work was not great. Other heavier-type units were fitted with moil points or spades and used as pavement breakers, etc. These latter machines are hand-held and their blow energy is limited to about 140 J (100ft-lb) or less, because humans are unable to react to the forces developed by the piston if this level is exceeded.

While great strides have been made in the development of mining machines in general, there still remained, until comparatively recently, many areas of application where a powerful selective mining machine was needed. Some types of selective mining could, of course, be handled by the ripping or milling-type roadheaders, despite their attendant dust problem, but still there remained many applications such as in a solid heading where these machines proved ineffective, and the only other alternative available was the use of an explosive with all its inherent disadvantages.

During the late 1960s it was recognized by many in the mining industry in both the U.K. and the U.S.A. that because of the rising cost of labour, the dust problem associated with roadheaders, and the detrimental effects of blast-hole work, a new and better method of high-energy selective mining was required. This was particularly true with regard to vein mining.

British and Continental Developments

Lothians Impact Plough

In 1951 the N.C.B.'s Mining Research Establishment began developmental work on a machine capable of delivering high-energy blows of the order of approximately 1400 J (1000 ft-lb). By 1956 an experimental pneumatic impulse rig had been assembled. It was tested underground in the Garw coal seam at Cwmtillery colliery. While this prototype unit was not, naturally, entirely satisfactory, it nevertheless gave sufficient indication of its

potential as a breaker unit to encourage further research work and development. Consequently a machine, known as a 'Lothians impact plough' was designed and built in collaboration with the Lothians Area, Scottish Division. In 1961 it was installed underground at Easthouses colliery in Scotland where it performed satisfactorily. This machine and all subsequent impact units designed by the Central Engineering Establishment (now M.R.D.E.) were operated hydraulically and most incorporated nitrogen springs.

The Mark I Lothians impact plough was followed in 1962 by a more powerful Mark II impact unit, and work was also commenced that year on the construction of the Mark III Lothians impact plough which was expected to be able to give an impact capacity of approximately 2700 J (2000 ft-lb). It was completed in 1964.

Although the Lothians impact plough 'cut anything in its path' it was, according to H. M. Hughes (M.R.D.E.) finally abandoned because of problems associated with the steering.

In 1965 one of these impact mechanisms was fitted to a specially constructed rig in the Bretby workshops. It was capable of vertical, longitudinal, and transverse motion and was tested on a ripping face in Cadley Hill colliery. The following year another was mounted on a drill boom and tested at Bulwell limestone quarry, Nottinghamshire, and also in the Middleton limestone mine in Derbyshire. By 1968 a prototype 'impact ripper'* made at Bretby was performing satisfactorily in the Lea Hall colliery.

It should be noted, however, that while the impact mechanisms of these prototype machines performed fairly well, the drill booms on which they were mounted did not. These booms were not designed to accept the continuous high-impact loadings to which they were subjected and under such loadings rapidly deteriorated.

Developmental work on the Bretby impact rippers ceased following commercial development of the machines in 1969 as the Board is only concerned with research and developmental work and not with general

---

*This name was coined at M.R.D.E. to cover the N.C.B. patented ripping machine using an impact tool.

Figure 531. The Belgian cutter loader showing the loading and conveying mechanism. The impact tool was mounted on top of this unit. (Courtesy Inichar, Belgium.)

Figure 532. Belgian cutter loader with extension loader mechanism. The loader mechanism can be fitted for right- or left-hand loading. (Unfortunately no photograph or drawing of the impact tool appeared to be available.) (Courtesy Inichar, Belgium.)

manufacturing. The patent for the Bretby impact ripper is held by the N.C.B. and licence has been granted to various manufacturers in Britain for manufacturing purposes.

Towards the end of 1969 and the beginning of 1970 the N.C.B. began purchasing hydraulically powered impact hammers made by various manufacturers both in Britain and abroad. Some of these machines were roof mounted (i.e. suspended from the tunnel crown rib supports) others were fitted to a boom and mounted on a walking base structure. Both the Mining Research and Development Establishment (M.R.D.E.) of the N.C.B. and Gullick Dobson produced floor-mounted units (Figures 533 and 534). Other machines tested by the N.C.B. at that time were produced by Ingersoll-Rand (the Hobgoblin), Krupp (Model HM 400), Macol-Lemand (roof mounted),

Eimco Ltd., Anderson Mavor Ltd., (in conjunction with the M.R.D.E.), and Comp-Air Construction and Mining Ltd. (Holman), making a total of 18 machines in all. At that date most of the machines (with the exception, perhaps, of one or two) performed satisfactorily and according to an article which appeared in the publication *World Construction* dated October 1974,[4] the N.C.B. were expecting to install a further 15 machines. By 1977 some 50 impact breakers were being used by the N.C.B. in their mines.

The Belgian Cutter Loader

Steps towards the development of a machine for such work were also taken by the Institut Nationale de l'Industrie Charbonnière de Belgique in 1955. The machine was designed

580

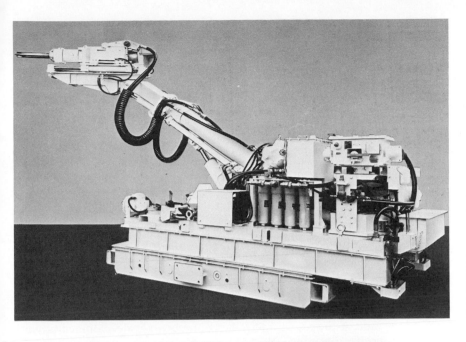

Figure 533. Gullick Dobson mounted impact breaker ripping machine. (Courtesy Gullick Dobson Ltd.)

Figure 534. Gullick Dobson mounted impact breaker with bucket loader attachment in use underground. (Courtesy Gullick Dobson Ltd.)

and built by the Institute (Inichar) with the assistance of various Belgian mining engineers and manufacturers. It was exhibited in Paris in 1955 and was basically a self-propelled cutter loader. The loader mechanism (Figures 531 and 532) comprised a transversely positioned frontal screw or auger revolving over a transverse scoop blade which curved up behind the screw to provide a trough for conveying the coal to the side. The screw was rotated by means of a lateral chain. This chain was fitted with picks, so that in addition to rotating the screw-type loader mechanism, it also trimmed the side wall. The transmission chain and shearing mechanism could be fitted to either side of the machine.

Where necessary the frontal loader mechanism could be lengthened by attaching an extension piece. It could also be adapted for right- or left-hand loading by changing the pitch of the screw. The unit, which was designed mainly for work in soft coals, was mounted on caterpillar tracks which were independently driven. This made the machine highly manoeuvrable, even on fairly irregular surfaces, and Inichar has reported that it has been operated fairly successfully on slight slopes. The coal was loaded onto a bottom belt conveyor which carried it to the rear.

Traction and loading power were provided by three compressed-air motors having a combined output of 12.7 kW (17 hp). The coal-cutting machine was about 1.82 m (6 ft) (long) × 1.21 m (4 ft) (wide) by 40.6 cm (16 in) (high) and weighed 1.50 t.

Of particular significance was a unique type of 'gun' or 'impact tool' which was mounted on top of the unit. The impact tool was fitted with a moil point and was powered by compressed air. It could be moved from side to side of the unit by means of a rack and pinion mechanism. The sledgehammer of the striking ram weighed 25 kg (55 lb) and it delivered three blows per second. According to Inichar the gun or impact tool performed satisfactorily in rock formations, but in soft coal the tip tended to thrust in and become jammed.

The impact unit was developed from the Dufrasne gun which was built at Winterslag in Belgium in 1946. The unit which was adapted for use on the loader was extensively modified and improved by the National Factory of Liège (Fabrique Nationale de Liège). The Dufrasne gun was similar in concept to the compressed-air-powered puncher machines or pneumatic hammers which were developed originally in America and later found their way into many British coal mines during the early 1880s.

American Developments

The 'Hobgoblin'

In 1966 Ingersoll-Rand began development work on a mounted impact breaker. This work was stimulated when the company became acquainted with Jack Ottestad (President of Impulse Products). Apparently Ottestad had at one time been involved with the aero-space industry where he had carried out various

experiments with gas-oil impulse devices which produced high acceleration rates for missile-testing purposes. According to John Adams (Manager of Ingersoll-Rand) when the company

> first came across his activities in 1966 Jack (Ottestad) was attempting to apply the gas-oil actuators to the construction industry. Heavy blunt weights were hurled relatively long strokes by gas-oil impulse devices to impact high energies into whatever they hit. After Ingersoll-Rand and Impulse began working together, field demonstrations indicated that impacting on the surface was inefficient and dangerous. The design was then changed to a short-stroke, relatively low-velocity piston striking a moil point and driving it into the work.

Ingersoll-Rand is a licensee under certain Impulse Products patents, and for years Impulse has done breaker developmental work under contract for Ingersoll-Rand.

While Ingersoll-Rand and Impulse Products were carrying out this developmental work towards the end of 1966 and the beginning of 1967, escalating costs forced the Anaconda Company to have a good inward look at its mining system with the idea of improving production, cutting costs, and producing a better product for the consumer. With this end in view it directed its Mining Research Department in Butte to carry out investigations which covered its entire range of mining activities such as exploration, development, stoping, transportation, and hoisting. Most particularly it was realized that a new and better method of rock-breaking was needed, in other words 'selective mining', so that the quality and the quantity per man-shift of the end product could be improved. After examining various rock-breaking methods currently available the company asked Ingersoll-Rand to supply it with a high-energy impact breaker which was to be mounted on a Gradall. Anaconda's decision to try this type of machine was largely influenced by Bailey and Dean's paper on some work carried out during the period 1955-57 by the U.S. Army in the Camp Tuto tunnel in Greenland where various methods of picking and prying solid ice were tried. This work was thoroughly discussed by Bailey and Dean in their paper which was submitted to the

Eighth Rock Mechanics Symposium in Minnesota during September 1966.

The first impact breaker produced by Ingersoll-Rand for the Anaconda Company was called the 'Hobgoblin', so named because it was introduced in October (Hallowe'en) 1969).

That same month another impact breaker made its appearance. It was the Krupp 400 unit, which was advertised in *Engineering News Record* by the Krupp Company of Germany. The Krupp machine was reputed to be able to produce about 542 J (400 ft-lb) of energy per blow while the Hobgoblin was capable of producing about 1360 J (1000 ft-lb) at approximately 600 blows per minute.

The first mounted impact breaker supplied by Ingersoll-Rand was tested in an old tunnel leading into the Anaconda Company's Berkeley pit stope and was installed at the back of the tunnel for cut and fill stoping work. The success of these trials induced the company to order another four machines which were supplied to carry out further tests underground. These tests were divided into several phases and each new machine, as it was progressively introduced to succeeding phases carried improvements and modifications suggested by the preceding test phase. For instance as a result of work carried out in phase II the machine used in phase III was designed as a 'stoper' and was capable of breaking and transporting waste and ore separately. In subsequent phases the machines carried improvements to the hydraulic system and a move was made towards giving the man in the cab complete push-button and lever control for all mining operations carried out by the machine.

Unfortunately, the Anaconda Company closed their trial operations before the five machines were fully tested and, since then, most of these units have been sold to other mining companies. Some Hobgoblins subsequently built by the Ingersoll-Rand Company have been mounted on Lee Norse mining machines and these were sold to the United States Steel and Consolidated Coal company for use as boomheaders, while other units have been delivered to the South African Chamber of Mines for selective mining of thin gold seams, and a few have been put to work in the Prague subway cutting out the full face of the tunnels.

Initially all of Ingersoll-Rand's machines were hydraulically powered, the reason for this being that the company felt the mining and construction industry were heavily pneumatically oriented at the time the first breakers were introduced and the introduction of all-hydraulic units gave the development of hydraulics a better chance. However, during the early 1970s the company produced both hydraulically powered and pneumatically powered breakers of 1400 and 700 J (1000 and 500 ft-lb) blow energy. Of interest is the fact that when an air-breaker is supplied for use with a backhoe and compressor unit, Ingersoll-Rand recommend that it should only be considered for jobs of short duration (i.e. two months or less a year). Apparently, this restriction does not apply to the hydraulic units.

Description

While there are, naturally, minor differences between the design of one manufacturer's unit and another the basic concept of the machine remains the same. Some machines such as the Krupp are fitted with interchangeable sealed accumulators, in others, i.e. the Shand, CompAir, Gullick Dobson, and the Ingersoll-Rand, the accumulators are built into the unit and need periodic recharging to compensate for the loss of nitrogen gas during operation.

The Montabert BRH 250 and 501 hydraulic rock breakers (Figure 535) operate in much the same manner as do their hydraulic drifter drills. Both are fitted with an external nitrogen-filled accumulator (positioned at the side in the case of the drifter and at the top or head in the case of the breaker). The oil flow through the various high- and low-pressure ports is controlled by a distributor or slide which is raised by the piston on its upward stroke and forced down by the oil pressure which enters the rear chamber when the piston uncovers a port on its downward journey.

So far as the Ingersoll-Rand unit (Figure 536) is concerned, the high-energy impacting action of the breaker is achieved by the compression

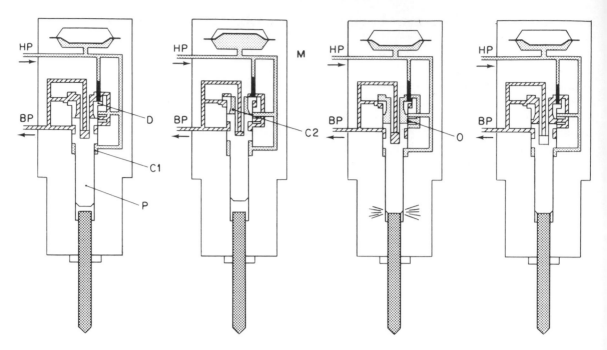

Figure 535. Operating diagram of Montabert BRH 250 and BRH 501.L hydraulic rock breakers. (HP) High-pressure oil line; (C1) high-pressure oil is fed into chamber C1; (P) piston; (D) distributor; (M) membrane in the accumulator; (O) orifice O is opened by P on its downward journey, thus allowing oil pressure to force D down; (C2) D cuts off supply of oil to chamber C2 thus opening the outlets to the low-pressure line; (BP) low-pressure line. (Courtesy ETS Montabert, S.A.)

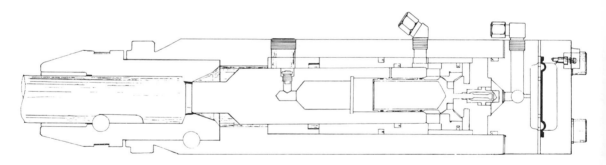

Figure 536. Sectional drawing of Ingersoll-Rand mounted impact breaker showing incapsulated nitrogen gas cylinder. (Courtesy Ingersoll-Rand.)

and expansion of nitrogen gas in a cylinder. The cylinder is encapsulated within the impact head of the breaker which operates as follows.

Hydraulic oil enters a chamber in the impact head above the cylinder and raises the head to the top of its stroke. At the same time the nitrogen gas within the cylinder is compressed by a piston which is forced down into the cylinder by the fluid. At a preset pressure a valve in the head opens and allows the

hydraulic fluid to escape, thus releasing the head which had been held at the top of its stroke by the oil while the pressure was being built up. The head is released and is immediately forced violently downwards by the driving force of the hydraulic oil plus the explosive energy of the nitrogen gas as it expands and thrusts the piston in its cylinder upwards. Using the combined kinetic energy of the expanding nitrogen gas and the pressure of the hydraulic fluid to propel it, the head is impacted forcibly against the shank end of the tool. The tool in turn strikes the rock or other material being attacked.

Attached to the top plate of the drill is a surge chamber holding air at a pressure of 0.1 MPa (15 lb/in²). At the bottom of the surge chamber is a flexible neoprene diaphragm which separates the chamber from the hydraulic fluid. The purpose of this chamber is to dampen or absorb the pressure impulses which would otherwise be transmitted through the return line to the hydraulic reservoir.

By increasing or decreasing the pressure in the nitrogen chamber and by controlling the rate of flow of the hydraulic fluid respectively, both the impact energy and the blow frequency of the breaker tool may be varied to suit the special requirements of the job in hand.

The Ingersoll-Rand tool can be changed on site by removing an external 'O' ring and pushing out a retainer pin. This operation allows the tool or bit to slip out.

Apart from the conventional 'moil point' tool, a variety of tools of various shapes is offered by manufacturers of these units. These include the chisel tool, the blunt tool, the tamping tool, the spade, and the sheet driver (the latter being for pile-driving).

Shovels, Backhoes, and Bucket Excavators[7-10]

William Otis designed and developed his first steam shovel during the mid-1830s. He patented the design in 1837. Before the advent of this revolutionary machine, strip mining in America and elsewhere was largely a manual operation. Picks, shovels, and wheelbarrows were used to dig and remove the overburden in order that the coal seam or ore might be exposed.

However, 'Necessity is the mother of invention', and this obviously laborious and time-consuming method of soil removal called for improvements.

In 1866 Kirkland, Blackeney, and Groves began stripping the overburden from their newly opened mine at Grape Creek, Danville, Illinois, by means of ploughs and scrapers. The overburden was drawn to one side and then removed in wheelbarrows and carts. Similar excavation methods were employed by Michael Kelly of Hungry Hollow, Illinois, some nine years later. According to Hollingsworth of Bucyrus-Erie, Kelly removed the overburden during the summer and mined the coal during the winter.

Figure 537. Original Otis steam shovel. (Courtesy Bucyrus-Erie Company.)

Figure 538. Otis steam shovel at work about 1870. (Courtesy Bucyrus-Erie Company.)

By this time the first Otis steam shovel (Figures 537 and 538) had been built and tested. The prototype unit was powered by a vertical single-cylinder steam-engine mounted on a vertical boiler. The boom was constructed of timber. It was centrally mounted on a cast-iron jib which allowed it to pivot through 90° in the vertical plane and swing 180° horizontally.

In 1877 it occurred to two gentlemen — J. N. Hodges and A. J. Armil — who had been using the horse, plough, and scraper method of stripping, that an Otis-type shovel might be used for this work. Accordingly a unit was procured and put to work on their Pittsburg property. Unfortunately, however, as the coal seam was at least 90 cm (3 ft) deep in most areas and the overburden some 240 to 370 cm (8 to 12 ft) thick, the short boom of the Otis shovel was unable to handle it and the machine was ultimately abandoned.

Nevertheless, a way had been shown whereby strip mining could be mechanized and gradually, over the succeeding years, these machines were further developed by various people or companies throughout the world (Figure 539).[7]

In 1911 at the instigation of two men — Rant Holmes and W. G. Hartshorn, the Marion Company built a rail-mounted, revolving, steam stripping shovel with a 20 m (65 ft) boom. It was designated Model 250 (Figure 540) and it carried a 2.7 cu. m (3½ cu. yd) dipper on a 12 m (40 ft) arm which was centrally pivoted inside the long lattice-type boom. This machine was the true forerunner of the modern stripping shovel and was recognised as being the largest of its type in the world at that time. To prevent the lower frame from becoming twisted — a problem common to excavators of that period — Marion fitted the 250 with hydraulic levelling jacks — one on each corner.[7]

By the 1930s the crane navvy or power shovel, the skimmer scoop and the back-acting trencher were in general use for soil-removal purposes.

586

Figure 539. First Bucyrus steam shovel was called a 'Thomson' after its designer. It was shipped in 1882. (Courtesy Bucyrus-Erie Company.)

Crane Navvy[10]

The crane navvy had a dipper or bucket fitted with steel cutting teeth or tines. This bucket was mounted on the end of a boom which, in turn, was pivoted on a jib which extended from the front end of the base of the main structure. The dipper or bucket was raised and lowered by an operating rope and its contents were discharged from a hanging door at the back of the bucket opened by another rope. The upper structure of the unit was mounted on a roller ring base which allowed it to revolve. Traction was provided by caterpillar or crawler tracks.*[9]

---

*The 'caterpillar' or crawler track system was developed simultaneously during the period 1904-9 by Benjamin Holt in California and David Roberts in England. Holt combined with Best (a competitor in this field) and the first caterpillar tracks were produced. In England Roberts and the Hornsby Company developed a steerable track system capable of running non-stop for a distance of 60 km (40 miles) and, as a result, Roberts was awarded £1000 as a prize by the War Department. Holt promptly purchased the patent rights from Roberts. Thus in 1915 when Rusten was asked by Churchill to manufacture the first tank, the company was forced to repurchase the manufacturing rights for the caterpillar track (see also Chapter 17, Military use of the tunnelling machine — Churchill's excavator).

In operation the bucket or dipper made an upward cut, working away from the machine, and for this reason it was used extensively for working against relatively steep faces. The depth of the cut made could be controlled by altering the angle of the bucket at the point where it made contact with the jib arm or boom.

Skimmer Scoop[10]

The skimmer scoop was a special-purpose machine basically similar in concept to the crane navvy. It was designed specifically for removing shallow overburden, or for other projects where a shallow horizontal cut was needed, such as road ripping.

The jib of the navvy or excavator was set parallel with the ground and the dipper was mounted beneath the navvy jib in such a way that it could run back and forth beneath it. It was hauled towards the machine by ropes during the filling or stripping operation and then discharged, or the navvy jib was raised and the dipper was run out to the end of the arm and emptied.

Figure 540. Marion shovel — Model 250. (Courtesy Marion Power Shovel Co. Inc.)

Figure 541. JCB 'Major loader' (1949). (Courtesy JCB Sales Ltd.)

Figure 542. JCB 'Load-over' (1953). (Courtesy JCB Sales Ltd.)

Figure 543. JCB 'Mark I back-actor' (1953). (Courtesy JCB Sales Ltd.)

Back-acting Trencher[10]

The back-acting trencher machines were similar to the skimmer scoops described above In that the navvy jib was also set parallel with the ground but, instead of the dipper running back and forth beneath the main jib, it was supported by an arm which was pivoted at the top end of the supporting navvy jib. During operation the bucket was drawn towards the trench-digging machine in much the same way as the modern backhoe machine operates.

Hydraulic Shovels and Excavators[8]

The modern backhoe, excavator or digger-type machines which have recently been used in conjunction with shields for tunnelling work (described in the 'Hybrid' section of this book) evolved from the early back-acting trencher-type machines operated by rope and chain. Later with the advent of hydraulics came the sophisticated modern hydraulically operated backhoes and front-end loader excavators of today. Important forerunners of this type of machine were the JCB (J. C. Bamford) 'Major loader' (1949) (Figure 541), the JCB 'load-over

(1953), the JCB (Figure 542) 'Mark I 'back-actor' (1953) (Figure 543) and the JCB 'Hydra-digger' (1954) (Figure 544). According to J. C. Bamford the Major loader is reputed to be 'the first hydraulic loader in Europe'. The hydraulically powered Hydra-digger featured a glass fibre cab and a boom and digger combination. Apparently the glass fibre cabin was 'difficult to produce at first'.

Today backhoes, shovels, and excavator machines come in a wide range of sizes, the larger units being mainly designed for surface excavation work while many of the smaller excavators or shovel machines have been put to work underground either with or without a protective shield (Figure 545).

A powerful excavator (Model RH300) (Figure 546) produced by O & K Orenstein & Koppel AG of West Germany and exhibited at the A.M.C. International Mining Show, Las Vegas, has a 23 cu. m (30 cu. yd) shovel. A 31 cu. m (40 cu. yd) shovel is available for loading coal or other lightweight material, while a heavy-duty 18 cu. m (23 cu. yd) shovel could be fitted, if required, for heavier work.

The unit's heavy-duty hydraulic cylinders enable the RH300 to exert some 200 MPa

589

(30,000 lb/in²) of thrust at the shovel teeth. The shovel can penetrate into the face up to a height of 12 m (39 ft) while maximum reach is 16.2 m (53 ft). Backhoe attachments include 16 and 12 cu. m (21 and 16 cu. yd) buckets which are capable of digging to depths of 12.8 m (43 ft) and 15.8 m (52 ft) respectively. The unit is crawler mounted and can travel at a maximum speed of 2.75 km/h (1.7 mph). (A pontoon-mounted version equipped with a backhoe can reach depths of some 25 m (82 ft) below water level.)

Of interest is O & K's claim that assembly time has been reduced to seven days, whereas

Figure 546. The world's largest hydraulic mining shovel at AMC International Mining Show, Las Vegas, 1978. This new machine, the RH 300 has been developed by O & K Orenstein & Koppel AG, West Germany. The gigantic mining shovel weighs some 420 metric tonnes (930,000 lb) and has a standard shovel capacity of 22 cu.m (30 cu. yd). For loading coal or other lightweight material a 30 cu.m (40 cu. yd) shovel is available. (Courtesy O & K Orenstein & Koppel AG.)

rope shovel machines of comparable bucket size take some three to four months to assemble on site. This time-saving feature has been made possible by the special design of individual modules which do not exceed 3.6 m (12 ft) in width or 3 m (10 ft) in height, thus facilitating transport to, and assembly at, the mine site.

While the RH300 was designed primarily for surface excavation work, these latter features suggest a possible additional use underground during the excavation of, say, large storage caverns.

Correspondence

Ingersoll-Rand Company Ltd. — J. W. Adams, Phillipsburg, U.S.A.
Ingersoll-Rand (Aus.) Ltd. — J. H. Whitehead, S. Melbourne, Australia.
Voest-Alpine A G, Vienna, Austria.
Voest-Alpine (Aus.) Pty. Ltd. — McGuinn, N. Sydney, Australia.

Institut Nationale de l'Industrie Charbonnière de Belgique (INICHAR), Brussels, Belgium
Paurat GmbH, Voerde, West Germany.
Demag AK. (now Mannesman Demag Bergwerktechnik) Duisburg, West Germany.
H. B. Zachry Co. — B. Cloud, San Antonio, U.S.A.
Eickhoff Maschinenfabrik u. Eisengiesserei GmbH, Bochum, West Germany.
AEC Inc. — J. Kogelmann, State College, U.S.A.
Bucyrus-Erie Company, South Milwaukee, U.S.A.
Ruston-Bucyrus Limited, Lincoln, U.K.
Marion Power Shovel Division of Dresser Pty. Ltd. Ind. U.S.A.
Dosco Overseas Engineering Ltd. — B. Reid, Tuxford, U.K.
Westfalia Lünen, Lünen, West Germany.

References

1. Dr. Z. Ajitay, 20 Jahre-Gewinnungs-Lademaschine, Bauart "F" Mitteilungen Des Ungarischen Forschungsinstitutes Fur Bergbau 1970, Voest-Alpine, Austria.

2. Martin Hunt (Editor), A Titan at Gateshead, *Tunnels and Tunnelling,* **9,** No. 2, 35-36, March/April 1977.

3. Andre Borie, *Marche Pour La Fourniture. De Quatre Machines Eickhoff Type EVR 200* (Dénommées dans le texte Les Machines), Société des Entreprises de Travaux Publics, Paris, France.

4. A. H. Morris and I.G. Rodford, Impact ripping underground, *World Construction,* Oct. 1974, 59-62, New York.

5. Wm. Ross Wayment and Innes P. Grantmyre, Development of a high blow energy hydraulic impactor, *Proc. R.E.T.C.,* 1976, Las Vegas, A.I.M.E., New York, pp. 611-625.

6. P. J. G. Du Toit, Mechanical Rock Pick, *Mining Congress Journal,* April 1973, 60-63, Washington, D.C.

7. John A. Hollingsworth, Jr. (Bucyrus-Erie Company), *History of Development of Strip Mining Machines,* South Milwaukee, Wisconsin, U.S.A.

8. Profile of JCB, *Contract Journal,* Centenary Supplement, 12 April 1973, JCB Sales Ltd., Rochester, U.K.

9. Ruston-Bucyrus Cranes & Excavators. Lincoln, England, 1874-1974, *Construction News,* London.

10. *Encyclopaedia Britannica,* 1929, Vol. 8, pp. 943-945.

11. Gregory B. Sparks, Application of hard rock continuous miners to cut-and-fill slot stoping, *Mining Congress Journal,* May, 1980, **66,** No. 5, 29-33, Washington, D.C.

12. Hermann Boldt, *Use of drivage techniques and proving of the take to ensure best mining conditions in the West German coalmining industry,* (translation) Glückauf. Annual Set 114 (1978), No. 3, 55/60.

TWENTY-THREE

Hydraulic Mining

The Hydrominer[1,2]

High-pressure water excavation has been in use for many hundreds of years. It was used fairly extensively from the mid-1850s onwards by miners in California and Alaska to sluice out gold-bearing deposits. However, apart from an isolated hydraulic operation utilizing a fairly low-pressure head at a distance of about 6 m (20 ft) for washing down gold-bearing material, this method of mining is now no longer used there.

In 1915 the Russians began using hydraulic mining for extracting coal and this mining method was further developed by D. Muchnik during the 1930s. After the successful development of the Soviet Union's first water jet mine in 1939 other larger mines were established by the Russians, making the present total about 11 mines. Other countries such as Japan, Germany, Poland, Czeckoslovakia and mainland China with similar coal formations amenable to this type of mining, soon followed suit. In 1970, the Kaiser mine in Sparwood, British Columbia, also tried the method in their Balmer coal seam. Some 3100 t of coal in a 6 hour shift were reputed to have been extracted by a three-man crew at the Kaiser mine.

The water-jet extraction method is particularly advantageous in coal-mining because both the dust and the spark problem are eliminated. It is most suited to thick, sharply dipping coal seams where the water and coal mixture flow naturally away from the face or working areas; so that the stability of the roof and floor shales is not threatened, as could occur if the seams lay in horizontal planes.

Generally speaking, water jets flowing at the rate of some 5700 litres/min (1500 gal/min) at 14 MPa (2000 lb/in^2) pressure are directed at the coal seam through a monitor nozzle which may be from 12 to 21 m (40 to 70 ft) away from the face, while the operator is stationed even further back in a control cab.

The cut coal and water mixture is then channelled via a feeder breaker into a dewatering plant for coal separation and water-clarification and recycling.

In view of the current interest in hydraulic mining the Rock Mechanics and Explosives Research Centre, University of Missouri-Rolla, Missouri, U.S.A., undertook some studies on the subject, and a paper dealing with this work was presented by associate professors David A. Summers and Clark R. Barker.

Because the longwall mining machine (which operates along a face up to 180 m (200 yd) long) can produce over 12,000 t of coal a day, compared with about 700 to 1000 t produced by a continuous miner in a board and pillar operation, the Research Center chose a longwall unit for their experiments.

Both the Meco-Moore cutter loader and the Huwood slicer loader machines were examined as potential basic hydraulic units. The final machine which evolved was called a 'Hydrominer' and it incorporated features from both the above-mentioned units.

As described more fully in the section on coal-mining in Britain, the Huwood slicer loader cuts a web of coal from the back of the face which is then pushed sideways on to the conveyor by the wedge head of the plough blade. The two cutting jibs on the Meco-Moore cutter loader cut two horizontal slots or grooves at the bottom and middle (or top) of the seam. This is followed by a vertical shearing jib which makes a third cut at the back of the seam.

The Hydrominer featured five sets of high-pressure oscillating water jets in place of the cutting teeth and picks on the Huwood and Meco-Moore machines. Three oscillating jets of water moving in a vertical plane cut a slot at the back of the seam in much the same manner as did the Huwood slicer's oscillating teeth on the leading edge of its plough blade, while two jets moving in a horizontal plane cut the top and bottom of the seam respectively, much like the cutting chains and picks on the Meco-Moore's twin cutting jibs. The coal was then broken and loaded on to the conveyor by a plough blade similar to that utilized by the Huwood slicer loader. A 61 cm (2 ft) and a 91 cm (3 ft) web of coal was thus taken by the Hydrominer's five cutting jets at cutting speeds of 3.05 m/min (10 ft/min) and 1.52 m/min (5 ft/min) respectively.

When the head was initially designed, prior experiments (carried out in the laboratory under an earlier Bureau of Mines contract) indicated that a pressure of 70 MPa (10,000 lb/in²) with an estimated flow rate of 230 litres/min) (50 gal/min) would probably give better results than if the higher pressures and lower flow rates advocated elsewhere were used. However, in a letter to the author, Dr. Summers advised that during the surface trials the top two cutting arms (i.e. the top vertical and the top horizontal) were not used and the machine was therefore operated at only 115 litres/min (25 gal/min).

In operation the head slides along the bottom cut, using it as a guide. The back cut facilitated the breaking and removal of the coal web by the vertical wedge, and the top cut was designed to provide a smooth roof line and also remove any coal left behind by the wedge. The head was powered by a hydraulic haulage unit.

It was anticipated that if (as originally planned) the various jet arms were oscillated by means of hydraulic cylinders, small pieces of coal bouncing back from the cutting surface would damage the drive pistons. To avoid this the arms were eccentrically connected to flywheels driven by a chain and sprocket. The chain and sprocket were in turn powered by a hydraulic drive system supplied by an externally located pump unit.

However, as there was no room at the bottom for a flywheel, the lowest vertical arm was inflexibly joined to the upper vertical arm.

During operation the middle arm oscillated '180 deg out of phase'[2] with the top and bottom vertical arms. According to Summers et al,[2] this action appeared to be more effective in cutting the coal.

In their paper, 'Experimentation in hydraulic coal mining',[2] Summers et al concluded by commenting that:

1. The Hydrominer mines coal with effectively no dust.
2. The size of product is on average larger than that from a shearer with substantially less fine coal produced. By choice of jet angle in the nozzles this size could be controlled to minimize large, i.e. plus 6 in. coal.
3. The haulage forces required to move the unit are lower than those required to pull a shearer down the face under equivalent conditions. Concurrently, there is much less vibration of the cutting unit as it advances since the actual coal cutting occurs ahead of the Hydrominer.
4. The Hydrominer is able to cut webs of increasing depth, in the tests from 1.5 to 3 ft, with little increase in unit horsepower. The head, although designed for this operation to mine only 32 in of coal, was able to successfully cut coal to the maximum seam height of 54 in.
5. The cutting head was able to penetrate, without loss in speed and no wear on the machine, pyrite layers 2 in thick.
6. Coal coming from the Hydrominer was infused with water so that no dust will be generated by later degradation of the coal as it is transported out of the mine.

In a letter to the author dated October 1978 Professor Summers commented that since their research began they have received information that three German companies (Westfalia, GHH, and Eickhoff) have all undertaken hydraulic mining machine research.

References

1. Assoc. Professors David A. Summers and Clark R. Barker, The development of a water jet mining machine for coal extraction, Paper presented at R.E.T.C. Las Vegas, Nevada, June 1976,

Canadian Inst. of Mining and Metallurgy, Eng. Inst. of Canada.

2. David A. Summers, Clark R. Barker and Marian Muzurkiewicz, Experimentation in hydraulic coal mining, A.I.M.E. Annual Meeting, Atlanta, Georgia, New York, March 1977.

Correspondence

University of Missouri-Rolla — Prof. D. A. Summers, Missouri, U.S.A.

TWENTY-FOUR

Hydraulic Conveyor Systems

The effectiveness of a tunnelling machine, no matter how perfectly it functions, is largely governed by the efficiency of the muck-removal system. As has been noted on numerous occasions throughout this book, many tunnelling machines performed well, but progress was hindered by the inability of the muck-removal system to cope with the large amounts of cuttings produced by the unit, the machine, perforce, having to mark time while this was being disposed of.

In an effort to find a faster, safer, and more reliable continuous haulage system, which was flexible enough to keep pace with the excavator, advancing and receding with it, the Consolidation Coal Company, through the Mining Research Division of Continental Oil (its parent company), commenced a research and development programme in 1969 to establish the feasibility of installing a 'coarse material hydraulic transportation system'.

Over a period of some seven years numerous experiments were undertaken with different types and sizes of pipe and coal, and varying pressures and water levels to determine the best method of transporting the material from the face to the preparation plant in one system.

Finally two 25.4 cm (10 in) diameter flexible, extensible rubber hoses mounted on wheels and capable of advancing a maximum distance of 300 m (1000 ft), were selected to carry the coal from the various faces in the mine to a central collecting point where it was channelled into rigid steel pipelines which took it via the main haulage ways to the surface and thence overland to the preparation plant.

The coal broken from the face by the continuous miners, varied in size, with some lumps as large as 30 cm (12 in) or more.

However, the limit size for the transport system was 10 cm (4 in) and this problem was overcome by installing an injection hopper in which was mounted a submersible roll crusher. The crusher accepted the larger pieces of coal and crushed it into 10 cm (4 in) pieces or less so that it could be mixed with water and pumped to the surface in the form of a coarse slurry.

To acquire the requisite flow rates with the volumes available in the hopper, the particular size of the injection hopper chamber and the water level controls were important considerations, necessitating careful design. Other factors such as the size of the equipment, i.e. centrifugal pumps, crusher, hopper, etc. were also significant because these needed to be handled and transported in the confined spaces of an underground mine. However, tests proved that these specifications were adequately met.

The system was, in effect, an automated closed-circuit operation (including start-up and shut-down) with the coal and water moving to the surface as a slurry and the filtered water returning to the face afterwards for recirculation.

After all major research and development had been completed and the system proved satisfactory, the Consolidation Coal Company began work on a full-scale commercial installation at the Loveridge mine which was expected to be in operation by May 1978.

As pointed out by Dahl and Petry in their article[1] on the above project, hydraulic transportation of coarse material offers many advantages which may be enjoyed by operators of civil engineering equipment as well as by miners. These include greater safety, less moving equipment and better roof control because it opens the way for the development of narrower mining machinery. It would be particularly useful in conjunction with water-jet-assisted tunnelling machines and equipment and hydraulic mining machines.

References

1. H. Douglas Dahl and Eston F. Petry, Update on slurry transportation from face to cleaning plant, *Mining Congress Journal,* Dec. 1977, Washington, D.C.

Correspondence

Consolidation Coal Co. — Messrs. Dahl & Petry, Norfolk, U.S.A.

Hybrid Machines

TWENTY-FIVE

Boom-type Hybrid Machines

A 'hybrid' machine is defined as one which combines two or more different types of machine in a single unit, i.e. a shield with a boomheader unit or a rock machine combined with a soft ground shield, etc.

Thomas Thomson 'Ladder Excavator'

The first practical application of this concept was made during the late 1890s when a 'Thomson ladder excavator' (patented in London by T. Thomson on 12 January 1897 under No. 865) was used in conjunction with an ordinary Greathead-type shield, although as a completely independent unit. This machine is of particular interest in the evolutionary chain because it presages the modern boom-type continuous miner of the early 1950s.

Mainly because of the frequent failure of the electric motor and ancillary equipment, Thomson's machine was not considered particularly successful when tried in 1897 on the Central London Railway. However, Copperthwaite[1] commented that if further developmental work on the machine had been carried out at the time it might, perhaps, have fared better.

The excavator (Figure 547) consisted of a dredger ladder[c] mounted on a swinging upper frame[a] which revolved on a bottom frame, and in this connection the reader's attention is drawn to the similarity between the muck-loading system of Thomson's unit and that of the Westfalia Lünen boom cutter loader (Figures 524-529).

The bottom frame of Thomson's ladder excavator (in effect a carriage on wheels) was designed to run on a track, laid as wide apart as was practicable in the tunnel, in order that as much room as possible could be provided between the rails for the movement of workmen and waggons.

The machine was advanced and retracted from the face by means of chains [d] bolted to the tunnel iron-work and led through the top of the bottom frame and then over a double drum or barrel.

In order to enable the cutting jib or dredger ladder to slew from side to side across the working face, the top frame revolved on the bottom frame. The dredger ladder was also supported by a projecting jib [e] so that it could be raised or lowered as desired.

In operation the machine was advanced to the face and the cutting jib lowered to the floor of the tunnel. After sumping into the face for a certain distance the cutter jib was slowly raised until the dredger ladder had reached the top of its stroke. As soon as one complete cut was made, the top frame was swung to one side and the arm lowered to the floor for the commencement of another cut.

When the excavator had advanced about 50 cm (20 in) into the clay face (or enough space was made for the erection of a ring of cast-iron tunnel lining) the unit was run back approximately 3 m (10 ft) in order to enable the workers to erect the ring. After this the excavator would again be advanced to the face and the entire cycle recommenced.

The dredger ladder consisted of a series of buckets [f] each of which carried four picks arranged along its forward edge. These buckets were joined together by the separating links of the bucket chain, one on either side.

Regarding this machine Copperthwaite[1] remarks that

the lifting cable of the arm carrying the buckets, which passes over the pulley between the

601

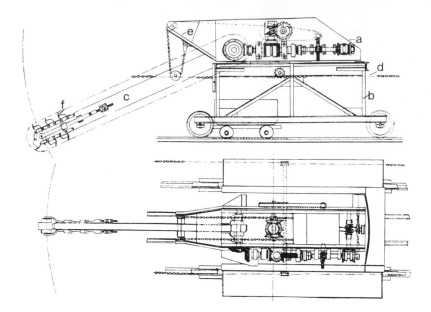

Figure 547. T. Thomson 'Ladder excavator', patented on 12 Jan. 1897 under No. 865 in London

cantilevers is obviously attached much too near the fulcrum of the arm to lift this latter save at an expenditure of power compared with the actual pressure required to cut the clay of some three to one; and this cannot be improved without limiting the vertical movement of the arm.

Copperthwaite speaks of a mechanical disadvantage of 'some three to one'. Yet it is interesting to realize that many modern machines of the roadheader type, i.e. Dosco's Mark 2A or Anderson Mavor's RH1 etc. operate efficiently at much higher disadantage ratios — an indication of the change in thinking brought about by the advent of hydraulics.

DeMAG Aktiengesellschaft (now Mannesmann Demag Bergwerktechnik)

Walking Prop Support Unit — VS1

The first of the modern hybrid machines was the walking prop support unit introduced by Demag in 1962 and designated 'Face heading machine Model VS1 "Unicorn"'. (Figures 548 and 549). Basically it consisted of a boom-heading component and walking supports. The Unicorn was designed specifically for driving gate-end roads* in mining operations and for rise headings.** It served both to loosen and load the material while the operating crew within the vicinity of the machine were protected by the walking and support system. This system consisted of seven hydraulic props and cantilever roof supports which faced the direction of travel of the machine.

The walking supports, which were capable of a total forward thrust of 120 kN (27,000 lb), could be steered so that any desired heading angle could be followed.

During operation the rotating cutting head was first fed about 50 cm (20 in) into the coal by the (clamped) walking support. The centre props mounted on the machine were set, allowing the walking support props to be moved up to the machine and reset. The face was then worked by the cutting head, the loose coal being transferred to the side conveyor by a transverse conveyor. The side conveyor transported the coal to the gate-end conveyor. When the face was worked off the machine was repositioned in the sequence described above.

*Gate-end roads — roads used to provide access to the main and tail-gate sections of a longwall installation.
**Headings or stableholes — access roads driven slightly ahead of the working face.

Figure 548. Demag face heading machine VS1 — prototype model. (Courtesy Mannesmann Demag Bergwerktechnik.)

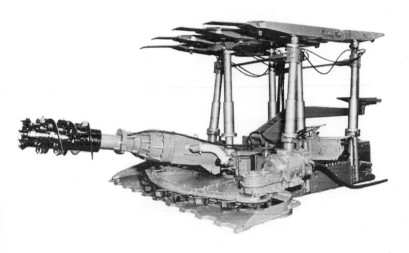

Figure 549. Demag face heading machine VS1 — second-generation unit. (Courtesy Mannesmann Demag Bergwerktechnik.)

Face Heading Machine — VS2E

The VS2E unit, (Figures 550 and 551) a more sophisticated version of the original VS1 was produced by Demag in 1967/68. Designed for handling larger cross-sections the VS2E combination walking support and boomheader machine was capable of cutting rock with a compressive strength up to 98 MPa (14,200 lb/in²). Several versions of the VS2E were built, some suitable for rectangular cross-sectioned tunnels and others for tunnels with arched profiles.

The pivot joint of the cutting jib (which could be slewed in all directions) was mounted on the machine frame, which in turn was connected with the laterally arranged walking supports. The cutting head was similar to the Demag VS3 model head described in the section entitled, 'Boom-type machines — Demag face heading machine VS3'.

A recessed scraper chain conveyor on a front-loading ramp collected the material and carried it along the side of the machine to the rear. The machine was moved by the walking supports which also assisted in anchoring the unit firmly between the floor and the roof during the cutting operation. The four cantilever roof support sections (which were mounted on the walking supports and also on four props on the central machine frame) were hinged to cater for undulations in the floor and roof of the tunnel. Each roof support section could be individually extended as a forepole, varying distances up to 1000 mm, (39 in) to provide additional protection at the face.

The unit was advanced by releasing the props on the machine frame while the walking support jacks were still extended. The main frame and cutting jib were advanced until the rotating cutting head was biting into the face.

Figure 550. Demag walking prop support face heading machine Model VS2E for rectangular tunnel cross-sections. (Courtesy Mannesmann Demag Bergwerktechnik.)

Figure 551. Demag face heading machine Model VS2E for arched tunnel cross-sections. (Courtesy Mannesmann Demag Bergwerktechnik.)

The machine props were then extended. With their props retracted the two walking supports were simultaneously drawn up to the machine and their props again extended, anchoring the unit firmly in the tunnel. After the cutting cycle had been completed the machine was repositioned close to the face by repeating the above sequence.

Although the walking prop machines filled an urgent need at the time they were introduced, these units have to a large extent been superseded by the crawler-mounted VS3 units with their lining erection system and by the T.S.S.M. machines. According to Demag the VS2E units are now only made for very special applications.

Boomheader and Shield Machines

A hybrid range of machines which combined a boomheading unit with a shield was introduced by Demag from 1967 onwards. These fell into the T.S.S.M. series and, depending upon the particular model selected, incorporated various types of 'shield feed'. Model T.S.S.M. (Figure 552) used thrust jacks which reacted against newly erected tunnel lining. If no lining was needed, but tunnel conditions warranted support at the face, the shield (Model T.S.S.M.-S) came equipped with wall grippers or 'clamping shields'. Model T.S.S.M.-R was designed for pipe-jacking operations and the shield was moved forward by the jacks in the thrust pit or by interjacking stations (see also section on pipe jacking).

Rigid or telescopic cutting booms are available (Figures 553 and 554). The telescopic boom is useful in varied ground conditions. If the ground is unstable the telescopic jib can be retracted into the shield, but if the ground is competent it can be extended a distance of 1 to 1½ m (3 to 5 ft) beyond the cutting edge of the shield. A choice of circular, arched, or rectangular shield is offered by Demag.

Digger-Type Shields[2-4]

Another important hybrid development is the digger-type shield (Figures 555 to 559). These machines are somewhat similar to the Demag boomheader and shield units. They consist of a shield, within the protection of which an

Figure 552. Demag 2.5 m-4.5 m (8 ft 3 in-14 ft 9 in) diameter shield, Model TSSM (hybrid machine). (Courtesy Mannesmann Demag Bergwerktechnik.)

Figure 553. Komatsu Ltd. mechanical shield Model TM507S (hybrid). (Courtesy Komatsu, Ltd.)

Figure 554. Priestley/Anderson Strathclyde semi-mechanized shield prior to leaving the works. This was the first of the boom-shield type of unit to be manufactured in Britain. (Courtesy Robert L. Priestley Ltd.)

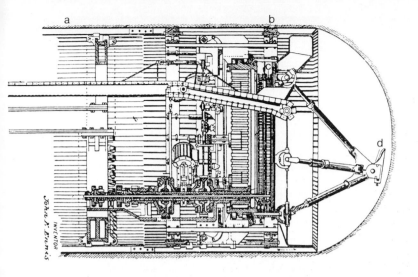

Figure 555. J. E. Ennis ripper scraper. U.S. Pat. No. 690,137 dated 31 Dec. 1901 (application filed 6 April 1897.)

articulated digging arm, hoe, or bucket, etc. operates (Figures 560-564). This type of unit is particularly effective in soft or caving ground in which harder inclusions are encountered.

J. E. Ennis Digger Shield

The idea for a digger shield (Figure 555) was first mooted by John E. Ennis who was granted U.S. Patent No. 690,137 for the design on 31 December 1901 (U.S.). Its modern counterpart may be recognized in the Robbins ripper scraper machine (Figures 556 and 557). However, while the shield section of the modern digger unit is constructed along conventional lines, i.e. in one piece, the Ennis unit consisted of a supporting frame composed of approximately 120 longitudinally placed I-beams or rails, arranged in a circle, their external diameter corresponding with the diameter of the unlined tunnel. (This concept was tested in practice by Hochtief AG of Essen, Germany, with the horseshoe-shaped Westfalia blade shield) (Figures 561 and 562).

The rails [a] of the Ennis machine, which could be advanced independently as required, were supported by a main frame [b] situated within the shield. Also attached to the main frame was a supplementary frame which in turn supported the digging and gearing mechanism ancillary thereto.

The gear devices in the structure were so arranged that at each complete revolution of the master cog-wheel [c], 12 of the independently movable rails were advanced, one at a time. (Thus 10 revolutions of the master cog-wheel advanced the entire set of 120 rails a predetermined distance.)

In operation, after all the rails had been satisfactorily advanced by the master cog-wheel, additional revolutions were imparted to this same wheel which then carried the digger operating mechanism and digger-plough [d] forward to the face. The digger-operating gear mechanism was so arranged as to be able to impart either a spiral movement to the shovel-carrying frame or a direct thrust. Alternatively, the shovel or digger could be made to operate in a reverse direction, i.e. following a spiral path from the centre to the circumference of the cut and then back from the circumference of the cut to the centre. According to Ennis this would enable the cutting-shovel or digger arm to clear away material from around the sides of a boulder and then, by utilizing the outward spiral movement, pull the boulder away from the face.

The digger arm was manoeuvred by differentially adjustable screws which in turn produced a telescopic motion to the tripod legs. The reader may care to compare this with the Robbins machine, Model 184S-169 (Figures 556 and 557) in which three hydraulic cylinders perform a similar function (the Robbins machine was used in 1975 to excavate the branch route of the Washington Metro).

607

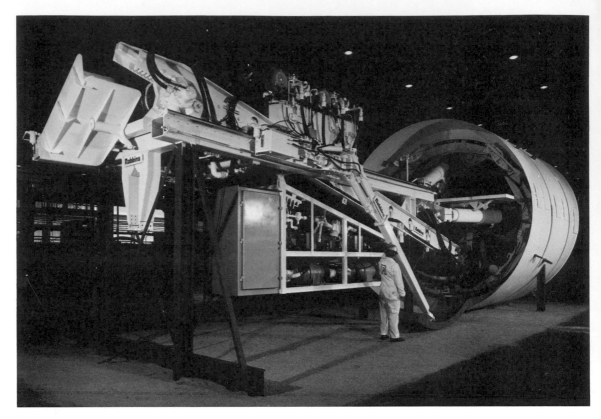

Figure 556. Rear view of Robbins ripper scraper Model 184S-169/170 showing the hydraulic rams and conveyor system. (Courtesy The Robbins Company.)

Figure 557. Front view of Robbins ripper scraper Model 184S-169/170. (Courtesy The Robbins Company.)

Figure 558. Robbins ripper scraper — Model 143S-141. (Courtesy The Robbins Company.)

Figure 559. Zokor digger shield with rotating bucket excavator arm ready for shipment to Teheran, Iran. (Courtesy Zokor International Ltd. U.K.)

Figure 560. One of two Markham shields being constructed for use in the second Dartford tunnel under the river Thames. These 11 m (36 ft) diameter shields weighed 355 t each and one unit was equipped with four boom-mounted impact breaker units for mechanical excavation. The drive under the Thames proved to be extremely hazardous as many problems were encountered due to water, necessitating at times the use of compressed air up to 0.21 MPa (30 lb/in²) despite previous ground treatment. (Courtesy Markham & Company.)

The muck was disposed of by two buckets situated opposite each other and behind the digger-arm mechanism. During operation the buckets lifted the spoil and carried it until they had reached their highest point, when a trip arm device dropped the bottom of the bucket, thus allowing its contents to be spilled into a chute. The chute then guided the material on to an endless chain of buckets which carried it to the rear of the shield.

Robbins Ripper Scraper and MEMCO Mining Machine

Amongst the first of the modern digger-type shields used were the MEMCO mining machine* and the Robbins ripper scraper unit (Model 221S-132) (1970) (Figure 558) which made their debut in the difficult ground

---

*See Chapter 11, Mechanized shields — California State Water Project — Castaic tunnels.

609

Figure 561. Horseshoe-shaped Westfalia blade shield similar to that used in the Frankfurt/Main single-tract rapid transit system tunnels by Hochtief AG of Essen, Germany. Two Westfalia 'Fuchs' (fox) boom cutter loaders are installed on the lower platform. (Courtesy Westfalia Lünen.)

conditons of the Castaic and San Fernando tunnels, respectively.

Iñ the case of the Robbins 'ripper scraper' the articulated digging arm formed an integral part of the unit itself, which was also equipped with powerful breasting doors. Caving of the face was thus effectively controlled by the hydraulically operated doors which were able to withstand considerable pressures at the crown and upper sections of the face. The shield was advanced by hydraulic jacks capable of exerting thrusts of approximately 1,500 kg/lin. cm. (8500 lb/lin. in.) of cutting edge. The articulated digging boom could be fitted with a variety of tools, depending upon the demands of the ground conditions in the tunnel. Thus not only could harder inclusions such as boulders and floaters be dealt with individually, but difficult or awkward obstructions extending beyond the gauge limits of the shield could also be handled effectively.

Longitudinally positioned guide rails allowed the MEMCO unit axial movement within the shield. However, freely moving wheel-type or crawler excavators such as boomheader units or mounted hydraulic impact ripper machines are now frequently used in conjunction with shields. The advantage of this latter method is that the boom unit may be advanced to the face or retracted within the shield, according to the demands of the strata at the heading. The disadvantage is that there is inadequate face control in difficult non-cohesive ground and over-excavation may occur when these conditions are encountered.

Zokor Digger-type Shields

A recently formed company operating as the Zokor Corporation in the United States and as Zokor International Ltd. in the U.K. and

Figure 562. Front view of the Westfalia blade shield showing the chain flight conveyor on one of the boom cutter loaders. (Courtesy Westfalia Lünen.)

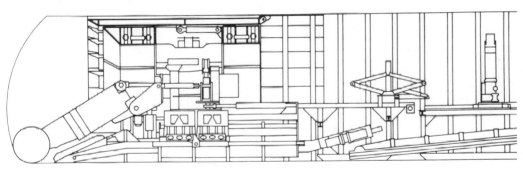

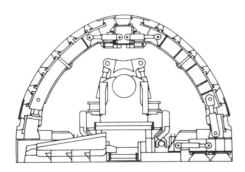

Figure 563. Photograph and
sectional drawings of a
horseshoe-shaped Westfalia blade
shield equipped with a Westfalia
'Bison' WAV 200 selective
cutting machine. (Courtesy
Westfalia Lünen.)

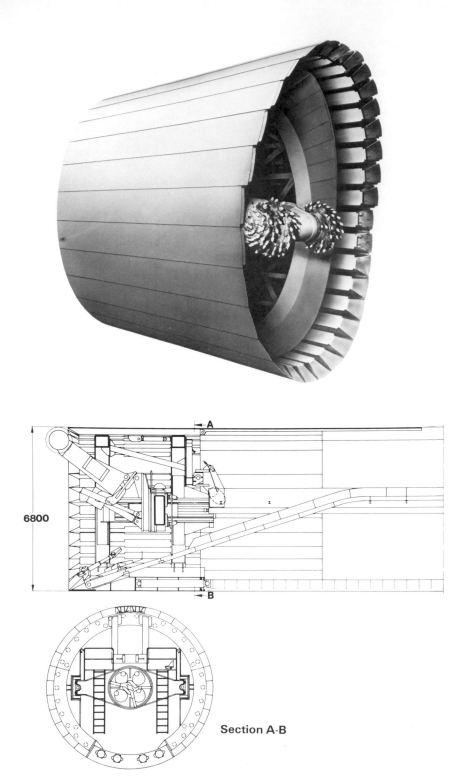

6800

Section A-B

Figure 564. Photograph and sectional drawings of a circular Westfalia blade shield equipped with a Westfalia selective cutting machine. (Courtesy Westfalia Lünen.)

elsewhere, has now produced a digger shield which, in effect, has evolved from both the Memco mining machine with its 'Big John' excavator bucket and the Robbins ripper scraper with its flower-petal-like breasting doors. Like the Memco unit, the Zokor machine (Figure 559) also incorporates hydraulically powered forepoling plates which provide support from the crown to the springline of the tunnel. These forepoling plates, which have a stroke of 1.5 m (59 in), form an integral part of the shield itself. Seven breasting plates provide face protection when necessary.

Two basic types of excavator bucket arm are available. The first is an extensible swing-type excavator arm which enables the operator to cut further ahead and also outside the periphery of the tunnel. When used in its retracted position the extensible boom provides greater breakout force over the tunnel face area. The manufacturers recommend this type of excavator arm for a wide range of ground conditions from soft to hard clay, including sand with boulder intrusions. The second type of excavator bucket arm (Figure 559), shown fitted to a unit which was shipped to Teheran, Iran, is recommended where harder types of ground are expected. Instead of the swing-hinge type action of the former, the articulated boom of this excavator arm rotates on a turntable, thus enabling the bucket to exert maximum breakout force in all its digging directions.

The rotating boom excavator arm does not extend telescopically as the manufacturers feel 'this would reduce the maximum breakout force that could otherwise be obtained'.

Bade & Theelen Breasting Shield[5,6]

To meet the demands of President Sadat for the swift excavation and construction of the Ahmed Hamdi road tunnel, beneath the Suez Canal at El Shallufa, Bade & Theelen GmbH, together with Tarmac Overseas Ltd. and Sir William Halcrow & Partners, designed a new type of breasting shield with unusual segment erection and support equipment.

Soft ground tunnelling is normally carried out in two stages. In the first operation the entire tunnel is excavated and lined. When this work has been completed the second stage of construction work commences. In the case of rail or road tunnels, this secondary phase of the operation consists in laying the traffic deck and frequently proves to be as time consuming as the initial excavation and lining work has been.

In an attempt to cut this lengthy procedure, a 120 m (393 ft) long trailer (figure 565) incorporating a high deck was constructed which was pulled behind the shield. The trailer allowed for the simultaneous excavation of the tunnel, erection of lining segments, and the laying of the road deck platform. The trailer was mounted on wheels which ran on rails.

Figure 565. General view of the 'open cut' working site on the west side of the Suez Canal showing the shield trailer in the starting position. (Courtesy Bade & Theelen GmbH.)

Figure 566. Front view of the Bade & Theelen breasting shield with 'closed' breast panels. The bottom shovel is discernible below the lowest platform. (Courtesy Bade & Theelen GmbH.)

Figure 567. View of the liner segment conveyor on the high deck of the trailer. One segment is discernible at the rear end of the conveyor which is capable of carrying one complete liner ring consisting of 15 segments and the keystone. (Courtesy Bade & Theelen GmbH.)

The forward end of the 11.8 m (38.7 ft) diameter shield (Figure 566) consisted of three work platforms with hydraulically operated cutting edges. Hydraulically operated, independently controlled, hinged breast panels were fitted to the under side of the front edge of each platform, and to the crown section of the shield. These breast plates numbered 32 in all and could be raised or lowered to meet the demands of the ground being excavated

A shovel loader was located behind the breast plates on each platform. The shovels were mounted on rotatable telescopic booms which enabled them, together, to cover the entire face. Because they were capable of rotation, conventional cutting heads could also be mounted on the end of the booms, if desired. Each head was capable of a maximum ripping out force of some 50 MPa (7300 lb/in^2).

Thirty double-acting hydraulic thrust rams moved the shield forward. A special segment conveyor (Figure 567), capable of carrying an entire tunnel lining ring (consisting of 15 segments and the keystone) ran along rails on the upper deck of the trailer.

Material cut from the face dropped to the invert where it was moved by a shovel to a scraper conveyor. With the aid of a series of rubber belt conveyors the debris was lifted to a belt conveyor running on the high deck of the trailer. This high-deck conveyor carried the dirt

Figure 568. View of the bottom area of the trailer before entering the tunnel. Above left on the high deck the bottom of the rubber belt conveyor may be seen. (Courtesy Bade & Theelen GmbH.)

to the rear and dropped it into loading-out vehicles. Thus, as lining segments and dirt were being transported to and from the face on the high deck of the trailer, the area below (Figure 568) was left free for the erection of the road deck with on-site precast concrete sections.

The tunnel was being driven through firm blue-clay mudstone.

Summary

While this unique arrangement of shield and digger arm enables the operator to deal fairly effectively with mixed faces of soft and medium hard ground, including soft ground interspersed with floaters, this type of machine cannot, generally speaking, be used in fractured and weathered rock strata lying in the upper compressive strength ranges beyond the capabilities of boom-type machines. However, an exception to this is the Zokor rotary excavator which, according to the manufacturers has been found to be most effective in very hard fractured and weathered strata. In such ground the bucket is reputed to be capable of handling material beyond the capabilities of most boom-type machines.

Boom-type Trench-Cutting Machine

Jaws I and Jaws II — H. B. Zachry Company

An unusual boom-type trench-cutting machine has recently been developed by the H. B. Zachry Company, General Contractors from Texas, U.S.A. Two models, viz. Jaws I and Jaws II have so far been produced.

The Zachry units are somewhat similar to the early boomheading machines produced by the Hungarians during the 1940s in that they feature two parallel cutting wheels at the end of a boom. However, the Hungarians used contra-rotating discs, whereas the cutter blades on the Zachry units rotate in the same direction.

The machines were built specifically for working some 120 km (75 miles) of sewer and storm drain trenches in Riyadh, Saudi Arabia. They were developed to cut ditches in rock areas where blasting was prohibited and the ground was beyond the capabilities of the conventional backhoe.

An excavator superstructure with a telescoping boom was mounted on the undercarriage of a crawler-type grading machine. The front boom and cutting wheels, designed by the Zachry Company, were then added to the telescoping boom.

Jaws I (Figure 569) has a four-track undercarriage and weighs 31,700 kg (69,820 lb) while Jaws II (Figure 570) (a smaller version of the unit) has a two-track carrier and weighs 21,500 kg (47,500 lb). Power is provided by a 220 to 240 hp (164 to 179 kW) supercharged Caterpillar D-333 engine. Hydraulic systems drive the cutting wheels and provide the power for the tramming or forward action.

The cutter blades which rotate transversely are rigged to the front end of the articulated hydraulically controlled telescopic boom and

Figure 569. Jaws I. (Courtesy H. B. Zachry Co.)

Figure 570. Jaws II. (Courtesy H. B. Zachry Co.)

are fitted with tungsten-carbide teeth. These teeth, which number 90 per wheel, are similar to those used on conventional underground mining equipment.

Spacers on each unit allow wheel width adjustment from a minimum of 53 cm (21 in) out to 183 cm (72 in) in 10 cm (4 in) (Jaws II) and 15 cm (6 in) (Jaws I) increments. The minimum cutting width of the wheels is set far enough apart to allow the boom and gear box to follow them into the trench when deep cuts are made. The depth of a single pass is governed by the distance between the outside diameter of the wheel and its hub (i.e. 91 cm (36 in) and 76 cm (30 in) in the case of Jaws I and II

respectively). The cutting depth is controlled by extending the telescopic boom and by adding 122 cm (4 ft) extension pieces as necessary. (Jaws I has a digging depth of 8.5 m (28 ft) and Jaws II a digging depth of 6.4 m (21 ft) without extensions.)

Each wheel has ten segments with nine tooth sockets (per segment) to hold the teeth.

The lacing pattern of each segment is so arranged that the first tooth socket is positioned in the centre and the second, third, fourth, etc., are spaced consecutively further from the centre until the ninth tooth socket is cutting gauge. That is the effective cutting width of the wheel is 11.4 cm (4½ in).

616

Figure 571. Side view of Jaws I. (Courtesy H. B. Zachry Co.)

Figure 572. Cutting the trench. (Courtesy H. B. Zachry Co.)

When the outer or gauge teeth wear sufficiently to reduce the width of the cut to 10 cm (4 in) or less, they are replaced by new teeth and the worn gauge teeth are refitted to the central cutting section (i.e. in the first three or four sockets) where their continued use does not affect hole width.

The 234 cm (92 in) diameter cutting wheels on Jaws I and the 208 cm (82 in) diameter cutting wheels on Jaws II normally rotate at 105 rpm with cutter tip speed at 820 metres (2700 ft) per minute. A water spray which operates at the rate of 6.8 litres (1.5 gal) per minute is provided for each wheel for cooling and dust control purposes.

The boom can be rotated and shifted sideways 91 cm (36 in) either side of the centre line of the base machine for ditch alignment purposes.

Each crawler track is independently powered by a hydraulic motor for travel in both directions and either side of the machine may be raised or lowered separately for stability purposes to maintain the unit on an even keel on sloping or uneven ground.

According to the manufacturers, digging speeds vary from 46 to 61 cm (18 to 24 in)/min when cutting through rock with compressive strengths in the 103 MPa (15,000 lb/in^2) to 138 MPa (20,000 lb/in^2) range and from 122 to 244 cm (4 to 8 ft)/min when cutting through say limestone rock in the 48 MPa (7000 lb/in^2) to 69 MPa (10,000 lb/in^2) range.

The manufacturers also advise that when the cutting wheels are run at the foregoing speeds they are capable of cutting through steel rebar and metal obstructions with ease and in

617

addition can cut cleanly through concrete or asphalt pavement.

During operation the sides of the trench are cut first, after which the loosened material is removed by means of a backhoe.

Travelling speeds may be varied from 15 cm (6 in) to 38.1 m/min (125 ft/min), the 38.1 m (125 ft) rate being the maximum for on-the-road travel.

According to Ed. Marten, Zachry's Plant Superintendent, the company has only experienced one minor teething problem since the installation of Jaws I in Saudi Arabia. This concerned the failure of a rigid coupling used from the hydraulic motor to the gearbox. The problem was field corrected by replacing the rigid coupling with a double universal joint.

References

1. William Charles Copperthwaite, *Tunnel Shields and the Use of Compressed Air in Subaqueous Works,* Archibald Constable & Co., London, 1906.
2. D. B. Sugden, Ground support in mechanically bored tunnels, *Journal Inst. of Engineers, Australia,* Aug. 1973, Barton, A.C.T., Australia.
3. D. B. Sugden, Tunnel boring machines and systems — a survey. *Journal Inst. of Engineers, Australia,* Dec. 1975. Barton, A.C.T.
4. D. B. Sugden, Tunnel boring machines — their advantages and disadvantages, *Contracting and Con. Eng.,* Aug. 1973, Sydney, Australia.
5. David Martin (Ed.) Tunnelling under the Suez Canal, *Tunnels and Tunnelling,* Sept. 1979, London.
6. *Bade & Theelen Main Shield, Model ADS-1180-LS/BV for Ahmed Hamdi Tunnel Project in Suez/Egypt* (technical data), Bade & Theelen GmbH, Lehrte, West Germany.

Correspondence

Demag Aktiengesellschaft (now Mannesman Demag Bergwerktechnik), Duisberg, West Germany.
The Robbins Company, Seattle, U.S.A.
Memco (Mining Equipment Mnf. Corp.), Racine, U.S.A.
Zokor International (U.K.) Ltd., London.
Bade & Theelen GmbH, Lehrte, West Germany.
H. B. Zachry Company, Texas, U.S.A.

TWENTY-SIX

Full-face Hybrid Machines

The Melbourne Type T.B.M. (Melbourne, Australia) 1966-1976[1,3]

The variable ground conditions encountered in the California State water project tunnels influenced the development of shield excavators which had some degree of flexibility built into their basic design. This enabled them to cope more readily with anomalous ground conditions. In the same way, in their turn, did the Melbourne & Metropolitan Board of Works (M.M.B.W.) south-eastern trunk sewer tunnels Melbourne, Australia, inspire the design of the first hard-rock hybrid machine. However, there was one important difference, the M.M.B.W. had, to some extent, a forewarning of the type of country which would be met along their tunnel routes.

In the initial design stages of the project, 110 holes were bored. The Board's laboratory staff then subjected the core samples from these holes to numerous tests which attempted to measure, amongst other important properties, such qualities as the unconfined compressive strength of the rock, its hardness (using Mohs's scratch test), moisture content, etc.

In addition three test and three permanent shafts were excavated. The results of these tests indicated, with some degree of accuracy, the various types of ground conditions which could be expected. They showed that water-bearing Tertiary sandy sediment over Silurian bedrock lay in the southern section of the Kew to Moorabbin tunnel route, while mudstone and sandstone strata, ranging in compressive strength between 14 and 69 MPa (2000 and 10,000 lb/in²) were to be expected in other sections. There were, too, indications of badly faulted and fractured siltstone, blocky sandstone and plastic clay (the result of igneous

dykes which had weathered). Lightly fractured sections of fairly good cohesive material (mudstone and sandstone) were interspersed with areas of soft running ground and bands of heavily fissured rock combined with soft sticky clay. To say the least, the picture was hardly encouraging. The Board were then faced with the question of choice of method of driving the tunnel, that is, should it be by conventional means or by machine?

Method of Tunnelling

So far as conventional tunnelling methods were concerned, some local experience had been gained in smaller diameter tunnels where partly mechanized systems had been used. This experience, coupled with the geological report, appeared to indicate that if conventional methods were used in the larger diameter tunnels, support would need to be extensive and costly, and average penetration rates of as low as 3 m (10 ft) per shift could be expected most of the way. In addition, blasting in the metropolitan area was definitely undesirable for obvious reasons.

By 1966 enough machine-bored tunnels had been driven to show that substantial savings in such areas as primary and permanent ground support systems could be made, if the conditions were right. Very little was known at that stage, however, of the likely behaviour of hard-rock machines in such country. Moreover, such reports as were available of European attempts to machine-bore similar ground were somewhat discouraging.

To offset this gloomy picture was the precedent set by the Tasmanian Hydro-Electric Commission (see Chapter 5 on the Poatina

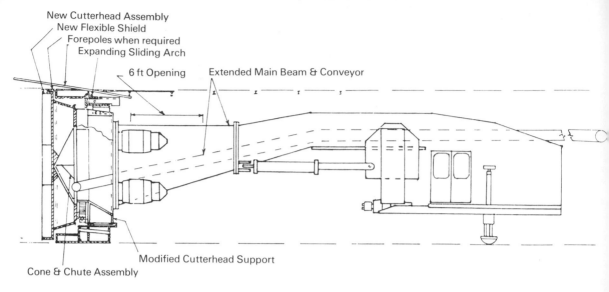

Figure 573. Modifications to M.M.B.W. T.B.M. (Robbins) Melbourne, Australia. (Courtesy Melbourne & Metropolitan Board of Works.)

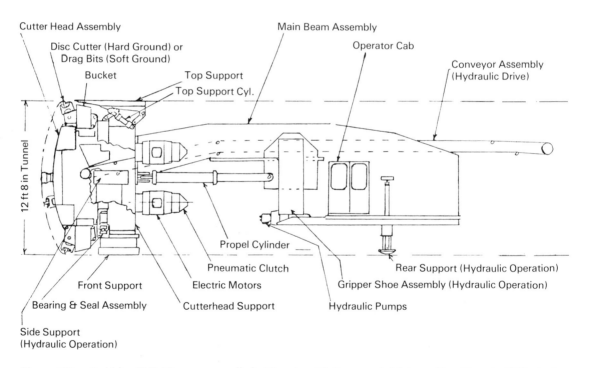

Figure 574. Robbins T.B.M. — as supplied. (Courtesy Melbourne & Metropolitan Board of Works.)

tunnel). To gain as much data as possible on current boring technology, an eight-week study tour of European and American methods was undertaken by three engineers from the M.M.B.W. (i.e. Brack, Jordan, and Neyland).

As a result of this tour and a study commissioned from the Californian consulting firm of Jacobs Associates, a decision was finally made to call for tenders for a suitable hard-rock tunnelling machine.

620

The Basic Machine

The tunnel borer (Model 132-123) (Figures 574 to 575) which was eventually selected was built by the Robbins Company of Seattle and, apart from certain additional components which had been included by the manufacturers in order to conform with the Board's performance specifications, was basically similar in design to the Robbins 161-108 unit used at Poatina. It was capable of exerting a total thrust of 2442 kN (550,000 lb) and a torque of 426 kNm (310,000 lb ft). The domed cutter head was 3.86 m (12 ft 8 in) in diameter. Provision had, however, been made for the later modification of the head to 4.39 m (14 ft 5 in) and to 3.4 m (11 ft 2 in).

In order to cope with the performance specifications the design for the T.B.M. allowed for the incorporation of certain unique features. The most significant of these was the provision of a canopy or shield which was mounted over the cutter head and extended as far as the tunnel springline. The shield consisted of three hydraulically controlled and radially arranged rigid metal supports. As a further means of protection a short tail skirt was extended behind each of the three supports to allow for the safe erection of tunnel ribs.

To enable the thrust system to be used in relatively soft ground, including badly fractured sections, the front shoe and side grippers were larger than those normally supplied with this type of machine. In addition it was intended that the design of the muck disposal system would be able to cope with blocky or sticky clay conditions when these were encountered.

Eight loading buckets were placed on the periphery of the cutter head. These lifted the excavated material and then dropped it into a chute at the top of the machine immediately behind the cutter head.

The cutter head was dressed with a central 25 cm (10 in) diameter tri-cone cutter and 25 Robbins disc-type cutters (which could be removed and replaced with tungsten-carbide-tipped drag bits if necessary). Expected penetration rates were approximately 2.13 m (7 ft)/hour.

Table 14 shows the programme of work for which the T.B.M. was required:

Table 14

Tunnel	Length (m)	(ft)	Diameter m	(ft in)
S.E.T.S., Section 3A	3460	11,350	3.86	(12 8)
S.E.T.S., Section 3B	3450	11,320	4.39	(14 5)
S.E.E.O., Frankston[a]	3800	12,470	4.01	(13 2)
S.E.T.S., Section 4	9260	30,380	3.4	(11 2)

[a]This was not an initial requirement, but was slotted into the original programme as a result of reprogramming later on

Figure 575. Robbins 'Melbourne' machine Model 132-123 as supplied. Note the solid domed head and the canopy. (Courtesy The Robbins Company.)

621

The machine commenced work in mid-1968 and in the ensuing year a distance of 1340 m (4400 ft) was driven.

The numerous difficulties and problems which were met with during this initial shakedown period strongly emphasized the need for several drastic changes in the design of the machine.

The front end of the canopy was located 90 cm (2 ft 11 in) behind the central cutter and this left an unsupported area 90 cm (2 ft 11 in) wide. In the highly unstable ground this distance proved too great. Moroever, the domed boring head with its protruding cutters was found to be unsatisfactory when blocky material was present. Instead of crushing and removing the spoil as they had been designed to do, the cutters tended to grab it in chunks and pull it across the face. The net result of this was that the unstable ground ahead of the machine was disturbed in much the same manner as had occurred in the first section of the Newhall Tunnel (see 'California State water project'). Large pieces of material (which had been picked up by the side entry buckets) clogged the bucket openings. This problem was further compounded by cavities which frequently formed when the jammed rock was forced into open jointed material in the wall on the left of the T.B.M. Often before the cutters and buckets could be freed the entire face would collapse and the machine would stall (as had also occurred in the Newhall tunnel).

When caving occurred ahead of the machine it became necessary to carry out a certain amount of hand mining in order to stabilize the ground. This work was greatly hampered by the fact that access to the front of the machine was difficult.

The design of the hydraulically actuated shield supports also proved to be inadequate in that when operators attempted to lower ram pressures which had built up because of high rock loadings, this could only be done by allowing the shield to subside. As the shield had a smaller radius than the bored tunnel the crown of the tunnel and part of the walls was unevenly supported. This led to the movement (and occasionally the complete collapse) of the rock mass above the machine. These factors were compounded by the rigidity of the shield

supports and their inadequate size. Again the 'shields' were held parallel to the machine by the support linkage. This meant that when too abrupt vertical steering was applied, shields would tend either to 'dig in' or to leave a tapering space — either condition leading to further difficulties.

To overcome these problems engineers of the M.M.B.W. (in association with the Robbins Company and their consulting engineer, D.B. Sugden) made several important modifications to the shield, cutter head and the conveyor system of the machine.

Modifications

The modifications to the T.B.M. were carried out in two stages.

Stage I

As mentioned above, experience had shown that when a rigid component such as a shield, canopy, or similar device was mounted on top of a T.B.M. such a device frequently caused strain to both the machine and the tunnel structure during steering operations. In order that steering operations should not be inhibited by these factors, the new roof support was made of flexible steel. In effect it consisted of three short segments to each of which was attached a number of trailing flexible fingers measuring 15 cm (6 in) in width and about 1.83 m (6 ft) in length.

Primary tunnel support could then be erected in comparative safety beneath the flexible fingers. During a working cycle, temporary support for the trailing fingers was provided by another unusual innovation which was developed by engineers of the M.M.B.W. This was an expandable cantilever arch which was connected to the tunnel support ribs by means of chains. Three supporting beams were longitudinally arranged on spring mountings and the expanding arch was carried on rollers which tracked along these beams.

In operation the arch worked as follows. When unstable ground conditions warranted it, it was possible to drive forepoles (i.e. bars, pipes or rails, etc.) through the gaps (11 cm (4½ in) wide) left between the trailing fingers. The forepoles were supported cantilever fashion on the movable arch which expanded and simultaneously braced the forepoles and supported the trailing fingers of the shield as the T.B.M. advanced. Other modifications which were carried out at that time were changes to the spacing of the loading buckets and to the buckets themselves, in an effort to prevent the blocking which had occurred. In addition extensions were made to the side supports (steering shoes) to attempt to improve ground support in that area.

Stage II

Despite the above improvements, performance of the machine was far from satisfactory and the Board decided in February to invite tenders for a more suitable machine for the conditions. At the same time Robbins were asked to submit a proposal for whatever changes they considered would be likely to make the machine more effective.

In the event both steps were taken. A Calweld 'windscreen wiper' machine was ordered and Sugden (joint consultant at that time to both the Board and the Robbins Company), prepared a proposal for radical modifications to the Board's 'Robbins' machine. These proposals were designed to incorporate a form of the Board's patented forepoling system. In addition, drawing on experience gained on the job up to that time and taking into account several previous consultations with the Board's staff — in particular F. G. Watson, Workshops and Plant Services Engineer, and A. J. Neyland, Project Engineer, many other changes which seemed necessary were suggested.

The Robbins Company's Chief Engineer came to Melbourne and the new proposals were submitted to the Board. A wooden full-scale mock-up was made in a matter of days by the Board and the suggestions of all job-site staff obtained for the final design. These proposals were accepted and the final design was carried out by the Robbins Company in Seattle.

So far as the actual modifications were concerned, these were complex and numerous, but basically included the replacement of the domed head by a new revolving cutter head and a full circle flexible slotted shield which wrapped completely around the cutter-head support and replaced the sliding support and steer shoes for the front of the machine. The new 'Melbourne-type' head, as it came to be known, was manufactured in Australia by Marfleet and Weight. The flexible shield design was a further stage of an evolutionary process (used later on subsequent rock tunnelling machines). It consisted of trailing flexible fingers which extended from the shield for a distance of up to 2.7 m (9 ft). These fingers were fitted to the upper part of the shield down as far as the springline. The new shield now provided complete protection around the entire circumference (3.86 m (12 ft 8 in) diameter) of the machine at its most vulnerable point, namely around and immediately behind the cutter head, and also provided extra cover under which to erect the primary tunnel supports.

The cantilever arch and forepoling system were retained. This new cutter-head design reduced to a minimum the unsupported ground area which lay over the cutter head between the front edge of the shield and the tunnel face. In addition, drag bit cutters (which did not extend as far forward as disc cutters) were used to ensure that this distance was kept to a minimum.

Robbins acquired rights to the Board's patented forepoling system conceived by F. G. Watson and patented the entire design of cutter head and shield with forepoling arrangement as the joint inventions of Watson and Sugden.

Essential to the machine's improved performance was the modified ground support. This was achieved by three-piece expanding steel ribs and prefabricated steel mesh panels designed so that they could be quickly assembled 'on the run' with simple preformed 'pins'. Frequently the finely fractured ground would necessitate the lining of these panels with hessian sacking to retain the fines.

In addition in order to allow for easier access so that hand-mining operations could be carried out ahead of the machine, or to facilitate the changing of cutters, etc. a permanent opening was provided in the cutter head. Formerly, restricted access had been available through a bulkhead-type door in the centre of the original domed cutter head.

After the 'Melbourne head' had been fitted the remaining 2130 m (7000 ft) of Section 3A was driven. When the T.B.M. was put to work in Section 3B, the head and shield were modified to enable to it bore a 4.39 m (14 ft 5 in) diameter tunnel. The head and shield were again modified for the 4.0 m (13 ft 2 in) diameter south-eastern effluent outfall tunnel at Frankston. Though disc cutters were retained in case they were needed, drag bit cutters were mainly used throughout the length of the afore-mentioned tunnels.

So far as the last section was concerned (Section 4 — diameter 3.4 m (11 ft) a newly constructed head, dressed with narrow 28 cm (11 in) diameter disc cutters designed by the Robbins Company, was used. Later, further improvements, which were outlined in a letter to the author from H. L. Reid (Resident Engineer, Group 'C') of the M.M.B.W., greatly increased the efficiency of the T.B.M. and its rate of progress. He commented that:

All these factors contributed to a general improvement in production. Below is a table of production figures taken over the period of use of the T.B.M. (see Table 15).

In conclusion Reid commented that:

Two hand-mined sections of S.E.T.S. Section 4, were done using 2 shields. Two old river valleys crossed the line of the tunnel and the Silurian mudstone was overlain by Quaternary sands and clay. Railway lines were laid in concrete in the liner plates and the machine moved through. The speed of the machine through this section was only limited by the time to erect services behind the machine.

After completing its programme of work the T.B.M. cutter head and shield were modified to 3.66 m (12 ft) diameter and later moved to the Dandenong valley trunk sewer tunnel where it began a drive through 16,800 m (55,000 ft) of Silurian mudstone — country very similar to that encountered in the S.E.T.S. project.

The Robbins/Grandori Tandem T.B.M.[2]
(Model 144-151 for Orichella and Timpagrande tunnels, Sila, Calabria, Southern Italy)

An interesting hybrid machine (Figure 576) was developed and built by the Robbins Company of Seattle for use in two 4.3 m (14 ft) diameter pressure tunnels, required by the Italian Power Board for their hydro-electric scheme. The basic design of the machine was in accordance with specification proposals submitted by Carlo Grandori, Managing Director of the Italian company S.E.L.I., s.r.l., Rome, Italy.

The tunnel routes lay through 8000 m (26,000 ft) of varied strata comprising Sila granite which fell within the 69 to 207 MPa (10,000 to 30,000 lb/in²) compressive strength range, and badly fractured and faulty strata which included breccias, etc. It was thus anticipated that two types of ground condition would be encountered:

(a) sections needing full primary tunnel lining support in the form of precast segments or tunnel ribs etc.; and
(b) sections which would be capable of standing without such support.

Where ground conditions were good and the rock fell into the higher compressive strength range, the most efficient machine to use would be a rock T.B.M. fitted with side-wall grippers which provided the thrust reaction for the unit's advance rams. On the other hand, where the ground was unstable, necessitating the provision of primary tunnel supports and lining, the obvious choice of machine would be a mechanized shield where the unit was

Table 15

	Best shift		Best day		Best week		Best month	
	(m)	(ft)	(m)	(ft)	(m)	(ft)	(m)	(ft)
S.E.T.S. Section 3A	20.7	68	45.7	150	229	751	—	
S.E.T.S. Section 3B	26.2	86	56.4	185	258	846	813	2667
S.E.E.O. Frankston	17.4	57	43.6	143	201	659	534	1752
S.E.T.S. Section 4	27.7	91	66.4	218	318	1043	945	3100

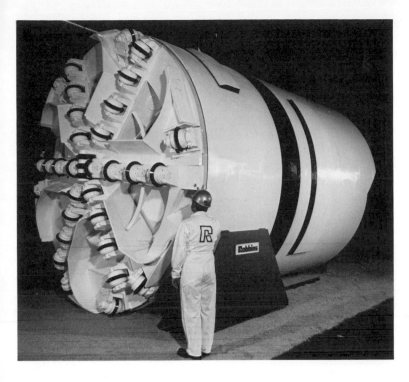

Figure 576. The Robbins 'Grandori' tandem mechanized shield Model 144-151 used in the Orichella and Timpagrande Tunnels in southern Italy in 1972. (Courtesy The Robbins Company.)

advanced by the machine's rams reacting against the precast tunnel lining or support system.

This seemed to indicate a need for two distinct types of machine, namely a mechanized shield and a rock T.B.M. However, apart from the not inconsiderable factor of expense in purchasing two such units, there was also the practical aspect which involved changing from one machine to another each time ground conditions warranted it. An inconvenient and time-consuming operation which in the long run would probably cost the contractor more to execute than it would save.

To meet the situation Grandori suggested that the Robbins Company should design and build for them a special prototype machine embodying, in one unit, as many of the essential characteristics of the two types of machine as was practicable. The Robbins Company accepted the challenge and produced their 144-151 unit. It was completed by 1972 and delivered to Italy.

Basically it consisted of two shields in tandem joined axially by 12 thrust jacks, which reacted against the grippers in the rear shield. The cutter head and its support as well as the unit's main drive motors and gear reduction

system, etc. were contained in the front shield. Behind this came the second shield, the forward end of which partially telescoped into the rear end of the front shield. The rear shield housed an auxiliary set of 12 thrust jacks and the unit's main grippers. The grippers, which consisted of a section of the second shield's skin on each side, extended radially against the tunnel wall.

When ground conditions rendered the unit's main grippers inoperable, the auxiliary jacks in the rear shield were used to advance the unit. These jacks were also used to thrust each newly erected segment ring firmly back against the previous set during the advance and repositioning phase of the gripper system and rear shield. Forward thrust was transmitted through the main thrust jacks and the auxiliary jacks which, in turn, reacted against the recently installed tunnel lining. The auxiliary jacks also served to hold the lining segments in place when the machine was being advanced.

Two tangentially arranged hydraulic cylinders operated between the trailing edge of the front shield and the leading edge of the rear shield. These jacks maintained the relative angular positions of the two shields and counteracted cutter head torque and roll drag. Two additional grippers were installed in the

625

front shield to provide reaction when the rear shield was being pulled forward for repositioning. These grippers were arranged at 45° on either side of the vertical radius of the top of the shield and also provided supplementary reaction to cutter head torque during the boring operation.

Following the Melbourne & Metropolitan Board's recent experiences in Australia with the Robbins Company's 132-123 unit, the Italian machine was equipped with a 'Melbourne-type' cutter head, to enable it to cope with the anticipated sections of fractured and blocky rock strata which lay ahead. It featured an open six-spoked wheel-type cutter head fitted with one tri-disc and 28 disc cutters measuring 30 cm (1 ft) in diameter.

The unit's rams were capable of exerting a maximum forward thrust of 6326 kN (1,400,000 lb) and the cutter head which was powered by 448 kW (600 hp), could develop a torque of 762 kNm (560,000 lb ft). A muck conveyor and a segment erector system were provided by a jumbo structure which followed behind the machine.

Steering was effected through the main thrust jacks.

Teething Problems

The expression, 'one man's meat is another man's poison', might perhaps be aptly applied to the 'Melbourne-type' cutter head which was fitted to the tandem shield. In both cases badly fractured and blocky ground was anticipated and encountered. However, in Melbourne this consisted mainly of Silurian mudstone and sandstone, whereas in Italy it was composed of Sila granite — a much harder substance, and perhaps herein lay the difference.

In the first 3 km (2 mile) of tunnel, Grandori reported that hard boulders were found to be embedded in a softer matrix in some sections of the badly fractured Sila granite formations. Large pieces of rock frequently fell from the face in these areas. They passed through between the spokes and jammed themselves firmly behind the cutter arms, where they caused extensive damage to the muck-loading and conveying system. When this occurred the wedged rocks had to be manually broken and

removed, a time-consuming and costly operation. The situation was tolerated until the machine entered a particularly bad area at 'Chainage 3 + 107'. At this point a major face collapse jammed the cutter head securely and prevented further progress.

The mishap induced Grandori and his team to remove the cutter head and effect suitable modifications to it which, hopefully, would enable the machine to proceed more efficiently than had hitherto proved possible.

Basically the cutter-head modifications consisted of closing up the large gaps which existed between the cutting arms with heavily welded steel bars arranged concentrically between the arms. These formed a strong grill or screen which prevented rock fragments measuring more than 20 cm (8 in) in diameter from passing through into the muck-handling and conveying equipment. The completed grillwork was situated only 3 cm (1.2 in) from the tunnel face which, in Grandori's opinion, was the key to the success of the modification, as this distance represented the average depth of rock spall produced by the disc cutters.

Problems relating to the hydraulic steering system became manifest soon after the machine commenced work, but after certain modifications had been carried out to the hydraulic circuits this difficulty was overcome.

Other problems concerned the grippers and the main thrust cylinders. In practice it was found that occasionally when the rear shield needed to be pulled forward it became wedged. At such times the auxiliary front grippers proved to be inadequate as an anchor reaction. This was partly due to the fact that the rear shield was somewhat longer than the front one. To overcome this difficulty, the front shield was lengthened by about 0.6 m (2 ft) and in addition three extra grippers were installed on the front shield as a safety precaution. These grippers were so positioned that they automatically wedged against the tunnel wall if the machine commenced backsliding. The extra grippers were also useful during normal boring operations for steering and as stabilizers for the front shield.

The hard Sila granite formation caused undue stress to the main thrust cylinders which, in practice, proved to be too small for the job.

Another benefit derived from the new cylinders was that more room was available in the hitherto extremely crowded front shield.

The upper tunnel was holed through in October 1975. Before the machine was put to work in the lower tunnel the machine was given a general overhaul in the site workshop and the modification work, which had been partially carried out after the tunnel face collapse, was completed.

Work on the lower tunnel was commenced early in 1976.

The Buckskin Water Tunnel T.B.M. (Robbins T.B.M. 233-172)

The development of the Robbins 233-172 T.B.M. for the Buckskin water tunnel in Arizona came as a direct result of Grandori's experiences with the tandem shield in Italy.

Basically the Buckskin machine (Figure 577 to 579) was a 7.16 m (23 ft 6 in) diameter completely shielded rock-boring machine consisting of two shields which telescoped inside one another for the boring stroke and yet were flexible enough to permit normal steering control.

Within the front shield were housed the cutter head, the cutter-head support and the six

They were therefore replaced with eight new cylinders with larger bores. These were capable of providing greater thrust, and the risk of cylinder failure was reduced because the new jacks were not required to work to capacity.

Secondary breasting skin

Figure 579. Model 233-172 depicting part of the secondary breasting skin arrangement tried in the Buckskin Tunnel. (Courtesy The Robbins Company.)

main drive motors, the main thrust jacks, and the forward end of the machine conveyor.

Within the rear gripper shield were housed the segment erector, the rear end of the machine conveyor, and of course the grippers. The unit's four main hydraulic thrust jacks were mounted between the cutter-head support and the mid-section of the gripper shield.

It was anticipated that the compressive strength of the rock in the Buckskin tunnel would range from 69 to 274 MPa (10,000-40,000 lb/in²). To deal with this extremely hard stratum the cutter head on the Buckskin unit was a dome-shaped, robustly built, steel plate weldment. It was supported on a pair of large tapered main bearings and was equipped with 39 cm (15 in) diameter disc cutters. The peripherally mounted muck buckets were fitted with grille bars which acted

as a screen designed to prevent large pieces of rock, which may have fallen from the face or roof, from entering the muck-handling and conveying systems.

The new gripper mechanism on the Buckskin unit was a significant departure from the simpler type of clamp or gripper normally fitted to rock-tunnelling machines. Based somewhat on the design of the thrust expansion ring used on certain mechanized shields (see section on mechanized shields — McAlpine tunnelling machines, etc.) the grippers expanded circumferentially against the tunnel wall to provide thrust reaction for the unit's propulsion jacks. Their mode of operation may, perhaps, be more simply explained by comparing them with the conventional drum brakes fitted to a great many modern vehicles.

The grippers were also used to steer the unit

628

via the main thrust jacks. Between the main beam carrier and the gripper shield was a horizontal cylinder to control horizontal steering. When the original designs were drawn an additional steering aid for use on a constant horizontal curve was planned. It was to be provided by the cutter head and cutter-head support which would have been eccentrically positioned in relation to the front shield. (This method of steering was first mooted by Whitaker in 1917.) However, in the final analysis this latter method was not used.

Another important feature of the Buckskin unit concerned the installation of two segment clamp cylinders which were used to position and hold each newly erected segment ring firmly back against the previous set during the advance and repositioning phase of the gripper system and rear shield in much the same manner as did the auxiliary thrust jacks on the Grandori tandem shield.

The Buckskin unit commenced work towards the last quarter of 1976 and it was expected that areas of blocky fractured strata similar to that encountered in Australia and Italy would be met along the proposed 10,700 m (35,000 ft) tunnel route which penetrated layers of andesite, agglomerates, and tuff (comminuted rock debris ejected from a volcano).

According to Eugene G. Murphy* (Project Manager for the Buckskin Mts. tunnel project), the tunnel was started in a well-cemented andesite rock formation which varied from 2 m (6 ft) cube blocks to crushed material, visicular andesite, agglomerate, and tuff. However, as the tunnel progressed the calcite and gypsum cementation between the blocks was replaced by clay or the joints became completely devoid of any adhesive material.

Through this difficult formation (460 m (1500 ft) of tunnel were bored, the last 150 m (500 ft) being the worst, because of the total lack of cementation and because of the gaps between the large blocks, these gaps at times measuring as much as 5 to 8 cm (2 or 3 in) in width. This caused a large degree of overbreak at the crown (from 180 to 240 cm (6 to 8 ft) on both sides of the tunnel borer and in front of the cutter head.

At the 460 m (1500 ft) mark the overbreak

---

*J. F. Shea Co. Inc. (General Contractors).

was some 6 m (21 ft) high. The collapse of large blocks caused structural damage to the machine, constantly plugged the muck hopper, and affected the general stability of the T.B.M., making direction and grade difficult to maintain. Its ability to propel was all but nil, thus rendering the machine incapable of further progress.

Field observations during the difficult phases of boring through the blocky ground indicated that the existing face was some 150 to 240 cm (5 to 8 ft) ahead of the cutter head and that the cutters were, for all practical purposes, merely stirring up the blocky rock which, in turn, was churning the ground ahead. As Murphy so aptly described it, the T.B.M. was, in effect, acting as a giant horizontal blender, no work being performed by the cutters other than the mixing action. Although cutter wear was minimal the cutters were nevertheless failing structurally due to their impact with the blocks of rock.

To overcome these problems Murphy reported that the following steps were taken:

1. The cutter-head face was advanced a distance of 30 cm (1 ft) and the consequent gap between the new plate and the original cutter head was filled with a cement grout so that the thrust against the new cutter head could be handled.

2. A ring made from 61 cm (24 in) diameter pipe, with 2.5 cm (1 in) thick walls, was welded around each cutter, the edge of the 61 cm (24 in) pipe being trimmed to fit it to the spherical shape of the cutter head.

3. Three circumferential rings made of pre-bent 2.5 cm (1 in) plate and radial stiffeners made of precut 2.5 cm (1 in) plate were welded to the cutter head between the outer gauge cutters and the inner circle of cutters, 2.5 cm (1 in) facing plate was then pre-bent to conform to the spherical shape of the head and was welded over the circumferential rings and stiffeners. (The inner circle of cutters was left uncovered so that the cutter head could be removed to install seals when necessary.)

4. Blockouts were also fitted around the gauge cutters and the muck buckets as a protective measure.

629

The modifications were carried out at the 460 m (1510 ft) mark in the chamber formed by the 6 m (19 ft 8 in) high overbreak. The entire operation was conducted through a 91 × 122 cm (3 × 4 ft) opening in the side of the machine.

All the pieces were specially made to size, to enable them to pass through this opening. Each component needed for the modification work had therefore to be hauled the 460 m (1510 ft) along the tunnel and passed through the opening before it could be fitted to the new cutter head. The entire job took some three months (of three shifts per day, six days per week) to complete.

Boring was recommenced on 3 January 1977.

While the problem of overbreak and blocking of the conveyor system appeared to have been solved, a new difficulty became apparent. The shield skin plate which was designed with a horizontal joint 90 cm (3 ft) below the springline was now being deflected by the blocky ground at the crown and sides. This deflection was severe enough to hinder the erection of segments within the tail shield section. On 7 January 1977 the T.B.M. was once again stopped so that five stiffener beams could be installed in the tail shield skin plate, while the plate itself was increased in thickness from 25 to 37 mm (1 to 1½ in).

After these modifications were carried out the T.B.M. successfully handled blocky, soft and various grades of ground in between. Thus while the first 460 m (1500 ft) of tunnel took nine months to drive, the subsequent 2000 m (7000 ft) were bored through in less than a year.

Progress as at 26 October 1977 was as follows:

Best single shift	15 m	(50 ft)
Best single day	33 m	(110 ft)
Best single week	130 m	(425 ft)
Best month	530 m	(1735 ft)

Summary

The evolutionary developmental pattern of this particular type of hybrid machine is clearly discerned from a study of its history. Beginning with the Melbourne unit it can be seen that the troublesome closed dome head on the original

Figure 580. Cutter head of Robbins Model 151-191 machine built for use in the Baikal-Amur mainline service tunnel, Nijne Angarsk, Siberia, U.S.S.R. Note the secondary breasting plates. (Courtesy The Robbins Company.)

T.B.M. was replaced with the open-type 'Melbourne head' which in those particular conditions appeared to work more efficiently. The rigid hydraulic canopy support system gave way to the full circle flexible shield with its unique arrangement of trailing flexible finger supports and cantilever forepoling arch mechanism.

In the tandem shield can be seen the first daring attempt by the Robbins Company and Grandori to combine the best attributes of both mechanized shield and rock tunneller in one workable unit. So far as Grandori was concerned, his experiences indicated the unsuitability of the open 'Melbourne-type' cutter head for his particular ground conditions, so he partly closed the gaps with a grillework and penetration rates immediately showed an improvement.

While the Buckskin unit did not have an auxiliary set of jacks in its second shield for thrusting off the primary lining of the tunnel (considered to be unnecessary in the Arizona tunnel) it nevertheless incorporated many of the important innovations which were made in Italy. These included the retention of the double shield and the provision of two segment clamp cylinders to hold the segments in place during the regripping phase. The grillework protective screen over the peripheral buckets was also an idea which evolved as a result of Grandori's experiences. Once again a change in the physical aspects of the ground challenged the ingenuity of the personnel concerned. To overcome the problem of the open-jointed blocky andesite with joints measuring as much as 8 cm (3 in) in width, unusual and unique modifications were effected to the T.B.M. used on the Buckskin Mts. project.

As a result of the experiences gained on the Buckskin project the cutter head of the Robbins 151-191 fully shielded hard-rock machine (Figure 580) destined for work in Russia during 1979-82 featured cutters protected by breasting plates and rings.

References

1. A. J. Neyland, R. F. Murrell, F. G. Watson and A. J. Cusworth, *Tunnel boring in fractured and weathered sedimentary rock,* Australian Geomechanics Society, Raise and Tunnel Boring Symposium, Melbourne, Aug. 1970.
2. Carlo Grandori, *Fully mechanized tunnelling machine and method to cope with the widest range of ground conditions – experiences with a hard rock prototype machine, Proc. R.E.T.C.,* Las Vegas, 1976, American Inst. of Mining, Met. and Pet. Engs., New York.
3. D. B. Sugden, *Ground support in mechanically bored tunnels, Journal Inst. of Engineers, Australia,* Dec. 1975. Barton, A.C.T. Australia.

Correspondence

The Robbins Company, Seattle, U.S.A.

Melbourne & Metropolitan Board of Works, Melbourne, Australia.

Buckskin Mts. Tunnel Project — Eugene G. Murphy, Shea Co. Inc., Utah, U.S.A.

Cutters

TWENTY-SEVEN

Cutters

Both fixed cutters or drag teeth and rolling cutters i.e. disc, kerf, milled tooth, and tungsten-carbide button insert roller cutters, etc. have changed and developed over the years, to some extent keeping pace with the development of the machines themselves (Figures 581 to 586).

Although fixed cutters, i.e. drag teeth, are generally used for softer materials while roller cutters, from the disc to the tungsten-carbide insert, are used in progressively harder rock formations, specially designed drag teeth with tungsten-carbide tips have been used very successfully by Atlas Copco on units with high-speed revolving heads in fairly hard sandstone formations. In addition, drag bits have also been used by the Chamber of Mines of South Africa Research Organization on their drag-bit rock cutter units which have successfully cut through quartzite in the gold-mines.

By and large, however, the fixed or drag tooth tends to wear out a lot faster than does the roller cutter when put to work in hard or abrasive rock formations.

One of the most difficult problems still unsolved is that of finding some standard measure whereby the drillability of rocks may be accurately predicted so that the correct type of cutter can be chosen for the project. At present measurements are made of the unconfined compressive strength of the rock and its abrasive properties are assessed.

Sometimes very hard rock which is not abrasive may be cut with discs or milled tooth cutters, whereas, conversely, fairly soft rock in the lower compressive strength range which is highly abrasive, will result in early disc cutter failure due to wear. In addition, some medium-strength types of rock have exhibited plastic or resilient properties which resist the pressure and cutting force exerted on them by the tools, the rock tending to yield instead of break out as it normally does. This type of rock is extremely difficult to penetrate and will resist the action of the cutters more effectively than would brittle rock in the higher compressive strength ranges.

Below is a description of the disc and roller cutters now in use. However, it should be noted that this description is intended only as a general guide and that roller cutters are made in a wide variety of shapes and sizes which only roughly conform to the various groups described (Figures 587 to 597).

Disc Cutter

This is a cutter with a single circular disc or cutter blade of hardened steel alloy which revolves freely about its axis as it rolls around the face. Some disc cutters carry double or triple disc blades on a single cutter mounting, and sometimes the peripheral edge of the disc is fitted with a tungsten-carbide insert.

Kerf Cutter

The kerf cutter is a form of disc cutter with several discs on one mounting. Tungsten-carbide button inserts are pressed into the periphery of the discs so that the inserts project beyond the softer metal of the disc.

Milled Tooth Cutter

Special heat-treated alloy steels are used to manufacture milled tooth cutters. Rows of teeth of varying patterns and depth are cut on

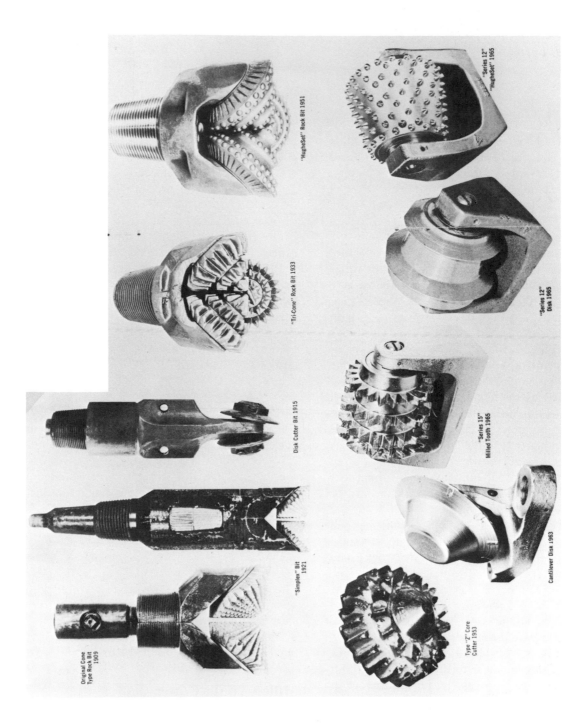

Figure 581. Evolutionary progress of Hughes Tool Company rock bit cutters from 1909 to 1965. (Courtesy Hughes Tool Company.)

Original Cone Type Rock Bit 1909

"Simplex" Bit 1921

Disk Cutter Bit 1915

Type "Z" Core Cutter 1953

"Series 15" Milled Tooth 1965

Cantilever Disk 1963

"Tri-Cone" Rock Bit 1933

"HughesSet" Rock Bit 1951

"Series 12" Disk 1965

"Series 12" "HughesSet" 1965

Figure 582. Ingersoll-Rand one-row replaceable tungsten carbide button ring face cutter mounted in steel bracket. (Courtesy Ingersoll-Rand.)

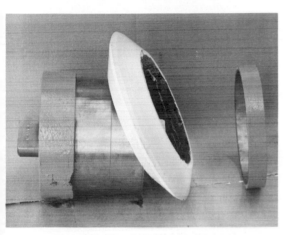

Figure 583. Ingersoll-Rand face cutter body, replaceable disc ring and ring keeper. (Courtesy Ingersoll-Rand.)

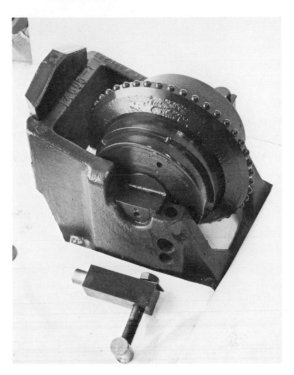

Figure 584. Ingersoll-Rand face cutter showing retainer assembly. (Courtesy Ingersoll-Rand.)

Figure 585. Ingersoll-Rand gauge cutter assembly. (Courtesy Ingersoll-Rand.)

the outside surface of the cutter cone. While this tool is used mainly for shaft drilling, it is sometimes fitted to raise drills or full-face tunnelling machines and has proved fairly effective in comparatively hard material which is not too abrasive.

Tungsten-Carbide Button Insert Roller Cutter

This cutter is generally used in very hard rock which is highly abrasive. It is made by inserting a multitude of button-shaped tungsten-carbide bits into the external surface of a cone-shaped roller cutter.

Figure 586. Ingersoll-Rand gauge cutter bracket and scraper. Note: the scraper is spring-loaded to maintain constant contact with the rock. This is to assist in clearing the path immediately in front of the cutter. (Courtesy Ingersoll-Rand.)

Figure 587. Ingersoll-Rand tungsten carbide button ring cutter. (Courtesy Ingersoll-Rand.)

Most manufacturers have developed their own methods and formulae for the composition, carburization, pressing, shaping, and sintering of the tungsten steel into the various insert shapes required; and this

Figure 588. Ingersoll-Rand tungsten carbide button ring cutter. (Courtesy Ingersoll-Rand.)

information is regarded by individual manufacturers as highly confidential.

Roller cutters (cone-shaped cutters with milled teeth) were used by the oil industry for drilling purposes from their introduction in 1909 by the Hughes Tool Company. Since then Hughes and other manufacturers of cutters have made significant contributions towards their development.

Its small size resulted from its method of evolution. When tunnelling machines, raise drills, and other large-diameter drilling equipment first began appearing on the scene these small cutters were used because they were available at that time. The cutters were small enough to be held and bolted into position by one man. While this may have been of some advantage, especially in confined spaces where cutter changing is a problem, the cutters themselves were too small to provide adequate room for bearings, cone shell and mountings, etc. No seals were incorporated in the earlier roller cutters as there was no room for them. Early bearing failure was a constant problem because of their small size and also because the lack of an adequate seal meant that grit and dirt entered the bearing area. Moreover, because

638

Figure 589. Ingersoll-Rand disc cutters. (Courtesy Ingersoll-Rand.)

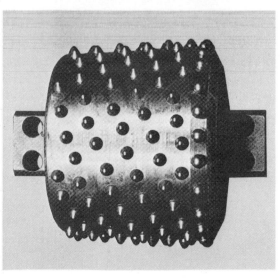
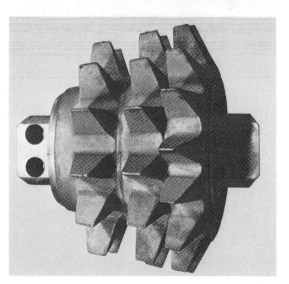

Figure 590. Single and double disc, carbide insert and toothed cutters. (Courtesy Wirth GmbH)

639

Figure 591. Wirth 1-ring-disc-cutter used on shaft boring machine SB-VI-500/600. (Courtesy Thyssen Schachtbau.)

Figure 593. Wirth 1-ring-disc-element for multiple disc-centre-cutter used on shaft boring machine SB-VI-500/600. (Courtesy Thyssen Schachtbau.)

Figure 592. Wirth 5-ring-disc-centre-cutter used on shaft boring machine SB-VI-500/600. (Courtesy Thyssen Schachtbau.)

Figure 594. Robbins 30 cm (12 in) standard cutter assembly. (Courtesy The Robbins Company.)

Figure 596. Robbins 30 cm (12 in) stinger abutted cutter assembly. (Courtesy The Robbins Company.)

Figure 595. Robbins 30 cm (12 in) narrow cutter assembly. (Courtesy The Robbins Company.)

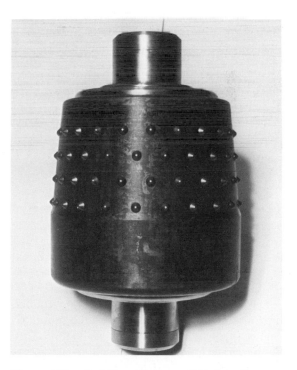

Figure 597. Robbins 24.4 cm (9⅝ in) diameter roller and carbide cutter assembly. (Courtesy The Robbins Company.)

the cutters were small a large number was needed to cover a given area.

So far as the drilling industry was concerned this meant that the drill string was continually being raised to change the cutters, a time-consuming and costly operation.

According to J. M. Glass (Hughes Tool Company) the early use of rolling cutters for shafts apparently involved the sectioning of 'tri-cone' rock bits for attachment by welding to a plate. The next step called for the manufacture of a cutter specifically designed for this purpose which would fit on the same bit size. The 'bolt-on' cantilever cutter was the result. However, due to the geometry of the cantilever cutter and its size at that time, little room was available for the inclusion of an adequate seal capable of withstanding the high mud pressures encountered down a deep hole. This problem was later alleviated by the installation of a patented pressure compensator during the late 1950s.

The disc cutter, pioneered by the Robbins Company evolved along different lines.

Over the years discs were used first as 'cutters' by Wilson and Brunton in 1856 and 1866, respectively, and then later by McKinlay during the early 1920s as 'wedges' or 'bursting wheels' to cut the remaining concentric ring or ridge left behind after the drag teeth or fixed cutters had made a groove.

During the early 1950s James Robbins (senior) used fixed cutters and discs in this manner on his first full-face tunnelling machine, but later it was found that the disc cutter alone would cut the rock and, moreover, it lasted longer than the fixed cutter had done.

Later, as cutters increased in size seals were incorporated. More room meant improved bearing designs, while mountings became stronger and better able to withstand the high loads imposed on them. New types of metal seals were introduced which, in turn, prolonged the life of the bearing and thus the cutter.

During the driving of the Poatina tunnel, bearing failure reached a maximum of about 60 per cent of all cutter failures, but due to improved techniques introduced by engineers of the Tasmanian Hydro-Electric Commission, this figure was reduced to about 10 per cent, despite increasing rock hardness.

Disc cutters with metal-faced seals consisting of two matched hardened alloy-steel sealing rings and two resillient rubber toric rings were used by the Robbins Company on their 161 Poatina unit. The rubber torics positioned the metal seals and absorbed any assembly end play or deflection, providing a constant sealing face pressure. This type of seal was originally developed for caterpillar crawler tractors, and their adaptation as a disc cutter seal was pioneered by the Robbins Company on their 161 Poatina unit. Because of its effectiveness this type of seal has since been used extensively by manufacturers of roller cutters throughout the world.

As time has progressed numerous experiments have been carried out to determine the influence of various factors such as cutter edge geometry, cutter configuration and true roll, cutter type, rotary speed of cutter head and cutter, load, angular orientation, lubrication, etc. on the tool's ability to penetrate and break the material being attacked, and its wearing capabilities.

Tungsten-carbide insert cutters capable of cutting very hard rock have been available for some years, but high replacement cost has precluded their use in most standard tunnelling projects where blasting still offers a cheaper alternative. They are, however, used extensively today in drilling and raiseboring work. No doubt as ways and means are found to produce these cutters more cheaply they will find greater application in other areas.

References

1. R. L. Dixon and E. P. Worden, Rolling cutters design, Raise and Tunnel Boring Symposium, Australian Geomechanics Society, Melbourne, June, 1970.
2. I. M. Ogilvy and C. M. Perrott, Design faults in cutters used for raise boring, *Technical Report 31/ME* (Aug. 1974), C.S.I.R.O. Division of Tribophysics, University of Melbourne, Australia.

Correspondence

Hughes Tool Company — J. M. Glass, Houston, U.S.A.
The Robbins Company, Seattle, U.S.A.

Humour Underground

Figure 598. Boring the first tunnel with an early type of rotary excavator Heath Robinson, *Heath Robinson Railway Ribaldry*. (Courtesy Duckworth and Company Ltd., London.)

Figure 599. Sectional view of the excavations for the Severn tunnel, showing the hard and fossiliferous nature of the ground to be penetrated. Heath Robinson, *Heath Robinson Railway Ribaldry*. (Courtesy Duckworth and Company Ltd., London.)

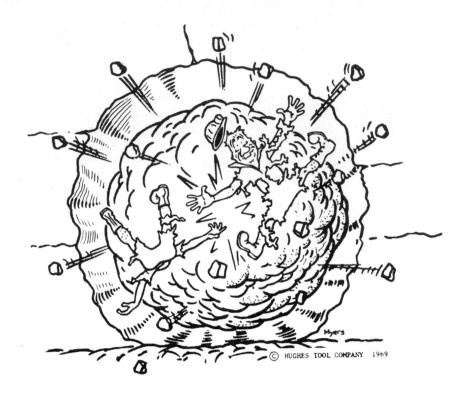

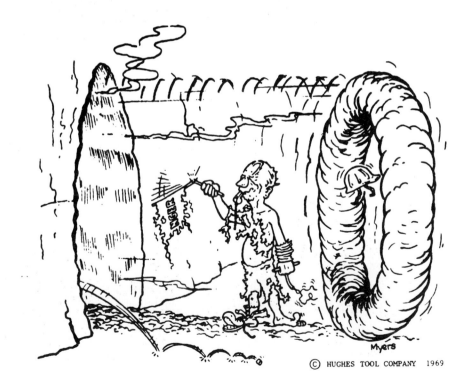

Figure 600. Problems associated with explosives. (Courtesy Hughes Tool Company.)

646

Figure 601. Early attempts at mechanization. (Courtesy Hughes Tool Company.)

Figure 602. Lack of a muck-removal system would appear to have its problems. (Kevin Baily, cartoonist for the *Mercury* newspaper, Hobart, Tasmania.)

Figure 603. Problems arising from an inefficient muck-removal system. (Courtesy Hughes Tool Company.)

Figure 604. Suggested muck removal system. (Courtesy Hughes Tool Company.)

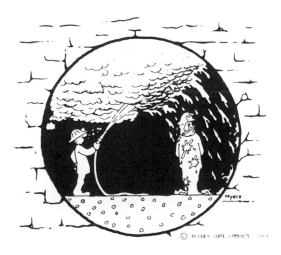

Figure 605. Problems associated with an inadequate ventilation system. (Courtesy Hughes Tool Co.)

648

Figure 606. This ventilation system appears to be *too* efficient! (Courtesy Hughes Tool Co.)

Figure 607. Guidance control problems. (Courtesy Hughes Tool Co.)

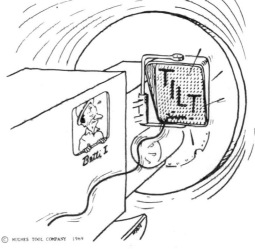

Figure 608. The solution — 'laser guidance'. (Courtesy Hughes Tool Co.)

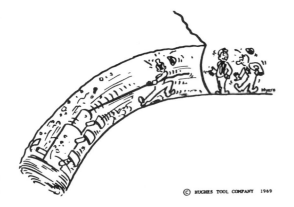

Figure 609. Maintaining grade! (Courtesy Hughes Tool Co.)

Figure 610. Rock masher and bone crusher! (David B. Sugden.)

Figure 611. New pneumatic conveyor turns coal to dust (see Figure 463 — Colonel O'Toole machine). (Bruce G. Stack.)

Appendix

Horizontal rock drill patents

R, rotary; P, percussion

1845 *Kranner* (Prague) (R).
1849 *J. J. Couch* Pat. 6237, 27 Mar. (U.S.A.)
    (P).
1851 *J. W. Fowle* Pat. 7972. 11 Mar. (Caveat
    filed 9 May 1849) Pat. reissued 5 June
    1866. Pat. 2,275 (U.S.A.) (P).
    Cavé (Paris) 15 Oct. (P)
1852 *L. P. Jenks.* Pat. 9,379, 2 Nov. Two
    cross-heads connected together. Feed
    automatic. Pat. claims changing rate of
    rotation and rate of feed. (U.S.A.).
    J. J. Couch. Pat. 9,415, 23 Nov.
    Improvement on original patent (U.S.A.)
    (P).
1854 *Schumann.* Also 17 Feb. 1857, 1 Feb.
    1860 and in 1862 (tested but not patented
    — valve motion by special auxiliary
    cylinder, rotary motion by screw on fly-
    wheel shaft; advance by hand (Germany)
    (P).
1855 *T. Bartlett.* (First trials 1854) Pat. 1,913,
    23 Aug. Sardinian patent 30 June, 1855.
1856 *G. H. Wood.* Pat. 15,540, 12 Aug. Drill
    mounted on frame and driven forward by
    a spring (U.S.A.).
1857 *L. P. Jenks.* Pat. 17,765, 7 July. Drill
    forced back against rubber spring, the
    recoil of which forced the tool forward
    (U.S.A.) (P).
1858 *Schwartzkopff and Philippson.* Pat.
    2,477, 5 Nov. Rotary borer applied to a
    steam-drill. Mounted by ball and socket.
    Hand-feed. Rocking valve driven by an
    arm on the piston rod. Rotation by a ball
    driven by valve stem (invented 1857) (R).

1859 *J. H. Johnson. (Germain Sommeiller).*
    Pat. 50, 6 Jan. (Sardinian Pat. 25 June,
    1857 and March 1861). This is the
    renowned Sommeiller drill which was
    used extensively in the Mont Cenis tunnel.
    Drill was attached to the piston rod and
    the valve, rotation and feed were operated
    by a small engine placed in the rear of the
    machine (three types) (Sardinian) (P).
1861 *W. Harsen.* Pat. 31,430, 12 Feb. Reissued
    11 Aug. 1874, Pat. No. 6,009. Cylinder
    and valve motion similar to a steam-
    engine. Hollow piston. Drill bar of any
    required length passing through hollow
    piston and was moved with the piston.
    Drill bar held by means of four wedges or
    cams on each end of the piston, these
    cams being held on the drill bar by means
    of sliding collars. Drill bar and piston
    rotated together by means of ratchet
    operated by spiral groove in the shield of
    machine. A novel characteristic of this
    drill was a tappet bar for operating the
    valve. The bar had an inclined slot in
    which the pin from the valve stem worked
    (U.S.A.).
1861 *Poole, Wright, Hemming, and Searby.*
    Pat. 373, 13 Feb. (U.K.). Cylinder recipro-
    cates and strikes drill like a hammer (P).
1862 *E. S. Crease.* Pat. 464, 21 Feb.
1863 *G. Low.* Pat. 903, 9 April. U.K. (P).
    E. S. Crease. Pat. 2,624, 24 Oct.
    De La Haye. 27 Feb. (France).
    Chr. G. Barthelson. Hammer machine.
    Hammer is driven against drill by pressure
    of spring. 31 Mar. (Sweden).
    Sachs. Improvement on Schumann drill.
    (This was made after Sachs had experi-
    mented for some time with Schumann's

drill at Altenberg Moresnet, making frequent alterations to Schumann's drill during this trial period.)

1864 *Hipp*. A modification of Schwarzkopff's drill. Drill attached to piston of working cylinder. Used at Bingen-on-the-Rhine.

R. H. Lamborn. Pat. No. 42,669, 10 May. Compressed-air drill for mining (U.S.A.).

S. Gwynn. Pat. 44,722, 18 Oct. Piston struck the drill like a hammer (U.S.A.).

S. Gwynn. Pat. 44,862, 1 Nov. Hollow piston rod of the Couch type. A spiral ratchet was attached to the piston rod for rotating the tool (U.S.A.).

1865 *H. Haupt*. Pat. 46,668, 7 Mar. Combination of mining machinery (U.S.A.).

A. Shiland. Pat. 46,949, 21 Mar. Spiral grooves on drill bar for rotation (U.S.A.).

J. D. Butler. Pat. 47,390, 25 April. Hollow piston rod of the Couch type. The tool was griped by means of steam pressure (U.S.A.).

H. Haupt. Pat. 47,541, 2 May. Couch type. Drill mounted between two columns (U.S.A.).

H. Haupt. Pat. 47,819, 23 May. Drill had griper-box for regulating feed. Momentum feed (U.S.A.).

J. L. Smith. Pat. 47,870, 23 May. Tool having three or more radial cutting edges (U.S.A.).

W. Bickel. Pat. 48,785, 18 July. A drill bit which had an ordinary bit, and also at right angles to the cutting edge were two chisels or reamers, the object being to make a round hole (U.S.A.).

J. H. Johnson. Pat. 981, 6 April. (U.K.). Identical with American patent Haupt's drill, 1865, No. 47,819.

G. Low. Pat. 1,778, 5 July. Valve in Low's earlier drill was operated by a bell-crank, but in the later one it was operated by a spiral bar (U.K.).

Bergström. 4 Dec. Bergström's drill is an improvement on Schumann's machine (Sweden).

Fontenay. Sliding valve moved one way by compressed air, pushed back by three-cornered eccentric. Rotary motion by ratchet and pawl.

Stapff (invented 1865 but no patent issued). Working piston hollow.

J. M. May. Pat. No. 49,129, 1 Aug. Improvement in the drill bit. The drill is beveled like a chisel, one wing being beveled one way and the other the opposite way, so that as it strikes it will cause the tool to rotate, the object being to dispense with all the rotating mechanism in the machine (U.S.A.).

1866 *G. Haseltine*. Pat. No. 3065, 22 Nov. Similar to Burleigh's American Patent No. 59,960.

W. Brooks, S. F. Gates and C. Burleigh. Pat. No. 52,960, 6 Mar. Hollow piston (Couch type). (P) (U.S.A.).

C. Burleigh. Pat. 52,961, 6 Mar. (P) (U.S.A.)

W. R. King. Pat. 53,305, 20 Mar. Reissued 26 Oct. 1875. Telescopic legs for tripod. Guide-plates for drill. (U.S.A.).

S. F. Gates. Pat. No. 55,277, 5 June. Drill bit. Four cutting edges radiate from the centre (U.S.A.).

L. P. Jenks and G. A. Gardner. Pat. No. 55,307, 5 June. Drill mounted on column. Raised and lowered by means of a screw (U.S.A.).

R. Nutty. Pat. No. 58,175, 8 Sept. Hollow piston rod. Frame for mounting drills (U.S.A.).

C. Burleigh. Pat. 59,960, 27 Nov. (U.S.A.) (P).

G. F. Case. Pat. 59,963, 27 Nov. Feed device (U.S.A.).

C. D. Foote. Pat. 60,497, 18 Dec. (U.S.A.).

Jordan & Darlington. Pat. 3,395, 26 Dec. Valve and feed are worked by hand using the same crank. The rotation is performed by a spiral bar and ratchet.

1867 *F. B. Doering*. Pat. No. 43, 7 Jan. Also 9 Nov. 1866 and 7 June 1867. Also Swedish pat. 5 Mar. 1868. This inventor made several drills and carriages of different design (U.K.).

E. S. Crease. Pat. 296, 2 Feb. A spiral bar for rotation. Improved column for tunnelling. Same valve movement as his patent of 1864, No. 2624.

F. B. Doering. Pat. 1704, 10 June. The

valve was operated directly by steam without metallic connections (U.K.).

James Asbury McKean. Pat. 2,607, 16 Sept. Mounted on two columns joined at bottom. Valve driven directly by a projecting arm on piston rod. Machine stationary, drill fed by screw. Successor to English patent of Haupt's drill.

Jordan & Darlington. Pat. 3,386, 29 Nov. Rotation by two ratchets, having a spiral for one and a straight groove for the other.

De La Roche Tolay. Diamond bit. Bit pressed against rock by piston in spiral cylinder. Rotary motion effected by Perret's hydraulic pressure engine. Piston-rod is six-cornered and is hollow; water passes through it to wash away dust in the bore-hole (R).

S. W. Robinson and De V. Wood. Reissued 18 Feb. 1873. Direct-action drill of Fowle type. Pat. 71,329, 26 Nov. (U.S.A.) (P).

F. B. Doering. Pat. 72,465, 24 Dec. USA Pat. (See UK Patents 7 Jan., 1867, No. 43, and 10 June, 1867, No. 1,704.

1868 *Beaumont and Appleby.* Pat. 1,682. Application of diamond drill to tunnelling (R) (U.K.).

C. Schumann. Pat. 73,053, 7 Jan. Valve rotation and feed operated by levers etc. (similar to Saxe's drill of Germany) (U.S.A.).

De V. Wood and S. W. Robinson. Pat. 76,131. 31 Mar. Momentum-valve movement by which the valve was reversed after the blow was struck (U.S.A.).

F. B. Doering. Pat. 77,597, 5 May. (U.S.A.).

De V. Wood and S. W. Robinson. Pat. 78,853, 9 June. Improvement in the forward head (U.S.A.).

C. Burleigh. Pat. 80386, 28 July. (U.S.A.).

W. Hall. Jr. Pat. 80,406. 28 July. Chuck in which the forward end is split or divided and the two parts are forced against the shank by means of two bolts (U.S.A.).

R. Gidly. Pat. 84,543, 1 Dec. Peculiar frame for adjusting in all directions (U.S.A.).

William Robert Lake. Pat. No. 1,183. 8 April. Improvements on Johnson and McKean rotary valve. (some slight modifications in feed) (U.S.A.).

F. B. Doering. Pat. 2,965, 20 Sept. Feed operated by means of a piston driven by water (U.K.).

1869 *Edward Thomas Hughes.* Pat. 2,387, 10 Aug. Drill mounted by ball-and-socket joint; valve moved by a piston. rotation by a spiral bar; feed by hand (U.K.).

R. Nutty. Pat. 87,061, 16 Feb. Inner and outer steam cylinders, peculiar rotating device at the forward end (U.S.A.).

J. P. Frizell. Pat. 94,097, 24 Aug. Automatic feed device, etc. (U.S.A.).

Thomas Brown. Pat. 3,411, 25 Nov. Reissue of Burleigh's patent.

1870 *W. R. Lake.* Pat. 1,104. 14 April. Improvement on McKean's rotation. Supplementary spiral wheel. Tool feeds through centre of piston rod. Rotary valve operated by wing tappets fixed in the valve rod.

H. Osterkamp. Pat. 1,466, 20 May. (tried 1869). Piston valve operated without metallic connections.

F. B. Taylor. Pat. 3,131, 29 Nov. Improvement on Lake's or McKean's drill.

De V. Wood. Pat. 98,901, 10 Jan.

A. Blatchly. Pat. 100,252, 1 Mar.

H. Osterkamp. Pat. 106,197, 9 Aug. U.S.A. Pat. (U.K. Pat. 20 May, 1870, No. 1466).

C. Peck. Pat. 110,280, 20 Dec.

1871 *S. Ingersoll.* Pat. 112,254, 28 Feb, reissued 16 Feb. 1875. Tripod consists of two ordinary legs and one forked one, the legs of the tripod having a telescopic adjustment. Automatic feeding device (see No. 115,478) (U.S.A.).

A. Ball. Pat. 112,885, 21 Mar. (U.S.A.).

C. Burleigh. Pat. 113,850, 18 April. Pawl to regulate feed.

C. S. Pattison. Pat. 114,193, 25 April. Reinforcing rods for fastening the forward head (U.S.A.).

Simon Ingersoll. Pat. 115,478, 30 May.

Reissued 16 Feb. 1875, No. 6,193. Internal Tappets, improvement on Ingersoll's patent of 28 Feb. 1871. (U.S.A.).

G. Phillips. Pat. 117,678. 1 Aug. Spring fork for preventing a rebound of the valve in the Burleigh drill of Pat. 59,960 (U.S.A.).

S. Ingersoll. Pat. 120,279. 24 Oct. Spiral bar for rotating which is cut with radial grooves in combination with the nut (U.S.A.).

M. Ball and J. A. Stansbury. Pat. 121,315, 28 Nov. Drill standard having adjustable legs (U.S.A.).

J. Darlington. Pat. 2,657, 7 Oct. Rotation by spiral bar, the ratchet being in the piston. Spiral bar also operates valve.

1872 *P. Savage.* Pat. 130,246, 6 Aug. (U.S.A.).

J. Cody. Pat. 130,412, 13 Aug. (U.S.A.).

J. G. Cranston. Pat. 143, 17 June. Valve operation similar to Burleigh's.

F. D. Watteau (Dubois-Francois) Pat. 1,398, 7 May.

M. A. Soul (Sachs drill). Automatic feed. Pat. 1,980, 1 July. Feed is automatic. The feed, rotation and valve are all operated by a bell-crank, which is rocked by the reciprocating movement of the piston.

E. Le Gros. Pat. 2,008, 3 July. Similar to Ingersoll's American Pat. 112,254.

J. H. Johnson (Rand & Waring). Improved tripod. Pat. 3,125, 23 Oct.

C. J. Ball. Pat. 3,491, 21 Nov. The main piston forms the valve. (These are called valveless machines.)

Brydon, Davidson & Warrington. Pat. 3,507, 23 Nov.

Brydon, Davidson & Warrington. Pat. 3,921, 26 Dec. Improved tripod.

1873 *Azzolina Del Acqua.* Pat. 52,960. The valve is a cock valve similar to the one first used on the Burleigh drill. The feed and rotation are on the same general principle as the Brooks, Gates, and Burleigh drill, the drill being attached to a central screw and rotation and feed performed by rotating the screw (U.S.A.) (P).

De V. Wood. Pat. 138,777, 13 May. (U.S.A.).

H. C. Sergeant. Pat. 140,596, 8 July.

Drill-carriage. A single upright column which may move horizontally on a bed-plate. Clamp and elastic cushion (U.S.A.).

G. E. Nutting and J. C. Githens. Pat. 140,637, 8 July. Piston-valve. Pawls of rotating device held by friction, so that they may slip if necessary. Improved chuck (U.S.A.) (P).

J. Doty. Pat. 140,767. 15 July. Improved feeding device for the Burleigh rock-drill. (U.S.A.).

H. C. Sergeant. Pat. 143,261, 10 July. Improvement in valve movement. Tappets in the steam-ports. (U.S.A.).

D. Kennedy. Pat. 143,355, 30 Sept. Piston within the main piston, for operating the valve and feeding device (U.S.A.).

W. Roberts, Jr. Pat. 145,364, 9 Dec. Chuck. Solid head in which are half-boxes, said boxes being forced against the shank of the tool by means of wedge-shaped bolts (U.S.A.).

E. Edwards. Pat. 541, 13 Feb. Valveless. Section of spiral rod is triangular. Ratchet-click for rotating. Is without pins or trunnions, and is placed in a recess in the head of piston (U.K.).

E. Edwards. Pat. 1,280, April. Feeds with piston and pawl. Piston worked by steam from the main cylinder and is regulated by poppet opened by the piston at the end of the stroke.

J. G. Cranston. Pat. 1,455, 22 April.

J. Darlington. Pat. 1,734, 13 May.

Brydon & Davidson. Pat. 1,991, 3 June.

W. R. Lake-McKean's. Pat. 2,263, 30 June. The McKean drill is the result of several patents granted under different names: Lake, Taylor, and others.

1874 *F. E. B Beaumont.* Pat. 1,149. Valve is a piston worked without tappets; for rotation, a spiral bar passes into a square mortise in the piston (U.K.) (P).

Warsop & Hill. Pat. 1,181. Valve operated by spiral bar.

W. Manson. Pat. 1,608. (Similar to Wood's U.S.A. Pat. 98,901.)

A. M. Clark. (Chenot). Pat. 1,676. (See Bartlett's UK. Pat. 1,913, 1855.

J. H. Johnson-(Waring). Pat. 1,716. (see Waring's U.S.A. Pat. No. 178,214.)

Sturgeon & White. Pat. 2,085. Long screw goes through piston rod for feeding and drill, similar to Brooks, Gates, and Burleigh U.S.A. Pat. 52,960, 1866. Rotary valve.

G. Haseltine (Thurston). See Wood's U.S.A. Pat. 138,777.

Hosking & Blakewell. Pat. 2,760.

H. B. Barlow (Turretini & Colladon). Pat. 3,038. Hydraulic pressure to feed. Annular piston.

E. Edwards. Pat. 3,342.

F. E. B. Beaumont. Pat. 4,402. Small trunk engine for valve and rotation. Feed is automatic; feed-click is operated one way by pawl and the other way by a cone on the piston rod.

S. Ingersoll. Pat. 147,402, 10 Feb. Improvement to Pat. 112,254. Feeding device (U.S.A.) (P).

S. Ingersoll, G. R. Cullingworth, H. C. Sergeant, and A. H. Elliott. Pat. 147,403, 10 Feb. Cam-shaped tappets and rubber buffer in front end (U.S.A.).

E. S. Winchester. Pat. 148,273, 3 Mar. (U.S.A.), Flutes cut on face of piston, so that a jet of steam will cause rotation.

R. Brydon, J. S. Davidson and T. A. Warrington. Pat. 148,924, 24 Mar. (See U.K. Pat. 3 June, 1873, No. 1991). (U.S.A.).

J. B. Waring. Pat. 152,712, 30 June. (U.S.A.) Tripod. Legs attached to frame at upper ends by ball-and-socket joints. Reissued 19 Oct. 1875, No. 6705.

J. B. Waring. Pat. 156,003, 13 Oct. Tripod. Rear leg in two pieces which can be adjusted laterally. (U.S.A.).

C. S. Pattison. Pat. 157,133, 24 Nov. (U.S.A.).

E. S. Winchester. Pat. 158,009, 22 Dec. (U.S.A.).

J. C. Githens. Pat. 158,060, 22 Dec. A V-shaped disc attached to the spiral rod for rotating and a V-shaped ring on a piston, fitting into the disc, and steam passages for forcing the piston and ring together and apart (U.S.A.).

1875 *W. L. Wise.* Pat. 3.

Haseltine — (Winchester & Phelps). Pat. 485 (see Winchester's U.S.A. Pats. 165,646, etc.

F. E. B. Beaumont. Pat. 829.

Brown-(Burleigh's) (see Burleigh's U.S.A. Pat. No. 162,528).

E. Edwards. Pat. 2013.

G. H. Reynolds (see Reynold's U.S.A. Patents). U.K. Pat. 2,287.

Blake. Pat. 4,207. (See Ingersoll's U.S.A. Pat. 4 July, 1876. 179,561.)

J. Hanrahan. Pat. 158,704, 12 Jan. (U.S.A.).

E. S. Winchester. Pat. 159,241, 26 Jan. (U.S.A.).

E. S. Winchester. Pat. 159,242, 26 Jan. (U.S.A.).

W. F. Tallman and J. N. Mandeville. Pat. 159,471, 2 Feb. (U.S.A.).

H. P. Bell. Pat. 159,885, 16 Feb. (U.S.A.).

Turretini. Similar to 'Darlington' drill. Also Wood & Robinson's drill, U.S.A. Pat. 71,329.

N. W. Horton. Pat. 161,616, 6 April. (U.S.A.).

G. E. Nutting and J. C. Githens. Pat. 161,631, 6 April. Valve moved by a three-armed rocking lever (U.S.A.).

W. W. Goodwin. Pat. 161,948, 13 April. (U.S.A.).

G. E. Nutting and J. C. Githens. Pat. 162,302, 20 April. Tripod for supporting a drill (U.S.A.).

G. H. Reynolds. Pat. 162,419, 20 April (U.S.A.).

C. Burleigh. Pat. 162,528, 27 April (U.S.A.).

G. H. Reynolds and W. Teft. Pat. 163,257, 11 May (U.S.A.).

D. Kennedy. Pat. 163,785, 25 May (U.S.A.).

J. H. Mandeville. Pat. 164,315, 8 June. Tripod improvement on Pat. 159,471. Combination of ball and socket with rotating shaft (U.S.A.).

G. H. Reynolds. Pat. 164,395, 15 June (U.S.A.).

G. H. Reynolds. Pat. 163,395, 15 June (U.S.A.).

G. H. Reynolds. Pat. 164,396, 15 June (U.S.A.).

J. C. Githens. Pat. 164,990, 29 June (U.S.A.).

J. C. Githens. Pat. 164,991, 29 June (U.S.A.).

E. S. Winchester. Pat. 165,646, 13 July. (The Union drill is a combination of the patents of E. S. Winchester and G. H. Reynolds.) (U.S.A.).

J. C. Githens. Pat. 166,273, 3 Aug. (U.S.A.).

C. Ferroux. Pat. 167,324, 31 Aug. (U.S.A.).

H. Thomas. Pat. 168,938, 19 Oct. (U.S.A.).

G. B. Seddon and W. McFaul. Pat. 169,121, 26 Oct. (U.S.A.).

J. B. Waring. Pat. 169,389, 2 Nov. (U.S.A.).

1876 *J. B. Waring.* Pat. 172,529, 18 Jan. (U.S.A.).

J. Brandon. Pat. 174,768, 14 Mar. (U.S.A.).

L. W. Coe. Pat. 174,352, 7 Mar. (Re-issued 25 April 1876. No. 7,079.) (U.S.A.).

L. W. Coe. Pat. 175,931, 11 April. (U.S.A.).

J. B. Waring. Pat. 178,214, 30 May. (Used at Perkiomen Tunnel, Pennsylvania.) (U.S.A.).

S. Ingersoll. Pat. 179,561, 4 July (U.S.A.).

G. H. Reynolds. Pat. 179,818, 11 July (U.S.A.).

Roe and Tallman. Pat. 180,730, 8 Aug. (U.S.A.).

E. S. Winchester. Pat. 181,386, 22 Aug. (U.S.A.).

A. Herring. Pat. 181,576, 29 Aug. (U.S.A.).

Thos. Lawrie. Pat. 34, 4 Jan.

Johann Richard Schram. Pat. 119, 11 Jan.

Ernest De Pass. Pat. 408, 2 Feb.

Andrew B. Brown. Pat. 529, 10 Feb.

Jas. Grafton Jones. Pat. 793, 25 Feb.

Chas. J. Copeland. Pat. 2,189, 24 May.

Wm. W. Dunn. Pat. 2,208, 25 May.

James Mawson. Pat. No. 2,225, 26 May.

Miles Kennedy and Jos. Eastward. Pat. 2,233, 27 May.

John Darlington. Pat. 4,045, 19 Oct. (prov. protection only). Outer surface of the cylinder is screwed to permit of feed. Pressure is introduced through a revolving joint so that the cylinder may be rotated. Feed is by means of a lever struck by the piston (P).

Jas. Mawson. Pat. 4,537, 2 Nov.

W. R. Lake. Pat. 4,682, 4 Dec.

1877 *Adolf Mezger.* used in Rothschönberg tunnel.

R. Schram.

Champonnois. Rotary ratchet-drill.

Lisbet. Rotary drill.

Lisbet & Jacquet. Lisbet's machine simplified. Was used in French and Belgian coal-mines. Rejected at Saarbrücken.

Rziha. Rotary drill similar to Lisbet drill.

Abegg and Richards. Rotary drill. Simple ratchet-drill.

Villepigue. Rotary drill — worked by hand.

Wm. Wynn Kenrick. Pat. 162, 12 Jan. (U.K.).

Thomas B. Ford. Pat. 188,734, 27 Mar. (U.S.A.).

Jos. C. Githens. Pat. 188,045, 6 Mar. (U.S.A.).

Thos. B. Ford. Pat. 188,734, 27 Mar. (U.S.A.).

Thos. B. Ford. Pat. 189,853, 24 April. (U.S.A.).

A. J. Mershon. Pat. 190,332, 1 May. (U.S.A.).

O. B. Keeley and J. Fleming. Pat. 190,871, 15 May. (U.S.A.).

A. A. Goubert. Pat. No. 192,068. 19 June (U.S.A.).

W. W. Dunn. Pat. 194,419, 21 Aug. (U.S.A.).

Geo. Warsop and Henry Walker Hill. Pat. 3,218, 24 Aug. (U.S.A.).

Thos. N. Gaines. Pat. 196,574, 30 Oct. (U.S.A.).

Uriah Cummings. Pat. 196,788, 6 Nov. (U.S.A.).

J. A. Albright. Pat. 197,075, 13 Nov. (U.S.A.).

Prescott B. Buckminster. Pat. 198,086, 25 Dec. (U.S.A.).

Franklin Keenan. Pat. 198,625, 25 Dec. (U.S.A.).

P. S. Buckminster. Pat. 198,486, 25 Dec. (U.S.A.).

Wm. Wynn Kenrick. Pat. 162, 12 Jan. (U.K.).

Jno. Shaw and William Timbrel Clark. Pat. 619, 14 Feb.

Philip John Legros. Pat. 2,054, 26 Mar.

William Wallace Dunn. Pat. 1,485, 16 April.

Wm. Robert Lake. Pat. 1,985, 19 May.

Auguste Alexandre Goubert and Nathaniel Waterman Pratt. Pat. 2,863, 26 July.

Jos. Roseby and Wm. Balmer. Pat. 3,272, 28 Aug.

Thos. Brown Jordan. Pat. 3,510, 18 Sept. (U.K.).

Jno. Drysdale and Roderick Walter Bayner, and Jno Snawdon Stonehouse. Pat. 3,921, 23 Oct. (U.K.).

Ernest De Pass. Pat. 4,027, 30 Oct. (U.K.).

Theodore Frölich. Pat. 4,033, 30 Oct. (U.K.).

John Albert Reinhold Hildebrandt. Pat. 4,316, 17 Nov. (Germany).

Jas. Kergulland. Pat. 4,841, 20 Dec. (U.K.) (R).

1878 S. Winchester. Pat. 199,389, 22 Jan. (U.S.A.).

A. Brandt. Ebensee, Austria. (U.S.A. Pat.) 200,024, 5 Feb.

Charles Burleigh. Pat. 200,690, 26 Feb. (U.S.A.).

W. H. Elliot. Pat. 202,338, 16 April. (U.S.A.).

H. C. Sergeant. Pat. 202,060, 2 June (U.S.A.).

Stephen Morey. Pat. 204,990, 18 June. Hand drill. (U.S.A.).

Chas D. Pierce. Pat. 205,901, 9 July. Hand drill (U.S.A.).

Saml. G. Bryer. Pat. 205,998, 16 July (U.S.A.).

Robt. Allison. Pat. 206,067, 27 June (U.S.A.).

Edward S. Winchester. Pat. 208,448, 24 July (U.S.A.).

John Julien. Pat. 207,122, 20 Aug. (U.S.A.).

Sam'l G. Bryer. Pat. 207,162, 20 Aug. (U.S.A.).

Aaron J. Mershon. Pat. 207,885, 10 Sept. (U.S.A.).

Uriah Cummings. Pat. 210,189, 26 Nov. (U.S.A.).

Thos. B. Jordan and Thos. R. Jordan. Pat. 211,022, 17 Dec. (U.S.A.).

Thomas Brown Jordan and Thos. Rowland Jordan. Pat. 2,572, 26 June. Pneumatic and hydraulic drill. Improvement upon T.B.J. Pat. 1,877, dated 2 Feb, 1877 (U.K.).

Ernest De Pass. Pat. 629, 14 Feb. (U.K.) (P).

Jno. Albert Reinhold Hildebrandt. Pat. 1,973, 16 May. (U.K.).

Geo. Forsyth and Edmund Barnes Ulverston. Pat. 1,944, 18 May (U.K.) (P).

Jno. Brown. Sr. Pat. 2,165, 30 May (U.K.). (P).

Martin Macdermott. Pat. 2,598, 28 June. (U.K.) (P).

Carl Pieper. Pat. 2,696, 5 July. (Dresden, Saxony).

Peter Jensen. Pat. 3,012, 29 July. (U.K.).

1879 Henry Richmann and Uriah Kline Arnold. Pat. 104, 10 Jan.

Tom Bellingham. Pat. 923, 8 March (U.K.).

Jas A. Gulland. Pat. 1,809, 7 May. Diamond drill. (U.K.) (R).

Carl Pieper. Pat. 2,258, 7 June. Improvement to Pat. 2,696, 5 May. 1878 (Strasse, Berlin).

Wm. Lowber Neill. Pat. 2,258, 24 June. (U.K.) (P).

Jno. Imray. Pat. 4,097, 10 Oct. (U.K.) (P).

Max Blumenreich. Pat. 4,455, 1 Nov. Hydraulic (Berlin).

Geo. M. Githens. Pat. 212,598, 25 Feb. (U.S.A.).

J. B. Johnson. Pat. 213,663, 25 Mar. (U.S.A.).

Henry Richmann. Pat. 214,704, 22 April (U.S.A.).

Uriah Cummings. Pat. 215,101, 6 May (U.S.A.).

Thomas Murphy. Pat. 215,152, 6 May (U.S.A.).

Harris Morse. Pat. 223,529, 4 Aug. (U.S.A.) (R).

John Grey Cranston. Pat. 22,388, 19 Aug. (U.S.A.) (P).

E. J. Williams. Pat. 221,646, 11 Nov. (U.S.A.).

John B. Wheat. Pat. 222,972, 23 Dec. (U.S.A.).

1880 *John Brown.* Pat. 223,474, 13 Jan. (U.S.A.).

John Fleming. Pat. 224,412, 10 Feb. (U.S.A.).

Robert MacMullen. Pat. 9,072, 10 Feb. (U.S.A.).

John H. Parkinson. Pat. 225,161, 2 Mar. (U.S.A.).

John Atchinson. Pat. 229,074, 12 Mar. Diamond rock-drill (U.S.A.) (R).

Sam'l G. Bryer. Pat. 227,878, 12 Mar.

Improvement on Pat. 205,998 and 207,162, in manner of moving the pawl of the ratchet of the screw nut, to move the cylinder lengthwise in its arch (U.S.A.).

William L. Neill. Pat. 226,539. April 13 (U.S.A.).

Jas. Ward. Pat. 228,797, 15 June (U.S.A.).

Robert Magil. Pat. 227,908 (U.S.A.).

Geo. M. Githens. Pat. 228,056, 25 June (U.S.A.).

Geo. M. Githens. Pat. 235,080, 12 Dec. (U.S.A.).

1890 *C. H. Shaw.* Hammer-type drill (p). Denver (U.S.A.).

1893 *J. G. Leyner.* Compressed-air drill — piston type (p).

1899 *George Leyner.* Pat. 626,761, 13 June. Hammer-type drill (p) (U.S.A.).

Raise Drill Machines

Calweld-Smith Inc.

Model no.	Year of manufacture	Number of this model manufactured	Thrust kN (lb)	Nominal raise diameter m (ft)	Pilot drill diameter mm (in)
VTB	1967	12	356 (80,000)	To 1.32 (To 4' 4")	
BH-80-60	1974	4	1,668 (375,000) to 2,224 (500,000)		

Dresser Industries Inc. (Drilling Equipment Operations)

Model no.	Year of manufacture	Number of this model manufactured	Thrust kN (lb)	Nominal raise diameter m (ft)	Pilot drill diameter mm (in)
7200	1967		2,200 (500,000)	1.8 (6)	
480	1968-72	18	1,335 (300,000)	1.2-1.8 (4-6)	250.8-311.1 (9⅞-12¼)
800	1971-73	8	2,669 (600,000)	1.8-3.0 (6-10)	311.1 maximum (12¼)
500	1971-79	7	2,224 (500,000)	1.2-1.8 (4-6)	250.8-311.1 9⅞-12¼)
300	1972-79	7	800 (180,000)	0.91-1.2 (3-4)	250.8 nominal (9⅞)

Note: In 1964 Nichols Universal Drilling Co. built raise-boring machines Models RD1 and RD2 which were first used in the Cœur d'Alene mining district of Idaho and Anaconda in Butte, Montana. Subsequently Security Engineering of Dresser Industries purchased the Nichols assets in 1965. With previous experience in the manufacture of cutters and reamer heads the company was able to offer a complete raise drill equipment package. The company's name was then changed to Dresser Oil Field and Mining Equipment and later (approx 1971) to Dresser Mining Services and Equipment Division. Dresser's involvement in raise boring can be traced back to the application of cutters and the first raise-boring machine built by the Robbins Company.

Hughes Tool Company

Model no.	Year of manufacture	Number of this model manufactured	Thrust kN (lb)	Nominal raise diameter m (ft)	Pilot drill diameter mm (in)
MRD 100	1962		900 (200,000)	1.2 (4)	
MRD 100M	1965		900 (200,000)	1.2 (4)	
MRD 60	1966		900 (200,000)	1.2 (4)	
MRD 200	1967		900 (200,000)	1.2 (4)	
MRD 200M	1970		1,000 (225,000)	1.5 (5)	

Raise Drill Machines — continued

Ingersoll-Rand Company

Model no.	Year of manufacture	Number of this model manufactured	Thrust kN (lb)	Nominal raise diameter m (ft)	Pilot drill diameter mm (in)
RBM-7	1972	16	3,020 (680,000)	2.44 (8)	311 (12¼)
RBM-6	1974	8	2,130 (480,000)	1.83 (6)	280 (11)
RBM-211	1977	2	4,444 (1,000,000)	3.66 (12)	349 (13¾)

Note: Ingersoll-Rand Drill Division purchased by the Robbins Company, November 1979, now called 'Robbins Machine Incorporated & Products' sold under Robbins name.

Koken Boring Machine Company Limited

Model no.	Year of manufacture	Number of this model manufactured	Thrust kN (lb) (reaming pull)	Nominal raise diameter m (ft)	Pilot drill diameter mm (in)
BM-1	1967	1			
BM-100	1968	8			
BM-40	1969	2			
BM-200	1969	2			
Current models					
BM-50N	1970	18	441 (99,200)	0.5-0.8 (1.64-2.62)	200.0 (7⅞)
BM-100N	1971	14	1,570 (353,000)	1.15-1.45-1.75 (3.77-4.76-5.74)	250.8 (9⅞)
BM-150N	1971		2,160 (485,000)	1.75-1.83-2.13 (5.74-6.00-7.00)	270.0 (10⅝)
BM-200N	1970	2	2,750 (617,000)	1.83-2.13-2.43 (6.0-7.0-8.0)	349.3 (13¾)

Komatsu Limited (Robbins Company Licensee)

Model no.	Year of manufacture	Number of this model manufactured	Thrust kN (lb)	Nominal raise diameter m (ft)	Pilot drill diameter mm (in)
6ZRK	1973	1	1,372 (308,000)	1.5 (4-9)	Boring down 250 (0-10) Boring up 381 (1-3)

Raise Drill Machines — continued

The Robbins Company

Model no.	Year of manufacture	Number of this model manufactured	Thrust kN (lb)	Nominal raise diameter m (ft) (hard rock)	Pilot drill diameter mm (in)
31R (1101)	1962	1	n.a.	0.9 (3)	171.4 (6¾)
41R	1963	20	890 (200,000)	1.2 (4)	228.6 (9)
61R	1967	58	1,400 (315,000)	1.8 (6)	279.4 (11)
81R	1971	2	5,560 (1,250,000)	2.4-3.6 (8-12)	352.4 (13⅞)
71R	1972	23	2,520 (567,000)	2.4 (8)	279.4 (11)
11D (11MD)	1973	17	220 (49,308)	Rotary drill	200.0-250.8 (7⅞-9⅞)
32R	1973	4	800 (180,000)	1.2 (4)	228.6 (9)
23R	1974	2	340 (77,000)	0.9 (3)	200.0 (7⅞)
52R	1974	13	1,560 (350,000)	Blind hole 1.5 (5)	
72R	1974	7	2,730 (615,000)	2.1 (7)	279.4 (11)
82R	1974	5	3,340 (750,000)	2.4 (8)	311.1 (12¼)
84R	1974	3	3,340 (750,000)	2.4 (8)	311.1 (12¼)
33R	1975	1	3,400 (77,000)	Blind hole 0.9 (3)	
85R	1975	4	4,450 (1,000,000)	3.0-3.6 (1-12)	349.2 (13¾)
63R	1976	1	1,400 (315,000)	1.8 (6)	279.4 (11)
34R	1977	1	800 (180,000)	1.2 (4)	228.6 (9)
121R	1978	1	8,900 (2,000,000)	3.6 (12)	349.2 (13¾)
121BR	1978	1	5,560 (1,250,000)	Raise drill/ Blind-hole drill 4.5 (15)	558.8 (22)
80BR	1978	1	4,450 (1,000,000)	Raise drill/ Blind-hole drill 4.5 (15)	406.4 (16)
43R	1979	3	1,555 (350,000)	1.2-1.5 (4-5)	228 (9)

Raise Drill Machines — continued

Subterranean Tools Inc. (Division of Kennametal Inc.) now Subterranean Equipment Company

Model no.	Year of manufacture	Number of this model manufactured	Thrust kN (lb)	Nominal raise diameter m (ft)	Pilot drill diameter mm (in)
003	1971	6	400 (100,000)	0.9 (3)	200 (7⅞)
007	1971	3	2,200 (500,000)	2.1 (7)	
004	1972	12	900 (200,000)	1.2-1.5 (4-5)	228 (9)
005	1972	7	1,300 (300,000)	1.5-1.82 (5-6)	279 (11)
009	1973	7	3,100 (700,000)	2.13-4.57 (7-15)	279 (11) or 311 (12¼)
006	1974	n.a.	1,800 (400,000)	1.82 (6)	
010	1974	n.a.	4,400 (1,000,000)	3.7 (12)	
UR-60	1973	3		1.5 (5)	
UR-36	1975	1		0.9 (3)	

Note: One of the first machines built by Subterranean Tools Inc. in 1971 (Model 007) was installed in the Molybdenum mine in Colorado in September 1971. In 1973 Subterranean Tools Inc. became a wholly owned subsidiary of Kennametal Inc. Complete raise-boring systems were offered until 15 September 1978. That year Kennametal Inc. decided it would no longer participate in the rock-cutting machine market and a new company, Subterranean Equipment Company, was formed to service the customers of Subterranean Tools Inc. Although not affiliated with Kennametal Inc., Subterranean Equipment Company were given a non-exclusive distributorship for Kennametal cutters.

Tampella-Tamrock

Model no.	Year of manufacture	Number of this model manufactured	Thrust kN (lb)	Nominal raise diameter m (ft)	Pilot drill diameter mm (in)
1000E	1973-79	5	3,200 (719,000)	2.4[a]-3.7[b] (8-12)	280-311 (11-12¼)

Note: Initially Tamrock acted as the agent for Dresser Industries in Scandinavian countries but later started its own raise-boring machine programme.
[a]Hard rock
[b]Soft rock

Raise Drill Machines — continued

TURMAG Turbo-Maschinen AG — Sprockhovel/West Germany

Model no.	Year of manufacture	Number of this model manufactured	Thrust kN (lb)	Nominal raise diameter m (ft)	Pilot drill diameter mm (in)
PVI/120120	to 1965	240	120 (27,000)	0.6 (1.9)	143 (5.63)
P 600	from 1965	42	120 (27,000)	0.6 (1.9)	143 (5.63)
P 30	to 1968	108	250 (56,200)	1.2 (3.9)	193 (7.60)
P 1200	from 1968	60	250 (56,200)	2.4 (7.9)	193 (7.60)
EH 6000	from 1976	3	500 (112,000)	6.0 (19.7)	216 (8.50)

Raise and Box Drill Machines

Wirth Maschinen und Bohrgerate Fabrik GmbH
(Wirth Maschinen GmbH)

Model no.	Year of manufacture	Number of this model manufactured	Thrust kN (lb)	Nominal raise diameter m (ft)	Pilot drill diameter mm (in)
HG 160			863 (194,000)		
HG 210			1,569 (353,000)		
HG 250			2,648 (595,000)		
HG 170S					

Other manufacturers of raise drill machines not included in raise drill machine list: U.S.S.R.

Hard-rock Tunnelling Machines

Atlas Copco Maschinen AG (obtained Patent Rights from Habegger in 1968). *Medium to hard-rock circular and rectangular full-face and mini machines* (see also 'Habegger, Limited' machines)

Model type and year	Project location	Rock strength kg/cm² (MPa) lb/in²	Tunnel length m (ft)	Machine diameter m (ft in)	Cutter head kW (hp)	Thrust kg (lb) kN	Torque kg m (lb ft) kN m	Type cutter and machine weight (tons)
Circular FF 340 4/1967- 11/1967	Julia 1 Switzerland	Flysch 1,835-2,039 (180-200) 26,000-29,000	295 (968)	3.40 (11 2)				
FF 340 5/1968- 10/1968	Julia 2 Switzerland	Metam. schist 612 (60) 8,700	880 (2,887)	3.40 (11 2)				
FF 340 6/1969- 10/1971	Rorschach Switzerland	Sandstone 612-1,835 (60-180) 8,700-26,000	4,530 (14,862)	3.40 (11 2)				
FF 400 8/1973- 7/1975	Elikon -A Greece	Limestone 1,224-1,631 (120-160) 17,000-23,000	5,509 (18,075)	4.25 (13 11½)				
Circular FF 836 FF 840 FF 845 1967-74	Seikan Japan (machines manufactured by IHI — licence holder)		approx. 4,900 (16,076)	3.60 (11 10) 4.00 (13 2) 4.50 (14 9)				
FF 945	Seikan Japan			4.50 (14 9)				
FF 340				3.40 (11 2)				
Rectangular FF 4826 (Prototype) 10/1971- 6/1972	White Pine, U.S.A., copper mine		310 (1,017)	4.80 × 2.60 (15 9 × 8 6)				
Rectangular Mini FF Prototype 8/1971-4/1972	Rorschach, Switzerland	Sandstone 1,020-1,631 (100-160) 15,000-23,000	350 (1,148) 4 tunnels	1.30 × 2.10 (4 3 × 6 11)				
Mini FF 1524 7/1973- 10/1973 (Mini-0)	Innsbruck, Austria	Breccia limestone 306-510 (30-50) 4,400-7,200 1,427-1,835 (140-180) 20,000-26,000	400 (1,312)	1.50 × 2.40 (4 11 × 7 10½)				
Mini FF 1524 12/1973- 1/1974	Neuchâtel, Switzerland	Jura limestone 1,224-1,427 (120-140) 17,000-20,000	115 (377)	1.50 × 2.40 (4 11 × 7 10½)				
Mini FF 1524 3-5/1974	Rochester, U.S.A.	Dolomite limestone 1,276-1,428 (125-140) 18,000-20,000	204 (669)	1.50 × 2.40 (4 11 × 7 10½)				
Rect. Mini FF 1524 (Mini-0) 1/1976-3/1977	Trento, Italy (tunnel flooded 3/1977)	Dolomitic limestone	1,339 (4,390)	1.50 × 2.40 (4 11 × 7 10½)		Not applicable		
12/1977- 2/1978	Balmholz II, Switzerland (conveyor-tunnel also tool test)	Limestone 3,060 (300) 44,000	90 (295)	1.50 × 2.40 (4 11 × 7 10½)		Not applicable		

Hard-rock Tunnelling Machines

Atlas Copco Maschinen AG. Medium to hard-rock circular and rectangular full-face and mini machines

Model type and year	Project location	Rock strength kg/cm² (MPa) lb/in²	Tunnel length m (ft)	Machine diameter m (ft in)	Cutter head kW (hp)	Thrust kg (lb) kN	Torque kg m (lb ft) kN m	Type cutter and machine weight t (tons)
5/1978	Laufenburg, Switzerland (Test)	Granite 3,000 (295) 43,500	11 (36)	1.50 × 2.40 (4 11 × 7 10½)	Not applicable			
(Mini-1) 10/1974	Barden Road, Sydney, Australia	Sandstone 408-612 (40-60) 5,800-8,700	425 (1,394)	1.50 × 2.40 (4 11 × 7 10½)	Not applicable			
2/1975	Como, Sydney, Australia	Sandstone 408-612 (40-60) 5,800-8,700	598 (1,962)	1.50 × 2.40 (4 11 × 7 10½)	Not applicable			
11/1975- 1/1976	Forbes Creek, Sydney, Australia	Sandstone 408-612 (40-60) 5,800-8,700	490 (1,610)	1.50 × 2.40 (4 11 × 7 10½)	Not applicable			
3/1976- 5/1976	New Port, Sydney, Australia	Sandstone, shale 408-612 (40-60) 5,800-8,700	330 (1,083)	1.50 × 2.40 (4 11 × 7 10½)	Not applicable			
6/1976	Avalon, Sydney, Australia	Siltstone mixed with sandstone 300-750 (30-74) 4,350-10,800	795 (2,607)	1.50 × 2.40 (4 11 × 7 10½)	Not applicable			
(Mini-1) 12/1977	Drummoyne, Sydney, Australia	Sandstone 408-612 (40-60) 5,800-8,700	1,182 (3,875)	1.50 × 2.40 (4 11 × 7 10½)	Not applicable			
6/1978	Mill Creek, Sydney, Australia	Sandstone 408-612 (40-60) 5,800-8,700	462 (1,514)	1.50 × 2.40 (4 11 × 7 10½)	Not applicable			
11/1978	Forbes Creek II, Sydney, Australia	Sandstone 408-612 (40-60) 5,800-8,700	511 (1,675)	1.50 × 2.40 (4 11 × 7 10½)	Not applicable			
7/1979	Engadine III, Sydney, Australia	Shale, sandstone 300-500 (30-50) 4,350-7,250	647 (2,121)	1.50 × 2.40 (4 11 × 7 10½)	Not applicable			
Rect. Mini FF 1524 (Mini-2) 8/1974- 10/1974	Washington Metro 1, U.S.A.	Shale 2,039 (200) 29,000	200 (656)	1.50 × 2.40 (4 11 × 7 10½)	Not applicable			
11/1974	Washington Metro 2, U.S.A.	Shale 2,039 (200) 29,000	50 (164)	1.50 × 2.40 (4 11 × 7 10½)	Not applicable			
11/1975- 11/1976	Quincy, Ill., U.S.A.	Metamorphic limestone with layers of chert 1,630-1,840 (160-180) 23,000-26,000	2,385 (7,825)	1.50 × 2.40 (4 11 × 7 10½)	Not applicable			

Hard-rock Tunnelling Machines

Atlas Copco Maschinen AG. Medium to hard-rock circular and rectangular full-face and mini machines — continued

Model type and year	Project location	Rock strength kg/cm² (MPa) lb/in²	Tunnel length m (ft)	Machine diameter m (ft in)	Cutter head kW (hp)	Thrust kg (lb) kN	Torque kg m (lb ft) kN m	Type cutter and machine weight t (tons)
Rect. Mini FF 1524 (Mini-3) 8/1974	Radenthein, Austria (test boring in magnesite mine)		15 (49)	1.50 × 2.40 (4 11 × 7 10½)		Not applicable		
11/1974-2/1975	Innsbruck, Austria	Dolomite limestone approx. 1,224 (120) 17,000	520 (1,706)	1.50 × 2.40 (4 11 × 7 10½)		Not applicable		
17-26 March 1975	Feldkirch, Austria	Limestone, approx. 1,780 (175) 25,000	112 (367)	1.50 × 2.40 (4 11 × 7 10½)		Not applicable		
8-9/1975	Balmholz, Thun, Switzerland	Limestone 3,060 (300) 44,000	50 (164)	1.50 × 2.40 (4 11 × 7 10½)		Not applicable		
(Mini-3) 5/1976	Penarroya, France	Sandstone 1,120-1,330 (110-130) 16,000-19,000	190 (623)	1.50 × 2.40 (4 11 × 7 10½)		Not applicable		
6/1976	Aarburg, Switzerland	Limestone, marl 612-816 (60-80) 8,704-1,200 200-400 (20-40) 2,800-5,700	64 (210)	1.50 × 2.40 (4 11 × 7 10½)		Not applicable		
12/1976	Kiruna, Sweden	Test bore in magnetite ore	27 (89)	1.50 × 2.40 (4 11 × 7 10½)		Not applicable		
1/1977	Trondheim, Norway		120 (394)	1.50 × 2.40 (4 11 × 7 10½)		Not applicable		
6/1977	Lausen, Switzerland	Limestone 1,000-1,300 (100-130) 14,800-19,000	251 (823)	1.50 × 2.40 (4 11 × 7 10½)		Not applicable		
5/1978	Skoumksa, Norway		629 (2,062)	1.50 × 2.40 (4 11 × 7 10½)		Not applicable		
(Mini-4) 1/1976	Inclined shaft of 30°. Duge Power Station, Norway (test bore)	Granite, gneiss		1.50 × 2.40 (4 11 × 7 10½)		Not applicable		
3/1976	Round Hill, England	Sandstone 306-510 (30-50) 4,300-7,200	376 (1,240)	1.50 × 2.40 (4 11 × 7 10½)		Not applicable		
1/1977	Trieste, Italy	Sandy limestone 2,550 (250) 36,000	170 (558)	1.50 × 2.40 (4 11 × 7 10½)		Not applicable		
6/1977	Trenlo II, Italy	Dolomite, sandstone	738 (2,420)	1.50 × 2.40 (4 11 × 7 10½)		Not applicable		
9/1977-2/1979	Simssee, Germany	Very soft sandstone 150-300 (15-30) 2,115-4,350	948 (3,110)					

Hard-rock Tunnelling Machines

Atlas Copco Maschinen AG. Medium to hard-rock circular and rectangular full-face and mini machines — continued

Model type and year	Project location	Rock strength kg/cm² (MPa) lb/in²	Tunnel length m (ft)	Machine diameter m (ft in)	Cutter head kW (hp)	Thrust kg (lb) kN	Torque kg m (lb ft) kN m	Type cutter and machine weight t (tons)
(Mini-5) 6/1976-9/1977	Engadine, Sydney, Australia	Sandstone 408-612 (40-60) 5,800-8,700	3,018 (10,190)	1.50 × 2.40 (4 11 × 7 10½)	Not applicable			
12/1977-3/1978	Bonnet Bay, Sydney, Australia	Sandstone 408-612 (40-60) 5,800-8,700	511 (1,675)	1.50 × 2.40 (4 11 × 7 10½)	Not applicable			
8/1978	Engadine, Sydney, Australia	Sandstone 408-612 (40-60) 5,800-8,700	1,229 (4,030)	1.50 × 2.40 (4 11 × 7 10½)	Not applicable			
11/1979	Deep Creek, Sydney, Australia	Sandstone 408-612 (40-60) 5,800-8,700	1,349 (4,422)	1.50 × 2.40 (4 11 × 7 10½)	Not applicable			
(Mini-6) 12/1975-1/1976	Beaver project, Montreal, Canada	Shale 700-1,500 (70-150) 10,150-21,750	152 (498)	1.50 × 2.40 (4 11 × 7 10½)	Not applicable			
(Mini-6) 2/1977	Fitzpatrick, Montreal, Canada	Shale 700-1,500 (70-150) 10,150-21,750	135 (442)	1.50 × 2.40 (4 11 × 7 10½)	Not applicable			
11/1979	Pewanhee, Canada		45 (147)	1.50 × 2.40 (4 11 × 7 10½)	Not applicable			
(Mini-7) 1977	Tusco, U.S.A.		537 (1,760)	1.50 × 2.40 (4 11 × 7 10½)	Not applicable			
1979	Columbia, U.S.A.		114 (373)	1.50 × 2.40 (4 11 × 7 10½)	Not applicable			
12/1979	St. Louis, U.S.A.		140 (459)	1.50 × 2.40 (4 11 × 7 10½)	Not applicable			
(Mini-8) 4/1978	Schwarzenburg, Switzerland	Sandstone 300-500 (30-50) 4,350-7,250	145 (475)	1.50 × 2.40 (4 11 × 7 10½)	Not applicable			
6/1978	Bremgarten, Switzerland	Sandstone	351 (1,150)	1.50 × 2.40 (4 11 × 7 10½)	Not applicable			
11/1978	Chinon I, France		176 (577)	1.50 × 2.40 (4 11 × 7 10½)	Not applicable			
11/1978	Chinon II, France		178 (583)	1.50 × 2.40 (4 11 × 7 10½)	Not applicable			
11/1978	Chinon III, France		175 (573)	1.50 × 2.40 (4 11 × 7 10½)	Not applicable			
1/1979	Isla Bella, Italy		120 (393)	1.50 × 2.40 (4 11 × 7 10½)	Not applicable			
5/1979	Sonceboz, Switzerland	Tura, limestone	331 (1,085)	1.50 × 2.40 (4 11 × 7 10½)	Not applicable			
11/1979	Lausanne, Switzerland	Limestone	120 (393)	1.50 × 2.40 (4 11 × 7 10½)	Not applicable			
(Mini-9) 9/1978	Pennant Hills, Sydney, Australia	Sandstone 408-612 (40-60) 5,800-8,700	1,079 (3,540)	1.50 × 2.40 (4 11 × 7 10½)	Not applicable			
5/1979	Hornsby Heights, Sydney, Australia	Sandstone 408-612 (40-60) 5,800-8,700	Section I 1,301 (4,265)	1.50 × 2.40 (4 11 × 7 10½)	Not applicable			

667

Hard-rock Tunnelling Machines

Atlas Copco Maschinen AG. Medium to hard-rock circular and rectangular full-face and mini machines — continued

Model type and year	Project location	Rock strength kg/cm² (MPa) lb/in²	Tunnel length m (ft)	Machine diameter m (ft in)	Cutter head kW (hp)	Thrust kg (lb) kN	Torque kg m (lb ft) kN m	Type cutter and machine weight t (tons)
12/1979/1980	Hornsby Heights, Sydney, Australia	Sandstone 408-612 (40-60) 5,800-8,700 (total length of project 8,000 m)	Section II 650 (2,131)	1.50 × 2.40 (4 11 × 7 10½)	Not applicable			
(Mini-10) 1979	Balmholz, Switzerland Roller bit test	Limestone 3,060 (300) 44,000	100 (1,450)	1.50 × 2.40 (4 11 × 7 10½)	Not applicable			
MIDI FF 2131								
(Midi-1) 8/1978	Buffalo, U.S.A.	Limestone with intrusions of chalk 400 (40) 5,700	452 (1,481)	2.15 × 3.30 (7 1 × 10 10)				
(Midi-2) 1978	St. Etienne, France		181 (593)	2.15 × 3.30 (7 1 × 10 10)	Not applicable			

Bade & Co. GmbH (now Bade & Theelen GmbH) Rock Machines

Model type and year	Project location	Rock strength kg/cm² (MPa) lb/in²	Tunnel length m (ft)	Machine diameter m (ft in)	Cutter head kW (hp)	Thrust kg (lb) kN m	Torque kg m (lb ft) kN m	Type cutter and machine weight t (tons)
SVM-33-RM 1956/57	Potash mine, Hangsen, Hanover			3.30 (10 10)	210 (282)	81,000 (179,000) 794		Cutters and breakers 28 (27.5)
SVM-40-RM 1961	Coal-mine, Essen			4.0 (13 1½)	550 (737)	457,000 (1,008,800) 4,482		Gear-tooth type rock bit rollers 105 (103)

Calweld (Division of Smith International Pty. Ltd.) Soft ground, medium and hard-rock tunnelling machines

Model type and year	Project location	Rock strength kg/cm² (MPa)	Tunnel length m (ft)	Machine diameter m (ft in)	Cutter head kW (hp)	Thrust kg (lb) kN	Torque kg m (lb ft) kN m	Type cutter and machine weight t (tons)[a]
TBM-1 2/1962	Chicago sewer, Kenny, Gibson Roberts, Drummond & Bronneck	Clay, gravel, boulders, sands	1,680 (5,500)	2.29 (7 6)	22.4 (30)	54,400 (120,000) 534	6,290 (45,500) 62	Spade 9.07 (8.93)
TBM-2 2/1962	Storm drain, Fulton	Caliche	2,070 (6,800)	2.13 (7 0)	29.8 (40)	54,400 (120,000) 534	6,290 (45,500) 62	Spade 10.7 (10.5)
TBM-3 4/1963	Chicago sewer, Healy	Clay, gravel, boulders, sands, water	5,490 (18,000)	2.74 (9 0)	112 (150)	160,000 (353,000) 1,569	6,290 (45,500) 62	Spade and ripper 20.4 (20.1)
TBM-4 2/1964	Chicago sewer, Kenny	Clay, gravel, boulders, sands, water	3,660 (12,000)	2.74 (9 0)	74.6 (100)	224,000 (494,000) 2,197	8,750 (63,000) 86	Spade and ripper
TBM-5 1/1964	Chicago sewer, Healy	Clay, gravel, boulders, sands, water	2,440 (8,000)	5.79 (19 0)	224 (300)	256,000 (565,000) 2,511	68,600 (496,000) 673	Spade 65.7 (64.7)

[a]American ton = 2,000 lb

Calweld. Soft ground, medium and hard-rock tunnelling machines

Model type and year	Project location	Rock strength kg/cm² (MPa) lb/in²	Tunnel length m (ft)	Machine diameter m (ft in)	Cutter head kW (hp)	Thrust kg (lb) kN	Torque kg m (lb ft) kN m	Type cutter and machine weight t (tons)[a]
TBM-6 9/1964	Chicago sewer, Healy	Clay, gravel, boulders, sands, water	1,580 (5,200)	7.92 (26 0)	224 (300)	256,000 (565,000) 2,511	68,600 (496,000) 673	Spade 65.7 (64.7)
TBM-7 7/1964	Toledo, Peirce	Sticky clay	1,830 (6,000)	2.13 (7 0)	351 (470)	160,000 (353,000) 1,569	6,360 (46,000) 62	Spade and ripper 13.6 (13.4)
TBM-8 8/1964	Cleveland, Kassouf	Soft blue clay, shale	2,130 (7,000)	2.57 (8-5)	351 (470)	192,000 (424,000) 1,883	6,290 (45,500) 62	Spade 18.2 (17.9)
TBM-9 8/1964	Chicago, Kenny	Sandy clay	3,050 (10,000)	3.45 (11 4)	112 (150)	320,000 (706,000) 3,138	12,600 (90,900) 124	Spade 40.8 (40.2)
TBM-10 1/1965	Texas, Gibraltar Co.	Soft clay, sands, unstable	1,050 (3,460)	4.95 (16 3)	112 (150)	385,000 (848,000) 3,776	20,600 (148,800) 202	Spade 27.2 (26.8)
TBM-11 10/1965	Chicago, Kenny	Clay, gravel, boulders, sands, water	2,010 (6,600)	3.45 (11 4)	224 (300)	449,000 (989,000) 4,403	36,500 (264,000) 358	Spade 60.8 (59.8)
TBM-12 1/1966	Munich subway, Wayss & Freytag	Limestone, chert, alluvium	1,830 (6,000)	6.71 (22 0)	373 (500)	2,070,000 (4,560,000) 20,300	68,600 (496,000) 673	Spade 61.3 (60.3)
TBM-13 3/1966	Portsmouth, Streeters Ltd.	Silty, sands, clay	2,290 (7,500)	3.12 (10 3)	209 (280)	320,000 (706,000) 3,138	17,100 (124,000) 168	Spade 43.1 (42.4)
TBM-14 4/1966	Japan Water Supply, Hazama-Gumi	Sandstone, hard clay	7,240 (23,760)	3.84 (12 7)	298 (400)	224,000 (494,000) 2,197	36,500 (264,000) 358	Spade and ripper 79.3 (78.1)
TBM-15 5/1966	Coventry sewer, Streeters Ltd.	Red sandstone	3,050 (10,000)	3.45 (11 4)	298 (400)	224,000 (494,000) 2,197	36,500 (264,000) 358	Disc 60.8 (59.8)
TBM-16 7/1966	Chatham sewer, Arnold & Nathan	Chalk with layers of flint	n.a.	2.13 (7 0)	149 (200)	192,000 (424,000) 1,883	11,300 (82,000) 111	Spade 25.0 (24.6)
TBM-17 6/1966	Minneapolis storm drain, American Structures	Sandstone	2,040 (6,700)	4.88 (16 0)	298 (400)	445,000 (982,000) 4,364	36,500 (264,000) 358	Disc 73.5 (72.3)
TBM-18 8/1966	Anaheim storm drain, Baker-Anderson, Solum	Silty clay, water, sand	3,050 (10,000)	2.95 (9 8)	149 (200)	320,000 (706,000) 3,138	47,400 (343,000) 465	Oscillator S-type 21.7 (21.4)
TBM-19 1/1967	Newhall tunnel, California State water project	Sandstone, siltstone, mudstone	5,470 (17,950)	7.80 (25 7)	597 (800)	3,420,000 (7,536,000) 33,540	110,000 (793,000) 1,079	Spade and disc 109 (107)
TBM-20 12/1966	Pleasant Hills	Sandy clay, sandstone	2,740 (9,000)	4.06 (13 4)	298 (400)	314,000 (692,000) 3,079	36,500 (264,000) 358	Spade and disc 50.0 (49.1)
TBM-21 2/1967	Lompoc missile site	Sandstone		3.05 (10 0)	187 (250)	156,000 (343,000) 1,530	6,290 (45,500) 62	Spade 31.8 (31.3)
TBM-22 4/1967	Munich subway, Zueblin-Hochtief	Limestone, chert, flint	2,740 (9,000)	6.86 (22 6)	373 (500)	572,000 (1,260,000) 5,610	68,600 (496,000) 673	Spade and ripper 67.2 (66.1)
TBM-23 6/1967	Montreal storm drain, Spino Con.	Glacial till, clay, sands, gravel	2,210 (7,250)	2.44 (8 0)	244 (300)	256,000 (564,000) 2,511	47,400 (343,000) 465	Oscillator diamond 30.4 (29.9)

[a] American ton = 2,000 lb

669

Calweld. Soft ground, medium and hard-rock tunnelling machines — continued

Model type and year	Project location	Rock strength kg/cm² (MPa) lb/in²	Tunnel length m (ft)	Machine diameter m (ft in)	Cutter head kW (hp)	Thrust kg (lb) kN	Torque kg m (lb ft) kN m	Type cutter and machine weight t (tons)[a]
TBM-24 8/1967	Sewer lines, Mitchell Bros.	Chalk	2,260 (7,400)	3.45 (11 4)	298 (400)	352,000 (777,000) 3,452	36,500 (264,000) 358	Spade 45.3 (44.6)
TBM-25 12/1967	San Francisco subway	Flowing watery sand	2,130 (7,000)	5.49 (18 0)	783 (1,050)	2,720,000 (6,000,000) 26,675	597,000 (4,320,000) 5,854	Oscillator diamond 136 (134)
TBM-26 6/1967	Milwaukee, Grange Con.	Running material, glacial till	1,370 (4,500)	3.20 (10 6)	223 (300)	753,000 (1,659,000) 7,385	89,100 (644,000) 874	Oscillator sq. bar 34.0 (33.5)
TBM-33 4/1969	Rome subway, S.A.C.O.P.	Tuff clay, mudstone, gravel, marl, hard silt	4,420 (14,500)	6.15 (20 2)	522 (700)	1,140,000 (2,512,000) 11,180	68,600 (496,000) 673	Disc, spade and ripper 114 (112)
TBM-34 4/1969	Rome subway, S.A.C.O.P.	Tuff clay, mudstone, gravel, marl, hard silt	4,420 (14,500)	6.15 (20 2)	522 (700)	1,140,000 (2,512,000) 11,180	68,600 (496,000) 673	Disc, spade and ripper 114 (112)
TBM-35 7/1968	Munich subway, Gruen & Bilfinger	Marl, limestone, flint, alluvium	1,280 (4,200)	7.75 (25 5½)	671 (900)	5,650,000 (12,452,000) 55,410	196,000 (1,420,000) 1,922	Oscillator diamond 135 (133)
TBM-36 5/1968	Calumet sewer, Kiewit	Clay, gravel, rocks	7,620 (25,000)	2.59 (8 6)	250 (335)	1,200,000 (2,656,000) 11,768	24,200 (175,000) 237	Spade 38.5 (37.9)
TBM-37 6/1968	Calumet sewer, American Structures, Kiewit	Clay, gravel, rocks	3,050 (10,000)	2.69 (8 10)	224 (300)	422,000 (930,000) 4,139	27,700 (200,000) 272	Oscillator S-type 38.5 (37.9)
TBM-39 8/1968	Edmonton sewer, Edmonton City	Clay	1,220 (4,000)	2.13 (7 0)	56 (75)	28,100 (62,000) 276	4,150 (30,000) 41	Spade 9.53 (9.38)
TBM-40 10/1969	Climax mine tunnel, Climax Molybdenum	Granitic porphyry, gneiss	610 (2,000)	3.96 (13 0)	597 (800)	512,000 (1,128,000) 5,021	48,000 (347,000) 471	GT 90.7 (89.3)
TBM-41 3/1969	Sepulveda sewer tunnel, California State water project, Drummond & Bronneck	Sandstone	2,230 (7,300)	3.66 (12 0)	224 (300)	186,000 (1,800,000) 1,825	72,500 (524,000) 711	Oscillator diamond 77.1 (75.9)
TBM-42 1/1969	Edmonton sewer, Alta-West Const.	Clay, sand, gravel	1,830 (6,000)	2.84 (9 4)	224 (300)	680,000 (1,500,000) 6,669	44,300 (320,000) 434	Oscillator spade 49.9 (49.1)
TBM-43 2/1969	Tongariro water tunnel, Codelfa Cogefar	Sand, gravel conglomerate	6,000 (19,700)	4.04 (13 3)	298 (400)	953,000 (2,100,000) 9,346	55,300 (400,000) 542	Oscillator diamond 78.5 (77.3)
TBM-44 2/1969	Edmonton sewer, City of Edmonton	Clay	3,050 (10,000)	2.13 (7 0)	56 (75)	28,100 (62,000) 276	4,150 (30,000) 41	Spade 9.53 (9.38)
TBM-45 4/1969	Winnipeg sewer, Earthworm Ltd.	Sand, clay, till, conglomerate	1,520 (5,000)	2.79 (9 2)	224 (300)	680,000 (1,500,000) 6,669	44,300 (320,000) 434	Oscillator diamond 46.7 (46.0)
TBM-46 3/1969	Bunker Hill utility, Artukovich	Clay, silty sand	274 (900)	2.03 (6 8)	56 (75)	28,100 (62,000) 276	4,150 (30,000) 41	Spade 9.07 (8.93)

[a]American ton = 2,000 lb

670

Calweld. Soft ground, medium and hard-rock tunnelling machines — continued

Model type and year	Project location	Rock strength kg/cm² (MPa) lb/in²	Tunnel length m (ft)	Machine diameter m (ft in)	Cutter head kW (hp)	Thrust kg (lb) kN	Torque kg m (lb ft) kN m	Type cutter and machine weight t (tons)[a]
TBM-47 5/1969	Newhall water project, California state water project, Kiewit	Sandstone, siltstone, mudstone	2,740 (9,000)	7.87 (25 10)	1,340 (1,800)	2,750,000 (6,060,000) 26,969	55,300 (4,000,000) 542	Oscillator diamond 272 (268)
TBM-48 4/1969	Chicago sewer tunnel, May Company	Clay	1,980 (6,500)	2.39 (7 10)	75 (100)	192,000 (424,000) 1,883	8,160 (59,000) 80	Spade and ripper 20.0 (19.7)
TBM-49 7/1969	Melbourne water tunnel, Melbourne & Met. Board of Works	Clay, silt, sandstone	10,060 (33,000)	3.40 (11 2)	224 (300)	748,000 (1,650,000) 7,336	72,500 (524,000) 711	Oscillator diamond 74.9 (73.7)
TBM-50 8/1969	Barcelona subway, M.Z.O.V.-CYT	Clay, sand	8,230 (27,000)	5.94 (19 6)	522 (700)	1,130,000 (2,500,000) 11,082	68,600 (496,000) 673	Spade and ripper 114 (112)
TBM-51 5/1969	Edmonton sewer, Edmonton City	Clay	4,570 (15,000)	2.13 (7 0)	56 (75)	28,100 (62,000) 276	4,150 (30,000) 41	Spade 9.53 (9.38)
TBM-52 6/1969	Munich subway, Gruen & Bilfinger	Alluvium, clay, sand, limestone	4,330 (14,200)	7.77 (25 6)	671 (900)	5,650,000 (12,452,000) 55,410	138,000 (1,000,000) 1,353	Drag ripper 136 (134)
TBM-53 11/1969	Mather 'B' mine, Cleveland Cliffs	Hematite	61 (200) Cross cuts	2.97 (9 9)	187 (250)	136,000 (300,000) 1,334	166,000 (1,200,000) 1,628	Oscillator ripper 34.0 (33.5)
TBM-55 10/1969	Lake de Smet, Wyoming, Eagle-Western	Lignite and shale	2,560 (8,400)	3.17 (10 5)	112 (150)	45,400 (100,000) 445	8,300 (60,000) 81	Spade and ripper 27.2 (26.8)
TBM-56 stock				3.66 (12 0)	224 (300)	56,200 (124,000) 551	18,000 (130,000) 177	Disc 40.8 (40.2)
TBM-57 5/1970	Corridor Constructors, Detroit sewer tunnel	Clay, silt, sand	4,450 (14,600)	5.03 (16 7)	373 (500)	1,190,000 (2,660,000) 11,670	113,000 (820,000) 1,108	Spade 90.7 (89.3)
TBM-58 5/1970	Lawrence Ave. No. 2, Reliance Underground Const. Co.	Stiff clay	2,590 (8,500)	4.04 (13 3) 2.97 (9 9)	112 (150)	56,200 (124,000) 551	8,570 (62,000) 84	Spade and ripper 27.2 (26.8)
TBM-59 4/1970	Tauranga, New Zealand, Canadian Constructors	Volcanic tuff	3,660 (12,000)	2.29 (7 6)	75 (100)	76,700 (169,000) 752	7,610 (55,000) 75	Spade and ripper 13.6 (13.4)
TBM-60 6/1970	Edmonton sewer, City of Edmonton	Clay, sand	3,050 (10,000)	4.27 (14 0)	112 (150)	448,000 (988,000) 4,394	12,900 (93,000) 127	Spade 72.5 (71.4)
TBM-63 2/1971	Upper Salt Creek, Chicago, Kenny	Sand, silt	4,880 (16,000)	2.74 (9 0)	149 (200)	399,000 (880,000) 3,913	13,100 (95,000) 128	Spade and ripper 8.61 (8.48)
TBM-64 10/1971	AMAX Coal Co., Ayrshire Coal — 17° incline, McGuire Shaft and Tun. Corp.	Limestone, sandstone, shale	914 (3,000)	5.18 (17 0)	522 (700)	544,000 (1,200,000) 5,335	144,000 (1,044,000) 1,412	Discs 74.9 (73.7)

[a]American ton = 2,000 lb

671

Calweld. Soft ground, medium and hard-rock tunnelling machines — continued

Model type and year	Project location	Rock strength kg/cm² (MPa) lb/in²	Tunnel length m (ft)	Machine diameter m (ft in)	Cutter head kW (hp)	Thrust kg (lb) kN	Torque kg m (lb ft) kN m	Type cutter and machine weight t (tons)[a]
TBM-65 4/1972	Madrid subway, Pacifico-Oporto. F.O.C.-S.I.C.O.P.	Stiff clay, penuela blackish gray clay, gypsum or soft dolemite	5,110 (16,750)	8.48 (27 10)	746 (1,000)	6,420,000 (10,176,000) 62,961	144,000 (1,044,000) 1,412	Discs 364 (358)
TBM-66 5/1972	Eastern Suburbs R.R. Sydney, Australia. Codelfa Construction	Sandstone and basalt	3,960 (13,000)	5.08 (16 8)	597 (800)	544,000 (1,200,000) 5,335	72,200 (522,000) 708	Discs 140 (138)
TBM-69 8/1972	San Donato section Rome-Florence R.R. Vianini	Shale, sandstone, marl, limestone	11,000 (36,150)	11.20 (36 8³/₁₆)	1,490 (2,000)	6,530,000 (14,400,000) 64,040	415,000 (3,000,000) 4,070	Kerf 725 (714)
TBM-70 9/1972	Tonner tunnels, J. F. Shea Co. Inc.	Sandstone, shale, limestone, siltstone	6,960 (22,850)	3.43 (11 3)	298 (400)	6,530,000 (14,400,000) 64,040	41,500 (300,000) 407	Kerf 61.3 (60.3)
TBM-71 11/1972	Crosstown Interceptor sewer, Austin, Texas, Peter Kiewit Sons	Shale, limestone	9,200 (30,200)	3.20 (10 6)	298 (400)	6,530,000 (14,400,000) 64,040	41,500 (300,000) 407	Kerf 58.9 (58.0)

[a]American ton = 2,000 lb

Mannesman Demag AG (Formerly Demag Aktiengesellschaft, Germany) Medium to hard-rock machines

Model type and year	Project location	Rock strength kg/cm² (MPa) lb/in²	Tunnel length m (ft)	Machine diameter m (ft in)	Cutter head kW (hp)	Thrust kg (lb) kN	Torque kg m (lb ft) kN m	Type cutter and machine weight t (tons)[a]
21H 1966 (1)	Sewerage tunnel, Dortmund	Green Sandstone 300-800 (29-78) 4,300-11,000	2,800 (9,190)	2.10 (6 11)	110 (147)	100,000 (220,000) 981		18-23 ring-tooth rollers 40 (39.4)
20-23H 1967 (2)	Water mains, Lake Constance-Stuttgart Veringendorf	Jurassic limestone to 2000 (196) 28,000	3,135 (10,300)	2.14 (7 0)	220 (295)	120,000 (265,000) 1,177	22,400 (162,000) 220	18-23 deposit-weld discs 55 (54.1)
20-23H 1970 (2)	Freshwater tunnel, Stuttgart	Siltstone and sandstone 800-1,500 (78-147) 11,000-21,000	2,400 (7,870)	2.30 (7 7)	220 (295)	120,000 (265,000) 1,177	22,400 (162,000) 220	18-23 deposit-weld discs 55 (54.1)
20-23H 1972 (2)	Freshwater tunnel, Stuttgart	Marl and siltstone 400-600 (39-59) 5,700-8,500	300 984)	2.30 (7 7)	220 (295)	120,000 (265,000) 1,177	22,400 (162,000) 220	16-18 discs with TC buttons
20-23H 1972 (2)	Sewage tunnel, Trondheim, Norway	Greenstone and schist 1,500-2,000 (147-196) 21,000-28,000	4,500 (14,800)	2.30 (7 7)	220 (295)	120,000 (265,000) 1,177	22,400 (162,000) 220	7-10 cutters with/without T.C. buttons 55 (54)

[a]American ton = 2,000 lb

672

Mannesmen Demag AG. Medium to hard-rock machines

Model type and year	Project location	Rock strength kg/cm² (MPa) lb/in²	Tunnel length m (ft)	Machine diameter m (ft in)	Cutter head kW (hp)	Thrust kg (lb) kN m	Torque kg m (lb ft) kN m	Type cutter and machine weight t (tons)[a]
20-23/24H 1973 (2)	Sewage tunnel, Waiblingen	Keuper marl 50-600 (5-59) 700-8,500 Shell limestone 1,500 (147) 21,000	1,000 (3,281)	2.40 (7 10)	220 (295)	120,000 (265,000) 1,177	22,400 (162,000) 220	8-11 cutters with/without T.C. buttons
20-23H 1977 (2)	Water tunnel, Harz	Silicious slate, granite, diabase 500-2,000 (49-196) 7,000-28,000	47,000 (15,419)	2.30 (7 7)	220 (295)	120,000 (265,000) 1,177	22,400 (162,000) 220	8-11 cutters with/without T.C. buttons
20-23H 1968 (3)	Sewerage tunnel, Ramscheid	Greywacke schist 800-1,200 (78-118) 1,100-1,700	300 (984)	2.30 (7 7)	220 (295)	120,000 (265,000) 1,177	22,400 (162,000) 220	18-23 deposit-weld discs 55 (54.1)
20-23H 1968 (3)	Sewerage tunnel, Wuppertal	Greywacke and schist 2,940 (288) 42,000	990 (3,250)	2.30 (7 7)	220 (295)	120,000 (265,000) 1,177	22,400 (162,000) 220	18-23 discs with T.C. buttons 55 (54.1)
20-12H 1968 (3)	Experimental gallery, Drensteinfurt	Marl 600 (59) 8,500	200 (656)	2.30 (7 7)	220 (295)	120,000 (265,000) 1,177	22,400 (162,000) 220	18-23 discs with T.C. buttons 55 (54.1)
20-23H 1970 (3)	Outfall tunnel, Dortmund	Green Sandstone 300-800 (29-78) 4,300-11,000	460 (1,510)	2.30 (7 7)	220 (295)	120,000 (265,000) 1,177	22,400 (162,000) 220	18-23 discs with T.C. buttons 55 (54.1)
20-23H 1971 (3)	Sewer tunnel, Kohlfurt	Schist, limestone and sandstone 200-1,400 (20-137) 2,800-20,000	1,400 (4,590)	2.30 (7 7)	220 (295)	120,000 (265,000) 1,177	22,400 (162,000) 220	7-10 discs with T.C. buttons
20-23H 1973 (3)	Sewage tunnel, Dortmund	Green Sandstone 300-800 (30-78) 4,300-11,000	800 (2,624)	2.30 (7 7)	220 (295)	120,000 (265,000) 1,177	22,400 (162,000) 220	8-11 cutters with/without T.C. buttons
20-23H 1975 (3)	Sewage tunnel, Dortmund	Green Sandstone 300-800 (30-78) 4,300-11,000	800 (2,625)	2.40 (7 10)	220 (295)	120,000 (265,000) 1,177	22,400 (162,000) 220	8-11 cutters with/without T.C. buttons 55 (54)
20-23H 1975 (3)	Sewage tunnel, Dortmund	Green Sandstone 300-800 (30-78) 4,300-11,000	600 (1,969)	2.40 (7 10)	220 (295)	120,000 (265,000) 1,177	22,400 (162,000) 220	8-11 cutters with/without T.C. buttons 55 (54)
28-31H 1968 (4)	Connection Oker-Grane dams, Harz	Schist and sandstone 1,200-2,500 (118-245) 17,000-36,000	6,187 (20,300)	3.15 (10 4)	375 (503)	250,000 (551,000) 2,452	22,400 (162,000) 220	18-23 discs with T.C. buttons 115 (113)
28-31H 1971 (4)	Sewer tunnel, Heiligenhaus	Siltstone and limestone 850-1,400 (83-137) 12,000-20,000	800 (2,630)	2.85 (9 4)	375 (503)	250,000 (551,000) 2,452	22,400 (162,000) 220	8-14 discs with T.C. buttons 105 (103)

[a]American ton = 2,000 lb

Model type and year	Project location	Rock strength kg/cm² (MPa) lb/in²	Tunnel length m (ft)	Machine diameter m (ft in)	Cutter head kW (hp)	Thrust kg (lb) kN m	Torque kg m (lb ft) kN m	Type cutter and machine weight t (tons)[a]
28-31H 1974 (4)	Sewage tunnel, Büsnau	Sandstone 700-1,000 (69-198) 10,000-14,000 Slipstone 250-500 (25-49) 3,600-7,000	2,800 (9,186)	2.80 (9 2)	375 (503)	250,000 (551,000) 2,452	22,400 (162,000) 220	11-17 cutters with/without T.C. buttons 90 (89)
28-31H 1978 (4)	Water tunnel, Wuppertal-Barmen	Sandstone, siltstone 3,000 (294) 43,000	3,200 (10,500)	3.15 (10 4)	375 (503)	250,000 (551,000) 2,452	22,400 (16,100) 220	11-17 cutters with/without T.C. buttons 90 (89)
20-23/24H 1968 (5)	Water tunnel, Loch Lomond, Scotland	Whinstone, schist, limestone, quartz 300-3,000 (29-294) 4,300-43,000	1,200 (3,940)	2.40 (7 10)	220 (295)	120,000 (265,000) 1,177	22,400 (162,000) 220	18-23 discs with T.C. buttons 55 (54.1)
20-23/24H 1978 (5)	Water tunnel, NaBfeld	Gneis	6,100 (20,013)	2.30 (7 7)	220 (295)	120,000 (265,000) 1,177	22,400 (162,000) 220	18-23 discs with T.C. buttons 55 (54.1)
20-23H 1969 (6)	Sewer tunnel, Stockholm, Sweden	Granite 3,000-4,000 (294-392) 43,000-57,000	2,200 (7,220)	2.30 (7 7)	220 (295)	160,000 (353,000) 1,569	22,400 (162,000) 220	18-23 discs with T.C. buttons 55 (54.1)
20-23H 1971 (6)	Freshwater tunnel, Brac, Yugoslavia	Limestone dolomite 1,200-1,400 (118-137) 17,000-20,000	8,600 (28,200)	2.30 (7 7)	220 (295)	160,000 (353,000) 1,569	22,400 (162,000) 220	7-10 discs with T.C. buttons
20-23H 1978 (6)	Böckstein	1,000-1,600 (98-157) 1,400-22,700	930 (3,051)	2.15 (7 0)	220 (295)	160,000 (353,000) 1,569	22,400 (162,000) 220	Discs with T.C. buttons 55 (54.1)
24-27H 1970 (7)	Freshwater tunnel, Erzgebirge, Czechoslovakia	Gneiss 1,000-2,500 (98-245) 14,000-36,000	6,300 (20,700)	2.67 (8 9)	285 (382)	200,000 (441,000) 1,961	22,400 (162,000) 220	8-14 discs with T.C. buttons 77 (85.8)
24-27H 1973 (7)	Freshwater tunnel, Rusova II, Czechoslovakia	Gneiss 1,000-2,500 (98-245) 14,000-36,000	3,100 (10,171)	2.67 (8 9)	285 (382)	200,000 (441,000) 1,961	22,400 (162,000) 220	11-15 cutters with/without T.C. buttons 80 (79)
24-27H 1975 (7)	Freshwater tunnel, Karlsbad, Czechoslovakia	Granite 1,500-2,500 (147-245) 21,000-36,000	1,290 (4,232)	2.67 (8 9)	285 (382)	200,000 (441,000) 1,961	22,400 (162,000) 220	11-15 cutters with/without T.C. buttons 80 (79)
24-27H 1975 (7)	Teplice, Czechoslovakia	Sandstone	1,100 (3,609)	2.67 (8 9)	285 (382)	200,000 (441,000) 1,961	22,400 (162,000) 220	11-15 cutters with/without T.C. buttons 80 (79)
24-27/30H 1972 (8)	Sewer tunnel, Durban, South Africa	Schist and sandstone 1,000-1,800 (98-177) 14,000-26,000	2,050 (6,730)	3.10 (10 2)	285 (382)	200,000 (441,000) 1,961	2,240 (162,000) 220	16 discs with T.C. buttons 82 (80.7)
24-27/30H 1977 (8)	Freshwater tunnel, Damascus, Syria	Limestone marly limestone 500-1,500 (49-147) 7,000-20,000	6,000 (19,685)	2.93 (9 7)	285 (382)	200,000 (441,000) 1,961	2,240 (162,000) 220	11-15 cutters with/without T.C. buttons 80 (79)

[a] American ton = 2,000 lb

Mannesmen Demag AG. Medium to hard-rock machines — continued

Model type and year	Project location	Rock strength kg/cm^2 (MPa) lb/in^2	Tunnel length m (ft)	Machine diameter m (ft in)	Cutter head kW (hp)	Thrust kg (lb) kN m	Torque kg m (lb ft) kN m	Type cutter and machine weight t (tons)[a]
34-38H 1973 (9)	Headrace tunnel, Albula, Switzerland	Limestone and schist 900-1,700 (88-167) 13,000-24,000	4,500 (14,800)	3.80 (12 6)	440 (590)	320,000 (705,000) 3,138	47,900 (346,000) 470	12-19 cutters with/without T.C. buttons 125 (123)
34-38H 1978 (9)	Pilot tunnel, Blaubeuren	1,500 (147) 20,000	2 × 270 (2 × 885)	3.80 (12 6)	440 (590)	320,000 (705,000) 3,138	47,900 (346,000) 470	12-19 cutters with/without T.C. buttons 125 (123)
24-27H 1973 (10)	Sewage tunnel, Halifax, Canada	Greywacke	1,000 (3,281)	2.42 (7 11)	285 (382)	200,000 (441,000) 1,961	22,400 (162,000) 220	11-15 cutters with/without T.C. buttons 125 (123)
24-27H 1978 (10)	Headrace tunnel, Böckstein		4,200	2.70 (8 10)	285 (382)	200,000 (441,000) 1,961	22,400 (162,000) 220	11-15 cutters with/without T.C. buttons 125 (123)
24-27H 1977 (10)	Cable tunnel Prague, Czechoslovakia	Schist 500-1,000 (49-98) 7,000-14,000 Quarzite 2,000 (196) 28,000	700 (2,296) bored so far	2.67 (8 9)	285 (382)	200,000 (441,000) 1,961	22,400 (162,000) 220	11-15 cutters with/without T.C. buttons 125 (123)
38/42HS 1973 (11)	Inclined shaft (70%) headrace tunnel, Mapragg, Switzerland	Limestone 800-1,300 (78-127) 11,000-18,000	1,400 (4,600)	4.20 (13 9)	440 (590)	320,000 (705,000) 3,138	47,900 (346,000) 470	17-25 cutters with/without T.C. buttons 150 (148)
54-58/60H 1973 (12)	Pilot heading, B.A.G. Niederrhein	Schist, sandy-shale, sandstone 400-1,600 (39-157) 5,700-23,000	2,640 (8,700)	6.00 (19 8)	880 (1,180)	640,000 (1,410,000) 6,276	112,000 (809,800) 1,100	18-34 cutters with/without T.C. buttons 300 (295)
54-58/60H 1974 (12)	Pilot heading, 1-0N B.A.G., Niederrhein	Schist, sandy shale, sand-sandstone 400-1,600 (39-157) 5,700-23,000	2,890 (9,482)	6.00 (19 8)	880 (1,180)	640,000 (1,410,000) 6,276	112,000 (809,800) 1,100	18-34 cutters with/without T.C. buttons 300 (295)
54-58/60H 1977 (12)	B.A.G., Niederrhein	Schist, sandy shale and sandstone 400-1,600 (39-157) 5,700-23,000	2,000 (6,561)	6.00 (19 8)	880 (1,180)	640,000 (1,410,000) 6,276	112,000 (809,800) 1,100	18-34 cutters with/without T.C. buttons 300 (295)
54-58/60H 1979 (12)	B.A.G., Niederrhein	Schist, sandy shale and sandstone 400-1,600 (39-157) 5,700-23,000	1,700 (5,577)	6.00 (19 8)	880 (1,180)	640,000 (1,410,000) 6,276	112,000 (809,800) 1,100	18-34 cutters with/without T.C. buttons 300 (295)
36HS 1975 (13)	Pilot heading, Dawdon, England	Schist, sandy shale 600 (59) 8,500	1,000 (3,281)	3.65 (12 0)	440 (590)	320,000 (705,000) 3,138	47,900 (346,000) 470	18-22 cutters with/without T.C. buttons 85 (84)

[a]American ton = 2,000 lb

Model type and year	Project location	Rock strength kg/cm² (MPa) lb/in²	Tunnel length m (ft)	Machine diameter m (ft in)	Cutter head kW (hp)	Thrust kg (lb) kN m	Torque kg m (lb ft) kN m	Type cutter and machine weight t (tons)[a]
34-38H 1975 (14)	Freshwater tunnel, Kielder, England	Sandstone, limestone, marl 2,000 (196) 28,000	12,000 (39,370)	3.50 (11 6)	440 (590)	320,000 (705,000) 3,138	33,600 (243,000) 330	12-19 cutters with/without T.C. buttons 130 (128)
34-38H 1977 (14)	Cable tunnel, Prague, Czechoslovakia	Schist 500-1,000 (49-98) 7,000-14,000 Quartzite 2,000 (196) 28,000	10,000 (32,808)	3.50 (11 6)	440 (590)	320,000 (705,000) 3,138	33,600 (243,000) 330	12-19 cutters with/without T.C. buttons 130 (128)
34-38H 1975 (15)	Freshwater tunnel, Kielder, England	Sandstone, limestone, 2,000 (196) 28,000 Dolorite 4,500 (441) 64,000	7,000 (22,966)	3.50 (11 6)	440 (590)	320,000 (705,000) 3,138	33,600 (243,000) 330	12-19 cutters with/without T.C. buttons 130 (128)
54-58/60H 1976 (16)	Pilot heading, B.A.G., Niederrhein	Schist, sandy shale and sandstone 400-1,600 (39-157) 5,700-23,000	2,720 (8,924)	6.00 (19 8)	880 (1,180)	640,000 (1,410,000) 6,276	112,000 (810,000) 1,100	18-34 cutters with/without T.C. buttons 300 (295)
54-58/61H 1977 (17)	Pilot heading, B.A.G., Dortmund	Schist, sandy shale and sandstone 400-1,600 (39-157) 5,700-23,000	7,000 (22,966)	6.10 (20 00)	880 (1,180)	640,000 (1,410,000) 6,276	112,000 (810,000) 1,100	19-34 cutters with/without T.C. buttons 300 (295)
54-58/60H 1979 (17)	Pilot heading, B.A.G., Westfalen A.R.G.E., Victoria	Schist, sandy shale and sandstone 400-1,600 (39-157) 5,700-23,000	1,600 (5,249)	6.10 (20 00)	880 (1,180)	640,000 (1,410,000) 6,276	112,000 (810,000) 1,100	19-34 cutters with/without T.C. buttons 300 (295)
24-27H 1978 (18)	Water tunnel, Kühtai	Gneiss	5,000 (16,404)	2.70 (8 10)	285 (382)	200,000 (441,000) 1,961	22,400 (162,000) 220	11-15 cutters with/without T.C. buttons
45H 1979 (19)	Coal formations, Petrosani	Sandstone, sandy shale, schist	10,000 (32,808)	4.80 (15 9)	525 (708)	500,000 (1,100,000) 4,903	83,000 (600,000) 814	24-28 cutters with/without T.C. buttons
55H 1979 (20)	Pilot heading, coal formations, Saarbergwerke	Schist, sandstone and sandy shale 400-1,600 (39-157) 5,700-23,000	10,000 (32,808)	6.00 (19 8)	640 (858)	640,000 (1,410,000) 6,276	125,000 (904,000) 1,226	30-34 discs with/without T.C. buttons

[a]American ton = 2,000 lb

Dresser Industries Hard Rock Machines

Model type and year	Project location	Rock strength kg/cm² (MPa) psi	Tunnel length m (ft)	Machine diameter m (ft in)	Cutter head kW (hp)	Thrust kg (lb) kN	Torque kg m (lb ft) kN m	Type cutter and machine weight t (tons)[a]
205 5/1971	Navajo Irrigation Project Tunnel No. 3	70-420 (7-41) Sandstone, shale 1000-6000	4,800 (15,800)	6.25 (20 6)	537 (720)	718,000 (1,583,400) 7,041	146,000 (1,054,620) 1,432	36 double disc cutters (Dresser) and 32 conical picks (Kennamet) 245 (270)

[a]American ton = 2,000 lb

Dresser Industries Hard Rock Machines — continued

Model type and year	Project location	Rock strength kg/cm² (MPa) psi	Tunnel length m (ft)	Machine diameter m (ft in)	Cutter head kW (hp)	Thrust kg (lb) kN	Torque kg m (lb ft) kN m	Type cutter and machine weight t (tons)[a]
205	Navajo Irrigation Project Tunnel No. 3A	70-420 (7-41) Sandstone, shale 1000-6000	1,043 (3,423)	6.25 (20 6)	537 (720)	718,000 (1,583,400) 7,041	146,000 (1,054,620) 1,432	36 double disc cutters (Dresser) and 32 conical picks (Kennamet) 245 (270)

[a]American ton = 2,000 lb

Greenside/McAlpine T.B.M.s

Model Type and year	Project location	Rock strength kg/cm² (MPa) lb/in²	Tunnel length m (ft)	Tunnel diameter m (ft in)	Cutter head kW (hp)	Thrust kg (lb) kN	Torque kg m (lb ft) kN m	Type cutter and machine weight t (tons)
Single head 1968	Hinkley Nuclear Power Station	Limestone/ mudstone 780-2,360 (76-230) 11,000-34,000	1,100 (3,600)	3.1-4.1 (10 2-13 5½)	138 (185)	12,200 (27,000) (119) for sumping		Two-edged picks (tungsten carbide tipped)
1969	Hunterston Nuclear Power Station	Sandstone 640-1,100 (63-108) 9,100-15,600	850 (2,800)	3.6-4.0 (11 9½-13 1)	138 (185)			Two-edged picks (tungsten carbide tipped)
1970	St. Maximin, Provence	Limestone 1,280-1,430 (126-140) 18,200-20,300	1,100 (3,600)	4.1-4.7 (13 5½-15 5)	138 (185)			Two-edged picks (tungsten carbide tipped)
Double head 1970	St. Maximin, Provence	Limestone 1,280-1,430 (126-140) 18,200-20,300		4.1-4.7 (13 5½-15 5)	224 (300)			Two-edged picks (tungsten carbide tipped)
1970	Severn-Wye cable tunnel	Limestone/ sandstone 1,400-2,500 (137-245) 19,900-35,600	1,815 (5,950)	3.5 (11 6)	224 (300)		4,180 (30,000) 41 on the drum 17,538 (127,000) 172 on the arm	Two-edged picks (tungsten carbide tipped)

Habegger Limited Rock machines

Model type and year	Project location	Rock strength kg/cm² (MPa) lb/in²	Tunnel length m (ft)	Machine diameter m (ft in)	Cutter head kW (hp)	Thrust kg (lb) kN	Torque kg m (lb ft) kN m	Type cutter and machine weight t (tons)
836 1966	Pilot railway, tunnel, Japan	352-2,390 (35-234) Tuff, andesite 5,000-34,000	20,100 (66,000)	3.51 (11 6)	485 (650)	234,000 (515,000) 2,295	69,200 (500,000) 679	Picks 86.3 (85)
836 1967	Hydroelectric, Chur, Switzerland	984-1,480 (97-145) Shale, quartz, limestone 14,000-21,000	5,790 (19,000)	3.51 (11 6)	485 (650)	234,000 (515,000) 2,295	69,200 (500,000) 679	Picks 86.3 (85)
829 1967	Water tunnel, Stuttgart, Germany			2.90 (9 6)	410 (550)			Picks 66 (65)
836 1968	Pilot rail tunnel, Japan			3.51 (11 6)	485 (650)	234,000 (515,000) 2,295	69,200 (500,000) 679	Picks 86.3 (85)
840 1968	Pilot rail tunnel, Japan			4.04 (13 3)				

Hughes Tool Company Medium and hard-rock machines

Model type and year	Project location	Rock strength kg/cm² (MPa) lb/in²	Tunnel length m (ft)	Machine diameter m (ft in)	Cutter head kW (hp)	Thrust kg (lb) kN	Torque kg m (lb ft) kN m	Type cutter and machine weight t (tons)[a]
40 in 1958	Hugh B. Williams Various Texas stone quarries	Chalk, sandstone, limestone, granite 176, 1,480, 1,550 and 2,110 (17, 145, 152 and 207) 2,500, 21,000, 22,000, 30,000	Approx. 3,230 (10,600)	1.01 (3 4)	78.3 (105)	36,200 (80,000) 355 40,800 (90,000) 400 77,100 (170,100) 756	3,312 (24,000) 32	9 milled tooth rollers and carbide rollers
1959	National Coal Board, England	Chalk, sandstone, limestone, granite 176, 1,480, 1,500 and 2,110 (17, 145, 152 and 207) 2,500, 21,000, 22,000, 30,000	Approx. 3,230 (10,600)	1.01 (3 4)	78.3 (105)	36,200 (80,000) 355 40,800 (90,000) 400 77,100 (170,100) 756	3,312 (24,000) 32	9 milled tooth rollers and carbide rollers
1960	American Gilsonite Co., Utah	Chalk, sandstone, limestone, granite 176, 1,480, 1,550 and 2,110 (17, 145, 152 and 207) 2,500, 21,000, 22,000, 30,000	Approx. 3,230 (10,600)	1.01 (3 4)	78.3 (105)	36,200 (80,000) 355 40,800 (90,000) 400 77,100 (170,100) 756	3,312 (24,000) 32	9 milled tooth rollers and carbide rollers
40 in Mod. to 54 in 1961-64	American Gilsonite Co., Utah	Siltstone, sandstone 352-1,480 (35-145) 5,000-21,000	Approx. 4,820 (15,800)	1.32 (4 4)	78.3 (105)	40,800 (90,000) 400	3,312 (24,000) 32	13 milled tooth rollers
100 in 1961	Peter Kiewit Sons	Shale, sand-stone, limestone 141, 562 and 1,970 (14, 55 and 193) 2,000, 8,000, 28,000		2.54 (8 4)	96.0 (130)	45,400 (100,000) 445		22 milled tooth rollers and drag teeth
80 in 1963	Morrison Knudsen — Arizona, Michigan. Mogollon Rim, Wirth, West Germany	Sandstone, granite (79-101) 11,000-15,000 (79-101) 1,340-2,110 (131-207) 19,000-30,000	2,290 (7,500)	2.03 (6 8)	216 (290)	272,000 (600,000) 2,668	17,664 (128,000) 173	(Ser. 12) 24 milled tooth rollers and carbide rollers
80 in mod. to 84 in 1963	Morrison Knudsen Co., White Pine, White Pine Copper Co., Michigan	Sandstone, shale 914,844 and 1,480 (90, 83 and 145) 13,000, 12,000 and 21,000	805 (2,640)-3,230 (10,600)	2.13 (7 0)	216 (290)	83,000 (600,000) 814	17,664 (128,000) 173	(Ser. 12) 24 and 14 milled tooth rollers, carbide rollers and disc cutter
238 in to 254 in Betti I 1965	Fenix & Scisson Inc., New Mexico Navajo Irrigation	Sandstone 387-492 (38-48) 5,500-7,000	3,050 (10,600)	6.04-6.45 (19 10-21 2)	858 (1,150)	635,000 (1,400,000) 6,227	156,630 (1,135,000) 1,536	(Ser. 12 and 15) 44 milled tooth rollers and disc cutter 236 (260)
102 in	Reynolds Electrical & Engineering Co., Las Vegas			2.59 (8 6)	224 (300)	272,000 (600,000) 2,668	17,700 (128,000) 174	(Ser. 12) 21 milled tooth or carbide cutters

[a]American ton = 2,000 lb

Jarva Inc. Rock machines

Model type and year	Project location	Rock strength kg/cm² (MPa) lb/in²	Tunnel length m (ft)	Machine diameter m (ft in)	Cutter head kW (hp)	Thrust kg (lb) kN	Torque kg m (lb ft) kN m	Type cutter and machine weight t (tons)[a]
Mark 14 (101) 1308 (1) 10/1965- 5/1966	22 St. relief sewer, Philadelphia, Pa.	Mica schist hornblende 422-1,758 (41-172) 6,000-25,000	497 (1,632)	4.17 (13 8)	373 (500)	393,000 (866,000) 3,854	38,000 (275,000) 373	Carbide insert kerf 73 (80)
1006 10/1966- 4/1967 (1)	St. Louis sewer, St. Louis Metro. sewer dist.	Limestone 984-1,336 (97-131) 1,400-19,000	1,249 (4,100)	3.20 (10 6)	373 (500)	393,000 (866,000) 3,854	38,000 (275,000) 373	Carbide insert kerf 73 (80)
1300 9/1967- 9/1968 (1)	Development drift, Cleveland Cliffs Iron Company	Hematite, greywacke shale 703 (69) 10,000	325 (1,070)	3.96 (13 0)	373 (500)	393,000 (866,000) 3,854	38,000 (275,000) 373	Carbide insert kerf 73 (80)
1400 1/1970- 6/1970 (1)	16° incline shaft, Oak Park mine, Hanna Coal Div.	Sandstone, shale 492 (48) 7,000	549 (1,800)	4.27 (14 0)	373 (500)	393,000 (866,000) 3,854	38,000 (275,000) 373	Carbide insert kerf 73 (80)
Mark 8 (102) 800 6/1965- 12/1965 (2)	St. Louis sewer, St. Louis Metro. Sewer Dist.	Limestone 984-1,195 (97-117) 14,000-17,000	945 (3,100)	2.44 (8 0)	224 (300)	254,000 (560,000) 2,491	18,000 (130,000) 177	Carbide insert 27 (30)
800 1/1968- 4/1968 (2)	St. Louis sewer, St. Louis Metro. Sewer Dist.	Limestone 984-1,195 (97-117) 14,000-17,000	858 (2,814)	2.44 (8 0)	224 (300)	254,000 (560,000) 2,491	18,000 (130,000) 177	Carbide insert 27 (30)
800 3/1968- 7/1968 (2)	Sewer tunnel, Metro. Sanitary Div. of Chicago	Limestone 703-1,125 (69-110) 10,000-16,000	701 (2,300)	2.44 (8 0)	224 (300)	254,000 (560,000) 2,491	18,000 (130,000) 177	Carbide insert 27 (30)
800 8/1969- 8/1969 (2)	St. Louis sewer, St. Louis Metro. Sewer Dist.	Limestone 984-1,195 (97-117) 14,000-17,000	149 (490)	2.44 (8 0)	224 (300)	254,000 (560,000) 2,491	18,000 (130,000) 177	Carbide insert 27 (30)
1000 4/1970- 7/1970 (2)	Milwaukee sewer, Milwaukee, Sewerage Comm.	Limestone, shale 1,055 (103) 15,000	1,313 (4,309)	3.05 (10 0)	280 (375)	192,000 (424,000) 1,882	24,000 (170,000) 235	Carbide insert 41 (45)
806 6/1972- 7/1973 (2)	N. Branch Interceptor, N. Heading, N.Y.C. Dept. of Public Works	Mica, schist 1,406-2,109 (138-207) 20,000-30,000	2,705 (8,875)	2.59 (8 6)	280 (375)	312,000 (688,000) 3,059	22,000 (157,000) 215	43 (47)
900 11/1974- 2/1975 (2)	Water diversion tunnel, Alabama Power Co.	845 (83) 12,000	245 (805)	2.74 (9 0)	280 (375)	312,000 (688,000) 3,059	22,000 (157,000) 215	43 (47)
802 5/1975- 7/1975 (2)	Sanitary sewer, McCandless T'ship Sanitary Authority	Sandstone 563 (55) 8,000	195 (640)	2.49 (8 2)	280 (375)	312,000 (688,000) 3,059	22,000 (157,000) 215	43 (47)
802 2/1976- 4/1976 (2)	Water Div., Amos Power Plant, American Electric Power	Sandstone 563 (55) 8,000	198 (650)	2.44 (8 0)	280 (375)	312,00 (688,000) 3,059	22,000 (157,000) 215	43 (47)

[a]American ton = 2,000 lb

Model type and year	Project location	Rock strength kg/cm² (MPa) lb/in²	Tunnel length m (ft)	Machine diameter m (ft in)	Cutter head kW (hp)	Thrust kg (lb) kN	Torque kg m (lb ft) kN m	Type cutter and machine weight t (tons)[a]
802 5/1976- 7/1976 (2)	Mine development drift, Grace Mine, Bethlehem Steel	Chlorite magnetite 563 (55) 8,000	183 (600)	2.49 (8 2)	280 (375)	312,000 (688,000) 3,059	22,000 (157,000) 215	43 (47)
1000 7/1978- 12 /1978 (2)	− 30% incline slope, Westmoreland Coal Co.	Shale, coal, sandstone, limestone, fireclay 845-1,680 (83-165) 12,000-165,000	319 (1,045)	3.05 (10 0)	280 (375)	312,000 (688,000) 3,059	22,000 (157,000) 215	45 (50)
1000 5/1979 (2)	+ 25% incline slope, United Pocahontas Coal Co.	Shale, sandstone, limestone 563-1,125 (55-110) 8,000-16,000	945 (3,100)	3.05 (10 0)	280 (375)	312,000 (688,000) 3,059	22,000 (157,000) 215	45 (50)
Mark 8 (103) 800 11/1965- 12/1965 (3)	St. Louis sewer, St. Louis Metro. Sewer Dist.	Limestone 984-1,195 (97-117) 14,000-17,000	272 (891)	2.44 (8 0)	223 (300)	254,000 (560,000) 2,490	18,000 (132,000) 177	27 (30)
800 4/1966- 7/1966 (3)	St. Louis sewer, St. Louis Metro. Sewer Dist.	Limestone 984-1,195 (97-117) 14,000-17,000	937 (3,075)	3.05 (10 0)	223 (300)	254,000 (560,000) 2,490	18,000 (132,000) 177	27 (30)
1000 9/1967- 12/1967 (3)	St. Louis sewer, St. Louis Metro. Sewer Dist.	984-1,336 (97-131) 14,000-19,000	1,003 (3,290)	3.05 (10 0)	223 (300)	254,000 (560,000) 2,490	18,000 (132,000) 177	32 (35)
1000 3/1968- 7/1968 (3)	St. Louis sewer, St. Louis Metro. Sewer Dist.	984-1,336 (97-131) 14,000-19,000	792 (2,598)	3.05 (10 0)	223 (300)	254,000 (560,000) 2,490	18,000 (132,000) 177	32 (35)
900 10/1968- 12/1969 (3)	Development drift, Hecla Mining Co.	Quartzite 2,109-3,023 (198-296) 30,000-43,000	152 (500)	2.74 (9 0)	223 (300)	254,000 (560,000) 2,490	18,000 (132,000) 177	32 (35)
900 11/1971- 2/1972 (3)	Sewer tunnel, City of Dunkirk, New York	Shale 211 (21) 3,000	699 (2,294)	2.74 (9 0)	179 (240)	192,000 (424,000) 1,880	12,000 (88,000) 118	32 (35)
800 7/1972- 9/1972 (3)	Utility R.R. crossing, Ohio Bell Telephone Co.	Shale 211 (21) 3,000	183 (600)	2.44 (8 0)	179 (240)	192,000 (424,000) 1,880	12,000 (88,000) 118	32 (35)
801 3/1974- 11/1974 (3)	Contract No. 2, French Creek Sanitary District	Sandstone, shale 181-984 (18-97) 4,000-14,000	975 (3,200)	2.48 (8 1½)	279 (375)	183,000 (403,000) 1,790	22,000 (159,000) 216	43 (47)
1006 9/1975- 2/1976 (3)	Milwaukee sewer tunnel, Milwaukee Sewerage Comm.	Limestone 1,055 (103) 15,000	940 (3,085)	3.05 (10 0)	279 (375)	227,000 (500,000) 2,220	22,000 (159,000) 216	50 (55)
800 8/1976- 11/1976 (3)	Sewer tunnel, Davenport, Iowa	Limestone 844-985 (83-97) 12,000-14,000	917 (3,010)	2.44 (8 0)	279 (375)	183,000 (403,000) 1,790	22,000 (159,000) 216	50 (55)

[a]American ton = 2,000 lb

Model type and year	Project location	Rock strength kg/cm² (MPa) lb/in²	Tunnel length m (ft)	Machine diameter m (ft in)	Cutter head kW (hp)	Thrust kg (lb) kN	Torque kg m (lb ft) kN m	Type cutter and machine weight t (tons)[a]
907 2/1977- 3/1977 (3)	Sewer tunnel, Little Blue Valley Sewer Dist.	Sandstone 703 (69) 10,000	251 (825)	2.92 (9 7)	279 (375)	183,000 (403,000) 1,790	22,000 (159,000) 216	50 (55)
907 10/1977- 12/1977 (3)	Sewer tunnel, Mill Creek Interceptor	Shale 563 (55) 8,000	145 (475)	2.44 (8 0)	279 (375)	183,000 (403,000) 1,790	22,000 (159,000) 216	50 (55)
800 6/1979 (3)	Water Main W-80, Washington Suburban Sanitary Comm.	Quartz gneiss 1,055-2,460 (103-241) 15,000-35,000	1,402 (4,600)	2.44 (8 0)	279 (375)	183,000 (403,000) 1,790	22,000 (159,000) 216	50 (55)
Mark 11 (104) 1000 4/1967- 11/1967 (4)	27° inclined shaft, Republic Steel Corp.	Magnetite, hornblende, biotite and grey granite gneiss 703-2,461 (69-241) 10,000-35,000	234 (768)	3.05 (10 0)	223 (300)	281,000 (620,000) 2,760	24,000 (170,000) 235	50 (55)
1200 9/1968- 6/1969 (4)	River Mt. tunnel, U.S. Bureau of Reclamation	Tuffs, rhyolite, rhyodocite 70-352 210-703 281-1,617 (6.9-35 21-69 28-159) 1,000-5,000 3,000-10,000 4,000-23,000	6,096 (20,000)	3.66 (12 0)	223 (300)·	281,000 (620,000) 2,760	24,000 (170,000) 235	59 (65)
1102 10/1969- 3/1970 (4)	Milwaukee sewer, Milwaukee Sewerage Comm.	Limestone 1,055 (103) 15,000	1,171 (3,841)	3.40 (11 2)	223 (300)	281,000 (620,000) 2,760	24,000 (170,000) 235	59 (65)
102 1/1971- 4/1971 (4)	Milwaukee Sewerage Comm.	Limestone 1,055 (103) 15,000	849 (2,784)	3.40 (11 2)	223 (300)	281,000 (620,000) 2,760	24,000 (170,000) 235	59 (65)
1102 7/1971- 10/1971 (4)	Contract 817 Milwaukee Sewerage Comm.	Limestone 1,055 (103) 15,000	1,343 (4,407)	3.40 (11 2)	223 (300)	281,000 (620,000) 2,760	24,000 (170,000) 235	59 (65)
1102 12/1971- 2/1972 (4)	Contract 843, Milwaukee Sewerage Comm.	Limestone 1,055 (103) 15,000	857 (2,813)	3.40 (11 2)	223 (300)	281,000 (620,000) 2,760	24,000 (170,000) 235	59 (65)
1200 5/1974- 8/1974 (4)	Mill Run sewer, city of Springfield, Ohio	Dolomite 1,130 (111) 16,000	701 (2,235)	3.66 (12 0)	279 (375)	257,000 (566,000) 2,520	25,000 (183,000) 245	59 (65)
1200 8/1975- 4/1976 (4)	Tijuana Aqueduct, Mexican government, Baja, Calif., Mexico	Migmatite 2,800-4,570 (275-448) 40,000-65,000	843 (2,766)	3.66 (12 0)	279 (375)	257,000 (566,000) 2,520	25,000 (183,000) 245	59 (65)
1200 9/1977- 4/1978 (4)	Mine development drift, Pachuca, Mexico	Shale 563 (55) 8,000	1,530 (5,000)	3.66 (12 0)	279 (375)	257,000 (566,000) 2,520	25,000 (183,000) 245	59 (65)

[a]American ton = 2,000 lb

Model type and year	Project location	Rock strength kg/cm² (MPa) lb/in²	Tunnel length m (ft)	Machine diameter m (ft in)	Cutter head kW (hp)	Thrust kg (lb) kN	Torque kg m (lb ft) kN m	Type cutter and machine weight t (tons)[a]
Mark 21 (105) 2000 10/1968- 10/1969 (5)	Bay Area Rapid Transit System, San Francisco, Calif.	Serpentine, greenstone, chert, breccia 70-2,810 (69-276) 1,000-40,000	2,104 (6,900)	6.10 (20 0)	559 (750)	771,000 (1,700,000) 7,560	91,000 (660,000) 892	195 (215)
Mark 21 (106) 1610 6/1969- 11/1970 (6)	Calumet 18E., Ext. A, Metro. Sanitary Dist. of Chicago, Ill.	Dolomite, limestone 984-2,742 (14,000-39,000) 97-269	5,424 (17,794)	5.13 (16 10)	746 (1,000)	771,000 (1,700,000) 7,560	123,000 (890,000) 1,210	195 (215)
1902 2/1976- 4/1978 (6)	Section K-2, Contract 1K0011, WMATA, Washington	Quartz, gneiss 1,055-2,460 (103-241) 15,000-35,000	1,768 (5,800)	6.33 (20 9)	746 (1,000)	816,000 (1,800,000) 8,000	91,000 (660,000) 892	227 (250)
Mark 12 (107) 1100 2/1971- 8/1971 (7)	Queen Lane raw water conduit, city of Philadelphia, Pa.	Mica schist, quartz 422-1,758 (42-172) 6,000-25,000	1,763 (5,784)	3.35 (11 0)	373 (500)	544,000 (1,200,000) 5,430	33,000 (240,000) 324	82 (90)
1403 5/1972- 10/1972 (7)	Moss Point Drainage System, city of Euclid, Ohio	Shale 211 (21) 3,000	1,122 (3,682)	4.34 (14 3)	373 (500)	544,000 (1,200,000) 5,430	33,000 (240,000) 324	82 (90)
1302 2/1974- 7/1974 (7)	Northwest Area inter- ceptor sewer, Contract No. 3, Cleveland Sewer Dist.	Shale 281 (28) 4,000	1,478 (4,850)	4.00 (13 2)	373 (500)	544,000 (1,200,000) 5,430	33,000 (240,000) 324	82 (90)
1202 3/1975- 12/1975 (7)	Northwest Area inter- ceptor sewer, Contract No. 6, Cleveland Sewer Dist.	Shale 281 (28) 4,000	1,981 (6,500)	3.71 (12 2)	373 (500)	544,000 (1,200,000) 5,430	33,000 (240,000) 324	82 (90)
1206 11/1978- 3/1979 (7)	Sanitary sewer, City of Willoughby, Ohio	Shale 280 (27) 4,000	1,524 (5,000)	3.81 (12 6)	373 (500)	544,000 (1,200,000) 5,430	33,000 (240,000) 324	82 (90)
Mark 18 (108) 480 10/1971- 3/1972 (8)[b]	Sakawa river water tunnel, Tokyo, Japan	Tuffs, sand- stone, con- glomerate 563-1,266 (55-124) 8,000-18,000	1,000 (3,280)	4.80 (15 9)	559 (750)	850,000 (1,875,000) 8,340	73,000 (525,000) 716	204 (225)
Mark 12 (109) 1100 2/1972- 3/1973 (9)	N. Branch Interceptor S. Heading, N.Y.C. Dept. of Public Works	Mica schist 1,406-2,109 (138-207) 20,000-30,000	2,863 (9,392)	3.35 (11 0)	373 (500)	544,000 (1,200,000) 5,430	33,000 (240,000) 324	82 (90)
1102 11/1973- 6/1975 (9)	South Ottawa collector sewer, Phase 1	Shale, limestone 210-420 (21-41) 3,000-6,000	3,977 (13,049)	3.40 (11 2)	373 (500)	544,000 (1,200,000) 5,430	33,000 (240,000) 324	82 (90)
Mark 12 (109) 1300 3/1976- 3/1977 (9)	Water inlet tunnel, Municipality of Metro. Toronto	Shale 422 (41) 6,000	3,048 (10,000)	3.96 (13 0)	373 (500)	544,000 (1,200,000) 5,430	33,000 (240,000) 324	86 (95)

[a]American ton = 2,000 lb   [b]Kawasaki/Jarva

Jarva Inc. Rock machines — continued

Model type and year	Project location	Rock strength kg/cm² (MPa) lb/in²	Tunnel length m (ft)	Machine diameter m (ft in)	Cutter head kW (hp)	Thrust kg (lb) kN	Torque kg m (lb ft) kN m	Type cutter and machine weight t (tons)[a]
1300 11/1977- 5/1978 (9)	Sewer tunnel, Sayerville, Contract 33F, N.J.	Shale 563 (55) 8,000	1,280 (4,200)	3.96 (13 0)	373 (500)	544,000 (1,200,000) 5,430	33,000 (240,000) 324	86 (95)
Mark 22 (111) 2100 2/1972- 6/1976 (11)	Kaimai R.R. deviation tunnel, New Zealand Ministry of Works	Welded tuff, ignimbrite, andesite 14-1,055 (1.3-103) 200-15,000	4,780 (15,684)	6.40/5.94 (21 0/19 0)	746 (1,000)	1,000,000 (2,200,000) 9,810	128,000 (921,000)	227 (250)
Mark 24 (112) 2303 11/1974- 1/1977 (12)	Contract No. 313, Melbourne Underground Rail Loop Authority, Australia	Sandstone, siltstone, clay 7-1,756 (0.6-172) 100-25,000	5,200 (17,000)	7.10 (23 3½)	746 (1,000)	454,000 (1,000,000) 4,450	194,000 (1,400,000) 1,900	182 (200)
Mark 12 (113) 1400 9/1975- 2/1976 (13)	8% Urling No. 3 slope, R & P Coal Company, near Indiana, Pa.	Shale, coal, sandstone, limestone, fireclay 140-845 (14-83) 2,000-12,000	762 (2,500)	4.27 (14 0)	373 (500)	544,000 (1,200,000) 5,340	34,000 (244,000) 333	122 (134)
1400 9/1975- 2/1976 (13)	10% Urling No. 4 slope, R & P Coal Company, near Indiana, Pa.	Shale, coal, sandstone, limestone, fireclay 140-845 (14-83) 2,000-12,000	396 (1,300)	4.27 (14 0)	373 (500)	544,000 (1,200,000) 5,340	34,000 (244,000) 333	122 (134)
1400 9/1979 (13)	Conveyor Haulage tunnel, R & P Coal Co., near Indiana, Pa.	Shale, sandstone 563-845 (55-83) 8,000-12,000	427 (1,400)	4.27 (14 0)	373 (500)	544,000 (1,200,000) 5,340	34,000 (244,000) 333	122 (134)
Mark 12 (114) 1200 6/1975- 1/1978 (14)	Project A-206 108″ water-main, City of Montreal	Limestone, shale, Gabbro intrusive 984 (97) 14,000	7,976 (26,138)	3.66 (12 0)	373 (500)	544,000 (1,200,000) 5,340	34,000 (244,000) 333	96 (106)
1206 (14)	Bi-County water tunnel, Washington Suburban Sanitary Comm., Md.	Quartz, granite, schist 1,266 (124) 18,000	5,477 (18,000)	3.8 (12 6)	373 (500)	544,000 (1,200,000) 5,340	34,000 (244,000) 333	96 (106)
Mark 30 (115) 3001 4/1977 (15)	Contract 72-049-2H Addison to Wilmette Metro. Sanitary Dist. of Greater Chicago	Limestone 984-2,742 (97-269) 14,000-39,000	8,458 (27,750)	9.17 (30 1)	1,790 (2,400)	1,361,000 (3,000,000) 13,300	436,000 (3,150,000) 4,280	454 (500)
Mark 22 (116) 2201 10/1976- 11/1978 (16)	Contract 72-049-2H, Addison to Wilmette Metro. Sanitary Dist. of Greater Chicago	Limestone 984-2,742 (97-269) 14,000-39,000	7,315 (24,000)	6.73 (22 1)	895 (1,200)	907,000 (2,000,000) 8,890	160,000 (1,156,000) 1,570	318 (350)

[a]American ton = 2,000 lb

683

Model type and year	Project location	Rock strength kg/cm² (MPa) lb/in²	Tunnel length m (ft)	Machine diameter m (ft in)	Cutter head kW (hp)	Thrust kg (lb) kN	Torque kg m (lb ft) kN m	Type cutter and machine weight t (tons)[a]
Mark 10 (117) 800 12/1975- 5/1976 (17)	Snyder storm relief sewer, Contract No. 3, town of Amherst, N.Y.	Sandstone 563 (55) 8,000	1,308 (4,292)	2.44 (8 0)	279 (375)	227,000 (500,000) 2,220	22,000 (162,000) 216	50 (55)
907 12/1976- 4/1977 (17)	Cont. 1005, Sect. 1.3, Montreal Urban Community, Quebec, Canada	Limestone, shale, gabbro intrusive 984, 280, 2,040 (97, 27, 200) 14,000, 4,000, 29,000	3,350 (11,000)	2.92 (9 7)	279 (375)	227,000 (500,000) 2,220	22,000 (162,000) 216	54 (60)
Mark 12 (118) 1307 12/1976- 10/1977 (18)	Nashville Ave. sewer, Chicago Dept. of Public Works	Limestone 984-2,742 (97-269) 14,000-39,000)	3,700 (12,139)	4.14 (13 7)	447 (600)	544,000 (1,200,000) 5,340	41,000 (293,000) 402	104 (115)
1307 7/1979 (18)	Recycled mill water tunnel, Inland Steel, East Chicago	Limestone 1,125 (110) 16,000	610 (2,000)	4.14 (13 7)	447 (600)	544,000 (1,200,000) 5,340	41,000 (293,000) 402	104 (115)
Mark 6 (119) 606 5/1977- 1/1979 (19)	Sewer tunnel, Davenport, Iowa	Limestone 845-984 (83-97) 12,000-14,000	1,219 (4,000)	1.98 (6 6)	149 (200)	148,000 (325,000) 1,450	11,000 (84,000) 108	25 (27)
706 8/1979 (19)	Hales Corners interceptor sewer, Milwaukee, Wis.	Limestone 1,055 (103) 15,000	1,950 (6,400)	2.29 (7 6)	149 (200)	148,000 (325,000) 1,450	11,000 (84,000) 108	25 (27)
Mark 12 (120) 1202 3/1978 10/1978 (20)	Intake and discharge water tunnel, Perry Nuclear Power Plant, Ohio	Shale 703 (69) 10,000	1,730 (5,678)	3.19 (12 2)	373 (500)	544,000 (1,200,000) 5,340	34,000 (239,000) 333	95 (105)
1206 7/1979 (20)	Bi-County water tunnel, Washington Suburban Sanitary Comm., Rockville, Md.	Quartz, gneiss, schist 1,266 (124) 18,000	4,572 (15,000)	3.81 (12 6)	447 (600)	544,000 (1,200,000) 5,340	34,000 (239,000) 333	95 (105)
Mark 30 (121) 3203 4/1979 (21)	Contract 75-125-2H, Metro. Sanitary Dist. of Greater Chicago	Limestone 984-2,742 (97-269) 14,000-39,000	7,498 (24,600)	9.83 (32 3)	1,790 (2,400)	1,361,000 (3,000,000) 13,300	392,000 (2,833,000) 3,840	765 (850)
Mark 15 (122) 1503 3/1979 (22)	Contract 75-124-2H, Metro. Sanitary Dist. of Greater Chicago	Limestone 984-2,742 (97-269) 14,000-39,000	1,981 (6,500)	4.65 (15 3)	559 (750)	544,000 (1,200,000) 5,340	57,000 (419,000) 559	162 (180)
Mark 10T (123) 1006 1/1980 (23)	Three Rivers water quality management program, Atlanta, Ga.	Gneiss, schist, granite 914-1,547 (90-152) 13,000-22,000	8,473 (27,800)	3.20 (10 6)	335 (450)	418,000 (920,000) 4,100	26,000 (189,000) 255	64 (70)

[a] American ton = 2,000 lb

684

Jarva Inc. Rock machines — continued

Model type and year	Project location	Rock strength kg/cm^2 (MPa) lb/in^2	Tunnel length m (ft)	Machine diameter m (ft in)	Cutter head kW (hp)	Thrust kg (lb) kN	Torque kg m (lb ft) kN m	Type cutter and machine weight t (tons)[a]
Mark 6 (124) 606 6/1980 (24)	Lemont intercepting sewer, 1, 2, and 3 M.S.D.G.C., Chicago	Limestone 1,125 (110) 16,000	5,377 (17,640)	1.98 (6 6)	149 (200)	192,000 (422,000) 1,880	12,000 (87,000) 118	31 (35)
Mark 6 (125) 702 8/1980 (25)	Bi-County water tunnel, Washington Suburban Sanitary Comm. Rockville, Md.	Quartz, gneiss, schist 1,266 (124) 18,000	305 (1,000)	2.18 (7 2)	149 (200)	192,000 (422,000) 1,880	12,000 (87,000) 118	33 (38)

Jarva Inc. Mechanised shields and hybrid machines

Model type and year	Project location	Rock strength kg/cm^2 (MPa) lb/in^2	Tunnel length m (ft)	Machine diameter m (ft in)	Cutter head kW (hp)	Thurst kg (lb) kN	Torque kg m (lb ft) kN m	Type cutter and machine weight t (tons)[a]
SM-1500 Hard-rock slot machine 1979	Mine entry — haulage system, Westmoreland Coal Co., MacAlpin mine to East Gulf mine	Bottom rock	5.6 km (3½ m)	4.57 wide (15 0) 107 cm high (42 in)	224 (300)	68,000 (150,000) 667	20,000 (147,000) 200	Disc cutters 65 (72)
S0911W Mechanized shield 4/1971- 1/1972	Calumet Intercept sewer, No. 17G, Chicago Metro. San. Dist., Chicago, Ill.	Clay	2,118 (6,950)	3.02 (10 0)	134 (180)	435,000 (959,000) 4,266	17,000 (123,000) 167	Trencher teeth 27 (30)
S1205W 4/1971- 5/1972 Mechanized shield	Inner City relief sewer, Ft. Wayne, Ind.	Clay	1,167 (3,830)	3.79 (12 5)	112 (150)	680,000 (1,500,000) 6,669	26,000 (188,000) 255	Trencher teeth 36 (40)
S2209BJ 9/1971- 12/1971 Mechanized shield	Flint River Improv. Atlanta Airport City	Clay	272 (891)	6.95 (22 9½)		2,722,000 (6,000,000) 26,690		159 (175)
S1907-BH 3/1974- 10/1974 Hybrid machine	Northwest Area interceptor sewer, Contract No. 2, Cleveland Sewer Dist.	Stiff grey clay	366 (1,200)	5.97 (19 7)	Shield 45 (60) Backhoe 150 (110)	839,000 (1,840,000) 8,179		38 (42)
S1800BJ 8/1974- 4/1975 Mechanized shield	Section D-4a Contract 1D0041, W.M.A.T.A., Washington D.C.	Stiff clay with traces of wet sandy silt	381 (1,250)	5.50 (18 0)	185 (250)	2,268,000 (5,000,000) 22,242		86.6 (95)
S1800BJ 1/1975- 4/1975 Mechanized shield	Section D-4a Contract 1D0041, W.M.A.T.A., Washington D.C.	Stiff clay with traces of wet sandy silt	381 (1,250)	5.50 (18 0)	185 (250)	2,268,000 (5,000,000) 22,242		86.6 (95)

[a]American ton = 2,000 lb

Jarva Inc. Mechanised shields and hybrid machines

Model type and year	Project location	Rock strength kg/cm² (MPa) lb/in²	Tunnel length m (ft)	Machine diameter m (ft in)	Cutter head kW (hp)	Thrust kg (lb) kN	Torque kg m (lb ft) kN m	Type cutter and machine weight t (tons)[a]
M/F (110) S1907W 4/1972- 2/1973 (10)	Moss Point drainage system, city of Euclid	Shale, clay 141 (14) 2,000	1,335 (4,380)	5.97 (19 7)	690 (925)	2.268,000 (5,000,000) 22,200	111,000 (800,000) 1,090	Rotary discs 159 (175)
S2205W 12/1973- 10/1974 (10)	Capitol Hill relief sewer, Sect. 4, Dept. of Environmental Services	Stiff brown clay, silty clay	594 (1,948)	6.85 (22 5½)	690 (925)	2,268,000 (5,000,000) 22,200	111,000 (800,000) 1,090	182 (200)

[a]American ton = 2,000 lb

Komatsu Limited (Komatsu-Robbins) Rock Machines

Model type and year	Project location	Rock strength kg/cm² (MPa) lb/in²	Tunnel length m (ft)	Machine diameter m (ft in)	Cutter head kW (hp)	Thrust kg (lb) kN	Torque kg m (lb ft) kN m	Type cutter and machine weight t (tons)[a]
TM230G 1964	Nihama tunnel and others, Japan	Rock 598-1,620 (59-159) 8,500-23,000	2,000 (6,560)	2.3 (7 7)	112 (150)	136,000 (300,000) 1,334	27,700 (87,000) 272	Discs 24.5 (27)
TM320G 1967	Matsushima mine, Japan	Rock 984 (97) 14,000	Contin.	3.2 (10 6)	298 (400)	249,000 (550,000) 2,442	79,600 (250,000) 781	Discs 69.8 (77)
TM445G 1967	Enasan tunnel, Japan	Rock 1,410 (138) 20,000	8,500 (27,900)	4.45 (14 7)	477 (640)	499,000 (1,100,000) 4,894	191,000 (600,000) 1,873	Discs 180 (198)
TM430G 1967	Inuyama tunnel, Japan	Rock 1,760 (173) 25,000	1,800 (5,910)	4.3 (14 1)	477 (640)	499,000 (1,100,000) 4,894	191,000 (600,000) 1,873	Discs 120 (132)
1970	Kagawa water supply tunnel, Japan	Rock 2,110 (207) 30,000	8,000 (26,200)	4.3 (14 1)	477 (640)	499,000 (1,100,000) 4,894	191,000 (600,000) 1,873	Discs 120 (132)
TM450G 1968	Taiheizan pilot tunnel for railway	Soft shale, sandstone	800 (2,620)	4.5 (14 9)	500 (670)	499,000 (1,100,000) 4,894	191,000 (600,000) 1,873	Discs 145 (160)
TM350G 1969	Okinawa water tunnel, Japan	Rock 70.3-2,110 (7-207) 1,000-30,000	2,690 (8,820)	3.5 (11 6)	395 (530)	369,000 (814,000) 3,619	57,300 (180,000) 561	Discs 81.6 (90)
TM480G 1970	Kanagawa Pref. wide area water supply tunnel Japan	Rock 400-1,400 (39-138) 6,000-20,000	1,600 (5,250)	4.8 (15 9)	500 (670)	499,000 (1,100,000) 4,894	85,600 (270,000) 839	Discs 154 (170)
TM40G 1972	Gunma Pref. water tunnel, Japan	984-2,110 (97-207) 14,000-30,000	275 (900)	4.4 (14 5)	500 (670)	499,000 (1,100,000) 4,894	85,600 (270,000) 839	Discs 152 (168)
TM340G 1977	Isfahan water tunnel, Iran	Rock 100-2,000 (98-196) 1,400-28,000	5,600 (18,370)	3.4 (11 2)	400 (540)	370,000 (814,000) 3,630	57,300 (180,000) 561	Discs 90 (100)

[a]American ton = 2,000 lb

Komatsu Limited (Komatsu-Robbins) Rock Machines — continued

Model type and year	Project location	Rock strength kg/cm² (MPa) lb/in²	Tunnel length m (ft)	Machine diameter m (ft in)	Cutter head kW (hp)	Thrust kg (lb) kN	Torque kg m (lb ft) kN m	Type cutter and machine weight t (tons)[a]
TM490G 1978	Kouhrang development project, Iran	Rock 100-1,800 (98-176) 1,400-26,000	2,000 (6,560)	4.9 (16 1)	500 (670)	600,000 (1,320,000) 5,900	85,600 (270,000) 839	Discs 160 (176)
TM370G 1979	Kouhrang development project, Iran	Rock 100-1,800 (98-176) 1,400-26,000	6,500 (21,320)	3.7 (12 2)	400 (540)	370,000 (814,000) 3,630	57,300 (180,000) 839	Discs 95 (105)
TM370G 1979	Kouhrang development project, Iran	Rock 100-1,800 (98-176) 1,400-26,000	3,400 (11,150)	3.7 (12 2)	400 (540)	370,000 (814,000) 3,630	57,300 (180,000) 839	Discs 95 (105)

[a]American ton = 2,000 lb

Fried. Krupp GmbH. Medium to semi-hard rock machines

Model type and year	Project location	Rock strength kg/cm² (MPa) lb/in²	Tunnel length m (ft)	Machine diameter m (ft in)	Cutter head kW (hp)	Thrust kg (lb) kN	Torque kg m (lb ft) kN m	Type cutter and machine weight t (tons)
KTF 280 5/1967- 11/1968	Talheim Swabian Alb., S.W. Germany	Brown Jurassic	9,500 (31,000)	2.9 (9 6)	240 (321)	35,000 (77,000) 343	25,000 (181,000) 245	76.2 (75) Toothed discs with hardened inset teeth
KTF 340 1967	Mine, West Germany	Shale, limestone 1,800 (177) 26,000	At 12/1968 1,350 (4,429)	3.7 (12 1½)	240 (321)	35,000 (77,000) 343	25,000 (181,000) 245	81.3 (80) Toothed discs with hardened inset teeth

Lawrence Manufacturing Company (Subsidiary of Ingersoll-Rand Limited) Alkirk-Lawrence rock machines

Model type and year	Project location	Rock strength kg/cm² (MPa) lb/in²	Tunnel length m (ft)	Machine diameter m (ft in)	Cutter head kW (hp)	Thrust kg (lbs) kN	Torque kg m (lb ft) kN m	Type cutter and machine weight t (tons)[a]
HRT-12 1964	Water tunnel, Richmond, U.S.A., New York	Shale, pegmatite with intrusions. Hard Manhattan schist 211-1,970 (21-193) 3,000-28,000	122 (400)	3.66 (12 0)	537 (720)	680,000 (1,500,000) 6,669	48,400 (350,000) 475	Button rollers 71.1 (70)
HRT-13 1968/9	Sewage tunnel, Chicago	Dolomitic limestone 1,122-2,600 (110-255) 1,600-37,000	1,970 (6,463)	4.17 (13 8)	448 (600)	680,000 (1,500,00) 6,669	65,700 (475,000) 644	Button discs Tri-cone button
007 1969	Sewage tunnel, Chicago	Dolomitic limestone 1,122-2,600 (110-255) 1,600-37,000	4,950 (16,240)	4.1 (13 5)	448 (600)	680,000 (1,500,000) 6,669	65,700 (475,000) 644	Button discs Tri-cone button

[a]American ton = 2,000 lb

Lawrence Manufacturing Company. Alkirk-Lawrence rock machines — continued

Model type and year	Project location	Rock strength kg/cm² (MPa) lb/in²	Tunnel length m (ft)	Machine diameter m (ft in)	Cutter head kW (hp)	Thrust kg (lb) kN	Torque kg m (lb ft) kN m	Type cutter and machine weight t (tons)[a]
006 1969	Freshwater supply, Port Huron tunnel, Mich.	Antrim shale and limestone boulders 245-1,060 (24-104) 3,500-15,000	9,620 (33,000)	5.6 (18 4)	560 (750)	680,000 (1,500,000) 6,669	79,500 (575,000) 780	Carbide button discs
006R 1972	Sewage tunnel, Rochester, New York	Sandstone and shale 214-2,182 (21-214) 3,000-31,000	8,994 (29,508)	5.61 (18 5)	671 (900)	907,000 (2,000,000) 8,900	145,000 (1,050,000) 1,420	Carbide button disc rollers
008 1969	Dorchester tunnel, Boston, Mass.	Argillite and andesite 1,693-3,569 (166-350) 24,000-50,000	10,000 (33,000)	3.8 (12 6)	448 (600)	680,000 (1,500,000) 6,669	48,400 350,000 475	Carbide button rollers
009 1969	Long haulage tunnel, Magma copper mine, Ariz.	Dacite, quartzite, limestone and conglomerate 846-3,447 (83-338) 1,200-49,000	2,592 (9,400)	3.8 (12 6)	448 (600)	680,000 (1,500,000) 6,669	65,700 (475,000) 644	Carbide button single row
010 1970	Cookhouse tunnel, Dept. of Water Affairs, South Africa	Sandstone, siltstone, mudstone and dolorite 1,693-3,304 (166-324) 24,000-47,000	10,976 (43,000)	5 (16 6)	597 (800)	680,000 (1,500,000) 6,669	79,500 (575,000) 780	Carbide button single row
Nov 1972	Albula tunnel	Schist, limestone, and shale 280-844 (27-83) 4,000-12,000	6,100 (20,000)	3.81 (12 6)	392 (525)	680,000 (1,500,000) 6,669	48,400 (380,000) 475	Single row carbide button
Dec 1972	Paris sewer tunnel	Marl, limestone 1,050 (103) 2,110 (207) 15,000-30,000	2,680 (8,800)	4.85 (15 11)	448 (600)	680,000 (1,500,000) 6,669	65,700 (475,000) 644	Single row carbide button

[a]American ton = 2,000 lb

Mitsubishi Heavy Industries Limited (Mitsubishi-Hughes) rock machines[b]

Model type and year	Project location	Rock strength kg/cm² (MPa) lb/in²	Tunnel length m (ft)	Machine diameter m (ft in)	Cutter head kW (hp)	Thrust kg (lb) kN	Torque kg m (lb ft) kN m	Type cutter and machine weight t (tons)[a]
126″ 1968	Japanese National Railway and others, Japan	Sandstone, tuff, andesite and granite 844-1,690 (83-166) 12,000-24,000	1,000 (3,280)	3.2 (10 6)	298 (400)	349,000 (770,000) 3,423		(Ser. 12) Milled tooth rollers and carbide rollers
177″ 1969-73	Japanese National Railway, Japan	Andesite, granite 844-2,110 (83-207) 12,000-30,000	2,000 (6,560)	4.5 (14 9)	560 (750)	454,000 (1,000,000) 4,452		(Ser. 12 and 15) 34 Milled tooth rollers, carbide rollers and disc cutters

[a]American ton = 2,000 lb   [b]Licensee of Hughes Tool Company

National Coal Board, United Kingdom. Rock machines

Model type and year	Project location	Rock strength kg/cm² (MPa) lb/in²	Tunnel length m (ft)	Machine diameter m (ft in)	Cutter head kW (hp)	Thrust kg (lb) kN	Torque kg m (lb ft) kN m	Type cutter and machine weight t (tons)
Bretby 1961-62	Trial site — limestone quarry, Breedon-on-the-Hill, Leicestershire, U.K.	Dolomitized limestone 633-2,530 av. strength 1,410 (62-248) av. strength (138) 9,000-36,000 av. strength 20,000	62.3 (205)	5.49 (18 0)	560 (750)	447,000 (986,000) 4,384		88 roller cutters 305 (300)
1965	Iron ore mine, Dragonby, U.K., United Steel Co. Ltd.	Limestone, iron ore 281-914 (28-90) 4,000-13,000	Approx. 3,230 (10,600)	5.49 (18 0)	560 (750)	447,000 (986,000) 4,384		20 disc cutters 11 roller cutters 305 (300)
1968	Experimental	Limestone 1,550 (152) 22,000		1.83 (6 0)	70.9 (95)			Discs

The Robbins Company. Rock machines

Model type and year	Project location	Rock strength kg/cm² (MPa) lb/in²	Tunnel length m (ft)	Machine diameter m (ft in)	Cutter head kW (hp)	Thrust kg (lb) kN	Torque kg m (lb ft) kN m	Type cutter and machine weight t (tons)[a]
910-101 1953	Oahe Dam, Pierre, S. Dak.	Shale, faulted and jointed, bentonite 14-28 (1-3) 200-400	6,858 (22,500) six tunnels total by machines 910 and 930	8 (26 3)	298 (400)	45,400 (100,000) 445	38,900 (281,000) 381	Fixed and disc 113 (125)
930-102 1954 1955	Oahe Dam, Pierre, S. Dak.	Shale, faulted and jointed, bentonite 14-28 (1-3) 200-400	6,858 (22,500) six tunnels total by machines 910 and 930	8 (26 3)	298 (400)	45,400 (100,000) 445	38,900 (281,000) 381	Fixed and disc 113 (125)
101-103 1956	Sewer tunnel, Pittsburgh, Pa.	Tough shale 352-844 (35-83) 5,000-12,000		2.44 (8 0)	231 (310)	53,100 (117,000) 521	19,100 (138,000) 187	Fixed and disc 15.4 (17)
102-104 1956	Sewer tunnel, Pittsburg, Pa.	Shale and limestone interbedded 352-1,050 (35-103) 5,000-15,000		2.59 (8 6)	231 (310)	53,100 (117,000) 521	14,800 (107,000) 145	Fixed and disc 15.4 (17)
103-105 1956	Sewer tunnel, Chicago, Ill.	Hard Chicago limestone 1,266-1,760 (124-173) 18,000-25,000		2.74 (9 0)	231 (310)	53,100 (117,000) 521	19,100 (138,000) 187	Fixed 15.4 (17)
1957	Steep Rock iron mines, Atikokan, Ontario	Limonite 35-352 (3-35) 500-5,000	304.8 (1,000)	2.74 (9 0)	231 (310)	53,100 (117,000) 521	19,100 (138,000) 187	Fixed 15.4 (17)

[a]American ton = 2,000 lb

Model type and year	Project location	Rock strength kg/cm² (MPa) lb/in²	Tunnel length m (ft)	Machine diameter m (ft in)	Cutter head kW (hp)	Thrust kg (lb) kN	Torque kg m (lb ft) kN m	Type cutter and machine weight t (tons)ᵃ
1957	Thetford mines, Asbestos, Quebec	Greenstone 352-1,050 (35-103) 5,000-15,000	30 (100)		231 (310)	53,100 (117,000) 521		Fixed 15.4 (17)
103-105 1960	Grants, N.M.	Sandstone 84-176 (8-17) 1,200-2,500		2.74 (9 0)	231 (310)	53,100 (117,000) 521	19,100 (138,000) 187	Fixed 15.4 (17)
131-106 1956	Humber river sewer, Toronto, Ontario	Sandstone, shale, crystalline limestone 562-1,900 (55-186) 8,000-27,000	4,510 (14,800)	3.28 (10 9)	254 (340)	142,000 (314,000) 1,393	24,300 (314,000) 238	Discs — 24 59 (65)
1961	Sewer tunnel, Ohio	Shale 352-703 (35-69) 5,000-10,000	4,510 (14,800)	3.28 (10 9)	254 (340)	142,000 (314,000) 1,393	24,300 (314,000) 238	Discs — 24 59 (65)
351-107 1960	Oahe dam, Pierre, S. Dak.	Shale, faulted and jointed bentonite seams 14-28 (1-3) 200-400	seven tunnels 2,380 (7,800)	9.0 (29 6)	507 (680)	118,000 (260,000) 1,157	94,600 (684,000) 928	Discs — 44 159 (175)
261 (Converted from 351) 1961	Gardiner dam div. tunnels, South Saskatchewan river, Canada	Soft shale, badly faulted and caving	5,580 (18,302)	7.82 (25 8)	671 (900)	118,000 (260,000) 1,157	484,000 (3,500,000) 4,747	Fixed 159 (175)
161-108 1961-73	Poatina tunnel hydro project, Tasmania, Australia	Massive mudstone, sandstone 703-1,200 (69-118) 1,000-17,000	6,860 (22,500)	4.93 (16 2)	448 (600)	279,000 (615,000) 2,736	105,000 (762,000) 1,030	33 discs 104 (115)
1964	Rhyndaston Expan., Hobart, Tasmania	Massive mudstone, sandstone 703-1,200 (69-118) 1,000-17,000	945 (3,100)	4.93 (16 2)	448 (600)	279,000 (615,000) 2,736	105,000 (762,000) 1,030	33 discs 104 (115)
1973	Shoalhaven C5 tunnel, Sydney		2,440 (8,000)	4.93 (16 2)	448 (600)	279,000 (615,000) 2,736	105,000 (762,000) 1,030	33 discs 104 (115)
71-109 1962	Sewer, McLean, Va.	Shale 352-1,050 (35-103) 5,000-15,000		2.18 (7 2)	74.6 (100)	178,000 (393,000) 1,746		14 discs 31.7 (35)
71A-109 1963	Homer-Wauseca Iron Mine, Iron River, Mich.	Limestone, iron ore 703-1,050 (69-103) 10,000-15,000	354 (1,163)	2.13 (7 0)	74.6 (100)	178,000 (393,000) 1,746		14 discs 31.7 (35)
71B-109 1963	Sewer, Chicago, Ill.	Limestone 352-1,050 (35-103) 5,000-15,000		2.13 (7 0)	74.6 (100)	178,000 (393,000) 1,746		14 discs 31.7 (35)
371-110 1963-64	Mangla dam, West Pakistan	Soft sandstone, clay with hard sandstone, lime 70-562 (7-55) 1,000-8,000	4,270 (14,000)	11.2 (36 8)	746 (1,000)	272,000 (600,000) 2,668	726,000 (5,250,000) 7,120	53 discs 32 fixed tools 290 (320)

ᵃAmerican ton = 2,000 lb

Model type and year	Project location	Rock strength kg/cm² (MPa) lb/in²	Tunnel length m (ft)	Machine diameter m (ft in)	Cutter head kW (hp)	Thrust kg (lb) kN	Torque kg m (lb ft) kN m	Type cutter and machine weight t (tons)[a]
1968-72	Twin tube, 2nd Mersey tunnel, Liverpool, England			10.6 (34 11)	746 (1,000)	272,000 (600,000) 2,668	726,000 (5,250,000) 7,120	53 discs 32 fixed tools 290 (320)
72-112 1965-67	Sumitoma Metal Mining Co. Ltd., Japan	Chlorite, schist 703-1,410 (69-138) 10,000-20,000		2.13 (7 0)	112 (150)	178,000 (393,000) 1,746	11,800 (85,000) 116	Discs 31.7 (35)
81-113 1965	Sooke Lake-Goldstream water tunnel, Victoria, B.C.	Hard and soft schist 1,050-1,410 (103-138) 15,000-20,000	914 (3,000)	2.59 (8 6)	149 (200)	178,000 (393,000) 1,746	15,200 (110,000) 149	18-20 discs 1 tri-cone 40.8 (45)
1968	Water tunnel, Starvation, Ut.	Sandstone and shale 70-562 (7-55) 1,000-8,000	1,620 (5,300)	2.59 (8 6)	149 (200)	178,000 (393,000) 1,746	15,200 (110,000) 149	18-20 discs 1 tri-cone 40.8 (45)
1968-69	Alb Tun. Water Supply, Stuttgart, West Germany	Limestone 352-1,050 (35-103) 5,000-15,000	4,940 (16,200)	2.90 (9 6)	149 (200)	178,000 (393,000) 1,746	15,200 (110,000) 149	18-20 discs 1 tri-cone 40.8 (45)
1970	Water supply, Vienna, Austria	Limestone 352-1,050 (35-103) 5,000-15,000	905 (2,970)	2.90 (9 6)	149 (200)	178,000 (393,000) 1,746	15,200 (110,000) 149	18-20 discs 1 tri-cone 40.8 (45)
1972	Sewer tunnel, Geneva, Switzerland	Molasse 633-914 (62-90) 9,000-13,000	5,000 (16,405)	2.90 (9 6)	149 (200)	178,000 (393,000) 1,746	15,200 (110,000) 149	18-20 discs 1 tri-cone 40.8 (45)
1977	Drainage tunnel, Beckenreid		500 (1,640)	2.90 (9 6)	149 (200)	178,000 (393,000) 1,746	15,200 (110,000) 149	18-20 discs 1 tri-cone 40.8 (45)
1977	Drainage tunnel, Oeffingen	1,200 (117) 17,000		2.59 (8 6)	149 (200)	178,000 (393,000) 1,746	15,200 (110,000) 149	18-20 discs 1 tri-cone 40.8 (45)
73-114 1964	Water tunnel, Payson, Ariz.	Sandstone 210-490 (20-48) 3,000-7,000	8.5 (28)	2.13 (7 10)	112 (150)	178,000 (392,700) 1,750	11,755 (85,000) 115	16 discs 31.7 (35)
73-114-1 1975	Water jet test, Skykomish, Washington	Granite-gneiss 1,400-3,160 (137-309) 20,000-45,000	12 (40)	2.13 (7 10)	112 (150)	178,000 (392,700) 1,750	11,755 (85,000) 115	16 discs 31 jets 31.7 (35)
73-114-2 1980	Buffalo, N.Y.	Limestone 910-2,100 (89-205) 13,000-30,000		2.13 (7 10)	112 (150)	178,000 (392,700) 1,750	11,755 (85,000) 115	
74-115 1965-66	Water tunnel pilot bore, Bessans, France	Schist 703-1,270 (69-125) 10,000-18,000	300 (984)	2.18 (7 2)	112 (150)	178,000 (393,000) 1,746	11,800 (85,000) 116	14 discs 31.7 (35)
1966	Pilot tunnel, Met. Paris	Limestone 70-1,050 (7-103) 1,000-1,500		2.18 (7 2)	112 (150)	178,000 (393,000) 1,746	11,800 (85,000) 116	14 discs 31.7 (35)
121-116 1965 Hybrid	Azotea Water tunnel, Chama, N.M.	Shale and sandstone 70-562 (7-55) 1,000-8,000	20,400 (67,000)	3.81 (12 6) 4.03 (13 3)	298 (400)	218,000 (480,000) 2,138	40,400 (292,000) 396	27-29 discs 1 tri-cone 68 (75)

[a]American ton = 2,000 lb

691

Model type and year	Project location	Rock strength kg/cm² (MPa) lb/in²	Tunnel length m (ft)	Machine diameter m (ft in)	Cutter head kW (hp)	Thrust kg (lb) kN	Torque kg m (lb ft) kN m	Type cutter and machine weight t (tons)[a]
1968-69	Water hollow tunnel, Heber City, Ut.	Sandstone and conglomerate 352-1,050 (35-103) 5,000-15,000	4,820 (15,800)	3.94 (12 11)	298 (400)	218,000 (480,000) 2,138	40,400 (292,000) 396	27-29 discs 1 tri-cone 68 (75)
111-117 1965-66	Sewer tunnel, Baden, Switzerland	Soft sandstone 70-352 (7-35) 1,000-5,000	2,060 (6,768)	3.71 (12 2)	224 (300)	209,000 (460,000) 2,050	44,000 (318,000) 432	26-28 discs 1 tri-cone 65.3 (72)
1966	Sondier test tunnel, Julia, Sweden	Shale, conglomerate 492-1,410 (48-138) 7,000-20,000	154 (505)	3.71 (12 2)	224 (300)	209,000 (460,000) 2,050	44,000 (318,000) 432	26-28 discs 1 tri-cone 65.3 (72)
1966	Sewer tunnel, Lucerne, Switzerland	Sandstone conglomerate 633-984 (62-97) 9,000-14,000	563 (1,847)	3.71 (12 2)	224 (300)	209,000 (460,000) 2,050	44,000 (318,000) 432	26-28 discs 1 tri-cone 65.3 (72)
1967	Sewer tunnel, Gaislingen, West Germany	Limestone, marl 1,195-1,476 (117-145) 17,000-20,000	1,440 (4,718)	3.71 (12 2)	224 (300)	209,000 (460,000) 2,050	44,000 (318,000) 432	26-28 discs 1 tri-cone 65.3 (72)
1968-69	Conveyor tunnel, Blaubeuren, West Germany	Limestone 281-1,830 (28-179) 4,900-26,000	742 (2,434)	3.71 (12 2)	224 (300)	209,000 (460,000) 2,050	44,000 (318,000) 432	26-28 discs 1 tri-cone 65.3 (72)
1969-70[b]	Sonnenberg pilot tunnel, Lucerne, Switzerland	Sandstone 773-1,200 (76-118) 11,000-17,000	2,810 (9,229)	3.71 (12 2)	224 (300)	209,000 (460,000) 2,050	44,000 (318,000) 432	26-28 discs 1 tri-cone 65.3 (72)
111-117 1971	Test bore, Sondier tunnel, Seelisberg, Switzerland	Marl 703 (69) 10,000	654 (2,146)	3.71 (12 2)	224 (300)	209,000 (460,000) 2,050	44,000 (318,000) 432	26-28 discs 1 tri-cone 65.3 (72)
1971	Utility tunnel, Lugano Switzerland	Limestone 1,200 (117) 17,000	1,300 (4,248)	3.71 (12 2)	224 (300)	209,000 (460,000) 2,050	44,000 (318,000) 432	26-28 discs 1 tri-cone 65.3 (72)
1971	Mine Addit, Schelklingen, West Germany	1,410 (138) 20,000	1,240 (4,068)	3.71 (12 2)	224 (300)	209,000 (460,000) 2,050	44,000 (318,000) 432	26-28 discs 1 tri-cone 65.3 (72)
1972	Pipeline tunnel, Obergestein, Switzerland	Schist 1,200-1,800 (117-177) 17,000-26,000	1,840 (6,037)	3.71 (12 2)	224 (300)	209,000 (460,000) 2,050	44,000 (318,000) 432	26-28 discs 1 tri-cone 65.3 (72)
1973	Friedental tunnel, Lucerne, Switzerland		500 (1,640)	3.7 (12 2)	224 (300)	209,000 (460,000) 2,050	44,000 (318,000) 432	26-28 discs 1 tri-cone 65.3 (72)
1974-75	Lavtina-Stausee tunnel, Gigerwald, Switzerland		5,800 (19,030)	3.50 (11 6)	224 (300)	209,000 (460,000) 2,050	44,000 (318,000) 432	26-28 discs 1 tri-cone 65.3 (72)
1977	Inclined shaft ('emergency' cable car), Sunnegga, Zermatt, Switzerland		1,800 (5,900)	3.50 (11 6)	224 (300)	209,000 (460,000) 2,050	44,000 (318,000) 432	26-28 discs 1 tri-cone 65.3 (72)

[a]American ton = 2,000 lb   [b]Used in conjunction with Wirth Enlarging T.B.M.

Model type and year	Project location	Rock strength kg/cm² (MPa) lb/in²	Tunnel length m (ft)	Machine diameter m (ft in)	Cutter head kW (hp)	Thrust kg (lb) kN	Torque kg m (lb ft) kN m	Type cutter and machine weight t (tons)[a]
81-118 1965-66	Sewer tunnel, Frieburg, Switzerland	Soft sandstone 422-775 (41-76) 6,000-11,000	2,150 (7,041)	2.59 (8 6)	149 (200)	178,000 (393,000) 1,746	16,300 (118,000) 160	19 discs 40.8 (45)
1967-68	Zurichberg water supply, Zurich, Switzerland	Sandstone and marl 246-703 (24-69) 3,500-10,000	2,050 (6,716)	2.59 (8 6)	149 (200)	178,000 (393,000) 1,746	16,300 (118,000) 160	19 discs 40.8 (45)
1969	Hydro tunnel, Wattenback, Austria	Lime, philyte 773-1,195 (76-117) 11,000-17,000	979 (3,212)	2.59 (8 6)	149 (200)	178,000 (393,000) 1,746	16,300 (118,000) 160	19 discs 40.8 (45)
1969-70	Sewer tunnel, Flawil, Switzerland	Marl and conglomerate 633-703 (62-69) 9,000-10,000	663 (2,175)	2.59 (8 6)	149 (200)	178,000 (393,000) 1,746	16,300 (118,000) 160	19 discs 40.8 (45)
1970	Sewer tunnel, Berne, Switzerland	Sandstone 422-703 (41-69) 6,000-10,000	873 (2,864)	2.59 (8 6)	149 (200)	178,000 (393,000) 1,746	16,300 (118,000) 160	19 discs 40.8 (45)
1971-73	Hardhof water tunnel, Zurich, Switzerland	Sandstone 773 (76) 11,000	5,610 (18,393)	2.59 (8 6)	149 (200)	178,000 (393,999) 1,746	16,300 (118,000) 160	19 discs 40.8 (45)
1977 81-118	Water tunnel, Feickirch		700 (2,297)	2.59 (8 6)	149 (200)	178,000 (393,000) 1,746	16,300 (118,000) 160	19 discs 40.8 (45)
1977	Power tunnel, Scheubsberg		1,638 (5,374)	2.59 (8 6)	149 (200)	178,000 (393,000) 1,746	16,300 (118,000) 160	19 discs 40.8 (45)
1977	Rosenberg highway tunnel, Switzerland		500 (1,640)	2.59 (8 6)	149 (200)	178,000 (393,000) 1,746	16,300 (118,000) 160	19 discs 40.8 (45)
104-120 1966	Blanco water tunnel, Col.	Soft shale 70-352 (7-35) 1,000-5,000	12,800 (42,000)	3.02 (9 11)	224 (300)	172,000 (380,000) 1,687	28,400 (205,000) 279	20-25 discs 1 tri-cone 54.4 (60)
104-120-1 1968	Water supply reservoir, Pseux, Switzerland	Limestone 984-2,320 (97-228) 14,000-33,000	524 (1,719)	3.02 (9 11)	224 (300)	172,000 (380,000) 1,687	28,400 (205,000) 279	20-25 discs 1 tri-cone 54.4 (60)
1969	Mine vent tunnel, Muhlbach, Austria	Philyte, green rock 773-1,480 (76-145) 11,000-21,000	1,020 (3,334)	3.02 (9 11)	224 (300)	172,000 (380,000) 1,687	28,400 (205,000) 279	20-25 discs 1 tri-cone 54.4 (60)
104-120-1 1970	Electric cable tunnel, Berne, Switzerland	Sandstone 420-700 (41-69) 6,000-10,000	662 (2,172)	3.02 (9 11)	224 (300)	172,000 (380,000) 1,687	28,400 (205,000) 279	20-25 discs 1 tri-cone 54.4 (60)
1972	Stadtler sewer tunnel, Cham, Switzerland	Sandstone 800-1,100 (78-108) 11,000-16,000	3,336 (10,950)	3.02 (9 11)	224 (300)	172,000 (380,000) 1,687	28,400 (205,000) 279	20-25 discs 1 tri-cone 54.4 (60)
1977	Scheubsberg tunnel, Switzerland		822 (2,700)	3.22 (10 7)	224 (300)	172,000 (380,000) 1,687	28,400 (205,000) 279	20-25 discs 1 tri-cone 54.4 (60)

[a]American ton = 2,000 lb

Model type and year	Project location	Rock strength kg/cm² (MPa) lb/in²	Tunnel length m (ft)	Machine diameter m (ft in)	Cutter head kW (hp)	Thrust kg (lb) kN	Torque kg m (lb ft) kN m	Type cutter and machine weight t (tons)[a]
1978	Milchbuck tunnel, Zurich, Switzerland		2,700 (8,858)	3.22 (10 7)	224 (300)	172,000 (380,000) 1,687	28,400 (205,000) 279	20-25 discs 1 tri-cone 54.4 (60)
104-121A 1966-67	Oso water tunnel, Durango, Col.	Soft shale 70-352 (7-35) 1,000-5,000	8,530 (28,000)	3.10 (10 2)	224 (300)	172,000 (380,000) 1,687	28,400 (205,000) 279	20-25 discs 1 tri-cone 54.4 (60)
1976	Ottawa, Ontario	Dolomitic shale	4,429 (1,350)	3.22 (10 7)	224 (300)	172,000 (380,000) 1,687	28,400 (205,000) 279	20-25 discs 1 tri-cone 54.4 (60)
181-122 1968-72	Copper mine, White Pine, Mich.	Hard shale 1,050-2,110 (103-207) 15,000-30,000	2,590 (8,500)	5.49 (18 0)	895 (1,200)	717,000 (1,580,000) 7,031	238,000 (1,720,000) 2,334	47 discs 1 tri-disc 227 (250)
181-122-1 1979	Buffalo, N.Y.	Dolomitic shale	5,739 (18,830)	5.66 (18 7)	895 (1,200)	717,000 (1,580,000) 7,031	238,000 (1,720,000) 2,334	47 discs 1 tri-disc 227 (250)
132-123 1968-69 Hybrid	South-eastern trunk sewer, Melbourne, Australia	Blocky sandstone, hard siltstone, fat plastic clay 352-1,410 (35-138) 5,000-20,000	1,110 (3,655)	3.86 (12 8)	298 (400)	249,000 (550,000) 2,442	43,400 (314,000) 426	25 discs 1 tri-cone 67.1 (74)
132-123A 1969-70	South-eastern trunk sewer, Melbourne, Australia	Blocky sandstone, hard siltstone, fat plastic clay 352-1,410 (35-138) 5,000-20,000	2,390 (7,840)	3.86 (12 8)	298 (400)	249,000 (550,000) 2,442	43,400 (314,000) 426	25 discs 1 tri-cone 67.1 (74)
132-123A-2 1970	South-eastern trunk sewer, Melbourne, Australia	Blocky sandstone, hard siltstone, fat plastic clay 352-1,410 (35-138) 5,000-20,000	3,420 (11,223)	4.39 (14 5)	298 (400)	161,000 (356,000) 1,579	43,400 (314,000) 426	21 discs fixed 67.1 (74)
132-123A-3 1970-71	Frankston sewer, Melbourne, Australia	Blocky sandstone, hard siltstone, fat plastic clay 352-1,410 (35-138) 5,000-20,000	7,270 (23,862)	4.01 (13 2)	298 (400)	161,000 (356,000) 1,579	43,400 (314,000) 426	21 discs fixed 67.1 (74)
132-123-4 1972	South-eastern trunk sewer, Melbourne, Australia	Blocky sandstone, hard siltstone, fat plastic clay 352-1,410 (35-138) 5,000-20,000	10,100 (33,000)	3.40 (11 2)	298 (400)	161,000 (356,000) 1,579	43,400 (314,000) 426	21 discs fixed 67.1 (74)
132-123-5 1974-75	South-eastern trunk sewer, Melbourne, Australia	1,050-1,410 (103-138) 15,000-20,000	7,620 (25,000)	3.66 (12)	298 (400)	161,000 (356,000) 1,579	43,400 (314,000) 426	21 discs fixed 67.1 (74)
112-124 1968-69	Reussport highway tunnel, Lucerne, Switzerland	Medium hard sandstone 281-1,050 (28-103) 4,000-15,000	3,050 (10,000)	3.30 (10 10)	298 (400)	213,000 (470,000) 2,089	41,500 (300,000) 407	23 discs 63.5 (70)

[a] American ton = 2,000 lb

The Robbins Company. Rock machines — continued

Model type and year	Project location	Rock strength kg/cm² (MPa) lb/in²	Tunnel length m (ft)	Machine diameter m (ft in)	Cutter head kW (hp)	Thrust kg (lb) kN	Torque kg m (lb ft) kN m	Type cutter and machine weight t (tons)[a]
112-124 1969	By-pass tunnel des Laufenbach, Ruti, Switzerland	Sandstone, marl, conglomerate 773 (76) 11,000	573 (1,879)	3.30 (10 10)	298 (400)	213,000 (470,000) 2,089	41,500 (300,000) 407	23 discs 63.5 (70)
1972	Buonas sewage tunnel, Rotkreuz, Switzerland	Molasse 598 (59) 8,500	1,610 (5,280)	3.30 (10 10)	298 (400)	213,000 (470,000) 2,089	41,500 (300,000) 407	23 discs 63.5 (70)
1972	Lammschlucht tunnel, Fluel-Sorenberg, Switzerland	Conglomerate 1,600 (157) 23,000	1,540 (5,053)	3.30 (10 10)	298 (400)	213,000 (470,000) 2,089	41,500 (300,000) 407	23 discs 63.5 (70)
1973	Littau sewage tunnel, Schachenhof/ Emmenzopf, Switzerland		650 (2,132)	3.30 (10 10)	298 (400)	213,000 (470,000) 2,089	41,500 (300,000) 407	23 discs 63.5 (70)
1974-75	Kiemen sewage tunnel, Cham, Risch, Switzerland	Limestone and shale 1,200 (118) 17,000	1,850 (6,068)	3.30 (10 10)	298 (400)	213,000 (470,000) 2,089	41,500 (300,000) 407	23 discs 63.5 (70)
1977	Inclined tunnel, Sarelli, Switzerland		480 (1,575)	3.30 (10 10)	298 (400)	213,000 (470,000) 2,089	41,500 (300,000) 407	23 discs 63.5 (70)
112-124 1977	Industrial supply tunnel, Monaco		952 (3,123)	3.30 (10 10)	298 (400)	213,000 (470,000) 2,089	41,500 (300,000) 407	23 discs 63.5 (70)
1977	Water supply tunnel, Sines, Portugal		7,000 (22,966)	3.30 (10 10)	298 (400)	213,000 (470,000) 2,089	41,500 (300,000) 407	23 discs 63.5 (70)
82-125 1968-70	Water and power, Tasmania, Australia	Sandstone and mudstone 703-1,200 (69-118) 10,000-17,000	2,590 (8,500)	2.44 (8 0)	149 (200)	257,000 (565,500) 2,520	18,600 (134,600) 182	17 discs 1 tri-cone 40.8 (45)
82-125-1 1972	Rosedale connector Toronto sewer, Toronto, Canada	Banded limestone and shale 492-1,050 (48-103) 7,000-15,003	1,250 (4,100)	2.64 (8 8)	149 (200)	257,000 (565,500) 2,520	18,600 (134,600) 182	17 discs 1 tri-cone 40.8 (45)
1974	Toronto sewer tunnel		2,440 (8,000)	2.64 (8 8)	149 (200)	257,000 (565,500) 2,520	18,600 (134,600) 182	17 discs 1 tri-cone 40.8 (45)
1977	Robbins shop							
122-126 1968-69	Bermajales water tunnel, Granada, Spain	Shale and sandstone 70-703 (7-69) 1,000-10,000	7,925 (26,000)	3.81 (12 6)	298 (400)	213,000 (470,000) 2,089	41,500 (300,000) 407	26 discs 68.0 (75)
1970-71	Tajo-Segura water project, Albacete, Spain	Limestone with clay pockets and water, incline gallery 25% grade 844-1,050 (83-103) 12,000-15,000	476 (1,561)	3.81 (12 6)	298 (400)	213,000 (470,000) 2,089	41,500 (300,000) 407	26 discs 68.0 (75)

[a] American ton = 2,000 lb

Model type and year	Project location	Rock strength kg/cm² (MPa) lb/in²	Tunnel length m (ft)	Machine diameter m (ft in)	Cutter head kW (hp)	Thrust kg (lb) kN	Torque kg m (lb ft) kN m	Type cutter and machine weight t (tons)[a]
1975 122-126-1	Trasvase Ter — Liobregat	Clay with granite, pure granite schist, limestone	1,685 (5,527)	3.81 (12 6)	298 (400)	213,000 (470,000) 2,089	41,500 (300,000) 407	26 discs 68.0 (75)
141-127 1969-70	Southwest interceptor sewer — 13A Chicago, Ill.	Limestone approx. 1,050-1,760 (103-173) 15,000-25,000	5,300 (17,400)	4.22 (13 10)	448 (600)	404,000 (890,000) 3,962	85,500 (618,000) 838	27 discs 1 tri-cone 99.8 (110)
141-127-1 1971-72	Layout and current water tunnel, Heber City, Ut.	Shale and sandstone 703-1,410 (69-138) 10,000-20,000	7,930 (26,000)	3.94 (12 11)	448 (600)	404,000 (890,000) 3,962	85,500 (618,000) 838	27 discs 1 tri-cone 99.8 (110)
1975	Montreal, Canada	n.a.	8,534 (28,000)	3.94 (12 11)				
352-128 1970-72	Heitersberg rail tunnel, Zurich, Switzerland	Sandstone (molasse) 562-844 (55-83) 8,000-12,000	2,596 (8,500)	10.67 (35 0)	746 (1,000)	717,000 (1,580,000) 7,031	353,000 (2,550,000) 3,461	62 discs 317 (349)
352-128-1 1980	Gubrist tunnel, Switzerland					717,000 (1,580,000) 7,031	353,000 (2,550,000) 3,461	62 discs 317 (349)
182-129 1970-73	Altomira water diversion, S.E. of Madrid, Spain	Limestone 844-1,050 (83-103) 12,000-15,000	9,140 + (30,000 +)	5.49 (18 0)	597 (800)	499,000 (1,100,000) 4,894	135,000 (977,000) 1,324	36 discs 166 (183)
182-129-1 1974-75	Bilbao railroad tunnel, Bilbao, Spain	Limestone 562-984 (55-97) 8,000-14,000	1,070 (3,510)	5.79 (19 0)	597 (800)	499,000 (1,100,000) 4,894	135,000 (977,000) 1,324	36 discs 166 (183)
182-129-1 1975-76	Tunnel De Orthuela, Alicante, Spain	Limestone and marl	4,130 (13,550)	5.79 (19 0)	597 (800)	499,000 (1,100,000) 4,894	135,000 (977,000) 1,324	36 discs 166 (183)
1977-78	Llauset Moralets project, Huesca, Spain		1,421 (4,660)	5.79 (19 0)	597 (800)	499,000 (1,100,000) 4,894	135,000 (977,000) 1,324	36 discs 166 (183)
182-129-2 1975	Crevillente tunnel, Murcia, Spain		4,300 (14,110)	5.48 (18 0)	597 (800)	499,000 (1,100,000) 4,894	135,000 (977,000) 1,324	36 discs 166 (183)
n.a.	Tunnel de Orijuila, Alicante, Spain		4,130 (13,550)	5.48 (18 0)	597 (800)	499,000 (1,100,000) 4,894	135,000 (977,000) 1,324	36 discs 166 (183)
n.a.	Tunnel de Portugalete, Vizcaya, Spain		1,065 (3,490)	5.48 (18 0)	597 (800)	499,000 (1,100,000) 4,894	135,000 (977,000) 1,324	36 discs 166 (183)
1979	Tunnel de Villarejo, Cuenca, Spain		4,500 (14,765)	5.48 (18 0)	597 (800)	499,000 (1,100,000) 4,894	135,000 (977,000) 1,324	36 discs 166 (183)
162-130 1970-75	Tajo-Segura water project, S.E. of Madrid, Spain	Limestone 844-1,050 (83-103) 12,000-15,000	32,200 (105,524)	4.6 (15 0) 5.11 (16 9)	448 (600)	404,000 (890,000) 3,962	87,700 (633,800) 860	32 discs 113 (125)
162-130-1 1975-76	Water tunnel outlet, Yacambu, Venezuela	Hard sandstone with shale bands 301-211 (30-207) 4,300-30,000	5,000 (16,400)	4.6 (15 0) 5.11 (16 9)	448 (600)	404,000 (890,000) 3,962	87,700 (633,800) 860	32 discs 113 (125)

[a] American ton = 2,000 lb

The Robbins Company. Rock machines — continued

Model type and year	Project location	Rock strength kg/cm² (MPa) lb/in²	Tunnel length m (ft)	Machine diameter m (ft in)	Cutter head kW (hp)	Thrust kg (lb) kN	Torque kg m (lb ft) kN m	Type cutter and machine weight t (tons)[a]
162-131[b] 1970-75	Tajo-Segura water project, S.E. of Madrid, Spain	Limestone 844-1,050 (83-103) 12,000-15,000	11,000 (36,080)	4.6 (15 0) 5.11 (16 9)	448 (600)	404,000 (890,000) 3,962	87,700 (633,800) 860	32 discs 113 (125)
123-133 1970-71	Thompson-Yarra water diversion, Melbourne	Quartzite 1,500-1,970 (152-193) 22,000-28,000	2,400 (8,000)	3.66 (12 0)	448 (600)	422,000 (930,000) 4,139	66,800 (483,000) 655	27 discs 116 (128)
123-133-1 1972	Gas pipeline, Holland-Italy	Schist 1,200-1,760 (118-173) 17,000-25,000	1,025 (3,362)	3.66 (12 0)	448 (600)	422,000 (930,000) 4,139	66,800 (483,000) 655	27 discs 116 (128)
123-133-1	Gries tunnel, Ulrichen, Switzerland			3.66 (12 0)	448 (600)	422,000 (930,000) 4,139	66,800 (483,000) 655	27 discs 116 (128)
1973	Ferden tunnel, Ferden Hohtenn, Switzerland	Soft gneiss 984 (97) 14,000	1,665 (5,641)	3.66 (12 0)	448 (600)	422,000 (930,000) 4,139	66,800 (483,000) 655	27 discs 116 (128)
123-133-1 1975	Pfander highway tunnel, Bregenz, Austria	Molasse 600-800 (59-78) 8,500-11,000	4,645 (15,240)	3.66 (12 0)	448 (600)	422,000 (930,000) 4,139	66,800 (483,000) 655	27 discs 116 (128)
1977	Kielder tunnel, England		8,000 (26,247)	3.66 (12 0)	448 (600)	422,000 (930,000) 4,139	66,800 (483,000) 655	27 discs 116 (128)
124-134 1972	Fernheiz tunnel, Zurich, Switzerland	Sandy shale to 1,700 (to 167) 24,000	3,700 (12,136)	3.71 (12 2) 3.91 (12 10)	373 (500)	302,000 (665,000) 2,962	58,800 (425,000) 577	26 discs 68.0 (75)
1973	Escape tunnel, Aarau, Switzerland	Hard limestone	200 (656)	3.71 (12 2)	373 (500)	302,000 (665,000) 2,962	58,800 (425,000) 577	26 discs 68.0 (75)
1975	Nisellas Headrace-Tomils tunnel, Tomils, Switzerland	Lime schist with sandstone, layers quartz 300-700 (29-69) 4,300-10,000	6,200 (20,342)	3.71 (12 2)	373 (500)	302,000 (665,000) 2,962	58,800 (425,000) 577	26 discs 68.0 (75)
1975	Langenegg tunnel, Oesterreich, (Austria)	Marl	5,550 (18,209)	3.71 (12 2)	373 (500)	302,000 (665,000) 2,962	58,800 (425,000) 577	26 discs 68.0 (75)
125-135 1970-72	Nchanga water diversion and mining, Zambia	Granite 1,410-1,760 (138-173) 20,000-25,000	3,200 (10,500)	3.66 (12 0)	448 (600)	422,000 (930,000) 4,139	66,800 (483,000) 655	27 discs 113 (125)
163-136 1971-72	Ruhr coal Coal 16 ft drifts in West Germany	Shale and sandstone 703-1,410 (69-138) 10,000-20,000	6,800 (22,300)	4.88 (16 0)	448 (600)	408,000 (900,000) 4,001	88,000 (636,000) 863	36 discs 1 tri-disc 169 (186)
163-136-1 1976	Coal tunnel, Monopol, West Germany		7,000 (22,966)	5.4 (17 9)	448 (600)	408,000 (900,000) 4,001	88,000 (636,000) 863	36 discs 1 tri-disc 169 (186)
163-136-2 1976	Monopol coal pit, Kamen/ Westfalen, West Germany	Schist, sand schist, sandstone 301-1,055 4,300-15,000	8,300 (30-103) (27,231)	5.4 (17 9)	448 (600)	408,000 (900,000) 4,001	88,000 (636,000) 863	36 discs 1 tri-disc 169 (186)

[a]American ton = 2,000 lb   [b]Bored in conjunction with 162-130.

Model type and year	Project location	Rock strength kg/cm² (MPa) lb/in²	Tunnel length m (ft)	Machine diameter m (ft in)	Cutter head kW (hp)	Thrust kg (lb) kN	Torque kg m (lb ft) kN m	Type cutter and machine weight t (tons)[a]
126-137 1970-71	Mid-Toronto interceptor sewer, Toronto, Canada	Shale 492-1,050 (48-103) 7,000-15,000	5,850 (19,200)	3.66 (12 0)	298 (400)	249,000 (550,000) 2,442	44,900 (325,000) 440	26 or 24 discs 1 tri-cone 66.2 (73)
1973	Toronto sewer, Toronto, Ontario, Canada	Shale 492-1,050 (48-103) 7,000-15,000	3,660 (12,000)	3.66 (12 0)	298 (400)	249,000 (550,000) 2,442	44,900 (325,000) 440	26 or 24 discs 1 tri-cone 66.2 (73)
126-137-1 1974-75	Ottawa sewer Phase II, Ottawa, Canada	Limestone with vertical faults, shale, water 703-1,410 (69-138) 10,000-20,000	2,440 (8,000)	3.66 (12 0)	298 (400)	249,000 (550,000) 2,442	44,900 (325,000) 440	26 or 24 discs 1 tri-cone 66.2 (73)
211-138 1971-72	Brasimone-Suviana water tunnel, Italy	Sandstone and marl 984 (97) 14,000	4,544 (14,909)	6.40 (21 0)	671 (900)	717,000 (1,580,000) 7,032	136,000 (980,000) 1,334	41 discs 1 tri-disc 298 (328)
211-138-1 1973-75	Taloro tunnel, Sardinia	Granite, very sound to broken 703-2,460 (69-241) 10,000-35,000	2,200 (7,218)	6.65 (21 10)	671 (900)	717,000 (1,580,000) 7,032	136,000 (980,000) 1,334	41 discs 1 tri-disc 298 (328)
142-139 1971	Kitheron water tunnel, Athens, Greece	Limestone 1,050-1,410 (103-138) 15,000-20,000	1,200 (3,937)	4.27 (14 0)	448 (600)	404,000 (890,000) 3,962	71,900 (520,000) 705	32 discs 1 tri-disc 130 (143)
1972-75	Ghiona water tunnel, Ghiona, Greece	Limestone 1,055-1,406 (103-138) 15,000-20,000	14,500 (47,574)	4.27 (14 0)	448 (600)	404,000 (890,000) 3,962	71,900 (520,000) 705	32 discs 1 tri-disc 130 (143)
142-139-1 1977	Escape gallery, Yacambu, Venezuela	Sandstone and shale 301-2,110 (30-207) 4,300-30,000	61 (200)	4.27 (14 0)	448 (600)	404,000 (890,000) 3,962	71,900 (520,000) 705	32 discs 1 tri-disc 130 (143)
105-144 1972	Mt. Greenwood sewer tunnels (1) and (2) Chicago, Ill.	Limestone 1,050-2,110 (103-207) 15,000-30,000	(1) 1,660 (5,431) (2) 804 (2,638)	3.05 (10)	298 (400)	318,000 (700,000) 3,119	42,900 (310,000) 421	23 discs 68.0 (75)
105-144-1 1978	Miner's Ranch Tunnel, Oroville, California	Granite Amphibolite 2,100-3,150 (206-309) 15,000-45,000	1,270 (4,169)	3.5 (11 0)	373 (500)	404,000 890,000 3,960	71,900 (520,000) 705	24 discs 100 (110)
142-145 1972-75	Kirfi water tunnel, Athens, Greece	Limestone 1,050-1,410 (103-138) 15,000-20,000	13,000 (42,640)	4.27 (14)	448 (600)	404,000 (890,000) 3,962	71,900 (520,000) 705	32 discs 1 tri-disc 99.8 (110)
142-145-1 1975-77	Agios Nikolaos tunnel, Mornos Dam, Athens, Greece	Limestone	3,000 (9,843)	4.60 (15 6)	448 (600)	404,000 (890,000) 3,962	71,900 (520,000) 705	32 discs 1 tri-disc 99.8 (110)

[a]American ton = 2,000 lb

The Robbins Company. Rock machines — continued

Model type and year	Project location	Rock strength kg/cm² (MPa) lb/in²	Tunnel length m (ft)	Machine diameter m (ft in)	Cutter head kW (hp)	Thrust kg (lb) kN	Torque kg m (lb ft) kN m	Type cutter and machine weight t (tons)[a]
133-146 1973 Hybrid	Water tunnel, pipehead, Potts Hills, Sydney, Australia	Faults, crushed zones, high water volume, unweathered shales, siltstones and fine sandstone with dolerite dikes 141-562 (14-55) 2,000-8,000	7,800 (25,600)	3.93 (12 10)	373 (500)	316,000 (696,000) 3,099	74,700 (540,000) 733	29 discs 1 tri-disc 81.6 (90)
133-146-1 1974-75	Mt. Lyell mine, Tasmania, Australia	Andesite and schist 300-2,700 (29-265) 4,300-38,000	3,048 (10,000)	3.93 (12 10)	373 (500)	316,000 (696,000) 3,099	74,700 (540,000) 733	29 discs 1 tri-disc 81.6 (90)
164-147 1972	Atomic Laboratory Main Ring (CERN), Geneva, Switzerland	Meers molasse 562-773 (55-76) 8,000-11,000	6,388 (21,000)	4.8-5.06 (15 9-16 6)	448 (600)	499,000 (1,100,000) 4,894	88,100 (637,000) 864	36 discs 1 tri-disc 145 (160)
164-147-1 1975-76	Water tunnel, Yacambu Inlet, Venezuela	Shale with thin quartzite intrusions 350-2,100 (34-206) 5,000-30,000	14,000 (45,930)	4.8-5.06 (15 9-16 6)	448 (600)	499,000 (1,100,000) 4,894	88,100 (637,000) 864	36 discs 1 tri-disc 145 (160)
107-149 1972	Tanes drinking water, Oviedo, Asturias, Spain	Dolomitic limestone, sandstone 1,050-1,760 (103-173) 1,500-25,000	10,700 (35,097)	3.04 (10 0)	298 (400)	301,000 (665,230) 2,952	54,200 (391,791) 522	23 discs 1 tri-disc 67.1 (74)
144-151 1972 Tandem Sh. (hybrid)	Orichella and Timpagrande tunnels. SILA/Calabria southern Italy	Altered granite 700-1,400 (68-137) 10,000-20,000	4,000 (13,124)	4.27 (14 0)	645 (600)	654,000 (1,400,000) 6,326	77,700 (561,700) 762	28 discs 1 tri-disc 100 (110)
144-151-1 1979	Tunjenta tunnel, Chevon project, Columbia	Mica-schist with quartz 490-2,810 (48-275) 7,000-40,000	5,791 (19,000)	4.27 (14 0)	645 (600)	654,000 (1,400,000) 6,326	77,700 (561,700) 762	28 discs 1 tri-disc 100 (110)
231-152 1973	R.A.T.P. subway of Paris, France	Limestone molasse 1,550-2,040 (152-200) 22,000-29,000	4,700 (15,400)	7.01 (23 0)	671 (900)	718,000 (1,583,000) 7,041	175,000 (1,266,000) 1,716	45 discs 1 tri-disc 263 (290)
231-152-1 1976	Bajina Basta, Yugoslavia	Limestone 984-1,406 (97-138) 14,000-20,000	8,000 (26,246)	7.01 (23 0)	671 (900)	718,000 (1,583,000) 7,041	175,000 (1,266,000) 1,716	45 discs 1 tri-disc 263 (290)
135-153 1973	Schwelme tunnel, Wuppertal, West Germany	Dolomite, limestone, lime-sandstone 1,050-1,760 (103-173) 1,500-25,000	2,600 (8,530)	3.96 (13 0)	298 (400)	301,000 (665,230) 2,952	71,600 (517,450) 702	29 discs 1 tri-disc 90.7 (100)

[a]American ton = 2,000 lb

Model type and year	Project location	Rock strength kg/cm² (MPa) lb/in²	Tunnel length m (ft)	Machine diameter m (ft in)	Cutter head kW (hp)	Thrust kg (lb) kN	Torque kg m (lb ft) kN m	Type cutter and machine weight t (tons)[a]
135-153-1 1977	Hydro tunnel, Soelk, Austria	Mica-schist with quartz 490-3,160 (48-309) 7,000-45,000	3,300 (10,827)	3.96 (13 0)	298 (400)	301,000 (665,230) 2,952	71,600 (517,450) 702	29 discs 1 tri-disc 90.7 (100)
108-154 1973	Simeri tunnel, Calabria, southern Italy	Altered granite gneiss schist 35-2,460 (3-241) 500-35,000	3,963 (13,000)	3.33 (10 11)	298 (400)	359,000 (792,000) 3,521	44,800 (324,000) 439	25 discs 1 tri-disc 68.0 (75)
108-154-1 1977	Sado-Morgavel tunnel, Sines, Portugal	Shale, sandstone inclusions 352-703 (35-69) 5,000-10,000	21,000 (68,901)	3.2 (10 7)	298 (400)	359,000 (792,000) 3,521	44,800 (324,000) 439	25 discs 1 tri-disc 68.0 (75)
91-155 1973	Crosstown waste-water interceptor Shoal Creek to Bull Creek, Austin, Tex.	Limestone shale 50-148 (5-15) 700-2,100	8,380 (27,500)	2.64 (8 8) 2.90 (9 6)	224 (300)	302,000 (665,000) 2,962	26,500 (191,599) 260	20 discs 1 tri-disc 63.5 (70)
91-155-1 1980	Calumet intercepting sewer, Chicago, Ill.	Limestone shale 50-148 (5-15) 700-2,100	5,790 (12,123)	2.6 (8 9)	224 (300)	302,000 (665,000) 2,962	26,500 (191,599) 260	20 discs 1 tri-disc 63.5 (70)
201-158 1973	Post Trasvase, Madrid, Spain	Limestone, marl, clay 352-1,050 (35-103) 5,000-15,000	11,200 (36,747) 5 tunnels	6.10 (20)	604 (810)	585,000 (1,289,000) 5,737	154,000 (1,114,000) 1,510	43 discs 279 (308)
201-158-1 1977	Padrum tunnel, Asturias, Spain	Limestone and impure limestone 562-2,390 (55-234) 8,000-34,000	3,000 (9,843)	6.10 (20)	604 (810)	585,000 (1,289,000) 5,737	154,000 (1,114,000) 1,510	43 discs 279 (308)
191-161 1974-75	Rockville Route, Section A6a, Washington, D.C.	Granite gneiss, chorite schist 703-1,760 (69-173) 10,000-25,000	5,768 (18,924)	5.79 (19 0)	671 (900)	839,000 (1,850,000) 8,228	145,000 (1,050,000) 1,422	45 discs 259 (285)
191-161-1 1975-76	Rockville Route, Section A6a, Washington, D.C.	Granite gneiss chorite schist 703-1,760 (69-173) 10,000-25,000	4,512 (14,806)	5.79 (19 0)	671 (900)	839,000 (1,850,000) 8,228	145,000 (1,050,000) 1,422	45 discs 259 (285)
1977	Rockville Route, Section Alla, Washington, D.C.	Granite gneiss chorite schist 703-1,760 (69-173) 10,000-25,000	6,707 (22,000)	5.79 (19 0)	671 (900)	839,000 (1,850,000) 8,228	145,000 (1,050,000) 1,422	45 discs 259 (285)
191-161-2 1980	Culver-Goodman, Rochester, N.Y.	Sandstone, limestone 1,800-3,280 (176-321) 25,600-46,600	8,628 (28,000)	5.79 (19 0)	671 (900)	839,000 (1,850,000) 8,228	145,000 (1,050,000) 1,422	45 discs 259 (285)
114-163 1974	Libanon gold-mine, Gold-fields of South Africa	Quartzite 1,270-2,110 (125-207) 18,000-30,000	Mine development	3.35 (11 0)	418 (560)	406,000 (896,000) 3,982	52,600 (380,000) 516	28 discs 63.5 (70)

[a]American ton = 2,000 lb

Model type and year	Project location	Rock strength kg/cm² (MPa) lb/in²	Tunnel length m (ft)	Machine diameter m (ft in)	Cutter head kW (hp)	Thrust kg (lb) kN	Torque kg m (lb ft) kN m	Type cutter and machine weight t (tons)[a]
105-165 1974	Oslo sewer tunnel, Oslo, Norway	Calciferous shale with cyenite and diabase intrusions 352-2,110 (35-207) 5,000-30,000	5,000 (16,400)	3.15 (10 4)	373 (500)	318,000 (700,000) 3,119	42,900 (310,000) 421	23 discs 1 tri-disc 68.0 (75)
1977	Fosdalen mine, Norway	Greenstone and magnetite 1,550-3,030 (152-297) 22,000-43,000	670 (2,200)	3.15 (10 4)	373 (500)	318,000 (700,000) 3,119	42,900 (310,000) 421	23 discs 1 tri-disc 68.0 (75)
105-165-1 1977	Sulitjelma, Norway	Granite 1,690-2,890 (153-283) 24,000-41,000	360 (1,180)	3.15 (10 4)	373 (500)	318,000 (700,000) 3,119	42,900 (310,000) 421	23 discs 1 tri-disc 68.0 (75)
1978	Eidford Hydro-electric Scheme, Rwanda, W. Africa	280-1,050 (27-102) 3,000-15,000	2,800 (9,240)	3.15 (10 4)	373 (500)	318,000 (700,000) 3,119	42,900 (310,000) 421	23 discs 1 tri-disc 68.0 (75)
105-165-2 1979	Mukungwa Power Scheme, Rwanda, W. Africa	Sandstone, schist and quartzite	2,000 (6,562)	3.08 (10 1)	373 (500)	318,000 (700,000) 3,119	42,900 (310,000) 421	23 discs 1 tri-disc 68.0 (75)
262-166 1974	Bramefarine tunnel, French Alps, France	Limestone 141-422 (14-41) 2,000-6,000	3,810 (12,500)	8.08 (26 6)	671 (900)	721,000 (1,590,000) 7,071	212,000 (1,530,000) 2,079	56 discs 308 (340)
145-168 1975 45° incline slope	Hydro Penstock, Grimsel Oberaar, Switzerland	Alaskite, gneiss 703-1,760 (69-173) 10,000-25,000	812 (2,664)	4.3 (14)	634 (850)	862,000 (1,900,000) 8,454	97,900 (708,183) 960	15½ in discs
1978	Negro Ruico tunnel, Chivor project, Columbia	Schist and quartzite 700-2,460 (68-241) 10,000-35,000	10,000 (32,810)	4.3 (14)	634 (850)	862,000 (1,900,000) 8,454	97,900 (708,183) 960	15½ in discs
232-171 1976	Hydro tunnel, Split, Yugoslavia	Hard limestone 1,000-2,000 (98-196) 14,000-28,000	9,144 (30,000)	7.125 (24.3)	895 (1,200)	1,603,000 (3,534,000) 15,700	19,660 (1,706,000) 193	15½ in discs 330 (363)
233-172 1976 Shielded TBM (Hybrid)	Buckskin water tunnel, Ariz.	Andesite aglomerate, tuff 700-2,800 (69-274) 10,000-40,000	10,700 (35,000)	7.16 (23 5)	895 (1,200)	1,202,000 (2,650,000) 11,800	215,000 (1,550,000) 2,110	15½ in discs 320 (350)
212-173 1977	Seabrook cooling tunnel, N.H.	Dolomite and gneiss 1,400-2,800 (137-274) 20,000-40,000	3,964 (13,000)	6.43 (21 1)	358 (480)	907,030 (2,000,000) 8,897	183,940 (1,330,000) 1,800	15½ in discs 320 (350)
212-174 1977	Seabrook cooling tunnel, N.H.	Dolomite and gneiss 1,400-2,800 (137-274) 20,000-40,000	3,964 (13,000)	6.43 (21 1)	358 (480)	907,030 (2,000,000) 8,897	183,940 (1,330,000) 1,800	15½ in discs 320 (350)

[a]American ton = 2,000 lb

Model type and year	Project location	Rock strength kg/cm² (MPa) lb/in²	Tunnel length m (ft)	Machine diameter m (ft in)	Cutter head kW (hp)	Thrust kg (lb) kN	Torque kg m (lb ft) kN m	Type cutter and machine weight t (tons)[a]
61-176 1977 Reef raiser	Blyvoor mine, Blyvoor, S. Africa	Quartzite 1,750 (171) 25,000		1.84 (6 ½)	149 (200)	201,900 (445,300) 1,981	182,000 (122,000) 1,787	14 (15)
61-177 1977 Reef raiser	East Driefontein mine, Carltonville, S. Africa	Quartzite 1,750 (171) 25,000		1.84 (6 ½)	149 (200)	201,900 (445,300) 1,981	182,000 (122,000) 1,787	14 (15)
185-178 1977	Upper Des Plaines 21, Contract 73-320-2S, Chicago, Ill.	Limestone and Racine formations, Brandon Bridge, Markgraf, and Romeo members of the Joliet formations	3,313 (10,870)	5.5 (18 0)	670 (900)	867,000 (1,911,000) 8,501	128,000 (924,000) 1,254	42 discs 213 (235)
185-178-1 1980	Subway tunnel, Buffalo, N.Y.	Limestone 1,120-2,460 (109-241) 16,000-35,000	2,134 (7,000)	5.95 (18 7)	670 (900)	867,000 (1,911,000) 8,501	128,000 (924,000) 1,254	42 discs 213 (235)
1010-179 1976 Shielded TBM	Vat tunnel, Ut.	Sandstone, limestone, shale 0-1,400 (0-140) 0-20,000	11,817 (38,760)	3.25 (10 8)	298 (400)	907,000 (2,000,000) 8,900	46,000 (334,000) 453	26 discs 68 (75)
116-181 1977-78	Aurland tunnel, Lysverken, Norway	Chlorite schist with calcite 420-1,050 (41-103) 6,000-15,000	6,200 (20,340)	3.5 (11 6)	447 (600)	506,000 (1,115,000) 5,000	59,490 (430,000) 580	100 (110)
1979	Lier water tunnel, Lier, Norway		3,600 (11,811)	3.5 (11 6)	447 (600)	506,000 (1,115,000) 5,000	59,490 (430,000) 580	100 (110)
129-182 1977	Sellrain-Silz tunnel, Salzburg, Austria	Hard schist, gneiss, amphibolite 980-1,550 (96-152) 14,000-22,000	4,500 (14,765)	3.9 (12 9½)	373 (500)	595,000 (1,312,000) 5,840	65,000 (469,000) 636	32 14 in discs 95 (105)
1979	Buers tunnel, Bludenz, Austria		1,550 (5,085)	3.9 (12 9½)	373 (500)	595,000 (1,312,000) 5,840	65,000 (469,000) 636	32 14 in discs 95 (105)
222-183 1977	Weller Creek, Contract 73-217-2S, Chicago, Ill.	Limestone and Racine forma- tions, Brandon Bridge, Markgraf and Romeo members of the Joliet formation 1,000-2,100 (98-200) 14,000-30,000	6,706 (22,000)	6.70 (22)	1,194 (1,600)	1,060,000 (2,340,000) 10,410	259,000 (1,870,000) 2,540	52 15 in discs 272 (300)
1210-187	Soelk project, Salzburg, Austria	Mica-schist with quartz 490-3,160 (48-309) (7,000-44,000)	12,000 (38,290)	3.52 (11 7)	448 (600)	505,700 (1,115,000) 4,960	60,000 (430,000) 583	29 discs 91 (100)

[a] American ton = 2,000 lb

The Robbins Company. Rock machines — continued

Model type and year	Project location	Rock strength kg/cm² (MPa) lb/in²	Tunnel length m (ft)	Machine diameter m (ft in)	Cutter head kW (hp)	Thrust kg (lb) kN	Torque kg m (lb ft) kN m	Type cutter and machine weight t (tons)[a]
1210-187 1979	Bodendorf project, Austria	n.a.	9,000 (12,000)	3.52 (11 7)	448 (600)	505,700 (1,115,000) 4,960	60,000 (430,000) 583	29 discs 91 (100)
116-188 1978	Oslo sewer, Oslo, Norway	Conglomerate 700-1,000 (69-98) 10,000-14,000	7,200 (23,950)	3.50 (11 5⅞)	448 (600)	450,000 (1,000,000) 4,400	60,600 (438,000) 594	14 in discs 86 (96)
116-189 1978-81	Oslo sewer, Oslo, Norway	Conglomerate 700-1,000 (69-98) 10,000-14,000	7,200 (23,950)	3.50 (11 5⅞)	448 (600)	450,000 (1,000,000) 4,400	60,600 (438,000) 594	14 in discs 86 (96)
213-190 1978	Chicago underflow, Crawford-Calumet, Chicago, Ill.	Dolomitic limestone 1,050-2,800 (103-274) 1,900-40,000	13,000 (42,000)	6.5 (21 3)	1,600 (1,194)	910,000 (2,000,000) 8,900	205,000 (1,480,000) 2,010	15½ in discs 295 (325)
151-191 1979-82 Shielded hard rock (hybrid)	Baikal-Amur mainline service tunnel, Nijne Angarsk, Siberia, U.S.S.R.	Limestone, sandstone, shale 780-1,770 (76-174) 11,000-25,000	20,000 (65,600)	4.56 (14 11½)	597 (800)	1,226,000 (2,702,900) 12,000	106,000 (766,700) 1,000	34 14 in discs 210 (230)
92-192 1978	Stillwater tunnel, Ut.	Sandstone and shale 350-1,050 (34-103) 5,000-15,000	12,875 (42,240)	2.91 (9 6½)	298 (400)	454,000 (1,000,000) 4,452	30,953 (223,807) 2,194	24 12 in discs 112 (123)
146-193 1978-80	Thomson Yarra, Victoria, Australia	Sandstone and shale 1,400-3,600 (140-355) 20,000-51,000	6,070 (19,910)	4.1 (13 6)	559 (750)	567,000 (1,250,000) 5,560	108,816 (787,000) 1,067	160 (145)
242-195 Shielded hard rock (hybrid) 1978	Park river diversion, Hartford, Conn.	Sandstone and shale, basalt 490-1,340 (48-131) 6,970-19,000	2,755 (9,040)	7.36 (24 2)	895 (1,200)	857,304 (1,890,000) 8,400	209,525 (1,515,000) 2,000	50 × 15½ in discs 4 × 12 in centre discs 320 (350)
353-196 1978	Chicago underflow, Central-Damen, Chicago, Ill.	Hard dolomitic limestone and shale 1,200-1,750 (117-171) 17,000-25,000	7,925 (26,000)	10.8 (35 3)	1,790 (2,400)	1,250,000 (2,760,000) 12,000	486,000 (3,510,000) 4,800	65 × 15½ in discs 4 × 12 in discs (centre) 740 (810)
353-197 1979	Chicago underflow, 59th-Central, Chicago, Ill.	Dolomitic limestone 350-2,260 (34-222) 4,980-32,000	5,700 (18,804)	10.8 (35 3)	1,790 (2,400)	1,250,000 (2,760,000) 12,000	486,000 (3,510,000) 4,800	65 × 15½ in discs 4 × 12 in discs (centre) 740 (810)
1011-198 1978	V.E.P.C.O. drainage tunnels, Bath County, Va.	Shale, limestone, siltstone and mudstone 1,050-2,450 (103-240) 15,000-35,000	7,315 (24,000)	3.25 (10 8)	298 (400)	386,467 (852,000) 3,800	48,405 (350,000) 470	23 × 14 in discs 4 × 12 in discs (centre) 63 (70)
321-199 1979	Chicago underflow, Roosevelt-Ogden, Chicago, Ill.	Dolomitic limestone 840-2,250 (82-221) 12,000-32,000	4,200 (13,780)	9.85 (32 4)	1,790 (2,400)	1,165,752 (2,570,000) 11,000	458,000 (3,304,00) 4,500	60 × 15½ discs 4 × 12 in discs (centre) 613 (675)

[a] American ton = 2,000 lb

Model type and year	Project location	Rock strength kg/cm² (MPa) lb/in²	Tunnel length m (ft)	Machine diameter m (ft in)	Cutter head kW (hp)	Thrust kg (lb) kN	Torque kg m (lb ft) kN m	Type cutter and machine weight t (tons)[a]
321-200 1979	Chicago underflow, Ogden-Addison, Chicago, Ill.	Dolomitic limestone 840-2,250 (82-221) 12,000-32,000	6,893 (22,614)	9.85 (32 4)	1,790 (2,400)	1,165,752 (2,570,000) 11,000	458,000 (3,304,000) 4,500	60 × 15½ discs 4 × 12 in discs (centre)
202-201 1979	Westfalen coal-mine, Ruhr, West Germany	Shale and sandstone 350-1,400 (34-137) 5,000-20,000	12,700 (41,670)	6.1 (20 0)	537 (720)	635,000 (1,400,000) 6,200	131,000 (950,000) 1,300	39 × 14 in discs 4 × 12 in discs (centre) 200 (220)
214-202 1979	Blumenthal coal-mines, Ruhr, West Germany	Shale and sandstone 350-1,400 (34-137) 5,000-20,000	10,600 (34,780)	6.5 (21 4)	716 (960)	635,040 (1,400,000) 6,200	174,535 (1,262,000) 1,700	42 × 14 in discs 4 × 12 in discs (centre) 200 (220)
93-203 1979-80	Scajaquada sewer, Buffalo, N.Y.	Limestone 910-2,100 (89-206) 13,000-30,000	9,146 (30,000)	2.8 (9 6)	447 (600)	326,592 (720,000) 3,200	4,840 (35,000) 47	18 × 14 in discs 4 × 12 in discs (centre) 70 (77)
136-204 1979	Haeusling tunnel project, Zillgrtal, Austria	Dolomite and phylite 1,200 (118) 17,000	1,700 (5,580)	4.2 (13 9)	447 (600)	458,136 (1,010,000) 4,500	8,436 (61,000) 83	30 × 14 in discs 4 × 12 in discs (centre) 71 (80)
203-205 1980	East 63rd Street subway, New York	Granite-schist 280-2,430 (27-238) 4,000-34,000	1,524 (5,000)	6.15 and 6.7 (20 2 and 22 0)	895 (1,200)	889,056 (1,960,000) 8,700	162,641 (1,176,000) 1,600	42 × 15½ in discs 4 × 12 in discs (centre) 445 (490)
186-206 1979-80	Buffalo subway, Buffalo, N.Y.	Dolomitic shale 1,320-2,140 (129-210) 19,000-30,000	3,200 (10,500)	5.6 (18 6)	895 (1,200)	762,048 (1,680,000) 7,500	150,221 (1,086,200) 1,500	38 × 15½ in discs 4 × 12 in discs (centre) 213 (235)
186-207 1979-80	Buffalo subway, Buffalo, N.Y.	Dolomitic shale 1,320-2,140 (129-210) 19,000-30,000	3,200 (10,500)	5.6 (18 6)	895 (1,200)	762,048 (1,680,000) 7,500	150,221 (1,086,200) 1,500	38 × 15½ in discs 4 × 12 in discs (centre) 213 (235)
187-208 1980	Ruilmare, Bucharest, Romania	Granite-gneiss, quartzitic schist 860-1,970 (84-193) 12,000-28,000	12,000 (39,360)	5.5 (18 0)	895 (1,200)	884,520 (1,950,000) 8,700	156,970 (1,135,000) 1,500	38 × 15½ in discs 4 × 12 in discs (centre) 215 (237)
94-209 1980	Turrach project, Austria	Schist-gneiss 420-2,400 (41-235) 6,000-34,000	10,000 (32,810)	3.02 (9 10 ¾)	298 (400)	335,665 (740,000) 3,300	32,915 (238,000) 320	20 × 14 in discs 4 × 12 in discs (centre) 71 (80)
147-210 1980	Calumet system, Chicago underflow, Chicago, Ill.	Dolomitic limestone 840-2,250 (82-221) 12,000-32,000	10,975 (36,000)	4.32 (14 2)	671 (900)	551,579 (1,216,000) 5,400	87,267 (631,000) 850	31 × 15½ in discs 4 × 12 in discs (centre) 113 (125)
251-211 1980	Grand Maison, Geneve, France	Granite, granite gneiss 3,480 (341) 50,000	5,865 (19,242)	7.7 (25 3)	1,491 (2,000)	1,163,484 (2,565,000) 11,000	276,600 (2,000,000) 2,700	53 × 15½ in discs 4 × 12 in discs (centre) 380 (418)

[a]American ton = 2,000 lb

The Robbins Company. Rock machines — continued

Model type and year	Project location	Rock strength kg/cm² (MPa) lb/in²	Tunnel length m (ft)	Machine diameter m (ft in)	Cutter head kW (hp)	Thrust kg (lb) kN	Torque kg m (lb ft) kN m	Type cutter and machine weight t (tons)[a]
148-212 1980	Brattset hydropower project, Norway	Quartz-diorite intrusive; phylite 280-2,710 (27-266) 4,000-38,000	12,000 (39,372)	4.5 (14 9)	783 (1,050)	635,040 (1,400,000) 6,200	63,504 (689,000) 630	31 × 15½ in discs 4 × 12 in discs (centre) 140 (154)
148-213 1980	Brattset hydropower project, Norway	Quartz-diorite intrusive; phylite 280-2,710 (27-266) 4,000-38,000	12,000 (39,372)	4.5 (14 9)	783 (1,050)	635,040 (1,400,000) 6,200	63,504 (689,000) 630	31 × 15½ in discs 4 × 12 in discs (centre) 140 (154)
193-214 1981	Selby project, United Kingdom	Calcareous sandstone 13,000-1,476 (127-145) 18,000-21,000	13,600 (44,620)	5.8 (19)	671 (900)	578,144 (1,274,230) 5,700	130,693 (945,000) 1,300	43 × 14 in discs 182 (200)

The Robbins Company Liner-thrust shields, digger/ripper scraper shields and shaft reamers, etc.

Model type and year	Project location	Rock strength kg/cm² (MPa) lb/in²	Tunnel length m (ft)	Machine diameter m (ft in)	Cutter head kW (hp)	Thrust kg (lb) kN	Torque kg m (lb ft) kN m	Type cutter and machine weight t (tons)[a]
341-111 1964 Liner-thrust shield pressurized face (hybrid)	Paris Metro subway, Paris, France	Mixed limestone, sandstone and clay 0-1,050 (103) 15,000	2,870 (9,430)	10.26 (33 8)	746 (1,000)	7,290,000 (16,080,000) 71,493	726,000 (5,250,000) 7,120	Discs and 180 fixed tools 454 (500)
221S-132 1970 Ripper-scraper shield (hybrid)	San Fernando water diversion, near Los Angeles, Calif.	Soft soil formations	8,230 (27,000)	6.71 (22 0)	634 (850)	3,180,000 (7,000,000) 31,186		Ripper teeth and scraper 259 (285)
113-140 1972 Liner-thrust TBM (hybrid)	Tajo-Segura water project, Albacete, Spain	Limestone 846-1,050 (83-103) 12,000-15,000	6,100 + (20,000) +	3.51 (11 6)	298 (400)	204,000 (450,000) 2,001	55,900 (404,000) 548	24 discs 68.0 (75)
143S-141 1971 Ripper-scraper shield (hybrid)	Sewer tunnel, Detroit, Mich.	Silty sand, highly compacted dry clay, to soft and sticky clay, occasional gravel, full and mixed face	3,048 (10,000)	4.41 (14 6)	448 (600)	1,270,000 (2,800,000) 12,455		Ripper scraper 97.1 (107)
143S-141 1973/4 Ripper-scraper shield (hybrid)	Sewer tunnel, Little Rock, Ark.	Sandy clay with small to medium cobble stones, gravel	1,219 (4,000)	4.41 (14 6)	448 (600)	1,270,000 (2,800,000) 12,455		Ripper scraper 97.1 (107)
143S-142 1971 Ripper-scraper shield (hybrid)	Sewer tunnel, Detroit, Mich.	Same as 143S-141 1971	3,780 (12,500)	4.41 (14 6)	448 (600)	1,270,000 (2,800,000) 12,455		Ripper scraper 97.1 (107)

[a]American ton = 2,000 lb

Model type and year	Project location	Rock strength kg/cm² (MPa) lb/in²	Tunnel length m (ft)	Machine diameter m (ft in)	Cutter head kW (hp)	Thrust kg (lb) kN	Torque kg m (lb ft) kN m	Type cutter and machine weight t (tons)[a]
143S-142 1973 Ripper-scraper shield (hybrid)	Mt. Clemens PCI 14, Mt. Clemens, Mich.	Clay, sand, gravel, hardpan	2,740 (9,000)	4.41 (14 6)	447 (600)	1,270,000 (2,800,000) 12,455		Ripper/scraper
361S-143 1973-5 Ripper-scraper shield (hybrid)	Castiglione rail tunnel between Rome and Florence, Italy	Consolidated slabby dry clay, and sand	7,400 (24,272)	Horseshoe 10.97 wide 9.45 high (36 wide 31 high)	1,027 (1,377)	5,440,000 (12,000,000) 53,350		Ripper scraper 558 (615)
106-148 Shielded TBM (hybrid)	La Coche hydro-electric tunnel, French Alps, France	Schist, brecchia, flysch, lime-stone, sandstone 703-1,760 (69-173) 10,000-25,000	7,700 (25,200)	3.02 (9 11⅛)	298 (400)	301,000 (665,000) 2,952	54,000 (392,000) 532	23 discs 1 tri-disc 67.1 (74)
75-150 1973 Oscillating	Sewer tunnel, West Lane Cove, Sydney, Australia	Hawkesbury sandstone 148-394 (15-39) 2,100-5,600	152 (500)	Rectangular with radius top and bottom 2.1 high 1.5 wide (7 high 5 wide)	104 (140)	45,400 (100,000) 445	9,220 (66,600) 90	4 discs 18.1 (20)
183S-156 1973 Ripper-scraper shield (hybrid)	New Carroltown Route D4a, Washington Metro, Washington, D.C.	Sand, clay	945 (3,100)	5.49 (18)	n.a.	2,270,000 (5,000,000) 22,262		Ripper scraper 200 (220)
183S-157 1973 Ripper-scraper shield (hybrid)	New Carroltown Route D4a, Washington Metro, Washington, D.C.	Sand, clay	945 (3,100)	5.49 (18)	n.a.	2,270,000 (5,000,000) 22,262		Ripper scraper 200 (220)
281S-159 1973 Ripper-scraper shield (hybrid)	Madrid Metro, Madrid, Spain	Clay with sand	3,500 (11,480)	8.59 (28 2)	n.a.	41,900,000 (92,400,000) 410,913		Ripper scraper 300 (331)
1977	Madrid Metro, Madrid, Spain		1,050 (3,444) 10/77 840 (2,755)	8.59 (28 2)	n.a.	41,900,000 (92,400,000) 410,913		Ripper scraper 300 (331)
109S-160 1973 Ripper-scraper shield (hybrid)	East Branch intercepting sewer, Port Richmond, N.Y.	Compact glacial till silt, sand boulders, water fill	5,180 (17,000)	3.05 (10)	261 (350)	771,000 (1,700,000) 7,561		Ripper scraper 43.5 (48)
109S-160 ex. sh. 1976 Hybrid	Eltingville sewer, Staten Island, New York	Sand, clay and fill. 1st tunnel 2nd tunnel	1,149 (3,771) 775 (2,543)	3.05 (10)	261 (350)	770,950 (1,700,000) 7,560		Ripper/scraper
109-160-2 1980	Red Hook tunnel, Brooklyn, New York	Glacial till	2,620 (8,600)	3.18 (10 5)	298 (400)	770,950 (1,700,000) 7,560		Ripper/scraper
165-162 1974 Shielded TBM (hybrid)	Channel service tunnel, Calais, France	Chalk (project delayed)	2,000 (6,560)	4.88 (16)	615 (825)	471,000 (1,039,000) 4,619	109,000 (787,800) 1,069	30 discs and 5 drag bits 245 (270)

[a]American ton = 2,000 lb

Model type and year	Project location	Rock strength kg/cm² (MPa) lb/in²	Tunnel length m (ft)	Machine diameter m (ft in)	Cutter head kW (hp)	Thrust kg (lb) kN	Torque kg m (lb ft) kN m	Type cutter and machine weight t (tons)[a]
127S-164 1974 Ripper-scraper shield (hybrid)	Berne sewer tunnel, Berne, Switzerland	Gravel with sand and clay	1,550 (5,080)	3.68 (12 1)	336 (450)	1,550,000 (3,420,000) 15,200		Ripper scraper 54.4 (60)
127-164-1 1977	Refalzaft, Sicily	Limestone and clay	3,200 (10,496) 10/77 2,200 (7,816)	3.68 (12 1)	336 (450)	1,550,000 (3,420,000) 15,200		Ripper scraper 54.4 (60)
128S-167 1975 Liner-thrust shield (hybrid)	Detroit Metro Contract PCI 24, Detroit, Mich.	Compacted sand, silty sand, clay and gravel	2,200 (7,200)	3.66 (12)	336 (450)	1,090,000 (2,400,000) 10,690		72.6 (80)
184S-169 1975 Liner-thrust shield (hybrid)	Branch Route — Section F2a, Washington Metro, Washington, D.C.	Sand and sandy clay, clay river cobbles and boulders, gravel, some water	1,372 (4,500)	5.5 (18)	358 (480)	2,720,000 (6,000,000) 26,675		200 (220)
184S-170 1975 Liner-thrust shield (hybrid)	Branch Route — Section F2a, Washington Metro, Washington, D.C.	Sand and sandy clay, clay river cobbles and boulders, gravel, some water	1,372 (4,500)	5.5 (18)	358 (480)	2,720,000 (6,000,000) 26,675		200 (220)
115S-175 1977 Digger shield (hybrid)	Yacambu water tunnel, Venezuela	Fault zone	1,000 (3,273)	3.6 (12 0)	254 (340)	206,840 (456,000) 2,028		80 (88) Probe/scraper
263S-180 Liner-thrust shield (hybrid)	Condotte d'Acqua, Rome, Italy	Soft plastic phylitic material	4,000 (13,124)	8.2 (27 1)	970 (1,300)	6,713,000 (14,800,000) 65,830		425 (468)
192S-185 1977 Liner-thrust shield (hybrid)	Caracas subway, Caracas, Venezuela	40% mica-schist, 60% clay and muddy sand	2,200 (7,218)	5.79 (19 0)	433 (580)	2,573,846 (5,675,000) 25,240		209 (230)
192S-186 1977 Liner-thrust shield (hybrid)	Caracas subway, Caracas, Venezuela	80% mica-schist, 20% clay 490-700 (48-69) 7,000-10,000	2,200 (7,218)	5.79 (19 0)	433 (580)	2,573,846 (5,675,000) 25,240		209 (230)
241S-184 1977 Shaft borer	U.S. Steel, Alabama	Sandstone, shale 0-1,400 (140) 20,000	340 (1,100) Shaft	7.44 (24 5)	500 (750)	680,000[b] (1,500,000) 6,670	226,000 (1,630,000) 2,210	56 13 in discs 227 (250)
1211SR-194 Shaft reamer 1978	Chicago underflow, Chicago, Ill.	Dolomitic limestone 1,050-2,110 (103-207) 15,000-30,000	Shaft 76.2 (250)	3.7 (12)	112 (150)	170,100 (375,000) 1,668	30,426 (220,000) 298	16 12 in discs 45 (50)

[a]American ton = 2,000 lb; [b]Includes dead weight

Subterranean Tools Inc. (Division of Kennametal Inc.)

Model type and year	Project location	Rock strength kg/cm^2 (MPa) lb/in^2	Tunnel length m (ft)	Machine diameter m (ft in)	Cutter head kW (hp)	Thrust kg (lb) kN	Torque kg m (lb ft) kN m	Type cutter and machine weight t (tons)
TB-11 1978	Anglo-American Corporation of South Africa, Vaal Reefs gold-mine	Variable up to 4,220 (414) 60,000	Approx. 2 km	3.5 (11 6)	450 (600)	499,000 (1,100,000) 4,890	55,310 400,000 542.4	T.C. button disc cutters 90 (88)

Thyssen (Great Britain) Limited. Rock machine

Model type and year	Project location	Rock strength kg/cm^2 (MPa) lb/in^2	Tunnel length m (ft)	Machine diameter m (ft in)	Cutter head kW (hp)	Thrust kg (lb) kN	Torque kg m (lb ft) kN m	Type cutter and machine weight t (tons)
FLP35 1975	Dawdon colliery, Co. Durham	302-1,012 (30-99) 4,300-14,000 Siltstone 264-1,617 (26-159) 3,800-23,000 Sandstone	Three tunnels 3,500 ea. (11,482) ea.	3.65 (12)	193.96 (260)	406,000 (895,000) 3,982	42,000 (304,000) 412	20/12 twin disc cutters 91.44 (90)

Union Industrielle Blanzy-Ouest. (now taken over by Bouygues, France)

Model type and year	Project location	Rock strength kg/cm^2 (MPa) lb/in^2	Tunnel length m (ft)	Machine diameter m (ft in)	Cutter head kW (hp)	Thrust kg (lb) kN	Torque kg m (lb ft) kN m	Type cutter and machine weight t (tons)
3 M 1968	R.A.T.P. (Paris Underground) R.E.R. (Regional Express System) Exploration drift	102-1,020 (10-100) Lime soil 1,500-15,000	500 (1,640)	3 to 4 (10 10-13 2)		150,000 (301,000) 1,470		4 discs 4 arms
5 M 1968	Drift, Borie, Vianden. (Grand Duchy of Luxemburg)	816-1,224 (80-120) Schist 12,000-17,000	800 (2,630)	4-5.5 (13 2-18 0)		165,000 (364,000) 1,617		4 discs or 8 discs 4 arms

Alfred Wirth & Co. Kg. Medium to hard-rock machines — (now Wirth Maschinen GmbH)

Model type and year	Project location	Rock strength kg/cm² (MPa) lb/in²	Tunnel length m (ft) inclin.	Machine diameter m (ft in)	Cutter head kW (hp)	Thrust kg (lb) kN	Torque kg m (lb ft) kN m	Type cutter and machine weight t (tons)
TB 1-214 E 2/1967 to 6/1967	Water tunnel, Zemmkraft- werke, Ginzling, Austria	Granite gneiss	263 (863)	2.14 (7 0)				Carbide insert
TB 1-240 8/1968	Grand Emosson, Châtelard/ Switzerland	Granite	60 (197)	2.40 (7 10)				Insert
TB 11-300 E 10/1968 to 9/1969	Grand Emosson Pentshaft 'Corbes', Châtelard/ Switzerland	Granite	1,145 (3,766) 29.3°	3.00 (9 10)	460 (617)	455,000 (1,003,000) 4,462	60,500 (437,000) 593	Insert
TB 1-214 10/1968 to 1/1969	Grand Emosson 'Barberine', Châtelard/ Switzerland	Granite	415 (1,362)	2.14 (7 0)				Insert
TB 1-214 3/1969 to 12/1969	Grand Emosson Pentshaft 'Barbarine', Châtelard/ Switzerland	Granite	1,019 (3,343) 40.50°	2.25 (7 5)				Carbide insert
TB 1-240 H 7/1969 to 12/1969	HAGA tunnel, sewer tunnel, Stockholm, Sweden	Granite	450 (1,476)	2.40 (7 10)				Insert
TBE 350/770 H 770/1046 H 9/1970	Road tunnel, Lucerne, Switzerland	Sandstone and marl	2,700 (8,858)	7.70 (25 3) 10.46 (34 4)	760 (1,019)	680,000 (1,500,000) 6,669 604,000 (1,330,000) 5,923	75,000 (542,000) 736 96.000 (694,000) 941	Discs
TB 11-300 E TBE 300/600 5/1970- 9/1972	Pentshaft, Wehr/Black Forest	Granite and gneiss	1,400 (4,593) 23.90°	3.00 (9 10) 6.30 (20 8)	460 (617)	455,000 (1,003,000) 4,462	60,500 (437,000) 593	Carbide insert
TB V-540 E-Sch 3/1972 5/1972	Test drilling horizontal. Colliery, Sophia Jacoba	Slate, schistous sandstone, sandstone		5.30 (17 5) 5.60 (18 4)				Discs
TB 11-300 H TBE 300/530 H 10/1971 to 9/1972	Main gallery in a colliery	Carbon, sandstone	1,600 (5,249)	5.30 (17 5)	460 (617)	440,000 (970,000) 4,315 455,000 (1,003,000) 4,462	26,000 (188,000) 255 60,500 (437,000) 593	Discs
TB 11-300 H 11/1970 7/1973	Water tunnel, Rocky Mountain, Col., U.S.A.	Granite	4,000 (13,123)	2.98 (9 9)	460 (617)	440,000 (970,000) 4,315	26,000 (188,000) 255	Insert and discs
TB 1-214 E 3/1971 to 7/1971	Tube tunnel, Meilen, Zürich	Shale	600 (1,973)	2.25 (7 5)				Discs
TB 1-240 H 10/1971 to 1/1972	Water tunnel, Orsières, Switzerland	Lime shale	1,600 (5,249)	2.40 (7 10)				Discs

Model type and year	Project location	Rock strength kg/cm^2 (MPa) lb/in^2	Tunnel length m (ft) inclin.	Machine diameter m (ft in)	Cutter head kW (hp)	Thrust kg (lb) kN	Torque kg m (lb ft) kN m	Type cutter and machine weight t (tons)
TB 1-240 E 9/1971	Hydropower Station, Hirzbachwater-tunnel, Austria	Calcareous mica slate	4,845 (15,937)	2.40 (7 10)				Disc cutters and carbide insert cutters
TB V-580 H 1/1972 8/1973	Hydropower Station, Echaillon, France	Slate, crystalline slate	4,357 (14,332)	5.80 (19 0)	760 (1,019)	635,000 (1,402,000) 6,227	76,000 (551,000) 745	Discs carbide insert
TB 11-300 E 7/1972	Ventilation shaft, St. Gotthard, Motto di Dentro/ Switzerland	Slate gneiss	850 (2,796) 40.5°	3.00 (9 10) 6.40 (21 00)	460 (617)	455,000 (1,003,000) 4,462	60,500 (437,000) 593	Discs
TB IV510-H 1972	Hydro-electric Power Station, Mapragg, Sarganzer Land, Switzerland	Calcareous sandstone	6,400 (21,052)	5.10 (16 9)	620 (831)	500,000 (1,102,000) 4,903	55,000 (139,700) 54	Discs
TB 11-330 H 2/1972 3/1973	Gas tunnel, Rothorn-stollen, Switzerland	Calcareous sandstone	3,000 (9,868)	3.30 (10 10)	460 (617)	440,000 (970,000) 4,315	26,000 (188,000) 255	Discs
TB 11-328/360 H 8/1972 7/1973	Glacier-Metro, Kaprun, Austria	Crystalline slate	3,200 (10,499) 23.9°	3.60 (11 10)	460 (617)	440,000 (970,000) 4,315	26,000 (188,000) 255	Discs and carbide inserts
TB 11-300 H 11/1972	Tschingelmad, Switzerland		1,200 (3,937)	3.10 (10 2)	460 (617)	440,000 (970,000) 4,315	26,000 188,000) 255	Discs
TB 1-253 H 3/1973	Inclined shaft, Hohtenn-Lotschental	Gneiss	1,100 (3,609) 1,103 (3,619) 29.3°	2.53 (8 4)	380 (509)	280,000 (617,000) 2,746	17,500 (127,000) 172	Carbide insert cutters
TB 11-300 H 2/1973 10/1974	Pressure gallery, Hohtenn-Lotschental	Crystalline slate	4,300 (14,144)	3.00 (9 10)	460 (617)	440,000 (970,000) 4,315	26,000 (188,000) 255	Disc cutters Carbide insert cutters
TB 11-370 H 7/1973 12/1974	Pressure gallery, Sarelli Sarganser Land	Calcareous sandstone	4,800 (15,789)	3.70 (12 2)	460 (617)	440,000 (970,000) 4,315	26,000 (188,000) 255	Disc cutters
TB V 580 H 2/1974	Pressure water gallery, Arc Isère, Spie-Batignolles	Sandstone, slate	9,675 (31,742)	5.80 (19 0)	760 (1,019)	635,000 (1,400,000) 6,227	76,000 (551,000) 745	Disc cutters
TB11-300E 3/1974 9/1974	Ventilation shaft pilot hole, St. Gotthard, Bäzberg	Granite	476 (1,565) 37.8°	3.00 (9 10)	460 (617)	455,000 (1,003,000) 4,462	60,500 (437,000) 593	Carbide inserts
TB11-359H 6/1975	Express motor road tunnel, pilot drill, Pfänder, near Bregenz	Lime, sandstone	3,350 (11,019)	3.59 (11 10)	460 (617)	440,000 (9,700,000) 4,315	26,000 (188,000) 255	Discs
TB1-253H 8/1975	Inclined gallery, Chiosta Piastra, Italy	Crystalline slate (gneiss)	1,080 (3,552)	2.53 (8 4)	380 (509)	280,000 (617,000) 2,746	17,500 (127,000) 172	Carbide inserts

Alfred Wirth & Co. Kg. Medium to hard-rock machines — continued

Model type and year	Project location	Rock strength kg/cm² (MPa) lb/in²	Tunnel length m (ft) inclin.	Machine diameter m (ft in)	Cutter head kW (hp)	Thrust kg (lb) kN	Torque kg m (lb ft) kN m	Type cutter and machine weight t (tons)
TB11-346H 9/1974	Pressure gallery, Rotlech, Heiterwang	Limestone	4,600 (15,131)	3.46 (11 5)	460 (617)	440,000 (970,000) 4,315	26,000 (188,000) 255	Discs
TB11-300E 9/1975	Pressure shaft, Rovina, Entracque	Crystalline slate (gneiss)	1,080 (3,540) 42.0°	3.00 (9 10)	460 (617)	455,000 (1,003,000) 4,462	60,500 (437,000) 593	Carbide inserts
TB1-280H 12/1975	Feed gallery, Lavtina	Limestone	6,000 (19,736)	2.80 (9 3)	380 (509)	280,000 (617,000) 2,746	17,500 (127,000) 172	Discs
TB11-325H 10/1975	Feed gallery, Acquedotto delle Capore, Vianini	Limestone	7,000 (23,026)	3.25 (10 8)	460 (617)	440,000 (970,000) 4,315	26,000 (188,000) 255	Discs
TBS-11-340H 1976	Tunnel in gold-mine, Anglo-American, South Africa	Quartzite, granite		3.40 (11 2)				Carbide inserts
TB-11 300E 6/1976	Pressure shaft, Chiotas-Entracque	Crystalline slate (gneiss)	1,080 (3,552) 42.0°	3.00 (9 10)	460 (617)	455,000 (1,003,000) 4,462	60,500 (437,000) 593	Carbine inserts
TB-11 330H 1/1977	Feed gallery, Guamacan, Vianini, Venezuela	Sandstone	13,000 (42,763)	3.30 (10 10)	460 (617)	440,000 (970,000) 4,315	26,000 (188,000) 255	Discs

Japanese Licensee Rock Machine Manufacturers*

Ishikawajima Harima (Habegger/Atlas Copco)   4 machines 1967-73;
Kawasaki Heavy Industries (Jarva/Atlas Copco)   2 machines 1971 and 1973;
Komatsu (Robbins)   11 machines 1964-79;
Mitsubishi Heavy Industries (Hughes Tool)   2 machines 1967 and 1968.

Other manufacturers of rock tunnelling machines not included in general list: U.S.S.R. and China.

*Richard J. Robbins, Lecture 1: History of rock boring. Economics and management of underground rock boring, The South African Institute of Mining and Metallurgy, 4-8 Feb., 1980.

Mechanized Shield Machine Manufacturers

Markham & Company Limited, U.K.
Sir Robert McAlpine & Sons Ltd, U.K.*
Robert L. Priestley Ltd., U.K.
Jarva Inc., U.S.A.
Kinnear Moodie & Company Ltd., U.K.*
Bade & Theelen GmbH, West Germany
Tunnelling Equipment (London) Ltd., U.K.
Calweld Inc. — Division of Smith International, U.S.A.*
W. Lawrence & Sons (London) Ltd., U.K.
Westfalia Lunen, West Germany
Arthur Foster, U.K.*
Lovat Tunnel Equipment Inc., Ontario
M & H Tunnel Equipment, U.S.A.*
Machinoexport U.S.S.R.
Aubrey Watson Ltd., U.K.*
Zokor International Ltd., U.S.A.
Decker Mnf. Co., U.S.A.

Memco, U.S.A.*
Robbins Company, U.S.A.
Elgood Mayo Corp., U.K.
Milwaukee Boiler Manufacturing Co., U.S.A.
Stelmo Ltd., U.K.
Marcon Ltd., U.K.
Martin Herrenknecht GmbH, West Germany
Komatsu Ltd., Japan
Nihon Koki Co., Ltd., Japan
Mitsubishi Heavy Industries Ltd., Japan
Mitsui Engineering & Shipbuilding Co., Ltd., Japan
Hitachi Construction Machinery Company, Ltd., Japan
Ishikawajima-Harima Heavy Industries Co. Ltd., Japan
Hitachi Shipbuilding & Engineering Co. Ltd., Japan
Kawasaki Heavy Industries Co. Ltd., Japan
Iseki Poly-Tech., Japan
R. Schäfer & Urbach GmbH, West Germany

*These companies have either stopped manufacturing soft-ground machines or have ceased trading.

Slurry and Earth Pressure Balanced Machine Manufacturers

Markham & Co. Ltd., U.K.
Robert L. Priestley Ltd., U.K.
Orenstein & Koppel, AG, West Germany
Gardner Engineering Corp., U.S.A.*
Wayss & Freytag, AK, West Germany
Komatsu Ltd., Japan
Mitsubishi Heavy Industries Ltd., Japan

Mitsui Engineering & Shipbuilding Co. Ltd., Japan
Hitachi Construction Machinery Co. Ltd.
Ishikawajima-Harima Heavy Industries Co., Ltd., Japan
Hitachi Shipbuilding & Engineering Co. Ltd., Japan
Kawasaki Heavy Industries Co. Ltd., Japan
Iseki Poly Tech., Japan

* These companies have either stopped manufacturing soft-ground machines or have ceased trading.

712

Boomheader Machines

AEC Inc. (formerly Alpine Equipment Corp.)

Model no.	Year of manufacture	Cutter-head type	Loading system	Number of this model manufactured	Cutter-head power kW (hp)	Total installed machine power kW (hp)	Machine weight t (tons)
Super ROC-MINER, Model 330	1977	Field-interchangeable milling or ripping head	Gathering arms	27	123-205 (165-275)	290-373 (390-500)	41.65 (41)
AEC Gantry Miner (Jumbo)	1978	Milling or ripping head	Gathering arms or L.H.D. vehicle	2	Max. 2 × 205 (Max. 2 × 275)	597 (800)	70 (68.9)
ROC-MINER E-Series and H-Series	1979	Ripping head	Gathering arms	31	75-112 (100-150)	150-187 (200-250)	24 (23.63)

Anderson Boyes

Model no.	Year of manufacture	Cutter-head type	Loading system	Number of this model manufactured	Cutter-head power kW (hp)	Total installed machine power kW (hp)	Machine weight t (tons)
Bretby roadheader Mark IIA (Dosco tracks)	1964	Milling	Gathering arm	2	37 (50)	74 (100)	26.4 (26)

Anderson Strathclyde Limited

Model no.	Year of manufacture	Cutter-head type	Loading system	Number of this model manufactured	Cutter-head power kW (hp)	Total installed machine power kW (hp)	Machine weight t (tons)
RH I	1969	Milling	Gathering arm		50 (65)	100 (130)	27.4 (27)
RII 10	1970-71	Milling	Gathering arms		50 (65)	100 (130)	25.4 (25)
RH 20 (crawler or walking base)	1975	Milling	Twin-flight loading chains		50 (65)	100 (130)	24.3 (24)
Boom miner (crawler or walking base)	1975-76	Milling	Encircling conveyor		60 (80)	120 (160)	22 (21.5)
RH 22 (crawler or walking base)	1975-76	Milling	Gathering arm		90 (120)	140 (185)	35 (34.4)
RH1/3 (crawler or walking base)	1975-76	Milling	Gathering arm		90 (120)	180 (240)	50 (49.2)

Distington Engineering Company Limited

Model no.	Year of manufacture	Cutter-head type	Loading system	Number of this model manufactured	Cutter-head power kW (hp)	Total installed machine power kW (hp)	Machine weight t (tons)
Bretby roadheader (Dosco tracks)	1964	Milling	Gathering arm	2	37 (50)	74 (100)	26.4 (26)

Dosco Overseas Engineering Limited

Model no.	Year of manufacture	Cutter-head type	Loading system	Number of this model manufactured	Cutter-head power kW (hp)	Total installed machine power kW (hp)	Machine weight t (tons)
Roadway cutter	1953-69	Milling	Encircling conveyor		37 (50)	85 (115)	18.28 (18)
Mark 2A	1970	Milling	Encircling conveyor		49 (65)	104 (140)	23.36 (23)
Twin-boom miner TB 600	1971	Two milling heads	Gathering arms (flight conveyor folding on return)		90 (120)	650 (900)	73.15 (72)
Mark 3		Milling	Gathering arms		143 (190)	300 (400)	60.96 (60)
SL 120	1971	Milling	Gathering arms		75 (100)	164 (220)	23.36 (23)
U.T.R. crawler mounted	1973	Milling	Bulldozer blade		37 (50)	86 (115)	18.28 (18)
LH 100	1976	Milling	Encircling single-strand conveyor		49 (65)	105 (140)	30.48 (30)

Gebr. Eickhoff Maschinenfabrik (Gebr. Eickhoff)

Model no.	Year of manufacture	Cutter-head type	Loading system	Number of this model manufactured	Cutter-head power kW (hp)	Total installed machine power kW (hp)	Machine weight t (tons)
EV2		Ripping	Gathering arms		80 (107)	173 (232)	33.52 (33)
EVA 160		Ripping	Gathering arms		160 (215)	310 (416)	52.83 (52)
EVR 160		Milling	Gathering arms		160 (215)	340 (456)	81.28 (80)
EVR 200		Milling	Gathering arms				
EV 100A					160 (215)	340 (456)	81.28 (80)
EV 100B		Milling	Gathering arms		160 (215)	340 (456)	81.28 (80)
EVR 120		Milling	Gathering arms		160 (215)	340 (456)	81.28 (80)

Machinoexport (U.S.S.R.)

Model no.	Year of manufacture	Cutter-head type	Loading system	Number of this model manufactured	Cutter-head power kW (hp)	Total installed machine power kW (hp)	Machine weight t (tons)
PK3 and PK-3M (Evolved from Hungarian F-4)	1953	Milling	Encircling conveyor		32 (43)	78 (104)	10.8 (10.8)
PK-9		Milling	Gathering arms		90 (118)	173 (232)	36 (35.4)

714

Mannesman Demag AG (formerly Demag AK)

Model no.	Year of manufacture	Cutter-head type	Loading system	Number of this model manufactured	Cutter-head power kW (hp)	Total installed machine power kW (hp)	Machine weight t (tons)
'Unicorn' VS1 (Hybrid)	1962	Milling — spiral pick arrangement	Encircling flight conveyor	21	25.3 (34)	47.7 (64)	12.5 (12.7)
Articulated Boomheader Excavator	1971	Milling		8	119.3 (160)	160.3 (215)	60 (60.96)
VS2E (Hybrid)	1968-72	Milling — spiral pick arrangement	Encircling flight conveyor	7	55.95 (75)	86.53 (116)	42.67 (42)
VS3	1978	Milling	Twin flight conveyor	5	160 (215)	264 (354)	71.12 (70)

Mavor & Coulson Ltd.

Model no.	Year of manufacture	Cutter-head type	Loading system	Number of this model manufactured	Cutter-head power kW (hp)	Total installed machine power kW (hp)	Machine weight t (tons)
Bretby roadheader Mark IIB (Lee-Norse tracks)	1964	Milling	Gathering arm	2	37 (50)	74 (100)	26.4 (26)

National Coal Board

Model no.	Year of manufacture	Cutter-head type	Loading system	Number of this model manufactured	Cutter-head power kW (hp)	Total installed machine power kW (hp)	Machine weight t (tons)
Bretby roadheader Mark I	1962	Milling	Gathering arms	1	30 (40)	60 (80)	20 (20)
Bretby roadheader Mark II	1963	Milling	Gathering arms	8	30 (40)	78 (105)	20 (20)
MRDE Boom ripper	1969	Milling	None	2	30 (40)	87.5 (130)	20 (20)

Paurat GmbH (under Dosco licence)

Model no.	Year of manufacture	Cutter-head type	Loading system	Number of this model manufactured	Cutter-head power kW (hp)	Total installed machine power kW (hp)	Machine weight t (tons)
SVM	1967	Milling	Encircling conveyor	15	50 (67)	88 (118)	20 (20)

Model no.	Year of manufacture	Cutter-head type	Loading system	Number of this model manufactured	Cutter-head power kW (hp)	Total installed machine power kW (hp)	Machine weight t (tons)
E124	1973	Milling	Flight conveyor	1	55 (74)	85 (141)	21 (21)
E141	1974	Milling	Gathering arms	10	55 (74)	85 (141)	24.4 (24)
Twin-boom miner	1970	Twin booms milling	None (separate loader)	1	110 (150)	242 (324)	50.8 (50)
E134 U.K. (Titan) West Germany (Roboter)	1975	Milling spiral pick arrangement	Two single-chain conveyors around deflector sheaves	31	200 (268)	300 (402)	62.9 (62)
E169	1978	Milling	Gathering	9	100 (134)	187 (206)	39.6 (39)

Paurat GmbH

Model no.	Year of manufacture	Cutter-head type	Loading system	Number of this model manufactured	Cutter-head power kW (hp)	Total installed machine power kW (hp)	Machine weight t (tons)
Trench-cutting machine (crawler mounted)	1975	Articulated boom — milling — spiral pick arrangement	Separate excavator	9	200 (268)	400 (536)	101.6 (100)

Paurat Special-Purpose Machines

GKL	1969	Cutting unit in walking frame with two booms.
PTF70	1970	With two booms.
PTF70A	1971	With telescoping boom for maximum cross-section of 144.5 m² to be cut from one position.
E124	1973	For small cross-sections, special loader for sticky material.
E128	1973	For large cross-sections without loader.
E130	1973	Similar to the E128, with loader.
E133	1973	Light machine mounted on an excavator chassis.
E135	1974	For circular cross-sections.
E136	1974	For railway tunnels in abrasive rock.
E149	1974	With impact ripper.
E141	1975	With gathering arm loaders.

Voest-Alpine (formerly Alpine Montan) (licence granted by Nikex in conjunction with Orszago Bányagépgyártó Vállalat)

Model no.	Year of manufacture	Cutter-head type	Loading system	Number of this model manufactured	Cutter-head power kW (hp)	Total installed machine power kW (hp)	Machine weight t (tons)
F-6A	1964	Ripping (twin head)	Gathering arms (single-chain conveyor)		30 (40)	60 (80)	12.19 (12)

Alpine Montan/Schäffer & Urbach

Model no.	Year of manufacture	Cutter-head type	Loading system	Number of this model manufactured	Cutter-head power kW (hp)	Total installed machine power kW (hp)	Machine weight t (tons)
Circular cutting equipment	1977	Ripping (twin head — with different diameters)	Chain conveyor	1	Cutterheads on twin booms driven by one motor		

Voest-Alpine

Model no.	Year of manufacture	Cutter-head type	Loading system	Number of this model manufactured	Cutter-head power kW (hp)	Total installed machine power kW (hp)	Machine weight t (tons)
Alpine miner AM50	1971	Ripping (twin head)	Gathering arms		100 (136)	155 (211)	24.38 (24)
Alpine miner AM100	1976	Ripping (twin head)	Gathering arms		228 (306)	456 (612)	75.18 (74)

Voros Csillag (Red Star) Tractor Works

Model no.	Year of manufacture	Cutter-head type	Loading system	Number of this model manufactured	Cutter-head power kW (hp)	Total installed machine power kW (hp)	Machine weight t (tons)
	1949	Ripper chain-driven cutter head with contra-rotating discs fitted with picks					
F-Type 1	1949	Ripper chain-driven cutter head with contra-rotating discs fitted with picks					
F-Type 2	1949	Ripper cutter-head with hemi-spherical discs fitted with picks			Motor at base of cutting boom	Separate motor for traction	
F-3, F-4, F-5, Serial 'Zero', 'Serial-1'	1949	Ripper cutter-head with hemi-spherical discs fitted with picks			Motor at base of cutting boom	Separate motor for traction	
F-Type	1950-51	Ripper cutter-head with hemi-spherical discs fitted with picks			Motor at base of cutting boom	Separate motor for traction	
F-6	1956-64	Ripping (first unit with twin heads)	Gathering arms		30 (40)	60 (81)	10.7 (10.5)

Model no.	Year of manufacture	Cutter-head type	Loading system	Number of this model manufactured	Cutter-head power kW (hp)	Total installed machine power kW (hp)	Machine weight t (tons)
Büffel WAV 170 211 100-211 119	1972-79	Radial cutting head	Platform with loading arm	20	200 (268)	290 (390)	53.8 (53)
WAV 178 211 120-211 122	1975-79	Radial cutting head	Platform with loading arm	3	200 (268)	290 (390)	73.15 (72)
Bison WAV 200 211 200-211 203	1975-78	Radial cutting head	Platform with loading arm	4	200 (268)	310 (416)	74.16 (73)
Frettchen FL-34	1976	Radial cutting head	Conveyor boom	1	13 (17.5)	22 (30)	3.04 (3)
FL-1-20 40 001-40 008	1978	Radial cutting head	Conveyor boom	8	20 (27)	25 (34)	3.55 (3.5)
Wühlmaus FL-13 13 001-13 027	1967-73	Radial cutting head	Conveyor boom	27	14 (19)	37 (50)	5.08 (5)
FL-31 31 001-31 014	1973-77	Radial cutting head	Conveyor boom	14	14 (19)	52 (71)	6.60 (6.5)
FL-2-25 39 001-39 003	1978-79	Radial cutting head	Conveyor boom	3	25 (34)	45 (61)	5.89 (5.8)
Fuchs (Prototype FL-SO1) ı	1963	Radial cutting head	Conveyor boom	1	15 (21)	23.8 (32)	5.08 (5)
FL-SO1 1001-1014	1965-67	Radial cutting head	Conveyor boom	11	15 (21)	23.8 (32)	5.08 (5)
FL-RO2 2001-2012	1965-66	Radial cutting head	Conveyor boom	10	15 (21)	23.8 (32)	5.08 (5)
FL-RO4 4001	1968	Radial cutting head	Conveyor boom	1	33 (45)	53 (72)	6.40 (6.3)
FL-3R-33 4002-4021	1972-78	Radial cutting head	Conveyor boom	20	33 (45)	53 (72)	6.40 (6.3)
FL-SO5 5001	1967	Radial cutting head	Conveyor boom	1	24 (33)	37.5 (51)	6.09 (6)
FL-35-33 5002	1967-79	Radial cutting head	Conveyor boom	42	24 (33)	37.5 (51)	6.09 (6)
FL-RO6 6002-6047	1966-78	Radial cutting head	Conveyor boom	44	24 (33)	37.5 (51)	5.58 (5.5)
FL-14 14 001	1978	Radial cutting head	Conveyor boom	1	33 (45)	55 (75)	6.09 (6)
FL-20 20 001	1978	Radial cutting head	Conveyor boom	1	22 (30)	33 (45)	9.14 (9)
FL-21 21 001-21 006	1969-70	Radial cutting head	Conveyor boom	6	24 (33)	37 (50)	3.55 (3.5)
FL-22 22 001	1969	Radial cutting head	Conveyor boom	1	35 (47)	86 (117)	6.60 (6.5)
FL-S25 25 001-25 002	1969-76	Radial cutting head	Conveyor boom	2	14 (19)	29.5 (40)	4.57 (4.5)
FL-35 35 001-35 002	1976	Radial cutting head	Conveyor boom	2	33 (45)	45 (61)	6.60 (6.5)
FL-3R-40 37 001-37 007	1977-79	Radial cutting head	Conveyor boom	7	40 (54)	73 (99)	9.14 (9)
Dachs FL-R12 12 001-12 021	1967-75	Radial cutting head	Conveyor boom	21	30 (41)	70 (96)	13.20 (13)
FL-R23 23 001	1969	Radial cutting head	Conveyor boom	1	53 (72)	101 (137)	13.20 (13)
FL-4R-53 23 002-23 046	1969-79	Radial cutting head	Conveyor boom	45	53 (72)	101 (137)	13.20 (13)

Model no.	Year of manufacture	Cutter-head type	Loading system	Number of this model manufactured	Cutter-head power k W (hp)	Total installed machine power k W (hp)	Machine weight t (tons)
FL-24 24 001-24 004	1972-76	Radial cutting head	Conveyor boom	4	35 (47)	86 (117)	12.19 (12)
FL-S26 26 001	1970	Radial cutting head	Conveyor boom	1	30 (47)	70 (96)	14.22 (14)
FL-R29 29 001	1971	Radial cutting head	Conveyor boom	1	53 (72)	101 (137)	13.20 (13)
FL-38 38 001-38 002	1976	Radial cutting head	Conveyor boom	2	35 (47)	75 (102)	6.09 (6)
Luchs FL-5R-90	1975-78	Radial cutting head	Conveyor boom		110 (150)	182 (248)	25.4 (25)
32 001-32 004 42 005-42 006 46 007-46 010	1975-78 1979			4 2 4			
FL-6R-110 43 001-43 002	1978-79	Radial cutting head	Conveyor boom	2	110 (150)	202 (275)	38.60 (38)
Firstenfreise FF-5R-90 36 001-36 004	1976-79	Radial cutting head	Conveyor boom	4	90 (122)	137 (186)	23.36 (23)

Mounted Impact Breakers — continued

Manufacturer	Model no.	Blow energy kg m (ft lb)	Blows per minute	Type of accumulator or spring	Total energy per unit time kg m/min (ft lb/min)	Hydraulic or pneumatic	Weight kg (lb)	Tool shank size (dia) mm (in)
Allied Steel & Tractor Products, U.S.A.	Rapid Ram Model 33	27.66 (200)	1,250		35,000 (250,000)	H	317 (700)	63.5 (2¼)
Allied Steel & Tractor Products, U.S.A.	Ho-Ram* Super 79	415 (3,000)	250	none	103,000 (750,000)	P	1,632 (3,600)	146 (4¾)
Allied Steel & Tractor Products, U.S.A.	Ho-Ram 7000B	138 (1,000)	400	none	55,000 (400,000)	P	499 (1,100)	101.6 (4)
Allied Steel & Tractor Products, U.S.A.	Ho-Ram Model 250	138 (1,000)	450	none	62,100 (450,000)	P	431 (950)	89 (3½)
Manufactured by Allied for Krupp in U.S.A.	Hy-Ram* Model 65	104 (750)	550		57,000 (412,500)	H	453 (1,000)	82.5 (3¼)
Manufactured by Allied for Krupp in U.S.A.	Hy-Ram Model 77	104 (750)	550		57,000 (412,500)	H	408 (900)	82.5 (3¼)
Manufactured by Allied for Krupp in U.S.A.	Hy-Ram Model 88	180 (1,300)	450		80,905 (585,000)	H	726 (1,600)	101.6 (4)
Manufactured by Allied for Krupp in U.S.A.	Hy-Ram Model 99	277 & 138 (2,000 & 1,000) at 450 & 900 blows per minute respectively	450-900		124,400 (900,000)	H	451 (3,200)	133.3 (5¼)
Anderson/Mavor, U.K.**	MC3A/4R	150 (1,085)	600		90,000 (651,000)	H	n.a.	
Champion**	H-16	2,212 (16,000)	110		243,000 (1,760,000)	H	n.a.	
Compair, U.K.**	Holbuster	248 (1,800)	180		45,000 (324,000)	H	635 (1,402)	
Contech**	HD-7	331 (2,400)	200		66,000 (480,000)	H	544 (1,200)	

Mounted Impact Breakers — continued

Manufacturer	Model no.	Blow energy kg m (ft lb)	Blows per minute	Type of accumulator or spring	Total energy per unit time kg m/min (ft lb/min)	Hydraulic or pneumatic	Weight kg (lb)	Tool shank size (dia) mm (in)
Contech**	HD-10	318 (2,300)	350		11,000 (805,000)	H	789 (1,740)	
Contech**	HD-5	138 (998)	450		62,000 (450,000)	H	489 (1,080)	
Contech**	HD-3	69 (500)	650		45,000 (325,000)	H	329 (727)	
Contech**	Miniram 125	48 (350)	750	none	36,000 (262,000)	P	235 (520)	
C.T.I. West Germany**	Nutcracker HD7	297 (2,150)	200		59,000 (430,000)	H	499 (1,100)	
Demag, West Germany**	VR40	1,300 (9,403)	138	none	179,000 (1,300,000)	P	5,499 (12,125)	
Demag, West Germany**	VR15	366 (2,647)	215	none	79,000 (569,000)	P	2,300 (5,071)	
Demag, West Germany**	DKB-750	276 (1,996)	600	none	166,000 (1,200,000)	P	870 (1,918)	
Demag, West Germany**	DKB-375	138 (998)	600	none	83,000 (600,000)	P	446 (984)	
Eimco**	Impactor	207 (1,500)	400		83,000 (600,000)	H	n.a.	
Furukawa**	1200	240 (1,736)	460	none	110,000 (800,000)	P	1,139 (2,513)	
Furukawa**	750	170 (1,230)	450	none	76,000 (550,000)	P	749 (1,653)	
Gardner Denver, U.S.A.	CB99A		101	none	n.a.	P	150 (330)	76 (3)
Guest**	125	96 (700)	500	none	48,000 (350,000)	P	204 (450)	
Gullick Dobson Ltd, U.K.	G.D. 3000	295 (2,220)	Variable 0-600	Internal nitrogen	177,000 (1,330,000)	H	820 (1,804)	102 (4)
Gullick Dobson Ltd, U.K.	G.D. 2000	204 (1,475)	Variable 0-600	Internal nitrogen	122,000 (885,000)	H	465 (1,025)	89 (3.5)
Hausherr**	HNM-1	210 (1,519)	400		84,000 (610,000)	H	n.a.	
Hausherr**	HNL-1	90 (651)	550		49,000 (360,000)	H	n.a.	
HED**	HB-500	69 (500)	500		35,000 (250,000)	H	244 (538)	
Hughes Tool Co., U.S.A.	Impactor AA-750Wg	17.3 (125)	1,000	Internal mechanical spring	17,300 (125,000)	H	105 (233)	47.63 (1-7/8)
Ingersoll-Rand, U.S.A.	Goblin G500	69 (500)	600	Sliding piston accumulator	41,000 (300,000)	H	527 (720)	76.2 (3)
Ingersoll-Rand, U.S.A.	Goblin G900	124 (900)	420	Sliding piston accumulator	52,000 (378,000)	H	363 (800)	76.2 (3)
Ingersoll-Rand, U.S.A.	Goblin G1100B	165 (1,100)	588	Sliding piston accumulator	97,000 (646,000)	H	500 (1,100)	76.2 (3)
Ingersoll-Rand, U.S.A.	Air Goblin ABM500	96.6 (698)	600	none	57,000 (420,000)	P	291 (640)	76.2 (3)
Ingersoll-Rand, U.S.A.	Air Goblin ABM1000	165.6 (1,200)	600	none	99,000 (720,000)	P	454 (1,000)	101 (4)

Mounted Impact Breakers — continued

Manufacturer	Model no.	Blow energy kg m (ft lb)	Blows per minute	Type of accumulator or spring	Total energy per unit time kg m/min (ft lb/min)	Hydraulic or pneumatic	Weight kg (lb)	Tool shank size (dia) mm (in)
I.P.H.**	202B	95 (687)	400		38,000 (270,000)	H	249 (551)	
Joy Manufacturing Co., U.S.A.	411 HEFTI	829 (6,000)	65	Nitrogen	54,000 (390,000)	H	1,179 (2,600)	
Joy Manufacturing Co., U.S.A.	514 HEFTI	2,765 (20,000)	25	Nitrogen	69,000 (500,000)	H	1,089 (2,400)	127 (5)
Joy Manufacturing Co., U.S.A.	206 HEFTI	138 (1,000)	200	Nitrogen	27,600 (200,000)	H	159 (350)	51 (2)
Kent Air Tool Co., U.S.A.	999	138 (1,000)	600	none	83,000 (600,000)	P	376 (829)	89 (3½)
Kent Air Tool Co., U.S.A.[a]	KB-999-H	138 (1,000)	600	none	83,000 (600,000)	P	567 (1,250)	89 (3½)
Kent Air Tool Co., U.S.A.[b]	KB-999-S	138 (1,000)	600	none	83,000 (600,000)	P	590 (1,300)	89 (3½)
Kent Air Tool Co., U.S.A.	555	69 (500)	600	none	44,490 (300,000)	P	220 (485)	64 (2½)
Kent Air Tool Co., U.S.A.	2000	276 (2,000)	600	none	165,960 (1,200,000)	P	744 (1,640)	133 (5¼)
Krupp & Co., West Germany[c]	HM 401	55 (400)	550		30,000 (220,000)	H		
Krupp & Co., West Germany	HM 110	46 (332)	700-1,000		45,900 (331,000)	H	170 (374)	65 (2.55)
Krupp & Co., West Germany	HM 200	82 (590)	480-650		53,000 (383,000)	H	530 (1,168)	80 (3.14)
Krupp & Co., West Germany	HM 600	194 (1,400)	360-500		96,700 (699,000)	H	925 (2,039)	100 (3.93)
Krupp & Co., West Germany[d]	HM 800	316 (2,280)	300-450 (600-900)		142,000 (1,020,000)	H	1,480 (3,263)	135 (5.31)
Krupp & Co., West Germany[d]	HM 900	377 (2,730)	300-450 (600-900)		169,500 (1,220,000)	H	1,480 (3,263)	135 (5.31)
Krupp & Co., West Germany[d]	HM 1200	459 (3,320)	250-400		183,000 (1,320,000)	H	1,650 (3,637)	150 (5.90)
Lee Norse**	Hard Nose-1	165 (1,200)	600		99,000 (720,000)	H		
Lemand**	Shand-Macol	272 (1,970)	200		545,000 (394,000)	H	522 (1,151)	
Lemand**	Overarm	240 (1,736)	180		43,200 (312,000)	H		
Lemand**	Shand Hammer	239 (1,730)	180		43,000 (311,000)	H	521 (1,150)	
McDowell**	Powa Ram	1,383 (10,000)	85		117,000 (850,000)	H	1,530 (3,375)	
Mindev**	BR 120	159 (1,155)	1,250		199,000 (1,440,000)	H	306 (675)	
Mindev**	380	134 (976)	500		67,000 (488,000)	H		
Mindev**	BR-40	117 (850)	2,000		235,000 (1,700,000)	H	204 (450)	
Mindev**	Woodpecker	8 (65)	1,500		13,000 (97,000)	H	39 (88)	
MKT**	RB-8	575 (4,158)	225	none	129,000 (935,000)	P	2,751 (6,065)	
MKT**	RB-4	345 (2,500)	275	none	95,000 (687,000)	P	1,757 (3,975)	

Mounted Impact Breakers — continued

Manufacturer	Model no.	Blow energy kg m (ft lb)	Blows per minute	Type of accumulator or spring	Total energy per unit time kg m/min (ft lb/min)	Hydraulic or pneumatic	Weight kg (lb)	Tool shank size (dia) mm (in)
MKT**	RB-2	138 (1,000)	300	none	41,000 (300,000)	P	929 (2,050)	
Montabert, France	BRH125	70 (506)	400-1,000	External nitrogen	70,000 (500,000)	H	272 (600)	70 (2.7)
Montabert, France	250	140 (1,012)	230-600	External nitrogen	84,000 (600,000)	H	550 (1,200)	95 (3.7)
Montabert, France	501	220 (1,590)	350-500	External nitrogen	110,000 (700,000)	H	1,000 (2,200)	114 (4.5)
Montabert, France	1000	450 (3,253)	275-450	External nitrogen	202,000 (1,000,000)	H	1,600 (3,500)	135 (5.3)
Montabert, France	2-96A	69 (500)	175	none	48,000 (350,000)	P	204 (450)	50.8 (2)
Muedon**	450	180 (1,302)	300	none	54,000 (390,000)	P	449 (992)	
National Coal Board, U.K.[e]	Bretby Impact Unit Mark I	85 (600)	180	Internal mechanical spring	15,000 (108,000)	H	450 (1,000)	76 (3)
National Coal Board, U.K.[f]	Bretby Impact Unit Mark II	140 (1,000)	120	Internal mechanical spring	16,800 (120,000)	H	450 (1,000)	76 (3)
NPK**	Dynamax 6000	600 (4,340)	150	none	90,000 (650,000)	P	n.a.	
NPK**	H-11X	387 (2,800)	500		193,000 (1,400,000)	H	1,202 (2,650)	
NPK**	802HB	280 (2,025)	400		112,000 (810,000)	H	729 (1,609)	
NPK**	H-9X	276 (2,000)	500		138,000 (1,000,000)	H	889 (1,962)	
NPK**	Dynamax 2500	270 (1,953)	200	none	53,900 (390,000)	P	749 (1,653)	
NPK**	602HB	210 (1,519)	400		84,000 (600,000)	H	499 (1,102)	
NPK**	H-6X	207 (1,500)	570		110,000 (850,000)	H	600 (1,323)	
NPK**	IPH600	170 (1,230)	310	none	52,000 (380,000)	P	624 (1,376)	
NPK**	IPH400	73 (533)	320	none	23,000 (170,000)	P	405 (893)	
NPK**	H-3X	69 (500)	580		40,000 (290,000)	H	249 (551)	
Oy Tampella Ab TAMROCK Industriel, Finland	HB300	30 (220)	1,320	Internal nitrogen	40,000 (290,000)	H	115 (250)	53-63 (2-2.48)
Quincy**	BBH36	13 (100)	1,000		13,000 (100,000)	H	39 (88)	
Quincy**	BBH31	13 (100)	500		6,000 (50,000)	H	29 (66)	
Racine Federated Inc., U.S.A. (formerly manufactured by Worthington Compressors, U.S.A.)	MB-600	69 (500)	600	Bladder	41,000 (300,000)	H	430 (950)	83 (3¼)
Rammer Oy**	S800	120 (886)	840		100,000 (740,000)	H	1,250 (2,756)	

Mounted Impact Breakers — continued

Manufacturer	Model no.	Blow energy kg m (ft lb)	Blows per minute	Type of accumulator or spring	Total energy per unit time kg m/min (ft lb/min)	Hydraulic or pneumatic	Weight kg (lb)	Tool shank size (dia) mm (in)
Roxon**	B-200	132 (959)	560		74,000 (530,000)	H	600 (1,323)	
Roxon**	B-700	299 (2,169)	400		119,000 (860,000)	H	1,000 (2,205)	
Schramm Inc., U.S.A.	B1100	207 (1,500)	500	none	83,000 (600,000)	P	499 (1,100)	101 (4)
Schramm Inc., U.S.A.**	B450	76 (550)	1,100	none	83,000 (600,000)	P	200 (441)	
Shand**		228 (1,650)	180		41,000 (297,000)	H	n.a.	
Shand**	Fluicon	228 (1,650)	200		45,600 (330,000)	H	521 (1,150)	
Stennuick**	BR150	n.a.	200	none	n.a.	P	650 (1,433)	
Theiss**	DCB 2500	268 (1,945)	1,250	none	336,000 (2,430,000)	P	1,059 (2,336)	
Worthington Compressors, U.S.A.[g]	Vanquisher 1000	138 (1,000)	500	none	69,000 (500,000)	P	498 (1,100)	101 (4)
Worthington Compressors, U.S.A.[g]	Vanquisher T500	69 (500)	500	none	35,000 (250,000)	P	297 (655)	76 (3)

*Obsolete.

**Data not obtained direct through company but indirectly through other sources.

[a]With heavy-duty side plates.
[b]With silencing muffler.
[c]Discontinued.
[d]Blows per minute may be doubled by operator pressing a button on the operator's panel during operation.

Note: The National Coal Board impact units were designed at the Mining Research and Development Establishment and made to their order:
[e](Mark I) Partially built in M.R.D.E. Workshops with components made by MacTaggart Scott & Co. Ltd, Scotland.
[f](Mark II) built by Universal Fisher Engineering Ltd, England.
[g]These models discontinued. Rights for the manufacture of Worthington hydraulic mounted impact breakers sold to Racine Federated Inc. several years ago.

Index

727

728

DEFINITIONS FOR MORE COMMONLY USED TUNNELLING MACHINES

SHIELD

Simple construction in a tunnel producing a working area which is protected against collapse either of the walls or roof of that section of the tunnel which has recently been excavated and in which no tunnel lining or other means of support has yet been installed. These machines are advanced by jacks or propel cylinders which thrust against a thrust-ring or the recently erected tunnel lining.

MECHANIZED SHIELD

A shield fitted with either a rotary, or oscillatory cutting head; or with active horizontal or vertical cutting grids or slump shelves, where the *gauge of the tunnel profile is cut by the cutting edge of the shield.* These machines are advanced by jacks or propel cylinders which thrust against a thrust-ring or the recently erected tunnel lining.

TUNNEL BORING MACHINE (T.B.M.)
or ROCK TUNNELLING MACHINE

A machine where the *gauge of the tunnel profile is cut by the outer or peripheral cutters* and *not* by the cutting edge of a shield. These machines are advanced by jacks or propel cylinders reacting against grippers which are expanded laterally against the tunnel walls.

BOOM-TYPE MACHINES

A machine fitted with a boom on which is mounted either a rotary cutting head (boomheader)[a] with ripping- or milling-type cutting head; a ripping or scraping implement; a digger implement; an impact breaker[b]; or an excavator- or knuckling-bucket (i.e. backhoe). These machines may be sledge mounted (now generally obsolete); wheel mounted; or crawler-track mounted.

HYBRID MACHINES

Hybrid Machine: A machine embracing two or more different types of machine in a single unit, i.e. a shield with a boomheader; or a rock tunnelling machine or T.B.M. with a full soft-ground shield, etc.

---

[a]These machines are frequently referred to as 'Roadheaders', a name coined by the Mining Research and Development Establishment (MRDE) of the National Coal Board (NCB) when the prototype machines were being developed and patented. Strictly speaking, therefore, the word 'Roadheader' should only be used when referring to the boomheader machines manufactured by Anderson Strathclyde Ltd for 12 years under NCB licence (which terminated in 1980) and by Dosco Overseas Engineering Ltd. (Dosco machines did not use gathering arms and were only licensed for the telescopic boom).

[b]Similarly, the word 'Impact Ripper' was a name coined by the MRDE to describe the prototype mounted impact breaker machines developed and patented by the NCB. Both Dosco Overseas Engineering Ltd and Guillick Dobson Ltd currently produce 'Impact Rippers' under NCB licence.

SHIELDED T.B.M.

A fully shielded T.B.M. where the outer or peripheral cutters cut the gauge of the tunnel profile, *fitted with propel cylinders or jacks which react against grippers* expanded against the tunnel walls, or with similar reacting jacks and ancilliary propel cylinders which react against a thrust-ring or the tunnel lining. (i.e. *all* 'Shielded T.B.Ms.' are equipped with *grippers* which expand against the tunnel walls and which provide reaction for the thrust forces of the propel cylinders. Such machines may or may not also be equipped with ancilliary propel cylinders which react against the tunnel lining or a thrust-ring.

LINER-THRUST SHIELD T.B.M.
or LINER-THRUST T.B.M.

A fully shielded T.B.M. where the outer or peripheral cutters cut the gauge of the tunnel profile, fitted *only* with propel or thrust jacks which react against a thrust-ring or the recently erected tunnel lining. No lateral grippers are used with this type of machine because of the soft nature of the tunnel walls.

BOOMSHIELD MACHINES

A shield within which is mounted a *boom* fitted with either an impact breaker; a rotary cutting head (boomheader); a ripping, scraping or digging implement; or an excavator- or knuckling-bucket.

SHIELD BOOMHEADER
or BOOMHEADER SHIELD

A shield within which is mounted a *boomheading* unit with either a milling- or ripping-type cutter head.

BOOMSHIELD IMPACT BREAKER
or BREAKER BOOMSHIELD
or BREAKER SHIELD

A shield within which is mounted an *impact breaker* unit.

SHIELD RIPPER-SCRAPER
or DIGGER SHIELD
or RIPPER-SCRAPER SHIELD

A shield within which is mounted a boom fitted with a *ripping, scraping or digging implement*.

BOOMSHIELD EXCAVATOR
or EXCAVATOR BOOMSHIELD
or EXCAVATOR SHIELD

A shield within which is mounted a boom fitted with an *excavating-* or *knuckling-bucket* (i.e. a backhoe-type machine)

STONE BRONZE AND IRON AGE DIGGING TOOLS

TUNNELLING MACHINES

ROCK AND HARD GROUND TUNNELLING MACHINES [25] [17]
[22] [16] [1]

- Maus 1846 (P)
- [2] Talbot (Discs) 1853 (R)
- [13] Wilson (Discs) 1856 (R)
- Beaumont (Discs) 1864 (P)
- Brunton (Discs) 1866 (R)
- Beaumont (Picks) 1875 (R)
- Beaumont/English (Sliding upper frame) 1880 (R)
- Whitaker 1921 (R)
- [3] Robbins (Snr) 1953 (R)
- [10] Hughes Tool Co 1958 (R)
- [4]

SHIELDS [26] [18] [16]

- Brunel 1818
- Beach 1869
- Greathead 1869
- [15] Jennings Nobbs (Needle Shield) (Bar Shield) 1889 and 1890
- Whitaker 1917
- Memco (Hydraulic Fore-poling) 1962
- [15] Westfalia Lünen (Blade Shield) 1978

MECHANIZED SHIELDS [18]

- Price 1896
- Markham (Price design) 1901

PRESSURIZED PLENUM CHAMBER MACHINES

- Campenon Bernard 1961
- Robbins Co. (Etoile T.B.M.) 1964
- Bade (MDS 550-GS) 1965

SLURRY MACHINES

- Gardner Eng. 1958/9
- Markham/Mitchell Con. - Kinnear Moodie (Universal Soft Ground T.B.M.) 1967
- Mitsubishi Heavy Industries 1970
- Bartlett/Nuttal-Priestley (Bentonite Shield) 1971
- Tekken Kensetsu

HYBRID MACHINES

FULL-FACE T.B.Ms. [25] [16]

- Robbins Co. (Melbourne-head) 1968
- Robbins Co. (Grandori Tandem Shield) 1972
- Robbins Co. (Buckskin T.B.M.) 1976

BOOM-TYPE MACHINES [26] [12]

- T. Thomson (Ladder Excavator) 1890
- [15] J.E. Ennis (Shielded Digger) 1901
- [24] Memco (Big John) (Shielded Excavator) 1967
- Robbins (Ripper Scraper) (Shielded Digger)